M. Herrmann S. Ikeda U. Orbanz

Equimultiplicity and Blowing up

An Algebraic Study

With an Appendix by B. Moonen

Springer-Verlag
Berlin Heidelberg New York
London Paris Tokyo

Manfred Herrmann
Ulrich Orbanz
Mathematisches Institut der Universität zu Köln
Weyertal 86−90, D-5000 Köln 41, FRG

Shin Ikeda
Mathematical Department Gifu College of Education
2078 Takakuwa, Gifu, Japan

With 11 Figures

The figure on the cover illustrates Theorem (20.5) of Chapter IV.
The geometry of this is elaborated in Chapter III,2.2 of the Appendix,
see in particular Theorem (2.2.2) and (2.2.32).

Mathematics Subject Classification (1980): 13 H 10, 13 H 15,
14 B 05, 14 B 15, 32 B 05, 32 B 30

ISBN 3-540-15289-X Springer-Verlag Berlin Heidelberg New York
ISBN 0-387-15289-X Springer-Verlag New York Berlin Heidelberg

Libary of Congress Cataloging-in-Publication Data. Herrmann, Manfred, 1932-.
Equimultiplicity and blowing up. Bibliography: p. Includes index. 1. Multiplicity
(Mathematics) 2. Blowing up (Algebraic geometry) 3. Local rings. I. Ikeda, S. (Shin),
1948-. II. Orbanz, Ulrich, 1945-. III. Title. QA251.38.H47 1988 512 88-4660
ISBN 0-387-15289-X (U.S.)

© Springer-Verlag Berlin Heidelberg 1988
Printed in Germany

Printing and binding: Druckhaus Beltz, Hemsbach/Bergstraße
2141/3140-543210

To Gerdi 1988
from the first author

Preface

This book is intended as a special course in commutative algebra and assumes only a general familiarity with topics on commutative algebra and algebraic geometry included in textbooks. We treat two kinds of problems. One of them consists in controlling Hilbert functions after blowing up convenient centers. This question arises directly from the resolution of algebraic and complex-analytic singularities. The other problem is to investigate Cohen-Macaulay properties under blowing up. We begin with some remarks on the background.

1) In the case of plane curves desingularization means classification of singularities, since by blowing up points we get a multiplicity sequence which describes the topological type of the singularity. For the case of higher dimensions and codimensions Zariski and Hironaka suggested to blow up regular centers D contained in the singular locus of the given variety X. In this case the hierarchy of numerical conditions on D is as follows:

(i) all points of D have the same multiplicity

(ii) all points of D have the same Hilbert polynomial

(iii) all points of D have the same Hilbert functions.

These three conditions coincide for hypersurfaces but they differ in general. For each condition there is an algebraic description, namely by reductions of ideals for (i) and by flatness conditions on the associated graded ring for (ii) and (iii).
Hironaka's inductive resolution of algebraic schemes over fields of characteristic zero makes use of numerical conditions arising from the Hilbert functions. The approach to the problem by a non-inductive procedure is open and one is still far from the knowledge of complexity and classification of singularities of dimension bigger than one.

Besides blowing up regular centers there are also approaches to desingularization which amount to blowing up singular centers; for example:

a) Zariski-Jung's desingularization of surfaces in characteristic zero, using generic projections and embedded resolution of the diseriminant locus, induces blowing ups at singular centers on the surface.

b) Also in the way of desingularization by blowing up non-regular centers one can mention M. Spivakovsky's resolution of two-dimensional complex-analytic singularities by Nash transformations and normalizations.

In order to control singularities under blowing up singular centers one can ask for generalizations of the numerical conditions (i), (ii) or (iii) and their algebraic descriptions. That we do by using generalized Hilbert functions and multiplicities. This allows to extend many classical results to a more general situation, and it leads us to three essential types of numerical conditions as three possibilities to make precise the naive idea of "equimultiplicity".

Note that the role of multiplicities and Hilbert functions in geometry is that they furnish some way of measuring and comparing singularities. The concept of multiplicity is older than that of the Hilbert function, but since Samuel has shown how to compute multiplicities via Hilbert functions, many results on multiplicities are consequences of the corresponding results for Hilbert functions. Still there are some results on multiplicities which are not parallel to Hilbert functions, due to the following facts: a) The multiplicity of a local ring is always the degree of a generic projection, which means algebraically that there is a system of parameters giving the same multiplicity as the maximal ideal. b) For multiplicities there is a projection formula for finite morphisms. (There is nothing similar to a) and b) for Hilbert functions, of course.) c) To derive a relation between multiplicities from Hilbert functions, one needs to know something on the dimensions, which occur as degrees of the Hilbert polynomials. In particular, lower dimensional components do not enter into the multiplicity. Therefore already Chevalley assumed his local rings to be quasi-unmixed. In numerous papers, Ratliff has developped a fairly complete theory for quasi-unmixed rings, and it is now clear that the notion of quasi-unmixedness gives the correct frame for the study of multiplicities.

2) Let X' be a blowing up of a variety X with center Y. In general the Cohen-Macaulay properties of X and X' are totally unrelated. But if Y is locally a complete intersection and if the local cohomology modules of the affine vertex over X (or the local cohomology of the vertex of the conormal cone of Y) are finitely generated in all orders $\leq \dim X$ (or $< \dim X$ respectively) then X' may become Cohen-Macaulay. This gives a motivation to study arithmetic properties as well as geometric ones of X' and its exceptional divisor.

The main purpose of the Appendix by B. Moonen is to provide a geometric description of the notion of multiplicity and a geometric interpretation of the notion of an equimultiple ideal within the realm of complex analytic geometry.

Now we give a detailed description of the contents of the book. Chapter I - III contain the basic techniques except local duality which is treated in Chapter VII. In Chapter I we recall all the basic facts about multiplicities, Hilbert functions and reductions of ideals. The second Chapter contains some general facts about graded rings that arise in connection with blowing up. We also recall the theory of standard bases. In Chapter III several characterizations of quasi-unmixed local rings are given. A very useful tool for these characterizations is the theory of asymptotic sequences which has been mainly developed by Ratliff and Katz. Our treatment follows closely the way of Katz.

Chapter IV presents various notions of equimultiplicity. For a hypersurface and a regular subvariety there exists a "natural" notion of equimultiplicity, and there are different directions of generalization: a) to the non-hypersurface case, b) to non-regular subvarieties. In these more general situations there are weaker and stronger notions, all of which specialize to equimultiplicity in the original case. We mention three essentially different algebraic generalizations of equimultiplicity together with a numerical description of each condition. Then we describe the hierarchy among these conditions. Finally we investigate these conditions concerning openess and transitivity properties.

Chapter V shows that these conditions are of some use to investigate Cohen-Macaulay properties under blowing up.

In Chapter VI we indicate that the new conditions of equimultiplicity are useful in the study of the numerical behaviour of singularities under blowing up singular centers. In this context we consider two essential topics: blowing up and semicontinuity. To prove semicontinuity one has to desingularize curves by blowing up points, and conversely for inequalities of Hilbert functions under blowing up other centers one has to use semicontinuity.

Chapter VII, presupposing the following Chapters VIII and IX, discusses local cohomology and duality over graded rings. For local rings, the theory of local duality can be found in textbooks. For the corresponding results over graded rings we give detailed proofs because they are not all out available in the literature.

Chapter VIII studies local rings (A,m) with finite local cohomology $H^i(A)$ for $i < \dim A$. If X is an irreducible non-singular projective variety over a field then the local ring at the vertex of the affine cone over X has always this property. We present the results on these rings in a unified manner according to S. Goto and N.V. Trung.

In Chapter IX the results of Chapter V are partially extended and rephrased in a different context by using cohomological methods. The main result is a general criterion of the Cohen-Macaulay property of Rees rings. Then we investigate Rees rings of certain equimultiple ideals. Finally we give special applications to rings with low multiplicities. In this context we also prove the equivalence between the "monomial property" and the "direct summand-property" in the sense of M. Hochster.

The Appendix consists of three parts. Part I treats the fundamentals of local complex analytic geometry in a fairly complete way, for the sake of reference, and convenience for the reader. Some emphasis is laid on effective methods, and so consequent use of the general Division Theorem, due to Grauert-Hironaka, is made. Part II exposes the geometric description of the multiplicity of a complex spacegerm as the local mapping degree of a generic projection. To handle the nonreduced case, the notion of compact Stein neighbourhoods is introduced, which allow a systematic transition from the algebraic to the analytic case. The connection with Samuel multiplicity is described. Part III develops the theory of compact Stein neighbourhoods further and thus deduces the properties of normal flatness in the analytic case from the algebraic case. Then the geometry of equimultiplicity along a smooth subspace is developed in some detail in § 2 with geometric proofs. Finally, § 3 treats the geometric content of the equimultplicity results of Chapter IV; these follow from the algebraic results via the method of compact Stein neighbourhoods.

We have to make some acknowledgments. First we would like to express our gratitude in particular to J. Giraud, J. Lipman, R. Sharp and J.L. Vicente for suggestions and encouragements during the preparation of this work. Furthermore we thank deeply D. Katz, L. Robbiano, O. Villamayor and K. Yamagishi for their careful reading of various positions of the manuscript or for their detailed suggestions and improvements. K. Yamagishi also worked out the main part of the appendix to Chapter V. Finally we have to emphasize the help of our students F. Bienefeldt, D. Rogoss, M. Ribbe and M. Zacher. Their stimulating questions in the seminars and their special contributions to the Chapters VII, VIII and IX (besides reading carefully the manuscript) have essentially improved the first version of the last three chapters. Chapter V contains the main results of the thesis of Dr. U. Grothe who worked out the main part of this chapter. Last not least we owe thanks to Mrs. Pearce from the Max-Planck-Institute of Mathematics in Bonn for typing services and for patience and skill. The third author has received support and great hospitality by the Max-Planck-Institute of Mathematics in Bonn by the Department of Mathematics of the University in Genova and the Department of Mathematics of the University of Kansas. The acknowledgements of the author of the Appendix are stated in the introduction there.

Table of contents

Chapter I. REVIEW OF MULTIPLICITY THEORY

In this chapter we collect all the basic facts about multiplicities, Hilbert functions and reductions of ideals. At the same time we will introduce the notations to be used throughout the book.

§ 1. The multiplicity symbol

Here we recall the definition and the main properties of the multiplicity symbol $e(\underline{x};M)$ introduced by D.J. Wright and D.G. Northcott. This section contains no proofs, since these may be found in full detail in the excellent book by Northcott ([7]). At the end of the third section we give some comments on the relation to the geometric idea of multiplicity.

For the rest of this section we will fix a noetherian ring R and a finitely generated R-module M. By $\mathrm{Ann}_R(M)$ or simply $\mathrm{Ann}(M)$ we will denote the annihilator ideal of M in R. By $\dim M = \dim R/\mathrm{Ann}(M)$ we mean the Krull-dimension of M. If $\mathfrak{a}$ is any ideal in R, we write $\mathrm{Min}(\mathfrak{a})$ to denote the set of those prime ideals which are minimal among all prime ideals of R containing $\mathfrak{a}$. If $x_1,\ldots,x_r$ are elements of R, we often write $\underline{x}$ for the sequence $(x_1,\ldots,x_r)$, and in this case we put $\underline{x}M = x_1M+\ldots+x_rM$. Finally, if M is an R-module of finite length this length will be denoted by $\lambda_R(M)$ or simply $\lambda(M)$.

<u>(1.1) Definition</u>. Let $x_1,\ldots,x_r \in R$. The sequence $\underline{x} = (x_1,\ldots,x_r)$ will be called a multiplicity system for M if $M/\underline{x}M$ is an R-module of finite length.

So $\underline{x} = (x_1,\ldots,x_r)$ is a multiplicity system for M if and only if $\mathrm{Min}(\mathrm{Ann}(M) + \underline{x}R)$ consists of finitely many maximal ideals of R, and in this case if $\underline{x}R \subset J(R)$, the Jacobson radical, then $r \geq \dim(M)$. In particular $\dim(M)$ is finite. If S is another noetherian ring and $\varphi : S \twoheadrightarrow R$ a surjective homomorphism, and if $y_1,\ldots,y_r$ are any elements of S such that $\varphi(y_i) = x_i$, $i = 1,\ldots,r$, then the sequence $\underline{y} = (y_1,\ldots,y_r)$ is a multiplicity system for M as an S-module via φ if and only if $\underline{x}$ is a multiplicity system for M as an R-module.

Let $\underline{x} = (x_1,\dots,x_r)$ be a multiplicity system for $M, r \geq 1$. Then clearly $(x_1,\dots,x_{r-1})$ is a multiplicity system for M/x_1M. Moreover $x_1R + \mathrm{Ann}(M) \subset \mathrm{Ann}(0:_M x_1)$, where $(0:_M x_1) = \{a \in M \mid x_1 a = 0\}$, and therefore $(x_1,\dots,x_{r-1})$ is also a multiplicity system for $(0:_M x_1)$. This allows to define the multiplicity symbol $e(\underline{x};M)$ inductively in the following way.

(1.2) Definition. Let $\underline{x} = (x_1,\dots,x_r)$ be a multiplicity system for M. We define the multiplicity symbol $e_R(\underline{x};M)$ by induction on r as

$$e_R(\underline{x};M) = \begin{cases} \lambda(M) & \text{if } r = 0 \\ e(x_1,\dots,x_{r-1};M/x_rM) - e(x_1,\dots,x_{r-1};(0:_M x_r)) & \text{if } r > 0 . \end{cases}$$

If $\underline{x}$ is not a multiplicity system for M we put

$$e_R(\underline{x};M) = \infty .$$

If $\varphi : S \longrightarrow R$ is a homomorphism of noetherian rings and $\underline{x} = (x_1,\dots,x_r)$ is a sequence of elements of S we put

$$e_R(\underline{x};M) = e_R(\varphi(x_1),\dots,\varphi(x_r));M) .$$

We also write e instead of e_R if there is no doubt about the ground ring. For $r = 0$ we also write $e(\underline{x};M) = e(\emptyset;M)$.

Using induction on r, one can show easily that $e(\underline{x};M) \leq \lambda(M/\underline{x}M)$. It can also be shown, although if is not obvious form the definition, that $e_R(\underline{x};M) \geq 0$. To prove this statement (by induction on r), one uses another important property of the multiplicity symbol, which we state next:

(1.3) Proposition. Let

$$0 \longrightarrow M' \longrightarrow M \longrightarrow M'' \longrightarrow 0$$

be an exact sequence of finitely generated R-modules , and let $\underline{x} = (x_1,\ldots,x_r)$ be a sequence of elements of R . Then $\underline{x}$ is a multiplicity system for M if and only if $\underline{x}$ is a multiplicity system for M' and M'' , and we have

$$e(\underline{x};M) = e(\underline{x};M') + e(\underline{x};M'') \quad .$$

We note that this definition of multiplicity is basically the same as given by Serre ([8], first published by Auslander-Buchsbaum [10]). We recall this definition. For $\underline{x} = (x_1,\ldots,x_r)$ and M as above, we denote by $K.(\underline{x};M)$ the Koszul complex and by $H_n(\underline{x};M)$ the n-th homology module of this complex. It is well-known that the annihilator of $H_n(\underline{x};M)$ contains $\underline{x}R$ and $\mathrm{Ann}(M)$. Therefore, if $\underline{x}$ is a multiplicity system for M , these modules have finite length $h_n(\underline{x};M) = \lambda_R(H_n(\underline{x};M))$, and we can define

$$\chi(\underline{x};M) = \sum_n (-1)^n h_n(\underline{x};M) \quad .$$

Now $H_0(\underline{x};M) = M/\underline{x}M)$ and $H_r(\underline{x};M) = (0:_M\underline{x}R)$. Therefore if $r = 1$, we see that

$$\chi(x_1;M) = \lambda(M/x_1M) - \lambda(0:_Mx_1) = e(x_1;M) \quad ,$$

and in fact we also have for arbitrary r (see [7], p. 370)

$$\chi(\underline{x};M) = e(\underline{x};M)$$

<u>(1.4) Definition</u>. $\underline{x} = (x_1,\ldots,x_r)$ is called a regular sequence on M

if

a) $\underline{x}M \neq M$ and
b) x_1 is a non-zero-divisor on M and
c) x_i is a non-zero-divisor on $M/x_1M+\ldots+x_{i-1}M$ for all $i \in \{2,\ldots,r\}$.

It is well-known that regular sequences can be characterized by the vanishing of the Koszul-homology, provided that $\underline{x}R$ is contained in the Jacobson radical of R . Moreover, if $\underline{x}$ is a multiplicity system for M , then regularity of $\underline{x}$ can be expressed by $e(\underline{x};M)$. We will state the corresponding results also for the graded case, to which we want to apply it later on.

<u>(1.5) Proposition.</u> Given R,M and $\underline{x} = (x_1,\ldots,x_r)$ as before, assume that $M \neq 0$ and $\underline{x}R$ is contained in the Jacobson radical of R . Then the following conditions are equivalent:

(i) $\underline{x}$ is a regular sequence on M .
(ii) $H_n(\underline{x};M) = 0$ for all $n > 0$.
(iii) $H_1(\underline{x};M) = 0$.

If, moreover, $\underline{x}$ is a multiplicity system for M , then these conditions are equivalent to

(iv) $e(\underline{x};M) = \lambda(M/\underline{x}M)$.

The proof is given in [7], Theorem 9, page 312. For the graded case see Proposition (11.9) in Chapter II.

<u>(1.6) Proposition.</u> ([7], p. 333) Let $\underline{x}$ be multiplicitiy system for M and let Max(R) denote the set of maximal ideals of R . Then

$$e_R(\underline{x};M) = \sum_{\mathrm{Max}(R)} e_{R_{\mathfrak{m}}}(\underline{x};M_{\mathfrak{m}}) .$$

Since $e_{R_{\mathfrak{m}}}(\underline{x};M_{\mathfrak{m}}) = 0$ if $\mathfrak{m} \not\supset \mathrm{Ann}(M) + \underline{x}R$, the above sum is actually finite. By this formula, most of the considerations on multiplicities may be restricted to modules over local rings. In this local case, we have the following result:

<u>(1.7) Proposition.</u> ([7], p. 335/336). Assume that R is local with maximal ideal $\mathfrak{m}$ and let $\underline{x} = (x_1,\ldots,x_r)$ be a multiplicity system for M such that $\underline{x}R \subset \mathfrak{m}$. Then

$$e(\underline{x};M) = 0 \Longleftrightarrow \dim(M) < r .$$

Therefore in Proposition (1.6) the sum can be restricted to those $\mathfrak{m} \in \mathrm{Max}(R)$ such that $\mathfrak{m} \supset \mathrm{Ann}(M) + \underline{x}R$ and $\dim(M_{\mathfrak{m}}) = r$, and then all summands will be strictly positive.

The most useful tool on multiplicity for our applications later on is the following result, which is called the Associative Law for Multiplicities. For a neat statement of this result we will use the following notation. If $\underline{y} = (y_1,\ldots,y_s)$ and $\underline{z} = (z_1,\ldots,z_t)$ are sequences of elements of R, we will use the symbol $(\underline{y},\underline{z})$ to denote the sequence $(y_1,\ldots,y_s,z_1,\ldots,z_t)$.

(1.8) Theorem. Let $\underline{y}$ and $\underline{z}$ be sequences of elements of R and let M be a finitely generated R-module such that $(\underline{y},\underline{z})$ is a multiplicity system for M. Then for any ideal $\mathfrak{a} \subset \mathrm{Ann}(M)$ such that $(\underline{y},\underline{z})$ is a multiplicity system for $R/\mathfrak{a}$ we have

$$e_R((\underline{y},\underline{z});M) = \sum_{\mathfrak{p} \in \mathrm{Min}(\mathfrak{a}+\underline{z}R)} e_{R/\mathfrak{p}}(\underline{y};R/\mathfrak{p})\, e_{R_{\mathfrak{p}}}(\underline{z};M_{\mathfrak{p}}) \quad .$$

(Here we have denoted the image of the sequence $\underline{y}$ in $R/\mathfrak{p}$ by $\underline{y}$ again, and similarly for $\underline{z}$.)

Proof. The formula is unchanged if we replace R by $R/\mathfrak{a}$, i.e. we may assume that $(\underline{y},\underline{z})$ is a multiplicity system for R itself. For this case the proof is given in [7], p.342 Theorem 18.

If in the above Theorem we choose $\underline{y} = (y_1,\ldots,y_s)$ in the Jacobson radical of R, then we know by Proposition (1.6) and (1.7) that $e_{R/\mathfrak{p}}(\underline{y};R/\mathfrak{p}) \neq 0 \Leftrightarrow \dim(R/\mathfrak{p}) = s$, and of course we have $\dim(R/\mathfrak{p}) \leq s$ for all $\mathfrak{p} \in \mathrm{Min}(\underline{z}R)$.

(1.9) Definition. Let M be an R-module of finite Krull dimension. Then we put

$$\mathrm{Assh}(M) = \left\{\mathfrak{p} \in \mathrm{Min}(\mathrm{Ann}(M)) \mid \dim(R/\mathfrak{p}) = \dim(M)\right\} \quad .$$

If $\mathfrak{a} \subset R$ is an ideal for which $R/\mathfrak{a}$ has finite Krull dimension we let

$$\mathrm{Assh}(\mathfrak{a}) = \mathrm{Assh}(R/\mathfrak{a}) \quad .$$

(1.10) Remark. By the above observations, we can rewrite Theorem (1.8) as

$$e_R((\underline{y},\underline{z});M) = \sum_{\mathfrak{p}\in \mathrm{Assh}(\mathfrak{a}+\underline{z}R)} e_{R/\mathfrak{p}}(\underline{y};R/\mathfrak{p})\, e_{R_\mathfrak{p}}(\underline{z};M_\mathfrak{p}) \quad ,$$

where now all factors $e_{R/\mathfrak{p}}(\underline{y};R/\mathfrak{p})$ appearing in this sum are all zero or all strictly positive. Of special importance is the case where $M = R$ and where $\underline{z}$ is the empty sequence. Assume that $\underline{x}$ is a multiplicity system for R contained in $J(R)$ consisting of $\dim(R)$ elements, so that $e(\underline{x};R) \neq 0$. Then

$$(1.10.1) \qquad e(\underline{x};R) = \sum_{\mathfrak{p}\in \mathrm{Assh}(R)} e(\underline{x};R/\mathfrak{p})\,\lambda(R_\mathfrak{p}) \quad ,$$

where now all summands on the right hand side are non-zero. We will call this equation the Reduction Formula.

We remark that the name of the Associative Law for Multiplicities is derived from its role in the theory of intersection multiplicities, where it can be used to show that the intersection product defines a multiplication of cycles which is associative (see Chevalley [12], Fulton [2], and App. II, 5.1.5, 4.1.8, 4.3.1.

§ 2. Hilbert functions

In this section we recall the classical results by Hilbert [13] and Samuel [17] about Hilbert functions. Proofs may be found in Zariski-Samuel [9] or Serre [8] . Before this we make some remarks on integer valued functions, and at the end we will describe the relation to the multiplicity defined in §1, see also App. II, § 4.

(2.1) Definition. Let $\mathbb{N}_0$ be the set of non-negative integers and let $F : \mathbb{N}_0 \longrightarrow \mathbb{Z}$ be any mapping. F will be called a polynomial function if there exists a polynomial $\Phi(X) \in \mathbb{R}[X]$ such that $F(n) = \Phi(n)$ for all large values of $n \in \mathbb{N}_0$.

Since F takes integer values, it is clear, e.g. by taking derivatives, that Φ has rational coefficients. Moreover Φ is uniquely determined by F , and we will write it by $P(F)$, the polynomial of

F . Since the polynomials $\binom{X}{i} = \frac{1}{i!}X(X-1)\ldots(X-i+1)$ are a $\mathbb{Q}$-basis of $\mathbb{Q}[X]$, every $\Phi(X) \in \mathbb{Q}[X]$ may uniquely be written as $\Phi(X) = \sum_{i=0}^{d} a_i \binom{X}{i}$, $d = \deg \Phi$. Moreover, if Φ takes integer values on an infinite set of integers, then $a_i \in \mathbf{Z}$ for all i .

<u>(2.2) Definition</u>. Let F be a polynomial function and assume

$$P(F) = \sum_{i=0}^{d} a_i \binom{X}{i} \neq 0 \ , \quad d = \deg P(F) \quad .$$

Then we define the degree $\deg F$ of F to be d , and a_d will be called the leading term of F .

Note that if $P(F) = \sum_{i=0}^{d} b_i X^i$, $d = \deg F$, then $a_d = d!b_d$, or equivalently, $a_d = \lim_{n \to \infty} \frac{d!P(F)(n)}{n^d}$.

From now on it will be convenient to consider only functions F defined on $\mathbb{N}_0$, the set of non-negative integers , and we will view such an F as a function on $\mathbf{Z}$ by assigning the value zero to negative integers. For two such functions F,G we write $F \leq G$ if $F(n) \leq G(n)$ for all n .

<u>(2.3) Definition</u>. For any function $f : \mathbb{N}_0 \longrightarrow \mathbf{Z}$ we put

$$(\Delta F)(n) = F(n) - F(n-1)$$

and

$$(IF)(n) = \sum_{k=0}^{n} F(k) \quad ,$$

and for any $r \geq 1$ we put

$$\Delta^r F = \Delta(\Delta^{r-1}F) \ , \quad I^r F = I(I^{r-1}F) \quad .$$

We list some properties of the operators I and Δ , which are easily verified or even obvious (compare [7], p. 322).

(2.4) Proposition. For any function $F : \mathbb{N}_0 \longrightarrow \mathbb{Z}$ we have

a) $\Delta(IF) = F = I(\Delta F)$.

b) F is polynomial $\Leftrightarrow$ ΔF is polynomial $\Leftrightarrow$ IF is polynomial.

c) If F is polynomial and $P(F) \neq 0$, then $\deg(IF) = \deg(F)+1$ (hence $\deg(I^r F) = \deg(F)+r)$, and if $P(\Delta F) \neq 0$, then $\deg(\Delta F) = \deg(F)-1$ (hence $\deg(\Delta^r F) = \deg(F)-r$ as long as $P(\Delta^r F) \neq 0)$.

d) If F is polynomial and $P(F) \neq 0$, then

$$\deg F = \sup\{r \mid P(\Delta^r F) \neq 0\} .$$

e) If F is polynomial and $P(F) \neq 0$, then F and IF have the same leading term. If moreover $P(\Delta F) \neq 0$ then also ΔF has the same leading term as F and IF .

f) If F is polynomial, then $P(\Delta F) = \Delta P(F)$.

For an arbitrary polynomial function F we will have $P(IF) \neq IP(F)$ in general. But $P(IF)$ and $IP(F)$ will have the same $d+1$ highest coefficients, where $d = \deg F$, and the same is true for $P(I^r F)$ and $I^r P(F)$.

Now we fix a noetherian ring R and a finitely generated R-module M again.

(2.5) Definition. Let $\mathfrak{q}$ be an ideal of R such that $\dim(M/\mathfrak{q}M) = 0$. Then we define a function by

$$H^{(0)}[\mathfrak{q},M](n) = \lambda_R(\mathfrak{q}^n M/\mathfrak{q}^{n+1}M)$$

and we put

$$H^{(r)}[\mathfrak{q},M] = I^r H^{(0)}[\mathfrak{q},M] .$$

The functions $H^{(r)}[\mathfrak{q},M]$ will be called the Hilbert functions of M with respect to $\mathfrak{q}$.

We note that in fact the modules $\mathfrak{q}^nM/\mathfrak{q}^{n+1}M$ have finite length, since $\dim(M/\mathfrak{q}M) = 0$. This also implies that $\mathrm{Min}(\mathfrak{q}+\mathrm{Ann}(M))$ is a finite set of maximal ideals of R. Putting $S = R \setminus \bigcup_{\mathfrak{m}\in W} \mathfrak{m}$, where $W = \mathrm{Min}(\mathfrak{q}+\mathrm{Ann}(M))$, we have $\mathfrak{q}^nM/\mathfrak{q}^{n+1}M \simeq \mathfrak{q}^nM/\mathfrak{q}^{n+1}M \otimes_R R_S$, and therefore

$$H^{(r)}[\mathfrak{q},M] = H^{(r)}[\mathfrak{q}R_S,M_S] \quad \text{for all} \quad r \geq 0 \quad .$$

So we see that the theory of these Hilbert functions is actually a theory over semi-local rings (and so it is treated in the text books), although it is sometimes convenient to have the definition in the more general case.

<u>(2.6) Theorem</u>. (Hilbert [13]) The functions $H^{(r)}[\mathfrak{q},M]$ are polynomial functions.

<u>(2.7) Theorem</u>. (Samuel [17]) a) Assume that $\mathfrak{q}$ is contained in the Jacobson radical of R and that $M \neq 0$. Then

$$\deg H^{(1)}[\mathfrak{q},M] = \dim(M) \quad .$$

b) Let $\underline{x}$ be a multiplicity system for M and assume that $e(\underline{x};M) \neq 0$. Then $e(\underline{x};M)$ is the leading term of each Hilbert function $H^{(r)}[\underline{x}R,M]$. ("Limit formula of Samuel", [7], Theorem 13, p. 329).

<u>(2.8) Remark</u>. We make the convention that the zero module has no dimension. We note that under the assumptions of (2.5) we have for any $\mathfrak{q}$

$$M = 0 \iff H^{(0)}[\mathfrak{q},M] = 0 \quad .$$

<u>(2.9) Definition</u>. Let $\mathfrak{q}$ be an ideal contained in the Jacobson radical of R such that $\dim M/\mathfrak{q}M = 0$, and let M be a finitely generated R-module. The common leading term of the Hilbert functions

$H^{(r)}[\mathfrak{q},M]$ will be called the multiplicity of M with respect to $\mathfrak{q}$, and will be denoted by $e(\mathfrak{q},M)$. We put $e(\mathfrak{q},M) = 0$ if $M = 0$. For a local ring R with maximal ideal $\mathfrak{m}$ we write $e(R)$ instead of $e(\mathfrak{m},R)$.

With this definition, the second part of the theorem of Samuel (i.e., Theorem (2.7),b)) may be restated as

$$(2.9.1) \qquad e(\underline{x};M) = e(\underline{x}R,M) ,$$

provided that $e(\underline{x};M) \neq 0$ and $\underline{x}R$ is contained in the Jacobson radical of R. Note that $e(\mathfrak{q},M) = 0$ if and only if $M = 0$.

§ 3. Generalized multiplicities and Hilbert functions

Again R will denote any noetherian ring.

(3.1) Definition. Given an ideal $\mathfrak{a} \subset R$, a finitely generated R-module M and a sequence $\underline{x}$ in R, we define

$$H^{(0)}[\underline{x},\mathfrak{a},M](n) = e_R(\underline{x};\mathfrak{a}^nM/\mathfrak{a}^{n+1}M)$$

and

$$H^{(r)}[\underline{x},\mathfrak{a},M] = I^rH^{(0)}[\underline{x},\mathfrak{a},M] .$$

Usually we will assume that $\underline{x}$ is a multiplicity system for $M/\mathfrak{a}M$. In that case, using the fact that $\mathrm{Ann}(M/\mathfrak{a}M)$ and $\mathfrak{a}+\mathrm{Ann}(M)$ coincide up to radical, we see that $\underline{x}$ is a multiplicity system on each $\mathfrak{a}^nM/\mathfrak{a}^{n+1}M$ and therefore the above definition is meaningful (in the sense that $H^{(0)}[\underline{x},\mathfrak{a},M]$ has only finite values). Also, if the image of the sequence $\underline{x}$ in $R/\mathfrak{a}$ is denoted by $\underline{y}$ then clearly

$$e_R(\underline{x};\mathfrak{a}^nM/\mathfrak{a}^{n+1}M) = e_{R/\mathfrak{a}}(\underline{y};\mathfrak{a}^nM/\mathfrak{a}^{n+1}M) .$$

Therefore we can apply the Associative Law for Multiplicities (Theorem (1.8)) to the ground ring $R/\mathfrak{a}$ to obtain the following result:

<u>(3.2) Proposition</u>. Let $\mathfrak{a} \subset R$ be an ideal, M a finitely generated R-module and $\underline{x},\underline{y}$ sequences of elements of R such that $(\underline{x},\underline{y})$ is a multiplicity system for $M/\mathfrak{a}M$. Then for every $i \geq 0$ we have

$$H^{(i)}[(\underline{x},\underline{y}),\mathfrak{a},M] = \sum_{\mathrm{Assh}(\mathrm{Ann}(M/\mathfrak{a}M)+\underline{y}R)} e(\underline{x};R/\mathfrak{p})H^{(i)}[\underline{y},\mathfrak{a}R_{\mathfrak{p}},M_{\mathfrak{p}}] .$$

In particular, if $\underline{x}$ is a multiplicity system for $M/\mathfrak{a}M$, then

$$H^{(i)}[\underline{x},\mathfrak{a},M] = \sum_{\mathfrak{p}\in\mathrm{Assh}(M/\mathfrak{a}M)} e(\underline{x};R/\mathfrak{p})H^{(i)}[\mathfrak{a}R_{\mathfrak{p}},M_{\mathfrak{p}}] .$$

<u>(3.3) Definition</u>. $\dim(\mathfrak{a},M) = \sup\{\dim M_{\mathfrak{p}} \mid \mathfrak{p} \in \mathrm{Assh}(M/\mathfrak{a}M)\}$.

<u>(3.4) Corollary</u>. Suppose $\underline{x}$ is a multiplicity system for $M/\mathfrak{a}M$. The functions $H^{(i)}[\underline{x},\mathfrak{a},M]$ are polynomial functions, and if $H^{(1)}[\underline{x},\mathfrak{a},M]$ is not identically zero then its degree equals $\dim(\mathfrak{a},M)$.

From Proposition (3.2) we can also derive conditions for the vanishing of $H^{(i)}[\underline{x},\mathfrak{a},M]$. For this purpose we write $\underline{x} = (x_1,\ldots,x_r)$, and then clearly $H^{(0)}[\underline{x},\mathfrak{a},M]$ is identically zero if $r > \dim(M/\mathfrak{a}M)$. If $r = \dim(M/\mathfrak{a}M)$ and $\mathfrak{a} + \underline{x}R \subset J(R)$ then $H^{(0)}[\underline{x},\mathfrak{a},M]$ vanishes if and only if $M/\mathfrak{a}M = 0$. (Note that $\mathrm{Assh}(M/\mathfrak{a}M) \subset \mathrm{Supp}(M/\mathfrak{a}M)$ if $M/\mathfrak{a}M \neq 0$.)

<u>(3.5) Corollary</u>. If $\underline{x}$ is a multiplicity system for $M/\mathfrak{a}M$ such that $e(\underline{x};M/\mathfrak{a}M) \neq 0$, then

$$H^{(i)}[\underline{x},\mathfrak{a},M] = \sum_{\mathfrak{p}\in\mathrm{Assh}(M/\mathfrak{a}M)} e(\underline{x};R/\mathfrak{p})H^{(i)}[\mathfrak{a}R_{\mathfrak{p}},M_{\mathfrak{p}}] ,$$

and all summands on the right hand side are non-zero if $\underline{x}R \subset J(R)$.

<u>(3.6) Corollary</u>. If $\mathfrak{a}$ is any ideal in R and if $\underline{x} = (x_1,\ldots,x_r)$ is a multiplicity system for $R/\mathfrak{a}$ contained in $J(R)$, then the functions $H^{(i)}[\underline{x},\mathfrak{a},R]$ vanish identically if and only if $r > \dim(R/\mathfrak{a})$.

We will frequently make use of the following notation: For any ideal $\mathfrak{a} \subset R$ and an arbitrary sequence $\underline{x}$ of elements of R we write $\mathfrak{a}(\underline{x}) = \mathfrak{a} + \underline{x}R$.

(3.7) Proposition. Let $\mathfrak{a} \subset R$ be any ideal, let M be a finitely generated R-module and let $\underline{x} = (x_1,\dots,x_r)$ be a multiplicity system for $M/\mathfrak{a}M$. We put $\mathfrak{b} = \mathfrak{a} + x_1R$ and $\underline{y} = (x_2,\dots,x_r)$. Then we have

$$H^{(1)}[\underline{x},\mathfrak{a},M] \leqq H^{(0)}[\underline{y},\mathfrak{b},M] \quad .$$

Proof. By the definition of the multiplicity symbol we know that

$$e(\underline{x};L) \leqq e(\underline{y};L/x_1L)$$

for any L for which $\underline{x}$ is a multiplicity system. Applying this remark to $L = \mathfrak{b}^nM/\mathfrak{a}^{n+1}M$ and using the fact that here

$$L/x_1L = \mathfrak{b}^nM/\mathfrak{a}^{n+1}M + x_1\mathfrak{b}^nM = \mathfrak{b}^nM/\mathfrak{b}^{n+1}M$$

(since $\mathfrak{b}^{n+1} = \mathfrak{a}^{n+1} + x_1\mathfrak{b}^n$), we see that

$$e(\underline{x};\mathfrak{b}^nM/\mathfrak{a}^{n+1}M) \leqq e(\underline{y};\mathfrak{b}^nM/\mathfrak{b}^{n+1}M) = H^{(0)}[\underline{y},\mathfrak{b},M](n) \quad .$$

Now we apply $e(\underline{x};-)$ to the exact sequence

$$0 \longrightarrow \mathfrak{b}^nM/\mathfrak{a}^{n+1}M \longrightarrow M/\mathfrak{a}^{n+1}M \longrightarrow M/\mathfrak{b}^nM \longrightarrow 0 \quad .$$

By [7] (Prop. 5 on p.307) we have $e(\underline{x};M/\mathfrak{b}^nM) = 0$ and therefore

$$e(\underline{x};\mathfrak{b}^nM/\mathfrak{a}^{n+1}M) = e(\underline{x};M/\mathfrak{a}^{n+1}M) = H^{(1)}[\underline{x},\mathfrak{a},M](n) \quad ,$$

which completes the proof.

(3.8) Corollary. With the same notation as above we have

$$H^{(r)}[\underline{x},\mathfrak{a},M] \leqq H^{(0)}[\mathfrak{a}(\underline{x}),M] \quad .$$

We point out that, if $\dim M/\mathfrak{a}M = 0$, then the empty set is a multiplicity system for $M/\mathfrak{a}M$, and in this case the classical Hilbert function $H^{(i)}[\mathfrak{a},M]$ may also be viewed as the generalized Hilbert function

$$H^{(0)}[\emptyset,\mathfrak{a},M](n) = e(\emptyset,\mathfrak{a}^nM/\mathfrak{a}^{n+1}M) = \lambda(\mathfrak{a}^nM/\mathfrak{a}^{n+1}M) \quad .$$

<u>(3.9) Definition</u>. For any ideal $\mathfrak{a} \subset R$, any finitely generated R-module M and any multiplicity system $\underline{x}$ for $M/\mathfrak{a}M$ we put

$$e(\underline{x},\mathfrak{a},M) = \begin{cases} \text{leading term of } H^{(1)}[\underline{x},\mathfrak{a},M] & \text{if this is } \not\equiv 0 \\ 0 \quad \text{if} \quad H^{(1)}[\underline{x},\mathfrak{a},M] \equiv 0 & . \end{cases}$$

By Corollary (3.5) we see that $e(\underline{x},\mathfrak{a},M) \neq 0$ if and only if $e(\underline{x};M/\mathfrak{a}M) \neq 0$ (and $\mathrm{Supp}(M) \cap \mathrm{Assh}(M/\mathfrak{a}M) \neq 0$) . We also note that, if we let $\underline{x} = (x_1,\ldots,x_r)$ and assume that $\underline{x}R + \mathfrak{a} \subset J(R)$, then

$$e(\underline{x};R/\mathfrak{a}) \neq 0 \iff e(\underline{x},\mathfrak{a},R) \neq 0 \iff r = \dim(R/\mathfrak{a}) \quad .$$

Again, if $\dim M/\mathfrak{a}M = 0$, then the multiplicity $e(\mathfrak{a},M)$ introduced in (2.9) may be viewed as the generalized multiplicity $e(\emptyset,\mathfrak{a},M)$, and we note that in this case $e(\mathfrak{a},M) = 0$ if and only if $M/\mathfrak{a}M = 0$.

We now want to derive some formulas about multiplicities analogous to those given for Hilbert functions above. To do so we will need some information about the degrees of the polynomial functions occuring in these formulas, since multiplicities are given by the highest degree term. Therefore, for varying ideals $\mathfrak{a}$, it is reasonable to impose some chain condition on R , and it will become clear later on that for our purposes it fits best to assume R universally catenarian and equidimensional (see Chapter III). For a fixed ideal $\mathfrak{a}$, however, it is often sufficient to assume that $\dim(R/\mathfrak{a}) + \mathrm{ht}(\mathfrak{a}) = \dim(R)$, where $\mathrm{ht}(\mathfrak{a}) = \inf\{\mathrm{ht}(\mathfrak{p}) \mid \mathfrak{p} \in \mathrm{Min}(\mathfrak{a})\}$, of course. We will collect some consequences of this condition in the following.

<u>(3.10) Remark</u>. Assume that an ideal $\mathfrak{a} \subset R$ satisfies the equality

$$\dim(R/\mathfrak{a}) + \mathrm{ht}(\mathfrak{a}) = \dim(R) \quad .$$

For any $\mathfrak{p} \in \mathrm{Min}(\mathfrak{a})$ we will have

$$ht(\mathfrak{a}) \leqq ht(\mathfrak{p}) \quad \text{and} \quad \dim(R/\mathfrak{p}) + ht(\mathfrak{p}) \leqq \dim R$$

in general. If we now assume in addition that $\dim(R/\mathfrak{a}) = \dim(R/\mathfrak{p})$, then we conclude that

$$ht(\mathfrak{a}) = ht(\mathfrak{p}) = \dim(R) - \dim(R/\mathfrak{p}) \quad .$$

This shows that the height function is constant on $Assh(\mathfrak{a})$.

We note also, without giving explicit statements, that given any finitely generated R-module M , similar remarks can be applied to the ideal $\mathfrak{a} + Ann(M)$.

(3.11) Proposition. Let M be a finitely generated R-module and let $\underline{x} = (x_1,\dots,x_r)$ is a multiplicity system for $M/\mathfrak{a}M$. Then we have

a) $$e(\underline{x},\mathfrak{a},M) = \sum_{\substack{\mathfrak{p}\in Assh(M/\mathfrak{a}M)\\ \dim(M_{\mathfrak{p}}) = \dim(\mathfrak{a},M)}} e(\underline{x};R/\mathfrak{p})\, e(\mathfrak{a}R_{\mathfrak{p}},M_{\mathfrak{p}}) \quad .$$

In particular, if $\mathfrak{a} = \mathfrak{p}$ is prime and $\mathfrak{p} \supset Ann(M)$, then

$$e(\underline{x},\mathfrak{p},M) = e(\underline{x};R/\mathfrak{p}) \cdot e(\mathfrak{p}R_{\mathfrak{p}},M_{\mathfrak{p}}) \quad .$$

b) If $\dim(M/\mathfrak{a}M) + ht(\mathfrak{a} + Ann(M)) = \dim R$, then

$$e(\underline{x},\mathfrak{a},M) = \sum_{\mathfrak{p}\in Assh(M/\mathfrak{a}M)} e(\underline{x},R/\mathfrak{p})\, e(\mathfrak{a}R_{\mathfrak{p}},M_{\mathfrak{p}}) \quad .$$

c) Assume that $\sqrt{\mathfrak{a}(\underline{x})}$ is a maximal ideal of height d and that $\sup\{ht(\mathfrak{p}) \mid \mathfrak{p} \in Assh(\mathfrak{a})\} = d - \dim R/\mathfrak{a}$. Then

$$e(\underline{x},\mathfrak{a},R) \leqq e(\mathfrak{a}(\underline{x}),R) \quad .$$

d) If N is a homomorphic image of M and $\mathfrak{b} \supset \mathfrak{a}$ another ideal such that $\dim(\mathfrak{a},M) = \dim(\mathfrak{b},N)$, then

$$e(\underline{x},\mathfrak{a},M) \geqq e(\underline{x},\mathfrak{b},N) \quad .$$

Proof. a) is clear by Corollary (3.5) and b) follows from a). If in c) $e(\underline{x},\mathfrak{a},R) = 0$, then there is nothing to prove. If $e(\underline{x},\mathfrak{a},R) \neq 0$, then $r = \dim(R/\mathfrak{a})$ and the assumptions assure that the functions $H^{(i+r)}[\underline{x},\mathfrak{a},R]$ and $H^{(i)}[\mathfrak{a}(\underline{x}),R]$ are of the same degree $(i \geq 1)$. Therefore the assertion is a consequence of Corollary (3.8). Finally d) follows from the corresponding inequality for the Hilbert functions (see Proposition (1.3)).

<u>(3.12) Corollary</u>. Let $\mathfrak{p}$ be any prime ideal of R and let $\underline{x}$ be a multiplicity system for $R/\mathfrak{p}$. Then

$$e(\underline{x},\mathfrak{p},R) = e(\underline{x};R/\mathfrak{p})e(R_{\mathfrak{p}}) \quad .$$

If moreover $\mathfrak{p}(\underline{x})$ is a maximal ideal of R and $e(\underline{x},\mathfrak{p},R) \neq 0$, then

$$e(\underline{x},\mathfrak{p},R) = e(R_{\mathfrak{p}}) \quad .$$

Let us give some geometric meaning to the inequality of Proposition (3.11), c). Assume that X is an algebraic variety, $Y \subset X$ is a subvariety of dimension r, y is a point of Y and $R = \mathcal{O}_{X,y}$. We assume that Y is given in $\mathcal{O}_{X,y}$ by a prime ideal $\mathfrak{p}$. Assume moreover that the multiplicity system $\underline{x} = (x_1,\ldots,x_r)$ for $R/\mathfrak{p}$ can be extended to a system of parameters of R. Then (after passing to completion) $\underline{x}$ defines a projection $f : X \longrightarrow \mathbf{A}^r$ onto an r-dimensional space such that $f|Y$ is a finite covering of degree $e(\underline{x};R/\mathfrak{p})$ (see explanation in §6). By definition, this degree is also the multiplicity of the fibre of $f|Y$ at the origin of $\mathbf{A}^r$. As a subscheme of X, this fibre is defined by the ideal $\mathfrak{p}(\underline{x})$. Now this fibre comes with two multiplicities, namely one given by $e(\mathfrak{p}(\underline{x}),R)$, and another one given by its multiplicity on Y, $e(\underline{x};R/\mathfrak{p})$, times $e(R_{\mathfrak{p}})$, the generic multiplicity of Y on X. Now Proposition (3.11) states that for any "projection" $\underline{x}$, the second number is always smaller than the first. Later on (see Chapter IV) we will study the case where both numbers are equal, and it will turn out that the condition of equality is independent of the projection, i.e. of the choice of $\underline{x}$. For a precise geometric description of $e(\underline{x},\mathfrak{a},R)$ see App. III, § 3.

§ 4. Reductions and integral closure of ideals

As before, R will denote a neotherian ring, although some of the statements are true for any commutative ring. We will make use of some classical results on integral closures of noetherian rings and valuations. Of particular importance are the following two facts which are recalled explicitly:

(4.1). If R is a reduced ring with finitely many minimal prime ideals, then the integral closure of R in its total ring of fractions is (isomorphic to) a finite product of integrally closed domains, ([1], Chapter 5, §1, n°2, Proposition 9. Here R need not be noetherian.)

(4.2). If R is a noetherian domain, then the integral closure of R in its field of fractions is a Krull domain, which is the intersection of all valuation rings $V \supset R$ belonging to discrete rank one valuation of the quotient field of R. ([1], Chapter 7, §1, n°8, Proposition 12.)

(4.3) Definition. Let $\mathfrak{a}$ be any ideal of R. An element $x \in R$ will be called integral over $\mathfrak{a}$ if there are elements $a_1, \ldots, a_n (n > 0)$ such that

$$x^n + a_1 x^{n-1} + \ldots + a_n = 0 \quad \text{and} \quad a_i \in \mathfrak{a}^i, \ i = 1, \ldots, n \quad .$$

Obviously, if x is integral over $\mathfrak{a}$ and $c \in R$ is arbitrary then cx is integral over $\mathfrak{a}$. Also any element integral over $\mathfrak{a}$ belongs to the radical of $\mathfrak{a}$. The integral dependence on the ideal $\mathfrak{a}$ can be translated into an integral dependence on a certain ring which we are going to introduce next.

(4.4) Definition. For any ideal $\mathfrak{a}$ of R we put

$$B(\mathfrak{a}, R) = R[\mathfrak{a}t] \quad ,$$

the subring of the polynomial ring R[t] generated by R and $\mathfrak{a}t$. Moreover, for any ideal $\mathfrak{b}$ of R we define an ideal of $B(\mathfrak{a}, R)$ by

$$B(\mathfrak{a}, \mathfrak{b} \subset R) = \bigoplus_{n \geq 0} (\mathfrak{b} \cap \mathfrak{a}^n) t^n = \mathfrak{b} \cdot R[t] \cap B(\mathfrak{a}, R) \quad .$$

If $\mathfrak{a}$ and R are fixed, we simply write $B(\mathfrak{b})$ instead of $B(\mathfrak{a},\mathfrak{b}\subset R)$.

The letter B in the above definition stands for blowing up, and the relation to this process will be explained in the next chapter. The ring $B(\mathfrak{a},R)$ is sometimes called the Rees ring or Rees algebra of R with respect to $\mathfrak{a}$, and various different notations are used for it in the literature. A more detailed study of this ring will be given in Chapters II, V and IX.

We note that $B(\mathfrak{a},\mathfrak{b}\subset R)$ is the unique ideal of $B(\mathfrak{a},R)$, for which there is a canonical isomorphism

$$B(\mathfrak{a},R)\,/\,B(\mathfrak{a},\mathfrak{b}\subset R)\;\simeq\;B(\mathfrak{a}+\mathfrak{b}/\mathfrak{b},R/\mathfrak{b}) \quad .$$

We now list some simple properties of the function B .

(4.5) Lemma. Let $\mathfrak{a},\mathfrak{b},\mathfrak{c},\ldots$ be ideals of R , and let $B(\mathfrak{b}) = B(\mathfrak{a},\mathfrak{b}\subset R)$ etc.. Then we have:

a) $B(\mathfrak{b})\cdot B(\mathfrak{c})\subset B(\mathfrak{b}\cdot\mathfrak{c})$ and $B(\bigcap_{j\in J}\mathfrak{b}_j) = \bigcap_{j\in J} B(\mathfrak{b}_j)$.

b) If $\mathfrak{p}\subset R$ is prime, then $B(\mathfrak{p})$ is prime in $B(\mathfrak{a},R)$.

c) $\mathfrak{b}\subset\mathfrak{c} \iff B(\mathfrak{b})\subset B(\mathfrak{c})$.

d) $B(\sqrt{\mathfrak{b}}) = \sqrt{B(\mathfrak{b})}$.

e) The minimal prime ideals of $B(\mathfrak{a},R)$ are exactly the ideals $B(\mathfrak{a},\mathfrak{p}\subset R)$, where $\mathfrak{p}$ is a minimal prime ideal of R .

Proof. a) follows from $(\mathfrak{b}\cap\mathfrak{a}^n)\cdot(\mathfrak{c}\cap\mathfrak{a}^m)\subset\mathfrak{b}\cdot\mathfrak{c}\cap\mathfrak{a}^{n+m}$ and $(\bigcap_{j\in J}\mathfrak{b}_j)\cap\mathfrak{a}^n = \bigcap_{j\in J}(\mathfrak{b}_j\cap\mathfrak{a}^n)$ for any n . b) is clear, since $B(\mathfrak{a},R)/B(\mathfrak{p}) \simeq B(\mathfrak{a}+\mathfrak{p}/\mathfrak{p},R/\mathfrak{p})$ is a domain. For c) we observe that $\mathfrak{b} = B(\mathfrak{b})\cap R$. To prove d), note first that $B(\sqrt{\mathfrak{b}})$ is a radical ideal by a) and b), containing $B(\mathfrak{b})$, so $\sqrt{I(\mathfrak{b})}\subset B(\sqrt{\mathfrak{b}})$. Conversely, if $xt^n (x\in\mathfrak{a}^n)$ is a homogeneous element of $B(\sqrt{\mathfrak{b}})$ then $x^m\in\mathfrak{b}$ for some m , and consequently $(xt^n)^m\in B(\mathfrak{b})$. Finally e) follows from a), c) and d) applied to $\mathfrak{b} = 0$.

The next result is easy but very useful, as we will see in a moment.

(4.6) Proposition. Let $\mathfrak{a}$ be any ideal in R and x an element of R. Then x is integral over $\mathfrak{a}$ if and only if the element xt of $R[t]$ is integral over the subring $B(\mathfrak{a},R)$.

Proof. Let

$$x^n + a_1 x^{n-1} + \ldots + a_n = 0 , \quad a_i \in \mathfrak{a}^i$$

be an equation of integral dependence of x over $\mathfrak{a}$. Multiplying by t^n we obtain

$$(4.6.1) \qquad (xt)^n + (a_1 t)(xt)^{n-1} + \ldots + (a_n t^n) = 0 ,$$

where the coefficients $a_i t^i$ belong to $B(\mathfrak{a},R)$. Conversely, if xt is integral over $B(\mathfrak{a},R)$, there is a homogeneous equation of integral dependence like (4.6.1) above, and cancelling t^n we get the result.

(4.7) Corollary. If x and y are integral over $\mathfrak{a}$, then so is $x+y$.

(4.8) Definition. For any ideal $\mathfrak{a}$ of R, the set

$$\{x \in R \mid x \quad \text{integral over} \quad \mathfrak{a}\}$$

is called the integral closure of the ideal $\mathfrak{a}$. Sometimes it will be denoted by $\overline{\mathfrak{a}}$, if no confusion is possible. An ideal $\mathfrak{b}$ such that $\mathfrak{a} \subset \mathfrak{b} \subset \overline{\mathfrak{a}}$ is called integral over $\mathfrak{a}$.

(4.9) Corollary. a) The integral closure $\overline{\mathfrak{a}}$ of an ideal $\mathfrak{a}$ is an ideal with the same radical as $\mathfrak{a}$ and containing the nilradical of R.

b) If $\mathfrak{b}$ is integral over $\mathfrak{a}$ and $\mathfrak{c}$ is integral over $\mathfrak{b}$ then $\mathfrak{c}$ is integral over $\mathfrak{a}$.

c) For any multiplicatively closed subset S of R we have $\overline{\mathfrak{a}R_S} = \overline{\mathfrak{a}}R_S$.

Now we turn to reductions of ideals, and we will show that the concepts of reductions and of integral closure are equivalent, at least for noetherian rings (see (4.13) below). Reductions have been introduced by Northcott and Rees in their fundamental paper [14], which also contains most of the basic facts on this notion, including the relation to integral dependence. Moreover they showed that, by using reductions, one can associate a certain number to any ideal of a local ring, namely its analytic spread. This number plays a dominant role in our book, and it will be defined and discussed in Chapter II, §10. Here we restrict to some results on reductions which do not involve the analytic spread.

(4.10) Definition. Let $\mathfrak{a}, \mathfrak{b}$ be ideals of R . $\mathfrak{a}$ is called a reduction of $\mathfrak{b}$ if $\mathfrak{a} \subset \mathfrak{b}$ and

$$\mathfrak{b}^n = \mathfrak{a}\mathfrak{b}^{n-1} \quad \text{for some } n .$$

If $\mathfrak{b}^n = \mathfrak{a}\mathfrak{b}^{n-1}$ as above, then of course $\mathfrak{b}^m = \mathfrak{a}\mathfrak{b}^{m-1}$ for all $m \geqq n$, and also $\mathfrak{a}^m\mathfrak{b}^{n-1} = \mathfrak{b}^{m+n-1}$ for all $m \geqq 1$. Using this it is clear that if $\mathfrak{a}$ is a reduction of $\mathfrak{b}$ and $\mathfrak{b}$ is a reduction of $\mathfrak{c}$ then $\mathfrak{a}$ is a reduction of $\mathfrak{c}$. The next Lemma will give the link to integral dependence.

(4.11) Lemma. For any ideal $\mathfrak{a}$ and any element x of R , x is integral over $\mathfrak{a}$ if and only if $\mathfrak{a}$ is a reduction of $\mathfrak{a} + xR$.

Proof. Assume first that

$$x^n + a_1 x^{n-1} + \ldots + a_n = 0 , \; a_i \in \mathfrak{a}^i .$$

Then $x^n \in \mathfrak{a} \cdot (\mathfrak{a} + xR)^{n-1}$ and consequently

$$(\mathfrak{a}+xR)^n = x^n R + \mathfrak{a}(\mathfrak{a}+xR)^{n-1} = \mathfrak{a}(\mathfrak{a}+xR)^{n-1} .$$

Conversely, if $\mathfrak{a}$ is a reduction of $\mathfrak{a} + xR$, then $x^n \in \mathfrak{a}(\mathfrak{a}+xR)^{n-1}$ for some $n > 0$, and hence $x = \sum_i a_i b_i$, $a_i \in \mathfrak{a}$, $b_i \in (\mathfrak{a}+xR)^{n-1}$. Now each b_i can be written as

$$b_i = \sum_{j=0}^{n-1} a_{ij} x^{n-1-j} , \qquad a_{ij} \in \mathfrak{a}^j$$

and therefore

$$x^n = \sum_{j=0}^{n-1} (\sum_i a_i a_{ij}) x^{n-1-j} , \qquad \sum_i a_i a_{ij} \in \mathfrak{a}^{j+1} ,$$

which gives an equation of integral dependence for x over $\mathfrak{a}$.

<u>(4.12) Corollary</u>. For any faithfully flat homomorphism $R \longrightarrow S$ of noetherian rings and any proper ideal $\mathfrak{a}$ in R we have

$$\overline{\mathfrak{a}S} \cap R = \overline{\mathfrak{a}} .$$

Proof. We have to show that any $x \in \overline{\mathfrak{a}S} \cap R$ is integral over $\mathfrak{a}$. By Lemma (4.11) there is an integer n such that

$$(\mathfrak{a} + xR)^n S = (\mathfrak{a}S + xS)^n = (\mathfrak{a}S)(\mathfrak{a}S + xS)^{n-1} = \mathfrak{a}(\mathfrak{a} + xR)^{n-1} S ,$$

and hence

$$(\mathfrak{a} + xR)^n = \mathfrak{a}(\mathfrak{a} + xR)^{n-1}$$

by faithful flatness.

<u>(4.13) Proposition</u>. Let $\mathfrak{a}, \mathfrak{b}$ be ideals of R such that $\mathfrak{a} \subset \mathfrak{b}$. Then $\mathfrak{a}$ is a reduction of $\mathfrak{b}$ if and only if $\mathfrak{b} \subset \overline{\mathfrak{a}}$, i.e. $\mathfrak{b}$ is integral over $\mathfrak{a}$.

Proof. If $\mathfrak{a}$ is a reduction of $\mathfrak{b}$, then $\mathfrak{a}$ is a reduction of $\mathfrak{a} + xR$ for any $x \in \mathfrak{b}$, and therefore x must be integral over $\mathfrak{a}$ by Lemma (4.11), showing that $\mathfrak{b} \subset \overline{\mathfrak{a}}$. To prove the converse, assume that $\mathfrak{b} = \mathfrak{a} + x_1 R + \ldots + x_n R$. If $\mathfrak{b} \subset \overline{\mathfrak{a}}$, then again by Lemma (4.11) we see that $\mathfrak{a} + x_1 R + \ldots + x_i R$ is a reduction of $\mathfrak{a} + x_1 R + \ldots + x_{i+1} R$ for $i = 0, \ldots, n-1$. Therefore $\mathfrak{a}$ is a reduction of $\mathfrak{b}$ by the transitivity property of reductions mentioned above.

We point out that here we used for the first time that R is noetherian, and actually the result is still valid if we only assume $\mathfrak{b}/\mathfrak{a}$ to be finitely generated.

(4.14) Proposition. Let $\mathfrak{b}$ be a reduction of $\mathfrak{a}$ in R, let $\underline{x}$ be a sequence of elements of R and let M be a finitely generated R-module. Then $\underline{x}$ is a multiplicity system for $M/\mathfrak{a}M$ if and only if it is a multiplicity system for $M/\mathfrak{b}M$, and in this case we have $\dim(\mathfrak{a},M) = \dim(\mathfrak{b},M)$ and

$$e(\underline{x},\mathfrak{a},M) = e(\underline{x},\mathfrak{b},M) \quad .$$

Proof. The first assertion is clear, since the property of being a multiplicity system for $M/\mathfrak{a}M$ depends only on the radical of $\mathfrak{a}$. By the formula (see (3.11))

$$e(\underline{x},\mathfrak{a},R) = \sum_{\substack{\mathfrak{p}\in \mathrm{Assh}(M/\mathfrak{a}M)\\ \dim M_{\mathfrak{p}}=\dim(\mathfrak{a},M)}} e(\underline{x};R/\mathfrak{p})\,e(\mathfrak{a}R_{\mathfrak{p}},M_{\mathfrak{p}}) \quad ,$$

for the second assertion we are reduced to the case that R is local and $\mathfrak{a}$ and $\mathfrak{p}$ are $\mathfrak{m}$-primary, where $\mathfrak{m}$ is the maximal ideal of R. Since $\mathfrak{a}\subset\mathfrak{b}$, we have $H^{(1)}[\mathfrak{b},M] \leq H^{(1)}[\mathfrak{a},M]$. On the other hand $\mathfrak{a}\mathfrak{b}^s = \mathfrak{b}^{s+1}$ for some fixed s by assumption, which implies $\mathfrak{b}^{n+1} = \mathfrak{a}^{n-s+1}\mathfrak{b}^s \subset \mathfrak{a}^{n-s+1}$ for every $n \geqq s$, showing that

$$H^{(1)}[\mathfrak{a},M](n-s) \leq H^{(1)}[\mathfrak{b},M](n) \quad \text{for all} \quad n \geqq s \quad .$$

Comparing the leading term in both inequalities above we see that

$$e(\mathfrak{a},M) \leqq e(\mathfrak{b},M) \leqq e(\mathfrak{a},M) \quad .$$

In the case M = R, this result has a very important converse, first proved by D. Rees [15] and later on extended by E. Böger [11]. We will give proofs of the theorems of Rees and Böger in Chapter III. These results will show that the notions of reduction resp. integral closure determine exactly the range in which an ideal can be moved without changing its multiplicity.

<u>(4.15) Remark</u>. In Chapter II we will show that if R is a local ring with an infinite residue field and if $\mathfrak{q}$ is an ideal primary to the maximal ideal of R, then $\mathfrak{q}$ has a reduction consisting of $d = \dim R$ elements. Since the radical of the reduction is $\sqrt{\mathfrak{q}}$, these elements are a system of parameters of R. If we assume this result, then Proposition (4.14) shows that for any such $\mathfrak{q}$ there is a system of parameters $\underline{x} = (x_1,\ldots,x_d)$ of R such that

$$e(\mathfrak{q},R) = e(\underline{x}R,R) = e(\underline{x};R) \quad .$$

<u>(4.16) Lemma</u>. Let $\mathfrak{a}$ and $\mathfrak{b}$ be two ideals of R such that $\mathfrak{a} \subset \mathfrak{b}$. Then $\mathfrak{a}$ is a reduction of $\mathfrak{b}$ if and only if $\mathfrak{a}R_{red}$ is a reduction of $\mathfrak{b}R_{red}$. In particular we have $\overline{\mathfrak{a}\cdot R_{red}} = \bar{\mathfrak{a}}\cdot R_{red}$.

Proof. Assume that $\mathfrak{a}R_{red}$ is a reduction of $\mathfrak{b}R_{red}$. Then for any $x \in \mathfrak{b}$ there are $a_i \in \mathfrak{a}^i$, $i = 1,\ldots,n$, such that the element

$$z = x^n + a_1 x^{n-1} + \ldots + a_n$$

is nilpotent. Now if $z^m = 0$, then this gives an equation of integral dependence for x over $\mathfrak{a}$, showing that $\mathfrak{b} \subset \bar{\mathfrak{a}}$. The converse is obvious.

<u>(4.17) Proposition</u>. Let $\mathfrak{a}$ and $\mathfrak{b}$ be ideals of R such that $\mathfrak{a} \subset \mathfrak{b}$. Then the following conditions are equivalent:

(i) $\mathfrak{a}$ is a reduction of $\mathfrak{b}$.

(ii) For any minimal prime ideal $\mathfrak{p}$ of R we have that $\mathfrak{a}+\mathfrak{p}/\mathfrak{p}$ is a reduction of $\mathfrak{b}+\mathfrak{p}/\mathfrak{p}$.

Proof. Clearly (i) implies (ii), so assume for the converse that (ii) holds. By Lemma (4.16) we may assume that R is reduced. Consequently $B(\mathfrak{a},R)$ is reduced by Lemma (4.5), d) and moreover every minimal prime ideal of $B(\mathfrak{a},R)$ can be written as $B(\mathfrak{p})$ for some minimal prime ideal $\mathfrak{p}$ of R. Therefore, if we denote the integral closure in the total ring of fractions by a bar, we have a canonical isomorphism

$$\overline{B(\mathfrak{a},R)} \simeq \prod_{\mathfrak{p}} \overline{B(\mathfrak{a}+\mathfrak{p}/\mathfrak{p},R/\mathfrak{p})} \quad ,$$

by (4.1), where $\mathfrak{p}$ runs over the minimal prime ideals of R. Now the assertion is clear from Propositions (4.6) and (4.13).

We close this section by characterizing the integral closure of an ideal by valuations, and the result will be analogous to the description of the integral closure of a noetherian domain as an intersection of discrete (rank one) valuation rings. For this purpose we need two definitions.

(4.18) Definition. A discrete valuation of R is a function $v : R \longrightarrow \mathbf{Z} \cup \{\infty\}$ satisfying

a) $v(xy) = v(x) + v(y)$ for any $x,y \in R$;

b) $v(x+y) \geqq \min\{v(x),v(y)\}$ for any $x,y \in R$.

(For a) we use the convention that $a + \infty = \infty$ for any $a \in \mathbf{Z}$.) v will be called non-negative on R, if $v(x) \geqq 0$ for all $x \in R$.

It is clear that if v is a discrete valuation of R, and if

$$\mathfrak{p} = \{x \in R \mid v(x) = \infty\}$$

then either $\mathfrak{p} = R$, or $\mathfrak{p}$ is a prime ideal of R and v induces a (classical) discrete rank one valuation on the quotient field of $R/\mathfrak{p}$.

(4.19) Definition. Let v be a discrete valuation on R and let $\mathfrak{a}$ be any ideal of R. Then we put

$$v(\mathfrak{a}) = \inf\{v(x) \mid x \in \mathfrak{a}\} \quad .$$

(4.20) Proposition. Let $\mathfrak{a}$ be any ideal of R and $x \in R$. The integral closure of $\mathfrak{a}$ will be denoted by $\overline{\mathfrak{a}}$. Then the following conditions are equivalent:

(i) $x \in \overline{\mathfrak{a}}$.

(ii) For any non-negative discrete valuation v of R we have

$$v(x) \geqq v(\mathfrak{a}) \quad .$$

Proof. (i) $\Rightarrow$ (ii) . Note first that $v(\mathfrak{a}^i) = iv(\mathfrak{a})$ for all i . Let

$$x^n + a_1x^{n-1} + \ldots + a_n = 0 \ , \quad a_i \in \mathfrak{a}^i \quad .$$

If $v(x) = \infty$, then there is nothing to prove. So assume that $v(x) < \infty$ and that $v(x) < v(\mathfrak{a})$ if possible. Now for any $i \geq 1$ we have

$$v(a_ix^{n-i}) = v(a_i) + (n-i)v(x) \geq iv(\mathfrak{a}) + (n-i)v(x) > nv(x) \quad ,$$

which would imply $v(x^n + a_1x^{n-1} + \ldots + a_n) = v(x^n) = v(0) = \infty$, a contradiction.

(ii) $\Rightarrow$ (i) . If (ii) holds for R then it holds for $R/\mathfrak{p}, \mathfrak{p}$ any (minimal) prime of R . Therefore by Proposition (4.17) we may assume that R is a domain. Then also $B(\mathfrak{a},R)$ is a noetherian domain and $\overline{B(\mathfrak{a},R)}$ is a Krull domain by (4.2). If v is any discrete rank one valuation on the quotient field $Q(B(\mathfrak{a},R))$ of $B(\mathfrak{a},R)$ whose valuation ring V contains $B(\mathfrak{a},R)$, then $v \mid R$ is a nonnegative discrete valuation of R and consequently

$$v(x) \geq v(\mathfrak{a})$$

by assumption. This implies

$$v(xt) = v(x) + v(t) \geq v(\mathfrak{a}) + v(t) = v(\mathfrak{a}t) \geq 0 \quad ,$$

and therefore $xt \in V$. Since $\overline{B(\mathfrak{a},R)}$ is an intersection of such valuation rings V , we conclude that $xt \in \overline{B(\mathfrak{a},R)}$, i.e. x is integral over $\mathfrak{a}$ by Proposition (4.6).

(4.21) Remark. If R is not noetherian, then Proposition (4.20) remains true if one allows in (ii) more general than discrete valuation and if R has only finitely many minimal prime ideals. On the other hand, in case R is noetherian one may restrict the condition (ii) above to finitely many valuations v . Namely if v is a valuation on $B(\mathfrak{a},R)$ such that $v(t) \geq 0$, then clearly $v(xt) \geq 0$. So one needs only to check for those finitely many valuations for which $v(t) < 0$.

<u>(4.22) Corollary</u>. For any ideal $\mathfrak{a}$ of R there are finitely many non-negative discrete valuations $v_1,\dots,v_n$ of R such that

$$\overline{\mathfrak{a}} = \bigcap_{i=1}^{n} \{x \in R \mid v_i(x) \geq v_i(\mathfrak{a})\} \quad .$$

<u>(4.23) Corollary</u>. For any ideals $\mathfrak{a},\mathfrak{b}$ of R we have

$$\overline{\overline{\mathfrak{a}} \cdot \overline{\mathfrak{b}}} = \overline{\mathfrak{a} \cdot \mathfrak{b}} \quad .$$

Proof. $\overline{\mathfrak{a}} \cdot \overline{\mathfrak{b}} \subset \overline{\mathfrak{a} \cdot \mathfrak{b}}$ follows from

$$v(\overline{\mathfrak{a}} \cdot \overline{\mathfrak{b}}) = v(\overline{\mathfrak{a}}) + v(\overline{\mathfrak{b}}) = v(\mathfrak{a}) + v(\mathfrak{b}) = v(\mathfrak{a} \cdot \mathfrak{b})$$

for any non-negative discrete valuation v of R .

<u>§ 5. Faithfully flat extensions</u>

For this section we will assume that R is a local ring with maximal ideal $\mathfrak{m}$, and we denote by $\hat{R}$ the completion of R (for the $\mathfrak{m}$-adic topology). For computations with multiplicities it is often convenient to pass to completion, so we need to have some information of how these things behave under the homomorphism $R \longrightarrow \hat{R}$. On the other hand, for some results it is essential to have an infinite residue field. So if $R/\mathfrak{m}$ is finite, then R may be replaced by $R^* = R[X]_{\mathfrak{m}[X]}$, which has an infinite residue field, and similar to $R \longrightarrow \hat{R}$, the homomorphism $R \longrightarrow R^*$ is faithfully flat. Another common property of these homomorphisms is that the maximal ideals of both $\hat{R}$ and R^* are generated by the maximal ideal of R .

<u>(5.1) Proposition</u>. Let $(R,\mathfrak{m}) \longrightarrow (S,\mathfrak{n})$ be a flat, local homomorphism of local rings, let $\mathfrak{a}$ be an ideal of R , M a finitely generated R-module and $\underline{x}$ a sequence of elements of $\mathfrak{m}$. Then we have:

a) $\operatorname{ht}(\mathfrak{a}) = \operatorname{ht}(\mathfrak{a}S)$.

b) $e(\underline{x};M \otimes_R S) = \lambda_S(S/\mathfrak{m}S) \cdot e(\underline{x};M)$.

c) $H^{(i)}[\underline{x},\mathfrak{a}S,M \otimes_R S] = \lambda_S(S/\mathfrak{m}S) \cdot H^{(i)}[\underline{x},\mathfrak{a},M]$.

d) $e(\underline{x}, \mathfrak{a}S, M \otimes_R S) = \lambda_S(S/\mathfrak{m}S) \cdot e(\underline{x}, \mathfrak{a}, S)$.

e) if moreover $\mathfrak{n} = \mathfrak{m} \cdot S$, then $\dim M = \dim M \otimes_R S$, R is regular if and only if S is regular and M is Cohen-Macaulay over R if and only if $M \otimes_R S$ is Cohen-Macaulay over S .

Proof. Note first that

$$(5.1.1) \qquad \dim S = \dim R + \dim(S/\mathfrak{m}S)$$

(see [5], p. 79). Moreover, if M has finite length then by using induction on $\lambda_R(M)$ one shows easily that

$$(5.1.2) \qquad \lambda_S(M \otimes_R S) = \lambda_S(S/\mathfrak{m}S) \cdot \lambda_R(M) \quad .$$

Furthermore, if $\lambda_R(M) = \infty$ then also $\lambda_S(M \otimes_R S) = \infty$. We conclude that $\underline{x}$ is a multiplicity system for M (resp. $M/\mathfrak{a}M$) if and only if it is a multiplicity system for $M \otimes_R S$ (resp. $M/\mathfrak{a}M \otimes_R S$) and $\lambda_S(S/\mathfrak{m}S)$ is finite.

a) (see [5], p. 79) We give an outline of the proof. Let $\mathfrak{P} \subset S$ be any minimal prime ideal of S . Using the going down property of flat homomorphisms ([5], p. 33) we see that $\mathfrak{p} = \mathfrak{P} \cap R$ is minimal over $\mathfrak{a}$. This shows that $\dim S_{\mathfrak{P}}/\mathfrak{p}S_{\mathfrak{P}} = 0$, and since $R_{\mathfrak{p}} \longrightarrow S_{\mathfrak{P}}$ is flat again, we conclude from (5.1.1) that $\dim R_{\mathfrak{p}} = \dim S_{\mathfrak{P}}$. So we have $ht(\mathfrak{P}) = ht(\mathfrak{P} \cap R)$ for any minimal prime $\mathfrak{P}$ of $\mathfrak{a}S$. On the other hand, since $Spec(S) \longrightarrow Spec(R)$ is surjective ([5], p. 28), every minimal prime $\mathfrak{p}$ of $\mathfrak{a}$ is the contraction of a minimal prime $\mathfrak{p}$ of $\mathfrak{a}S$, which proves $ht(\mathfrak{a}) = ht(\mathfrak{a}S)$.

b) By the observations made earlier we may assume that all the members in the formula are finite. Let $\underline{x} = (x_1, \ldots, x_r)$. We use induction on r , and we note that the case $r = 0$ is true by (5.1.2). If $r \geq 1$ then $(M/x_rM) \otimes_R S = (M \otimes_R S)/x_r(M \otimes_R S)$ and also $(0 :_M x_r) \otimes_R S = (0 :_{M \otimes_R S} x_r)$ by flatness. Therefore the desired formula follows immediately from the definition of $e(\underline{x};M)$ by inductive assumption.

c) This is a consequence of b) since $\mathfrak{a}^nM/\mathfrak{a}^{n+1}M \otimes_R S \simeq \mathfrak{a}^n(M \otimes_R S)/\mathfrak{a}^{n+1}(M \otimes_R S)$, and d) follows from c) of course.

e) By (1) we have $\dim R/\mathrm{Ann}_R(M) = \dim S/\mathrm{Ann}_R(M)S$. But $\mathrm{Ann}_R(M)\cdot S = \mathrm{Ann}_S(M\otimes_R S)$, showing that $\dim M = \dim M\otimes_R S$. In particular we have $\dim R = \dim S$, and therefore the assertion about regularity follows from $\dim_{R/\mathfrak{m}}\mathfrak{m}/\mathfrak{m}^2 = \dim_{S/\mathfrak{n}}\mathfrak{n}/\mathfrak{n}^2$. For the last assertion we choose a system of parameters $\underline{y} = (y_1,\ldots,y_s)$ for M . Then $\underline{y}$ is also a system of parameter for $M\otimes_R S$, and from faithful flatness of $R \longrightarrow S$ we conclude that $\underline{y}$ is an M-regular sequence if and only if it is an $M\otimes_R S$ - regular sequence.

(5.2) Remark. In the situation e) above the given proof shows that more generally we have $\mathrm{depth}_R M = \mathrm{depth}_S(M\otimes_R S)$.

§ 6. Projection formula and criterion for multiplicity one

The projection formula is a very useful tool for computing multiplicities in concrete examples (see §7), and it can also be used for theoretical results. Here we will use it to show that the definitions of multiplicity given by Chavalley ([12]) and by Samuel ([17]) coincide, and also we will derive a criterion for multiplicity one.

(6.1) Lemma. Let R be a noetherian ring and $\mathfrak{a}$ an ideal of R . Let $N\subset M$ be finitely generated R-modules and assume that $\underline{x}$ is a multiplicity system on $M/\mathfrak{a}M$. Then we have:

a) There is a number k such that

$$H^{(1)}[\underline{x},\mathfrak{a},N](n-k) \leqq H^{(1)}[\underline{x},\mathfrak{a},M](n) \quad \text{for all } n \geqq k \quad .$$

b) Either $\dim(\mathfrak{a},N) < \dim(\mathfrak{a},M)$ or

$$e(\underline{x},\mathfrak{a},N) \leqq e(\underline{x},\mathfrak{a},M) \quad .$$

c) If there is an M-regular element $z\in R$ such that $zM\subset N$, then $\dim(\mathfrak{a},N) = \dim(\mathfrak{a},M)$ and

$$e(\underline{x},\mathfrak{a},N) = e(\underline{x},\mathfrak{a},M) \quad .$$

Proof. a) By the Artin-Rees Lemma ([5], p. 68) there is an integer k such that

$$\mathfrak{a}^n M \cap N = \mathfrak{a}^{n-k}(\mathfrak{a}^k M \cap N) \subset \mathfrak{a}^{n-k} N \quad \text{for} \quad n \geqq k \quad .$$

Using the additivity of $e(\underline{x},-)$ (see Proposition (1.3)) we know that $H^{(1)}[\underline{x},\mathfrak{a},M](n) = e(\underline{x};M/\mathfrak{a}^{n+1}M)$, and therefore

$$H^{(1)}[\underline{x},\mathfrak{a},N](n-k) \leqq e(\underline{x};N/\mathfrak{a}^{n+1}M \cap N) \leqq H^{(1)}[\underline{x},\mathfrak{a},M](n)$$

for all $n \geqq k$. b) is a direct consequence of a), and c) follows from a), b) and the fact that zM is isomorphic to M .

<u>(6.2) Definition</u>. For any noetherian ring R we put

$$\mathrm{Maxh}(R) = \{\mathfrak{m} \in \mathrm{Max}(R) \mid \dim R_{\mathfrak{m}} = \dim R\}$$

<u>(6.3) Theorem</u>. (Projection formula) Let $R \subset S$ be noetherian rings, $\mathfrak{a}$ an ideal of R and $\underline{x}$ a multiplicity system for $R/\mathfrak{a}$. Assume that

1) S contains a finitely generated free R-module F and

2) R contains an S-regular element z such that $zS \subset F$.

Then $\underline{x}$ is a multiplicity system for $S/\mathfrak{a}S$ and we have

$$(\mathrm{rank}\, F)\cdot e(\underline{x},\mathfrak{a},R) = \sum_{\substack{\mathfrak{p}\in\mathrm{Assh}(\mathfrak{a})\\ \dim S_{\mathfrak{p}} = \dim(\mathfrak{a},S)}} e(\underline{x};R/\mathfrak{p}) \cdot \sum_{\mathfrak{n}\in\mathrm{Maxh}(S_{\mathfrak{p}})} [S_{\mathfrak{p}}/\mathfrak{n}:R_{\mathfrak{p}}/\mathfrak{p}R_{\mathfrak{p}}]\, e(\mathfrak{a}S_{\mathfrak{n}},S_{\mathfrak{n}})$$

where $S_{\mathfrak{p}} = S \otimes_R R_{\mathfrak{p}}$, and where $e(\mathfrak{a}S_{\mathfrak{n}},S_{\mathfrak{n}}) = 0$ if $\mathfrak{a}S \not\subset \mathfrak{n}$.

Proof. First note that by 1) and 2) S is a finitely generated R-module, $\dim(\mathfrak{a},S) = \dim(\mathfrak{a},R)$ and $\underline{x}$ is a multiplicity system for $S/\mathfrak{a}S$. First we treat the case that $\underline{x} = \emptyset$, so $\dim R/\mathfrak{a} = 0$. Then consequently $\dim S/\mathfrak{a}^n S = 0$ for all n , and by the Chinese Remainder Theorem we know that

$$S/\mathfrak{a}^n S \simeq \prod_{\substack{\mathfrak{n}\in \mathrm{Max}(S)\\ \mathfrak{a}S \subset \mathfrak{n}}} S_{\mathfrak{n}}/\mathfrak{a}^n S_{\mathfrak{n}} \quad .$$

Taking the Hilbert function of the finitely generated R-module S with respect to $\mathfrak{a}$ we have

$$H^{(1)}[\mathfrak{a},S] = \sum_{\mathfrak{n}\in \mathrm{Max}(S)} H^{(1)}[\mathfrak{a}S_{\mathfrak{n}},S_{\mathfrak{n}}] \cdot [S/\mathfrak{n} : R/\mathfrak{n}\cap R] \quad ,$$

where we have used the fomula

$$\lambda_R(S_{\mathfrak{n}}/\mathfrak{a}^n S_{\mathfrak{n}}) = [S/\mathfrak{n} : R/\mathfrak{n}\cap R]\, \lambda_{S_{\mathfrak{n}}}(S_{\mathfrak{n}}/\mathfrak{a}^n S_{\mathfrak{n}}) \quad .$$

Comparing the leading terms we obtain

$$e_R(\mathfrak{a},S) = \sum_{\substack{\mathfrak{n}\in \mathrm{Maxh}(S)\\ \dim S_{\mathfrak{n}} = \dim(\mathfrak{a},S)}} [S/\mathfrak{n} : R/\mathfrak{n}\cap R]\; e_{S_{\mathfrak{n}}}(\mathfrak{a}S_{\mathfrak{n}},S_{\mathfrak{n}}) \quad .$$

But by Lemma (6.1) we know that $e_R(\mathfrak{a},S) = (\mathrm{rank}\, F)\cdot e_R(\mathfrak{a},R)$, so the proof is complete in the case $\underline{x} = \emptyset$. For the general case we note that $\mathrm{Ann}_R(S) = (0)$ and therefore by Proposition (3.11)

$$e(\underline{x},\mathfrak{a},S) = \sum_{\substack{\mathfrak{p}\in \mathrm{Assh}(\mathfrak{a})\\ \dim S_{\mathfrak{p}} = \dim(\mathfrak{a},S)}} e(\underline{x};R/\mathfrak{p}) \;\; e(\mathfrak{a}R_{\mathfrak{p}},S_{\mathfrak{p}}) \quad .$$

Now $S_{\mathfrak{p}}$, as a module over $R_{\mathfrak{p}}$, satisfies again the assumptions 1) and 2), so from the first case we know that

$$\begin{aligned} e(\mathfrak{a}R_{\mathfrak{p}},S_{\mathfrak{p}}) &= (\mathrm{rank}\ F)\cdot e(\mathfrak{a}R_{\mathfrak{p}},R_{\mathfrak{p}}) \\ &= \sum_{\substack{\mathfrak{n}\in \mathrm{Maxh}(S_{\mathfrak{p}})\\ \dim S_{\mathfrak{n}} = \dim(\mathfrak{a}R_{\mathfrak{p}},S_{\mathfrak{p}})}} [S_{\mathfrak{p}}/\mathfrak{n} : R_{\mathfrak{p}}/\mathfrak{n}\cap R_{\mathfrak{p}}]\cdot e(\mathfrak{a}S_{\mathfrak{n}},S_{\mathfrak{n}}) \quad . \end{aligned}$$

and moreover by Lemma (6.1) again we have

$$e(\underline{x},\mathfrak{a},S) = (\mathrm{rank}\ F)\cdot e(\underline{x},\mathfrak{a},R) \quad .$$

To complete the proof in the general case we note that if $\mathfrak{p} \in \mathrm{Assh}(\mathfrak{a})$ and $\dim S_{\mathfrak{p}} = \dim(\mathfrak{a},S)$, then $\dim(\mathfrak{a}R_{\mathfrak{p}},S_{\mathfrak{p}}) = \dim S_{\mathfrak{p}}$, which shows that $\{\mathfrak{n} \in \mathrm{Max}(S_{\mathfrak{p}}) \mid \dim S_{\mathfrak{n}} = \dim(\mathfrak{a}R_{\mathfrak{p}},S_{\mathfrak{p}})\} = \mathrm{Max}\,h(S_{\mathfrak{p}})$, and of course we have $\mathfrak{n} \cap R_{\mathfrak{p}} = \mathfrak{p}R_{\mathfrak{p}}$ for any $\mathfrak{n} \in \mathrm{Max}(S_{\mathfrak{p}})$, q.e.d.

(6.4) Remarks. Of course there are special cases in which the projection formula of Theorem (6.3) simplifies considerably, and we will restate this formula in some of these cases in the corollaries below. For example, if R is local with maximal ideal $\mathfrak{m}$ and $\mathfrak{a}$ is $\mathfrak{m}$-primary, then $\dim(\mathfrak{a},S) = \dim R = \dim S$ (as was used in the proof of Theorem (6.3)). More important is to give conditions under which the assumptiones 1) and 2) of the Theorem are satisfied, and the most important cases for this are the following ones:

a) R is a domain and S is a finitely generated torsion-free R-module. In this case we may choose $s_1,\ldots,s_n \in S$ such that these elements are a basis of the $Q(R)$-vector space $S \otimes_R Q(R)$, so $n = [S \otimes_R Q(R) : Q(R)]$. Putting $F = \sum_{i=1}^{n} Rs_i$, it is clear that F is free and $zS \subset F$ for some nonzero $z \in R$.

b) S is finitely generated over R and the total quotient rings $Q(R)$ and $Q(S)$ coincide. In this case we may simply take $F = R$.

(6.5) Corollary. Let R be a local domain with quotient field K and let S be an overring of R which is a finitely generated torsion free R-module. For any ideal $\mathfrak{q}$ of R which is primary to the maximal ideal $\mathfrak{m}$ of R we have

$$(\dim_K S \otimes_R K)\,e(\mathfrak{q},R) = \sum_{\mathfrak{n}\in\mathrm{Max}\,h(S)} [S/\mathfrak{n} : R/\mathfrak{m}] \cdot e(\mathfrak{q}S_{\mathfrak{n}},S_{\mathfrak{n}}) \quad .$$

More generally we have

(6.6) Corollary. Let R be a domain and let S be an overring of R which is a finitely generated torsion free R-module. Then for $\mathfrak{a}$ and $\underline{x}$ as in the Theorem (6.3) we have

$$[S \otimes_R Q(R):Q(R)] \cdot e(\underline{x},\mathfrak{a},R) = \sum_{\substack{\mathfrak{p}\in \mathrm{Assh}(\mathfrak{a}) \\ \dim S_{\mathfrak{p}} = \dim(\mathfrak{a},S)}} e(\underline{x},R/\mathfrak{p}) \cdot \sum_{\mathfrak{n}\in \mathrm{Maxh}(S_{\mathfrak{p}})} [S_{\mathfrak{p}}/\mathfrak{n}:R_{\mathfrak{p}}/\mathfrak{p}R_{\mathfrak{p}}] \cdot e(\mathfrak{a}S_{\mathfrak{n}},S_{\mathfrak{n}}) .$$

(6.7) Historical Remark. For a local ring R with maximal ideal $\mathfrak{m}$ there are essentially three different notions of "classical" multiplicity, and all these notions coincide in a sense which we want to make precise. Two notions have been introduced above, namely

a) the so called Samuel multiplicity $e(\mathfrak{q},R)$, which is defined for any ideal $\mathfrak{q}$ with $\sqrt{\mathfrak{q}} = \mathfrak{m}$ (see (2,9));

b) the multiplicity $e(\underline{x};R)$, which is defined if $(R/\underline{x}R)$ is finite, but which is different from zero only if $\underline{x}$ is a system of parameters of R (see (1.2) and (1.7)).

These two notions coincide in the following sense: If $\underline{x}$ is a system of parameters then $e(\underline{x};R) = e(\underline{x}R,R)$ by (2.7). Conversely, given any $\mathfrak{q}$ with $\sqrt{\mathfrak{q}} = \mathfrak{m}$, and assuming $R/\mathfrak{m}$ to be infinite, there is a system of parameters $\underline{x}$ of R such that $e(\mathfrak{q},R) = e(\underline{x}R,R)$ (see (4.14)).

We recall a third notion of multiplicity, which preceded the other ones and was introduced by Chevalley ([12]). For this purpose we need some well-known facts about complete local rings. Let R be a reduced, complete local ring containing a field. Then we may assume that R contains its residue field $k = R/\mathfrak{m}$ and we suppose that k is infinite. It is known that for any system of parameters $x_1,\dots,x_d$ of R the canonical homomorphism $k[[X_1,\dots,X_d]] \longrightarrow k[[x_1,\dots,x_d]] = R_\circ \subset R$ is an isomorphism, and moreover R is a finitely generated $R_\circ$-module. Now the definition of multiplicity of Chevalley is the following:

c) The multiplicity of $\underline{x} = (x_1,\dots,x_d)$ is defined to be the vector space dimension $[R \otimes_{k[[\underline{x}]]} k((\underline{x})) : k((\underline{x}))]$ (note that R is torsion free over $k[[x]]$ since it is reduced), and the multiplicity of R is defined to be the smallest multiplicity that can be obtained by all possible choices of $\underline{x}$.

Now by the projection formula (Corollary (6.6)) we see that for any $\underline{x}$

$$e(\underline{x}R,R) = [R \otimes_{k[[x]]} k((\underline{x})) : k((x))] .$$

Comparing with the Samuel multiplicity of other m-primary ideals $\mathfrak{q}$ we note that $e(\mathfrak{m},R) \leq e(\mathfrak{q},R)$, hence if we choose $\underline{x}$ to generate a reduction of $\mathfrak{m}$, then $[R \otimes_{k[[\underline{x}]]} k((x)) : k((x))]$ is minimal and is equal to $e(\mathfrak{m},R)$. This shows that whenever the multiplicity of Chevalley is defined then it is equal to the Samuel multiplicity $e(\mathfrak{m},R)$. See for this and the following App. II (Introduction).

We conclude these remarks by indicating some geometric background for the multiplicity $e(\mathfrak{m},R)$ and the definition of Chevalley as in c) above. Starting with the last one, we note that for any system of parameters $x_1,\ldots,x_d$ of R , the inclusion $k[[x_1,\ldots,x_d]] \subset R$ defines a finite surjective morphism $\mathrm{Spec}(R) \longrightarrow \mathrm{Spec}(k[[x_1,\ldots,x_d]])$, which may be viewed as a projection onto d-dimensional affine space. Then the multiplicity of Chevalley, i.e. the number $[R \otimes k((x)) : k((x))]$ may be viewed as the covering degree, which is the number of points in a "sufficiently general" fibre. So the multiplicity of R in the sense of Chevalley amounts to taking the least possible covering degree for various projections onto affine space.

To explain the Samuel multiplicity, let $V \subset \mathbb{P}^n$ be a projective variety, defined by a homogeneous ideal $I \subset k[X_0,\ldots,X_n]$, where k is an infinite field. The degree of V in $\mathbb{P}^n$ is defined to be the maximal number of intersection points of V with a linear subspace of dimension $n - \dim V$ in $\mathbb{P}^n$ (assuming that this intersection is finite), and one can prove that this degree is given by the leading term of the Hilbert function of the graded ring $k[X_0,\ldots,X_n]/I$. Now taking R to be the local ring of the affine cone $C(V) \subset \mathbb{A}^{n+1}$ at the vertex, the Hilbert function of R with respect to the maximal ideal is the same as the Hilbert function of $k[X_0,\ldots,X_n]/I$. So we conclude that the degree of V is equal to the multiplicity of $C(V)$ at its vertex.

We will see in the next section that computing multiplicities in explicit examples is sometimes easy using Hilbert functions but in many cases the only reasonable way is via the projection formula, so depending on the example, the notion a) or c) above may be more appropriate for computations, while the purely algebraic definition of b) is more suitable for theoretical considerations.

The very first example is that of a regular local ring R , for which (by any of the definitions) the multiplicity is one. Conversely, one might guess that for any local ring $(R,\mathfrak{m})$ the condition $e(\mathfrak{m},R) = 1$ implies regularity. There are two typical classes of counterexamples to this conjecture (see § 7), but if these are excluded, the conjecture is true as we will show next. Before we need a lemma.

(6.8) Theorem. Let $(R,\mathfrak{m})$ be a local ring for which the completion $\hat{R}$ satisfies $\mathrm{Ass}(\hat{R}) = \mathrm{Assh}(\hat{R})$. If $e(\mathfrak{m},R) = 1$, then R is regular.

Proof. By Proposition (5.1) we may assume that R is complete and has an infinite residue field. Then by Remark (4.15) we may choose $\underline{x} = (x_1,\ldots,x_d)$ such that $d = \dim R$ and $\underline{x}R$ is a reduction of $\mathfrak{m}$. Then $e(\mathfrak{m},R) = e(\underline{x},R)$ and hence the Reduction Formula (1.10.1) implies

$$\sum_{\mathfrak{p}\in \mathrm{Assh}(R)} e(\underline{x};R/\mathfrak{p}) \cdot e(\mathfrak{p}R_{\mathfrak{p}},R_{\mathfrak{p}}) = 1 \quad .$$

Since $e(\underline{x};R/\mathfrak{p}) \neq 0$ for every $\mathfrak{p} \in \mathrm{Assh}(R)$, and since $\mathrm{Assh}(R) = \mathrm{Ass}(R)$, we see that $\mathrm{Ass}(R)$ consists of one prime ideal $\mathfrak{p}$ for which $e(\mathfrak{p}R_{\mathfrak{p}},R_{\mathfrak{p}}) = \lambda(R_{\mathfrak{p}}) = 1$. This means that R is a domain. Now we proceed by induction on $\dim R$ to show that R is regular and $\underline{x}$ is a regular system of parameters of R . If $\dim R = 0$ then R is a field and hence regular. So assume that $\dim R > 0$. Let K be the quotient field of R and denote by $\overline{R}$ the integral closure of R in K . Then $\overline{R}$ is a complete local ring which is a finitely generated R-module (see e.g. [9], p. 283). Applying the Projection Formula (6.3) we get

$$1 = e(\underline{x};R) = [\overline{R}/\overline{\mathfrak{m}} : R/\mathfrak{m}] \cdot e(\underline{x};\overline{R}) \quad ,$$

where $\overline{\mathfrak{m}}$ denotes the maximal ideal of $\overline{R}$. It follows that

$$\overline{R}/\overline{\mathfrak{m}} = R/\mathfrak{m} \quad \text{and} \quad e(\underline{x};\overline{R}) = 1 \quad .$$

Now we pass to $S = \overline{R}/x_1\overline{R}$. S is a complete local ring. Moreover, since $\overline{R}$ is normal, the ideal $x_1\overline{R}$ is unmixed and consequently S satisfies $\mathrm{Ass}(S) = \mathrm{Assh}(S)$. Finally, using the fact that x_1 is not a zero-divisor, we have

$$e((x_2,\ldots,x_d);S) = e(\underline{x};\overline{R}) = 1 \quad .$$

If $\mathfrak{n}$ denotes the maximal ideal of S, then using (2.9.1) we have

$$0 < e(\mathfrak{n},S) \leq e((x_2,\ldots,x_d)S,S) = e((x_2,\ldots,x_d);S) = 1 \quad .$$

Therefore, by applying the inductive hypothesis to S, we conclude that S is regular, and the images of $x_2,\ldots,x_d$ in S form a regular system of parameters of S. Therefore $\overline{R}$ is regular and $\underline{x}$ is a regular system of parameters of $\overline{R}$. In particular we have $\overline{\mathfrak{m}} = \underline{x}\overline{R}$ and hence $\overline{R}/\mathfrak{m}\overline{R} = R/\mathfrak{m}$. So by Nakayama's Lemma it is clear that $R = \overline{R}$, which concludes the proof.

<u>(6.9) Remark</u>. The condition $\operatorname{Assh}(R) = \operatorname{Ass}(R)$ for a local ring R means that R has not embedded components and $\dim R/\mathfrak{p} = \dim R$ for each minimal prime ideal $\mathfrak{p}$ of R. Such rings are sometimes called unmixed (since the zero ideal is unmixed in the classical sense). A quasi-unmixed local ring R is (by Nagata's definition in [6], p. 124) a ring whose completion $\hat{R}$ satisfies $\operatorname{Assh}(\hat{R}) = \operatorname{Min}(\hat{R})$. A complete theory of quasi-unmixed rings (local and also nonlocal) will be given in Chapter III. Here we note only that Theorem (6.8) does not remain true if the condition " $\hat{R}$ is unmixed" is replaced by " R is quasi-unmixed". A counterexample is given in (7.1).

§ 7. Examples

Most of this section is devoted to explicit computation of multiplicities, and we will see that for some cases the Hilbert function is the best tool for this computation, whereas for other examples it is much better to use reductions and the projection formula since the computations of the Hilbert functions seems almost impossible.

(7.1) Non-regular local rings of multiplicity one. We take any infinite field k and we put

$$R_1 = k[[x,y]]/(x^2,xy), \quad R_2 = k[[x,y,z]]/(xy,xz) \quad .$$

If $\mathfrak{m}_1$ denotes the maximal ideal of R_1, then

$$\dim_{R_1/\mathfrak{m}_1}(\mathfrak{m}_1^n/\mathfrak{m}_1^{n+1}) = \begin{cases} 1 & \text{if} \quad n = 0 \\ 2 & \text{if} \quad n = 1 \\ 1 & \text{if} \quad n \geqq 2 \end{cases}$$

and therefore $e(\mathfrak{m}_1, R_1) = 1$, although R_1 is not regular. Here the reason is that R_1 has an embedded component. Note that $\mathrm{Assh}(R_1) = \mathrm{Min}(R_1) = \{xR_1\}$.

For R_2 let us choose a system of parameters, namely x_1 = residue class of $y - x$ and x_2 = residue class of $z - x$. Putting $\underline{x} = (x_1, x_2)$ we have

$$\begin{aligned} e(\underline{x}; R_2) &= \sum_{\mathfrak{p} \in \mathrm{Assh}(0)} e(\underline{x}; R_2/\mathfrak{p}) \cdot e((R_2)_{\mathfrak{p}}) = e(\underline{x}; R_2/xR_2) = \\ &= e((y,z); k[[y,z]]) = 1 \end{aligned}$$

since $\mathrm{Assh}(0) = \{x R_2\}$ and therefore certainly $e(\mathfrak{m}_2, R_2) = 1$ ($\mathfrak{m}_2$ = maximal ideal of R_2). Here of course R_2 is not equidimensional, i.e. $\mathrm{Assh}(R_2) \neq \mathrm{Min}(R_2)$. Geometrically, the equation $xy = xz = 0$ define the union of the x-axis and the y-z-plane, and R_2 is the local ring at the point of intersection.

(7.2) The Hilbert function of a hypersurface. Let

$$R = k[[x_1, \ldots, x_d]]/(f(x_1, \ldots, x_d)), \quad k \text{ any field},$$

where $f(x_1, \ldots, x_d)$ is neither zero nor a unit. Put $S = K[[x_1, \ldots x_d]]$ and let $\mathfrak{m}$ be the maximal ideal of S. Then it is well known (and easily proved by induction on d) that

$$H^{(0)}[\mathfrak{m}, S](n) = \binom{n+d-1}{d-1} \quad .$$

Write $f = \sum_{\nu=e}^{\infty} f_\nu$, where f_ν is a homogeneous plynomial of degree ν and $f_e \neq 0$. If $\mathfrak{m}$ denotes the maximal ideal of R, then

$$\mathfrak{m}^n/\mathfrak{m}^{n+1} = \mathfrak{m}^n + fS/\mathfrak{m}^{n+1} + fS \simeq \mathfrak{m}^n/\mathfrak{m}^{n+1} + \mathfrak{m}^n \cap fS$$

and we deduce

$$H^{(0)}[\mathfrak{m},R](n) = \begin{cases} \binom{n+d-1}{d-1} & \text{if } n < e \\ \binom{n+d-1}{d-1} - \binom{(n-e)+d-1}{d-1} & \text{if } n \geq e \end{cases},$$

where we have used the fact that $\mathfrak{m}^n \cap fS = \mathfrak{m}^{n-e} \cdot f$ if $n \geq e$. The value given for $n \geq e$ also gives the plynomial associated to $H^{(0)}[\mathfrak{m},R]$, and its leading term is e. We conclude that

$$e = \mathrm{ord}_S(f) = e(\mathfrak{m},R) \qquad ,$$

and moreover the Hilbert function $H^{(0)}[\mathfrak{m},R]$ is completely determined by the numbers $e = e(\mathfrak{m},R)$ and $d-1 = \dim(R)$.

(7.3) Verqnese vaieties. Again we choose an infinite field k and we put $S = k[X_0,\ldots,X_d] = \bigoplus_{n\geq 0} S_n$, where S_n denotes the homogeneous part of degree n. Then for any $r \geq 1$ we put

$$S_n^{(r)} = S_{nr} \quad \text{and} \quad S^{(r)} = \bigoplus_{n\geq 0} S_n^{(r)} \qquad .$$

Finally we define $R^{(r)}$ to be the localization of $S^{(r)}$ at the homogeneous maximal ideal, and we denote the maximal ideal of $R^{(r)}$ by $\mathfrak{m}(R^{(r)})$. Then

$$\mathfrak{m}(R^{(r)})^n/\mathfrak{m}(R^{(r)})^{n+1} \simeq S_{nr}$$

which shows that

$$H^{(0)}[R^{(r)}](n) = \dim_k S_{nr} = \binom{nr+d}{d} \qquad .$$

Therefore we see that $e(R^{(r)}) = r^d$.

Let us also compute this number by using the projection formula and reductions. If we put $y_0 = X_0^r,\ldots,y_d = X_d^r$, then $\underline{y} = (y_0,\ldots,y_d)$ is a system of parameters of $R^{(r)}$ and we have

$$\underline{y} \cdot \mathfrak{m}(R^{(r)})^{d-1} = \mathfrak{m}(R^{(r)})^d \qquad .$$

This shows that $e(R^{(r)}) = e(\underline{y};R^{(r)})$ by Proposition (4.14). By the projection formula applied to $k[y_0,\ldots,y_d]_{(y_0,\ldots,y_d)} \subset R^{(r)}$ we have

$$e(R^{(r)}) = [Q(R^{(r)}) : k(y_0,\ldots,y_d)] \quad .$$

It is easy to check that $k(X_0,\ldots,X_d) = Q(R^{(r)})(X_0)$ and therefore $[k(X_0,\ldots X_d) : Q(R^{(r)})] = r$. On the other hand we have $[k(X_0,\ldots,X_d) : k(y_0,\ldots,y_d)] = r^{d+1}$, which proves again that $e(R^{(r)}) = r^d$. We want to note that the rings $R^{(r)}$ are Cohen-Macaulay. In the case $r = d = 2$, one easily sees that

$$\lambda(R^{(r)}/(y_0,y_1,y_2)R^{(r)}) = 4 = e(R^{(r)}) \quad ,$$

so $\underline{y}$ is a regular sequence on $R^{(r)}$ by Proposition (1.5). In the general case we use a different argument, which we state separately.

<u>(7.4) Lemma</u>. Let $A \subset B$ be rings such that $B = A \oplus M$ for some A-submodule M of B . If $\underline{x} = (x_1,\ldots,x_r)$ is a sequence of elements of A which is a regular sequence on B , then $\underline{x}$ is a regular sequence on A .

Proof. Choose $a, a_j \in A$ such that

$$ax_i = a_1x_1 + \ldots + a_{i-1}x_{i-1} \quad .$$

Thus $a \in x_1B + \ldots + x_{i-1}B$ by assumption, which means that

$$a = x_1b_1 + \ldots + x_{i-r}b_{i-1} \quad .$$

Writing $b_j = c_j + m_j$ with $c_j \in A$, $m_j \in M$, we conclude that

$$a = x_1c_1 + \ldots + x_{i-1}c_{i-1} \quad ,$$

as desired.

<u>(7.5) Corollary</u>. The rings $R^{(r)}$ of example (7.3) are Cohen-Macaulay.

Proof. The elements $y_0,\dots,y_d$ are a regular sequence in S, and $S = S^{(r)} \oplus M$ where

$$M = \bigoplus_{n \notin r\cdot\mathbf{Z}} S_n \quad .$$

Since M is an $S^{(r)}$-module, Lemma (7.4) tells us that $y_0,\dots,y_d$ is a regular sequence on $S^{(r)}$, hence also on the localization $R^{(r)}$.

We note here that instead of the local rings $R^{(r)}$ we could have used the graded rings $S^{(r)}$. Denoting the maximal homogeneous ideal of $S^{(r)}$ by $S_+^{(r)}$ we have

$$H^{(i)}\left[S_+^{(r)}, S^{(r)}\right] = H^{(i)}\left[R^{(r)}\right] \quad \text{for all } i \; .$$

The change in multiplicity by passing from R to $R^{(r)}$ (or from S to $S^{(r)}$) corresponds to the well known geometric fact that different embeddings of the same projective variety may have different degrees.

More generally than (7.3), for any positively graded noetherian ring $A = \bigoplus_{n\geq 0} A_n$ for which A_0 is artinian we have

$$e(A_+^{(r)}, A^{(r)}) = e(A_+, A) \cdot r^{\dim A - 1} \quad .$$

(7.6) Segre embeddings. Usually the Segre embedding is defined to be the morphism $\mathbb{P}^r \times \mathbb{P}^s \longrightarrow \mathbb{P}^{(r+1)(s+1)-1}$ given by the homomorphism

$$\varphi : k[Z_{ij} \mid 0 \leq i \leq r,\ 0 \leq j \leq s] \longrightarrow k[z_{ij}] \subset k[x_0,\dots,x_r,y_0,\dots,y_s]$$

where $\varphi(Z_{ij}) = z_{ij} = x_i y_j$. We will treat this as a special case of the following more general situation. We are given two graded noetherian rings

$$A = \bigoplus_{n\geq 0} A_n \quad \text{and} \quad B = \bigoplus_{n\geq 0} B_n$$

where we assume that $A_0 = B_0$ is a field k and that both rings are generated over k by elements of degree one. Then we define another graded ring $C = \bigoplus_{n\geq 0} C_n$ be putting

$$C_n = A_n \otimes_k B_n \quad .$$

Let us denote the homogeneous maximal ideal of A (resp. B, C) by $\mathfrak{m}(A)$ (resp. $\mathfrak{m}(B), \mathfrak{m}(C)$). Then we have

$$H^{(0)}[\mathfrak{m}(C),C](n) = H^{(0)}[\mathfrak{m}(A),A](n) \cdot H^{(0)}[\mathfrak{m}(B),B](n) \quad .$$

Let $r = \dim A_{\mathfrak{m}(A)} - 1$ and $s = \dim B_{\mathfrak{m}(B)} - 1$. Then comparing the leading terms in the equation above we see that

$$e(\mathfrak{m}(C),C) = \binom{r+s}{r} \cdot e(\mathfrak{m}(A),A) \cdot e(\mathfrak{m}(B),B) \quad .$$

Moreover, we know that $\dim C_{\mathfrak{m}(C)} = r + s + 1$. (In the next chapter we will show that $\dim A_{\mathfrak{m}(A)} = \dim A$ and similarly for B and C.)

For the classical Segre embedding mentioned above we obtain that the multiplicity of the vertex of the affine case (= degree of the embedding) of $\mathbb{P}^r \times \mathbb{P}^s$ in $\mathbb{P}^{(r+1)(s+1)-1}$ is equal to $\binom{r+s}{r}$. In this case we have $C = k[z_{ij}]$, $z_{ij} = x_i y_i$, and we point out that there is no obvious choice for a system of parameters of $C_{\mathfrak{m}(C)}$. A simple computation shows that no z_{ij} is contained in the radical of the ideal generated by $\{z_{i'j'} \mid (i',j') \neq (i,j)\}$. If now $r > 0$ and $s > 0$, then the number of z_{ij}'s is strictly bigger than the dimension of C_1 and hence the set $\{z_{ij} \mid 0 \leq i \leq r,\ 0 \leq j \leq s\}$ does not contain a system of parameters of $C_{\mathfrak{m}(C)}$. Therefore we cannot use the projection formula to compute the multiplicity of C as we could in example (7.3).

(7.7) A two-dimensional non Cohen-Macaulay domain. We put

$$R = k[[s^2,s^3,st,t]] \subset k[[s,t]] \quad ,$$

k a field again. Thus s^2, t is a system of parameters of R, and for a maximal ideal $\mathfrak{m}$ we have $(s^2,t) \cdot \mathfrak{m} = \mathfrak{m}^2$ and therefore $e(R) = e((s^2,t);R)$. By the projection formula applied to $k[[s^2,t]] \subset R$ we see that

$$e((s^2,t);R) = [Q(R) : k((s^2,t))] = 2 \quad .$$

Now we write

$$R = k[[x,y,z,w]] = k[[X,Y,Z,W]]/\mathfrak{P} ,$$

where $x = s^2$, $y = s^3$, $z = st$ and $w = t$. It is not difficult to check that $\mathfrak{P}$ is generated by

$$F_1 = X^3 - Y^2 , F_2 = XZ - YW , F_3 = Z^2 - XW^2 , F_4 = YZ - X^2W ,$$

and therefore

$$R/(s^2,t)R \simeq k[[Y,Z]]/(Y^2,YZ,Z^2) .$$

It follows that $\lambda(R/(s^2,t)R) = 3 > e((s^2,t);R)$, so (s^2,t) is not a regular sequence and R is not Cohen-Macaulay. For this example it is much more difficult to compute the Hilbert function. Using techniques developed by Robbiano and Valla ([16], see Chapter II, § 13) one can show that

$$G(\mathfrak{m},R) = \bigoplus_{n\geq 0} \mathfrak{m}^n/\mathfrak{m}^{n+1} \simeq k[X,Y,Z,W]/(Y^2,XZ - YW,Z^2,YZ) ,$$

and from this one deduces that

$$H^{(0)}[R](n) = \begin{cases} 1 & \text{if} \quad n = 0 \\ 4 & \text{if} \quad n = 1 \\ 3\binom{n+1}{1} - \left[\binom{n}{1} - 1\right] = 2n + 4 & \text{if} \quad n \geq 2 \end{cases} ,$$

and this gives $e(R) = 2$ again.

Let us use the same ring to compute a generalized multiplicity. We put $\mathfrak{p} = (s^2,s^3,st)\cdot R$. Then t is a multiplicity system for $R/\mathfrak{p}$ and

$$e(t,\mathfrak{p},R) = e(t,R/\mathfrak{p}) \cdot e(R_{\mathfrak{p}}) .$$

Of course we have $e(t;R/\mathfrak{p}) = 1$. To determine $e(R_{\mathfrak{p}})$ we note that the completion of $R_{\mathfrak{p}}$ is isomorphic to $k((t))[[s^2,s^3,st]] = k((t))[[s]]$. So we get $e(R_{\mathfrak{p}}) = 1$ and also $e(t,\mathfrak{p},R) = 1$. Later on we will see that $e(R_{\mathfrak{p}}) < e(R)$ depends on the fact that $\mathfrak{p}$ does not contain any element generating a reduction of $\mathfrak{p}$.

(7.8) Integrally closed ideals. For any noetherian ring R and any ideal $\mathfrak{a}$ of R , let v denote the order function with respect to $\mathfrak{a}$, i.e.

$$v(x) = \sup \{n \mid x \in \mathfrak{a}^n\} \quad .$$

Assume that v is a valuation. Then we claim that for any n the ideal $\mathfrak{a}^n$ is integrally closed. Assume the contrary. Then there is an $x \in R$ integral over $\mathfrak{a}^n$ such that $v(x) < n$. If

$$x^m + a_1 x^{m-1} + \ldots + a_m = 0, \; a_i \in \mathfrak{a}^{in}$$

is an equation of integral dependence for x over $\mathfrak{a}^n$ then for any $i \geqq 1$ we have

$$v(a_i x^{m-i}) = v(a_i) - iv(x) + v(x^m) > v(x^m)$$

and therefore

$$v(x^m + a_1 x^{m-1} + \ldots + a_m) = v(x^m) \quad ,$$

contradicting the equation of integral dependence. We note that for any ideal $\mathfrak{a}$ the function v satisfies

$$v(x + y) \geqq \min \{v(x), v(y)\} \; .$$

So a necessary and sufficient condition for v to be a valuation is

$$v(xy) = v(x) + v(y) \quad ,$$

and an easy computation shows that this is equivalent to the condition

$$G(\mathfrak{a},R) = \bigoplus_{n \geqq 0} \mathfrak{a}^n/\mathfrak{a}^{n+1} \quad \text{is a domain.}$$

Therefore the above result on integrally closedness of $\mathfrak{a}^n$ applies in particular to the maximal ideal of a regular local ring, or more generally to any prime ideal generated by a regular sequence.

(7.9) Reductions of power of an ideal. For any ideal $\mathfrak{a}$ of a noetherian ring R let $x_1,\ldots,x_s$ generate some reduction of $\mathfrak{a}$. Then we will show that, for any n, the elements $x_1^n,\ldots,x_s^n$ generate a reduction of $\mathfrak{a}^n$. Obviously $(x_1,\ldots,x_s)^n$ is a reduction of $\mathfrak{a}^n$, and therefore it is enough to show that $(x_1^n,\ldots,x_s^n)$ is a reduction of $(x_1,\ldots,x_s)^n$. Let t be any integer such that $nt \geqq s(n-1)+1$. Then a typical generator of $(x_1,\ldots,x_s)^{nt}$ is a monomial in $x_1,\ldots,x_s$ of degree nt, which necessarily contains a factor x_i^n for some i. This shows that

$$[(x_1,\ldots,x_s)^n]^t = (x_1^n,\ldots,x_s^n) \circ [(x_1,\ldots,x_s)^n]^{t-1} .$$

(7.10) Principal ideals in normal domains are integrally closed. This follows from the fact that if $a \cdot R$ is a non-zero ideal in the domain R, then $B(aR,R) = R[at]$ is isomorphic to the polynomial ring in one variable over R, so if R is normal then the same holds for $B(aR,R)$.

(7.11) Analytically irreducible local domains of dimension one. Let R be a local domain which is analytically irreducible, i.e. the completion $\hat{R}$ of R is again a domain. Then it is well known that the integral closure $\bar{R}$ of R is a finite R-module and moreover it is local, so it is a discrete (rank one) valuation ring. We assume that $\bar{R}$ has the same residue field as R. Let v denote the valuation of $\bar{R}$. Then

$$e(R) = \inf \{v(x) \mid x \in \mathfrak{m}\} = v(\mathfrak{m})$$

where $\mathfrak{m}$ is the maximal ideal of R. In fact, for any $x \in \mathfrak{m}$ it is clear that

$$y \text{ is integral over } xR \iff y/x \in \bar{R} \iff v(y) \geqq v(x) .$$

In particular, if we choose $x \in \mathfrak{m}$ with $v(x) = v(\mathfrak{m})$, then the integral closure of xR is $\mathfrak{m}$. Therefore $e(R) = e(xR,R) = v(\mathfrak{m})$ by the projection formula (note that $v(\mathfrak{m}) = e(x\bar{R},\bar{R})$). In the above argument we have made use of the following general fact: For any domain R, if $\bar{R}$ denotes the integral closure of R in its quotient field, then for any $x \in R$ the integral closure of the ideal xR is given by $x\bar{R} \cap R$.

References - Chapter I

Books

[1] N. Bourbaki, Algèbre Commutative, Eléments de Mathe. 27,28,30, 31, Hermann, Paris 1961-65.

[2] W. Fulton, Intersection Theory, Ergebnisse der Mathematik und ihre Grenzgebiete, 3. Folge Band 2, Springer, Berlin-Heidelberg-New York-Tokyo 1984.

[3] R. Hartshorne, Algebraic Geometry, Graduate Texts in Mathematics vol. 52, Springer, New York-Heidelberg-Berlin 1977.

[4] M. Herrmann -R. Schmidt - W. Vogel, Theorie der normalen Flachheit, Teubner Texte zur Mathematik, Leipzig 1977.

[5] H. Matsumura, Commutative Algebra, Benjamin, New York 1970.

[6] M. Nagata, Local Rings, Interscience Tracts in Pure and Applied Mathematics 13, J. Wiley, New York 1962.

[7] D.G. Northcott, Lessons on Rings, Modules, and Multiplicities, Cambridge University Press 1968.

[8] J.P. Serre, Algèbre Locale - Multiplicités, Lecture Notes in Math. 11, Springer, Heidelberg 1965.

[9] O. Zariski - P. Samuel, Commutative Algebra II, Van Nostrand, Princeton 1960.

Papers

[10] M. Auslander - D. Buchsbaum, Codimension and Multiplicity, Annals of Math. 68 (1958), 625-657.

[11] E. Böger, Einige Bemerkungen zur Theorie der ganz-algebraischen Abhängigkeit von Idealen, Math. Ann. 185 (1979), 303-308.

[12] C. Chevalley, Intersections of algebraic and algebroid varieties, Trans. Amer. Math. Soc. 57 (1945), 1-85.

[13] D. Hilbert, Über die Theorie der algebraischen Formen, Math. Ann. 36 (1890), 473-534.

[14] D.G. Northcott - D. Rees, A note on reduction of ideals with application to the generalized Hilbert function, Proc. Camb. Phil. Soc. 50 (1954), 353-359.

[15] D. Rees, $\mathfrak{a}$-transforms of local rings and a theorem on the multiplicities of ideals, Proc. Camb. Phil. Soc. 57 (1961), 8-17.

[16] L. Robbiano - G. Valla, On the equations defining the tangent cones, Math. Proc. Camb. Phil. Soc. 88 (1980).

[17] P. Samuel, La notion de multiplicité en algèbre et en géométrie algébrique, J. Math. Pures Appl. 30 (1951), 159-275.

Chapter II. $\mathbb{Z}$-GRADED RINGS AND MODULES

This chapter contains some general facts about graded rings that arise in connection with blowing up. We compute the dimensions of these rings and for certain cases we construct special systems of parameters. We also relate the multiplicities and Hilbert functions of the original ring to those of the various graded rings derived from it. Then we recall the theory of standard bases, and finally we show how to translate some well known results on flatness to the graded case. Our presentation uses also ideas of the unpublished thesis of E.C. Dade [5] .

§ 8. Associated graded rings and Rees algebras

By a graded ring or module we will always mean a ring or module graded over $\mathbb{Z}$. If M is graded, M_n will denote the homogenous part of degree n of M .

(8.1) Definition. A graded ring $A = \bigoplus_{n \in \mathbb{Z}} A_n$ is called simple if $A \neq (0)$ and if it does not contain any homogeneous ideal different from A and (0) .

(8.2) Lemma. Let A be a graded ring. Then the following conditions are equivalent:

(i) A is simple .

(ii) A_0 is a field and either $A = A_0$ or $A = A_0[t,t^{-1}]$, where t is algebraically independent over A_0 .

(iii) Every graded A-module M is free.

Proof. (i) $\Rightarrow$ (ii). If A is simple then also A_0 must be simple, hence A_0 is a field. If $A_n = (0)$ for all $n > 0$ then $\bigoplus_{n<0} A_n$ is an ideal different from A and therefore $A = A_0$; similaritly if $A_n = (0)$ for all $n < 0$. If $A \neq A$, let $s > 0$ be minimal such that $A_s \neq (0)$. Then for any $t \in A_s \setminus \{0\}$ we have $t \cdot A = A$ and therefore

$$A_n = \begin{cases} A_0 \cdot t^k & \text{if} \quad n = k \cdot s \\ (0) & \text{if} \quad n \notin s \cdot \mathbf{Z} \end{cases} .$$

Since t is a unit, it is algebraically independent over A_0 .

(ii) ⇒ (iii) . Assume that $A = A_0[t,t^{-1}]$ where t has degree $s > 0$ and A_0 is a field. For any graded A-module M let $\{e_i / i \in I\}$ be a homogeneous basis of $M_\circ \oplus \ldots \oplus M_{s-1}$ over A_0 . We will show that $M = \bigoplus_{i \in I} Ae_i$. Let $x \in M_n$ and write $n = gs + r$, $0 < r < s$. Then $t^{-q} x \in M_r$ and therefore the e_i generate M over A . Assume now that

$$a_1 e_{i_1} + \ldots . + a_t e_{i_t} = 0 \ , \ \{i_1, \ldots, i_t\} \subset I \ , \quad a_j \in A \ .$$

We may assume that all a_j are homogeneous of the same degree, say $s \cdot q$. Multiplying the above equation by t^{-q} we conclude that $t^{-q} a_j = 0$ for all j and consequently $a_j = 0$ for all j .

(iii) ⇒ (i) is obvious.

(8.3) Definition. A homogeneous ideal $\mathfrak{m}$ of a graded ring A will be called a maximal homogeneous ideal of A if it is proper and maximal among the homogenes ideals of A .

(8.4) Corollary. If A is a graded ring and $\mathfrak{m}$ is a maximal homogeneous ideal of A , then $A_0 \cap \mathfrak{m}$ is a maximal ideal of A_0 and either $A/\mathfrak{m} = k$ or $A/\mathfrak{m} \simeq k[T,T^{-1}]$, where $k = A_0/A_0 \cap \mathfrak{m}$ and T is an indeterminate.

(8.5) Definition. For any graded ring A and any ideal $\mathfrak{a}$ of A we put

$$H(\mathfrak{a}) = \bigoplus_{n \in \mathbf{Z}} \mathfrak{a} \cap A_n \quad .$$

So $H(\mathfrak{a})$ is the ideal generated by the homogeneous elements of A . We note that if $\mathfrak{p} \subset A$ is a prime ideal then $H(\mathfrak{p})$ is prime too.

(8.6) Proposition. Let A be a graded ring and let $\mathfrak{q} \subset \mathfrak{p}$ be non-homogeneous prime ideals of A. If $H(\mathfrak{p}) = H(\mathfrak{q})$ then $\mathfrak{p} = \mathfrak{q}$.

Proof. Let S be the set of homogeneous elements of A not contained in $\mathfrak{p}$, put

$$B = A_S/H(\mathfrak{p})A_S$$

and let $f : A \longrightarrow B$ be the canonical homomorphism. B is a graded domain, and if $\mathfrak{a}$ is a non-zero homogeneous ideal of B, then $f^{-1}(\mathfrak{a})$ contains an element of S. This shows that $\mathfrak{a} = B$ and consequently B is simple. Since $H(\mathfrak{p}) = H(\mathfrak{q}) \subset \mathfrak{q}$, $\mathfrak{p}B$ and $\mathfrak{p}B$ are non-zero prime ideals of B. Now, by Lemma (8.2), B has dimension one and therefore $\mathfrak{p}B = \mathfrak{q}B$. It follows that $\mathfrak{p} = \mathfrak{q}$.

(8.7) Definition. For any ring R, any ideal $\mathfrak{a} \subset R$ and any R-module M we put

$$G(\mathfrak{a},M) = \bigoplus_{n \in \mathbb{N}_0} \mathfrak{a}^n M/\mathfrak{a}^{n+1}M ,$$

and

$$B(\mathfrak{a},M) = \bigoplus_{n \in \mathbb{N}_0} (\mathfrak{a}^n M)t^n ,$$

and

$$R(\mathfrak{a},M) = \bigoplus_{n \in \mathbb{Z}} (\mathfrak{a}^n M)t^n .$$

Here we use the convention that $\mathfrak{a}^n = R$ if $n \leq 0$, and we consider $B(\mathfrak{a},M)$ and $R(\mathfrak{a},M)$ to be subgroups of

$$M[t,t^{-1}] = M \otimes_R R[t,t^{-1}] .$$

Furthermore, for any submodule N of M we put

$$G(\mathfrak{a},N \subset M) = \bigoplus_{n \in \mathbb{N}_0} (\mathfrak{a}^n M \cap N) + \mathfrak{a}^{n+1}M/\mathfrak{a}^{n+1}M$$

and

$$B(\mathfrak{a},N\subset M) = \bigoplus_{n\in\mathbb{N}_0} (\mathfrak{a}^n M\cap N)t^n ,$$

and

$$\mathcal{R}(\mathfrak{a},N\subset M) = \bigoplus_{n\in\mathbf{Z}} (\mathfrak{a}^n M\cap N)t^n .$$

<u>(8.8) Remark</u>. $G(\mathfrak{a},R)$, $B(\mathfrak{a},R)$ and $\mathcal{R}(\mathfrak{a},R)$ are graded rings, $G(\mathfrak{a},M)$ is a graded $G(\mathfrak{a},R)$-module, and similarly for $B(\mathfrak{a},M)$ and $\mathcal{R}(\mathfrak{a},M)$. $G(\mathfrak{a},N\subset M)$ (resp. $B(\mathfrak{a},N\subset M)$, $\mathcal{R}(\mathfrak{a},N\subset M)$) is a submodule of $G(\mathfrak{a},M)$ (resp. $B(\mathfrak{a},M)$, $\mathcal{R}(\mathfrak{a},M)$) , and we have natural isomorphisms

(8.8.1) $\quad G(\mathfrak{a},M)/G(\mathfrak{a},N\subset M) \simeq G(\mathfrak{a},M/N)$ over $G(\mathfrak{a},R)$;

(8.8.2) $\quad B(\mathfrak{a},M)/B(\mathfrak{a},N\subset M) \simeq B(\mathfrak{a},M/N)$ over $B(\mathfrak{a},R)$;

(8.8.3) $\quad \mathcal{R}(\mathfrak{a},M)/\mathcal{R}(\mathfrak{a},N\subset M) \simeq \mathcal{R}(\mathfrak{a},M/N)$ over $\mathcal{R}(\mathfrak{a},R)$.

Furthermore, if we view $G(\mathfrak{a};M)$ as an R-module, we have a natural isomorphism

(8.8.4) $\quad B(\mathfrak{a},M) / \mathfrak{a}B(\mathfrak{a},M) \simeq G(\mathfrak{a},M)$;

and if we put $u = t^{-1}$, we have natural isomorphisms

(8.8.5) $\quad \mathcal{R}(\mathfrak{a},M)/u\cdot\mathcal{R}(\mathfrak{a},M) \simeq G(\mathfrak{a},M)$;

and

(8.8.6) $\quad \mathcal{R}(\mathfrak{a},M)/(u-1)\mathcal{R}(\mathfrak{a},M) \simeq M$.

Clearly, if M_1,M_2 are two R-modules, then

$$G(\mathfrak{a},M_1\oplus M_2) \simeq G(\mathfrak{a},M_1)\oplus G(\mathfrak{a},M_2) ;$$

and

$$B(\mathfrak{a},M_1\oplus M_2) \simeq B(\mathfrak{a},M_1)\oplus B(\mathfrak{a},M_2) ,$$

and

$$\mathcal{R}(\mathfrak{a},M_1 \oplus M_2) \simeq \mathcal{R}(\mathfrak{a},M_1) \oplus \mathcal{R}(\mathfrak{a},M_2) \quad .$$

In particular, if M is finitely generated over R, then $G(\mathfrak{a};M)$ (resp. $B(\mathfrak{a},M)$, $\mathcal{R}(\mathfrak{a},M)$) is finitely generated over $G(\mathfrak{a},R)$ (resp. $B(\mathfrak{a},R)$, $\mathcal{R}(\mathfrak{a},M)$). If $\mathfrak{a}$ is generated by $a_1,\dots,a_n$, then $G(\mathfrak{a},R)$, $B(\mathfrak{a},R)$ and $\mathcal{R}(\mathfrak{a},R)$ are finitely generated R-algebras, since then

$$B(\mathfrak{a},R) = R[a_1t,\dots,a_nt]$$

and

$$\mathcal{R}(\mathfrak{a},R) = R[a_1t,\dots,a_nt,u] \quad .$$

We conclude that if R and M are noetherian, then so are $G(\mathfrak{a},R)$, $B(\mathfrak{a},R)$ and $\mathcal{R}(\mathfrak{a},R)$, and $G(\mathfrak{a},M)$ (resp. $B(\mathfrak{a},M)$; $\mathcal{R}(\mathfrak{a},R)$) is a noetherian module over $G(\mathfrak{a},R)$ (resp. $B(\mathfrak{a},M)$, $\mathcal{R}(\mathfrak{a},M)$).

(8.9) Notation. If $R,\mathfrak{a}$ or M are varied in the definition (8.7), we will always use the letter t to denote the indeterminate, and we will write u for t^{-1}.

(8.10) Lemma. a) The elements u and u^{-1} are regular on $\mathcal{R}(\mathfrak{a},M)$. b) If $\mathfrak{m}$ is a maximal homogeneous ideal of $\mathcal{R}(\mathfrak{a},R)$ containing $\mathfrak{a}$, then $u \in \mathfrak{m}$.

Proof. a) is easy. For b) assume $u \notin \mathfrak{m}$. Then by Lemma (8.2) we know that $at \notin \mathfrak{m}$ for some $a \in \mathfrak{a}$, which leads to the contradiction $a = u(at) \notin \mathfrak{m} \cap R$.

(8.11) Corollary. If R is local and $\mathfrak{a}$ proper, then each of the rings $G(\mathfrak{a},R)$, $B(\mathfrak{a},R)$ and $\mathcal{R}(\mathfrak{a},R)$ contains exactly one maximal homogeneous ideal, and the residue class ring for this ideal is a field.

(8.12) Remark. Let R be a noetherian ring, $\mathfrak{a}$ an ideal of R,

M a finitely generated R-module and let $\underline{x} = (x_1,\ldots,x_r)$ be a multiplicity system for $M/\mathfrak{a}M$. Let $x_1^*,\ldots,x_r^*$ be the images of $x_1,\ldots,x_r$ in $R/\mathfrak{a}$ and put $\underline{x}^* = (x_1^*,\ldots,x_r^*)$. Denote by $\mathfrak{a}^*$ the ideal $\bigoplus_{n>0}\mathfrak{a}^n/\mathfrak{a}^{n+1}$ of $G(\mathfrak{a},R)$. Then we have a canonical isomorphism

$$G(\mathfrak{a}^*,G(\mathfrak{a},M)) \simeq G(\mathfrak{a},M) \quad ,$$

from which we derive the equalities

$$(8.12.1) \qquad H^{(i)}[\underline{x}^*,\mathfrak{a}^*,G(\mathfrak{a},M)] = H^{(i)}[\underline{x},\mathfrak{a},M] \quad \text{for } i \geqq 0 \quad ,$$

and

$$(8.12.2) \qquad e(\underline{x}^*,\mathfrak{a}^*,G(\mathfrak{a},M)) = e(\underline{x},\mathfrak{a},M) \quad .$$

This is a trivial but very useful obervation.

§ 9. Dimension

(9.1) Lemma. Let A be a noetherian graded ring and $\mathfrak{p}$ be a prime ideal of A. If $\mathfrak{p} \neq H(\mathfrak{p})$ then

$$ht(\mathfrak{p}) = ht(H(\mathfrak{p})) + 1 \quad .$$

Proof. We use induction on $n = ht(\mathfrak{p})$, and we note that $n \geqq 1$ since $\mathfrak{q}$ is not homogeneous. For $n = 1$ the result is clear, so we assume $n \geqq 2$. Then $\mathfrak{p}$ contains a prime ideal $\mathfrak{q}$ of height $n-1$, and $H(\mathfrak{q}) \subset H(\mathfrak{p})$. If $\mathfrak{q}$ is homogeneous, then since $H(\mathfrak{p})$ is prime we must have $\mathfrak{q} = H(\mathfrak{p})$ and therefore $ht(H(\mathfrak{q})) = n-1$ as asserted. If $\mathfrak{q}$ is not homogeneous then $ht(H(\mathfrak{q})) = n-2$ by inductive assumption, and therefore $ht(H(\mathfrak{p})) \geqq n-1$ by Proposition (8.6), which concludes the proof.

(9.2) Corollary. Let A be a graded noetherian ring and M a finitely generated A-module. If $\mathfrak{p}$ is a non-homogeneous prime ideal of M such that $M_{\mathfrak{p}} \neq 0$ then

$$\dim M_{\mathfrak{p}} = \dim M_{H(\mathfrak{p})} + 1 \quad .$$

Proof. Note that $H(\mathfrak{p}) \supset \mathrm{Ann}_A(M)$ and therefore $M_{H(\mathfrak{p})} \neq 0$. The result now follows from Lemma (9.1) applied to $A/\mathrm{Ann}_A(M)$.

(9.3) Corollary. Assume in addition that for any maximal homogeneous ideal $\mathfrak{m}$ of A the ring $A/\mathfrak{m}$ is a field. Then if M is any graded A-module we have

$$\dim M = \sup\{\mathrm{ht}(\mathfrak{m}/\mathrm{Ann}(M)) \mid \mathfrak{m} \subset A \quad \text{maximal homogeneous}\} \quad .$$

In particular, if A contains a unique maximal homogenous ideal $\mathfrak{m}$ and if $A/\mathfrak{m}$ is a field then $\dim A = \mathrm{ht}(\mathfrak{m})$.

Proof. Passing to $A/\mathrm{Ann}(M)$ we may assume that $M = A$. By the above Lemma we know that for any non-homogeneous prime ideal $\mathfrak{p}$ of A there is a maximal homogeneous ideal $\mathfrak{m}$ such that $\mathrm{ht}(\mathfrak{m}) \geq \mathrm{ht}(\mathfrak{p})$.

(9.4) Example. $\dim k[t,t^{-1}] = 1$ but $\mathrm{ht}(\mathfrak{m}) = 0$.

(9.5) Corollary. Let R be a local ring, $\mathfrak{a}$ a proper ideal of R and $M \neq 0$ finitely generated R-module. Then

$$\dim \mathfrak{R}(\mathfrak{a},M) = \dim G(\mathfrak{a},M) + 1 \quad .$$

Proof. By Lemma (8.10) and Corollary (8.11), $\mathfrak{R}(\mathfrak{a},R)$ contains a unique maximal homogeneous ideal $\mathfrak{m}$ and $\mathfrak{R}(\mathfrak{a},R)/\mathfrak{m}$ is a field. Since u is regular on $\mathfrak{R}(\mathfrak{a},M)$, the assertion follows from the isomorphism (8.8.5).

(9.6) Lemma. For any ideal $\mathfrak{a}$ of a ring R and any R-module M we have:

a) $B(\mathfrak{a},\mathrm{Ann}_R(M) \subset R) = \mathrm{Ann}_{B(\mathfrak{a},R)} B(\mathfrak{a},M)$.

b) $\mathfrak{R}(\mathfrak{a},\mathrm{Ann}_R(M) \subset R) = \mathrm{Ann}_{\mathfrak{R}(\mathfrak{a},R)} \mathfrak{R}(\mathfrak{a},M)$.

Proof. a) Since both ideals are homogeneous, it is enough to check the equality for homogeneous elements. If $x = at^n \in B(\mathfrak{a}, \mathrm{Ann}_R(M) \subset R)$ with $a \in \mathfrak{a}^n \cap \mathrm{Ann}_R(M)$, and if $z = bt^m \in B(\mathfrak{a},M)$ with $b \in \mathfrak{a}^m M$, then clearly $xz = 0$. Conversely, if $y = at^n \in \mathrm{Ann}_{B(\mathfrak{a},R)} B(\mathfrak{a},M)$ and $m \in M$, then $y \cdot (m \cdot t^0) = (am)t^n = 0$, and therefore $a \in \mathrm{Ann}_R(M)$. The proof of b) is literally the same if one allows the degrees m and n to be negative.

Next we want to determine the dimensions of $G(\mathfrak{a},M)$, $B(\mathfrak{a},M)$ and $\mathfrak{R}(\mathfrak{a},M)$ in terms of $\dim M$, and we do this first in the local case. For the analytic case see Corollary 1.4.6.

<u>(9.7) Theorem</u>. Let R be a local ring, $\mathfrak{a}$ a proper ideal of R and M a non-zero, finitely generated R-module. Then we have:

a) $\dim_{G(\mathfrak{a},R)} G(\mathfrak{a},M) = \dim_R M$.

b) $$\dim_{B(\mathfrak{a},R)} B(\mathfrak{a},M) = \begin{cases} \dim_R M + 1 & \text{if } \mathfrak{a} \not\subset \mathfrak{p} \text{ for some } \mathfrak{p} \in \mathrm{Assh}(M), \\ \dim_R M & \text{otherwise}. \end{cases}$$

c) $\dim_{\mathfrak{R}(\mathfrak{a},R)} \mathfrak{R}(\mathfrak{a},M) = \dim_R M + 1$.

Proof. By Corollary (9.5), the isomorphisms (8.8.2) and (8.8.3) and Lemma (9.6) we may assume that $M = R$. To simplify notations, we will write $\dim R$ instead of $\dim_R R$ etc.

a) and c). If $\mathfrak{b}_1, \mathfrak{b}_2$ are ideals of R, then one checks easily that

$$G(\mathfrak{a},\mathfrak{b}_1 \subset R) \cdot G(\mathfrak{a},\mathfrak{b}_2 \subset R) \subset G(\mathfrak{a},\mathfrak{b}_1 \cdot \mathfrak{b}_2 \subset R) .$$

Therefore, if $\mathfrak{p}_1,\ldots,\mathfrak{p}_n$ are minimal prime ideals of R and $(\mathfrak{p}_1 \cdot \ldots \cdot \mathfrak{p}_n)^s = 0$ then

$$\prod_{i=1}^{n} G(\mathfrak{a},\mathfrak{p}_i \subset R)^s = 0 .$$

From this we conclude that any minimal prime $\mathfrak{P}$ of $G(\mathfrak{a},R)$ contains some $G(\mathfrak{a},\mathfrak{p}_i \subset R)$ and therefore

$$\dim G(\mathfrak{a},R) \leq \sup\{\dim G(\mathfrak{a}\cdot R/\mathfrak{p}_i, R/\mathfrak{p}_i) \mid \mathfrak{p}_i \in \mathrm{Min}(R)\} \quad .$$

By the isomorphism (8.8.5), Lemma (8.10) and Corollary (9.5) we know that $\dim G(\mathfrak{a},R) = \dim \mathcal{R}(\mathfrak{a},R) - 1 \geq \dim R$. Therefore, to prove a) and b) it is enough to show that $\dim G(\mathfrak{a},R) \leq \dim R$ for any domain R . This last statement is a consequence of the isomorphism (8.8.4) and the assertion b), which we will prove next.

b). Assume first that R is a domain. If $\mathfrak{a} = 0$ then clearly $\dim B(\mathfrak{a},R) = \dim R$. For $\mathfrak{a} \neq 0$ we use induction on the number n of generators of $\mathfrak{a}$. For $n = 1$, $B(\mathfrak{a},R)$ is isomorphic to a polynomial ring over R in one variable, and therefore $\dim B(\mathfrak{a},R) = \dim R + 1$ (by Hilbert's syzygy theorem or simply by Corollary (9.3) applied to $A = R[t]$) . Assume now that $n > 1$ and let $\mathfrak{a} = a_1R + \ldots + a_nR$. Putting $\mathfrak{a}_0 = a_1R + \ldots + a_{n-1}R$ we have a canonical surjection

$$B(\mathfrak{a}_0,R)[T] \twoheadrightarrow B(\mathfrak{a},R) \quad , \quad T \longmapsto a_nt \quad ,$$

whose kernel contains the non-zero-divisor $a_1T - a_n(at)$. It follows that

$$\dim B(\mathfrak{a},R) \leq \dim B(\mathfrak{a}_0,R) = \dim R + 1 \quad .$$

On the other hand, if $\mathfrak{p}$ is any prime ideal of height 1 of R , then we take any nonzero element $a \in \mathfrak{p} \cap \mathfrak{a}$ and consider the ideal of $B(\mathfrak{a},R)$ generated by at . It is clear that $\dim B(\mathfrak{a},R) > \dim R$ since $R = B(\mathfrak{a},R)/a\cdot t\cdot B(\mathfrak{a},R)$. This completes the proof in the case that R is a domain. For the general case we note first that clearly $\dim B(\mathfrak{a},R) \geq \dim R$ (since any chain of prime ideals $\mathfrak{p}_i$ of R gives a corresponding chain in $B(\mathfrak{a},R)$ by taking $B(\mathfrak{a},\mathfrak{p}_i \subset R)$. Now from Lemma (4.5), e), we know that

$$\dim B(\mathfrak{a},R) = \sup\{\dim B(\mathfrak{a}(R/\mathfrak{p}),R/\mathfrak{p}) \mid \mathfrak{p} \in \mathrm{Min}(R)\} \quad .$$

Therefore, if $\mathfrak{a} \not\subset \mathfrak{p}$ for some $\mathfrak{p} \in \mathrm{Assh}(R)$ then $\dim B(\mathfrak{a},R) = \dim R + 1$ by the first case. So assume finally that $\mathfrak{a} \subset \mathfrak{p}$ for all $\mathfrak{p} \in \mathrm{Assh}(R)$. Then for any $\mathfrak{p} \in \mathrm{Min}(R)$ we get

$$\begin{cases} \dim B(\mathfrak{a}(R/\mathfrak{p}),R/\mathfrak{p}) = \dim R & \text{if } \mathfrak{p} \in \mathrm{Assh}(R) , \\ \dim B(\mathfrak{a}(R/\mathfrak{p}),R/\mathfrak{p}) \leqq \dim (R/\mathfrak{p}) + 1 \leqq \dim R & \text{otherwise.} \end{cases}$$

Hence in this last case we must have $\dim B(\mathfrak{a},R) = \dim R$.

(9.8) Lemma. Let $R \longrightarrow S$ be a flat ring homomorphism and let $\mathfrak{a}$ be an ideal of R . For any R-module M we have

a) $B(\mathfrak{a}S, M \otimes_R S) \simeq B(\mathfrak{a},M) \otimes_R S$.

b) $\mathfrak{R}(\mathfrak{a}S, M \otimes_R S) \simeq \mathfrak{R}(\mathfrak{a},M) \otimes_R S$.

c) $G(\mathfrak{a}S, M \otimes_R S) \simeq G(\mathfrak{a},M) \otimes_R S$.

Proof. Clear.

(9.9) Remark. Given R and $\mathfrak{a}$ as before, let $\mathfrak{p}$ be a prime ideal of R containing $\mathfrak{a}$. Put

$$\mathfrak{p}^* = \mathfrak{p}/\mathfrak{a} \oplus \left(\bigoplus_{n>0} \mathfrak{a}^n/\mathfrak{a}^{n+1} \right) \subset G(\mathfrak{a},R) .$$

Then clearly $\mathfrak{p}^*$ is a prime ideal of $G(\mathfrak{a},R)$. Conversely, for any homogeneous prime ideal $\mathfrak{P}$ of $G(\mathfrak{a},R)$ let $\check{\mathfrak{P}}$ be the unique prime ideal of R containing $\mathfrak{a}$ such that

$$\check{\mathfrak{P}}/\mathfrak{a} = \mathfrak{P} \cap (R/\mathfrak{a}) = \mathfrak{P}_0 .$$

Then $(\check{\mathfrak{P}})^* \subset \mathfrak{P}$ and $(\mathfrak{p}^*)^\vee = \mathfrak{p}$. Consequently $\mathfrak{P} \longmapsto \check{\mathfrak{P}}$ defines a surjection

$$\{\text{homogeneous prime ideals of } G(\mathfrak{a},R)\} \longrightarrow \{\text{prime ideals of } R \text{ containing } \mathfrak{a}\},$$

and this surjection induces a one-to-one mapping on the maximal elements on each set (by Lemma (8.2)).

Now let R be noetherian, let M be a finitely generated R-module and let $\mathfrak{p}$ be a prime ideal containing $\mathfrak{a} + \mathrm{Ann}_R(M)$. Then

$M \neq 0$ and hence $G(\mathfrak{a}R_{\mathfrak{p}}, M_{\mathfrak{p}}) = G(\mathfrak{a},M) \otimes_R R_{\mathfrak{p}} \neq 0$. This implies that $G(\mathfrak{a},M)_{\mathfrak{p}^*} \neq 0$, since $\mathfrak{p}^*$ extends to the unique homogeneous maximal ideal of $G(\mathfrak{a}R_{\mathfrak{p}}, R_{\mathfrak{p}})$. From this we conclude that the mapping $\mathfrak{P} \longrightarrow \check{\mathfrak{P}}$ defined above induces a one-to-one mapping

$$\{\text{maximal ideals of } R \text{ containing } \mathfrak{a} + \mathrm{Ann}_R(M)\}$$
$$\to \{\text{maximal homogeneous ideals of } G(\mathfrak{a},R) \text{ containing } \mathrm{Ann}(G(\mathfrak{a},M))\} .$$

<u>(9.10) Corollary</u>. Let R be a noetherian ring, $\mathfrak{a}$ an ideal of R and $\mathfrak{p}$ a prime ideal of R containing $\mathfrak{a}$. Then

a) $\operatorname{ht} G(\mathfrak{a}, \mathfrak{p} \subset R) = \operatorname{ht}(\mathfrak{p})$.

b) $\operatorname{ht} B(\mathfrak{a}, \mathfrak{p} \subset R) = \begin{cases} \operatorname{ht}(\mathfrak{p}) & \text{if } \mathfrak{a}R_{\mathfrak{p}} \subset \mathfrak{q} \text{ for all } \mathfrak{q} \in \mathrm{Assh}(R_{\mathfrak{p}}) , \\ \operatorname{ht}(\mathfrak{p}) + 1 & \text{otherwise.} \end{cases}$

c) $\operatorname{ht} \mathcal{R}(\mathfrak{a}, \mathfrak{p} \subset R) = \operatorname{ht}(\mathfrak{p}) + 1$.

This is a consequence of Theorem (9.7), Lemma (9.8) and Remark (9.9).

<u>(9.11) Proposition</u>. Let R be a noetherian ring, $\mathfrak{a}$ an ideal of R and let M be a finitely generated R-module. Then we have:

a) If $G(\mathfrak{a},M) \neq 0$ then $\dim G(\mathfrak{a},M) = \sup\{\operatorname{ht}(\mathfrak{p}) \mid \mathfrak{p} \in \operatorname{Spec} R , \mathfrak{p} \supset \mathfrak{a} + \mathrm{Ann}(M)\}$.

b) If $B(\mathfrak{a},M) \neq 0$ then $0 \leq \dim B(\mathfrak{a},M) - \sup\{\operatorname{ht}(\mathfrak{p}) \mid \mathfrak{p} \in \operatorname{Spec} R , \mathfrak{p} \supset \mathfrak{a} + \mathrm{Ann}(M)\} \leq 1$. If $\sup\{\operatorname{ht}(\mathfrak{p})\} = \infty$ then $\dim B(\mathfrak{a},M) = \infty$.

c) If $\mathcal{R}(\mathfrak{a},M) \neq 0$ then $\dim \mathcal{R}(\mathfrak{a},M) = \sup\{\operatorname{ht}(\mathfrak{p}) \mid \mathfrak{p} \in \operatorname{Spec} R , \mathfrak{p} \supset \mathfrak{a} + \mathrm{Ann}(M)\} + 1$.

Proof. This follows from Corollary (9.10) together with Remark (9.9).

§ 10. Homogeneous Parameters

For later applications it will be essential to make explicit computations for systems of parameters of the graded rings defined in § 1, and in many cases we want these systems of parameters to consist of homogeneous elements. In this section we prove the existence of such homogeneous parameters in special cases, and we apply this to reductions of ideals in local rings. We start by defining systems of parameters in the non-local case.

(10.1) Definition. Let R be a noetherian ring of finite Krull dimension. A subset $\{a_1,\ldots,a_d\}$ of R is called a system of parameters of R if

a) $d = \dim R$, b) $a_1R + .. + a_dR \neq R$

c) $\sqrt{a_1R + \ldots + a_dR}$ is the intersection of finitely many prime ideals of height d (necessarily maximal by a)) .

$\{a_1,\ldots,a_s\} \subset R$ is called a set of parameters if it can be extended to a system of parameters. If R is graded, then $\{a_1,\ldots,a_s\}$ will be called a homogeneous system of parameters (resp. homogeneous set of parameters) if it is a system of parameters (resp. set of parameters) consisting of homogeneous elements. If $\{a_1,\ldots,a_s\} \subset R$ is a set of parameters, we often will simply say that $a_1,\ldots,a_s$ are parameters of R , provided there is no danger of confusion.

If M is any finitely generated R-module of finite Krull dimension, we call $\{x_1,\ldots,x_r\}$ a system of parameters (resp. a set of parameters) of M if the images of $x_1,\ldots,x_r$ in $R/\mathrm{Ann}_R(M)$ are a system of parameters (resp. a set of parameters) of this ring.

(10.2) Lemma. Let R be a noetherian ring and let M be a finitely generated R-module of finite Krull dimension. Then M has a system of parameters.

Proof. Replacing R by $R/\mathrm{Ann}_R(M)$ we may assume that $M = R$.

We proceed by induction on $d = \dim R$, and we may assume that $d > 0$. Let $\mathfrak{m}$ be a maximal ideal of height d and let $\mathrm{Min}(R) = \{\mathfrak{p}_1,\ldots,\mathfrak{p}_r\}$. Then there is some $a_1 \in \mathfrak{m} \setminus (\mathfrak{p}_1 \cup \ldots \cup \mathfrak{p}_r)$. Let

$\overline{R} = R/a_1R$ and let $\overline{\mathfrak{p}}$ be any prime ideal of $\overline{R}$ and let $\mathfrak{p}$ be its inverse image in R. Then clearly $ht(\overline{\mathfrak{p}}) = ht(\mathfrak{p}) - 1$. We conclude that $\dim \overline{R} = d - 1$, so we know by induction hypothesis that $\overline{R}$ has a system of parameters $\{\overline{a}_2,\ldots,\overline{a}_d\}$. Taking inverse images a_i of $\overline{a}_i$ for $i = 2,\ldots,r$ it is now clear that $\{a_1,\ldots,a_d\}$ is a system of parameters of R.

<u>(10.3) Lemma</u>. Let R be a noetherian ring of finite Krull dimension. Then if $\{a_1,\ldots,a_s\} \subset R$ is a set of parameters of R we have

$$\dim R/a_1R + \ldots + a_sR = \dim R - s \quad .$$

Proof. Assume that $\{a_1,\ldots,a_d\}$, $d \geq s$, is a system of parameters of R. Putting $\mathfrak{a}_i = a_1R + \ldots + a_iR$ for $i = 1,\ldots,d$ and $\mathfrak{a}_0 = 0$ we have

$$\dim R/\mathfrak{a}_i \geqq \dim R/\mathfrak{a}_{i+1} \geqq \dim R/\mathfrak{a}_i - 1 \quad \text{for} \quad i = 0,\ldots,r-1$$

and

$$\dim R/\mathfrak{a}_d = 0 \quad .$$

It follows that $\dim R/\mathfrak{a}_i = \dim R - i$ for $i = 1,\ldots,d$ and in particular for $i = s$.

<u>(10.4) Corollary</u>. Let R be a noetherian ring of finite Krull dimension and let $\mathfrak{a}$ be an ideal of R containing a set $\{a_1,\ldots,a_s\}$ of parameters of R. Then $\dim R/\mathfrak{a} \leqq \dim R - s$.

<u>(10.5) Remark</u>. If $\varphi : R \longrightarrow \overline{R}$ is a surjective homomorphism with nilpotent kernel then $\{a_1,\ldots,a_s\} \subset R$ is a set (resp. system) of parameters of R if and only if $\{\varphi(a_1),\ldots,\varphi(a_s)\}$ is a set (resp. system) of parameters of $\overline{R}$.

<u>(10.6) Remark</u>. We note that our definition of a system of parameters differs from the one given by Nagata ([4], p. 77).

<u>(10.7) Lemma</u>. Let $A = \bigoplus_{n\in\mathbf{Z}} A_n$ be a graded ring and put $A_+ = \bigoplus_{n>0} A_n$. Let $\mathfrak{p}_1,\dots,\mathfrak{p}_r$ be homogeneous prime ideals of A such that $A_+ \not\subset \mathfrak{p}_i$ for $i = 1,\dots,r$. If $\mathfrak{a}$ is any homogeneous ideal of A such that

$$\mathfrak{a} \not\subset \mathfrak{p}_i \quad \text{for} \quad i = 1,\dots,r \quad ,$$

then there is a homogeneous element a (of positive degree) such that

$$a \in \mathfrak{a} \ , \quad a \notin \mathfrak{p}_1 \cup \dots \cup \mathfrak{p}_r \quad .$$

Proof. We may assume that $\mathfrak{p}_i \not\subset \mathfrak{p}_j$ for $i \neq j$, so we can choose homogeneous elements

$$p_i \notin \mathfrak{p}_i \ , p_i \in \mathfrak{p}_j \quad \text{for} \quad j \neq i \ , \ i,j = 1,\dots,r \quad .$$

Since we assumed $A_+ \not\subset \mathfrak{p}_i$ for all i , we may assume the p_i to be of arbitrary large positive degree. Next we may choose homogeneous elements

$$a_i \in \mathfrak{a} \smallsetminus \mathfrak{p}_i \quad , \quad i = 1,\dots,r \quad .$$

Denoting the degree of a_i by d_i and the degree of p_i by e_i , by the above remark we may assume that

$$d_i + e_i > 0 \quad \text{for} \quad i = 1,\dots,r \quad .$$

If now $m > 0$ is any common multiple of the $d_i + e_i$ and if we put $m_i = m/(d_i + e_i)$, then

$$a = (a_1p_1)^{m_1} + \dots + (a_rp_r)^{m_r}$$

is homogeneous of degree m and $a \in \mathfrak{a}$, $a \notin \mathfrak{p}_1 \cup \dots \cup \mathfrak{p}_r$.

<u>(10.8) Theorem</u>. Let $A = \bigoplus_{n\geq 0} A_n$ be a noetherian graded ring such that A_0 is an artinian local ring and let $\mathfrak{q}$ be a homogeneous ideal

of A such that $\dim A/\mathfrak{q} = 0$. Then we have:

a) A has a homogeneous system of parameters contained in $\mathfrak{q}$.

b) If A_0 has an infinite residue field and $\mathfrak{q}$ is generated by elements of degree 1, then A has a homogeneous system of parameters consisting of elements of degree one contained in $\mathfrak{q}$.

Proof. By Remark (10.5) we may assume that A_0 is a field. For the proof we use induction on $d = \dim A$, and there is nothing to prove if $d = 0$. For $d > 0$ let $\mathrm{Assh}(A) = \{\mathfrak{p}_1,\ldots,\mathfrak{p}_r\}$. Since $d > 0$ we know that

$$\mathfrak{q} \cap A_+ \not\subset \mathfrak{p}_i \quad \text{for} \quad i = 1,\ldots,r$$

(notation as in Lemma (10.7)). By Lemma (10.7) we can choose a homogeneous element $a \in \mathfrak{q}$ such that $a \notin \mathfrak{p}_1 \cup \ldots \cup \mathfrak{p}_r$. Since $\dim A/aA = d-1$, the assertion a) follows from the induction hypothesis. To prove b), note that $\mathfrak{q} \cap A_1 \not\subset \mathfrak{p}_i \cap A_1$, since $\mathfrak{q}$ is generated by $\mathfrak{q} \cap A_1$. Since A_0 is infinite, we conclude that $\mathfrak{q} \cap A_1 \not\subset (\mathfrak{p}_1 \cup \ldots \cup \mathfrak{p}_r) \cap A_1$, so the element a above may be chosen to be of degree 1.

Theorem (10.8) need not be true if A_0 is local ring of positive dimension, as we see from the following example.

<u>(10.9) Example</u>. Let $R = \mathbf{Z}_{(2)}$ and put $A = R[X]/(2X)$, so $\dim A = 1$. If a is any homogeneous nonunit of A , then either $a \in 2 \cdot A$ or $a \in X \cdot A$, so in both cases we have $\dim A/aA \doteq 1$. Therefore A cannot have a homogeneous system of parameters.

<u>(10.10) Definition.</u> Let R be a local ring with residue field k and let $\mathfrak{a}$ be an ideal of R . Then we define

$$s(\mathfrak{a}) = \dim G(\mathfrak{a},R) \otimes_R k \qquad .$$

$s(\mathfrak{a})$ is called the analytic spread of $\mathfrak{a}$.

<u>(10.11) Remark</u>. a) By Theorem (9.7) we know that $s(\mathfrak{a}) \leq \dim R$. Moreover, if $\mathfrak{m}$ is the maximal ideal of R and $\mathfrak{q}$ is an $\mathfrak{m}$-primary ideal of R, then $s(\mathfrak{q}) = \dim R$.

b) Let $a_1,\ldots,a_s \in \mathfrak{a}$ and assume that $a_i \in \mathfrak{a}^{d_i}$, $i = 1,\ldots,s$. Denote the class $a_i + \mathfrak{m}\mathfrak{a}^{d_i} \in G(\mathfrak{a},R) \otimes_R k$ by $\tilde{a}_i$. Then it is easy to see, using Nakayama's lemma, that $\{\tilde{a}_1,\ldots,\tilde{a}_s\}$ is a homogeneous system of parameters if and only if

$$\mathfrak{a}^n = \sum_{i=1}^{s} a_i \mathfrak{a}^{n-d_i} \quad \text{for some } n . \tag{10.11.1}$$

From this description it is clear that $s(\mathfrak{a})$ is the least number of elements in $\mathfrak{a}$ satisfying a relation like (10.11.1).

c) From the definition it is clear that for any homomorphism $R \longrightarrow S$ of local rings and any ideal $\mathfrak{a}$ in R we have

$$s(\mathfrak{a}) \geq s(\mathfrak{a}S) .$$

Moreover, if $R \longrightarrow S$ is flat and local then

$$s(\mathfrak{a}) = s(\mathfrak{a}S) .$$

<u>(10.12) Lemma</u>. Let R be a local ring with maximal ideal $\mathfrak{m}$ and let $\mathfrak{a},\mathfrak{b}$ be ideals of R such that $\mathfrak{b} \subset \mathfrak{a}$. Then $\mathfrak{b}$ is a reduction of $\mathfrak{a}$ if and only if $\mathfrak{b} + \mathfrak{a}\mathfrak{m}$ is a reduction of $\mathfrak{a}$.

Proof. Assume that $(\mathfrak{b}+\mathfrak{m}\mathfrak{a})\mathfrak{a}^n = \mathfrak{a}^{n+1}$ for some n. Then

$$\mathfrak{b}\mathfrak{a}^n + \mathfrak{m}\mathfrak{a}^{n+1} = \mathfrak{a}^{n+1}$$

and therefore $\mathfrak{b}\mathfrak{a}^n = \mathfrak{a}^{n+1}$ by Nakayama's lemma. The converse is obvious.

<u>(10.13) Definition</u>. Let R be a noetherian ring and $\mathfrak{a},\mathfrak{b}$ ideals of R such that $\mathfrak{b} \subset \mathfrak{a}$. $\mathfrak{b}$ is called a minimal reduction of $\mathfrak{a}$ if it is a reduction of $\mathfrak{a}$ and if no ideal properly contained in $\mathfrak{b}$ is a reduction of $\mathfrak{a}$.

(10.14) Theorem. Let R be a local ring with maximal ideal $\mathfrak{m}$ and residue field k and let $\mathfrak{a}$ be an ideal of R. For $a \in \mathfrak{a}$, denote by $a°$ the class of a in $\mathfrak{a}/\mathfrak{m}\mathfrak{a}$. Given $a_1,\ldots,a_s \in \mathfrak{a}$, the following conditions are equivalent:

(i) $a_1,\ldots,a_s$ generate a reduction of $\mathfrak{a}$.

(ii) $\dim(G(\mathfrak{a},R) \otimes_R k)/(a_1°,\ldots,a_s°) = 0$.

Proof. First we note that (ii) holds if and only if $[(G(\mathfrak{a},R) \otimes k)/(a_1°,\ldots,a_s°)]_n = 0$ for large n. Let $\mathfrak{b} = a_1R + \ldots + a_sR$ and let $\mathfrak{b}°$ denote the ideal of $G(\mathfrak{a},R) \otimes_R k$ generated by $a_1°,\ldots,a_s°$. Then

$$(\mathfrak{b}°)_n = \mathfrak{b}\mathfrak{a}^{n-1} + \mathfrak{m}\mathfrak{a}^n/\mathfrak{m}\mathfrak{a}^n \quad .$$

Assume (i). Then $\mathfrak{a}^{m-1} \cdot \mathfrak{b} = \mathfrak{a}^m$ for some m. It follows that

$$[G(\mathfrak{a},R) \otimes k/\mathfrak{b}°]_n = 0 \quad \text{for} \quad n \geq m \quad .$$

Conversely, if (ii) is satisfied, then for some n we have

$$\mathfrak{a}^n \subset \mathfrak{b}\mathfrak{a}^{n-1} + \mathfrak{m}\mathfrak{a}^n \quad ,$$

proving that $\mathfrak{a}^n = \mathfrak{b}\mathfrak{a}^{n-1}$.

(10.15) Corollary. Every reduction $\mathfrak{b}$ of $\mathfrak{a}$ (in particular $\mathfrak{a}$ itself) contains a minimal reduction of $\mathfrak{a}$. If $a_1,\ldots,a_s \in \mathfrak{b}$ are choosen such that

a) $a_1°,\ldots,a_s°$ are linearly independent over k,

b) $\dim(G(\mathfrak{a},R) \otimes k)/(a_1°,\ldots,a_s°) = 0$,

c) s is minimal with respect of b),

then $a_1,\ldots,a_s$ is a minimal system of generators of a minimal reduction of $\mathfrak{a}$ contained in $\mathfrak{b}$.

Proof. The first assertion is a consequence of the second one. To prove the second one we put $\mathfrak{b} = a_1R + \ldots + a_sR$ and first we observe that $\mathfrak{b} \cap \mathfrak{m}\mathfrak{a} = \mathfrak{m}\mathfrak{b}$ by a). If now $\mathfrak{b}' \subset \mathfrak{b}$ is any reduction of $\mathfrak{a}$

then $\mathfrak{b}' + \mathfrak{m}\mathfrak{a} = \mathfrak{b} + \mathfrak{m}\mathfrak{a}$ by c). Therefore

$$\mathfrak{b} \subset (\mathfrak{b}' + \mathfrak{m}\mathfrak{a}) \cap \mathfrak{b} = \mathfrak{b}' + \mathfrak{m}\mathfrak{a} \cap \mathfrak{b} = \mathfrak{b}' + \mathfrak{m}\mathfrak{b} ,$$

and we conclude that $\mathfrak{b}' = \mathfrak{b}$.

(10.16) Remark. If $a_1,\dots,a_s$ generate a minimal reduction of $\mathfrak{a}$ with s minimal, then by Theorem (10.14) we know that $a_1^\circ,\dots,a_s^\circ$ are linearly independent over k . In particular $a_1,\dots,a_s$ are part of a minimal system of generators of $\mathfrak{a}$.

(10.17) Proposition. Assume in addition that R has an infinite residue field. Then $a_1,\dots,a_s \in \mathfrak{a}$ are a minimal set of generators of a minimal reduction of $\mathfrak{a}$ if and only if $a_1^\circ,\dots,a_s^\circ$ are a (homogeneous) system of paramters of $G(\mathfrak{a},R) \otimes_R k$.

Proof. This follows from Corollary (10.15) and Theorem (10.14).

(10.18) Definition. For any ring R and any finitely generated R-module M we denote by $\mu(M)$ the minimal number of generators of M.

(10.19) Corollary. Let R be a local ring with infinite residue field and let $\mathfrak{a}$ be a proper ideal of R . Then for any minimal reduction $\mathfrak{b}$ of $\mathfrak{a}$ we have $\mu(\mathfrak{b}) = s(\mathfrak{a})$.

(10.20) Proposition. For any proper ideal $\mathfrak{a}$ of a local ring R we have

a) $\sup\{ht(\mathfrak{p}) \mid \mathfrak{p} \in Min(R/\mathfrak{a})\} \leq s(\mathfrak{a}) \leq \mu(\mathfrak{a})$

b) $\dim R \leq \dim R/\mathfrak{a} + s(\mathfrak{a})$.

c) If $ht(\mathfrak{a}) = s(\mathfrak{a})$ then $\dim R/\mathfrak{a} + ht(\mathfrak{a}) = \dim R$.

d) Assume that $\dim R/\mathfrak{p} + ht(\mathfrak{p}) = \dim R$ for all $\mathfrak{p} \in Min(R/\mathfrak{a})$. Then $ht(\mathfrak{a}) = s(\mathfrak{a})$ implies $Min(R/\mathfrak{a}) = Assh(R/\mathfrak{a})$.

Proof. a) Let $s = s(\mathfrak{a})$ and assume that

$$\mathfrak{a}^n = \sum_{i=1}^{s} \mathfrak{a}^{n-d_i} a_i \quad \text{for some } n \text{ and some } a_i \in \mathfrak{a}^{d_i}$$

(see Remark (10.11),b)). Then $\sqrt{\mathfrak{a}} = \sqrt{a_1R + \ldots + a_sR}$ and therefore every minimal prime $\mathfrak{p}$ of $\mathfrak{a}$ is minimal over $a_1R + \ldots + a_sR$. It follows now from Krull's principal ideal theorem that $\mathrm{ht}(\mathfrak{p}) \leqq s$. The inequality $s(\mathfrak{a}) \leqq \mu(\mathfrak{a})$ follows from the definition of $s(\mathfrak{a})$.

b) Let $x_1,\ldots,x_r \in R$ and denote the class of x_i in $R/\mathfrak{a}$ by $\overline{x}_i$. $\overline{x}_1,\ldots,\overline{x}_r$ is a system of parameters of $R/\mathfrak{a}$ then clearly

$$\dim G(\mathfrak{a},R) \otimes_R R/\mathfrak{m} = \dim G(\mathfrak{a},R)/(\overline{x}_1,\ldots,\overline{x}_r) \geqq \dim R - r ,$$

which proves b).

c) This follows from b) and the inequality $\dim R/\mathfrak{a} + \mathrm{ht}(\mathfrak{a}) \leqq \dim R$.

d) If $\mathrm{ht}(\mathfrak{a}) = s(\mathfrak{a})$ then we know from a) that

$$\mathrm{ht}(\mathfrak{p}) = \mathrm{ht}(\mathfrak{a}) \quad \text{for all} \quad \mathfrak{p} \in \mathrm{Min}(R/\mathfrak{a}) .$$

Hence by the assumption of d) we know that $\dim R/\mathfrak{p}$ is constant on $\mathrm{Min}(R/\mathfrak{a})$ by c), which concludes the proof.

(10.21) Remark. Later on (20.8) we will show that under certain assumptions $\mathrm{ht}(\mathfrak{a}) = s(\mathfrak{a})$ implies $s(\mathfrak{a}) = \mu(\mathfrak{a})$. As an example for $\mathrm{ht}(\mathfrak{a}) < s(\mathfrak{a}) = \mu(\mathfrak{a})$ we may take $R = k[[X,Y]]/(X \cdot Y)$ and $\mathfrak{a} = X \cdot R$. Then $G(\mathfrak{a},R) \otimes_R k \simeq k[X]$ has dimension 1, whereas $\mathrm{ht}(\mathfrak{a}) = 0$.

(10.22) Definition. Let $\mathfrak{a}$ be a proper ideal in a local ring R with maximal ideal $\mathfrak{m}$ and let M be a finitely generated R-module. For any nonzero $m \in M$ we put

$$\mathrm{ord}(\mathfrak{a},M)(m) = \sup\{n \mid m \in \mathfrak{a}^n M\}$$

and we define

$$\mathrm{ord}(\mathfrak{a},M)(0) = \infty .$$

If $m \neq 0$ and $\mathrm{ord}(\mathfrak{a},M)(m) = d$ we define the initial form $\mathrm{in}(\mathfrak{a},M)(m) \in G(\mathfrak{a},M)$ by

$$\mathrm{in}(\mathfrak{a},M)(m) = m + \mathfrak{a}^{d+1}M \in \mathfrak{a}^d M/\mathfrak{a}^{d+1}M \quad .$$

In the case $M = R$ we will write $\mathrm{ord}(\mathfrak{a})$ resp. $\mathrm{in}(\mathfrak{a})$ instead of $\mathrm{ord}(\mathfrak{a},R)$ resp. $\mathrm{in}(\mathfrak{a},R)$.

For any nonzero $a \in R$ with $\mathrm{ord}(\mathfrak{a})(a) = d$ we define the fibre form $F(\mathfrak{a})(a) \in G(\mathfrak{a},R) \otimes_R R/\mathfrak{m}$ by

$$F(\mathfrak{a})(a) = a + \mathfrak{m}\mathfrak{a}^d \in \mathfrak{a}^d/\mathfrak{m}\mathfrak{a}^d \quad .$$

Finally we put $\mathrm{in}(\mathfrak{a},M)(0) = 0$ and $F(\mathfrak{a})(0) = 0$.

<u>(10.23) Remark</u>. Using the notation of (10.22) we note that $m \neq 0$ implies $\mathrm{in}(\mathfrak{a},M)(m) \neq 0$. If $a \in R$ and $m \in M$ are nonzero elements then

$$\mathrm{in}(\mathfrak{a},R)(a) \cdot \mathrm{in}(\mathfrak{m},M)(m) = \begin{cases} \mathrm{in}(\mathfrak{a},M)(am) \neq 0 \quad \text{if} \quad \mathrm{ord}(\mathfrak{a},M)(am) = \\ \qquad = \mathrm{ord}(\mathfrak{a},R)(a) + \mathrm{ord}(\mathfrak{a},M)(m) \\ 0 \quad \text{otherwise} \end{cases}$$

and similarly for any nonzero $b \in R$ we have

$$F(\mathfrak{a})(a) \cdot F(\mathfrak{a})(b) = \begin{cases} F(\mathfrak{a})(ab) \quad \text{if} \quad \mathrm{ord}(\mathfrak{a})(ab) = \mathrm{ord}(\mathfrak{a})(a) + \\ \qquad\qquad\qquad\qquad + \mathrm{ord}(\mathfrak{a})(b) \\ 0 \quad \text{otherwise} \ . \end{cases}$$

Note that $F(\mathfrak{a})(a)$ may be zero.

<u>(10.24) Proposition</u>. For any local ring R and any proper ideal $\mathfrak{a}$ of R the following conditions are equivalent:

(i) $\dim R = \dim R/\mathfrak{a} + s(\mathfrak{a})$.

(ii) $G(\mathfrak{a},R)$ has a homogeneous system of parameters.

Proof. Let $\mathfrak{m}$ denote the maximal ideal of R . Assume (i) and let $r = \dim R/\mathfrak{a}$, $s = s(\mathfrak{a})$. We choose $x_1,\dots,x_r \in R$ and $a_1,\dots,a_s \in \mathfrak{a}$ such that

$$\begin{cases} x_1 + \mathfrak{a}, \dots, x_r + \mathfrak{a} \text{ is a system of parameters of } R/\mathfrak{a} \text{ and} \\ F(\mathfrak{a})(a_1), \dots, F(\mathfrak{a})(a_s) \text{ is a (homogeneous) system of parameters of} \\ G(\mathfrak{a},R) \otimes_R R/\mathfrak{m} \quad . \end{cases}$$

Then clearly

$$\dim G(\mathfrak{a},R)/(\mathrm{in}(\mathfrak{a})(x_1), \dots, \mathrm{in}(\mathfrak{a})(x_r), \mathrm{in}(\mathfrak{a})(a_1), \dots, \mathrm{in}(\mathfrak{a})(a_s)) = 0$$

and since

$$r + s = \dim R = \dim G(\mathfrak{a},R) \quad ,$$

(ii) follows. Conversely, assume that $G(\mathfrak{a},R)$ has a homogeneous system of parameters $x_1^*, \dots, x_r^*, a_1^*, \dots, a_s^*$ such that

$$\begin{cases} \deg(x_i^*) = 0 \ , & i = 1, \dots, r \ , \\ \deg(a_j^*) > 0 \ , & j = 1, \dots, s \ . \end{cases}$$

Then

$$\dim R/\mathfrak{a} = \dim G(\mathfrak{a},R)/\left(\bigoplus_{n>0} \mathfrak{a}^n/\mathfrak{a}^{n+1} \right) \leqq \dim G(\mathfrak{a},R) - s = \dim R - s$$

and

$$\dim G(\mathfrak{a},R) \otimes_R R/\mathfrak{m} = s(\mathfrak{a}) \leqq \dim G(\mathfrak{a},R) - r = \dim R - r \quad .$$

This gives

$$\dim R/\mathfrak{a} + s(\mathfrak{a}) \leqq 2 \dim R - r - s \quad .$$

Since $r + s = \dim G(\mathfrak{a},R) = \dim R$ we see that

$$\dim R/\mathfrak{a} + s(\mathfrak{a}) \leqq \dim R \quad ,$$

which proves (i) in view of Proposition (10.20), b).

Next we want to describe a system of parameters in $B(\mathfrak{a},R)$ for some special cases. The first to note is that we cannot expect a homogeneous system of parameters in general. More precisely we have:

<u>(10.25) Proposition.</u> Let (R,m) be a local ring and let $\mathfrak{a}$ be an ideal of R such that $\dim B(\mathfrak{a},R) = \dim R + 1$ (see Theorem (9.7)). Let $\mathfrak{M}$ be the unique maximal ideal of $B(\mathfrak{a},R)$ and let $h_1,\dots,h_r \in \mathfrak{M}$ be a system of parameters of $B(\mathfrak{a},R)$. Then the number of homogeneous elements of $\{h_1,\dots,h_r\}$ is at most $\dim R - s(\mathfrak{a}) + 2$.

Proof. Assume without loss of generality that

$h_1,\dots,h_m$ are homogeneous of degree 0,

$h_{m+1},\dots,h_{m+n}$ are homogeneous of positive degree,

$h_{m+n+1},\dots,h_r$ are inhomogeneous.

By Lemma (10.3) we see that

$$\dim B(\mathfrak{a},R)/\mathfrak{m}B(\mathfrak{a},R) \leqq \dim B(\mathfrak{a},R) - m$$

and

$$\dim B(\mathfrak{a},R)/(\mathfrak{a}t)B(\mathfrak{a},R) \leqq \dim B(\mathfrak{a},R) - n \quad .$$

Since $B(\mathfrak{a},M)/\mathfrak{m}B(\mathfrak{a},R) \simeq G(\mathfrak{a},R) \otimes_R R/\mathfrak{m}$ and $B(\mathfrak{a},R)/(\mathfrak{a}t)B(\mathfrak{a},R) \simeq R$, by adding the above inequality we conclude that

$$s(\mathfrak{a}) + \dim R \leqq 2\dim R + 2 - m - n$$

or

$$m + n \leqq \dim R - s(\mathfrak{a}) + 2 \quad .$$

<u>(10.26) Remark</u>. a) If $\mathfrak{a}=\mathfrak{m}$, then $s(\mathfrak{a}) = \dim R$ and consequently there are at most two homogeneous parmeters among any system of parameters of $B(\mathfrak{m},R)$ contained in $\mathfrak{M}$.

b) By the same proof as above we see that in the case $\dim B(\mathfrak{a},R) = \dim R$ there are at most $\dim R - s(\mathfrak{a})$ homogeneous parameters.

<u>(10.27) Lemma</u>. Let $A \subset B$ be graded rings such that $A_n = A \cap B_n$ for all $n \in \mathbf{Z}$. Assume that

a) $A_n = 0$ for $n < 0$ and $A = A_0[A_1]$,

b) B is integral over A.

Then the ideal $B_1 \cdot B$ is integral over $A_1 \cdot B$.

Proof. For any $b \in B_1$ we have an equation of integral dependence

$$b^n + a_1 b^{n-1} + \ldots + a_n = 0 , \quad a_i \in A \quad .$$

Since b is homogeneous of degree one, we may assume that

$$a_i \in (A_1 \cdot A)^i \subset (A_1 \cdot B)^i ,$$

which proves the assertion.

<u>(10.28) Remark</u>. The above Lemma applies in particular to the situation $B(\mathfrak{a},R) \subset B(\mathfrak{b},R)$ where $\mathfrak{b}$ is integral over $\mathfrak{a}$ (see Proposition (4.6)).

<u>(10.29) Proposition</u>. Let R be any ring and let $a_1,\ldots,a_s \in R$. We put $\mathfrak{a} = a_1R + \ldots + a_sR$, and in $B(\mathfrak{a},R)$ we define

$$z_1 = a_1 , \; z_2 = a_2 - a_1 t, \ldots , \; z_s = a_s - a_{s-1} t , \; z_{s+1} = a_s t \quad .$$

and $\underline{z} = (z_1,\ldots,z_s)$. Then the ideal $\underline{z}B(\mathfrak{a},R) + (\mathfrak{a}t)B(\mathfrak{a},R)$ is integral over $\underline{z}B(\mathfrak{a},R)$.

Proof. We use induction on s , the case $s = 1$ being obvious. For $s > 1$ we show that

$$\text{(10.29.1)} \qquad a_s^s = \sum_{i=0}^{s-1} z_{i-1} \, z_{s+1}^{s-i-1} a_s^i \quad .$$

In fact, putting $a_0 = 0$ so that $z_1 = a_1 - a_0 t$, we have

$$\sum_{i=0}^{s-1} z_{i+1} z_{s+1}^{s-1-i} a_s^i = \sum_{i=0}^{s-1} (a_{i+1}-a_i t) a_s^{s-1-i} t^{s-1-i} a_s^i =$$

$$a_s^{s-1} \left(\sum_{i=0}^{s-1} a_{i+1} t^{s-1-i} - \sum_{i=0}^{s-1} a_i t^{s-i} \right) = a_s^s \quad .$$

This shows that a_s , and hence $a_{s-1}t = a_s - z_s$, is integral over $\underline{z}B(\mathfrak{a},R)$. Putting $\mathfrak{a}' = a_1R + \ldots + a_{s-1}R$, we know from the inductive assumption that $a_1,\ldots,a_{s-1}$ are integrally dependent on $(z_1,\ldots,z_{s-1}, a_{s-1}t)$ in $B(\mathfrak{a}',R)$ and a forteriori in $B(\mathfrak{a},R)$. So the result follows from transitivity of integral dependence.

<u>(10.30) Proposition</u>. Let $(R,\mathfrak{m})$ be a local ring and let $x_1,\ldots,x_s, y_1,\ldots,y_r$ $s \geq 1$ be a system of parameters of R . Let $\mathfrak{a}$ be an ideal of R which is integrally dependent on $x_1R + \ldots + x_sR$. Then putting

$$z_1 = x_1,\ z_2 = x_2 - x_1t, \ldots, z_s = x_s - x_{s-1}t,\ z_{s+1} = x_st$$

and

$$\underline{y} = (y_1,\ldots,y_r)\ ,\ \underline{z} = (z_1,\ldots,z_s)\ ,\ \underline{x} = (x_1,\ldots,x_s)$$

we have:

a) $\underline{x}B(\mathfrak{a},R) + \underline{y}B(\mathfrak{a},R) + (\mathfrak{a}t)B(\mathfrak{a},R)$ is integrally dependent on $\underline{y}B(\mathfrak{a},R) + \underline{z}B(\mathfrak{a},R)$.

b) $(\underline{y},\underline{z})$ is a system of parameters of $B(\mathfrak{a},R)$.

Proof. Note first that $s \geq 1$ implies $\dim B(\mathfrak{a},R) = \dim R + 1$. Now obviously the radical of $\underline{x}B(\mathfrak{a},R) + \underline{y}B(\mathfrak{a},R) + (\mathfrak{a}t)B(\mathfrak{a},R)$ is the unique maximal homogeneous ideal of $B(\mathfrak{a},R)$, hence b) is a consequence of a). Now consider the integral extension

$$B(\underline{x}R,R) \subset B(\mathfrak{a},R) \quad .$$

By Lemma (10.27), $(\mathfrak{a}t)B(\mathfrak{a}R,R)$ is integral over $(\underline{x}Rt)B(\underline{x}R,R)$, so we are reduced to the case $\mathfrak{a} = \underline{x}R$. Finally in this case the assertion follows from Propositon (10.29).

§ 11. Regular sequences on graded modules

The purpose of this section is to extend the well-known facts about regular sequences, Cohen-Macaulay modules etc. to the graded case. We start with the graded version of classical results by Krull, Nakayama and Zariski.

(11.1) Theorem. Let $A = \bigoplus_{n \in \mathbb{Z}} A_n$ be a graded noetherian ring and let $\mathfrak{a} \subset A$ be a homogeneous ideal. Then the following conditions are equivalent:

(i) Every finitely generated graded A-module M is separated in the $\mathfrak{a}$-adic topology.

(ii) $\mathfrak{a}$ is contained in every maximal homogeneous ideal of A.

(iii) For every finitely generated graded A-module M, the equation $\mathfrak{a}M = M$ implies $M = 0$.

Proof. We may assume that $A \neq \mathfrak{a}$. For (i) ⇒ (ii) assume that $\mathfrak{a} \not\subset \mathfrak{m}$ for some maximal homogeneous ideal $\mathfrak{m}$ of A. Then $\mathfrak{a}^n + \mathfrak{m} = A$ for every $n \geq 1$, so $\mathfrak{m} = A$ by (i), which is a contradiction.

(ii) ⇒ (iii). Let M be generated by $z_1, \ldots, z_n$. If $\mathfrak{a}M = M$ then there are $a_{ij} \in \mathfrak{a}$, $i,j = 1,\ldots,n$, such that

$$z_i = \sum_{j=1}^{n} a_{ij} z_j , \quad i = 1,\ldots,n \quad .$$

Denoting by C the $n \times n$-matrix $(\delta_{ij} - a_{ij})$ we conclude that

$$\det(C) \cdot M = 0 \quad .$$

By (ii), $\det(C)$ is not contained in any maximal homogeneous ideal of A. Since $\operatorname{Ann}_A(M)$ is homogeneous, this shows that $\operatorname{Ann}_A(M) = A$ and consequently $M = 0$.

(iii) ⇒ (i). Given a finitely generated graded A-module M, let

$$N = \bigcap_{n \geq 0} \mathfrak{a}^n M \quad .$$

By the Artin-Rees lemma we know that

$$\mathfrak{a}^n M \cap N = \mathfrak{a}^{n-r}(\mathfrak{a}^r M \cap N) \subset \mathfrak{a}^{n-r} N$$

for some fixed r and for all $n \geqq r$. It follows that $\mathfrak{a}N = N$ and hence $N = 0$ by (iii).

<u>(11.2) Theorem</u>. Let R be a noetherian ring and M a finitely generated R-module. Let $\underline{x} = (x_1, \ldots, x_r)$ be a sequence of elements of R satisfying

a) $\underline{x}M \neq M$;

b) for every $i \in \{1, \ldots, r\}$ the module $M/x_1M + \ldots + x_{i-1}M$ is separated in the x_iR-adic topology;

c) for every $\mathfrak{p} \in \mathrm{Ass}(M/\underline{x}M)$, the sequence $\underline{x}_{\mathfrak{p}} = (\varphi_{\mathfrak{p}}(x_1), \ldots, \varphi_{\mathfrak{p}}(x_r))$ is regular on $M_{\mathfrak{p}}$, where $\varphi_{\mathfrak{p}} : R \longrightarrow R_{\mathfrak{p}}$ denotes the canonical homomorphism.

Then $\underline{x}$ is a regular sequence on M .

Proof. We use induction on r , the case $r = 0$ being trivial. So assume that $r \geqq 1$. Assume that for some $u \in M$

$$ux_r \in (x_1, \ldots, x_{r-1})M \quad .$$

We will first show that

(11.2.1) $\qquad u \in (x_1, \ldots, x_{r-1})M + x_r^n M$ for all n .

For $n = 0$ this is clear, so assume that

$$u = \sum_{i=1}^{r-1} x_i u_i + x_r^n u_r \ , \quad u_1, \ldots, u_r \in M$$

for some $n \geqq 0$. Given $\mathfrak{p} \in \mathrm{Ass}(M/\underline{x}M)$ we know from c) that

$$(\varphi_{\mathfrak{p}}(x_1), \ldots, \varphi_{\mathfrak{p}}(x_{r-1}), \ \varphi_{\mathfrak{p}}(x_r^{n+1})) \text{ is regular on } M_{\mathfrak{p}} \quad .$$

Since

$$ux_r = \sum_{i=1}^{r-1} x_i x_r u_i + x_r^{n+1} u_r \in (x_1,\dots,x_{r-1})M$$

we conclude that there is some $s \in R \setminus \bigcup_{\mathfrak{p} \in \mathrm{Ass}(M/\underline{x}M)} \mathfrak{p}$ such that

$$su_r \in (x_1,\dots,x_{r-1})M \subset (x_1,\dots,x_r)M \quad .$$

Since s is not a zero-divisor on $M/(x_1,\dots,x_r)M$, we have

$$u_r = \sum_{i=1}^{r} v_i x_i \ , \quad v_i \in M \quad .$$

It follows that

$$u = \sum_{i=1}^{r-1} x_i(u_i + x_r^n v_i) + x_r^{n+1} v_r \quad (x_1,\dots,x_{r-1})M + x_r^{n+1}M \quad .$$

This proves (11.2.1), and together with b) this shows that x_r is regular on $\overline{M} = M/(x_1,\dots,x_{r-1})M$. To conclude the proof, we will show that c) holds for M if $\underline{x}$ is replaced by $(x_1,\dots,x_{r-1})$. Once this is shown, the inductive hypothesis implies that $(x_1,\dots,x_{r-1})$ is a regular sequence on M , so the theorem follows. To prove the above assertion, we will show that

(11.2.2) every associated prime of $\overline{M} = M/(x_1,\dots,x_{r-1})M$ is contained in some associated prime of $M/\underline{x}M$.

So let y be any zero-divisor on $\overline{M}$ and choose $u \in M \setminus (x_1,\dots,x_{r-1})M$ such that $yu \in (x_1,\dots,x_{r-1})M$. Since $u \notin (x_1,\dots,x_{r-1})M$, by b) we may choose n maximal such that

$$u = u_1 + x_r^n u_2 \ , \quad u_1 \in (x_1,\dots,x_{r-1})M \quad .$$

Then clearly $u_2 \notin \underline{x}M$. But $yx_r^n u_2 \in (x_1,\dots,x_{r-1})M$, and since x_r^n is regular on $\overline{M}$, it follws that $yu_2 \in \underline{x}M$, i.e. y is a zero-divisor on $M/\underline{x}M$. This proves (11.2.2) and thereby completes the proof of the theorem.

(11.3) Corollary. Let A be a graded noetherian ring, $M \neq 0$ a finitely generated graded A-module and $\underline{x} = (x_1,\ldots,x_r)$ a sequence of homogeneous elements of A. Assume that every maximal homogeneous ideal $\mathfrak{m}$ of A satisfies

a) $\underline{x}A \subset \mathfrak{m}$;

b) $\underline{x}_{\mathfrak{m}}$ is a regular sequence on $M_{\mathfrak{m}}$ provided $M_{\mathfrak{m}} \neq 0$ (notation as in (11.2)).

Then $\underline{x}$ is a regular sequence on M.

Proof. By Theorem (11.1), (ii) $\Rightarrow$ (iii), we know that $\underline{x}M \neq M$ and moreover that any finitely generated graded A-module is seperated in the x_iA-adic topology for any $i = 1,\ldots,r$. If $\mathfrak{p}$ is any associated prime ideal of $M/\underline{x}M$, it is homogeneous and therefore it is contained in some maximal homogeneous ideal of A. Assumption b) implies therefore that $\underline{x}_{\mathfrak{p}}$ is regular on $M_{\mathfrak{p}}$ for any $\mathfrak{p} \in \mathrm{Ass}(M/\underline{x}M)$, so the corollary follows from Theorem (11.2).

Next we want to make use of the natural grading on the Koszul homology of a graded module with respect to a sequence of homogeneous elements. First we recall some standard notations.

(11.4) Definition. Let A be a graded ring, M a graded A-module and let d be any integer. Then we define $M(d)$ to be the graded A-module given by

$$M(d)_n = M_{d+n}\,, \quad n \in \mathbb{Z} \quad .$$

(11.5) Definition. Let A be a graded ring, M a graded A-module and let S be a multiplicatively closed subset of A consisting of homogeneous elements. Then we define

$$M_{(S)} = (M_S)_0 = \left\{ \frac{m}{s} \in M_S \;\middle|\; m \in M \text{ and } s \in S \text{ are homogeneous of the same degree} \right\} .$$

If $\mathfrak{p}$ is a homogeneous prime ideal, then we put $M_{(\mathfrak{p})} = M_{(S)}$, when S is the set of homogeneous elements of $A \smallsetminus \mathfrak{p}$.

(11.6) Definition. Let A be a graded ring, M a graded A-module and $\underline{x} = (x_1,\dots,x_r)$ a sequence of homogeneous elements of A. Let $d_j = \deg x_j$ for $j = 1,\dots,r$. The graded Koszul complex $K.(\underline{x},M)$ of $\underline{x}$ with respect to M is defined by

$$K_i(\underline{x},M) = \bigwedge^i \left(\bigoplus_{j=1}^{r} M(-d_j) \right), \quad i \in \mathbf{Z}$$

and

$$\partial_i(m_{j_1} \wedge \dots \wedge m_{j_i}) = \sum_{k=1}^{i} (-1)^{k+1} x_{j_k} \cdot (m_{j_1} \wedge \dots \wedge \hat{m}_{j_k} \wedge \dots \wedge m_{j_i}) .$$

(11.7) Remark. Except for the grading, the above definition of $K.(\underline{x},M)$ is the ordinary one, of course. It is easy to check that the homomorphisms $\partial_i : K_i(\underline{x},M) \longrightarrow K_{i-1}(\underline{x},M)$ are homogeneous (of degree zero). Therefore the grading of $K.(\underline{x},M)$ induces a natural grading on the homology, which we denote by $H.(\underline{x},M)$ as usual. So the symbol $H_i(\underline{x},M)_n$ denotes the degree n part of the graded A-module $\operatorname{Ker} \partial_i / \operatorname{Im} \partial_{i+1}$.

(11.8) Lemma. Let A be a graded ring and let M be a graded A-module. Then the following conditions are equivalent:

(i) $M = 0$.

(ii) $M_{(\mathfrak{m})} = 0$ for every maximal homogeneous ideal $\mathfrak{m}$ of A.

(iii) $M_{\mathfrak{m}} = 0$ for every maximal homogeneous ideal $\mathfrak{m}$ of A.

Proof. Clearly (i) ⇒ (ii) ⇒ (iii). Now (iii) signifies that $\operatorname{Ann}_A(M)$ is not contained in any maximal homogeneous ideal of A and therefor $M = 0$.

(11.9) Proposition. Let A be a graded noetherian ring, M a finitely generated graded A-module and $\underline{x} = (x_1,\dots,x_r)$ a sequence of homogeneous elements of A. If $\underline{x}A$ is contained in every maximal homogeneous ideal of A, then the following conditions are equivalent:

(i) $\underline{x}$ is a regular sequence on M.

(ii) $H_i(\underline{x},M) = 0$ for all i .

(iii) $H_1(\underline{x},M) = 0$.

If morover, $\underline{x}$ is a multiplicity system on M , then these conditions are equivalent to

(iv) $e(\underline{x};M) = \lambda_A(M/\underline{x}M)$.

Proof. The implication (i) ⇒ (ii) is well-known, and (ii) ⇒ (iii) is obvious. To prove (iii) ⇒ (i), let $\mathfrak{m}$ be any maximal homogeneous ideal of A for which $M_\mathfrak{m} \neq 0$. As before we denote by $\varphi_\mathfrak{m} : A \longrightarrow A_\mathfrak{m}$ the canonical homomorphism and we put $\underline{x}_\mathfrak{m} = (\varphi_\mathfrak{m}(x_1),\ldots,\varphi_\mathfrak{m}(x_r))$. Then

$$H_1(\underline{x}_\mathfrak{m},M_\mathfrak{m}) \simeq H_1(\underline{x},M) \otimes_A A_\mathfrak{m} = 0$$

by (iii), and therefore $\underline{x}_\mathfrak{m}$ is a regular sequence on $M_\mathfrak{m}$ by (1.5). Now (i) follows from Corollary (11.3). Assume now in addition that $\underline{x}$ is a multiplicity system on M , $\mathrm{Ass}(M/\underline{x}M)$ consists of a finite number of maximal homogeneous ideals of A , say $\mathfrak{m}_1,\ldots,\mathfrak{m}_n$. Let $S = A \setminus (\mathfrak{m}_1 \cup \ldots \cup \mathfrak{m}_n)$. Then $\lambda_A(M/\underline{x}M)$ does not change if A is replaced by A_S , and the same is true for $e(\underline{x};M)$ (Proposition (1.6)). If $\varphi_S : A \longrightarrow A_S$ denotes the canonical homomorphism again and $\underline{x}_S = (\varphi_S(x_1),\ldots,\varphi_S(x_r))$, we know from (1.5) that $\underline{x}_S$ is a regular sequence on M_S . If $\mathfrak{m}$ is any maximal homogeneous ideal of A , then $\underline{x}A \subset \mathfrak{m}$ implies that either $M_\mathfrak{m} = 0$ or $\mathfrak{m} \in \{\mathfrak{m}_1,\ldots,\mathfrak{m}_n\}$. It follows that, for any such $\mathfrak{m}$, either $M_\mathfrak{m} = 0$ or $\underline{x}_\mathfrak{m}$ is a regular sequence on $M_\mathfrak{m}$. This proves (iv) ⇒ (i) in view of Corollary (11.3). Finally, (i) implies that $\underline{x}_S$ is a regular sequence on M_S , which implies (iv) by (1.5) again.

<u>(11.10) Definition</u>. Let R be a noetherian ring and M a finitely generated non-zero R-module. M is said to be Cohen-Macaulay (over R) if for each prime ideal $\mathfrak{p}$ of R either $M_\mathfrak{p} = 0$ or $M_\mathfrak{p}$ is a Cohen-Macaulay $R_\mathfrak{p}$-module.

<u>(11.11) Theorem</u>. Let A be a graded noetherian ring and M a finitely generated graded A-module. Assume that $M \neq 0$ and that for each

maximal homogeneous ideal $\mathfrak{m}$ of A we have

$$\begin{cases} M_{\mathfrak{m}} = 0 \text{ or} \\ M_{\mathfrak{m}} \text{ is a Cohen-Macaulay } A_{\mathfrak{m}}\text{-module.} \end{cases}$$

Then M is a Cohen-Macauly A-module.

Proof. Let $\mathfrak{p}$ be any prime ideal of A such that $M_{\mathfrak{p}} \neq 0$. If $\mathfrak{p}$ is homogeneous, then $\mathfrak{p}$ is contained in some maximal homogeneous ideal of A and hence $M_{\mathfrak{p}}$ is Cohen-Macaulay by assumption. For non-homogeneous $\mathfrak{p}$ we prove the assertion by induction on $\dim M_{H(\mathfrak{p})}$, when $H(\mathfrak{p})$ denotes the prime ideal of A generated by all homogeneous elements of $\mathfrak{p}$. Since $\mathfrak{p}$ is non-homogeneous, $\mathfrak{p}$ is not an associated prime ideal of M and therefore clearly $\operatorname{depth} M_{\mathfrak{p}} > 0$. If now $\dim M_{H(\mathfrak{p})} = 0$ then $\dim M_{\mathfrak{p}} = 1$ by Corollary (9.2), showing that $M_{\mathfrak{p}}$ is Cohen-Macaulay. Assume now that $\dim M_{H(\mathfrak{p})} > 0$. Let S be the set of homogeneous elements of $A \setminus \mathfrak{p}$. Replacing A by A_S we may assume that $H(\mathfrak{p})$ is the unique maximal homogeneous ideal of A. Since $M_{H(\mathfrak{p})}$ is Cohen-Macaulay and $\dim M_{H(\mathfrak{p})} > 0$ we know that $H(\mathfrak{p})$ is not associated to M. By assumption we know that $A/H(\mathfrak{p}) \simeq k[T,T^{-1}]$ (see Lemma (8.2)), so putting $A_+ = \bigoplus_{n>0} A_n$ we know that $A_+ \not\subset H(\mathfrak{p})$. If now $\mathfrak{q}$ is any associated prime of M we have $\mathfrak{q} \subset H(\mathfrak{p})$ and hence $A_+ \not\subset \mathfrak{q}$. Using Lemma (10.7) we see that there is a homogeneous element $x \in H(\mathfrak{p})$ which is not contained in any associated prime of M, i.e. x is M-regular. Therefore $(M/xM)_{\mathfrak{p}} \cong M_{\mathfrak{p}}/xM_{\mathfrak{p}}$ is Cohen-Macaulay by inductive assumption, and since x is $M_{\mathfrak{p}}$-regular, we conclude that $M_{\mathfrak{p}}$ is Cohen-Macaulay.

<u>(11.12) Corollary</u>. Let A be a graded ring having a unique maximal homogeneous ideal $\mathfrak{m}$, and let M be a finitely generated graded A-module. Assume that M has a system of parameters $\underline{x} = (x_1,\ldots,x_r)$ such that $\sqrt{\operatorname{Ann}_A(M) + \underline{x}A} = \mathfrak{m}$. (Note that this implies that $A/\mathfrak{m}$ is a field). Then the following conditions are equivalent:

(i) M is Cohen-Macaulay.

(ii) $M_{\mathfrak{m}}$ is Cohen-Macaulay.

(iii) $e(\underline{x};M) = \lambda_A(M/\underline{x}M)$.

If, moreover, $x_1,\dots,x_r$ are homogeneous, then these conditions are equivalent to

(iv) $\underline{x}$ is a regular sequence on M.

Proof. (i)$\Longleftrightarrow$(ii) was shown in Theorem (11.11). Let $\varphi_{\mathfrak{m}} : A \longrightarrow A_{\mathfrak{m}}$ be the canonical homomorphism and put $\underline{x}_{\mathfrak{m}} = (\varphi_{\mathfrak{m}}(x_1),\dots,\varphi_{\mathfrak{m}}(x_r))$. Then $\underline{x}_{\mathfrak{m}}$ is a system of parameters of $M_{\mathfrak{m}}$, and moreover the assumption $\sqrt{\operatorname{Ann}_A(M) + \underline{x}A} = \mathfrak{m}$ implies

$$\lambda_{A_{\mathfrak{m}}}(M_{\mathfrak{m}}/\underline{x}_{\mathfrak{m}}M_{\mathfrak{m}}) = \lambda_A(M/\underline{x}M)$$

and

$$e(\underline{x}_{\mathfrak{m}};M_{\mathfrak{m}}) = e(\underline{x};M) \quad \text{(by Proposition (1.6))} \quad .$$

So from Proposition (1.5) we know that (iii) holds if and only if $\underline{x}_{\mathfrak{m}}$ is a regular sequence on $M_{\mathfrak{m}}$, proving (ii) $\Longleftrightarrow$ (iii). Finally if the x_i are homogeneous, then (iii) $\Longleftrightarrow$ (iv) by Proposition (11.9).

As an easy application of Corollary (11.12) we obtain the following well-known result:

<u>(11.13) Corollary</u>. If R is a local Cohen-Macaulay ring then the polynomial ring $R[T_1,\dots,T_n]$ is Cohen-Macaulay.

Proof. If $x_1,\dots,x_d$ is a system of parameters of R then $x_1,\dots,x_d, T_1,\dots,T_n$ is a homogeneous system of parameters of $R[T_1,\dots,T_n]$ which is a regular sequence.

<u>(11.14) Corollary</u>. Let M be a graded finitely generated Cohen-Macaulay A-module, where A is noetherian and A_0 local. If x is a homogeneous element of A such that $\dim M/xM = \dim M - 1$, the following statements hold:

(i) x is M-regular.

(ii) M/xM is Cohen-Macaulay.

(11.15) Remark. a) In the situation of Corollary (11.12), if $\underline{x} = (x_1,\dots,x_r)$ is any homogeneous system of parameters for M then necessarily $\sqrt{\mathrm{Ann}_A(M) + \underline{x}A} = \mathfrak{m}$ since $\underline{x}A$ is homogeneous.

b) The condition (iii) in Corollary (11.12) could be replaced by the same condition for all systems of parameters $\underline{y}$ of M satisfying $\sqrt{\mathrm{Ann}_A(M) + \underline{y}A} = \mathfrak{m}$ (provided there is at least one with this property); similarly for (iv).

c) Theorem (11.11) allows the following generalisation, to be proved in Chapter VII (see Cor. (33.26). If $\mathfrak{p}$ is any non-homogeneous prime ideal of A such that $M_{\mathfrak{p}} \neq 0$ then

$$\mathrm{depth}_{A_{\mathfrak{p}}}(M_{\mathfrak{p}}) = \mathrm{depth}_{A_{H(\mathfrak{p})}}(M_{H(\mathfrak{p})}) + 1 \quad .$$

(11.16) Proposition. Let R be a noetherian ring, $\mathfrak{a}$ a proper ideal of R and M a finitely generated R-module. Assume that $M \neq 0$ and that $\mathfrak{a}$ is contained in the Jacobson radical of R. Then:

a) $G(\mathfrak{a},M)$ is Cohen-Macaulay if and only if $\mathbf{R}(\mathfrak{a},M)$ is Cohen-Macaulay.

b) If $G(\mathfrak{a},M)$ is Cohen-Macaulay, then also M is Cohen-Macaulay.

Proof. Note first that $M \neq 0$ implies $G(\mathfrak{a},M) \neq 0$ and $\mathbf{R}(\mathfrak{a},M) \neq 0$. To prove a), let $\mathfrak{M}$ be any maximal homogeneous ideal of $\mathbf{R}(\mathfrak{a},R)$ and let $\mathfrak{m} = \mathfrak{M} \cap R$. Consider $G(\mathfrak{a},M) \simeq \mathbf{R}(\mathfrak{a},M)/u\cdot\mathbf{R}(\mathfrak{a},M)$ (see (8.8.5)) as an $\mathbf{R}(\mathfrak{a},R)$-module. As such we have

$$\sqrt{\mathrm{Ann}(G(\mathfrak{a},M))} = \sqrt{\mathrm{Ann}\ \mathbf{R}(\mathfrak{a},M) + uR(\mathfrak{a},R)}$$

and since $u \in \mathfrak{M}$ (Lemma (8.10), note that $\mathfrak{a} \subset \mathfrak{M}$ by assumption) we see that

$$\mathbf{R}(\mathfrak{a},M)_{\mathfrak{M}} = 0 \iff G(\mathfrak{a},M)_{\mathfrak{M}} = 0 \quad .$$

If $\mathbf{R}(\mathfrak{a},M)_{\mathfrak{M}} \neq 0$ then

$$\dim \mathbf{R}(\mathfrak{a},M)_{\mathfrak{M}} = \dim G(\mathfrak{a},M)_{\mathfrak{M}} + 1$$

by Corollary (9.5), and

$$\operatorname{depth} \mathcal{R}(\mathfrak{a},M)_{\mathfrak{m}} = \operatorname{depth} G(\mathfrak{a},M)_{\mathfrak{m}} + 1$$

by Lemma (8.10). This proves a). To show b), let $\mathfrak{m}$ be a maximal ideal of R such that $M_{\mathfrak{m}} \neq 0$. Then $G(\mathfrak{a} R_{\mathfrak{m}},M_{\mathfrak{m}}) \simeq G(\mathfrak{a},M) \otimes_R R_{\mathfrak{m}}$ by Lemma (9.8), and this module is different from zero since $\mathfrak{a} \in \mathfrak{m}$. By assumption, $G(\mathfrak{a} R_{\mathfrak{m}},M_{\mathfrak{m}})$ is Cohen-Macaulay, and the same is true for $\mathcal{R}(\mathfrak{a}R_{\mathfrak{m}},M_{\mathfrak{m}})$ by a). By (8.8.6) we know that

$$\mathcal{R}(\mathfrak{a}R_{\mathfrak{m}},R_{\mathfrak{m}}) / (u-1)\, \mathcal{R}(\mathfrak{a}R_{\mathfrak{m}},R_{\mathfrak{m}}) \simeq R_{\mathfrak{m}} \quad ,$$

so $\mathcal{R}(\mathfrak{a}R_{\mathfrak{m}},R_{\mathfrak{m}})$ contains a unique maximal ideal $\mathfrak{N}$ such that $u-1 \in \mathfrak{N}$ and $\mathfrak{N}/(u-1)\,\mathcal{R}(\mathfrak{a}R_{\mathfrak{m}},R_{\mathfrak{m}}) = \mathfrak{m}R_{\mathfrak{m}}$. Now $\mathcal{R}(\mathfrak{a}R_{\mathfrak{m}},M_{\mathfrak{m}})_{\mathfrak{N}}$ is Cohen-Macaulay by a) and $u-1$ is a non-zero divisor on this module by Lemma (8.10). We conclude that

$$M_{\mathfrak{m}} \simeq \mathcal{R}(\mathfrak{a}R_{\mathfrak{m}},M_{\mathfrak{m}}) / (u-1)\mathcal{R}(\mathfrak{a}R_{\mathfrak{m}},M_{\mathfrak{m}}) \simeq \mathcal{R}(\mathfrak{a}R_{\mathfrak{m}},M_{\mathfrak{m}})_{\mathfrak{N}} / (u-1)\mathcal{R}(\mathfrak{a}R_{\mathfrak{m}},M_{\mathfrak{m}})_{\mathfrak{N}}$$

is Cohen-Macaulay too.

(11.17) Remark. Chapter V will contain a detailed study of the interdependence of the Cohen-Macaulay property of M, $G(\mathfrak{a},M)$ and $B(\mathfrak{a},M)$. By means of examples we will see that $B(\mathfrak{a},M)$ may be Cohen-Macaulay without M being so. The last section of the present chapter will also contain an example of a Cohen-Macaulay module M for which $G(\mathfrak{a},M)$ is not Cohen-Macaulay. The reason why we have to postpone such an example lies in the fact that in explicit examples the computation of $G(\mathfrak{a},M)$ may be very difficult. More about this will be given in Section 13. But before we review the basic geometric concept of blowing up and the importance of $G(\mathfrak{a},R)$ for this concept.

§ 12. Review on blowing up

Blowing up is a very important kind of transformation in algebraic (and analytic) geometry. The most striking result using these transformations are Hironaka's proofs for desingularization of algebraic varieties of characteristic zero and of complex-analytic spaces. But also for other purposes, blowing up is a useful tool, e.g. for the classification of surfaces, for the elimination of the indetermination of a rational map etc. The interested reader is refered to Hartshorne's book [2] for these geometric aspects of blowing up.

The morphism obtained by blowing up a variety X involves a subvariety Y of X, called the center of blowing up. For general center Y, nothing can be said about the blowing up morphism except that it is proper and birational. Therefore, to study this kind of morphism means to describe how it depends on properties of X, Y and the embedding $Y \subset X$. For classical applications, Y is supposed to be non-singular and X is "equimultiple along Y" or even normally "flat along Y" (see Chapter IV). One of the main topics of this book is to study some algebraic and some numerical properties of X under blowing up suitable centers Y.

We do not intend to give any kind of complete description of blowing up. Instead we restrict ourselves to those aspects that we will need later on, and this means that we will give a purely algebraic description of blowing up and of the local homomorphism arising this way. We start by recalling the notion of Proj. For the corresponding analytic notion see App. III, 1.2.7 and 1.4.4.

<u>(12.1) Definition</u>. Let $A = \bigoplus_{n \geq 0} A_n$ be a positively graded ring, and let $f \in A$ be a homogeneous element of positive degree. We define

a) $\mathrm{Proj}(A) = \{\mathfrak{p} \mid \mathfrak{p}$ homogeneous prime ideal of A such that $A_n \not\subset \mathfrak{p}$ for some $n > 0\}$.

b) $D_+(f) = \{\mathfrak{p} \in \mathrm{proj}(A) \mid f \notin \mathfrak{p}\}$.

<u>(12.2) Remark</u>. a) The sets $D_+(f)$ form a basis for a topology on $\mathrm{Proj}(A)$. Denoting the topological space $\mathrm{Proj}(A)$ by X and a prime ideal $\mathfrak{p} \in \mathrm{Proj}(A)$ by x, we put $\mathcal{O}_{X,x} = A_{(\mathfrak{p})}$. This makes X a scheme (see [2], p. 76 for details). In our notation we will not distinguish between the point set, the topological space and the scheme $\mathrm{Proj}(A)$.

b) In the case that A is noetherian it is clear from the definition that $\mathrm{Proj}(A) = \emptyset$ if and only if $A_n = 0$ for all large n, and this is equivalent to saying that the ideal $\bigoplus_{n>0} A_n$ of A is nilpotent.

c) If $\mathfrak{p} \in \mathrm{Proj}(A)$ and $\mathfrak{p}_0 = \mathfrak{p} \cap A_0$, then by definition there is a canonical local homomorphism

$$\varphi_{\mathfrak{p}} : (A_0)_{\mathfrak{p}_0} \longrightarrow A_{(\mathfrak{p})} \quad .$$

This defines a morphism

$$\varphi : \mathrm{Proj}(A) \longrightarrow \mathrm{Spec}(A_0) \quad .$$

For most of our applications, we may identify the morphism φ with the family $\{\varphi_{\mathfrak{p}} \mid \mathfrak{p} \in \mathrm{Proj}(A)\}$ of local homomorphism. We will restrict ourselves to the case that A is noetherian, and then φ is proper [2], P. 100.

d) If A is generated over A_0 by homogeneous elements $f_1,\dots,f_n$ then obviously $\mathrm{Proj}(A)$ is covered by $D_+(f_1),\dots,D_+(f_n)$. Moreover the open subschemes $D_+(f_i)$ are affine with coordinate ring

$$A_{(f_i)} = \left\{ \frac{a}{f_i^n} \mid a \in A \text{ homogeneous, } \deg a = n \deg f_i \right\}$$

(see [2] again).

e) If $\mathfrak{a}$ is a homogeneous ideal of A, then $A \longrightarrow A/\mathfrak{a}$ defines a morphism $\mathrm{Proj}(A/\mathfrak{a}) \longrightarrow \mathrm{Proj}(A)$, making $\mathrm{Proj}(A/\mathfrak{a})$ a closed subscheme of $\mathrm{Proj}(A)$ .Conversely every closed subscheme of $\mathrm{Proj}(A)$ is defined by some homogeneous ideal $\mathfrak{a}$ of A and the homomorphism $A \longrightarrow A/\mathfrak{a}$. For the particular case $\mathfrak{a} = \bigoplus_{n>0} A_n$ we obtain a morphism $\psi : \mathrm{Proj}(A/\mathfrak{a}) = \mathrm{Spec}(A_0) \longrightarrow \mathrm{Proj}(A)$, and the composition of ψ with the morphism φ of c) is the identity on $\mathrm{Spec}(A_0)$, i.e. ψ is a section for the "structural morphism" φ.

<u>(12.3) Definition</u>. Let R be a ring and $\mathfrak{a}$ an ideal of R. We define the blowing up of R with center $\mathfrak{a}$ to be the morphism

$$\mathrm{Proj}(B(\mathfrak{a},R)) \longrightarrow \mathrm{Spec}(R) \quad .$$

The scheme $\mathrm{Proj}(B(\mathfrak{a},R))$ will also be denoted by $\mathrm{Bl}(\mathfrak{a},R)$.

(12.4) Remark. There is a variety of names given to what we call blowing up of R with center $\mathfrak{a}$, e.g. blowing up of $\mathrm{Spec}(R)$ with center $V(\mathfrak{a})$, blowing up of $\mathrm{Spec}(R)$ along $V(\mathfrak{a})$, blowing up $V(\mathfrak{a})$ inside $\mathrm{Spec}(R)$ etc. Of course there is a more general notion of blowing up a scheme X only in the case that X is affine, and the general case may be obtained by patching ([2]).

For the local description of the blowing up morphism we need the following

(12.5) Definition. Let R be a ring, $\mathfrak{a}$ an ideal of R and x an element of $\mathfrak{a}$. We denote by $R[\mathfrak{a}x^{-1}]$ or $R[\mathfrak{a}/x]$ the subring of R_x generated by the image of R and by the set $\{a/x \mid a \in \mathfrak{a}\}$.

(12.6) Proposition. Let R be a ring, $\mathfrak{a}$ an ideal of R and $x \in \mathfrak{a}$. Then there exists an unique R-algebra homomorphism

$$\alpha : R[\mathfrak{a}/x] \longrightarrow B(\mathfrak{a},R)_{(xt)}$$

and this is an isomorphism.

Proof. If α exists then $\alpha(1) = 1$ and α is determined by the images $\alpha(a/x)$, $a \in \mathfrak{a}$. Assume that

$$\alpha\left(\frac{a}{x}\right) = \frac{bt^n}{(xt)^n}\ , \ b \in \mathfrak{a}^n \ .$$

Then

$$x\alpha\left(\frac{a}{x}\right) = \alpha(a) = a \cdot \alpha(1) = \frac{a}{1}$$

and therefore

$$\frac{a}{1} = x \cdot \frac{bt^n}{(xt)^n} \ .$$

Hence there is some $m \geq 0$ such that

$$x^m t^m (ax^n t^n - xbt^n) = 0 \quad \text{in} \quad R[t] \ .$$

Multiplying by t we get

$$x^m t^m ((at)(xt)^n - (xt)bt^n) = 0 \ .$$

Viewing this as an equation in $B(\mathfrak{a},R)$ it follows that

$$\frac{at}{xt} = \frac{bt^n}{(xt)^n} = \alpha\left(\frac{a}{x}\right) .$$

So we see that α is uniquely determined by the rule $\alpha(a/x) = at/xt$ for each $a \in \mathfrak{a}$. Defining α this way, it is easy to check that α is an isomorphism.

<u>(12.7) Corollary</u>. Let $\mathfrak{q} \in \mathrm{Proj}(B(\mathfrak{a},R))$ and choose $x \in \mathfrak{a}$ such that $xt \notin \mathfrak{q}$. Then there is a unique prime ideal $\mathfrak{p}$ of $R[\mathfrak{a}/x]$ such that $R[\mathfrak{a}/x]_{\mathfrak{p}}$ and $B(\mathfrak{a},R)_{(\mathfrak{q})}$ are isomorphic as R-algebras, and moreover this isomorphism is unique.

<u>(12.8) Remark</u>. In $R[\mathfrak{a}/x]$ we have

$$\mathfrak{a} \cdot R[\mathfrak{a}/x] = x \cdot R[\mathfrak{a}/x] \quad ,$$

a principal ideal generated by a non-zerodivisor. It is easy to see that this defines a Cartier divisor on $Bl(\mathfrak{a},R)$. Moreover the blowing up morphism is universal with respect to making $\mathfrak{a}$ invertible (see [2] p.164 for a precise statement and proof).

Recall the following fact on graded rings. If $A = \bigoplus_{n \in \mathbf{Z}} A_n$ is graded and S a multiplicatively closed set of homogeneous elements of A , then any homogeneous ideal $\mathfrak{a}$ of A extends in a natural manner to $A_{(S)}$ namely to the ideal

$$\left\{ \frac{a}{b} \mid a \in \mathfrak{a}_n \, , \ b \in S \cap A_n \, , n \in \mathbf{Z} \right\} .$$

This extension has the same properties as the usual extension to localizations. Returning to the isomorphism α of Proposition (12.6) it is easy to check that $\mathfrak{a}R[\mathfrak{a}/x] \simeq \mathfrak{a}B(\mathfrak{a},R)_{(xt)}$ corresponds to the extension of the homogeneous ideal $\mathfrak{a}B(\mathfrak{a},R)$ to $B(\mathfrak{a},R)_{(xt)}$. Using (8.8.4) we see that $\mathrm{Proj}(G(\mathfrak{a},R))$, as a closed subscheme of $\mathrm{Bl}(\mathfrak{a},R)$, is a Cartier divisor.

(12.9) Definition. $\mathrm{Proj}(G(\mathfrak{a},R))$ is called the exceptional divisor of $\mathrm{Bl}(\mathfrak{a},R)$.

(12.10) Proposition. Let $\varphi : \mathrm{Bl}(\mathfrak{a},R) \longrightarrow \mathrm{Spec}(R)$ be the blowing up of R with center $\mathfrak{a}$ and let $E \subset \mathrm{Bl}(\mathfrak{a},R)$ be the exceptional divsor. Then φ is an isomorphism outside E .

Proof. Let p be a point of $\mathrm{Bl}(\mathfrak{a},R)$ outside E corresponding to $R[\mathfrak{a}/x]_{\mathfrak{p}}$ for some x and some $\mathfrak{p}$. Now $p \notin E$ means that $\mathfrak{p} \not\supset \mathfrak{a}$ and therefore $x \notin \mathfrak{p}$. Putting $\mathfrak{p}_0 = \mathfrak{p} \cap R$ it is clear that $R_{\mathfrak{p}_0} \simeq R[\mathfrak{a}/x]_{\mathfrak{p}}$. This shows that the structure sheaves at p and $\varphi(p)$ are isomorphic, and moreover the same holds for any point p' in the open subset $D(x)$ of $\mathrm{Spec}(R[\mathfrak{a}/x])$.

(12.11) Proposition. Let $R \longrightarrow S$ be a flat homomorphism of rings and let $\mathfrak{a}$ be an ideal of R . Then $\mathrm{Bl}(\mathfrak{a}S,S)$ and $\mathrm{Bl}(\mathfrak{a},R) \otimes_R S$ are canonically isomorphic over $\mathrm{Spec}(S)$.

Proof. This follows immediately from Lemma (9.8).

We note that (12.11) does not remain true without flatness assumption. A trivial example is the homomorphism $R \longrightarrow R/\mathfrak{a}$. Here $\mathrm{Bl}(\mathfrak{a},R) \otimes_R R/\mathfrak{a}$ is the exceptional divisor, whereas $\mathrm{Bl}(\mathfrak{a}(R/\mathfrak{a}),R/\mathfrak{a})$ is empty.

We are mostly interested in properties of the blowing up morphism which are local, at least on the base space $\mathrm{Spec}(R)$. For such properties we may restrict our attention to the case that R is local (by Proposition (12.11)). For the description of local properties of the blowing up morphism it is convenient to make the following

(12.12) Definition. Let $\alpha : R \longrightarrow R_1$ be a homomorphism of local rings and let $\mathfrak{a}$ be an ideal of R . α will be called a blowing up homomorphism of $(R,\mathfrak{a})$ if there exist $x \in \mathfrak{a}$ and a prime ideal $\mathfrak{p}$ in $R[\mathfrak{a}/x]$ such that R_1 and $R[\mathfrak{a}/x]_{\mathfrak{p}}$ are isomorphic as R-algebras. (We point out that α is not assumed to be local. If we want to assume that α is local, then we call it a local blowing up homomorphism of $(R,\mathfrak{a})$.)

In Proposition (12.10) we described the blowing up locally outside the exceptional divisor E. On E we have the following result which is parallel to Corollary (12.7):

<u>(12.13) Proposition</u>. Let $\alpha : R \longrightarrow R_1$ be a blowing up homomorphism of $(R,\mathfrak{a})$ such that $\mathfrak{a}R_1 \neq R_1$. Then there are unique homogeneous prime ideals Q of $\mathbf{R}(\mathfrak{a},R)$ and $\mathfrak{q}$ of $G(\mathfrak{a},R)$ such that

$$(12.13.1)\qquad \left.\begin{aligned} R_1 &\simeq \mathbf{R}(\mathfrak{a},R)_{(Q)} \\ R_1/\mathfrak{a}R_1 &\simeq G(\mathfrak{a},R)_{(\mathfrak{q})} \end{aligned}\right\} \text{ as R-algebras.}$$

Moreover, the R-algebra isomorphisms (12.13.1) are unique.

<u>(12.14) Proposition</u>. Let $\varphi : R \longrightarrow R_1$ be a blowing up homomorphism of $(R,\mathfrak{a})$. Then $\dim R_1 \leqq \dim R$. Hence $\dim \mathrm{Bl}(\mathfrak{a},R) \leqq \dim R$.

Proof. Let $\mathfrak{m}_1$ be the maximal ideal of R_1. If $\mathfrak{a} \not\subset \varphi^{-1}(\mathfrak{m})$ then $R_1 = R_{\mathfrak{p}}$ with $\mathfrak{p} = \varphi^{-1}(\mathfrak{m}_1)$, hence $\dim R_1 \leqq \dim R$. If $\varphi^{-1}(\mathfrak{m}_1) \supset \mathfrak{a}$, write $R_1/\mathfrak{a}R_1 = G(\mathfrak{a},R)_{(\mathfrak{q})}$. Then $\mathfrak{q}$ is different from the unique maximal homogeneous ideal of $G(\mathfrak{a},R)$ and therefore

$$\dim G(\mathfrak{a},R)_{(\mathfrak{q})} = \mathrm{ht}(\mathfrak{q}) < \dim G(\mathfrak{a},R) = \dim R \quad .$$

Since $R_1/\mathfrak{a}R_1 = R_1/xR_1$ for some non-zerodivisor x of R, we have $\dim R_1 = \mathrm{ht}(\mathfrak{q}) + 1$ and the result follows.

<u>(12.15) Remark</u>. We note that without further assumption, the above Proposition cannot be strengthened. For example, let $R = k[[x,y,z]]\,(xy,xz)$ and $\mathfrak{a} = xR$. Then $R_x = k((x))$, $R[\mathfrak{a}/x] = k[[x]]$

and hence $Bl(\mathfrak{a},R) = \operatorname{Spec}(k[[x]])$. So we get $\dim Bl(\mathfrak{a},R) = 1$ but $\dim R = 2$. In the next chapter we will see that for a special class of rings (quasi-unmixed rings), a lot more can be said about dimensions under blowing up.

Later on we want also to consider multiplicities and Hilbert functions of rings of the form $G(\mathfrak{a},R)_{(\mathfrak{q})}$. The result we need (Corollary (12.20)) is a special case of a more general result, for which we introduce some notation.

(12.16) Definition. Let A be a graded ring and S a multiplicatively closed subset of homogeneous elements of A containing 1. By giving $\deg X = 1$ to the variable X we define a graded ring $A_{[S]}$ by

$$A_{[S]} = A_{(S)}[X,X^{-1}] \quad .$$

For any homogeneous ideal $\mathfrak{a}$ of A we put

$$\mathfrak{a}_{(S)} = \left\{ \frac{a}{b} \in A_{(S)} \mid a \in \mathfrak{a} \right\} = \mathfrak{a}A_S \cap A_{(S)}$$

and

$$\mathfrak{a}_{[S]} = \mathfrak{a}_{(S)} \cdot A_{[S]} \quad .$$

If $\mathfrak{p}$ is a homogeneous prime ideal of A and S the set of homogeneous elements of A outside $\mathfrak{p}$, we write $A_{[\mathfrak{p}]}$, $\mathfrak{a}_{(\mathfrak{p})}$ and $\mathfrak{a}_{[\mathfrak{p}]}$ instead of $A_{[S]}$, $\mathfrak{a}_{(S)}$ and $\mathfrak{a}_{[S]}$ respectively.

(12.17) Proposition. Let A be a graded ring and let S be a multiplicatively closed subset of homogeneous elements of A containing 1. For any $x = s/t \in A_S$ such that $s,t \in S$ and $\deg x = 1$ there is a canonical isomorphism

$$\varphi : A_{[S]} \xrightarrow{\sim} A_S$$

for which $\varphi | A_{(S)}$ is the inclusion and $\varphi(X) = x$. Moreover, for any homogeneous ideal $\mathfrak{a}$ of A we have

$$\mathfrak{a}A_S = \varphi(\mathfrak{a}_{[S]}) \quad .$$

Proof. Clearly φ is well-defined by the above conditions, and it remains to show that φ is an isomorphism. Let $a/s \in A_S$ be such that $a \in A_d$, $s \in S \cap A_e$. Then mapping a/s to $\frac{a}{s} X^{e-d} . X^{d-e} \in A_{[S]}$ defines the inverse of φ, as one easily checks. For the second assertion, let $f = \sum \frac{a_i}{s_i} \cdot X^i \in \mathfrak{a}_{[S]}$. Then $a_i \in \mathfrak{a}$ for all i and therefore $\varphi(f) \in \mathfrak{a}A_S$. Finally, if $\frac{a}{s} \in A_S$ where $a \in \mathfrak{a}_d$ and $s \in S \cap A_e$, then

$$\varphi^{-1}\left(\frac{a}{s}\right) = \frac{a}{s} X^{e-d} X^{d-e} \in A_{[S]} ,$$

since $\frac{a}{s} X^{e-d} = \frac{a}{sX^{d-e}} \in \mathfrak{a}_{(S)}$.

<u>(12.18) Corollary</u>. Let A be a graded ring and $\mathfrak{p}$ a homogeneous prime ideal. Assume that $A \setminus \mathfrak{p}$ contains homogeneous elements s,t such that $\deg s - \deg t = 1$. Let $\mathfrak{p}' = \mathfrak{p}_{(\mathfrak{p})} \cdot A_{(\mathfrak{p})}[X]$. Then there is a canonical isomorphism

$$\varphi_{\mathfrak{p}} : A_{(\mathfrak{p})}[X]_{\mathfrak{p}'} \xrightarrow{\sim} A_{\mathfrak{p}}$$

sending X to s/t .

Proof. Let S denote the set of homogeneous elements of $A \setminus \mathfrak{p}$. Then clearly $A_{\mathfrak{p}} \simeq (A_S)_{\mathfrak{p}A_S}$. By Proposition (12.17) we know that

$$(A_S)_{\mathfrak{p}A_S} \simeq (A_{[S]})_{\mathfrak{p}_{[S]}}$$

and the conclusion follows by observing that clearly

$$(A_{[S]})_{\mathfrak{p}_{[S]}} \simeq A_{(S)}[X]_{\mathfrak{p}'}$$

since $X \notin \mathfrak{p}'$.

<u>(12.19) Corollary</u>. Assume in addition that A is noetherian. Then

a) $A_{(\mathfrak{p})}$ is regular if and only if A is regular.

b) $A_{(\mathfrak{p})}$ is Cohen-Macualay if and only if $A_{\mathfrak{p}}$ is Cohen-Macaulay. Moreover, if $\mathfrak{a}$ is a proper ideal in a local noetherian ring R then

c) $G(\mathfrak{a},R)$ Cohen-Macaulay implies $Bl(\mathfrak{a},R)$ Cohen-Macaulay.

Proof. a) and b) are a direct consequence of Proposition (5.1) and Corollary (12.18). To prove c) we note that $G(\mathfrak{a},R)$ Cohen-Macaulay implies R Cohen-Macaulay by Proposition (11.16), b). Hence it is enough to show that $Bl(\mathfrak{a},R)$ is Cohen-Macaulay on the exceptional divisor E. Since E is defined by a non-zerodivisor, the claim follows from b) above.

<u>(12.20) Corollary</u>. Given $A,\mathfrak{p}$ and $\varphi_{\mathfrak{p}}$ as in Corollary (12.18), let $\mathfrak{a}$ be any homogeneous ideal in A and let $x_1,\dots,x_r \in A_{(\mathfrak{p})}$. Then:

a) $x_1,\dots,x_r$ are a multiplicity system (resp. system of parameters) for $A_{(S)}/\mathfrak{a}_{(S)}$ if and only if $\varphi_{\mathfrak{p}}(x_1),\dots,\varphi_{\mathfrak{p}}(x_r)$ are a multiplicity system (resp. system of parameters) for $A_{\mathfrak{p}}/\mathfrak{a}A_{\mathfrak{p}}$.

b) For all $i \geq 0$ we have

$$H^{(i)}[\underline{x},\mathfrak{a}_{(\mathfrak{p})},A_{(\mathfrak{p})}] = H^{(i)}[\underline{y},\mathfrak{a}A_{\mathfrak{p}},A_{\mathfrak{p}}] \quad ,$$

where $\underline{x} = (x_1,\dots,x_r)$ and $\underline{y} = (\varphi_{\mathfrak{p}}(x_1),\dots,\varphi_{\mathfrak{p}}(x_r))$.

c) With $\underline{x}$ and $\underline{y}$ as above we have

$$e(\underline{x},\mathfrak{a}_{(\mathfrak{p})},A_{(\mathfrak{p})}) = e(\underline{y},\mathfrak{a}A_{\mathfrak{p}},A_{\mathfrak{p}}) \quad .$$

Proof. By Corollary (12.18), $A_{(\mathfrak{p})} \longrightarrow A_{\mathfrak{p}}$ is a faithfully flat extension and $\mathfrak{a}_{(\mathfrak{p})} \cdot A = \mathfrak{a}A_{\mathfrak{p}}$ by Proposition (12.17); in particular we have

$$\mathfrak{p}_{(\mathfrak{p})} \cdot A_{\mathfrak{p}} = \mathfrak{p}A_{\mathfrak{p}} \quad ,$$

so the result follows from Proposition (5.1).

We restate the above Corollary in the most important special case:

(12.21) Corollary. Let $A = \bigoplus_{n \geq 0} A_n$ be a graded ring, generated as an A_0-algebra by A_1. Then for any $\mathfrak{p} \in \mathrm{Proj}(A)$ we have

$$H^{(i)}[A_{(\mathfrak{p})}] = H^{(i)}[A_{\mathfrak{p}}] \quad , \quad i \geq 0$$

and

$$e(A_{(\mathfrak{p})}) = e(A_{\mathfrak{p}}) \quad .$$

We use the remaining part of this section to recall the notion of a strict transform of a closed subvariety under blowing up. This notion is very important in geometric applications, and the methods of the next section will give an algorithm to compute the equations of the strict transform locally from given equations of the subvariety. We note, however, that we will not make any geometric use of the notion of strict transforms, except in App.II 2.2.2.

(12.22) Definition. Let R be a ring and let $\mathfrak{a}, \mathfrak{h}$ be ideals of R. The closed subscheme $\mathrm{Proj}(B(\mathfrak{a}+\mathfrak{h}/\mathfrak{h}, R/\mathfrak{h}))$ of $\mathrm{Proj}(B(\mathfrak{a},R))$ is called the strict transform of $V(\mathfrak{h})$ in $\mathrm{Bl}(\mathfrak{a},R)$. If $R \longrightarrow R_1$ is a blowing up homomorphism of $(R,\mathfrak{a})$ and if $\mathfrak{h}_1 \subset R_1$ defines the strict transform of $V(\mathfrak{h})$ in $\mathrm{Bl}(\mathfrak{a},R)$ locally at R_1 then $\mathfrak{h}_1$ is called the strict transform of $\mathfrak{h}$ in R_1.

(12.23) Remark. a) Since $B(\mathfrak{a}+\mathfrak{h}/\mathfrak{h}, R/\mathfrak{h}) \simeq B(\mathfrak{a},R)/B(\mathfrak{a},\mathfrak{h} \subset R)$, the strict transform is indeed a closed subscheme of $\mathrm{Bl}(\mathfrak{a},R)$. A more general notion of strict transform, replacing the inclusion $V(\mathfrak{a}) \longrightarrow \mathrm{Spec}(R)$ by any morphism, can be found in [2], p.165 .

b) Let $X = \mathrm{Spec}(R)$, let $\pi : \mathrm{Bl}(\mathfrak{a},R) \longrightarrow X$ be the blowing up, let $D = V(\mathfrak{a})$ the center and put $Y = V(\mathfrak{h})$. If Y' denotes the strict transform of Y in $\mathrm{Bl}(\mathfrak{a},R)$, it is not difficult to see that Y' is the (scheme-theoretic) closure of $\pi^{-1}(Y \setminus D)$.

Let us turn to the problem of finding the equations of the strict transform of a subvariety. If locally for some homomorphism $R \longrightarrow R_1$ the strict transform is $\mathfrak{h}_1 \subset R_1$, then we want to find generators of the ideal $\mathfrak{h}_1$ in terms of generators of the original ideal $\mathfrak{h} \subset R$. Of particular interest is the case where the strict transform is (locally at R_1) contained in the exceptional divisor. For this case we have $\mathfrak{b}_1 \supset \mathfrak{a}R_1$ and it is enough to find generators for $\mathfrak{b}_1/\mathfrak{a}R_1$ in $R_1/\mathfrak{a}R_1$. Now

$$R_1/\mathfrak{a}R_1 \simeq G(\mathfrak{a},R)_{(\mathfrak{q})} ,$$

and on the exceptional divisor $E = \mathrm{Proj}(G(\mathfrak{a},R))$, the strict transform of $V(\mathfrak{b})$ is defined by $G(\mathfrak{a},\mathfrak{b}\subset R)$. Take any homogeneous generators

$$b_1^*,\ldots,b_n^* \in G(\mathfrak{a},\mathfrak{b}\subset R)$$

and assume that

$$\mathfrak{a}R_1 = a\cdot R_1 , \quad a\in\mathfrak{a} ,$$

which means that the class a^* of a mod $\mathfrak{a}^2$ is not contained in $\mathfrak{q}$. Then by definition $\mathfrak{b}_1/\mathfrak{a}R_1$ is generated by

$$b_1^*/(a^*)^{d_1},\ldots,b_n^*/(a^*)^{d_n} , \quad d_i = \deg b_i^* , \; i = 1,\ldots,n .$$

Now $b_i^* = \mathrm{in}(\mathfrak{a})(b_i)$ for some $b_i\in\mathfrak{b}$, $d_i = \mathrm{ord}(\mathfrak{a})(b_i)$ and

$$\frac{b_i^*}{(a^*)^{d_i}} = \frac{b_i}{a^{d_i}} .$$

Therefore the problem of finding generators for the strict transform may be reformulated in the following way. Given $\mathfrak{a}$ and $\mathfrak{b}$ as before, determine some generators $b_1,\ldots,b_n$ of $\mathfrak{b}$ such that $\mathrm{in}(\mathfrak{a})(b_1),\ldots,\mathrm{in}(\mathfrak{a})(b_n)$ generate the ideal $G(\mathfrak{a},\mathfrak{b}\subset R)$. This leads to the notion of a standard base, which will be discussed in the next section. There is a more refined notion leading to an effective algorithm, in particular Hilbert functions can be computed. See App. I, § 2.

§ 13. Standard bases

In this section we restrict ourselves to local rings, although some of the results may be generalized to the non-local case (see [12]).

(13.1) Definition. Let $(R,\mathfrak{m})$ be a local ring, M a finitely generated R-module and N a submodule of M. Let $\underline{m} = (m_1,\dots,m_t)$ be a sequence of elements of N. Then $\underline{m}$ is called an $(\mathfrak{a},M)$-standard-base of N, if the submodule $G(\mathfrak{a},N\subset M)$ is generated by $\{in(\mathfrak{a},M)(m_1),\dots,in(\mathfrak{a},M)(m_t)\}$.

The main part of this section consists in characterization of standard bases. For the computations the notation given in (8.7) and (10.22) is too heavy. Therefore in this section - and only here - we will use the following

(13.2) Notation. We fix a local ring $(R,\mathfrak{m})$, a proper ideal $\mathfrak{a}$ of R and a finitely generated R-module M. The order functions $ord(\mathfrak{a},R)$ and $ord(\mathfrak{a},M)$ will simply be denoted by 0, so

$$0(z) = \begin{cases} \sup\{n \mid z \in \mathfrak{a}^n\} & \text{if} \quad z \in R\,, \\ \sup\{n \mid z \in \mathfrak{a}^n M\} & \text{if} \quad z \in M\,. \end{cases}$$

Initial forms $in(\mathfrak{a},R)(z)$ and $in(\mathfrak{a},M)(z)$ will be denoted by z^*. So $0^* = 0$, and for $z \neq 0$ we have

$$z^* = \begin{cases} z + \mathfrak{a}^{0(z)+1} & \text{if} \quad z \in R\,, \\ z + \mathfrak{a}^{0(z)+1}M & \text{if} \quad z \in M\,. \end{cases}$$

We put $R^* = G(\mathfrak{a},R)$, $M^* = G(\mathfrak{a},M)$, and if N is a submodule of M, we put $\tilde{N} = G(\mathfrak{a},N\subset M)$, i.e.

$$\tilde{N} = \bigoplus_{n\in\mathbb{N}_0} (\mathfrak{a}^n M \cap N) + \mathfrak{a}^{n+1}M/\mathfrak{a}^{n+1}M \quad .$$

So $m_1,\dots,m_t \in N$ is an $(\mathfrak{a},M)$-standard-base if and only if

$$\tilde{N} = R^*m_1^* + \dots + R^*m_t^* \quad .$$

(13.3) Proposition. If N is a submodule of M and if $m_1,\dots,m_t \in N$ is an $(\mathfrak{a},M)$-standard-base of N, then

$$N = Rm_1 + \dots + Rm_t \quad .$$

Proof. Given $y \in N$ we will use induction on d to show that

$$(13.3.1) \quad \begin{cases} \text{for any } d \geq 0 \text{ there are } a_1(d),\dots,a_t(d) \in R \\ \text{such that } y - (a_1(d)m_1 + \dots + a_t(d)m_t) \in \mathfrak{a}^d M \end{cases}$$

There is nothing to prove if $d = 0$, so assume that $d > 0$. By assumption there are $a_1(d-1),\dots,a_t(d-1) \in R$ such that

$$z = y - (a_1(d-1)m_1 + \dots + a_t(d-1)m_t) \in \mathfrak{a}^{d-1}M \quad .$$

Since $m_1,\dots,m_t$ is an $(\mathfrak{a},M)$-standard-base of N, we can choose $b_1,\dots,b_t \in R$ such that

$$z^* = b_1^* m_1^* + \dots + b_1^* m_t^* \quad .$$

Since $0(z) \geq d-1$, we conclude that

$$z - (b_1 m_1 + \dots + b_t m_t) \in \mathfrak{a}^d M \quad .$$

So we complete the proof of (13.3.1) by putting $a_i(d) = a_i(d-1) + b_i$ for $i = 1,\dots,t$. Finally (13.3.1) implies that

$$y \in \bigcap_{d \geq 0} (Rm_1 + \dots + Rm_t + \mathfrak{a}^d M) = Rm_1 + \dots + Rm_t \quad .$$

(13.4) Definition. Let A be a graded ring, H a graded A-module and $d \in \mathbf{Z}$. We define a graded A-module $H(d) = \bigoplus_{n \in \mathbf{Z}} H(d)_n$ by putting

$$H(d)_n = H_{n+d} \quad .$$

(13.5) Definition. Given a sequence $\underline{m} = (m_1,\dots,m_t)$ of elements of M we define

a) a homomorphism $f(\underline{m}) : R^t \longrightarrow M$ by

$$f(\underline{m})(a_1,\dots,a_t) = a_1 m_1 + \dots + a_t m_t \quad ;$$

b) a graded homomorphism

$$f^*(\underline{m}) : \bigoplus_{i=1}^{t} R^*(-O(m_i)) \longrightarrow M^*$$

by

$$f^*(\underline{m})(a_1^*,\ldots,a_t^*) = a_1^* m_1^* + \ldots + a_t^* m_t^* \quad ;$$

c) a map $O(\underline{m}) : R^t \longrightarrow \mathbf{Z}$ by

$$O(\underline{m})(a_1,\ldots,a_t) = \min\{ O(a_i) + O(m_i) \mid i = 1,\ldots,t\} \quad ;$$

d) a map $h(\underline{m}) : R^t \longrightarrow \bigoplus_{i=1}^{t} R^*(-O(m_i))$ by

$$h(\underline{m})(a_1,\ldots,a_t) = (u_1,\ldots,u_t)$$

where

$$u_i = \begin{cases} a_i^* & \text{if} \quad O(a_i)+O(m_i) = O(\underline{m})(a_1,\ldots,a_t) \\ 0 & \text{otherwise} \end{cases} .$$

(13.6) Remark. Let $u_1,\ldots,u_t \in R^*$ be homogeneous elements. Then $(u_1,\ldots,u_t)$ is homogeneous in $\bigoplus_{i=1}^{t} R^*(-O(m_i))$ if and only if $\deg u_i + O(m_i)$ is a constant (independent of i). It follows that for any $(a_1,\ldots,a_t) \in R^t$, $h(\underline{m})(a_1,\ldots,a_t)$ is homogeneous of degree $O(\underline{m})(a_1,\ldots,a_t)$. It is important to note here that

$$h(\underline{m})(\operatorname{Ker} f(\underline{m})) \subset \operatorname{Ker} f^*(\underline{m}) \quad .$$

(13.7) Theorem. Let N be a submodule of M and let $\underline{m} = (m_1,\ldots,m_t)$ be a sequence of elements of N such that $N = Rm_1 + \ldots + Rm_t$. Then the following conditions are equivalent.

(i) $\underline{m}$ is an $(\mathfrak{a},M)$-standard-base of N.

(ii) For any $d \geqq 0$ we have

$$N \cap \mathfrak{a}^d M = \mathfrak{a}^{d-O(m_1)} m_1 + \ldots + \mathfrak{a}^{d-O(m_t)} m_t \quad .$$

(iii) For any $z \in N$ there are $a_1,\dots,a_t \in R$ such that $z = a_1m_1 + \dots + a_tm_t$ and $0(z) \leq 0(a_i) + 0(m_i)$ for all i.

(iv) $h(\underline{m})(\operatorname{Ker} f(\underline{m}))$ is the set of homogenous elements of $\operatorname{Ker} f^*(\underline{m})$.

(v) $h(\underline{m})(\operatorname{Ker} f(\underline{m}))$ generates $\operatorname{Ker} f^*(\underline{m})$.

Proof. The proof will be given according to the following diagram:

$$\begin{array}{ccccccc} (i) & \Rightarrow & (ii) & \Longleftrightarrow & (iv) & \Longleftrightarrow & (v) \\ \nwarrow & & \swarrow & & & & \\ & (iii) & & & & & \end{array}$$

(i) $\Rightarrow$ (ii). Let d be any nonnegative integer and let $z \in N \cap \mathfrak{a}^dM$. Choosing $a_1,\dots,a_t \in R$ such that

$$z^* = a_1^*m_1^* + \dots + a_t^*m_t^* \quad , \quad \deg(a_i^*m_i^*) = \deg(z^*)$$

we see that $0(a_i) = 0(z) - 0(m_i) \geq d - 0(m_i)$ for $a_i \neq 0$ and therefore

$$z \in \mathfrak{a}^{d-0(m_1)}m_1 + \dots + \mathfrak{a}^{d-0(m_t)}m_t + \mathfrak{a}^{d+1}M \quad .$$

This shows that

$$N \cap \mathfrak{a}^dM \subset \mathfrak{a}^{d-0(m_1)}m_1 + \dots + \mathfrak{a}^{d-0(m_t)}m_t + \mathfrak{a}^{d+1}M \quad \text{for all} \quad d \geq 0 .$$

Intersecting with N we obtain

$$\begin{aligned} N \cap \mathfrak{a}^dM &\subset \mathfrak{a}^{d-0(m_1)}m_1 + \dots + \mathfrak{a}^{d-0(m_t)}m_t + N \cap \mathfrak{a}^{d+1}M \\ &\subset \mathfrak{a}^{d-0(m_1)}m_1 + \dots + \mathfrak{a}^{d-0(m_t)}m_t + \mathfrak{a}^{d+2}M \quad . \end{aligned}$$

Proceeding inductively we conclude that

$$N \cap \mathfrak{a}^dM \subset \mathfrak{a}^{d-0(m_1)}m_1 + \dots + \mathfrak{a}^{d-0(m_t)}m_t + \mathfrak{a}^{d+n}M \quad \text{for all} \quad n \geq 0 ,$$

so (ii) follows from Krull's Intersection Theorem.

(ii) $\Rightarrow$ (iii) . This is obvious.

(iii) $\Rightarrow$ (i). For $z \in N$, $z \neq 0$, choose $a_1, \ldots, a_t \in R$ such that

$$z = a_1 m_1 + \ldots + a_t m_t \quad \text{and} \quad 0(a_i) + 0(m_i) \geqq 0(z) , \quad i = 1, \ldots, t .$$

Now $0(a_i m_i) \geqq 0(a_i) + 0(m_i) \geqq 0(z) \geqq \min\{0(a_i), 0(m_i)\}$ implies $z^* = b_1^* m_1^* + \ldots + b_t^* m_t^*$ where

$$b_i^* = \begin{cases} a_i^* & \text{if } 0(a_i m_i) = 0(z) \\ 0 & \text{otherwise} \end{cases} .$$

(ii) $\Rightarrow$ (iv) . Let $s^* = (s_1^*, \ldots, s_t^*) \in \operatorname{Ker} f^*(\underline{m})$ be homogeneous of degree d , so $0(s_i) = d - 0(m_i)$. Putting

$$z = s_1 m_1 + \ldots + s_t m_t$$

we get that

$$z \in N \cap \mathfrak{a}^{d+1} M = \mathfrak{a}^{d+1-0(m_1)} m_1 + \ldots + \mathfrak{a}^{d+1-0(m_t)} m_t .$$

Choosing $r_i \in \mathfrak{a}^{d+1-0(m_i)}$, $i = 1, \ldots, t$ such that $z = r_1 m_1 + \ldots + r_t m_t$, we see that

$$s = (s_1 - r_1, \ldots, s_t - r_t) \in \operatorname{Ker} f(\underline{m}) .$$

By construction it is clear that $h(\underline{m})(s) = s^*$, which proves (iv).

(iv) $\Rightarrow$ (ii) . For $d \geqq 0$ given let $z \in N \cap \mathfrak{a}^d M$. Using descending induction on e we will show that

$$z \in \mathfrak{a}^{d-e-0(m_1)} m_1 + \ldots + \mathfrak{a}^{d-e-0(m_t)} m_t , \quad e = 0, \ldots, d .$$

We assume $e < d$, since for $e = d$ there is nothing to prove. So by inductive assumption we may assume that

$$z = a_1 m_1 + \ldots + a_t m_t , \ a_i \in \mathfrak{a}^{d-e-1-0(m_i)} , \ i = 1, \ldots, t .$$

If $a_i \in \mathfrak{a}^{d-e-O(m_i)}$ for all i then we are done. So we may assume that $O(a_i) = d-e-1-O(m_i)$ for some i, which means that

$$O(\underline{m})(a_1,\ldots,a_t) = d-e-1 \quad .$$

Let $h(\underline{m})(a_1,\ldots,a_t) = (u_1,\ldots,u_t)$. Since $O(z) \geq d$, we see that $(u_1,\ldots,u_t) \in \operatorname{Ker} f^*(\underline{m})$, so by assumption we may choose $b_1,\ldots,b_t \in R$ such that

$$h(\underline{m})(b_1,\ldots,b_t) = (u_1,\ldots,u_t) \quad \text{and} \quad b_1m_1 + \ldots + b_tm_t = 0 \quad .$$

It follows that $a_i - b_i \in \mathfrak{a}^{d-e-O(m_i)}$ for $i = 1,\ldots,t$ and hence

$$z = (a_1-b_1)m_1 + \ldots + (a_t-b_t)m_t \in \mathfrak{a}^{d-e-O(m_1)}m_1 + \ldots + \mathfrak{a}^{d-e-O(m_t)}m_t \quad .$$

(iv)$\Leftrightarrow$(v). Trivially (iv) implies (v). For the converse let $s_1,\ldots,s_n \in \operatorname{Ker} f(\underline{m})$ be elements such that $h(\underline{m})(s_1),\ldots,h(\underline{m})(s_n)$ generate $\operatorname{Ker} f^*(\underline{m})$. If now $u \in \operatorname{Ker} f^*(\underline{m})$ is any homogeneous element, we may choose $a_i \in R$ such that

$$u = a_1^* \, h(\underline{m})(s_1) + \ldots + a_n^* \, h(\underline{m})(s_n)$$

and $O(a_i) = \deg u - \deg h(m)(s_i)$, $i = 1,\ldots,n$. Putting

$$s = a_1s_1 + \ldots + a_ns_n \quad ,$$

we have $s \in \operatorname{Ker} f(\underline{m})$, and it is immediate from the definition that $h(\underline{m})(s) = u$.

(13.8) Remark. For explicit computations (like in the examples in §14) the most important condition in the above Theorem is (v). For the equivalence of (i) and (ii) it is not necessary to assume R is local. For a proof in the general case, Krull's Intersection Theorem may be replaced by the Artin-Rees Lemma (see [13]). Theorem (13.7) is a special case of a quite general result which is published in [12].

While Theorem (13.7) is valid for any module M, in the applications we will always have $M = R$. This case has a special feature in connection with regular sequences.

(13.9) Lemma. Let $\underline{x} = (x_1,\dots,x_r)$ be a sequence of elements of R and let $y \in \mathfrak{m}$. Assume that

a) $(x_1,\dots,x_r,y)$ is an $(\mathfrak{a},R)$-standard base of $\underline{x}R + yR$,

b) y is a non-zero-divisor mod $\underline{x}R$.

Then $\underline{x}$ is an $(\mathfrak{a},R)$-standard base of $\underline{x}R$.

Proof. Since y is a non-zero-divisor mod $\underline{x}R$ we have

$$\underline{x}R \cap y\mathfrak{a}^n = y(\underline{x}R \cap \mathfrak{a}^n) \quad \text{for any} \quad n \geq 0 .$$

We will show by induction on n that

$$\underline{x}R \cap \mathfrak{a}^n = \sum_{i=1}^{r} \mathfrak{a}^{n-0(x_i)} x_i ,$$

the case $n = 0$ being trivial. For $n > 0$ we have by Theorem (13.7)

$$\begin{aligned}\underline{x}R \cap \mathfrak{a}^n &= \underline{x}R \cap (\mathfrak{a}^n \cap (\underline{x}R+yR)) = \underline{x}R \cap \left(\sum_{i=1}^{r} \mathfrak{a}^{n-0(x_i)} x_i + \mathfrak{a}^{n-0(y)} y\right) \\ &= \sum_{i=1}^{r} \mathfrak{a}^{n-0(x_i)} x_i + \underline{x}R \cap y\mathfrak{a}^{n-0(y)} \\ &= \sum_{i=1}^{r} \mathfrak{a}^{n-0(x_i)} x_i + y(\underline{x}R \cap \mathfrak{a}^{n-0(y)}) .\end{aligned}$$

If now $0(y) = 0$, then $\underline{x}R \cap \mathfrak{a}^n = \sum_{i=1}^{r} \mathfrak{a}^{n-0(x_i)} x_i$ by Nakayama's Lemma. If $0(y) > 0$, the inductive hypothesis implies

$$\underline{x}R \cap \mathfrak{a}^n = \sum_{i=1}^{r} \mathfrak{a}^{n-0(x_i)} x_i + y \cdot \sum_{i=1}^{r} \mathfrak{a}^{n-0(y)-0(x_i)} x_i = \sum_{i=1}^{r} \mathfrak{a}^{n-0(x_i)} x_i$$

again.

<u>(13.10) Theorem</u>. For any sequence $\underline{x} = (x_1,\ldots,x_r)$ of elements of R the following conditions are equivalent.

(i) $\underline{x}$ is a regular sequence and an $(\mathfrak{a},R)$-standard base of $\underline{x}R$.

(ii) $in(\mathfrak{a},R)(x_1),\ldots,in(\mathfrak{a},R)(x_r)$ is a regular sequence in $G(\mathfrak{a},R)$.

Proof. (i) $\Rightarrow$ (ii) . We use induction on r . In the case $r = 1$ assume that $a^* = in(\mathfrak{a},R)(x_1) = 0$ for some $a \in R$. If $a^* \neq 0$ let $d = 0(a)$ and $e = 0(x_1)$. Then

$$ax_1 \in x_1R \cap \mathfrak{a}^{d+e+1} = \mathfrak{a}^{d+1}x_1 \quad .$$

It follows that $a \in \mathfrak{a}^{d+1}$ in contradiction to $d = 0(a)$. For $r > 1$ we know by Lemma (13.9) that $x_1,\ldots,x_{r-1}$ is an $(\mathfrak{a},R)$-standard base of $x_1R + \ldots + x_{r-1}R$. Putting $\overline{R} = R/x_1R + \ldots + x_{r-1}R$, we obtain

$$G(\overline{\mathfrak{a}},\overline{R}) = G(\mathfrak{a},R)/(in(\mathfrak{a},R)(x_1),\ldots,in(\mathfrak{a},R)(x_{r-1})) \quad .$$

Denoting the image of x_r in $\overline{R}$ by $\overline{x}_r$ we will show that

(13.10.1) $$ord(\overline{\mathfrak{a}},\overline{R})(\overline{x}_r) = ord(\mathfrak{a},R)(x_r) \quad .$$

Let $ord(\mathfrak{a},R)(x_r) = d$ and assume that $\overline{x}_r \in \overline{\mathfrak{a}}^{d+1}$. Then there are $a_1,\ldots,a_{r-1} \in R$ such that

$$x_r + \sum_{i=1}^{r-1} a_ix_i \in \mathfrak{a}^{d+1} \cap \underline{x}R = \sum_{i=1}^{r} \mathfrak{a}^{d+1-0(x_i)} x_i \quad .$$

So

$$x_r + \sum_{i=1}^{r-1} a_ix_i = \sum_{i=1}^{r} b_ix_i \quad \text{for some} \quad b_i \in \mathfrak{a}^{d+1-0(x_r)} \quad .$$

In particular, since $0(x_r) > 1$, we have $b_r \in \mathfrak{m}$. Now

$$(1-b_r)x_r \in x_1R + \ldots + x_{r-1}R$$

and therefore $1-b_r \in x_1R + \ldots + x_{r-1}R \subset \mathfrak{m}$. This leads to the contradiction $1 \in \mathfrak{m}$, and therefore (13.10.1) holds. This means that $\mathrm{in}(\bar{\mathfrak{a}},\bar{R})(\bar{x}_r)$ is the image of $\mathrm{in}(\mathfrak{a},R)(x_r)$ under the canonical homomorphism

$$G(\mathfrak{a},R) \longrightarrow G(\bar{\mathfrak{a}},\bar{R}) \quad .$$

Clearly $\bar{x}_r$ is an $(\bar{\mathfrak{a}},\bar{R})$-standard base of $\bar{x}_r\bar{R}$, so by the case $r = 1$ we know that the image of $\mathrm{in}(\mathfrak{a},R)(x_r)$ in $G(\bar{\mathfrak{a}},\bar{R})$ is a non-zero divisor. Since $\mathrm{in}(\mathfrak{a},R)(x_1),\ldots,\mathrm{in}(\mathfrak{a},R)(x_{r-1})$ is a regular sequence in $G(\mathfrak{a},R)$ by inductive assumption, the proof of (i) ⇒ (ii) is complete.

(ii) ⇒ (i). We proceed by induction on r again. For $r = 1$, assume that $ax_1 = 0$ with $a \neq 0$. Then clearly

$$\mathrm{in}(\mathfrak{a},R)(a) \cdot \mathrm{in}(\mathfrak{a},R)(x_1) = 0 \quad ,$$

hence $\mathrm{in}(\mathfrak{a},R)(a) = 0$, which is impossible. Moreover, $\mathrm{in}(\mathfrak{a},R)(x_1)$ being a non-zero-divisor in $G(\mathfrak{a},R)$ means exactly

$$x_1R \cap \mathfrak{a}^n = \mathfrak{a}^{n-0(x_1)}x_1 \quad \text{for all} \quad n \geq 0 \quad ,$$

so x_1 is an $(\mathfrak{a},R)$-standard base of x_1R. For $r > 0$ we put $\bar{R} = R/x_1R$, $\bar{\mathfrak{a}} = \mathfrak{a}\bar{R}$. By the case $r = 1$ we know that x_1 is regular in R and

$$G(\bar{\mathfrak{a}},\bar{R}) = G(\mathfrak{a},R)/(\mathrm{in}(\mathfrak{a},R)(x_1)) \quad .$$

Let $g : G(\mathfrak{a},R) \longrightarrow G(\bar{\mathfrak{a}},\bar{R})$ be the canonical homomorphism. Then $g(\mathrm{in}(\mathfrak{a},R)(x_2)),\ldots,g(\mathrm{in}(\mathfrak{a},R)(x_r))$ is a regular sequence and in particular $g(\mathrm{in}(\mathfrak{a},R)(x_i)) \neq 0$. This implies

$$g(\mathrm{in}(\mathfrak{a},R)(x_i)) = \mathrm{in}(\bar{\mathfrak{a}},\bar{R})(\bar{x}_i) \ , \ i = 2,\ldots,r \quad ,$$

where $\bar{x}_i$ denotes the image of x_i in $\bar{R}$. Now the inductive assumption implies that

$$\bar{x}_2,\ldots,\bar{x}_r \quad \text{is a regular sequence in} \quad \bar{R} = R/x_1R$$

and

$$G(\mathfrak{a}(R/x_1R),R/x_1R) = G(\mathfrak{a},R)/(\mathrm{in}(\mathfrak{a},R)(x_1),\ldots,\mathrm{in}(\mathfrak{a},R)(x_r)) \quad ,$$

which concludes the proof.

We conclude this section by showing how our methods above can be used to prove a classical result due to D. Rees .

(13.11 Theorem). Let $\underline{x} = (x_1,\ldots,x_r)$ be a sequence of elements of $\mathfrak{m}$ and let

$$h : R/\underline{x}R[X_1,\ldots,X_r] \longrightarrow G(\underline{x}R,R)$$

be the canonical homomorphism defined by $h(X_i) = \mathrm{in}(\underline{x}R,R)(x_i)$. The following conditions are equivalent:

(i) $\underline{x}$ is a regular sequence.

(ii) h is an isomorphism.

Proof. Clearly (ii) ⇒ (i) by Theorem (13.10). For (i) ⇒ (ii) we use induction on r , the case $r = 0$ being trivial. If $r \geqq 1$, h induces by inductive assumption an isomorphism

$$\bar{h} : \bar{R}/(\bar{x}_1,\ldots,\bar{x}_{r-1})\bar{R}[X_1,\ldots,X_{r-1}] \longrightarrow G((\bar{x}_1,\ldots,\bar{x}_{r-1})\bar{R},\bar{R}) \quad ,$$

where $\bar{R} = R/x_r$. By (13.10), $\underline{x}$ is an $\underline{x}R$-standard base of $\underline{x}R$, hence by (13.9), x_r is an $\underline{x}R$-standard base of x_rR . Therefore

$$G(\underline{x}\bar{R},\bar{R}) \simeq G(\underline{x}R,R) \ / \ \mathrm{in}(\underline{x}R,R)(x_r) \quad ,$$

where $\underline{x}\bar{R} = (\bar{x}_1,\ldots,\bar{x}_{r-1})\bar{R}$, and $\mathrm{in}(\underline{x}R,R)(x_r)$ is a non-zero-divisor in $G(\underline{x}R,R)$. Hence the isomorphism $\bar{h}$ can be written in this way

$$\bar{h} : R/\underline{x}R[X_1,\ldots,X_{r-1}] \longrightarrow G(\underline{x}R,R) \ / \ (\mathrm{in}(\underline{x}R,R)(x_r)) \quad .$$

Assume now, if possible, that $\mathrm{Ker}\, h \neq 0$ and let $f(x_1,\ldots,x_r)$ be a non-zero homogeneous element of $\mathrm{Ker}\, h$ of minimal degree n . We write

$$f(X_1,\ldots,X_r) = \sum_{i=0}^{n} a_i X_r^i \quad , \quad a_i \in R/\underline{x}R[X_1,\ldots,X_{r-1}] \quad ,$$

where a_i is homogeneous of degree $n-i$. Then

$$\overline{h}(f(X_1,\ldots,X_{r-1},0)) = \overline{h}(a_\circ(X_1,\ldots,X_{r-1})) = 0$$

implies $a_\circ = 0$, and therefore $f = X_r\, g$. Applying h we get

$$\mathrm{in}(\underline{x}R,R)(x_r) \cdot h(g) = 0 \quad ,$$

which implies $g \in \operatorname{Ker} h$ in contradiction to the minimality of the degree of f.

<u>(13.12) Corollary</u>. Let $\underline{x} = (x_1,\ldots,x_r)$ be a regular sequence in R. Then $x_r \cdot R \cap (xR + \ldots + x_{r-1}R)^n = x_r(x_1R + \ldots + x_{r-1}R)^n$ for all $n \geq 0$.

Proof. Let $\underline{y} = (x_1,\ldots,x_{r-1})$. Since $G(\underline{y}R,R)$ is a polynomial ring over $R/\underline{y}R$ by Theorem (13.11), we know that $\mathrm{in}(\underline{y}R,R)(x_r)$ is a non-zero-divisor in $G(\underline{y}R,R)$. So by Theorem (13.10), x_r is a $(\underline{y}R,R)$-standard base of x_rR, which means that

$$x_rR \cap (\underline{y}R)^n = x_r \cdot (\underline{y}R)^n \quad \text{for all} \quad n \geq 0 \quad .$$

by Theorem (13.7).

At the end of the last section we raised the question of how to determine generators of the strict transform of an ideal under blowing up. Using Theorem (13.7) we can now give the following answer to this question:

<u>(13.13) Proposition</u>. Let $R \longrightarrow R_1$ be a blowing up homomorphism of $(R,\mathfrak{a})$ and let $\mathfrak{b}$ be a proper ideal of R. Let $f_1,\ldots,f_r$ be an $(\mathfrak{a},R)$-standard base of $\mathfrak{b}$ and let $d_i = \mathrm{ord}(\mathfrak{a},R)(f_i)$, $i = 1,\ldots,r$. If x is any element of $\mathfrak{a}$ such that $\mathfrak{a}R_1 = xR_1$, then

$$f_1' = f_1/x^{d_1},\ldots,f_r' = f_r/x^{d_r}$$

generate the strict transform $\mathfrak{b}_1$ of $\mathfrak{b}$ in R_1 (see Definition (12.22)).

Proof. Let $\mathfrak{p} \subset B(\mathfrak{a},R)$ be the homogeneous prime ideal for which

$$R_1 = B(\mathfrak{a},R)_{(\mathfrak{p})} \quad .$$

By assumption we have $xt \notin \mathfrak{p}$. From the definition of the strict transform it is clear that

$$\mathfrak{b}_1 = \left\{ \frac{a}{b} \mid a \in B(\mathfrak{a},\mathfrak{b} \subset R)_n \ , \ b \in B(\mathfrak{a},R)_n \setminus \mathfrak{p} \right\} \quad .$$

For any $b \in B(\mathfrak{a},R)_n \setminus \mathfrak{p}$, $b/(xt)^n$ is a unit in R_1 , hence

$$\mathfrak{b}_1 = \left\{ \frac{a}{(xt)^n} \mid a \in B(\mathfrak{a},\mathfrak{b} \subset R)_n \right\} \quad .$$

Therefore, to prove that $f_1',\ldots,f_r'$ generate $\mathfrak{b}_1$, it suffices to show that

$$f_1 t^{d_1},\ldots,f_r t^{d_r} \quad \text{generate} \quad B(\mathfrak{a},\mathfrak{b} \subset R) \quad .$$

Now any homogeneous element of degree d of $B(\mathfrak{a},\mathfrak{b} \subset R)$ has the form zt^d where

$$z \in \mathfrak{b} \cap \mathfrak{a}^d = \sum_{i=1}^{r} \mathfrak{a}^{d-d_i} f_i \qquad \text{(by Theorem (13.7))} \quad .$$

So $z = \sum_{i=1}^{r} a_i f_i$, $a_i \in \mathfrak{a}^{d-d_i}$ and therefore in $B(\mathfrak{a},R)$ we have

$$zt^d = \sum_{i=1}^{r} \left(a_i t^{d-d_i} \right) \left(f_i t^{d_i} \right) \quad .$$

§ 14. Examples

In Theorem (13.11), the structure of $G(\underline{x}R,R)$ is given in the case that $\underline{x}$ is a regular sequence in R . Using this result (or rather its Corollary (13.12)) we can describe $B(\underline{x}R,R)$ and $\mathcal{R}(\underline{x}R,R)$ in terms of generators and relations.

<u>(14.1) Proposition</u>. Let $\underline{x} = (x_1,\dots,x_r)$ be a regular sequence in the local ring R. Then we have:

a) The kernel of the canonical R-epimorphism

$$\beta : R[T_1,\dots,T_r] \twoheadrightarrow B(\underline{x}R,R) \ , \ \beta(T_i) = x_i t$$

is generated by $\{x_iT_j - x_jT_i \mid 1 \leq i \leq j \leq r\}$.

b) The kernel of the canonical R-epimorphism

$$\rho : R[T_1,\dots,T_rU] \twoheadrightarrow \mathfrak{R}(\underline{x}R,R)$$

is generated by $\{x_i - UT_i \mid 1 \leq i \leq r\}$.

Proof. a) Let H be the kernel of β and denote by H_0 the ideal of $R[T_1,\dots,T_r]$ generated by $\{x_iT_j - x_jT_i \mid 1 \leq i \leq j \leq r\}$. Clearly $H_0 \subset H$. We use induction on r to show that $H = H_0$, the cases $r = 0$ and $r = 1$ being trivial. So assume $r > 1$. Now H and H_0 are homomeneous ideals, and we will show that every non-zero homogeneous element of H belongs to H_0 . So let

$$F = \sum_{i=0}^{d} a_i(T_1,\dots,T_{r-1})T_r^i \in H$$

be homogeneous of degree d , $a_i(T_1,\dots,T_{r-1}) \in T[R_1,\dots,T_{r-1}]$ are homogeneous of degree $d-i$. For $d = 0$ nothing is to prove. For $d > 0$ we proceed by induction on d . By assumption we have

$$\beta(F) = t^d \cdot \sum_{i=0}^{d} a_i(x_1,\dots,x_{r-1})x_r^i = 0 \quad ,$$

hence $\sum_{i=0}^{d} a_i(x_1,\dots,x_{r-1})x_r^i = 0$ and therefore

$$a_\circ(x_1,\dots,x_{r-1}) \in x_rR \cap (x_1R+\dots+x_{r-1}R)^d = x_r \cdot (x_1R+\dots+x_{r-1}R)^d$$

by Corollary (13.12). It follows that there is a homogeneous polynomial $b_\circ(T_1,\dots,T_{r-1}) \in R[T_1,\dots,T_{r-1}]$ of degree d such that

$$a_\circ(x_1,\dots,x_{r-1}) = x_r \cdot b_\circ(x_1,\dots,x_{r-1}) \quad .$$

Note that $x_r b_\circ(T_1,\dots,T_{r-1}) = x_r \sum_{i=1}^{r-1} \alpha_i T_i = \sum_{i=1}^{r-1} \alpha_i (x_r T_i - x_i T_r) + \left(\sum_{i=1}^{r-1} \alpha_i x_i\right) T_r$, for some α_i . If we define

$$G = x_r b_\circ (T_1,\dots,T_{r-1}) + \sum_{i=1}^{d} a_i (T_1,\dots,T_{r-1}) T_r^i ,$$

then we see that $G \in H$. Therefore

$$F - G = a_\circ(T_1,\dots,T_{r-1}) - x_r b_\circ(T_1,\dots,T_{r-1}) \in H \cap R[T_1,\dots,T_{r-1}] ,$$

so our inductive assumption on r implies $F - G \in H_0$. Furthermore, since $d = \deg b_\circ > 0$,

$$G \equiv T_r G_1 (T_1,\dots,T_r) \mod H_0 \quad ,$$

where G_1 is homogeneous of degree $d-1$. Since $\beta(G) = 0 = \beta(T_r)\beta(G_1)$ and $\beta(T_r) = x_r t$ is a non-zero-divisor in $B(\underline{x}R,R)$, we have $G_1 \in H$. So our inductive assumption on d implies $G_1 \in H_0$ and therefore also $F \in H_0$.

b) Giving degree -1 to U , ρ is a graded homomorphism and $\operatorname{Ker}\rho$ is homogeneous again. Let K_0 be the ideal of $R[T_1,\dots,T_r,U]$ generated by $\{x_i - UT_i \mid 1 \leq i \leq r\}$. We will show that every homogeneous element of $\operatorname{Ker}\rho$ belongs to K_0 , and this will prove the assertion since $K_0 \subset \operatorname{Ker}\rho$. So let $F \in \operatorname{Ker}\rho$ be homogeneous of degree d , i.e.

$$F = \sum_{j=0}^{s} F_j (T_1,\dots,T_r) U^j \quad ,$$

when $F_j(T_1,\dots,T_r)$ are homogeneous of degree $d+j$. Then

$$F = F_0 (T_1,\dots,T_r) + \sum_{j=1}^{s} g_j (x_{i_j} - UT_{i_j}) + \sum_{j=1}^{s} g_j x_{i_j} \quad , \quad 1 \leq i_1 \leq \dots \leq i_s \leq 1 \quad .$$

Hence

$$F \equiv G(T_1,\dots,T_r) \mod K_0 \quad , \quad \deg G = d \quad ,$$

and since β is the restriction of ρ to $R[T_1,\dots,T_r]$, we have $G \in \operatorname{Ker}\beta$. Note now that

$$x_iT_j - x_jT_i = T_j(x_i - UT_i) - T_i(x_j - UT_j) \in K_0 \quad ,$$

and therefore $\operatorname{Ker}\beta \subset K_0$ by a). Now it is clear that also $F \in K_0$.

In connection with this example we make the following observation. If R is a local Cohen-Macaulay ring and $\underline{x}$ is a regular sequence in R then $G(\underline{x}R,R)$ and $\mathfrak{R}(\underline{x}R,R)$ are Cohen-Macaulay. For $G(\underline{x}R,R)$ this is clear from Theorem (13.11) and Corollary (11.13). For $\mathfrak{R}(\underline{x}R,R)$ it follows then from Proposition (11.16),a), but it can also be seen from the description of $\mathfrak{R}(\underline{x}R,R)$ above. Namely $R[T_1,\dots,T_r,U]$ is Cohen-Macaulay by Corollary (11.13) again, and an easy computation shows that the generators $x_1 - UT_1,\dots,x_r - UT_r$ of $\operatorname{Ker}\rho$ are a regular sequence in $R[T_1,\dots,T_r,U]$ (even without assuming R to be Cohen-Macaulay). Finally we note that our results of Chapter V (see Theorem (25.4)) imply that also $B(\underline{x}R,R)$ is Cohen-Macaulay, a fact which is not immediate from the description above.

(14.2) In Lemma (9.8) we noted that forming $G(\mathfrak{a},M)$, $B(\mathfrak{a},M)$ and $\mathfrak{R}(\mathfrak{a},M)$ is compatible with flat base change. This is no longer true if the flatness assumption is dropped. As a typical non-flat homomorphism $R \longrightarrow S$ take $k[[x,y]] \longrightarrow k[[x]] \simeq k[[x,y]]/y\cdot k[[x,y]]$. Choosing $\mathfrak{a} = xR + yR$ and using (14.1) we obtain:

a) $G(\mathfrak{a},R) \simeq k[x,y]$, so $G(\mathfrak{a},R) \otimes_R S \simeq k[x,y]$, but
   $G(\mathfrak{a}\cdot S, R\otimes_R S) \simeq k[x]$.

b) $B(\mathfrak{a},R) \simeq R[X,Y]/(xY - yX)$, so $B(\mathfrak{a},R)\otimes_R S \simeq S[X,Y]/(xY)$, but
   $B(\mathfrak{a}S, R\otimes_R S) \simeq S[X]$.

c) $\mathfrak{R}(\mathfrak{a},R) \simeq R[X,Y,U]/(x-UX, y-UY)$, so $B(\mathfrak{a},R)\otimes_R S \simeq$
   $\simeq S[X,Y,U]/(x - UX, UY)$, but $\mathfrak{R}(\mathfrak{a},S,R\otimes_R S) \simeq S[X,U]/(x - UX)$.

(14.3) Let A be a graded (noetherian) ring having a unique maximal homogeneous ideal $\mathfrak{m}$, and let $\underline{x}$ be a sequence of homogeneous elements in $\mathfrak{m}$. Then we know from Corollary (11.3) that $\underline{x}$ is a regular sequence in A if and only if it is a regular sequence in $A_{\mathfrak{m}}$. If $\underline{x}$ does not consist of homogeneous elements, then this equivalence is no longer valid. Take

$$A = k[x,y,z]$$

and $x_1 = x(z-1)$, $x_2 = y(z-1)$, $x_3 = z$, $\underline{x} = (x_1,x_2,x_3)$. Then $\underline{x}$ is not a regular sequence, but it becomes regular after localizing at the origin $xA + yA + zA$.

(14.5) There is a classical example, due to Samuel, that a graded ring over a finite field need not have a homogeneous system of parameters of degree one. For this let k be the prime field of characteristic 2 and let

$$A = k[X,Y] / (XY(X+Y)) = k[x,y] \quad .$$

Here $A_1 = \{x,y,x+y\}$ and hence every generator of A_1 is contained in a minimal prime ideal of A . Here $\dim A = 1$, and x^2+y^2+xy is a homogeneous parameter of degree 2 . If we replace the polynomials by power series, i.e. we put

$$R = k[[X,Y]]/(XY(X+Y)) = k[[x,y]] \quad , \quad \mathfrak{m} = xR + yR$$

then $G(\mathfrak{m},R) \simeq A$ and therefore $\mathfrak{m}$ is a minimal reduction of itself, although $s(\mathfrak{m}) = 1 < 2 = \mu(\mathfrak{m})$.

(14.6) Here is an easy example that different ideals in a local ring may define the same scheme by blowing up. Take

$$R = k[[X,Y,Z]]/(X^2,XY,XZ) = k[[x,y,z]] \quad ,$$

$$\mathfrak{a}_1 = \mathfrak{m} = xR + yR + zR \quad , \quad \mathfrak{a}_2 = yR + zR \quad .$$

Since $R_x = 0$, x does not contribut to the blowing up, so the morphisms

$$B\ell(\mathfrak{a}_1,R) \longrightarrow \operatorname{Spec} R \quad \text{and} \quad B\ell(\mathfrak{a}_2,R) \longrightarrow \operatorname{Spec} R$$

are the same. The difference is in the exceptional divisor, whose image is $\operatorname{Spec}(R/\mathfrak{a}_1)$ in the first case and $\operatorname{Spec}(R/\mathfrak{a}_2)$ in the second case.

The above scheme $B\ell(\mathfrak{a}_1,R) = B\ell(\mathfrak{a}_2,R)$ is covered by the two affine pieces

$$\operatorname{Spec}(k[[X,Y]][X/Y]) \quad \text{and} \quad \operatorname{Spec}(k[[X,Y]][Y/X]) \quad ,$$

glued in the obvious way. Now if we forget about the morphism $B\ell(\mathfrak{a}_1,R) \longrightarrow \operatorname{Spec} R$ and just consider the scheme $B\ell(\mathfrak{a}_1,R)$, then this is the same as blowing up $k[[X,Y]]$ at the origin. These examples should make clear that the blowing up is not just a scheme, but it is a pair consisting of a morphism and an exceptional divisor (or equivalently a morphism and a center of blowing up). We note however that in some rare cases the center is "determined" by the morphism $B\ell(\mathfrak{a},R) \longrightarrow \operatorname{Spec} R$ (see [1], (1.9.7).

<u>(14.7)</u> The classical geometric situation for blowing up is the following one. We put

$$S = k\left[X_1,\ldots,X_n\right]_{(x_1,\ldots,x_n)} \quad ,$$

we choose $f \in k[X_1,\ldots,X_n]$ such that $f(0,\ldots,0) = 0$ and we put $R = S/fS$. Let $\mathfrak{M}$ be the maximal ideal of S , $\mathfrak{m}$ the maximal ideal of R and

$$e = \operatorname{ord}(\mathfrak{M},S)(f) \quad ,$$

so e is the degree of the least homogeneous component of f . Then by the definition of a strict transform and by Proposition (13.13) we know that we may compute $B\ell(\mathfrak{m},R)$ in the following way. If $R \longrightarrow R_1$ is any blowing up homomorphism of $(R,\mathfrak{m})$, then there is a blowing up homomorphism $S \longrightarrow S_1$ of $(S,\mathfrak{M})$ such that R_1 is S-isomorphic to $S_1/f'S_1$, where

$$f' = f/x^e \quad , \quad xS_1 = \mathfrak{M}S_1 \quad .$$

Assume for example that $\mathfrak{M}S_1 = X_1S_1$ and write

$$f = f_e(X_1,\ldots,X_n) + f_{e+1}(X_1,\ldots,X_n) + \ldots + f_m(X_1,\ldots,X_n)$$

f_i homogeneous of degree i . Then

$$f' = f/X_1^e = f_e\left(1,\frac{X_2}{X_1},\dots,\frac{X_n}{X_1}\right) + X_1 f_{e+1}\left(1,\frac{X_2}{X_1},\dots,\frac{X_n}{X_1}\right) + \dots$$

$$\dots + X_1^{m-e}\, f_m\left(1,\frac{X_2}{X_1},\dots,\frac{X_n}{X_1}\right) \quad .$$

So writing $Y_i = X_i/X_1$, $i = 1,\dots,n$, we see that f' is obtained by substituting $X_i = X_1Y_i$ and cancelling the common factor X_1^e . We will give two explicit examples for plane curves, in which case it is convenient to write x,y for the variables.

First let

$$f = f(x,y) = x^3 - y^2 \quad ,$$

so $e = 2$. If the exceptional divisor at a point of the blowing up is defined by x , a strict transform is defined by f/x^2 , so we put $x' = x$, $y' = y/x$, and we have

$$f(x,y) = x^3 - x^2y'^2$$

and hence

$$f/x^2 = x - y'^2 \in k[x,y][\tfrac{y}{x}] = k[x,y']$$

(note that $y = xy'$) . If we look at this equation at a point of the exceptional divisor, then x is a non-unit, and f/x^2 is a non-unit if and only if y' is in the maximal ideal. This means that the only point on the exceptional divisor is given by $x = 0, y' = 0$, the origin of $\mathrm{Spec}(k[x,y'])$. We conclude that the corresponding ring S_1 is equal to $k[x,y']_{(x,y')}$. Now looking for $k[x,y][\frac{x}{y}] = k[x'',y]$ where $x'' = x/y$, we have

$$f(x,y) = y^3(x'')^3 - y^2 \ , \quad f/y^2 = y(x'')^3 - 1 \quad .$$

Again on the exceptional divisor y is a non-unit, so f/y^2 does not define any point there.

For the purpose of blowing up, the above example is the simplest one. Something different happens in the following one:

$$g = (x-y)^2 + x^3 \quad .$$

Again $e = 2$, and as above g is non-singular outside the origin. Let us again determine the points on the exceptional divisor of blowing up the origin. On $k[x,y][\frac{y}{x}] = k[x,y']$ we have the strict transform

$$g/x^2 = g_1 = (1-y')^2 + x^3 \quad ,$$

so on the exceptional divisor we get the point $x = 0,\ y' = 1$ of multiplicity 2 . On $k[x,y][\frac{x}{y}] = k[x",y]$ we get

$$g/y^2 = g_2 = (x"-1)^2 + y(x")^3 \quad .$$

Here the only point on the exceptional divisor is $x" = 1$, $y = 0$, and this is of multiplicity 2 again. But actually both points are the same. Namely, if we localize e.g. $k[x,y']$ at the maximal ideal $(x,y'-1)$, then $y' = y/x$ becomes a unit and hence g_1 and

$$g_2 = \frac{y^2}{x^2}\, g_1$$

define the same ideal. This is a general fact: If we blow up the local ring of a plane curve at a singularity, and if this blowing up contains a point of the same multiplicity, then this point is unique (see [4*]).

We also point out another important difference between the two above examples. In the first example, after blowing up there was just one point to look at, and the coordinates were given by x and y' . But in the second example instead of using x and y' as parameters, we had to choose x and $y'-1$. This translation of coordinates is an important phenomenon for the computations of resolution of singularities, because it means that looking for the singularities after blowing up one has to consider all possible changes of coordinates.

(14.8) In the example (14.7) given before, it is also important to know that the rings S_1 occuring there, i.e. the quadratic transforms of the regular local ring S , are regular again. More generally we have the following fact: Let S be a regular local ring, let $\mathfrak{p} \neq 0$ be a prime ideal of S such that $S/\mathfrak{p}$ is regular, and let $S \longrightarrow S_1$ be a blowing up homomorphism of $(S,\mathfrak{p})$. Then S_1 is regular.

One possible proof of this fact is like this: Since $\mathfrak{p}$ is generated by a regular sequence, we know from Theorem (13.11) and Corollary (12.19)a) that $B\ell(\mathfrak{p},S)$ is Cohen-Macaulay. In Chapter VI we will show that $e(S_1) \leqq e(S) = 1$, so by our criterion of multiplicity one (Theorem (6.8)), S_1 must be regular. Another method is to use that actually all localizations of $G(\mathfrak{p},S)$ are regular since $G(\mathfrak{p},S)$ is a polynomial ring over a regular local ring. From this it is immediate that also $B\ell(\mathfrak{p},S)$ is regular. A still different proof has been given by Abkyankar([1], (1.4.2)), who constructs explictly a regular system of parameters of S_1 .

It is not clear to us if there are also other ideals $\mathfrak{a}$ of a regular local ring S such that $B\ell(\mathfrak{a},S)$ is regular.

<u>(14.9)</u> Let $R = k[[X_1,\ldots,X_n]]$ or $R = k\left[X_1,\ldots,X_n\right]_{(X_1,\ldots,X_n)}$ and let $f_1,\ldots,f_m$ be any homogeneous polynomials in $X_1,\ldots,X_n$. If $\mathfrak{m}$ denotes the maximal ideal of R and $\mathfrak{a} = f_1R+\ldots+f_mR$, then $f_1,\ldots,f_m$ are an $(\mathfrak{m},R)$-standard base of $\mathfrak{a}$. In fact, the criterion of Theorem (13.7), (V) is trivially satisfied.

Now the remaining examples consist in explicit computations of standard bases in $R = k[[X,Y,Z,\ldots,]]$ with respect to the maximal ideal $\mathfrak{m}$ of R . We use the notation introduced in (13.2) (with $\mathfrak{a} = \mathfrak{m}$, $M = R$) .

<u>(14.10)</u> In $R = k[[X,Y,Z,W]]$ we consider $\mathfrak{a} = \underline{g}R$ where

$$g_1 = ZW - XY^2 \ , \ g_2 = Z^2 - XW^2 \ , \ \underline{g} = (g_1,g_2) \quad .$$

Clearly $\underline{g}$ is a regular sequence in R , but $g_1^* = ZW, g_2^* = Z^2$ is not a regular sequence in $G(\mathfrak{m},R) \simeq k[X,Y,Z,W]$. Hence by Theorem (13.10), $\underline{g}$ is not an $(\mathfrak{m},R)$-standard base of $\mathfrak{a}$. We have

$$(Z_1 - W) \in \operatorname{Ker} f^*(\underline{g}) \quad ,$$

but since

$$Zg_1 - Wg_2 = XW^3 - XY^2Z \neq 0 \quad ,$$

there is no element in $\operatorname{Ker} f(\underline{g})$, whose image under $h(\underline{g})$ is $(Z,-W)$.

Now we put $g_3 = Zg_1 - Wg_2 = XW^3 - XY^2Z \in \mathfrak{a}$ and claim that (g_1,g_2,g_3) is an $(\mathfrak{m},R)$-standard base of $\mathfrak{a}$. We have

$$g_3^* = XW^3 - XY^2Z \quad ,$$

and it is not difficult to check that $\operatorname{Ker} f^*(g_1,g_2,g_3)$ is generated by

$$v_1 = (Z,-W,0) \ , \ v_2 = (XW^2,-XY^2,-Z) \quad .$$

We have

$$0(g_1,g_2,g_3)(v_1) = 3 \ , \ 0(g_1,g_2,g_3)(v_2) = 5 \quad .$$

Clearly $u_1 = (Z,-W,-1) \in \operatorname{Ker} f(g_1,g_2,g_3)$, and since

$$0(-1) + 0(g_3) = 4 > 0(g_1,g_2,g_3)(u_1)$$

we have by definition

$$h(g_1,g_2,g_3)(u_1) = v_1 \quad .$$

Similarly $u_2 = (XW^2,-XY^2,-Z) \in \operatorname{Ker} f(g_1,g_2,g_3)$ and $h(g_1,g_2,g_3)(u_2) = v_2$. So we are done by Theorem (13.7)(v).

<u>(14.11)</u> In $R = k[[X,Y,Z,W]]$ we take $\mathfrak{a} = \underline{g}R$ where $\underline{g} = (g_1,\dots,g_4)$ and

$$g_1 = Y^2W^3 - Z^2 \ , \ g_2 = YZ - XW^3 \ , \ g_3 = XZ - Y^3 \ , \ g_4 = Y^4 - X^2W^3 \quad .$$

One can check that $g_1,\dots,g_3$, although they generate $\mathfrak{a}$, are not a standard base. Here $\operatorname{Ker} f^*(\underline{g})$ is generated by

$$v_1 = (X,0,Z,0) \quad , \quad v_2 = (Y,Z,0,0) \quad ,$$
$$v_3 = (0,-X,Y,0) \quad , \quad v_4 = (0,-Y^3,0,Z) \quad .$$

Putting

$$U_1 = (X,Y^2,Z,0) \quad , \quad U_2 = (Y,Z,W^3,0) \quad ,$$
$$U_3 = (0,-X,Y,1) \quad , \quad U_4 = (0,-Y^3,XW^3,Z)$$

we have $u_i \in \operatorname{Ker} f(\underline{g})$ for $i = 1,\dots,4$ and

$$h(\underline{g})(u_i) = v_i \quad , \quad i = 1,\dots,4 \quad .$$

So we obtain

$$G(\mathfrak{m},R)/G(\mathfrak{m},\mathfrak{a} \subset R) \simeq k[X,Y,Z,W]/(Z^2,YZ,XZ,Y^4)$$
$$= k[X,Y,Z,W]/(Z,Y^4) \cap (X,Y,Z)^2 \quad .$$

Going back to $R/\mathfrak{a}$, it is easy to check that the kernel of

$$\varphi : k[[X,Y,Z,W]] \longrightarrow k[[t^2,tu^3,tu^9,u^4]]$$
$$\varphi(X) = t^2 , \varphi(Y) = tu^3 , \varphi(Z) = tu^9 , \varphi(W) = u^4$$

is exactly $\mathfrak{a}$. Now applying the Projection formula to

$$k[[t^2,u^4]] \subset k[[t^2,tu^3,tu^9,u^4]]$$

and observing that

$$k((t^2,tu^3,tu^9,u^4)) = k((t^2,u^4))(tu)$$

defines an extension of degree 4, we see that $(t^2,u^4)\cdot R/\mathfrak{a}$ has multiplicity 4. But

$$R/\mathfrak{a} + XR + YR \simeq k[[Y,Z]]/(Y^3,YZ,Z^2)$$

has length 4, which shows that $R/\mathfrak{a}$ is Cohen-Macaulay. Considering the exceptional divisor of $B\ell(\mathfrak{m}/\mathfrak{a},R/\mathfrak{a})$, given by

$$G(\mathfrak{m}/\mathfrak{a},R/\mathfrak{a}) \simeq k[X,Y,Z,W]/(Z,Y^4) \cap (X,Y,Z)^2$$

at the affine piece $W \neq 0$, we obtain at the origin $X = Y = Z = 0$

a point with local ring

$$R_1 = k[X,Y,Z]_{(X,Y,Z)} / (Z,Y^4) \cap (X,Y,Z)^2 \quad .$$

Since R_1 has an embedded component, it is not Cohen-Macaulay. So we have shown that the blowing up of a Cohen-Macaulay scheme at a point need not be Cohen-Macaulay again.

(14.12) Without giving all the details we mention another example. In $R = k[[X,Y,Z,T]]$ we want to construct an $(\mathfrak{m},R)$-standard base of the ideal generated by

$$g_1 = X^2Y + X^4T + Z^6 \quad \text{and} \quad g_2 = XY^2 \quad .$$

Proceeding as before one sees immediately that

$$g_3 = Yg_1 - Xg_2 = X^4YT + YZ^6$$

must be added to g_1, g_2. Now $g_3^* = X^4YT$ is a multiple of g_1^*, but this does not mean that g_1, g_2, g_3 are a standard base. Using the same algorithm as in example (14.10) one shows that an $(\mathfrak{m},R)$-standard base is given by g_1, g_2, g_3 together with

$$g_4 = X^6T^2 + X^2Z^6T - YZ^6 \quad \text{and} \quad g_5 = X^8T^2 + 2X^4Z^6T + Z^{12} \quad .$$

After knowing that we have reached a standard base, g_3 can be omitted, so (g_1, g_2, g_4, g_5) is an $(\mathfrak{m},R)$-standard base of $g_1R + g_2R$, but for the construction of g_1, g_5 it is essential to keep g_3.

In example (7.7) we used that $(X^3 - Y^2, XZ - YW, Z^2 - XW^2, YZ - X^2W)$ is an $(\mathfrak{m},R)$-standard base of the corresponding ideal without giving a proof. This is easy, using the same method as above, and we leave it to the reader as an exercise.

Appendix

Homogeneous subrings of a homogeneous ring

Here we give an additional report on some of the topics concerning the more elementary parts of the study of homogeneous rings.

A noetherian (positively) graded ring $A = \bigoplus_{n \geq 0} A_n$ is called a homogeneous ring over a noetherian ring R , if A is generated over $A_0 = R$ by homogeneous elements of degree one. Recall that the classical parts of commutative algebra are mainly modelled by ideal theory. For studying homogeneous rings A we can investigate homogeneous subrings of A instead of homogeneous ideals. We want to sketch one branch of this way in this section. A graded subring of a homogeneous ring A over R is called a homogeneous subring when it is again a homogeneous ring over R . It is an obvious remark that there is a one-to-one correspondence between the homogeneous subrings of A and the R-submodules of A_1 . This concept was outlined by S. Abhyankar in [1], § 12 for investigating homogeneous domains over an algebraically closed field. One important result in Abhyankar's book may be seen in the fact that for any homogeneous domain A over an algebraically closed field, every homogeneous prime ideal $\mathfrak{p}$ of A with $\dim(A/\mathfrak{p}) = 1$ can be generated by linear forms. New aspects of Abhyankar's method were recently given by A. Ooishi [11]. We give a short survey of his work. First of all Ooishi extended the concept of the reduction of an ideal, given by D.G. Northcott and D. Rees [10] , in the following way. Suppose that R is a local ring with infinite residue field k . Let A be again a homogeneous ring over R and $M = \bigoplus_{n \in \mathbf{Z}} M_n$ a non-zero finitely generated graded A-module. Then a homogeneous subring B of A is called a reduction of A with respect to M if M is a finitely generated B-module. Moreover we call the Krull dimension of $M \otimes_R k$ the analytic spread, say $s(M)$, of M . These definitions are natural generalizations of the coresopnding notions for ideals in a local ring R in the sense of Northcott and Rees. In particular, if J is a reduction of an ideal I with respect to a finitely generated R-module E , then the Rees algebra $B(J,R)$ is a reduction of $B(I;R)$ with respect to $B(I,E)$, and $s(B(I,E)$ coincides with the analytic spread of I with respect to E in the sense of Northcott and Rees as defined in (10.10). Next Ooishi defined the socalled pseudo-flatness for homogeneous rings as follows: we say that M is a pseudo-flat A-module if

$s(M_{\mathfrak{p}})$ is constant for all $\mathfrak{p} \in \mathrm{Spec}(R)$. Then we have the following structure theorem for pseudo-flat modules.

Theorem 1. Let $A_0 = R$ be a reduced local ring with infinite residue field and let A be a homogeneous ring over R. Then M is a pseudo-flat A-module if and only if there is a polynomial R-subalgebra B of A so that M is a finitely generated faithful B-module.

The pseudo-flatness corresponds to the normal pseudo-flatness in the sense of Hironaka [8] (see [7], [9]). Note that for an ideal I of R, $G(I,R)$ is a pseudo-flat R/I-module if and only if R is normally pseudo-flat along I (or equivalently, I is equimultiple in the sense of our Chapter IV). Using the above structure theorem we can easily see that the set of prime ideals $\mathfrak{p}$ of R so that $M_{\mathfrak{p}}$ is a pseudo-flat $A_{\mathfrak{p}}$-module is an open set in $\mathrm{Spec}(R)$, and that for a locally pseudo-flat A-module M (i.e., $M_{\mathfrak{p}}$ is a pseudo-flat $A_{\mathfrak{p}}$-module for all $\mathfrak{p} \in \mathrm{Spec}(R)$) the function $\mathfrak{p} \overset{\psi}{\longmapsto} e(M \otimes_R k(\mathfrak{p}))$ satisfies upper semicontinuity, where e denotes the multiplicity with respect to the unique homogeneous maximal ideal of A.

If, in particular, $(R,\mathfrak{m})$ is a reduced local ring and $M/\mathfrak{m}M$ is a Cohen-Macaulay $A/\mathfrak{m}A$-module, then the pseudo-flatness of M and the constancy of the function ψ imply that M is R-free.

Ooishi also discussed the reduction exponent of graded modules: if R is a local ring with infinite residue field, then the reduction exponent of M, say $r_A(M)$, can be defined in the same way as in the classical case:

$$r_A(M) = \min\left\{ \gamma_B(M) \mid B \text{ is a minimal reduction of } A \text{ w.r.t. } M \right\},$$

where $\gamma_B(M) = \min\{t \mid B_1 M_n = M_{n+1} \text{ for all } n \geq t\}$. A reduction of A (w.r.t. M) which is minimal with respect to the relation of inclusion is called a <u>minimal reduction</u> of A (w.r.t. M). Putting $A = B(I,R)$ for an ideal I of R, one can immediately see that $r_A(A)$ coincides with the usual notion of a reduction exponent of the ideal I. If R contains an infinite field, then for a locally pseudo-flat A-module M the function $\mathfrak{p} \longmapsto r_{A_{\mathfrak{p}}}(M_{\mathfrak{p}})$ is upper semi-continuous. Moreover, we conclude for a homogeneous ring over a field the following theorem.

<u>Theorem 2</u>. Let A be a homogeneous ring over an infinite field k, but not regular, say $A = k[X_1,\ldots,X_v]/I$, where the X_i's are indeterminates over k of degree one and $v = \dim_k A_1$. Then we have

(1) $r_A(A) \geq \min\{n \mid I_n \neq (0)\} - 1$.

(2) If the equality holds in (1), then we get

$$e(A) \leq \binom{v+r}{r} - \binom{v+r-1}{r-1} \cdot \dim(A) ,$$

where $r = r_A(A)$.

We remark that we get in (2) the equality $e(A) = \binom{v+r}{r} - \binom{v+r-1}{r-1} \cdot \dim(A)$ if and only if A is Cohen-Macaulay. Moreover in this case A has a linear resolution, see [6], [14]. Concerning the inequality in (2) we know already by Abhyankar [1] that for a homogeneous domain A over an algebraically closed field we have

$$(*) \qquad e(A) \geq v + 1 - \dim(A) .$$

The appendix to Chapter V will be concerned with those homogeneous domains which satisfy the equality in (*).

References - Chapter II

Books

[1] S.S. Abhyankar, Resolution of singularities of embedded algebraic surfaces, Academic Press, New York and London, 1966.

[2] R. Hartshorne, Algebraic Geometry, Graduate Texts in Mathematics Vol. 52, Springer, New York - Heidelberg - Berlin, 1977.

[3] H. Matsumura, Commutative Algebra, 2nd. Ed., Benjamin, New York, 1980.

[4] M. Nagata, Local rings, Interscience Tracts in Pure and Appl. Math. Wiley, New York, 1962.

[4*] V. Cossart, J. Giraud and U. Orbanz, Resolution of surface singularities, Springer-Verlag, 1984.

Papers

[5] E.C. Dade, Multiplicity and monoidal transformations, Thesis Princeton, 1960.

[6] D. Eisenbud and S. Goto, Linear free resolutions and minimal multiplicity, J. Alg. 88, (1984), 89 - 133.

[7] M. Herrmann and U. Orbanz, On equimultiplicity, Math. Proc. Camb. Phil. Soc. 91, (1982), 207 - 213.

[8] H. Hironaka, Normal cones in analytic Whitney stratifications. Publ. Math. IHES 36, (1969), 127 - 138.

[9] J. Lipman, Equimultiplicity, reduction and blowing-up, Commutative Algebra, Analytic Methods, Dekker, New York, (1982), 111 - 147.

[10] D.G. Northcott and D. Rees, Reductions of ideals in local rings, Math. Proc. Camb. Phil. Soc. 50, (1954), 145 - 158.

[11] A. Ooishi, Reductions of graded rings and pseudo-flat graded modules, preprint.

[12] L. Robbiano, On the theory of graded structures, J. Symbolic Computation 2 (1986), 139 - 170.

[13] P. Valabrega and G. Valla, Form rings and regular sequences, Nagoya Math. J. 72 (1978), 93 - 101.

[14] P. Schenzel, Über die freien Auflösungen extremaler Cohen-Macaulay-Ringe, J. Algebra 64, (1980), 93 - 101.

References - Appendix - Chapter II

[1] S.S. Abhyankar, Reduction of singularities of embedded algebraic surfaces, Academic Press, New York and London, 1966.

[2] H. Hironaka, Normal cones in analytic Whitney stratifications, Publ. Math. IHES 36 (1969), 127 - 138.

[3] M. Herrmann - U. Orbanz, On equimultiplicity, Math. Proc. Camb. Phil. Soc. 91 (1982), 207 - 213.

[4] J. Lipman, Equimultiplicity, reduction and blowing-up, Commutative Algebra, Analytic Methods (ed. R. Draper), Marcel Dekker, New York 1982, 111 - 147.

[5] D.G. Northcott - D. Rees, Reductions of ideals in local rings, Proc. Camb. Phil. Soc. 50 (1954), 353 - 359.

[6] A. Ooishi, Reductions of graded rings and pseudo-flat graded modules, preprint 1986.

[7] D. Eisenbud - S. Goto, Linear free resolutions and minimal multiplicity, J. Algeba 88 (1984), 89 - 133.

[8] P. Schenzel, Über die freien Auflösungen extremaler Cohen-Macaulay-Ringe, J. Algebra 64 (1980), 93 - 101.

Chapter III. ASYMPTOTIC SEQUENCES AND QUASI-UNMIXED RINGS

In this chapter we give various characterizations of quasi-unmixed local rings. Most of the results are contained in papers by Ratliff, although with different proofs. Recently it has been recognized that a very useful tool for these characterizations are asymptotic sequences, which are somewhat analogous to regular sequences for the characterization of local Cohen-Macaulay rings. The theory of asymptotic sequences has been developed by Ratliff [10] and independently by Katz [8]. Our treatment follows closely the treatment by Katz [8]. We start by giving some auxiliary results.

§ 15. Auxiliary results on integral dependence of ideals

For this section we fix a noetherian ring R and an ideal $\mathfrak{a}$ of R. Recall from § 8 that

$$\mathcal{R}(\mathfrak{a},R) = \bigoplus_{n\in\mathbf{Z}} \mathfrak{a}^n t^n \subset R[t,u], \quad u = t^{-1}$$

and

$$\mathcal{R}(\mathfrak{a},\mathfrak{b}\subset R) = \bigoplus_{n\in\mathbf{Z}} (\mathfrak{b}\cap\mathfrak{a}^n)t^n = \mathfrak{b}R[t,u]\cap\mathcal{R}(\mathfrak{a},R)$$

for any further ideal $\mathfrak{b}$ of R.

(15.1) Proposition. Let $\mathfrak{b}_1,\mathfrak{b}_2,\mathfrak{p},\mathfrak{q}$, be ideals of R. Then:

a) $\mathcal{R}(\mathfrak{a},\mathfrak{b}_1\cap\mathfrak{b}_2\subset R) = \mathcal{R}(\mathfrak{a},\mathfrak{b}_1\subset R)\cap\mathcal{R}(\mathfrak{a},\mathfrak{b}_2\subset R)$;

b) if $\mathfrak{p}$ is prime, so is $\mathcal{R}(\mathfrak{a},\mathfrak{p}\subset R)$;

c) if $\mathfrak{q}$ is $\mathfrak{p}$-primary then $\mathcal{R}(\mathfrak{a},\mathfrak{q}\subset R)$ is $\mathcal{R}(\mathfrak{a},\mathfrak{p}\subset R)$-primary;

d) if $\mathfrak{b} = \mathfrak{q}_1\cap\ldots\cap\mathfrak{q}_n$

is an (irredundant) primary decomposition of $\mathfrak{b}$ then

$$\mathcal{R}(\mathfrak{a},\mathfrak{b}\subset R) = \mathcal{R}(\mathfrak{a},\mathfrak{q}_1\subset R)\cap\ldots\cap\mathcal{R}(\mathfrak{a},\mathfrak{q}_n\subset R)$$

is an (irredundant) primary decomposition of $\mathcal{R}(\mathfrak{a},\mathfrak{b}\subset R)$;

e) $\mathfrak{p} \longmapsto \mathcal{R}(\mathfrak{a},\mathfrak{p} \subset R)$ induces a bijection $\mathrm{Min}(R) \longrightarrow \mathrm{Min}(\mathcal{R}(\mathfrak{a},R))$;

f) $\mathcal{R}(\mathfrak{a},R)_{red} \simeq \mathcal{R}(\mathfrak{a}\cdot R_{red},R_{red})$.

Proof. a) and b) follow directly from the definition while c) is easy to prove. Then d), e) and f) are immediate consequences of a), b) and c).

Clearly for any n we have $\mathfrak{a}^n = u^n\mathcal{R}(\mathfrak{a},R) \cap R$, which has the following analogue for integral closures:

(15.2) Lemma. Given $x \in R$ and $n \in \mathbf{N}$, the following conditions are equivalent:

(i) x is integral over $\mathfrak{a}^n$;

(ii) xt^n is integral over $B(\mathfrak{a},R)$;

(iii) xt^n is integral over $\mathcal{R}(\mathfrak{a},R)$;

(iv) $x \in u^n\ \overline{\mathcal{R}(\mathfrak{a},R)}$;

(v) $x \in \overline{u^n\mathcal{R}(\mathfrak{a},R)}$.

Proof. (i) $\Rightarrow$ (ii) . Choose m and $a_i \in (\mathfrak{a}^n)^i$, $i = 1,\dots,m$, such that

$$x^m + a_1x^{m-1} + \dots + a_m = 0 \quad .$$

Then

$$(xt^n)^m + (a_1t^n)(xt^n)^{m-1} + \dots + a_mt^{nm} = 0$$

which proves (ii) since $a_it^{ni} \in B(\mathfrak{a},R)$. (ii) $\Rightarrow$ (iii) $\Rightarrow$ (iv) $\Rightarrow$ (v) are trivial. Finally, if (v) holds then

$$x^m + b_1x^{m-1} + \dots + b_m = 0$$

for some m and some $b_i \in u^{ni}\mathcal{R}(\mathfrak{a},R)$. Collecting terms of degree zero we may assume that $b_i = (b_it^{ni})u^{ni} \in \mathfrak{a}^{ni}$, showing that x is integral over $\mathfrak{a}^n$.

(15.3) Remark. In the following sections we want to study the primary decomposition or at least the associated prime of $\overline{\mathfrak{a}^n}$ for n varying, a problem that has been studied by many authors (see e.g. [3], [5], [6], [7], [8], [9], [10], [13]). The above Lemma indicates that this problem can be translated into a problem about associated primes of $\overline{u^n \mathcal{R}(\mathfrak{a},R)}$. The precise statement of the connection indicated here can be found in Theorem (16.9).

(15.4) Definition. For $\mathfrak{a}$ and R as above we define

$$W(\mathfrak{a}) = \{\mathfrak{p} \in \mathrm{Min}(R) \mid \mathfrak{a} + \mathfrak{p} \neq R\} .$$

We note here that clearly $W(\mathfrak{a}) = W(\mathfrak{a}^n)$ for any $n > 0$. Also, if R is local and $\mathfrak{a}$ is proper then $W(\mathfrak{a}) = \mathrm{Min}(R)$.

(15.5) Lemma. For $x \in R$, the following conditions are equivalent:

(i) x is integral over $\mathfrak{a}$;

(ii) the image of x in $R/\mathfrak{p}$ is integral over $\mathfrak{a} + \mathfrak{p}/\mathfrak{p}$ for all $\mathfrak{p} \in W(\mathfrak{a})$.

Proof. This is a direct consequence of the definition of $W(\mathfrak{a})$ together with Proposition (4.13) and Lemma (4.16).

Combining Lemmas (15.2) and (15.5), the natural object to study the integral dependence on $\mathfrak{a}$ or $\mathfrak{a}^n$ is

$$\prod_{\mathfrak{p}\in W(\mathfrak{a})} \overline{\mathcal{R}(\mathfrak{a} + \mathfrak{p}/\mathfrak{p}, R/\mathfrak{p})} \quad ;$$

and for notational purposes we give the following

(15.6) Definition. a) For $\mathfrak{a}$ and R as above we put

$$\overline{\mathcal{R}}(\mathfrak{a},R) = \prod_{\mathfrak{p}\in W(\mathfrak{a})} \overline{\mathcal{R}(\mathfrak{a} + \mathfrak{p}/\mathfrak{p}, R/\mathfrak{p})} \quad .$$

b) By $h(\mathfrak{a}) : R \longrightarrow \overline{\mathcal{R}}(\mathfrak{a},R)$ we denote the canonical homomorphism obtained by the composition

$$R \longrightarrow R_{red} \longrightarrow \overline{\mathcal{R}(\mathfrak{a}R_{red}, R_{red})} = \prod_{\mathfrak{p} \in Min(R)} \mathcal{R}(\mathfrak{a}+\mathfrak{p}/\mathfrak{p}, R/\mathfrak{p}) \longrightarrow \overline{\mathcal{R}}(\mathfrak{a},R) \quad ,$$

the last homomorphism being the canonical projection.

Since $\overline{\mathcal{R}}(\mathfrak{a},R)$ is a finite product of Krull domains, it makes sense to talk about essential valuations of $\overline{\mathcal{R}}(\mathfrak{a},R)$. Therefore the following definition will be meaningful:

(15.7) Definition. For $\mathfrak{a}$ and R as above, $Val(\mathfrak{a})$ will denote the set of valuations v of R of the form $v = w \circ h(\mathfrak{a})$, where w is some essential valuation of $\overline{\mathcal{R}}(\mathfrak{a},R)$ such that $w(h(\mathfrak{a})(\mathfrak{a})) > 0$.

(15.8) Remark. Observe that different (essential) valuations of $\overline{\mathcal{R}}(\mathfrak{a},R)$ induce different valuations on $\mathcal{R}(\mathfrak{a},R)$. Hence if $v = w \circ h(\mathfrak{a}) \in Val(\mathfrak{a})$, v induces a unique valuation on $\mathcal{R}(\mathfrak{a},R)$ that will be denoted by v again.

(15.9) Remark. Consider the canonical homomorphism

$$\mathcal{R}(\mathfrak{a},R) \longrightarrow \overline{\mathcal{R}}(\mathfrak{a},R)$$

and let us denote the image of u by u again. We already mentioned that $\overline{\mathcal{R}}(\mathfrak{a},R)$ is a finite product of Krull domains, which implies that

$$u^n \overline{\mathcal{R}}(\mathfrak{a},R) = \overline{u^n \overline{\mathcal{R}}(\mathfrak{a},R)} = \cap \left\{x \in \overline{\mathcal{R}}(\mathfrak{a},R) \mid w(x) \geq nw(u)\right\} ,$$

where the intersection is taken over all essential valuations w of $\overline{\mathcal{R}}(\mathfrak{a},R)$ for which $w(u) > 0$. In view of Lemma (15.2) we conclude that

$$\overline{\mathfrak{a}^n} = \left\{x \in R \mid v(x) \geq nv(u) \quad \text{for all} \quad v \in Val(\mathfrak{a})\right\} .$$

(15.10) Proposition. For $\mathfrak{a}$ and R as above and for any $n \geq 0$ we have

$$\overline{\mathfrak{a}^n} = \left\{x \in R \mid v(x) \geq nv(\mathfrak{a}) \quad \text{for all} \quad v \in Val(\mathfrak{a})\right\} .$$

Proof. If $x \in \overline{\mathfrak{a}^n}$ then $v(x) \geq v(\mathfrak{a}^n) = nv(\mathfrak{a})$ for any valuation v by Proposition (4.19). For the converse inclusion, observe that $\mathfrak{a} \subset u\mathfrak{R}(\mathfrak{a},R)$ and hence $v(\mathfrak{a}) \geq v(u)$ for any v. Therefore the proof follows from Remark (15.9) above.

(15.11) Lemma. $\bigcap_{n\geq 0} \overline{\mathfrak{a}^n} = \bigcap_{\mathfrak{p}\in W(\mathfrak{a})} \mathfrak{p}$.

Proof. Each $v \in \mathrm{Val}(\mathfrak{a})$ is induced by a valuation on $\mathfrak{R}(\mathfrak{a}+\mathfrak{p}/\mathfrak{p},R/\mathfrak{p})$ for some $\mathfrak{p} \in W(\mathfrak{a})$, and for this $\mathfrak{p}$ we have

(15.11.1) $$\mathfrak{p} = \{x \in R \mid v(x) = \infty\} .$$

Conversely, given any $\mathfrak{p} \in W(\mathfrak{a})$, (15.11.1) holds for a suitable $v \in \mathrm{Val}(\mathfrak{a})$. Therefore the Lemma is an immediate consequence of Proposition (15.10).

(15.12) Proposition. Let $R = \bigoplus_{\nu\in\mathbf{Z}^n} R_\nu$ be a graded noetherian ring and $\mathfrak{a}$ a homogeneous ideal of R. Then $\overline{\mathfrak{a}}$ is homogeneous.

Proof. By Proposition (4.17) it is enough to prove the claim in the case that R is a domain. Now $\mathfrak{R}(\mathfrak{a},R)$ is naturally $\mathbf{Z}^{n+1}$-graded, so the same is true for the integral closure $\overline{\mathfrak{R}(\mathfrak{a},R)}$ (see [1]). Now, by Lemma (15.2), (v), we know that

$$\overline{\mathfrak{a}} = u\cdot\overline{\mathfrak{R}(\mathfrak{a},R)} \cap R ,$$

and since u is homogeneous, $\overline{\mathfrak{a}}$ is a homogeneous ideal of R.

(15.13) Corollary. For R and $\mathfrak{a}$ as above, any associated prime of $\overline{\mathfrak{a}}$ is homogeneous.

(15.14) Corollary. Let R and $\mathfrak{a}$ be as above. Then for any indeterminates $X_1,\ldots,X_n$ over R we have

$$\overline{\mathfrak{a}\cdot R[X_1,\ldots,X_n]} = \overline{\mathfrak{a}}\cdot R[X_1,\ldots,X_n] .$$

Proof. Using induction we may assume that $n = 1$. We have to show that if $F(X_1) \in R[X_1]$ is integral over $\mathfrak{a} \cdot R[X_1]$ then the coefficients of F are integral over $\mathfrak{a}$. By Proposition (15.12) we may assume that F is homogeneous, in which case the assertion is immediate.

§ 16. Associated primes of the integral closure of powers of an ideal

For this section R denotes any noetherian ring, which need not be local unless otherwise stated. Recall that for any ideal $\mathfrak{a}$ of R, $\overline{\mathfrak{a}}$ denotes the integral closure of $\mathfrak{a}$ (see § 4). The purpose of this section is to characterise the associated primes of $\overline{\mathfrak{a}^n}$ as n varies. Before we recall some technical results to be used in later proofs.

(16.1) Lemma. Let $R \subset S$ be notherian rings, $\mathfrak{b}$ an ideal of S and $\mathfrak{a} = R \cap \mathfrak{b}$. For any $\mathfrak{p} \in \operatorname{Ass}(R/\mathfrak{a})$ there is some $\mathfrak{q} \in \operatorname{Ass}(S/\mathfrak{b})$ such that $\mathfrak{q} \cap R = \mathfrak{p}$.

Proof. We may assume that R is local with maximal ideal $\mathfrak{p}$. Let $\mathfrak{p} = (\mathfrak{a} :_R x)$ for some $x \in R \setminus \mathfrak{a}$. Then $x \notin \mathfrak{b}$, so $(\mathfrak{b} :_S x)$ is contained in some associated prime $\mathfrak{q}$ of $\mathfrak{b}$. It follows that $\mathfrak{q} \supset (I :_R x) = \mathfrak{p}$, and since $\mathfrak{p}$ is maximal, we must have $\mathfrak{q} \cap R = \mathfrak{p}$.

(16.2) Remark. Let $\mathfrak{a}$ be an ideal of R and let $\mathfrak{p}$ be any prime ideal of R. By Corollary (4.9) we know that $\overline{\mathfrak{a}^n}\, R_{\mathfrak{p}} = \overline{\mathfrak{a}^n R_{\mathfrak{p}}}$ for any n. Hence we see that $\mathfrak{p}$ is associated to $\overline{\mathfrak{a}^n}$ if and only if $\mathfrak{p}R_{\mathfrak{p}}$ is associated to $\overline{\mathfrak{a}^n R_{\mathfrak{p}}}$. We will make frequent use of this remark by assuming that R is local with maximal ideal $\mathfrak{p}$.

(16.3) Proposition. Let $\mathfrak{a}$ be any ideal of R. Then we have

$$\operatorname{Ass}(R/\overline{\mathfrak{a}^n}) \subset \operatorname{Ass}(R/\overline{\mathfrak{a}^{n+1}}) \quad .$$

Proof. Let $\mathfrak{p} \in \operatorname{Ass}(R/\overline{\mathfrak{a}^n})$. By Remark (16.2) we may assume that R is local with maximal ideal $\mathfrak{p}$. Now let

$$\mathfrak{p} = (\overline{\mathfrak{a}^n} : x) \quad \text{for some} \quad x \notin \overline{\mathfrak{a}^n} \quad .$$

First we show that $\bar{\mathfrak{a}}x \not\subset \overline{\mathfrak{a}^{n+1}}$. For this, recall from Proposition (15.10) that there is a finite set $\mathrm{Val}(\mathfrak{a})$ of valuations such that for any $m \in \mathbf{N}$

$$\overline{\mathfrak{a}^m} = \left\{ y \mid v(y) \geqq mv(\mathfrak{a}) \quad \text{for all} \quad v \in \mathrm{Val}(\mathfrak{a}) \right\} \quad .$$

Assume now if possible that $\bar{\mathfrak{a}}x \subset \overline{\mathfrak{a}^{n+1}}$ and fix any $v \in \mathrm{Val}(\mathfrak{a})$. Then

$$v(\bar{\mathfrak{a}}) + v(x) \geqq v(\overline{\mathfrak{a}^{n+1}})$$

and consequently

$$v(x) \geqq nv(\mathfrak{a}) \quad .$$

Since this holds for any $v \in \mathrm{Val}(\mathfrak{a})$, this leads to the contradiction $x \in \overline{\mathfrak{a}^n}$. So we may choose $a \in \bar{\mathfrak{a}}$ such that $ax \notin \overline{\mathfrak{a}^{n+1}}$. Then

$$ax\mathfrak{p} \subset \bar{\mathfrak{a}} \cdot \overline{\mathfrak{a}^n} \subset \overline{\mathfrak{a}^{n+1}} \quad , \qquad \text{(see Corollary (4.23))}$$

showing that $\mathfrak{p}$ consists of zero-divisors $\bmod \overline{\mathfrak{a}^{n+1}}$.

(16.4) Definition. For any ideal $\mathfrak{a}$ of R we define

$$A(\mathfrak{a}) = \bigcup_{n=1}^{\infty} \mathrm{Ass}(R/\overline{\mathfrak{a}^n}) \quad .$$

(16.5) Remark. In the following proposition we will consider the integral closure $\bar{R}$ of a reduced noetherian ring R . If $\mathfrak{p}_1,\ldots,\mathfrak{p}_n$ are the minimal primes of R and $s_i : R \longrightarrow R/\mathfrak{p}_i$ is the canonical surjection, we may write $\bar{R} = R_1 \times \ldots \times R_n$, where R_i is the integral closure of $s_i(R)$, and the morphism $s : R \longrightarrow \bar{R}$ is given by $s(y) = (s_1(y),\ldots,s_n(y))$. Since R is reduced, s is injective and we will identify R and $s(R)$. We will consider an element $x = (x_1,x_2,\ldots,x_n) \in \bar{R}$ and the finite extension $R[x]$. If now $x_1 \neq 0$

then the conductor $c(R[x]/R)$ contains some $b \in R$ such that $s_1(b) \neq 0$. This can be seen as follows. Writing $x_1 = s_1(a)/s_1(d)$ for some $a, d \in R$, $d \notin \mathfrak{p}$, and choosing m such that any element $z \in R[x]$ may be written as

$$z = \sum_{i=0}^{m} a_i x^i$$

we put $b = d^m$. Then $s_1(b) \neq 0$, and we have

$$dx = (s_1(d)x_1, 0, \ldots, 0) = (s_1(a), 0, \ldots, 0) \in s(R) \quad ,$$

so $d^m \left(\sum_{i=0}^{m} a_i x^i \right) = \sum_{i=0}^{m} a_i d^{m-i} (xd)^i \in R$, showing that $b \in t(R[x]/R)$.

Moreover, still in the situation $\overline{R} = R_1 \times \ldots \times R_n$, we will make use of the fact that any height one prime $\overline{\mathfrak{q}}$ of $\overline{R}$ can be written (after a permutation of the $\mathfrak{p}_i$) as

$$\overline{\mathfrak{q}} = \mathfrak{q}_1 \times R_2 \times \ldots \times R_n \;,\quad \mathfrak{q}_1 \text{ a height one prime of } R_1 \;,$$

and if v_1 is the valuation corresponding to $\mathfrak{q}_1$, $\overline{\mathfrak{q}}$ corresponds to a valuation $\overline{v}$ of $\overline{R}$ given by

$$\overline{v}(x_1, \ldots, x_n) = v_1(x_1) \quad .$$

Finally, if $a \in \overline{R}$ is a non-zerodivisor and $b \in \overline{R}$ is arbitrary, then $(a :_{\overline{R}} b)$ has associated primes of height one.

<u>(16.6) Lemma</u>. Let R be a reduced noetherian ring and $\overline{R}$ its integral closure. Let $\mathfrak{q}$ be a height one prime of $\overline{R}$. Then there exist a minimal prime ideal $\mathfrak{p}_\circ$ of $\overline{R}$ and $x \in \overline{R}$ such that

a) $\mathfrak{p}_\circ \subset \mathfrak{q}$

b) $x \notin \mathfrak{p}_\circ$

c) $ht(\mathfrak{q} \cap R[x]) = 1$

d) $v(x) \neq \infty$, where v is the valuation corresponding to $\mathfrak{q}$.

Proof. Putting $\overline{R} = R_1 \times \ldots \times R_n$ where $R_1, \ldots, R_n$ are normal domains, we may assume that

$$\mathfrak{q} = \mathfrak{q}_1 \times R_2 \times \ldots \times R_n \quad , \quad \mathfrak{q}_1 \text{ a height one prime of } R_1 .$$

We take $\mathfrak{p}_\circ = (0 \times R_2 \times \ldots \times R_n)$. Since the set

$$\{\mathfrak{q}' \in \operatorname{Spec} \overline{R} \mid \mathfrak{q}' \cap R = \mathfrak{q} \cap R\} = T$$

is finite and since there is no containment relation among the elements of T , we may choose $x \in \mathfrak{q}$ such that

$$x \notin \mathfrak{q}' \quad \text{for all} \quad \mathfrak{q}' \in T \setminus \{\mathfrak{q}\} \quad .$$

Writing $x = (x_1, \ldots, x_n)$, we may choose x such that $x_1 \neq 0$, so $x \notin \mathfrak{p}_\circ$. Now looking at $R[x]$, we see that any prime $\overline{\mathfrak{q}}$ of $\overline{R}$ for which $\overline{\mathfrak{q}} \cap R[x] = \mathfrak{q} \cap R[x]$ satisfies $\overline{\mathfrak{q}} \cap R = \mathfrak{q} \cap R$ and hence $\overline{\mathfrak{q}} \in T$. Since $x \in \overline{\mathfrak{q}}$, we conclude that $\overline{\mathfrak{q}} = \mathfrak{q}$ and so $\mathfrak{q}$ is the only prime ideal of $\overline{R}$ lying over $\mathfrak{q} \cap R[x]$. By going up it follows that $\operatorname{ht}(\mathfrak{q} \cap R[x]) = 1$.

<u>(16.7) Proposition</u>. Let R be a noetherian ring, $\overline{R}$ the integral closure of R_{red} and $s : R \longrightarrow \overline{R}$ the canonical homomorphism. Let $\mathfrak{a}$ be an ideal not contained in any minimal prime ideal of R . Then we have:

a) If $\mathfrak{a}\overline{R}$ is principal and $\mathfrak{p} \in A(\mathfrak{a})$ then there is a height one prime $\mathfrak{q}$ of $\overline{R}$ such that $s^{-1}(\mathfrak{q}) = \mathfrak{p}$.

b) Let $\mathfrak{q}$ be a height one prime of $\overline{R}$ such that $\mathfrak{a} \subset s^{-1}(\mathfrak{q})$. Then $s^{-1}(\mathfrak{q}) \in A(\mathfrak{a})$.

Proof. a) We may assume R reduced (see Lemma (4.16) and local with maximal ideal $\mathfrak{p}$. We may write

$$\mathfrak{p} = (\overline{\mathfrak{a}^n} : y) \quad \text{for some} \quad n \in \mathbf{N} \quad \text{and some} \quad y \in R \quad .$$

Let $\mathfrak{a}\overline{R} = a\overline{R}$. Then $a^n\overline{R}$ is integrally closed and hence $\overline{\mathfrak{a}^n} \subset a^n\overline{R}$ (note that s is the inclusion since R is reduced). This shows that

$$\mathfrak{p} \subset (a^n\overline{R} :_{\overline{R}} y) \quad .$$

Since $\mathfrak{a}$ is not contained in any minimal prime ideal of R, a is a non-zerodivisor in $\overline{R}$. So by Remark (16.5), any prime ideal $\mathfrak{q}$ of $\overline{R}$ minimal over $(a^n\overline{R} :_{\overline{R}} y)$ will have height one, and choosing one such $\mathfrak{q}$ we obtain $\mathfrak{q} \cap R = \mathfrak{p}$ since $\mathfrak{p}$ was maximal.

b) Let $\mathfrak{p} = s^{-1}(\mathfrak{q})$. Again we may assume R reduced and local with maximal ideal $\mathfrak{p}$. Let v be the valuation corresponding to $\mathfrak{q}$. By Lemma (16.6) we may choose $x \in \overline{R}$ such that

$$v(x) < \infty \quad \text{and} \quad ht(\mathfrak{q} \cap R[x]) = 1 \quad .$$

Moreover, by Remark (16.5) there is some $b \in R$ such that

$$bR[x] \subset R \quad \text{and} \quad v(b) < \infty \quad .$$

We put $v(b) = t-1$. Since $ht(\mathfrak{q} \cap R[x]) = 1$, $\mathfrak{q} \cap R[x]$ is minimal over $\mathfrak{a}^t R[x]$, so we may choose $k \in \mathbf{N}$ and $c \in R[x] \setminus \mathfrak{q}$ such that

$$c \cdot (\mathfrak{q} \cap R[x])^k \subset \mathfrak{a}^t R[x] \quad .$$

Now put $a = bc$ and note that $a \in R$ and $v(a) = t-1$. Since $v(\overline{\mathfrak{a}^t}) = v(\mathfrak{a}^t) = tv(\mathfrak{a}) \geq t > v(a)$, we see that $a \notin \overline{\mathfrak{a}^t}$. But

$$a\mathfrak{p}^k \subset \mathfrak{a}^t \cdot bR[x] \subset \mathfrak{a}^t \subset \overline{\mathfrak{a}^t} \quad ,$$

so $\mathfrak{p}$ consists of zero-divisors mod $\overline{\mathfrak{a}^t}$.

<u>(16.8) Proposition</u>. Let R be a noetherian ring, $\mathfrak{a}$ an ideal of R and $\mathfrak{q}$ a homogeneous prime ideal of $\mathcal{R}(\mathfrak{a},R)$. If $\mathfrak{q} \in A(u\mathcal{R}(\mathfrak{a},R))$ then $\mathfrak{q} \cap R \in A(\mathfrak{a})$.

Proof. Again we may assume R to be local with maximal ideal $\mathfrak{p} = \mathfrak{q} \cap R$. Let $\mathfrak{q}$ be associated to

$$(16.8.1) \qquad \overline{u^r\mathcal{R}(\mathfrak{a},R)} = \bigoplus_{n \in \mathbf{Z}} (\overline{\mathfrak{a}^{n+r}} \cap \mathfrak{a}^n)t^n \quad .$$

Writing $\mathfrak{q} = (\overline{u^r\mathcal{R}(\mathfrak{a},R)} : ct^k)$ for some $c \in \mathfrak{a}^k$, we see that

$$c \notin \overline{\mathfrak{a}^{r+k}} \quad .$$

But clearly $(ct^k)\mathfrak{p} = (c\mathfrak{p})t^k \subset \overline{u^r\mathcal{R}(\mathfrak{a},R)}$, which means that $c\mathfrak{p} \subset \overline{\mathfrak{a}^{r+k}}$ by (16.8.1). Hence $\mathfrak{p} \in A(\mathfrak{a})$.

<u>(16.9) Theorem</u>. Let R be a noetherian ring, $\mathfrak{a}$ an ideal of R and let $h = h(\mathfrak{a}) : R \longrightarrow \overline{\mathcal{R}}(\mathfrak{a},R)$ be the canonical homomorphism (15.6).

a) For any prime ideal $\mathfrak{p}$ of R the following conditions are equivalent:

(i) $\mathfrak{p} \in A(\mathfrak{a})$.

(ii) There is $\mathfrak{q} \in A(u\mathcal{R}(\mathfrak{a},R))$ such that $\mathfrak{q} \cap R = \mathfrak{p}$.

(iii) There is a height one prime $\overline{\mathfrak{q}} \subset \overline{\mathcal{R}}(\mathfrak{a},R)$ such that $u\overline{\mathcal{R}}(\mathfrak{a},R) \subset \overline{\mathfrak{q}}$ and $h^{-1}(\mathfrak{q}) = \mathfrak{p}$.

(iv) There is a valuation $v \in \mathrm{Val}(\mathfrak{a})$ such that $\mathfrak{p} = \{x \in R \mid v(x) > 0\}$.

b) If, for any ideal $\mathfrak{b}$, $s(\mathfrak{b}) : \mathrm{Spec}(R/\mathfrak{b}) \longrightarrow \mathrm{Spec}\, R$ denotes the canonical morphism, we have

$$A(\mathfrak{a}) = \bigcup_{\mathfrak{p} \in W(\mathfrak{a})} s(\mathfrak{p})(A(\mathfrak{a}\cdot(R/\mathfrak{p}))) \quad .$$

c) In the situation of a), if $\mathfrak{q}_\circ$ is any minimal prime ideal of $\mathcal{R}(\mathfrak{a},R)$ such that $\mathfrak{q}_\circ \cap R \subset \mathfrak{p}$, then the ideal $\mathfrak{q}$ in (ii) can be chosen to contain $\mathfrak{q}_\circ$.

Proof. a) (i) $\Rightarrow$ (ii) by Lemma (16.1) and (ii) $\Rightarrow$ (i) by Proposition (16.8). For the equivalence of (ii) and (iii) we may assume R reduced. Note that u is not contained in any minimal prime of $\mathcal{R}(\mathfrak{a},R)$, so we may apply Proposition (16.7), a) to deduce from (ii) that there is a height one prime $\mathfrak{q}'$ in $\overline{\mathcal{R}(\mathfrak{a},R)}$ such that $\mathfrak{q}' \cap R = \mathfrak{p}$. Now $\mathfrak{q}'$ contains a minimal prime $\mathfrak{p}_\circ$ of R such that $\mathfrak{p}_\circ + \mathfrak{a} \neq R$, i.e. $\mathfrak{p}_\circ \in W(\mathfrak{a})$ by definition, and hence $\mathfrak{q}'$ is the inverse image of a height one prime $\overline{\mathfrak{q}}$ of $\overline{\mathcal{R}}(\mathfrak{a},R)$ under the canonical surjection

$$\overline{\mathcal{R}(\mathfrak{a},R)} \longrightarrow \overline{\mathcal{R}}(\mathfrak{a},R) \quad .$$

Clearly $\overline{\mathfrak{q}}$ satisfies (iii). Conversely, given $\overline{\mathfrak{q}}$ as in (iii), the inverse image $\mathfrak{q}'$ of $\overline{\mathfrak{q}}$ in $\overline{\mathcal{R}(\mathfrak{a},R)}$ is a height one prime contracting to some $\mathfrak{q}$ in $\mathcal{R}(\mathfrak{a},R)$, which satisfies (ii) by Proposition (16.7), b). Finally the equivalence of (iii) and (iv) follows from the definition of Val(I) in (15.7).

b) By construction, $\overline{\mathfrak{R}}(\mathfrak{a},R)$ is a product of Krull domains which are canonically isomorphic to $\overline{\mathfrak{R}(\mathfrak{a}+\mathfrak{p}/\mathfrak{p},R/\mathfrak{p})}$ for $\mathfrak{p}\in W(\mathfrak{a})$. Therefore the assertion follows from the equivalence of (i) and (iii) in a).

c) Given $\mathfrak{q}_\circ$ we put $\mathfrak{p}_\circ = \mathfrak{q}_\circ \cap R$. By b) we have $\mathfrak{p}/\mathfrak{p}_\circ \in A(\mathfrak{a}(R/\mathfrak{p}_\circ))$ and hence, by a), $\mathfrak{p}/\mathfrak{p}_\circ$ is the inverse image of some height one prime $\bar{\mathfrak{a}}$ in $\overline{\mathfrak{R}(\mathfrak{a}+\mathfrak{p}_\circ/\mathfrak{p}_\circ,R/\mathfrak{p}_\circ)} \simeq \overline{\mathfrak{R}(\mathfrak{a},R)/\mathfrak{q}_\circ}$. Denoting by $\mathfrak{q}$ the inverse image of $\bar{\mathfrak{q}}$ under $\mathfrak{R}(\mathfrak{a},R) \longrightarrow \overline{\mathfrak{R}(\mathfrak{a},R)/\mathfrak{q}_\circ}$, clearly $\mathfrak{q}$ contains $\mathfrak{q}_\circ$ and $\mathfrak{q}$ satisfies (ii) by Proposition (16.7),b).

(16.10) Corollary. $A(\mathfrak{a})$ is finite.

(16.11) Remark. Trivially, if $\mathfrak{p}$ is any minimal prime of $\mathfrak{a}$ then $\mathfrak{p}\in A(\mathfrak{a})$. More generally, if $\mathfrak{p}_\circ \in W(\mathfrak{a})$ and $\mathfrak{p}$ is a minimal prime of $\mathfrak{a}+\mathfrak{p}_\circ$, then $\mathfrak{p}\in A(\mathfrak{a})$ by Theorem (16.9),b) above.

(16.12) Notation. If S is a domain and $R \to S$ a homomorphism, the transcendence degree of the quotient field of S over the quotient field of the image of R in S will be denoted by $\mathrm{tr.d.}_R(S)$. In particular, if $R\subset S$ are domains and $\mathfrak{q}$ is a prime ideal of S , then $\mathrm{tr.d.}_R(S/\mathfrak{q})$ denotes the transcendence degree of $Q(S/\mathfrak{q})$ over $Q(R/\mathfrak{q}\cap R)$.

(16.13) Definition. a) A noetherian domain R is said to satisfy the altitude formula if the following condition is satisfied: If $S\supset R$ is a domain that is a finitely generated R-algebra and if $\mathfrak{q}$ is any prime of S then

$$\mathrm{ht}(\mathfrak{q}) - \mathrm{ht}(\mathfrak{q}\cap R) = \mathrm{tr.d.}_R(S) - \mathrm{tr.d.}_R(S/\mathfrak{q}) \quad .$$

b) A noetherian ring R is said to satisfy the altitude formula if $R/\mathfrak{p}$ does for each minimal prime ideal $\mathfrak{p}$ of R .

(16.14) Remark. The most important cases of noetherian rings satisfying the altitude formula are complete local rings (see [2]) and algebras finitely generated over a field. The second example follows from the more general fact, which is immediate from the definition: If $R \subset S$ are domains such that S is a finitely generated R-algebra, and if R satisfies the altitude formula, so does S.

(16.15) Proposition. Let R be a noetherian domain satisfying the altitude formula. If an ideal $\mathfrak{a}$ of R can be generated by k elements, then any $P \in A(\mathfrak{a})$ satisfies $\mathrm{ht}(\mathfrak{p}) \leqq k$.

Proof. Assume first that $\mathfrak{a}$ is generated by one element a, which we may assume to be non-zero. Given $\mathfrak{p} \in A(\mathfrak{a})$, by Proposition (16.7),a) there is a height one prime $\mathfrak{q}$ in $\overline{R}$ such that $\mathfrak{q} \cap R = \mathfrak{p}$. By Lemma (16.6) we may choose $x \in \mathfrak{q}$ such that $\mathrm{ht}(\mathfrak{q} \cap R[x]) = 1$, and therefore the altitude formula implies $\mathrm{ht}(\mathfrak{p}) = 1$ (note that $\mathrm{tr.d.}_R(R[x]) = 0 = \mathrm{tr.d.}_R(R[x]/\mathfrak{q} \cap R[x])$ since $x \in \mathfrak{q}$). Now let $\mathfrak{a}$ be generated by $a_1,\ldots,a_k$ and choose $\mathfrak{p} \in A(\mathfrak{a})$. Lifting $\mathfrak{p}$ to $\mathfrak{q} \in A(u\mathfrak{R}(\mathfrak{a},R))$ such that $\mathfrak{q} \cap R = \mathfrak{p}$ (by Theorem (16.9)), we see that $\mathrm{ht}(\mathfrak{q}) = 1$ by the first case and by the fact that $\mathfrak{R}(\mathfrak{a},R)$ again satisfies the altitude formula (see Remark (16.14)). Now

$$\mathfrak{R}(\mathfrak{a},R) = R[a_1t,\ldots,a_kt,u] \quad \text{and} \quad u \in \mathfrak{q} \quad ,$$

and therefore $\mathrm{tr.d.}_R(\mathfrak{R}(\mathfrak{a},R)/\mathfrak{q}) \leqq k$. Since $\mathrm{tr.d.}_R(\mathfrak{R}(\mathfrak{a},R)) = \mathrm{ht}(\mathfrak{q})$, the altitude formula implies $\mathrm{ht}(\mathfrak{p}) = \mathrm{tr.d.}_R(\mathfrak{R}(\mathfrak{a},R)/\mathfrak{q})$, which concludes the proof.

(16.16) Example. The conclusion of Proposition (16.15) may fail to hold if R is not a domain. As the easiest example take

$$R = k[[x,y,z]] = k[[X,Y,Z]]/(XY,XZ) \quad , \quad \mathfrak{a} = (x-y)R \quad .$$

Then $\mathfrak{p}_\circ = (y,z)R$ is a minimal prime of R and the maximal ideal $\mathfrak{m}$ of R is minimal over $\mathfrak{a} + \mathfrak{p}_\circ$ (actually $\mathfrak{a} + \mathfrak{p}_\circ = \mathfrak{m}$). By Remark (16.11), $\mathfrak{m} \in A(\mathfrak{a})$, but $\mathrm{ht}(\mathfrak{m}) = \dim R = 2$.

(16.17) Corollary. Let R be a noetherian ring satisfying the altitude formula and let $\mathfrak{a}$ be an ideal of R generated by k elements.

Then any $\mathfrak{p} \in A(\mathfrak{a})$ contains a minimal prime $\mathfrak{p}_\circ \in W(\mathfrak{a})$ such that $ht(\mathfrak{p}/\mathfrak{p}_\circ) \leq k$.

Proof. By Theorem (16.9),b), $\mathfrak{p}$ contains $\mathfrak{p}_\circ \in W(\mathfrak{a})$ such that

$$\mathfrak{p}/\mathfrak{p}_\circ \in A(\mathfrak{a}\cdot(R/\mathfrak{p}_\circ)) \quad .$$

Since $R/\mathfrak{p}_\circ$ satisfies the altitude formula by assumption, the result follows from Proposition (16.15).

(16.18) Lemma. Let R be a noetherian ring, $\mathfrak{a}$ an ideal of R and $\mathfrak{p}_\circ$ a minimal prime ideal of R . If $\mathfrak{p}$ is a prime ideal minimal over $\mathfrak{a}+\mathfrak{p}_\circ$ then there exists an integer t with the following property: If $\mathfrak{b}$ is any ideal with $\mathfrak{b} \subset \mathfrak{a}^t$ and $\sqrt{\mathfrak{b}} = \sqrt{\mathfrak{a}}$ then $\mathfrak{p}$ is associated to $\mathfrak{b}$.

Proof. Since $\mathfrak{p}_\circ$ is minimal, there are an integer r and an element $x \in R \setminus \mathfrak{p}_\circ$ such that $x\cdot\mathfrak{p}_\circ^r = 0$. Since $x \notin \mathfrak{p}_\circ \in W(\mathfrak{a})$, we know from Lemma (15.11) that there is an integer t for which $x \notin \mathfrak{a}^t$. To show that t has the desired property we may assume R to be local with maximal ideal $\mathfrak{p} = \sqrt{\mathfrak{a}+\mathfrak{p}_\circ}$. Now given any $\mathfrak{b} \subset \mathfrak{a}^t$ with $\sqrt{\mathfrak{b}} = \sqrt{\mathfrak{a}}$, we have $\mathfrak{a}^n \subset \mathfrak{b}$ for some n and $\mathfrak{p}^k \subset \mathfrak{a}+\mathfrak{p}_\circ$ for some k . It follows that

$$(\mathfrak{p}^k)^{n+t-1} \subset (\mathfrak{a}+\mathfrak{p}_\circ)^{n+t-1} \subset \mathfrak{a}^n + \mathfrak{p}_\circ^t \subset \mathfrak{a} + \mathfrak{p}_\circ^t$$

and hence $x(\mathfrak{p}^k)^{n+t-1} \subset \mathfrak{b}$, which proves the claim since $x \notin \mathfrak{b}$.

(16.19) Theorem. Let $R \to S$ be a faithfully flat homomorphism of noetherian rings and let $\mathfrak{a}$ be an ideal of R . If S satisfies the altitude formula then for any prime ideal $\mathfrak{p}$ of R the following condtions are equivalent:

(i) $\mathfrak{p} \in A(\mathfrak{a})$.

(ii) There exists $\mathfrak{q} \in A(\mathfrak{a}\cdot S)$ such that $\mathfrak{q} \cap R = \mathfrak{p}$.

Proof. (i) $\Rightarrow$ (ii). Since $\overline{\mathfrak{a}^n S} \cap R = \overline{\mathfrak{a}^n}$ by (4.12), the result follows from Lemma (16.1).

(ii) $\Rightarrow$ (i). Let $\tilde{R} = \mathcal{R}(\mathfrak{a},R)$ and $\tilde{S} = \mathcal{R}(\mathfrak{a}S,S)$, so $\tilde{S} = \tilde{R} \otimes_R S$ and $\tilde{R} \to \tilde{S}$ is faithfully flat. Given $\mathfrak{q} \in A(\mathfrak{a}S)$, by Theorem (16.9) we can find $\tilde{\mathfrak{q}} \in A(u\tilde{S})$ such that $\tilde{\mathfrak{q}} \cap \tilde{S} = \mathfrak{q}$. By Corollary (16.17) $\tilde{\mathfrak{q}}$ contains a minimal prime $\tilde{\mathfrak{q}}_\circ$ such that $\tilde{\mathfrak{q}}$ is minimal over $u\tilde{S} + \tilde{\mathfrak{q}}_\circ$. By Lemma (16.18) there is an integer t such that $\tilde{\mathfrak{q}}$ is associated to any $\mathfrak{b} \subset \overline{u^t \tilde{S}}$ such that $\sqrt{\mathfrak{b}} = \sqrt{u\tilde{S}}$. Choosing $\mathfrak{b} = \overline{(u^t\tilde{R})} \cdot \tilde{S}$ we see by flatness that $\tilde{\mathfrak{q}} \cap \tilde{R}$ is associated to $\overline{u^t\tilde{R}}$, and hence $\tilde{\mathfrak{q}} \cap R \in A(\mathfrak{a})$. But $\tilde{\mathfrak{q}} \cap R = \mathfrak{p}$ by construction, so the proof is complete.

(16.20) Corollary. Let $\mathfrak{a}$ be an ideal in a local ring R with completion $\hat{R}$. For any prime ideal $\mathfrak{p}$ of R the following conditions are equivalent:

(i) $\mathfrak{p} \in A(\mathfrak{a})$.

(ii) There is $\hat{\mathfrak{p}} \in A(\mathfrak{a}\hat{R})$ such that $\hat{\mathfrak{p}} \cap R = \mathfrak{p}$.

(16.21) Lemma. Let $\mathfrak{a}$ be an ideal in a local ring R and let $\mathfrak{p},\mathfrak{q}$ be prime ideals such that $\mathfrak{q}$ is minimal in R and $\mathfrak{p} \in \mathrm{Min}(R/\mathfrak{a}+\mathfrak{q})$. Then there exists an n such that $\mathfrak{q} \in \mathrm{Ass}(R/\overline{\mathfrak{a}^n})$.

Proof. There is nothing to prove if $\mathfrak{a} \subset \mathfrak{q}$, so assume $\mathfrak{a} \not\subset \mathfrak{q}$. There are $x \in R \setminus \mathfrak{q}$ and $t \in \mathbf{N}$ such that $x^t \cdot \mathfrak{q} = 0$. Moreover there is $k \in \mathbf{N}$ such that $\mathfrak{p}^k \subset \mathfrak{a} + \mathfrak{q}$. Now choose n such that $x \notin \overline{\mathfrak{a}^n}$. Then

$$(\mathfrak{p}^k)^{t+n} \subset (\mathfrak{a} + \mathfrak{q})^{t+n} \subset \mathfrak{a}^n + \mathfrak{q}$$

and hence $x \cdot (\mathfrak{p}^k)^{t+n} \subset \mathfrak{a}^n$.

(16.22) Lemma. Let $\mathfrak{a}$ be an ideal in a noetherian ring R and let $\mathfrak{q}$ be a prime ideal in $\mathcal{R}(\mathfrak{a},R)$ associated to $u^n\mathcal{R}(\mathfrak{a},R)$ for some $n > 0$. Then there is a $k > -n$ such that $\mathfrak{q} \cap R$ is associated to $\mathfrak{a}^{n+k}$.

Proof. Let $\mathfrak{q} = (u^n\mathcal{R}(\mathfrak{a},R) : ct^k)$ where $c \in \mathfrak{a}^k$. Clearly $c \notin \mathfrak{a}^{n+k}$

but $c\cdot(\mathfrak{q}\cap R)\cdot t^k \subset u^n \mathcal{R}(\mathfrak{a},R)\cap \mathfrak{a}^k t^k = \mathfrak{a}^{n+k}t^k$, so $c(\mathfrak{q}\cap R)\subset \mathfrak{a}^{n+k}$. Finally $c \notin \mathfrak{a}^{n+k}$ implies $k > -n$.

(16.23) Theorem. Let R be a noetherian ring and $\mathfrak{a}$ an ideal of R . If $\mathfrak{p} \in A(\mathfrak{a})$ then $\mathfrak{p} \in \mathrm{Ass}(A/\mathfrak{a}^n)$ for some n .

Proof. First we treat the case that $\mathfrak{a}$ is principal, generated by a non-zerodivisor . We may assume R to be local with maximal ideal $\mathfrak{p}$ (see Remark (16.2)), and moreover we may assume that R is complete, since $\mathfrak{p}$ is associated to $\mathfrak{a}^n$ if and only if the same holds after completion. Now we know from Theorem (16.9),b), that there is a minimal prime $\mathfrak{p}_\circ$ of R such that $\mathfrak{p}/\mathfrak{p}_\circ \in A(\mathfrak{a}+\mathfrak{p}_\circ/\mathfrak{p}_\circ)$. By Proposition (16.15) we have $\mathrm{ht}(\mathfrak{p}/\mathfrak{p}_\circ) \leq 1$, and since $\mathfrak{a} = aR$ with $a \notin \mathfrak{p}_\circ$, it follows that $\mathfrak{p}$ is minimal over $\mathfrak{a}+\mathfrak{p}_\circ$. Therefore Lemma (16.21) implies $\mathfrak{p} \in \mathrm{Ass}(R/\mathfrak{a}^n)$ for some n .

In the general case, passing to $\mathcal{R}(\mathfrak{a},R)$ we may choose $\mathfrak{q} \in A(u\mathcal{R}(\mathfrak{a},R))$ such that $\mathfrak{q}\cap R = \mathfrak{p}$ (by Theorem (16.9),a)). By the first part of the proof we see that $\mathfrak{q}$ is associated to $u^m\mathcal{R}(\mathfrak{a},R)$ for some m , so $\mathfrak{p}$ is associated to $\mathfrak{a}^n$ for some n by Lemma (16.22).

(16.24) Remark. It can be shown without much effort that

$$\mathrm{Ass}(R/\mathfrak{a}^n) = \mathrm{Ass}(R/\mathfrak{a}^{n+1}) \quad \text{for large } n$$

for any ideal $\mathfrak{a}$ in a noetherian ring R (see [5] or [3]). An argument simular to the one used in Theorem (16.23) then shows that

$$A(\mathfrak{a}) \subset \mathrm{Ass}(R/\mathfrak{a}^n) \quad \text{for large } n \ .$$

We do not prove these facts here, since we will not use them, and moreover good proofs are available in [3]. The same text also treats the more general notion of an asymptotic sequence relative to an ideal.

(16.25) Lemma. Let R be a noetherian ring and $\mathfrak{a}$ an ideal of R. For the polynomial ring $S = R[x_1,\dots,x_n]$ we have

$$A(\mathfrak{a}\cdot S) = \{\mathfrak{p}\cdot S \mid \mathfrak{p} \in A(\mathfrak{a})\} \quad .$$

Proof. This is an immediate consequence of the fact that

$$\overline{(\mathfrak{a}S)^k} = \overline{\mathfrak{a}^k S} = \overline{(\mathfrak{a}^k)}\cdot S \quad \text{for any } k \quad (\text{see } (15.14)).$$

§ 17. Asymptotic sequences

Asymptotic sequences, to be defined and investigated in this section, have many properties analogous to regular sequences. In the next section we will use them to characterize quasi-unmixed local rings, which parallels the characterization of Cohen-Macaualy local rings by regular sequences.

(17.1) Definition. Let R be a noetherian ring and let $\underline{a} = (a_1,\dots,a_s)$ be a sequence of elements of R.

a) $\underline{a}$ is called an asymptotic sequence in R if $\underline{a}R \neq R$, a_1 is not contained in any minimal prime ideal of R and for each $i = 2,\dots,s$ we have $a_i \notin \mathfrak{p}$ for every $\mathfrak{p} \in A((a_1,\dots,a_{i-1})R)$.

b) $\underline{a}$ is called a maximal asymptotic sequence in R if $\underline{a}$ is an asymptotic sequence in R and there is no a_{s+1} in R such that $(a_1,\dots,a_{s+1})$ is an asymptotic sequence in R.

(17.2) Remark. If $\underline{a}$ is an asymptotic sequence in a local ring R then $\underline{a}$ is maximal if and only if $\mathfrak{m} \in A(\underline{a}R)$, $\mathfrak{m}$ being the maximal ideal of R.

(17.3) Lemma. Let R be a noetherian ring and for any $\mathfrak{p} \in \mathrm{Min}(R)$ let $f(\mathfrak{p}) : R \longrightarrow R/\mathfrak{p}$ denote the canonical surjection. Let $\underline{a} = (a_1,\dots,a_s)$ be a sequence of elements of R. Then we have:

a) The following conditions are equivalent:

(i) $\underline{a}$ is an asymptotic sequence in R .

(ii) $\underline{a}R \neq R$, and for each $\mathfrak{p} \in W(\underline{a}R)$ the sequence $(f(\mathfrak{p})(a_1),\ldots,f(\mathfrak{p})(a_s))$ is an asymptotic sequence in $R/\mathfrak{p}$.

b) If R is local then $\underline{a}$ is a maximal asymptotic sequence in R if and only if $(f(\mathfrak{p})(a_1),\ldots,f(\mathfrak{p})(a_s))$ is a maximal asymptotic sequence in $R/\mathfrak{p}$ for some $\mathfrak{p} \in \mathrm{Min}(R)$.

Proof. a) is an immediate consequence of Theorem (16.9),b). For b) note that $W(\underline{a}R) = \mathrm{Min}(R)$ in this case. Certainly, if $\underline{a}$ is not maximal then none of the sequences $(f(\mathfrak{p})(a_1),\ldots,f(\mathfrak{p})(a_s))$ will be maximal in $R/\mathfrak{p}$ by a), where $\mathfrak{p} \in \mathrm{Min}(R)$. Conversely, if $\underline{a}$ is maximal then $\mathfrak{m} \in A(\underline{a}R)$ by Remark (17.2), where $\mathfrak{m}$ is the maximal ideal of R , of course. By Theorem (16.9),b) again we see that $\mathfrak{m}/\mathfrak{p} \in A(\underline{a}\cdot(R/\mathfrak{p}))$ for some $\mathfrak{p} \in \mathrm{Min}(R)$ and hence $(f(\mathfrak{p})(a_1),\ldots,f(\mathfrak{p})(a_s))$ is a maximal asymptotic sequence in $R/\mathfrak{p}$.

<u>(17.4) Lemma</u>. Let R be a local ring with completion $\hat{R}$ and let $\underline{a}$ be a sequence of elements of R . Then $\underline{a}$ is an asymptotic sequence in R if and only if $\underline{a}$ is an asymptotic sequence in $\hat{R}$.

Proof. This is an immediate consequence of Corollary (16.20).

<u>(17.5) Remark</u>. a) Lemmas (17.3) and (17.4) reduce certain questions about asymptotic sequences to the case of complete local domains.

b) By Theorem (16.19), the completion $\hat{R}$ in Lemma (17.4) can be replaced by any faithfully flat extension of R satisfying the altitude formula.

c) Recall that any complete local domain R satisfies

$$\dim R = \dim R/\mathfrak{a} + \mathrm{ht}(\mathfrak{a}) \quad \text{for any ideal } \mathfrak{a} \text{ of } R .$$

<u>(17.6) Proposition</u>. Let R be a complete local domain and let $\underline{a} = (a_1,\ldots,a_s)$ be a sequence of elements in the maximal ideal of R . The following conditions are equivalent:

(i) $\underline{a}$ is an asymptotic sequence in R .

(ii) For any $j \in \{1,\dots,s\}$, $\mathrm{ht}(a_1R + \dots + a_jR) = j$.

(iii) $\mathrm{ht}(\underline{a}R) = s$.

(iv) $\dim R/\underline{a}R = \dim R - s$.

(v) For any $j \in \{1,\dots,s\}$, $\dim R/a_1R + \dots + a_jR = \dim R - j$

(vi) $\underline{a}$ is a sequence of parameters.

Proof. (iv) $\Longleftrightarrow$ (v) $\Longleftrightarrow$ (vi) is true for any local ring and (iii) $\Longleftrightarrow$ (iv) and (ii) $\Longleftrightarrow$ (v) by Remark (17.5),c). Assume (i) and let $j \in \{1,\dots,s\}$. By definition, a_j is not contained in any minimal prime of $a_1R + \dots + a_{j-1}R$, so (v) follows by induction. Finally we prove (ii) $\Rightarrow$ (i). For this let $\mathfrak{p} \in A(a_1R + \dots + a_{j-1}R)$ for some $j \in \{1,\dots,s\}$, where $a_1R + \dots + a_{j-1}R = 0$ for $j = 1$. Then $\mathrm{ht}(\mathfrak{p}) \leq j-1$ by Proposition (16.15), so $a_j \notin \mathfrak{p}$ by (ii) .

<u>(17.7) Remark</u>. The final step in the above proof can be phrased in the following way: If $\mathfrak{a}$ is an ideal of the principal class in a complete local domain then $A(\mathfrak{a}) = \mathrm{Min}(R/\mathfrak{a})$.

<u>(17.8) Theorem</u>. Let R be a local ring with completion $\hat{R}$ and let $\underline{a} = (a_1,\dots,a_s)$ be a sequence of elements in the maximal ideal of R . Then we have:

a) $\underline{a}$ is an asymptotic sequence in R if and only if for any minimal prime $\mathfrak{p}$ of $\hat{R}$, $\underline{a}(\hat{R}/\mathfrak{p})$ is an ideal of the principal class of height s .

b) $\underline{a}$ is a maximal asymptotic sequence in R if and only if it is an asymptotic sequence in R and for some $\mathfrak{p} \in \mathrm{Min}(\hat{R})$, the images of $a_1,\dots,a_s$ in $\hat{R}/\mathfrak{p}$ are a system of parameters of $\hat{R}/\mathfrak{p}$.

Proof. This follows from Lemma (17.4), Lemma (17.3),b) (applied to $\hat{R}$) and Proposition (17.6).

<u>(17.9) Corollary</u>. Let R be a local ring with completion $\hat{R}$, and let $(a_1,\dots,a_s)$ be any maximal asymptotic sequence in R . Then

$$s = \inf\{\dim \hat{R}/\mathfrak{p} \mid \mathfrak{p} \in \operatorname{Min}(\hat{R})\} \quad .$$

<u>(17.10) Proposition</u>. Let R be a local ring and let $\underline{a} = (a_1,\ldots,a_s)$ and $\underline{b} = (b_1,\ldots,b_s)$ be sequences of elements of the maximal ideal of R such that

$$\sqrt{\underline{a}R} = \sqrt{\underline{b}R} \quad .$$

Then $\underline{a}$ is an asymptotic sequence in R if and only if $\underline{b}$ is an asymptotic sequence in R .

Proof. Let $\hat{R}$ be the completion of R and let $\mathfrak{p} \in \operatorname{Min}(\hat{R})$. Then

$$\sqrt{\underline{a}(\hat{R}/\mathfrak{p})} = \sqrt{\underline{b}(\hat{R}/\mathfrak{p})} \quad ,$$

and hence $\underline{a}$ generates an ideal of the principal class of height s in $\hat{R}/\mathfrak{p}$ if and only if $\underline{b}$ does. Therefore the Proposition follows from Theorem (17.8),a).

<u>(17.11) Proposition</u>. Let R be a noetherian ring and let $\underline{x} = (x_1,\ldots,x_s)$ be a regular sequence in R . Then $\underline{x}$ is an asymptotic sequence.

Proof. Let $\mathfrak{a}_j = (x_1,\ldots,x_j)\cdot R$, and recall (see also Theorem (13.11)) that $\operatorname{Ass}(R/\mathfrak{a}_j) = \operatorname{Ass}(R/\mathfrak{a}_j^n)$ for all n . If now $\mathfrak{p} \in A(\mathfrak{a}_j)$ then $\mathfrak{p} \in \operatorname{Ass}(R/\mathfrak{a}_j)$ by Theorem (16.23). Hence if $j < s$ then $x_{j+1} \notin \mathfrak{p}$.

<u>(17.12) Question</u>. Let $\underline{a}$ be a sequence of elements of a noetherian ring R such that $\underline{a}R \neq R$. If $\underline{a}$ is an asymptotic sequence in $R_{\mathfrak{p}}$ for every $\mathfrak{p} \in A(\underline{a}R)$, does it follow that $\underline{a}$ is an asymptotic sequence in R ?

<u>(17.13) Lemma</u>. Let R be a noetherian ring and let $\underline{a}$ be a sequence of elements of R . Then $\underline{a}$ is an asymptotic sequence in R if and only if $\underline{a}$ is an asymptotic sequence in the polynomial ring $R[X]$.

Proof. This is immediate from Lemma (16.25).

<u>(17.14) Remark</u>. If R is a noetherian ring, $\underline{a}$ an asymptotic sequence in R and $\mathfrak{p}$ a prime ideal containing $\underline{a}R$, then the image of $\underline{a}$ in $R_{\mathfrak{p}}$ is an asymptotic sequence again. This follows from Remark (16.2).

§ 18. Quasi-unmixed rings

<u>(18.1) Definition</u>. Let R be a local ring. We put

$$a(R) = \sup\{n \mid \text{there exists an asymptotic sequence of length } n \text{ in } R\}.$$

If R is any noetherian ring and $\mathfrak{p}$ a prime ideal of R, we define $a(\mathfrak{p}) = a(R_{\mathfrak{p}})$.

<u>(18.2) Remark</u>. If R is local with completion $\hat{R}$ then

$$a(R) = \inf\{\dim \hat{R}/\mathfrak{p} \mid \mathfrak{p} \in \operatorname{Min} \hat{R}\} = a(\hat{R})$$

by Corollary (17.9).

<u>(18.3) Lemma</u>. Let R be a noetherian ring and $\mathfrak{q}$ a prime ideal in the polynomial ring $S = R[x]$. Putting $\mathfrak{p} = \mathfrak{q} \cap R$ we have

$$a(\mathfrak{q}) \geq a(\mathfrak{p}) + 1 - \operatorname{tr.d.}_R(S/\mathfrak{q}) \quad .$$

Proof. We may assume R local with maximal ideal $\mathfrak{p}$. If $\underline{a} = (a_1,\dots,a_s)$ is any asymptotic sequence in R then $\underline{a}$ is an asymptotic sequence in S by Lemma (17.13). So if $\operatorname{tr.d.}_R(S/\mathfrak{q}) = 1$ there is nothing to prove. If $\operatorname{tr.d.}_R(S/\mathfrak{q}) = 0$ then

$$\mathfrak{q} = \mathfrak{p}S + F\cdot S \quad , \quad F \text{ a monic polynomial} \quad .$$

Since F is monic, $F \notin \mathfrak{p}^*$ for any $\mathfrak{p}^* \in A(\underline{a}R)$ by Lemma (16.25), and hence $a_1,\dots,a_s, F$ is an asymptotic sequence in S and a forteriori in $S_{\mathfrak{q}}$. Therefore $a(\mathfrak{q}) \geq s+1$ as asserted.

(18.4) Proposition. Let R be a noetherian ring and $\mathfrak{p} \subset \mathfrak{q}$ prime ideals. Then

$$a(\mathfrak{q}) \leqq a(\mathfrak{p}) + ht(\mathfrak{q}/\mathfrak{p}) \quad .$$

Proof. Using induction on $ht(\mathfrak{q}/\mathfrak{p})$ we may assume that $ht(\mathfrak{q}/\mathfrak{p}) = 1$. (Note that $\mathfrak{p} \subset \mathfrak{q}' \subset \mathfrak{q}$ implies $ht(\mathfrak{q}/\mathfrak{p}) \geqq ht(\mathfrak{q}/\mathfrak{q}') + ht(\mathfrak{q}'/\mathfrak{p})$.) Moreover we may localize to assume that R is local with maximal ideal $\mathfrak{q}$, and hence $\dim R/\mathfrak{p} = 1$. Let $\underline{a} = (a_1,\ldots,a_s)$ be an asymptotic sequence with $a_i \in \mathfrak{p}$ and s maximal. Then $s \leqq a(\mathfrak{p})$, and we will show that $a(\mathfrak{q}) \leqq s+1$. If $\mathfrak{q} \in A(\underline{a}R)$ then $a(\mathfrak{q}) = s$, so assume $\mathfrak{q} \notin A(\underline{a}R)$. Now s maximal and $\dim R/\mathfrak{p} = 1$ imply $\mathfrak{p} \in A(\underline{a}R)$. Let $\hat{R}$ be the completion of R. By Corollary (16.20) we may choose $\hat{\mathfrak{p}} \in A(\underline{a}\hat{R})$ such that $\hat{\mathfrak{p}} \cap R = \mathfrak{p}$. By Proposition (17.6) there is a minimal prime $\mathfrak{p}_\circ$ of $\hat{R}$ such that $\mathfrak{p}_\circ \subset \mathfrak{p}$ and $ht(\mathfrak{p}/\mathfrak{p}_\circ) = s$. Therefore

$$\dim(\hat{R}/\mathfrak{p}_\circ) = \dim(\hat{R}/\hat{\mathfrak{p}}) + ht(\hat{\mathfrak{p}}/\mathfrak{p}_\circ) \leqq \dim(\hat{R}/\mathfrak{p}\hat{R}) + s = s+1 \quad ,$$

and consequently $a(\mathfrak{q}) = a(R) \leqq s+1$ (see Remark (18.2)).

(18.5) Remark. At the end of the above proof we have made use of the fact that a complete local domain R satisfies

(18.5.1) $\dim R = \dim(R/\mathfrak{p}) + ht(\mathfrak{p})$ for all prime ideals $\mathfrak{p}$ of R.

This is equivalent to saying that the first chain condition holds in R, where the first chain condition for a local ring means that all maximal chains of prime ideals in that ring have the same length. This first chain condition for an arbitrary local ring R is equivalent to (18.5.1) and

(18.5.2) $\dim R/\mathfrak{p} = \dim R$ for all $\mathfrak{p} \in \operatorname{Min} R$.

Recall also that a noetherian ring R is called catenary if for any prime ideals $\mathfrak{p}_1 \subset \mathfrak{p}_2 \subset \mathfrak{p}_3$ of R we have

$$ht(\mathfrak{p}_3/\mathfrak{p}_1) = ht(\mathfrak{p}_3/\mathfrak{p}_2) + ht(\mathfrak{p}_2/\mathfrak{p}_1) \quad .$$

Finally a noetherian ring R is called universally catenary if every finitely generated R-algebra is catenary.

(18.6) Lemma. If a local ring R satisfies (18.5.1), then

(18.6.1) $\quad \dim R = \dim(R/\mathfrak{a}) + \operatorname{ht}(\mathfrak{a})$ for any proper ideal $\mathfrak{a}$ of R .

Proof. Choosing $\mathfrak{p} \in \operatorname{Min}(R/\mathfrak{a})$ such that $\operatorname{ht}(\mathfrak{p}) = \operatorname{ht}(\mathfrak{a})$ we have

$$\dim R = \dim(R/\mathfrak{p}) + \operatorname{ht}(\mathfrak{p}) \leqq \dim(R/\mathfrak{a}) + \operatorname{ht}(\mathfrak{a}) \quad .$$

(18.7) Lemma. Let R be a local ring with completion $\hat{R}$. If $\hat{R}$ satisfies (18.6.1) so does R .

Proof. This is an immediate consequence of $\dim \hat{R}/\mathfrak{a}\hat{R} = \dim R/\mathfrak{a}$ and $\operatorname{ht}(\mathfrak{a}\hat{R}) = \operatorname{ht}(\mathfrak{a})$ for any ideal $\mathfrak{a}$ of R (see Proposition (5.1)).

(18.8) Definition. A local ring R is called equidimensional if

$$\dim R/\mathfrak{p} = \dim R \quad \text{for all} \quad \mathfrak{p} \in \operatorname{Min} R \quad .$$

(18.9) Lemma. Let R be a local ring with completion $\hat{R}$. If $\hat{R}$ is equidimensional, so is R .

Proof. If $\mathfrak{p} \in \operatorname{Min} R$ then $\operatorname{ht}(\mathfrak{p}\hat{R}) = 0$ by Proposition (5.1)). Choosing a minimal prime $\hat{\mathfrak{p}}$ of $\hat{R}$ containing $\mathfrak{p}\hat{R}$ we get

$$\dim R/\mathfrak{p} = \dim \hat{R}/\mathfrak{p}\hat{R} \geqq \dim \hat{R}/\hat{\mathfrak{p}} = \dim \hat{R} = \dim R \quad .$$

(18.10) Lemma. Let $R \to S$ be a flat local homomorphism of local rings. Let $\mathfrak{p}$ be a prime ideal of R and $\mathfrak{q} \in \operatorname{Min}(S/\mathfrak{p}S)$. Then

$$\operatorname{ht}(\mathfrak{q}) = \operatorname{ht}(\mathfrak{p}) \quad .$$

Proof. Note first that $\mathfrak{q} \cap R = \mathfrak{p}$ by going down. Now $R_{\mathfrak{p}} \longrightarrow S_{\mathfrak{q}}$ is flat again and we may assume that $\mathfrak{p}$ resp. $\mathfrak{q}$ is the maximal ideal of R resp. S. By [2], (13.B), Theorem 19

$$\dim S = \dim R + \dim S/\mathfrak{p}S \quad .$$

But $\mathfrak{q} \in \mathrm{Min}(S/\mathfrak{p}S)$ implies $\dim S/\mathfrak{p}S = 0$.

(18.11) Definition. A local ring is quasi-unmixed if its completion is equidimensional. A noetherian ring R is called quasi-unmixed if $R_{\mathfrak{m}}$ is quasi-unmixed for any maximal ideal $\mathfrak{m}$ of R .

(18.12) Proposition. A local ring R is quasi-unmixed if and only if every system of parameters is an asymptotic sequence of R .

Proof. By definition, R is quasi-unmixed if and only if $a(R) = \dim R$. Now given any system of parameters $x_1,\ldots,x_d$ of R and any minimal prime $\mathfrak{p}$ in the completion $\hat{R}$ of R , the images of $x_1,\ldots,x_d$ in $R/\mathfrak{p}$ form a system of parameters if and only if $\dim R/\mathfrak{p} = d$. Therefore the proposition follows from Theorem (17.8).

(18.13) Theorem. Let R be a quasi-unmixed noetherian ring. Then we have:

a) $R_{\mathfrak{p}}$ is quasi-unmixed for any prime $\mathfrak{p}$ of R .

b) The polynomial ring $R[x_1,\ldots,x_n]$ is quasi-unmixed.

c) R is universally catenary.

d) If R is local then the first chain condition holds in R .

Proof. Let $\mathfrak{p}$ be any prime ideal of R and let $\mathfrak{m}$ be a maximal ideal containing $\mathfrak{p}$. By assumption we have $a(\mathfrak{m}) = \mathrm{ht}(\mathfrak{m})$. By Proposition (18.4) we have

$$a(\mathfrak{p}) \geq a(\mathfrak{m}) - \mathrm{ht}(\mathfrak{m}/\mathfrak{p}) \geq a(\mathfrak{m}) - \mathrm{ht}(\mathfrak{m}) + \mathrm{ht}(\mathfrak{p}) = \mathrm{ht}(\mathfrak{p}) \quad .$$

This proves a) and

$$(18.13.1) \qquad \mathrm{ht}(\mathfrak{m}/\mathfrak{p}) = \mathrm{ht}(\mathfrak{m}) - \mathrm{ht}(\mathfrak{p}) \quad .$$

b) is a direct consequence of Lemma (18.3) using induction on n. Applying (18.13.1) to any localization of R we see that R is catenary, hence universally catenary by b). Finally, if R is local then R is equidimensional by Lemma (18.9), so the first chain condition holds by Remark (18.5).

(18.14) Corollary. Let R be a noetherian quasi-unmixed ring and let S be a multiplicatively closed subset of R. Then R_S is quasi-unmixed again.

Proof. This is a direct consequence of a) in Theorem (18.13).

(18.15) Remark. Let R be a noetherian domain. If R is universally catenary then the altitude formula holds in R ([2], (14.C), Theorem 23).

(18.16) Lemma. Let R be a local ring and let $\mathfrak{a}$ be a proper ideal of R satisfying $ht(\mathfrak{a}) = s(\mathfrak{a}) = s$. Then there are an integer n and elements $x_1,\ldots,x_s \in \mathfrak{a}^n$ such that

$$\overline{\mathfrak{a}^n} = \overline{(x_1,\ldots,x_s)R} .$$

Proof. Let $\mathfrak{m}$ be the maximal ideal of R. Using the notation of (10.22), we choose $x_1,\ldots,x_s$ such that $F(\mathfrak{a})(x_1),\ldots,F(\mathfrak{a})(x_s)$ are a homogeneous system of parameters of $G(\mathfrak{a},R) \otimes R/\mathfrak{m}$ consisting of elements of the same degree n. Then, for some $t \geqq n$, we will have

$$\mathfrak{a}^{m+n} = (x_1,\ldots,x_s)\mathfrak{a}^m \quad \text{for all} \quad m \geqq t ,$$

showing that $x_1,\ldots,x_s$ generate a reduction of $\mathfrak{a}^n$.

(18.17) Theorem. Let R be a local ring. The following conditions are equivalent:

(i) R is quasi-unmixed.

(ii) Every system of parameters of R is an asymptotic sequence.

(iii) R is equidimensional and universally catenary.

(iv) R is equidimensional and satisfies the altitude formula.

(v) R is equidimensional and for any $\mathfrak{p} \in \mathrm{Min}(R)$, $R/\mathfrak{p}$ is quasi-unmixed.

(vi) If $\mathfrak{a}$ is any ideal of the principal class in R then $\mathrm{ht}(\mathfrak{p}) = \mathrm{ht}(\mathfrak{a})$ for any $\mathfrak{p} \in A(\mathfrak{a})$.

(vii) If $\mathfrak{a}$ is any ideal of R satisfying $\mathrm{ht}(\mathfrak{a}) = s(\mathfrak{a})$ then $\overline{\mathfrak{a}}$ is unmixed (i.e. $\mathrm{ht}(\mathfrak{p}) = \mathrm{ht}(\overline{\mathfrak{a}})$ for any $\mathfrak{p} \in \mathrm{Ass}(R/\overline{\mathfrak{a}})$) .

(viii) If $\mathfrak{a}$ is any ideal of the principal class in R then $\overline{\mathfrak{a}}$ is unmixed.

Proof. The equivalence of (i) and (ii) has been established in Proposition (18.12), and (i) ⇒ (iii) by Theorem (18.13),c), and Lemma (18.9). (iii) ⇒ (iv) by Remark (18.14). Assume now (iv) and let $\mathfrak{p}$ be any minimal prime of R . Then any system of parameters $\underline{x}$ of R will remain a system of parameters in $R/\mathfrak{p}$, therefore Proposition (16.15) implies that the image of $\underline{x}$ in $R/\mathfrak{p}$ is an asymptotic sequence in $R/\mathfrak{p}$. So by Lemma (17.3),a), $\underline{x}$ is an asymptotic sequence in R , proving (iv) ⇒ (ii) and thereby the equivalence of (i), (ii), (iii) and (iv). The equivalence of (i) and (iii) shows that (i) ⟺ (v) . We proceed to show that (i) - (v) ⇒ (vi) ⇒ (vii) ⇒ (viii) ⇒ (ii) . Let $\mathfrak{p} \in A(\mathfrak{a})$ where $\mathfrak{a}$ is an ideal of the principal class. Then there is a minimal prime $\mathfrak{p}_o \subset \mathfrak{p}$ such that $\mathfrak{p}/\mathfrak{p}_o \in A(\mathfrak{a} + \mathfrak{p}_o/\mathfrak{p}_o)$ (by Theorem (16.9),b)). Therefore $\mathrm{ht}(\mathfrak{p}/\mathfrak{p}_o) \leqq \mathrm{ht}(\mathfrak{a})$ by Proposition (16.15). Now assuming (i) - (v) we know that R and $R/\mathfrak{p}_o$ satisfy (18.5.1) and therefore

$$\mathrm{ht}(\mathfrak{p}/\mathfrak{p}_o) = \dim R/\mathfrak{p}_o - \dim R/\mathfrak{p} = \dim R - \dim R/\mathfrak{p} = \mathrm{ht}(\mathfrak{p}) ,$$

which proves (vi). Now let $\mathfrak{a}$ be an ideal satisfying $\mathrm{ht}(\mathfrak{a}) = s(\mathfrak{a}) = s$. Then there are $x_1,\dots,x_s$ and an integer n such that

$$\overline{\mathfrak{a}^n} = \overline{(x_1,\dots,x_s)R} \quad \text{(see Lemma (18.16))}.$$

Hence $\mathrm{ht}(\mathfrak{p}) = s$ for any $\mathfrak{p} \in A(\mathfrak{a}^n) = A(\mathfrak{a})$ by (vi), and this holds in particular for all $\mathfrak{p} \in \mathrm{Ass}(R/\overline{\mathfrak{a}})$ so (vii) follows. (vii) ⇒ (viii) since any ideal $\mathfrak{a}$ of the principal class satisfies $\mathrm{ht}(\mathfrak{a}) = s(\mathfrak{a})$ (see Proposition (10.20),a)). Finally the implication (viii) $\Rightarrow$ (ii) is obvious.

(18.18) Remark. The proof given above for (i) - (v) $\Rightarrow$ (vi) actually proves the following generalization of Proposition (16.15): If $\mathfrak{a}$ is generated by s elements and R is quasi-unmixed then

$$\mathrm{ht}(\mathfrak{p}) \leqq s \quad \text{for any} \quad \mathfrak{p} \in A(\mathfrak{a}) \quad .$$

(18.19) Corollary. Let R be a quasi-unmixed local ring and $\mathfrak{a}$ a proper ideal of R. The following conditions are equivalent:

(i) $R/\mathfrak{a}$ is quasi-unmixed.

(ii) $R/\mathfrak{a}$ is equidimensional.

In particular $R/\mathfrak{p}$ is quasi-unmixed for any prime $\mathfrak{p}$ of R.

Proof. Use (iii) in Theorem (18.17).

(18.20) Corollary. Let R be a quasi-unmixed local ring and let $\mathfrak{a}$ be an ideal of the principal class of R. If $\mathfrak{b}$ is any ideal of R satisfying $\mathfrak{a} \subset \mathfrak{b} \subset \sqrt{\mathfrak{a}}$ then $R/\mathfrak{b}$ is quasi-unmixed.

(18.21) Corollary. Let R be a quasi-unmixed ring (not necessarily local). If $\underline{x}$ is a regular sequence of R then $R/\underline{x}R$ is quasi-unmixed.

(18.22) Theorem. Let A be a graded noetherian ring. If A is universally catenary, the following conditions are equivalent:

(i) A is quasi-unmixed.

(ii) For every maximal homogeneous $\mathfrak{m}$ of A, $A_{\mathfrak{m}}$ is quasi-unmixed.

Proof. Clearly (i) $\Rightarrow$ (ii) by definition. To prove (ii) $\Rightarrow$ (i) we need to show that $A_{\mathfrak{m}}$ is equidimensional for every inhomogeneous maximal ideal $\mathfrak{m}$ of A. Recall that

$$\mathrm{ht}(\mathfrak{m}) = \mathrm{ht}(H(\mathfrak{m})) + 1$$

(see Lemma (9.1)). Now let $\mathfrak{q}$ be any minimal prime ideal of A contained in $\mathfrak{m}$. Then $\mathfrak{q}$ is homogeneous and hence $\mathfrak{q} \subset H(\mathfrak{m})$. Since $A_{H(\mathfrak{m})}$ is equidimensional by assumption, we have

$$\mathrm{ht}(H(\mathfrak{m})/\mathfrak{q}) = \mathrm{ht}(H(\mathfrak{m})) \quad .$$

Since $\mathfrak{m}/\mathfrak{q}$ is inhomogeneous, we conclude that

$$\mathrm{ht}(\mathfrak{m}) \geqq \mathrm{ht}(\mathfrak{m}/\mathfrak{q}) = \mathrm{ht}(H(\mathfrak{m})/\mathfrak{q}) + 1 = \mathrm{ht}(H(\mathfrak{m})) + 1 = \mathrm{ht}(\mathfrak{m})$$

and therefore $\mathrm{ht}(\mathfrak{m}) = \mathrm{ht}(\mathfrak{m}/\mathfrak{q})$.

<u>(18.23) Theorem</u>. Let R be a local ring and let $\mathfrak{a}$ be a proper ideal of R. Consider the conditions:

(i) R is quasi-unmixed.

(ii) $B(\mathfrak{a},R)$ is quasi-unmixed.

(iii) $\mathfrak{R}(\mathfrak{a},R)$ is quasi-unmixed.

Then (i) and (iii) are equivalent, and if $\mathrm{ht}(\mathfrak{a}) > 0$, all three conditions are equivalent.

Proof. (i) $\Longleftrightarrow$ (ii). Note first that R is universally catenary if and only if $B(\mathfrak{a},R)$ is. Let $\mathfrak{n}$ be the unique maximal homogeneous ideal of $B(\mathfrak{a},R)$. We may assume $B(\mathfrak{a},R)$ to be universally catenary, so by Theorem (18.21) $B(\mathfrak{a},R)$ is quasi-unmixed if and only if $B(\mathfrak{a},R)_{\mathfrak{n}}$ is equidimensional. Now let $\mathfrak{q}$ be any minimal prime ideal of R and let $\mathfrak{q}^*$ be the corresponding minimal prime of $B(\mathfrak{a},R)$ (see Lemma (4.5),e)).

$$\dim B(\mathfrak{a},R)_{\mathfrak{n}}/\mathfrak{q}^* B(\mathfrak{a},R)_{\mathfrak{n}} = \dim R/\mathfrak{q} + 1$$

by Theorem (9.7). Hence $B(\mathfrak{a},R)_{\mathfrak{n}}$ is equidimensional if and only if the same holds for R. The proof of (i) $\Longleftrightarrow$ (ii) is literally the same, using the one-to-one correspondence of minimal primes of R and $\mathfrak{R}(\mathfrak{a},R)$ given in Proposition (15.1),e).

(18.24) Corollary. Let R be a quasi-unmixed local ring and let $\mathfrak{a}$ be a proper ideal of R. Then $G(\mathfrak{a},R)$ is quasi-unmixed.

Proof. Recall ((8.8.5)) that

$$G(\mathfrak{a},R) \cong \mathfrak{R}(\mathfrak{a},R)\,/\,u\cdot\mathfrak{R}(\mathfrak{a},R) \quad .$$

Since $u\cdot\mathfrak{R}(\mathfrak{a},R)$ is an ideal of the principal class, the Corollary follows from Theorem (18.23) and Corollary (18.20).

(18.25) Lemma. Let A be a graded ring and let S be a multiplicatively closed subset of homogeneous elements of A. If A is quasi-unmixed, so is $A_{(S)}$.

Proof. Recall ((12.17)) that

$$A_S \cong A_{(S)}[X,X^{-1}] \quad .$$

Now A_S is quasi-unmixed by (18.14). On the other hand there is a one-to-one correspondence of minimal primes of $A_{(S)}$ and those of $A_{(S)}[X,X^{-1}]$, and if $\mathfrak{q}$ in $A_{(S)}$ corresponds to $\mathfrak{q}^*$ in $A_{(S)}[X,X^{-1}]$ then

$$\dim A_{(S)}/\mathfrak{q} + 1 = \dim A_{(S)}[X,X^{-1}]/\mathfrak{q}^* \quad .$$

So if $A_{(S)}[X,X^{-1}]$ is quasi-unmixed, the same holds for $A_{(S)}$.

(18.26) Corollary. Let R be a quasi-unmixed local ring and let $\mathfrak{a}$ be a proper ideal of R. For any blowing up homomorphism $R \to R_1$ of $(R,\mathfrak{a})$, R_1 is quasi-unmixed.

Proof. By Proposition (12.10) and (12.13), R_1 is either a localization of R or

$$R_1 \simeq \mathfrak{R}(\mathfrak{a},R)_{(\mathfrak{q})}$$

for some homogeneous prime ideal $\mathfrak{q}$ of $\mathfrak{R}(\mathfrak{a},R)$. So R_1 is quasi-unmixed by Theorem (18.23) and Lemma (18.25).

§ 19. The theorem of Rees-Böger

The theorem of Rees (see [12]) is concerned with two $\mathfrak{m}$-primary ideals $\mathfrak{a} \subset \mathfrak{b}$ in a local ring R with maximal ideal $\mathfrak{m}$. It is the converse of Proposition (4.14) and states that if $\mathfrak{a}$ and $\mathfrak{b}$ have the same multiplicity then they have the same integral closure, provided the ring is quasi-unmixed. We give a short proof of this result and we will show that the property stated in Rees' theorem characterizes quasi-unmixed local rings. This was first observed by Ratliff ([11]). Finally we show how Böger's extension ([4]) of Rees' theorem can be simply obtained by localization; see also App. III 3.2.7.

(19.1) Lemma. Let R be a local ring and let $\underline{x} = (x_1,\ldots,x_d)$ be a system of parameters of R. Then

$$e(\underline{x}R,R) \leqq e(\underline{x}R/x_1R, R/x_1R)$$

with equality if and only if x_1 is a regular element in R.

Proof. By the Theorem of Samuel (see (2.9.1)) we have $e(\underline{x}R,R) = e(\underline{x};R)$ and $e(\underline{x}R/x_1R,R/x_1R) = e((x_2,\ldots,x_d);R/x_1R)$. So the Lemma is immediate from the definition of $e(\underline{x};R)$.

In Chapter VI we will study the general question of how dividing by a parameter affects multiplicities and Hilbert functions. To prove Rees' theorem we need a related result, which will be proved in (28.1), but for which we give an ad hoc proof here:

(19.2) Lemma. Let R be a local ring with maximal ideal $\mathfrak{m}$ and let $\mathfrak{q}$ be an $\mathfrak{m}$-primary ideal. For any $x \in \mathfrak{q}$ for which $\dim R/xR = \dim R - 1$ we have

$$e(\mathfrak{q},R) \leqq e(\mathfrak{q}/xR, R/xR) \quad .$$

Proof. By Proposition (5.1) we may assume that R has an infinite residue field. Therefore we may choose $x_2,\ldots,x_d \in \mathfrak{q}$, $d = \dim R$, such that the images of the x_i's generate a minimal reduction of $\mathfrak{q}/xR$, i.e. if we put $\mathfrak{q}' = (x,x_2,\ldots,x_d)R$ then $e(\mathfrak{q}/xR,R/xR) = e(\mathfrak{q}'/xR,R/xR)$. Then

$$e(\mathfrak{q},R) \leqq e(\mathfrak{q}',R) \leqq e(\mathfrak{q}'/xR,R/xR) = e(\mathfrak{q}/xR,R/xR) \quad ,$$

the first inequality since $\mathfrak{q}' \subset \mathfrak{q}$ (see Proposition (3.11),d)) and the second inequality by Lemma (19.1) above.

(19.3) Theorem. (Rees) Let R be a quasi-unmixed local ring with maximal ideal $\mathfrak{m}$ and let $\mathfrak{a} \subset \mathfrak{b}$ be $\mathfrak{m}$-primary ideals. The following conditions are equivalent:

(i) $e(\mathfrak{a},R) = e(\mathfrak{b},R)$;

(ii) $\overline{\mathfrak{a}} = \overline{\mathfrak{b}}$.

Proof. (ii) ⇒ (i) has been shown in Proposition (4.14). For the proof of (i) ⇒ (ii) we make some reductions first. Let $R^* = R[T]_{\mathfrak{m}R[T]}$. Then $e(\mathfrak{a}R^*,R^*) = e(\mathfrak{a},R)$ and $e(\mathfrak{b}R^*,R^*) = e(\mathfrak{b},R)$ by Proposition (5.1) and moreover $\overline{\mathfrak{a}} = \overline{\mathfrak{a}R^*} \cap R$ and $\overline{\mathfrak{b}} = \overline{\mathfrak{b}R^*} \cap R$ by Corollary (4.12). Hence we may assume that R has an infinite residue field, and we may choose $\underline{x} = (x_1,\ldots,x_d)$, $d = \dim R$, generating a minimal reduction of $\mathfrak{a}$. Then clearly $\mathfrak{a}$ may be replaced by $\underline{x}R$, so we assume $\mathfrak{a} = \underline{x}R$. Next we want to reduce to the case that R is a domain. For this we note that

$$e(\mathfrak{a},R) = \sum_{\mathfrak{p}\in \mathrm{Assh}(R)} e(\mathfrak{a}\cdot R/\mathfrak{p},R/\mathfrak{p})\cdot \lambda(R_{\mathfrak{p}})$$

by the reduction formula (1.10.1), and similarly

$$e(\mathfrak{b},R) = \sum_{\mathfrak{p}\in \mathrm{Assh}(R)} e(\mathfrak{b}R/\mathfrak{p},R/\mathfrak{p})\cdot \lambda(R_{\mathfrak{p}}) \quad .$$

(To prove the last formula, one has to replace $\mathfrak{b}$ by a minimal reduction again.) Since $e(\mathfrak{b}R/\mathfrak{p},R/\mathfrak{p}) \leqq e(\mathfrak{a}R/\mathfrak{p},R/\mathfrak{p})$ for all $\mathfrak{p}$, equality (i) implies

$$e(\mathfrak{a}R/\mathfrak{p},R/\mathfrak{p}) = e(\mathfrak{b}R/\mathfrak{p},R/\mathfrak{p}) \quad \text{for all} \quad \mathfrak{p} \in \mathrm{Assh}(R) \quad .$$

Now $\mathrm{Assh}(R) = \mathrm{Min}(R)$ since R is quasi-unmixed, so Proposition (4.17) tells us that it is enough to prove (i) ⇒ (ii) for domains. Now we proceed by induction on $d = \dim R$, assuming $d = 1$ first.

Then R is a Cohen-Macaulay domain and hence

$$e(\mathfrak{a},R) = e(x_1R,R) = \lambda(R/x_1R) \quad .$$

Moreover, for large n we have $\mathfrak{b}^n \subset x_1R$ and therefore

$$\begin{aligned} e(\mathfrak{b},R) &= \lambda(\mathfrak{b}^{n-1}/\mathfrak{b}^n) = \lambda(R/\mathfrak{b}^n) - \lambda(R/\mathfrak{b}^{n-1}) \\ &= \lambda(R/x_1R) + \lambda(x_1R/\mathfrak{b}^n) - \lambda(x_1R/x_1\mathfrak{b}^{n-1}) \\ &= e(\mathfrak{a},R) - \lambda(\mathfrak{b}^n/x_1\mathfrak{b}^{n-1}) \quad . \end{aligned}$$

By assumption (i) we conclude that $\mathfrak{b}^n = x_1\mathfrak{b}^{n-1}$ for large n as desired. Now let $d > 1$ and assume that the Theorem holds for local rings of dimension less that d (not only domains!). Let R be a d-dimensional domain. We will show that for any non-negative discrete valuation v of R we have $v(\mathfrak{b}) \geqq v(\mathfrak{a})$, and then the result will follow by Proposition (4.20). For the given v , let us assume that $v(x_1) \geqq v(x_2)$. We put

$$S = R[T] / (x_1 - x_2T) = R[x_1 / x_2]$$

and the proof will be complete if we can show that

(19.3.1) $$\overline{\mathfrak{a}S} = \overline{\mathfrak{b}S}$$

since v is non-negative on S . Let us define

$$S^* = R[T]_{\mathfrak{m}R[T]} \quad .$$

Then putting $x_1^* = x_1 - x_2T$ we see that

$$S^* = (x_1^*, x_2, \ldots, x_d)S^* \quad ,$$

showing that $(x_1^*, x_2, \ldots, x_d)$ is a system of parameters of S^* . Hence, using Lemmas (19.1) and (19.2), Proposition (5.1) and our assumption (i) we obtain

$$e(\mathfrak{b}S^*/x^*S^*, S^*/x^*S^*) \leqq e(\mathfrak{a}S^*/x^*S^*, S^*/x^*S^*)$$

$$= e(\mathfrak{a}S^*, S^*) = e(\mathfrak{a}, R) = e(\mathfrak{b}, R) = e(\mathfrak{b}S^*, S^*)$$

$$\leqq e(\mathfrak{b}S^*/x^*S^*, S^*/x^*S^*) \quad .$$

Note that S^*/x^*S^* is quasi-unmixed by Theorem (18.13) and Theorem (18.17), (vi). Therefore we can use our inductive assumption to conclude that $\mathfrak{a}S^*/x^*S^*$ and $\mathfrak{b}S^*/x^*S^*$ have the same integral closure. Since S^*/x^*S^* is the localization of S at $\mathfrak{m}S + x^*S$ which is the radical of both, $\mathfrak{a}S$ and $\mathfrak{b}S$, we have

$$\overline{\mathfrak{a}S} = \overline{\mathfrak{a}S^*/x^*S^*} \cap S$$

and

$$\overline{\mathfrak{b}S} = \overline{\mathfrak{b}S^*/x^*S^*} \cap S \quad ,$$

which proves (19.3.1) and thereby concludes the proof to the theorem.

<u>(19.4) Lemma</u>. Let R be a local ring and let $x \in R$ be a non-unit which is not nilpotent. Then R contains a prime ideal $\mathfrak{p}$ satisfying $\dim R/\mathfrak{p} = 1$ and $x \notin \mathfrak{p}$.

Proof. By assumption there is a minimal prime $\mathfrak{p}_o$ of R for which $x \notin \mathfrak{p}_o$. Now we use descending induction on i , $i = 1,\ldots,\dim R/\mathfrak{p}_o$, to show that R contains a prime ideal $\mathfrak{p}$ satisfying $\dim R/\mathfrak{p} \leqq i$, $\mathfrak{p} \neq \mathfrak{m}$ and $x \notin \mathfrak{p}$. Assume this holds for some i , $1 < i \leqq \dim R/\mathfrak{p}_o$. Then the image of x in $R/\mathfrak{p}$ is contained in only finitely many primes of height one. Since $R/\mathfrak{p}$ contains infinitely many primes of height one, we may choose a prime ideal $\mathfrak{p}' \neq \mathfrak{m}$ containing $\mathfrak{p}$ such that $x \notin \mathfrak{p}'$, and obviously $\dim R/\mathfrak{p}' \leqq i-1$.

<u>(19.5) Theorem</u>. Let R be a local ring with maximal ideal $\mathfrak{m}$. The following conditions are equivalent:

(i) R is quasi-unmixed.

(ii) For any two $\mathfrak{m}$-primary ideals $\mathfrak{a} \subset \mathfrak{b}$, $e(\mathfrak{a},R) = e(\mathfrak{b},R)$ implies $\overline{\mathfrak{a}} = \overline{\mathfrak{b}}$.

Proof. (i) $\Rightarrow$ (ii) is the Rees' Theorem (19.3). For (ii) $\Rightarrow$ (i) we first show that R is equidimensional. For this let $\mathfrak{p}_\circ$ be any minimal prime of R and let $x \notin \mathfrak{p}_\circ$ be a non-unit contained in all minimal primes of R different from $\mathfrak{p}_\circ$. By Lemma (19.4) above we may choose a prime $\mathfrak{p}$ such that $x \notin \mathfrak{p}$ and $\dim R/\mathfrak{p} = 1$. We define two $\mathfrak{m}$-primary ideals by putting

$$\mathfrak{a} = \mathfrak{p} + x^2R \ , \quad \mathfrak{b} = \mathfrak{p} + xR \quad .$$

Then certainly $\mathfrak{b}/\mathfrak{p}$ is not integral over $\mathfrak{a}/\mathfrak{p}$ and a forteriori $\mathfrak{b}$ is not integral over $\mathfrak{a}$. By assumption (ii) we must have

$$(19.5.1) \qquad e(\mathfrak{a},R) > e(\mathfrak{b},R) \quad .$$

Again, as in Theorem (19.3), the reduction formula (1.10.1) implies

$$e(\mathfrak{a},R) = \sum_{\mathfrak{q}\in \mathrm{Assh}(R)} e(\mathfrak{a}R/\mathfrak{q},R/\mathfrak{q})\ \lambda(R_{\mathfrak{q}})$$

and

$$e(\mathfrak{b},R) = \sum_{\mathfrak{q}\in \mathrm{Assh}(R)} e(\mathfrak{b}R/\mathfrak{q},R/\mathfrak{q})\ \lambda(R_{\mathfrak{q}}) \quad .$$

By construction we have

$$\mathfrak{a}\cdot(R/\mathfrak{q}) = \mathfrak{b}(R/\mathfrak{q}) \quad \text{for all} \quad \mathfrak{q} \in \mathrm{Assh}(R)\ ,\ \mathfrak{q} \neq \mathfrak{p}_\circ$$

and therefore (19.5.1) implies $\mathfrak{p}_\circ \in \mathrm{Assh}(R)$, i.e. $\dim R/\mathfrak{p}_\circ = \dim R$. To complete the proof we note that condition (ii) is inherited by the completion, which must be equidimensional as well by the above argument.

<u>(19.6) Theorem</u>. (Böger) Let R be a quasi-unmixed local ring and let $\mathfrak{a}$ be a proper ideal of R satisfying $\mathrm{ht}(\mathfrak{a}) = s(\mathfrak{a})$. Then for any proper ideal $\mathfrak{b} \supset \mathfrak{a}$ the following conditions are equivalent:

(i) $\mathfrak{a}$ is a reduction of $\mathfrak{b}$.

(ii) $\mathfrak{b} \subset \sqrt{\mathfrak{a}}$ and $e(\mathfrak{a}R_{\mathfrak{p}}) = e(\mathfrak{b}R_{\mathfrak{p}})$ for all $\mathfrak{p} \in \mathrm{Min}(\mathfrak{a})$.

Proof. Clearly (i) $\Rightarrow$ (ii). To prove (ii) $\Rightarrow$ (i), recall from Theorem (18.17), (vii) that $\overline{\mathfrak{a}}$ is unmixed since $\mathrm{ht}(\mathfrak{a}) = s(\mathfrak{a})$. Hence, by

Corollary (4.9),c) we conclude

$$\bar{\mathfrak{a}} = \bigcap_{\mathfrak{p}\in \mathrm{Min}(\mathfrak{a})} \overline{\mathfrak{a}R_{\mathfrak{p}}} \cap R = \bigcap_{\mathfrak{p}\in \mathrm{Min}(\mathfrak{a})} \overline{\mathfrak{a}R_{\mathfrak{p}}} \cap R \quad .$$

On the other hand, $\overline{\mathfrak{a}R_{\mathfrak{p}}} = \overline{\mathfrak{b}R_{\mathfrak{p}}}$ for any $\mathfrak{p} \in \mathrm{Min}(\mathfrak{a})$ by Rees' Theorem (19.3), and therefore

$$\bar{\mathfrak{a}} = \bigcap_{\mathfrak{p}\in \mathrm{Min}(\mathfrak{a})} \overline{\mathfrak{b}R_{\mathfrak{p}}} \cap R \supset \bar{\mathfrak{b}} \quad .$$

References

Books

[1] N. Bourbaki, Algèbre commutative, ch. VII.

[2] H. Matsumura, Commutative Algebra, Benjamin, New York 1970.

[3] S. McAdam, Asymptotic prime divisors, Lecture Notes in Mathematics 1023, Springer, Berlin-Heidelberg - New York 1983.

Papers

[4] E. Böger, Einge Bemerkungen zur Theorie der ganz-algebraischen Abhängigkeit von Idealen, Math. Ann. 185 (1970), 303 - 308.

[5] M. Brodmann, Asymptotic stability of $\mathrm{Ass}\, M/I^nM$, Proc. Amer. Math. Soc. 74 (1979), 16 - 18.

[6] S. Goto, Integral closedness of complete-intersection ideals, Preprint 1985.

[7] S. Goto and K. Yamagishi, Normality of blowing-up, Preprint 1984.

[8] D. Katz, Asymptotic primes and applications, Thesis, The University of Texas at Austin 1982.

[9] D. Katz, A note on asymptotic prime sequences, Proc. Amer. Math. Soc. 87 (1983), 415 - 418.

[10] L.J. Ratliff, Asymptotic sequences, J. Algebra 85 (1983), 337 - 360.

[11] L.J. Ratliff, On quasi-unmixed local domains, the altitude formula, and the chain condition for prime ideals II, Amer. J. Math. 92 (1970), 99 - 144.

[12] D. Rees, $\mathfrak{a}$-transforms of local rings and a theorem on multiplicities of ideals, Math. Proc. Camb. Phil. Soc. 57 (1961), 8 - 17.

[13] D. Rees, Rings associated with ideals and analytic spread, Math. Proc. Camb. Phil. Soc. 89 (1981), 423 - 432.

Chapter IV. VARIOUS NOTIONS OF EQUIMULTIPLE AND PERMISSIBLE IDEALS

§ 20. Reinterpretation of the theorem of Rees-Böger

(20.0) We reformulate the theorem of Rees-Böger (19.6) by use of the generalized multiplicity $e(\underline{x},\mathfrak{a},R)$ and give an application for complete intersections. Let $(R,\mathfrak{m})$ be a local ring and let $\mathfrak{p}$ be a prime ideal of R. Recall that, by definition (10.10), $s(\mathfrak{p})-1$ is the dimension of the fibre of the morphism

$$B\ell(\mathfrak{p},R) \longrightarrow \mathrm{Spec}(R)$$

at the closed point $\mathfrak{m}$ of $\mathrm{Spec}(R)$ (this fibre being $\mathrm{Proj}(G(\mathfrak{p},R)\otimes_R R/\mathfrak{m})$. Likewise, if $\mathfrak{q}$ is any prime ideal of R containing $\mathfrak{p}$, then $s(\mathfrak{p}R_{\mathfrak{q}})-1$ is the dimension of the fibre of the above morphism at the point $\mathfrak{q}$ (by flat base change). Now $s(\mathfrak{p}R_{\mathfrak{q}}) \leq s(\mathfrak{p})$ by (10.11), and $s(\mathfrak{p}R_{\mathfrak{p}}) = \dim R_{\mathfrak{p}} = \mathrm{ht}(\mathfrak{q})$ by Remark (10.11),a). This shows that $\mathrm{ht}(\mathfrak{p}) = s(\mathfrak{p})$ if and only if the fibre dimension of $B\ell(\mathfrak{p},R) \longrightarrow \mathrm{Spec}(R)$ is a constant function on $V(\mathfrak{p}) \subset \mathrm{Spec}(R)$.

The results of this section provide a link between multiplicities and dimensions, and in view of Chapter III it is not surprising that to obtain this link it is essential to assume R to be quasi-unmixed.

Recall that for an ideal $\mathfrak{a}$ in a ring R and a sequence $\underline{x} = (x_1,\ldots,x_r)$ of elements of R, $\mathfrak{a}(\underline{x})$ denotes the ideal generated by $\mathfrak{a}$ and $\underline{x}R$.

(20.1) Proposition. Let R be a local ring, $\mathfrak{a}$ an ideal of R and let $\underline{x}$ be a system of parameters of $(R,\mathfrak{a})$. If $\mathrm{ht}(\mathfrak{a}) = s(\mathfrak{a})$ then

$$e(\underline{x},\mathfrak{a},R) = e(\mathfrak{a}(\underline{x}),R) \quad .$$

Proof. We may assume that R has an infinite residue field, hence we may choose $\underline{z} = (z_1,\ldots,z_s)$ in R generating a minimal reduction of $\mathfrak{a}$, where $s = \mathrm{ht}(\mathfrak{a})$ by assumption. Then $s = \dim R - \dim R/\mathfrak{a}$ by Proposition (10.20),c), showing that $\underline{x} \cup \underline{z}$ is a system of parameters of R. Now $\mathfrak{a}$ is intergral over $\underline{z}R$ and $\mathfrak{a}(\underline{x})$ is integral over $\underline{x}R + \underline{z}R$, so we know from Proposition (4.14) and the Associativity Law that

$$e(\underline{x},\mathfrak{a},R) = e(\underline{x},\underline{z}R,R) = \sum_{\mathfrak{p}\in \mathrm{Assh}(\underline{z}R)} e(\underline{x};R/\mathfrak{p})e(\underline{z}R_{\mathfrak{p}})$$

$$= e(\underline{x}R + \underline{z}R,R) = e(\mathfrak{a}(\underline{x}),R) \quad .$$

(20.2) Remark. The above proof shows that, under the same assumptions $\underline{x}$ is not only a system of parameters of $(R,\mathfrak{a})$ but also a set of parameters of R . If we drop the assumption $\mathrm{ht}(\mathfrak{a}) = s(\mathfrak{a})$, this need not be true. But in the next Lemma we will show that a generic choice of the x's will give parameters of R again. This Lemma will be needed to prove the converse of Proposition (20.1) under the assumption that R is quasi unmixed; see also App. III 3.2.5, 3.2.6.

(20.3) Lemma. Let R be a local ring with infinite residue field, $\mathfrak{a}$ an ideal of R and $\underline{y} = \{y_1,\ldots,y_r\}$ a system of paramters of $(R,\mathfrak{a})$. If R has an infinite residue field, and $\mathrm{ht}(\mathfrak{a}) > 0$, then there is a set $\underline{x} = \{x_1,\ldots,x_r\}$ of elements of R such that

a) $\mathfrak{a}(\underline{x}) = \mathfrak{a}(\underline{y})$;

b) $e(\underline{x},\mathfrak{a},R) = e(\underline{y},\mathfrak{a},R)$;

c) $e(\mathfrak{a}(\underline{x}),R) = e(\mathfrak{a}(\underline{x})/\underline{x}R,R/\underline{x}R)$;

d) $\underline{x}$ is a set of parameters of R .

Proof. By induction we will construct elements $x_1,\ldots,x_i$ such that, upon letting $\underline{x} = \{x_1,\ldots,x_i\}$ and $\underline{z} = \{x_1,\ldots,x_i,y_{i+1},\ldots,y_r\}$, we have

(20.3.1) $\mathfrak{a}(\underline{z}) = \mathfrak{a}(\underline{y})$;

(20.3.2) $e(\underline{z},\mathfrak{a},R) = e(\underline{y},\mathfrak{a},R)$;

(20.3.3) $e(\mathfrak{a}(\underline{z}),R) = e(\mathfrak{a}(\underline{z})/\underline{x}R,R/\underline{x}R)$,

(20.3.4) $\underline{x}$ is a set of parameters of R .

There is nothing to prove if $i = 0$, so assume that $i > 0$ and that $x_1,\ldots,x_{i-1}$ are already constructed, satisfying the analogous conditions to (20.3.1) - (20.3.4). Then $\mathfrak{a} + x_1R + \ldots + x_{i-1}R \neq \mathfrak{a}(\underline{y})$, hence if $\mathfrak{p}_1,\ldots,\mathfrak{p}_s$ are the minimal primes of $\mathfrak{a} + x_1R + \ldots + x_{i-1}R$ then $\mathfrak{p}_j \not\supset \mathfrak{a}(y)$ for $j = 1,\ldots,s$. Moreover, denoting by $\mathfrak{p}_{s+1},\ldots,\mathfrak{p}_t$ the minimal primes of $x_1R + \ldots + x_{i-1}R$, we have $\mathrm{ht}(\mathfrak{p}_j) \leq i - 1 < \mathrm{ht}(\mathfrak{a}(\underline{y}))$ for $j = s + 1,\ldots,t$. So by later Corollary (22.9) applied to $R/x_1R + \ldots + x_{i-1}R$ we know that there is x_i satisfying

(20.3.5) the initial form of x_i in $G(\mathfrak{a}(y)/x_1R+\ldots+x_{i-1}R,\ R/x_1R+\ldots+x_{i-1}R)$ is weakly regular;

(20.3.6) $x_i \notin \mathfrak{a} + x_1R+\ldots+x_{i-1}R + \mathfrak{m}\mathfrak{a}(\underline{y})$;

(20.3.7) $x_i \notin \mathfrak{p}_j$ for $j = 1,\ldots,t$.

Now (20.3.1) follows by construction (note $\dim \mathfrak{a}(\underline{y})/\mathfrak{a} + \mathfrak{m}\mathfrak{a}(\underline{y}) = r$), and (20.3.3) is a consequence of [6*], VIII, § 10. (20.3.4) follows from (20.3.7), and finally (20.3.2) can be deduced from Proposition (3.11),a), since $zR + \mathfrak{p}/\mathfrak{p} = \underline{y}R + \mathfrak{p}/\mathfrak{p}$ for every $\mathfrak{p} \in \mathrm{Assh}(R/\mathfrak{a})$.

<u>(20.4) Lemma</u>. Let $(R,\mathfrak{m})$ be a local ring with infinite residue field and $\mathfrak{a}$ an ideal of height zero such that $\dim R = \dim R/\mathfrak{a} > 0$. For any system $\underline{y} = \{y_1,\ldots,y_d\}$ of parameters of $(R,\mathfrak{a})$ there is a system of parameters $\underline{x} = \{x_1,\ldots,x_d\}$ of R such that

a) $\mathfrak{a}(\underline{x}) = \mathfrak{a}(y)$;

b) $e(\mathfrak{a}(\underline{y}),R) = e(\underline{x}R,R)$;

c) $e(\underline{x},\mathfrak{a},R) = e(\underline{y},\mathfrak{a},R)$.

Proof. Let $x_1,\ldots,x_d$ generate a minimal reduction of $\mathfrak{a}(\underline{y})$ and let x_i^* denote the initial form of x_i in $G(\mathfrak{a}(\underline{y}),R)$. Then $x_1^*,\ldots,x_d^*$ are a homogeneous system of parameters of degree 1. Let

$$A = G(\mathfrak{a}(\underline{y})/\mathfrak{a}\ ,\ R/\mathfrak{a}) \otimes_R R/\mathfrak{m}$$

and let $\varphi : G(\mathfrak{a}(\underline{y}),R) \longrightarrow A$ be the canonical homomorphism. Now

$$\dim A = s(\mathfrak{a}(\underline{y})/\mathfrak{a}) = \dim R/\mathfrak{a} = \dim R$$

by assumption and hence A is a polynomial ring in d variables over $R/\mathfrak{m}$. Since $\varphi(x_1^*),\ldots,\varphi(x_d^*)$ is a homogeneous system of parameters of degree 1 and A, we conclude that

$$A_1 = \bigoplus_{i=1}^{d} R/\mathfrak{m}\cdot\varphi(x_i^*) \quad .$$

Since $A_1 \cong \mathfrak{a}(\underline{y})/\mathfrak{m}\mathfrak{a}(\underline{y})+\mathfrak{a}$, it follows that

$$A_1 \cong \mathfrak{m}\mathfrak{a}(\underline{y})+\mathfrak{a}+\underline{x}R/\mathfrak{m}\mathfrak{a}(y)+\mathfrak{a} = \mathfrak{a}(\underline{x})+\mathfrak{m}\mathfrak{a}(\underline{y})/\mathfrak{m}\mathfrak{a}(\underline{y})+\mathfrak{a} \quad .$$

Therefore $\mathfrak{a}(\underline{x})+\mathfrak{m}\mathfrak{a}(\underline{y}) = \mathfrak{a}(\underline{y})+\mathfrak{m}\mathfrak{a}(\underline{y})$ and hence $\mathfrak{a}(\underline{x}) = \mathfrak{a}(\underline{y})$ by Nakayama's Lemma. Since $\underline{x}$ was chosen to be a minimal reduction of $\mathfrak{a}(\underline{y})$ we have $e(\mathfrak{a}(\underline{y}),R) = e(\underline{x}R,R)$.

Now let $\mathfrak{p}\in \operatorname{Min}(R/\mathfrak{a})$. Then $\underline{x}$ will generate a (minimal) reduction of $\underline{y}(R/\mathfrak{p})$ in $R/\mathfrak{p}$. Hence $e(\underline{x};R/\mathfrak{p}) = e(\underline{y};R/\mathfrak{p})$ by Proposition (4.14), so c) follows from the Associativity Law (Proposition (3.11)).

<u>(20.5) Theorem</u>. Let $(R,\mathfrak{m})$ be a quasi-unmixed local ring, $\mathfrak{a}$ a proper ideal of R. Then the following conditions are equivalent:

(i) $\operatorname{ht}(\mathfrak{a}) = s(\mathfrak{a})$;

(ii) $e(\underline{x},\mathfrak{a},R) = e(\mathfrak{a}(\underline{x}),R)$ for any system of parameters $\underline{x}$ of $(R,\mathfrak{a})$.

(iii) There is a system of parameters $\underline{x}$ of $(R,\mathfrak{a})$ such that $e(\underline{x},\mathfrak{a},R) = e(\mathfrak{a}(\underline{x}),R)$.

Proof. (i) ⇒ (ii) has been shown in Proposition (20.1), while (ii) ⇒ (iii) is obvious. To prove (iii) ⇒ (i), let $\underline{x} = \{x_1,\ldots,x_r\}$ with $r>0$ (since there is nothing to prove for $r = 0$). By Proposition (5.1) and Remark (10.11) we may assume that $R/\mathfrak{m}$ is infinite. So by changing $\underline{x}$ if necessary according to Lemmas (20.3) and (20.4), we will have

$$(20.5.1.) \qquad e(\mathfrak{a}(\underline{x}),R) = \begin{cases} e(\mathfrak{a}(\underline{x})/\underline{x}R,R/\underline{x}R) & \text{if} \quad \operatorname{ht}(\mathfrak{a})>0 \\ e(\underline{x}R,R) & \text{if} \quad \operatorname{ht}(\mathfrak{a})=0 \end{cases} .$$

(Note that, since R is quasi-unmixed, we have $\dim R/\mathfrak{a} = \dim R$ in case $ht(\mathfrak{a}) = 0$). Now we choose $\underline{z} = \{z_1,\ldots,z_s\} \subset \mathfrak{a}$, $s = \dim R - r = ht(\mathfrak{a})$, such that $\underline{z}(R/\underline{x}R)$ is a (minimal) reduction of $\mathfrak{a}(R/\underline{x}R)$. If $r = \dim R$ (equivalently $ht(\mathfrak{a}) = 0$) then $\underline{z} = \emptyset$, of course. We will first prove that

(20.5.2) $$e(\underline{x},\mathfrak{a},R) = e(\underline{x},\underline{z}R,R) \quad .$$

In the case $ht(\mathfrak{a}) = 0$ this is simply (20.5.1). If $ht(\mathfrak{a}) > 0$ then

$$\begin{aligned} e(\underline{x},\mathfrak{a},R) &\leq e(\underline{x},\underline{z}R,R) && \text{by Proposition (4.13)} \\ &= e(\underline{x}R + \underline{z}R,R) && \text{by Proposition (20.1)} \\ &\leq e(\underline{z}(R/\underline{x}R),R/\underline{x}R) && \text{by Lemma (19.2)} \\ &= e(\mathfrak{a}(\underline{x})/\underline{x}R,R/\underline{x}R) && \text{by the choice of } \underline{z} \\ &= e(\mathfrak{a}(\underline{x}),R) && \text{by (20.5.1).} \end{aligned}$$

Therefore (20.5.2) is a consequence of our assumption (iii).

Now $\underline{z} \subset \mathfrak{a}$ implies $Assh(R/\mathfrak{a}) \subset Assh(R/\underline{z}R)$, from which we conclude

$$\begin{aligned} e(\underline{x},\mathfrak{a},R) &= \sum_{\mathfrak{p}\in Assh(R/\mathfrak{a})} e(\underline{x};R/\mathfrak{p})e(\mathfrak{a}R_{\mathfrak{p}},R_{\mathfrak{p}}) \\ &\leq \sum_{\mathfrak{p}\in Assh(R/\mathfrak{a})} e(\underline{x};R/\mathfrak{p})e(\underline{z}R_{\mathfrak{p}},R_{\mathfrak{p}}) \\ &= \sum_{\mathfrak{p}\in Assh(R/\underline{z}R)} e(\underline{x};R/\mathfrak{p})e(\underline{z}R_{\mathfrak{p}},R_{\mathfrak{p}}) \\ &= \sum_{\mathfrak{p}\in Assh(R/\underline{z}R)\setminus Assh(R/\mathfrak{a})} e(\underline{x};R/\mathfrak{p})e(\underline{z}R_{\mathfrak{p}},R_{\mathfrak{p}}) \\ &= e(\underline{x},\underline{z}R,R) - \sum_{\mathfrak{p}\in Assh(R/\underline{z}R)\setminus Assh(R/\mathfrak{a})} e(\underline{x};R/\mathfrak{p})e(\underline{z}R_{\mathfrak{p}},R_{\mathfrak{p}}) \\ &\leq e(\underline{x},\underline{z}R,R) \quad . \end{aligned}$$

Since $e(\underline{x};R/\mathfrak{p}) \neq 0$ for all $\mathfrak{p} \in Assh(R/\underline{z}R)$, equality (20.5.2) implies

(20.5.3) $$Assh(R/\mathfrak{a}) = Assh(R/\underline{z}R)$$

and

(20.5.4) $e(\underline{z}R_{\mathfrak{p}}, R_{\mathfrak{p}}) = e(\mathfrak{a}R_{\mathfrak{p}}, R_{\mathfrak{p}})$ for all $\mathfrak{p} \in \mathrm{Assh}(R/\mathfrak{a})$.

Since R is quasi-unmixed, $\mathrm{Assh}(R/\underline{z}R) = \mathrm{Min}(R/\underline{z}R)$, so necessarily

(20.5.5) $\mathrm{Assh}(R/\mathfrak{a}) = \mathrm{Min}(R/\mathfrak{a})$.

This implies

(20.5.6) $\sqrt{\mathfrak{a}} = \sqrt{\underline{z}R}$.

Now (20.5.4), (20.5.5) and (20.5.6) allow to apply the Theorem of Rees-Böger (19. 6), from which we conclude that $\underline{z}R$ is a reduction of $\mathfrak{a}$ and consequently $\mathrm{ht}(\mathfrak{a}) = s(\mathfrak{a})$.

<u>(20.6) Remark</u>. Let $\mathfrak{a}$, $\underline{x}$ and R be as in Theorem (20.5) and assume that $R/\mathfrak{m}$ is infinite and $\mathfrak{a}$ satisfies (iii). Then the proof given above has shown that any minimal reduction of $\mathfrak{a}(\underline{x})/\underline{x}R$ lifts to a minimal reduction of $\mathfrak{a}$.

<u>(20.7) Definition</u>. Let R be a local ring, $\mathfrak{a}$ a proper ideal of R . $\mathfrak{a}$ will be called a complete intersection ideal if it can be generated by a regular sequence.

<u>(20.8) Proposition</u>. Let R be a local ring and $\mathfrak{a}$ a proper ideal of R . Assume that

a) R is Cohen-Macaulay,

b) $\mathfrak{a}R_{\mathfrak{p}}$ is a complete intersection ideal for all $\mathfrak{p} \in \mathrm{Ass}(R/\mathfrak{a})$,

c) there is a system of parameters $\underline{x}$ of $(R,\mathfrak{a})$ such that $e(\underline{x},\mathfrak{a},R) = e(\mathfrak{a}(\underline{x}),R)$.

Then $\mathfrak{a}$ is a complete intersection ideal.

Proof. Assume first that R has an infinite residue field and use Theorem (20.5) to choose $\underline{z} = \{z_1,\ldots,z_s\}$, $s = \mathrm{ht}(\mathfrak{a})$, such that $\underline{z}R$ is a minimal reduction of $\mathfrak{a}$. Then $\underline{z}$ is a regular sequence by a), and for any $\mathfrak{p} \in \mathrm{Ass}(R/\mathfrak{a})$ we have

$$s = \mathrm{ht}(\mathfrak{a}) \leqq \mathrm{ht}(\mathfrak{a}R_{\mathfrak{p}}) = \mathrm{ht}(\underline{z}R_{\mathfrak{p}}) \leqq s$$

from which we conclude that $\mathrm{ht}(\mathfrak{a}R_{\mathfrak{p}}) = s$ and consequently, by assumption b),

(20.8.1) $\underline{z}R_{\mathfrak{p}} = \mathfrak{a}R_{\mathfrak{p}}$ for all $\mathfrak{p} \in \mathrm{Ass}(R/\mathfrak{a})$.

Since R is Cohen-Macaulay, $\underline{z}$ is a regular sequence in each $R_{\mathfrak{p}}$ and therefore

$$\mathrm{Ass}(R/\underline{z}R) = \mathrm{Ass}(R/\mathfrak{a}) \quad ,$$

which implies $\underline{z}R = \mathfrak{a}$ by (20.8.1).

If R has a finite residue field, by denoting by $\mathfrak{m}$ the maximal ideal of R we put

$$R^* = R[T]_{\mathfrak{m}[T]} \quad , \quad \mathfrak{m}^* = \mathfrak{m}R^* \ , \ \mathfrak{a}^* = \mathfrak{a}R^* \quad .$$

Then R^* is Cohen-Macaulay and $\mathrm{ht}(\mathfrak{a}) = \mathrm{ht}(\mathfrak{a}^*)$ by Proposition (5.1), and moreover $\mu(\mathfrak{a}) = \mu(\mathfrak{a}^*)$, where μ denotes the minimal number of generators. Our claim is equivalent to $\mu(\mathfrak{a}) = \mathrm{ht}(\mathfrak{a})$, so we need to show that $\mathfrak{a}^*$ is a complete intersection. Since $s(\mathfrak{a}) = s(\mathfrak{a}^*)$, $\mathfrak{a}^*$ satisfies c) (by Proposition (5.1) and Theorem (20.5)) and it remains to verify that $\mathfrak{a}^*$ satisfies b). For this let $\mathfrak{p}^* \in \mathrm{Ass}(R^*/\mathfrak{a}^*)$ and $\mathfrak{p} = \mathfrak{p}^* \cap R$. Then $\mathfrak{p}^* = \mathfrak{p}R^*$ and $R^*_{\mathfrak{p}^*} \simeq R_{\mathfrak{p}}[T]_{\mathfrak{p}R_{\mathfrak{p}}[T]}$. Since $\mathfrak{p} \in \mathrm{Ass}(R/\mathfrak{a})$ we know from b) that $\mathfrak{a}^*R^*_{\mathfrak{p}^*}$ is a complete intersection ideal. So the general case is reduced to the first case treated above and the proof is complete.

<u>(20.9) Theorem</u>. Let R be a quasi-unmixed local ring and let $\mathfrak{p}$ be a prime ideal of R for which $R/\mathfrak{p}$ is regular. Then the following conditions are equivalent:

(i) $e(R) = e(R_{\mathfrak{p}})$;

(ii) $\mathrm{ht}(\mathfrak{p}) = s(\mathfrak{p})$.

Proof. Choosing $\underline{x}$ to be a regular system of paramters of $(R,\mathfrak{p})$, we have

$$e(\underline{x},\mathfrak{p},R) = e(R_{\mathfrak{p}}) \quad \text{and} \quad e(\mathfrak{p}(\underline{x}),R) = e(R) \quad .$$

Therefore the Theorem is a direct consequence of Theorem (20.5).

(20.10) Example. Let k be an infinite field and let

$$R = k\,[[X,Y,Z]] \;/\; (X) \cap (Y,Z) = k\,[[x,y,z]] \quad .$$

Take $\mathfrak{p} = (y,z)$. Then $R/\mathfrak{p}$ is regular and $e(R) = e(R_{\mathfrak{p}}) = 1$. But on the other hand we have $0 = ht(\mathfrak{p}) < s(\mathfrak{p}) = 2$. This shows that the implication (i) $\Rightarrow$ (ii) of Theorem (20.9) is false if the local ring is not unmixed.

§ 21. Hironaka-Grothendieck homomorphism

Zariski suggested to get a desingularization of a given variety X by blowing up regular centers D contained in the singular locus of X . Of course one needs some condition for D which allows to conclude that blowing up X along D will actually improve the singularity. For surfaces (in characteristic 0 , embedded in a three dimensional non-singular variety) one condition for a regular curve to be a "permissible" center was the condition of equimultiplicity. In Hironaka's proof of resolution of singularities in characteristic 0 the notion of equimultiplicity was refined to normal flatness. One important aspect of normal flatness is that it can be translated into a numerical condition, using Hilbert functions. (As a result, Hilbert functions turned out to be useful numerical characters of singularities, in particular for desingularization.) The results relating normal flatness to Hilbert functions will be proved in § 22. Here we describe the main technical tool, a certain graded homomorphism first studied by Hironaka and later on generalized by Grothendieck and others.

(21.1) Definition. Let $(R,\mathfrak{m})$ be a local ring and I a proper ideal of R . Then R will be called normally flat along I if I^n/I^{n+1} is flat over R/I for all $n \geq 0$. I is called permissible, if R/I is regular and R is normally flat along I. Note that R is normally flat along I if and only if $G(I,R)$ is free over R/I.

In the next sections (in particular in Chapter VI) we want to choose a more general frame for this condition by using Cohen-Macauly properties of the graded module $G(I,R)$. The algebraic motivation for this can be seen as follows: recall (s. [5]), that if R is a regular local ring and M a finite R-module with $\dim M = \dim R$, then M is Cohen-Macaulay if and only if M is flat over R. Therefore if R/I is regular, normal flatness of R along I is equivalent to " I^n/I^{n+1} is Cohen-Macauauly over R/I with $\dim(I^n/I^{n+1}) = \dim(R/I)$ ". Hence, if R/I (= center of blowing up of $\mathrm{Spec}(R)$) is not regular, the natural generalization of normal flatness in some sense is the following Cohen-Macaulay property:

(21.2) Definition. Let (R,m) be a local ring and I a proper ideal of R. We call R normally Cohen-Macaulay along I if

$$\mathrm{depth}(I^n/I^{n+1}) = \dim(R/I) \quad \text{for all} \quad n \geq 0 .$$

(21.3) Remark. It is clear that if R/I is Cohen-Macaulay and R is normally flat along I then R is normally Cohen-Macaulay along I.

The interested reader is referred to § 23 where we will give a detailed explanation of how the various notions of this chapter are related. Moreover, Chapter VI contains some results showing how these notions can be used to study blowing ups with singular centers.

Our approach to the Hironaka-Grothendieck homomorphism and its connection to normal flatness is due to Robbiano [15]. We begin with the so called "local criterion of flatness", s. [5] and [1], III, § 5, no.2 and 4.

(21.4) Proposition. Let R be a local ring, I an ideal of R and M a finitely generated R-module. Then the follwoing statements are equivalent.

(1) M is free over R .

(2) M/IM is free over R/I and $\mathrm{Tor}_1^R(R/I,M) = 0$.

(3) M/IM is free over R/I and the canonical homomorphism $M/IM \otimes_R G(I,R) \xrightarrow{\pi} G(I,M)$ is an isomorphism.

For completeness we sketch the proof.

Proof. (1) ⇒ (2) and (1) ⇒ (3) are clear by the definition of flatness.

(2) ⇒ (1): Let $r = \mathrm{rk}_{R/I}\, M/IM$ be the rank of M/IM . Then by Nakayama's lemma we have an exact sequence

$$0 \longrightarrow K \longrightarrow R^r \longrightarrow M \longrightarrow 0 \quad .$$

Tensorizing with R/I , we get an exact sequence $0 \to \mathrm{Tor}_1^R(R/I,M) \to K/IK \to (R/I)^r \to M/IM \to 0$. From this we conclude $K/IK = 0$ and $K = 0$ by Nakayama's lemma.

(3) ⇒ (1): As before we get an exact sequence

$$0 \longrightarrow K \longrightarrow R^r \longrightarrow M \longrightarrow 0 \quad ,$$

which yields the following commutative diagram with exact rows and columns (where we put $F = R^r$):

$$\begin{array}{ccccccccc}
 & & 0 & & 0 & & 0 & & \\
 & & \downarrow & & \downarrow & & \downarrow & & \\
0 & \longrightarrow & K\cap I^{n+1}F & \longrightarrow & I^{n+1}F & \longrightarrow & I^{n+1}M & \longrightarrow & 0 \\
 & & \downarrow & & \downarrow & & \downarrow & & \\
0 & \longrightarrow & K\cap I^nF & \longrightarrow & I^nF & \longrightarrow & I^nM & \longrightarrow & 0 \\
 & & \downarrow & & \downarrow & & \downarrow & & \\
0 & \longrightarrow & K\cap I^nF/K\cap I^{n+1}F & \longrightarrow & I^nF/I^{n+1}F & \longrightarrow & I^nM/I^{n+1}M & \longrightarrow & 0 \\
 & & \downarrow & & \downarrow & & \downarrow & & \\
 & & 0 & & 0 & & 0 & &
\end{array}$$

Hence we get an exact sequence

$$\begin{array}{ccccccc} 0 \longrightarrow & \bigoplus_{n\geq 0} K\cap I^nF/K\cap I^{n+1}F & \longrightarrow & G(I,R)^r & \xrightarrow{f} & G(I,M) & \longrightarrow 0 \quad . \\ & & & \sigma \Big\uparrow \wr & \nearrow \pi & & \\ & & & M/IM \otimes_R G(I,R) & & & \end{array}$$

Here the isomorphism σ comes from the assumption that M/IM is free over R/I , and f corresponds to the given isomorphism π . From this we conclude that $K\cap I^nF = K\cap I^{n+1}F$ for all $n \geq 0$, i.e. $K \subset \bigcap_{n\geq 0} I^nF = (0)$.

We recall the following result with the line of proof for later reference:

<u>(21.5) Lemma</u>. Let R be a local ring, $\underline{x} = \{x_1,\ldots,x_r\}$ an R-sequence and M a finitely generated R-module. Then the following conditions are equivalent:

(i) $\underline{x}$ is an M-sequence.

(ii) $\mathrm{Tor}_1^R(R/\underline{x}R,M) = 0$.

Proof. Since $\underline{x}$ is a regular sequence, the Koszul complex $K.(\underline{x};R)$ gives a free resolution of $R/\underline{x}R$. Hence $\mathrm{Tor}_1^R(R/\underline{x}R,M) = H_1(\underline{x};M)$. Then the conclusion follows from Chapter II, Proposition (11.9).

<u>(21.6) Corollary</u>. With the notations as in Lemma (21.5) the following conditions are equivalent.

(i) M is free over R .

(ii) $M/\underline{x}M$ is free over $R/\underline{x}R$ and $\underline{x}$ is an M-sequence.

Following [15] we consider the following situation. Let $(R,\mathfrak{m},k)$ be a local ring, I an ideal of R and $\underline{x} = \{x_1,\ldots,x_r\}$ a set of elements of R . We put $L = \underline{x}R + I$. Note that

$$G(I,R) \otimes_R R/L \simeq \bigoplus_{n\geq 0} I^n/LI^n = \bigoplus_{n\geq 0} I^n/I^{n+1} + \underline{x}I^n \quad .$$

Moreover we have a canonical map

$$\bigoplus_{n\geq 0} I^n/I^{n+1} + \underline{x}I^n \longrightarrow G(L/\underline{x}R, R/\underline{x}R) \quad .$$

Hence we can define the following two surjective homomorphisms:

(21.7) Definition. $g(I,\underline{x})$ is the canonical homomorphism

$$g(I,\underline{x}): \quad G(I,R) \otimes_R R/L \longrightarrow G(L/\underline{x}R, R/\underline{x}R) \quad ,$$

and - for indeterminates $T_1,\ldots,T_r$ - $G(I,\underline{x})$ is the canonical homomorphism

$$G(I,\underline{x}): \quad (G(I,R) \otimes_R R/L)[T_1,\ldots,T_r] \longrightarrow G(L,R) \quad ,$$

such that $G(I,\underline{x})(T_i) := (x_i \bmod L^2) \in L/L^2$.

(21.8) Lemma. Let R , I and $\underline{x}$ be the same as in Definition (21.7). Then we have:

a) The following conditions are equivalent:

(i) $g(I,\underline{x})$ is an isomorphism;

(ii) $I^n \cap \underline{x}R = \underline{x}I^n$ for all $n \geq 0$;

(iii) $\mathrm{Tor}_1^R(R/\underline{x}R, R/I^n) = 0$ for all $n \geq 0$.

If one of these conditions is fulfilled then $x_1,\ldots,x_r \notin I$.

b) If $g(I,\underline{x})$ is an isomorphism, then $G(I,\underline{x})_0$ is injective, where $G(I,\underline{x})_0$ is the restriction of $G(I,\underline{x})$ to the subring $G(I,R) \otimes_R R/L$.

Proof. a) $g(I,\underline{x})$ is an isomorphism if and only if $I^{n+1} + I^n\underline{x} = I^{n+1} + (I^n \cap \underline{x}R)$ for all $n \geq 0$, hence if and only if $I^n \cap \underline{x}R \subseteq \underline{x}I^n + (I^{n+1} \cap \underline{x}R)$ for all $n \geq 0$, or equivalently: $I^n \cap \underline{x}R = \underline{x}I^n$. This proves (i)$\Longleftrightarrow$(ii) . To prove (ii)$\Longleftrightarrow$(iii) it is enough to note that

$$\mathrm{Tor}_1^R(R/\underline{x}R, R/I^n) \cong \underline{x}R \cap I^n/\underline{x}I^n \quad ,$$

which comes from the exact sequence $0 \to I^n \to R \to R/I^n \to 0$, if tensorized with $R/\underline{x}R$.

b) It is enough to show that $I^n \cap L^{n+1} \subset I^n L$. By a) we have:

$$I^n \cap L^{n+1} \subseteq I^n \cap (I^n L + \underline{x}R) = I^n L + I^n \cap \underline{x}R = I^n L .$$

Now we come to the Hironaka-Grothendieck-isomorphism, using the same notations as in (21.7).

<u>(21.9) Theorem</u>. The following conditions are equivalent.

(1) $g(I,\underline{x})$ is an isomorphism and $\underline{x}$ is an R-sequence.

(2) $\underline{x}$ is a regular sequence on $G(I,R)$.

(3) $\underline{x}$ is a regular sequence on R/I^n for every $n \geq 1$.

(4) $G(I,\underline{x})$ is an isomorphism.

Proof. (2) $\Longleftrightarrow$ (3): This follows from the exact sequence

$$0 \longrightarrow I^n/I^{n+1} \longrightarrow R/I^{n+1} \longrightarrow R/I^n \longrightarrow 0 .$$

(1) $\Rightarrow$ (3) is a consequence of Lemma (21.5) and Lemma (21.8).
(3) $\Rightarrow$ (1): Since condition (3) implies that $\underline{x}$ is an R-sequence, the conclusion follows again from Lemma (21.5) and Lemma (21.8).

(1) $\Rightarrow$ (4): By Lemma (21.8) we get in this case

$$\underline{x}R \cap L^n = \underline{x}R \cap (I^n + \underline{x}L^{n-1}) = \underline{x}L^{n-1} .$$

This shows that $\{x_1,\dots,x_r\}$ is an L-standard base of $\underline{x}R$ by Chapter II, § 13. Moreover $\underline{x}$ is an R-sequence, hence the initial forms $x_1^*,\dots,x_r^*$ form a $G(L,R)$-sequence by § 13.

For the next step we put $A = G(I,R) \otimes_R R/L$. By Lemma (21.8),b) A is a subring of $G(L,R)$ and $G(L,R) = A[x_1^*,\dots,x_r^*]$.
Claim: $A \cap \underline{x}^* A[\underline{x}^*] = (0)$.

In fact, if $A \cap \underline{x}^* A[\underline{x}^*] \neq (0)$, there is an element $a \in I^n - LI^n$ such that $a \in (\underline{x}R + L^{n+1}) \cap I^n$. On the other hand $(\underline{x}R + L^{n+1}) \cap I^n =$ $= (\underline{x}R + I^{n+1}) \cap I^n = \underline{x}I^n + I^{n+1} = LI^n$, hence $a \in LI^n$ which is a contradiction to the hypothesis. Now let $f(T_1,\dots,T_r) \in \operatorname{Ker} G(I,\underline{x})$. We may assume that f is homogeneous of degree $v > 0$ in $T_1,\dots,T_r$.

Since $\underline{x}^*$ is an $A[\underline{x}^*]$-sequence, the coefficients of f must belong to $A \cap \underline{x}^*A[\underline{x}^*] = (0)$. Hence $G(I,\underline{x})$ is an isomorphism.

(4) ⇒ (1): Since $G(I,\underline{x})$ is an isomorphism, the initial forms $x_1^*,\dots,x_r^*$ form a $G(L,R)$-sequence. Therefore we know, that $\underline{x}$ is an R-sequence and $\underline{x}R \cap L^n = \underline{x}L^{n-1}$ for all $n \geq 0$ by § 13. This gives an isomorphism

$$\tau_1 : G(L,R)/(x_1^*,\dots,x_r^*) \xrightarrow{\sim} G(L/\underline{x}R, R/\underline{x}R) \quad .$$

On the other hand, the isomorphism $G(I,\underline{x})$ induces an isomorphism

$$\tau_2 : G(I,R) \otimes_R R/L \xrightarrow{\sim} G(L,R)/(x_1^*,\dots,x_r^*) \quad .$$

Therefore $g(I,\underline{x}) = \tau_1 \circ \tau_2$ is an isomorphism, q.e.d.

As a consequence of Theorem (21.9) and Corollary (21.6) - applied to the graded pieces of $G(I,R)$ - we get the following theorem.

(21.10) Theorem. Let R, I and $\underline{x}$ be the same as in Definition (21.7). Assume in addition that $\underline{x}$ is a regular sequence on R/I. Then the following conditions are equivalent:

(1) $G(I,R)$ is free over R/I.

(2) $G(I,R) \otimes_R R/L$ is free over R/L and $\underline{x}$ is a regular sequence on $G(I,R)$.

(3) $G(I,R) \otimes_R R/L$ is free over R/L and $\underline{x}$ is a regular sequence on R/I^n for all $n \geq 0$.

(4) $g(I,\underline{x})$ is an isomorphism of free R/L-modules and $\underline{x}$ is an R-sequence.

(5) $G(I,\underline{x})$ is an isomorphism of free R/L-modules.

This theorem (21.10) has two important consequences, which we present in the Corollaries (21.11) and (21.12) (s. also [11], Chapter II, p. 184):

(21.11) Corollary. Let $(R,\mathfrak{m},k)$ be a local ring, $\mathfrak{p}$ a prime ideal of R such that $R/\mathfrak{p}$ is regular, and let $\underline{x} = x_1,\dots,x_r$ be a regular system of parameters mod $\mathfrak{p}$. Then the following conditions are equivalent:

(1) R is normally flat along $\mathfrak{p}$.

(2) $G(\mathfrak{p},\underline{x}) : (G(\mathfrak{p},R) \otimes_R k)[T_1,\ldots,T_r] \longrightarrow G(\mathfrak{m},R)$ is an isomorphism.

(21.12) Corollary. Let $(R,\mathfrak{m})$ be a local ring and $\mathfrak{p}$ a prime ideal of R such that

(i) $R/\mathfrak{p}$ is regular of dimension r ,

(ii) R is normally flat along $\mathfrak{p}$.

Then we have the following equality for Hilbert functions

$$H^{(0)}[R] = H^{(r)}[R_{\mathfrak{p}}] \quad .$$

Proof. This follows immediately from Corollary (21.11).

In the next section we will prove that for regular rings $R/\mathfrak{p}$ the normal flatness of R along $\mathfrak{p}$ can be even characterized by the numerical condition of Corollary (21.12). That will be called the criterion of permissibility,(s.(22.24) and App.III, 2.1.6, 2.1.7.

§ 22. Projective normal flatness and numerical characterization of permissibility

Let R be a (noetherian) local ring and I an ideal of R . In the first part of this section we study the flatness of the canonical morphism $\mathrm{Proj}(G(I,R)) \longrightarrow \mathrm{Spec}(R/I)$. In this case R is called "projectively normally flat along I ". The notation "projective" comes from the fact that the flatness of this morphism is equivalent to the property that I^n/I^{n+1} is flat for large n . We will show that projective normal flatness shares some essential properties with normal flatness.

We begin with some results on graded rings.

(22.1) Definition. Let $A = \bigoplus_{n \geq 0} A_n$ be a graded noetherian ring, generated by A_1 over A_0 . A is called projectively flat over A_0 if $\mathrm{Proj}(A) \longrightarrow \mathrm{Spec}(A_0)$ is a flat morphism.

(22.2) Remarks.

(1) If A is flat over A_0, then it is projectively flat over A_0.

(2) If $A^{(d)}$ denotes the graded ring $\bigoplus_{n \geq 0} A_{nd}$, then there exists a canonical isomorphism $\mathrm{Proj}(A) \simeq \mathrm{Proj}(A^{(d)})$ by sending $P \mapsto P \cap A^{(d)}$ for $P \in \mathrm{Proj}(A)$. In fact, let $P' \in \mathrm{Proj}(A^{(d)})$, put

$$P_{nd} = P' \cap A_{nd} \quad \text{and} \quad P_n = \left\{ x \in A_n \mid x^d \in P_{nd} \right\} .$$

Now it is clear, that there is a unique ideal $P \in \mathrm{Proj}(A)$ such that $P \cap A^{(d)} = P'$. Moreover this given bijection is an homomorphism of the topological spaces, where the basis open sets A_f and A_{f^d} for homogeneous $f \in A_+$ can be canonically identified. Therefore $\mathrm{Proj}(A)$ and $\mathrm{Proj}(A^{(d)})$ can be identified as schemes, s. [2], II, Proposition 2.4.

Let M be a graded A-module, P a homogeneous prime ideal of A and S the set of homogeneous elements of $A \setminus P$. Recall that $M_{(P)}$ denotes the homogeneous component of degree 0 of $S^{-1}M$, i.e.

$$M_{(P)} = \left\{ \frac{m}{s} \mid m \in M_n , \ s \in A_n \cap S , \ n \geq 0 \right\} .$$

One can easily see that for a finitely generated A-module M we have $M_{(P)} = 0$ for all $P \in \mathrm{Proj}(A)$ if and only if $M_n = 0$ for large n, s. Chapter II, Lemma (11.8).

(22.3) Proposition. The following conditions are equivalent:

(1) A is projectively flat over A_0.

(2) A_n is flat over A_0 for $n >> 0$.

(3) There exists an integer $d > 0$ such that $A^{(d)}$ is flat over A_0.

Proof. (2) $\Rightarrow$ (3) is trivial. (3) $\Rightarrow$ (1) is a consequence of $\mathrm{Proj}(A) = \mathrm{Proj}(A^{(d)})$.

(1) $\Rightarrow$ (2): For this we embed A into an exact sequence

$$(*) \qquad 0 \longrightarrow I \longrightarrow A_0[X] \longrightarrow A \longrightarrow 0 \quad ,$$

where $X = \{X_1,\dots,X_s\}$ are finitely many indeterminates which are sent to a system of generators of the A_0-module A_1. Let $P \in \operatorname{Proj}(A)$ and put $m := P \cap A_0$. We may assume that (A_0,m) is local. Moreover let $k := A_0/m$ and $T := \operatorname{Tor}_1^{A_0}(A_n,k)$. Then we have $T = \bigoplus_{n\geq 0} T_n$, where $T_n = \operatorname{Tor}_1^{A_0}(A_n,k)$.

Now we conclude from the exact sequence (*) that

$$T \simeq (I \cap m[X]) / (I \cdot m[X]) \quad ,$$

i.e. T is a finitely generated graded A-module. By assumption (1) we have $\operatorname{Tor}_1^{A_0}(A_{(P)},k) = 0$. But

$$\operatorname{Tor}_1^{A_0}(A_{(P)},k) \simeq \operatorname{Tor}_1^{A_0}(A,k)_{(P)} = T_{(P)} \quad ,$$

hence we know that $T_n = 0$ for large n. This proves (2).

(22.4) Definition. Let $A = \bigoplus_{n\geq 0} A_n$ be any positively graded ring and let $\underline{x} = \{x_1,\dots,x_r\}$ be a sequence of homogeneous elements of A. $\underline{x}$ is said to be weakly regular if there exists an integer $n_\circ$ with the property: for every $i \in \{1,\dots,r\}$ and every $a \in A_n$ with $n \geq n_\circ$, $ax_i \in x_1A + \dots + x_{i-1}A$ implies $a \in x_1A + \dots + x_{i-1}A$, where $x_\circ := 0$.

Recall that we denote by $H_i(\underline{x};A)_n$ the homogeneous component of degree n of the i-th Koszul homology. In the sequel we will always assume that A_0 is local and A is noetherian.

(22.5) Lemma. Consider the following conditions:

(1) $\underline{x}$ is a weakly regular sequence;

(2) $H_1(\underline{x};A)_n = 0$ for large n.

Then (1) ⇒ (2). If $\deg x_j = 0$ for $j = 1,\dots,r$, then (1) and (2) are equivalent.

Proof. The proof (1) ⇒ (2) is literally the same as for regular sequences (see e.g. [6]). If $\deg x_j = 0$ for $j = 1,\dots,r$, then we have $H_1(\underline{x};A_n) = H_1(\underline{x};A)_n$. Therefore $H_1(\underline{x};A)_n = 0$ if and only if

$\underline{x}$ is a regular sequence on A_n, which proves (2) $\Rightarrow$ (1) in this case.

(22.6) Corollary. If $\underline{x}$ is a weakly regular sequence and if $\deg x_j = 0$ for $j = 1,\ldots,r$, then any permutation of $\underline{x}$ is a weakly regular sequence.

Proof. Condition (2) of Lemma (22.5) is independent of the order of $\underline{x}$.

(22.7) Remark. Corollary (22.6) and Lemma (22.5) cannot be extended to a sequence $\underline{x}$ of elements with positive degrees, as the following example shows.

Take $A = k[[x]][Z]/(x\cdot Z^2)$ where the grading is taken with respect to Z. Then the sequence $\{Z^2, xZ\}$ is weakly regular, but xZ is not a weakly regular element.

(22.8) Proposition. With the same notations as before, assume that $\deg x_j = 0$ for $j = 1,\ldots,r$. Then the following conditions are equivalent.

(1) $\underline{x}$ is a weakly regular sequence on A.

(2) $H_1(\underline{x};A)_n = 0$ for $n \gg 0$.

(3) $H_1(\underline{x};A_{(P)}) = 0$ for all $P \in \operatorname{Proj}(A)$.

(4) $H_1(\underline{x};A_{(P)}) = 0$ for all $P \in \operatorname{Proj}(A)$ such that $\underline{x}A \subset P$.

Proof. The equivalence (1)$\Longleftrightarrow$(2) is clear by Lemma (22.5). The equivalence (3)$\Longleftrightarrow$(4) follows from the fact that $\underline{x}\cdot H_1(\underline{x};A_{(P)}) = 0$. Finally we get (2)$\Longleftrightarrow$(3) by Lemma (11.8).

Now let $(R,\mathfrak{m})$ be a local ring and I a proper ideal of R. We want to apply the results, obtained so far, to $A = G(I,R)$. We fix a sequence $\underline{x} = \{x_1,\ldots,x_r\}$ of elements in R, and we denote by $\underline{x}R$ the ideal generated by $\underline{x}$. Moreover we put again $I(\underline{x}) = I + \underline{x}R$. By $R \longrightarrow R_1$ we denote a local homomorphism obtained by the blowing up $f: X = B\ell(I,R) \longrightarrow \operatorname{Spec}(A)$, i.e. $R_1 = O_{X,x}$ for some $x \in f^{-1}(\{\mathfrak{m}\})$.

By abuse of language $\underline{x}$ will be sometimes considered as a sequence in $G(I,R)$ by which we mean the sequence of the images of $x_1,\dots,x_r$ in $R/I = G(I,R)_0$.

Using these notations we get as an immediate corollary of Proposition (22.8):

<u>(22.9) Corollary</u>. The following conditions are equivalent.

(1) $\underline{x}$ is weakly regular on $G(I,R)$.

(2) For any homomorphism $R \longrightarrow R_1$, obtained by the blowing up with center I , $\underline{x}$ is a regular sequence on R_1/IR_1 .

<u>(22.10) Definition</u>. R is called projectively normally flat along I if $G(I,R)$ is projectively flat over $R/I = G(I,R)_0$.

<u>(22.11) Remark</u>. By Proposition (22.3) R is projecitvely normally flat along I if and only if I^n/I^{n+1} is flat over R/I for large n .

<u>(22.12) Lemma</u>. Assume that $\underline{x}$ is a regular sequence on R/I . Then the follwing conditions are equivalent.

(1) R is projectively normally flat along I .

(2) $G(I,R) \otimes_R R/I(\underline{x})$ is projectively flat over $R/I(\underline{x})$, and $\underline{x}$ is a weakly regular sequence on $G(I,R)$.

Proof. (1) $\Rightarrow$ (2) : The first part of (2) is obvious. Note for the second part, that $H_1(\underline{x};R/I) = 0$ implies, for large n , $H_1(\underline{x};I^n/I^{n+1}) = 0$ as soon as I^n/I^{n+1} is flat over R/I . Therefore, $\underline{x}$ is weakly regular on $G(I,R)$ by Proposition (22.8).
(2) $\Rightarrow$ (1): Since $\underline{x}$ is regular on R/I , $H_1(\underline{x};I^n/I^{n+1}) = 0$ implies $\mathrm{Tor}_1^{R/I}(R/I(\underline{x}),I^n/I^{n+1}) = 0$ by Lemma (21.5). So we conclude by the local criterion of flatness.

<u>(22.13) Example</u>. Let $R = k[[X,Y]]/(XY,Y^2) = k[[x,y]]$, where k is any field. Take $I = y \cdot R$, $\underline{x} = \{x\}$. It follows that $G(I,R) \simeq k[[X]][T]/(XT,T^2)$ where T stands for the initial form of y in $G(I,R)$. Then x is regular on R/I , but x is a zero-divisor on $R = R/I^n$ for $n \geq 2$. On the other hand x is weakly regular on $G(I,R)$, since $I^n/I^{n+1} = 0$ for $n \geq 2$. This example shows that the property " $\underline{x}$ is weakly regular on $G(I,R)$ " is different from the property " $\underline{x}$ is a regular sequence on R/I^{n+1} for large n ". The next result shows that this difference in the example comes from the fact that the element x is a zero-divisor in R .

<u>(22.14) Proposition</u>. The following conditions are equivalent.

(1) $\underline{x}$ is a regular sequence on R/I^{n+1} for n large.

(2) $\underline{x}$ is a regular sequence on R and a weakly regular sequence on $G(I,R)$.

(3) $\underline{x}$ is a regular sequence on R and $I^n \cap \underline{x}R = \underline{x}I^n$ for $n >> 0$.

If one of these conditions is satisfied, then

$$x_i \notin I + x_1R + \ldots + x_{i-1}R + x_{i+1}R + \ldots + x_rR$$

for $1 \leq i \leq r$.

Proof. (1) $\Rightarrow$ (2): From the exact sequence

$$\text{(A)} \quad 0 \longrightarrow I^n/I^{n+1} \longrightarrow R/I^{n+1} \longrightarrow R/I^n \longrightarrow 0$$

and the induced exact sequences for the Koszul homology we get by Proposition (22.8) the second part of (2). To prove the first part, let $i \in \{1,\ldots,r\}$ and assume

$$ax_i \in x_1R + \ldots + x_{i-1}R \quad \text{for some} \quad a \in R \ .$$

From condition (1) we conclude that $a \in x_1R + \ldots + x_{i-1}R + I^n$ for large n , hence $a \in x_1R + \ldots + x_{i-1}R$ by Krull's intersection theorem.

(2) ⇒ (3): By (2) we find an integer k such that: $\underline{x}$ is regular on I^n/I^{n+1} for $n \geqq k$. Moreover using the Artin-Rees lemma we get an integer $c > 0$ with $I^{n+c} \cap \underline{x}R = I^n(I^c \cap \underline{x}R)$ for all $n \geqq 0$. To prove (3) it is enough to show that $I^{c+k} \cap \underline{x}R = \underline{x}I^{c+k}$.

Claim: This relation follows from the inclusion

$$\text{(B)} \qquad I^n \cdot \underline{x}R \cap I^{n+1} \subseteq I^{n+1} \cdot \underline{x}R \quad \text{for} \quad n \geqq k .$$

Proof of the claim: We have by the Artin-Rees lemma:

$$I^{c+k} \cap \underline{x}R = I^k(I^c \cap \underline{x}R) = I^{c+k} \cap I^k\underline{x}R \quad ;$$

hence we get for $c = 1$ by (B):

$$I^{k+1} \cap \underline{x}R \subseteq I^{k+1} \cdot \underline{x}R \; ; \quad \text{i.e.} \quad I^{k+1} \cap \underline{x}R = I^{k+1} \cdot \underline{x}R .$$

Using the relation for $c = 1$ and repeating this conclusion for $c = 2$ we obtain:

$$\begin{aligned} I^{k+2} \cap \underline{x}R &= I^{k+2} \cap I^k\underline{x}R = I^{k+2} \cap I^{k+1} \cap I^k\underline{x}R \\ &= I^{k+2} \cap I^{k+1}\underline{x}R \subseteq I^{k+2}\underline{x}R , \end{aligned}$$

hence $I^{k+2} \cap \underline{x}R = I^{k+2}\underline{x}R$. In this way one proves the claim for any $c > 0$.

Now from (A) we deduce the exact sequence

$$0 \longrightarrow \mathrm{Tor}_1^R(R/\underline{x}R, R/I^{n+1}) \longrightarrow \mathrm{Tor}_1^R(R/\underline{x}R, R/I^n)$$

for $n \geqq k$. This can also be written as $(n \geqq k)$

$$0 \longrightarrow (I^{n+1} \cap \underline{x}R) \; / \; (I^{n+1}\underline{x}R) \longrightarrow (I^n \cap \underline{x}R)/I^n\underline{x}R ,$$

which indeed implies (B).

(3) $\Rightarrow$ (1): Since $\underline{x}$ is regular on R, the Koszul complex is a finite free resolution of $R/\underline{x}R$, hence

$$H_1(\underline{x}; R/I^n) \simeq \mathrm{Tor}_1^R(R/\underline{x}R, R/I^n) \simeq I^n \cap \underline{x}R/I^n\underline{x}R$$

proving (1).

To prove the last claim of Proposition (22.14) observe that (1), (2), (3) are independent of the order of $\underline{x}$. So we may assume $i = r$. Now if $x_r \in I + x_1R + \ldots + x_{r-1}R$, then $x_r^n \in I^n + x_1R + \ldots + x_{r-1}R$ for all n, contradicting the assumption that $\underline{x}$ is a regular sequence on R/I^n for large n.

Recall that in § 21 we introduced two canonical homomorphisms of graded rings:

$$g(I,\underline{x}) : G(I,R) \otimes_R R/I(\underline{x}) \longrightarrow G(I(\underline{x})/\underline{x}R, R/\underline{x}R)$$

and

$$G(I,\underline{x}) : (G(I,R) \otimes_R R/I(\underline{x}))[T_1,\ldots,T_r] \longrightarrow G(I(\underline{x}),R) \quad .$$

In the following Theorem (22.15) we characterize weakly regular sequences via one of these homomorphisms:

(22.15) Theorem. With the notations as above the follwoing conditions are equivalent.

(1) $(\mathrm{Ker}\, g(I,\underline{x}))_n = 0$ for large n, and $\underline{x}$ is a regular sequence on R.

(2) $\underline{x}$ is a weakly regular sequence on $G(I,R)$ and a regular sequence on R.

(3) $\underline{x}$ is a regular sequence on R/I^{n+1} for large n.

Proof. For any n we have $(\mathrm{Ker}\, g(I,\underline{x}))_n = 0$ if and only if $I^n \cap \underline{x}R \subset I^n\underline{x}R + I^{n+1}$. From this, we see easily that $(\mathrm{Ker}\, g(I,\underline{x}))_n = 0$ for large n if and only if $I^n \cap \underline{x}R = I^n\underline{x}R$ for large n. So the theorem follows from Proposition (22.14).

We note that this theorem is parallel to Theorem (21.9), but with the important difference that for large degree it seems to be no reasonable equivalent property involving $G(I,\underline{x})$. On the other hand we will show in § 23 that if $\mathrm{Ker}(G(I,\underline{x}))$ is nilpotent for a system $\underline{x}$ of parameters modulo I , then we have equimultiplicity $\mathrm{ht}(I) = s(I)$.

<u>(22.16) Theorem</u>. With the same notations as in Theorem (22.15) assume in addition, that $\underline{x}$ is a regular sequence on R/I . Then the following conditions are equivalent:

(1) R is projectively normally flat along I , and $\underline{x}$ is a regular sequence on R .

(2) $G(I,R) \otimes_R R/I(\underline{x})$ is projectively flat over $R/I(\underline{x})$, and $\underline{x}$ is a weakly regular sequence on $G(I,R)$ as well as a regular sequence on R .

(3) $G(I,R) \otimes_R R/I(\underline{x})$ is projectively flat over $R/I(\underline{x})$, and $\underline{x}$ is a regular sequence on R/I^{n+1} for large n .

(4) $(\mathrm{Ker}\ g(I,\underline{x}))_n = 0$ for large n , $R/\underline{x}R$ is projectively normally flat along $I(\underline{x})/\underline{x}R$, and $\underline{x}$ is a regular sequence on R .

Proof. This follows from Lemma (22.12) and Theorem (22.15).

Before turning to some consequences of Theorem (22.16), we make the following simple observation:

<u>(22.17) Lemma</u>. Let $A = \bigoplus_{n \geq 0} A_n$ be a graded noetherian ring such that A_0 is local and $A = A_0[A_1]$, and let $K \subset A_+$ be a homogeneous ideal. Consider the conditions:

(1) $K = 0$;

(2) $K_n = 0$ for large n ;

(3) K is nilpotent.

Then we have the following implications.

(a) $(1) \Rightarrow (2) \Rightarrow (3)$.

(b) If A_+ contains a homogeneous non-zero-divisor, then $(1) \Longleftrightarrow (2)$.

(c) If Proj(A) is reduced, then (2)$\Longleftrightarrow$(3) .

Proof. (a) is clear. If $a \in A_+$ is a homogeneous non-zero-divisor and $k \in K_n$ is arbitrary, then (2) implies $a^n k = 0$ for large n . Therefore we have $k = 0$, which proves (b) . For (c) we note that if K is nilpotent and Proj(A) is reduced, then $K_{(\mathfrak{p})} = 0$ for all $\mathfrak{p} \in \mathrm{Proj}(A)$. So the assertion follows from Lemma (11.8).

(22.18) Corollary. Let $\mathfrak{p} \in \mathrm{Spec}(R)$ be a prime ideal such that $R/\mathfrak{p}$ is regular, and let $\underline{x}$ be a regular system of parameters modulo $\mathfrak{p}$. Assume that

(1) $\underline{x}$ is a regular sequence on R ,

(2) R is projectively normally flat along $\mathfrak{p}$,

(3) $\mathrm{depth}(G(\mathfrak{p},R) \otimes_R R/\mathfrak{m}) > 0$.

Then $\mathfrak{p}$ is permissible in R .

Proof. Since $R/\mathfrak{p}$ is regular and $\underline{x}$ is a regular system of parameters modulo $\mathfrak{p}$, R is normally flat along $\mathfrak{p}$ iff $\mathrm{Ker}(G(\mathfrak{p},\underline{x})) = 0$. This is equivalent to the condition $K := \mathrm{Ker}(g(\mathfrak{p},\underline{x})) = 0$ by assumption (1), see Theorem (21.9). Moreover $K = 0$ is equivalent to $K_n = 0$ for $n \gg 0$ by assumption (3) and Lemma (22.17), (b). Finally $K_n = 0$ for $n \gg 0$ follows from the assumptions (1) and (2) by Theorem (22.16), (1) $\Rightarrow$ (4).

(22.19) Corollary. Let $\mathfrak{p}$ be a prime ideal in R such that $R/\mathfrak{p}$ is regular and let $\underline{x}$ be a regular system of parameters modulo $\mathfrak{p}$. Assume that

(1) $R/\mathfrak{p}^n$ is Cohen-Macaulay for large n , and

(2) $\mathrm{depth}(G(\mathfrak{p},R) \otimes_R R/\mathfrak{m}) > 0$.

Then $\mathfrak{p}$ is permissible.

Proof. Using assumption (1) we obtain by Lemma (22.12) and Proposition (22.14) that $\underline{x}$ is a regular sequence on R , and R is projectively normally flat along $\mathfrak{p}$. Therefore the assertion follows from Corollary (22.18).

Now we turn to a numerical characterization of " R is normally Cohen-Macaulay along I " in terms of generalized Hilbert functions. From that we can finally obtain the announced criterion of permissibility (see also Corollary (21.12)).

We start with two technical lemmas:

(22.20) Lemma. Let I be an ideal in a local ring (R,m) and let $\underline{x} = \{x_1,\dots,x_r\}$ be a sequence of elements in m. Assume that $\underline{x}$ is a regular sequence on R/I^n for all $n > 0$. Then $x_2,\dots,x_r$ is a regular sequence on $R/(I,x_1)^n$ for all $n > 0$.

Proof. Note that the assumption of Lemma (22.20) is equivalent to the following two conditions (see [3], Lemma 1.35, p.17):

(1) $\underline{x}$ is a regular sequence on R,

(2) $I^n \cap \underline{x}R = I^n \cdot \underline{x}R$ for all $n > 0$.

Therefore it is enough to show that $(I,x_1)^n \cap (x_2,\dots,x_r) = (I,x_1)^n(x_2,\dots,x_r)$ for all $n > 0$. This will be done by induction on n. The assertion is clear for $n = 1$. For $r = 0$ we follow the usual convention that the empty set generates the zero ideal. So we assume $n > 1$.

Take $x \in (I,x_1)^n \cap (x_2,\dots,x_r)$. Since $(I,x_1)^n = I^n + x_1(I,x_1)^{n-1}$, one can write $x = a + x_1y$ with $a \in I^n$ and $y \in (I,x_1)^{n-1}$. Then we obtain:

$$a \in (x_1,\dots,x_r) \cap I^n = (x_1,\dots,x_r)I^n ,$$

i.e. we have $a = \sum_{i=1}^{r} a_ix_i$ with $a_i \in I^n$, hence

$$x = \sum_{i=2}^{r} a_ix_i + x_1(y + a_1) .$$

Using that $\underline{x}$ is a regular sequence on R we get by induction hypothesis:

$$y + a_1 \in (x_2,\dots,x_r) \cap (I,x_1)^{n-1} = (x_2,\dots,x_r)(I,x_1)^{n-1} ,$$

and therefore

$$x \in (x_2,\dots,x_r)\cdot(I^n + x_1(I,x_1)^{n-1}) = (x_2,\dots,x_r)\cdot(I,x_1)^n \quad ,$$

hence

$$(I,x_1)^n \cap (x_2,\dots,x_r) \subset (I,x_1)^n(x_2,\dots,x_r) \quad ,$$

which proves the claim.

(22.21) Remark. U. Orbanz and L. Robbiano tried to prove the following statement in [14], Theorem (1.16): "If I is an ideal in a local ring R and $\underline{x} = \{x_1,\dots,x_r\}$ is a regular sequence on R/I^n for large $n > 0$, then $x_2,\dots,x_r$ is a regular sequence on $R/(I,x_1)^n$ for large $n > 0$". However there is a gap in that proof and in fact S. Ikeda gave a counterexample to this special statement: Let $R = k[[X,Y,Z,W]]/(W^2,ZW,XZ-YW) = k[[x,y,z,w]]$, and put $I = ywR$ (k is any field). Then R is a two-dimensional Cohen-Macaulay ring, and $x+z$, y form a system of parameters of R. Since $I^2 = 0$, we have a regular sequence $x+z$, y on $R/I^n = R$ for $n \geq 2$. But y is a zero-divisor on $R/(I,x+z)^n$ for $n \geq 2$, because $y(x^{n-1}w) \in (I,x+z)^n = ((x+z)^n, yx^{n-1}w)$ and $x^{n-1}w \notin (I,x+z)^n$.

(22.22) Lemma. Let R, I and $\underline{x} = \{x_1,\dots,x_r\}$ be the same as in Lemma (22.20), let $I_1 = I + x_1R$ and $\underline{y} = x_2,\dots,x_r$. Assume that $\underline{x}$ is a system of parameters modulo I, $r > 0$, and that for some $n \geq 0$ we have

(1) $\operatorname{depth} I_1^n/I_1^{n+1} = r-1$,

(2) $H^{(0)}[\underline{y},I_1,R](n) = H^{(1)}[\underline{x},I,R](n)$.

Then x_1 is a non-zero-divisor on I_1^n/I^{n+1} .

Proof. Consider the exact sequence

$$0 \longrightarrow I_1^n/I^{n+1} \longrightarrow R/I^{n+1} \longrightarrow R/I_1^n \longrightarrow 0 \quad .$$

Since $\dim(R/I_1^n) < r$, we obtain

(A) $e(\underline{x};I_1^n/I^{n+1}) = e(\underline{x};R/I^{n+1}) = e(\underline{y};I_1^n/I_1^{n+1})$,

where the first equality follows from the additivity of the multiplicity, and the second equality comes from assumption (2). By definition we have

(B) $e(\underline{x};I_1^n/I^{n+1}) = e(\underline{y};I_1^n/I^{n+1} + x_1I_1^n) - e(\underline{y};N) =$

$= e(\underline{y};I_1^n/I_1^{n+1}) - e(\underline{y};N)$,

where $N := \mathrm{Ann}(x_1, I_1^n/I^{n+1}) = \left\{a \in I_1^n/I^{n+1} \mid x_1a = 0\right\}$. Comparing (A) and (B), we get

(C) $e(\underline{y};N) = 0$.

Hence if $r = 1$, then $\ell_R(N) = e(\underline{y}\ N) = 0$, i.e. x_1 is regular on I_1^n/I^{n+1} . Now assume that $r > 1$. We put $N_i := \mathrm{Ann}(x_1^i, I_1^n/I^{n+1})$ for $i > 1$, where $N_1 = N$. Then

$$0 =: N_0 \subseteq N_1 \quad N_2 \subseteq \ldots \subseteq I_1^n/I^{n+1} \quad ,$$

and there is a minimal integer $k \geq 0$ such that $N_k = N_{k+m}$ for all $m \geq 0$. The proof will be finished by showing that $k = 0$. First we conclude from (C) that

(D) $\dim N_1 < r - 1$.

Since N_1 is annihilated by x_1 and I^{n+1} , we have $I_1^{n+1} N_1 = 0$. Therefore there exists some $y \in R$ such that

(E) $yN_1 = 0$ and $y \notin \mathfrak{p}$ for all $\mathfrak{p} \in \mathrm{Assh}(R/I_1^{n+1})$.

For this y we show by induction on j that

(F) $y^jN_j = 0$.

This is clear for $j = 1$ by (E) . For $j > 1$ choose any element $z \in N_j$. By definition we have $x_1^jz = 0 = x_1^{j-1}(x_1z)$. So $x_1z \in N_{j-1}$, which implies $y^{j-1}x_1z = 0$ by induction hypothesis, i.e. $y^{j-1}z \in N_1$, hence $y(y^{j-1}z) = 0$ by (E), proving condition (F). Next we observe that

(G) y is regular on I_1^n/I_1^{n+1}.

Since I_1^n/I_1^{n+1} is a Cohen-Macaulay module of dimension $r-1$ by assumption (1) we have $\mathrm{Ass}(I_1^n/I_1^{n+1}) \subset \mathrm{Assh}(R/I_1^{n+1})$, so property (G) follows from (E). After these preparations assume that $k > 0$ and choose an element $z \in N_k \setminus N_{k-1}$. Since $y^k z = 0$, property (G) implies $z \in I_1^{n+1}/I^{n+1} = I^{n+1} + x_1 I_1^n / I^{n+1}$, and therefore $z = x_1 z_1$ for some $z_1 \in I_1^n/I^{n+1}$. Now $z = x_1^k z_1 \in N_k$ implies $x_1^{k+1} z_1 = 0$, i.e. $z_1 \in N_{k+1} = N_k$. But this implies $x_1 z_1^k = x_1^{k-1} z = 0$, which contradicts $z \notin N_{k-1}$. So k must be zero. This proves Lemma (22.22).

To state the main result we introduce the following notation: for $i \in \{0,1,\ldots,r-1\}$ we put

$$I_i = I + x_1R + \ldots + x_iR \quad \text{for} \quad i > 1 ,$$

$$I_0 = I ,$$

$$\underline{y}_i = \{x_{i+1},\ldots,x_r\} \quad \text{and} \quad \underline{y}_r = \emptyset .$$

(22.23) Theorem. Let R be a local ring and I a proper ideal of R. Assume that $x_1,\ldots,x_r$ is a system of parameters modulo I. Then the following conditions are equivalent.

(1) R/I^{n+1} is Cohen-Macaulay for all $n \geqq 0$.

(2) For all $i \in \{0,1,\ldots,r-1\}$ we have

$$H^{(0)}[\underline{y}_{i+1}, I_{i+1}, R](n) = H^{(1)}[\underline{y}_i, I, R](n)$$

for all $n \geqq 0$.

(3) $H^{(0)}[\underline{x}R + I, R] = H^{(r)}[\underline{x}, I, R]$.

(4) $\mathrm{depth}\, I^n/I^{n+1} = \dim R/I$ for all $n \geqq 0$, i.e. R is normally Cohen-Macaulay along I.

Proof. (1)$\Longleftrightarrow$(4): This is clear.

(1) $\Rightarrow$ (2): We make induction on r. Condition (2) is empty in the

case $r = 0$, so assume $r > 0$. By Lemma (22.20) we know that R/I_1^{n+1} is Cohen-Macaulay for all $n \geqq 0$, and therefore condition (2) is satisfied for $1 \leqq i \leqq r-1$ by induction hypothesis. Furthermore x_1 is regular on I_1^n/I^{n+1} , since R/I^{n+1} is Cohen-Macaulay. This last fact yields

$$e(\underline{x};I_1^n/I^{n+1}) = e(\underline{y}_1;I_1^n/I^{n+1} + x_1 I_1^n) =$$
$$= e(\underline{y}_1;I_1^n/I_1^{n+1}) = H^{(0)}[\underline{y}_1,I_1,R](n) \quad .$$

On the other hand we have

$$e(\underline{x};I_1^n/I^{n+1}) = e(\underline{x};R/I^{n+1}) = H^{(1)}[\underline{x},I,R](n) \quad .$$

This proves the remaining case $i = 0$.

(2) ⇒ (3): Using condition (2) for $i = 0$, we get

$$H^{(0)}[\underline{y}_1,I_1,R](k) = H^{(1)}[\underline{x},I,R](k) \quad .$$

Summarizing both sides over $k = 0,\ldots,n$ we obtain

$$H^{(1)}[\underline{y}_1,I_1,R](n) = H^{(2)}[\underline{x},I,R](n) \quad .$$

Applying again condition (2) to the left side, we have

$$H^{(0)}[\underline{y}_2,I_2,R](n) = H^{(2)}[\underline{x},I,R](n) \quad .$$

By successive summarizing both sides and then applying (2) to the left side, we get (3) .

(3) ⇒ (2): Note that in general (see Chapter II)

$$(*) \qquad H^{(1)}[\underline{y}_i,I_i,R] \leqq H^{(0)}[\underline{y}_{i+1},I_{i+1},R] \ , \quad i \in \{0,1,\ldots,r-1\} \quad .$$

Therefore condition (3) implies the equality in (*), which proves (2).

(2) ⇒ (4): We use induction on r . The case $r = 0$ is trivial since then condition (2) is empty. For $r > 0$ the inductive assumption implies $\operatorname{depth} I_1^n/I_1^{n+1} = r-1$ for all $n \geqq 0$, hence $x_2,\ldots,x_r$

is a regular sequence on I_1^n/I_1^{n+1} for all $n \geq 0$. Then x_1 is regular on I_1^n/I^{n+1} for all $n \geq 0$ by Lemma (22.22). Moreover since $(I_1^n/I^{n+1})/x_1 \cdot (I_1^n/I^{n+1}) \simeq I_1^n/I_1^{n+1}$, $\underline{x}$ is a regular sequence on I_1^n/I^{n+1} for all $n \geq 0$, in particular R/I is Cohen-Macaulay. This is condition (4) in the case $n = 0$. For $n > 0$ we consider the following exact sequences

$$\text{(S1)} \quad 0 \longrightarrow I_1^{n+1}/I^{n+1} \longrightarrow I_1^n/I^{n+1} \longrightarrow I_1^n/I_1^{n+1} \longrightarrow 0 \quad ,$$

and

$$\text{(S2)} \quad 0 \longrightarrow I^{n+1}/I^{n+2} \longrightarrow I_1^{n+1}/I^{n+2} \longrightarrow I_1^{n+1}/I^{n+1} \longrightarrow 0 \quad .$$

We know already that for all $n \geq 0$:

$$\operatorname{depth} I_1^n/I_1^{n+1} = r - 1 \quad \text{and} \quad \operatorname{depth} I_1^n/I^{n+1} = r \quad .$$

Therefore (S1) implies

$$\text{(S3)} \quad \operatorname{depth} I_1^{n+1}/I^{n+1} = r \quad \text{for all} \quad n \geq 0 \quad .$$

Using (S3) and $\operatorname{depth} I_1^{n+1}/I^{n+2} = r$, we conclude (4) from (S2), q.e.d.

As a consequence of Theorem (22.23) we obtain a numerical characterization of permissibility; see also [11], Theorem 3, where the socalled transitivity or normal flatness is used. This special property will be described in § 24.

<u>(22.24) Theorem</u>. Let $(R,\mathfrak{m})$ be a local ring and $\mathfrak{p}$ a prime ideal in R. Assume that $R/\mathfrak{p}$ is regular with $\dim(R/\mathfrak{p}) = r$. Then R is normally flat along $\mathfrak{p}$ if and only if

$$H^0[R] = H^{(r)}[R_{\mathfrak{p}}] \quad .$$

Proof. Since $R/\mathfrak{p}$ is regular, R is normally flat along $\mathfrak{p}$ if and only if R is normally Cohen-Macaulay along $\mathfrak{p}$. The last property is equivalent to $H^{(0)}[\mathfrak{p} + \underline{x}R, R] = H^{(r)}[\underline{x}, \mathfrak{p}, R]$ by Theorem (22.23). Then, choosing a regular system $\underline{x}$ of parameters modulo $\mathfrak{p}$, we get the desired equality of Hilbert functions.

§ 23. Hierarchy of equimultiplicity and permissibility

(23.0) In this section we give a detailed explanation of how the various notions of equimultiplicity are related. We will see that normal flatness along an ideal I implies $ht(I) = s(I)$. The same implication is true if we start with " R is normally Cohen-Macaulay along I ", provided $\dim(R) = \dim(R/I) + ht(I)$. In particular, if $R/\mathfrak{p}$ is regular then the condition $ht(\mathfrak{p}) = s(\mathfrak{p})$ is equivalent to $e(R) = e(R_{\mathfrak{p}})$, i.e. to equimultiplicity in the sense of Zariski. In quasi-unmixed rings R the condition $ht(I) = s(I)$ is equivalent to $e(\underline{x},I,R) = e(I+\underline{x}R,R)$, where $\underline{x}$ is a system of parameters modulo I. This indicates that the notion " R is normally Cohen-Macaulay along I " is the right counterpart to " R is normally flat along I " if R/I is not regular. Those "generalized" notions can be used in Chapter VI to study blowing ups with non-regular centers.

We start with a complete description of the relations between the various notions of equimultiplicity. For a geometric interpretation of (3) below see App. III, Them. 2.2.2.

Let $(R,\mathfrak{m},k)$ denote a local (noetherian) ring and let I be a proper ideal of R . We use the following notations:

(1) R nf $I \Longleftrightarrow R$ is normally flat along I

(1*) R nCM $I \Longleftrightarrow R$ is normally Cohen-Macaulay along I

(1**) R/I^{n+1} CM $\Longleftrightarrow R/I^{n+1}$ is a Cohen-Macaulay ring for $n \geq 0$

(2) R pnf $I \Longleftrightarrow R$ is projectively normally flat along I

(2*) R pnCM $I \Longleftrightarrow R$ is projectively normally Cohen-Macaulay along I ; i.e. $\operatorname{depth}(I^n/I^{n+1}) = \dim(R/I)$ for $n >> 0$

(2**) R/I^{n+1} CM , $n >> 0 \Longleftrightarrow R/I^{n+1}$ is Cohen-Macaulay ring for $n >> 0$

(3) R is equimultiple along $I \Longleftrightarrow ht(I) = s(I)$. [The name is motivated by Theorem (20.5).]

The relations between these conditions can be summarized in the following picture, where (*) means that $\dim(R) = \dim(R/I) + ht(I)$:

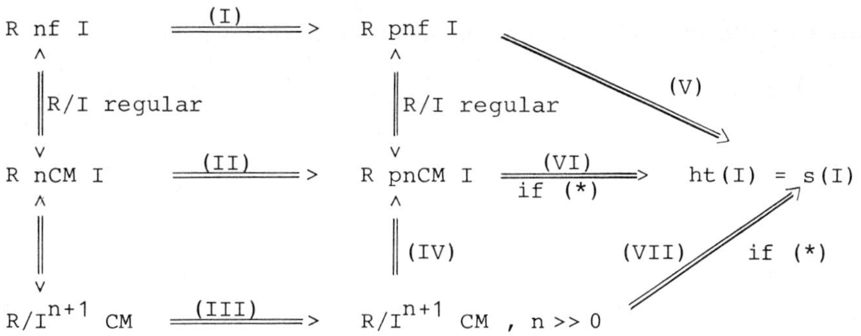

For more details (e.g. for the converse of (IV) under some additional assumptions) see [15], Theorem (1.10). In this drawing the implications (I), (II), (III) and (IV) are clear. Before proving the implications (V), (VI) and (VII) we give some examples to show that all implications (I) - (VII) are proper. Moreover we show that (VI) and (VII) do not hold if we drop the assumption $\dim(R) = \dim(R/I) + \mathrm{ht}(I)$. More precisely: without this dimension condition, even " R is normally Cohen-Macaulay along I " and " R/I^{n+1} is CM for $n \geq 0$ " don't imply $\mathrm{ht}(I) = s(I)$.

<u>(23.1) Example for (VI) and (VII).</u> Let $R = k[[X,Y,Z]]/(X) \cap (Y,Z) = k[[x,y,z]]$, where k is any field. Let $I = yzR$, so that $\mathrm{ht}(I) = 0$ and $\dim(R/I) = 1$, hence $\mathrm{ht}(I) + \dim(R/I) < \dim R$. On the other hand, it is not too hard to see that R/I^{n+1} is a Cohen-Macaulay ring and I^n/I^{n+1} is a Cohen-Macaulay module over R/I of dimension 1 for all $n \geq 0$, but $0 = \mathrm{ht}(I) < s(I) = 1$.

<u>(23.2) Example for (I), (II) and (III).</u> Let $R = k[[X,Y,Z,W]]/(W^2, WZ, WX, Z^2, XZ - YW) = k[[x,y,z,w]]$, let $\mathfrak{p} = (x,z,w)$. Then we have:

(1) $R/\mathfrak{p}^{n+1}$ is Cohen-Macaulay for $n \geq 2$;

(2) $R/\mathfrak{p}^2 \simeq k[[x,y,z,w]]/(x^2, xz, yw)$ is not Cohen-Macaulay;

(3) $R/\mathfrak{p} \simeq k[[Y]]$ is regular, i.e. R nf $\mathfrak{p}$ is equivalent to R nCM $\mathfrak{p}$.

It follows that $\mathfrak{p}/\mathfrak{p}^2$ is not Cohen-Macaulay, but $\mathfrak{p}^n/\mathfrak{p}^{n+1}$ is Cohen-Macaulay for $n \geq 2$. Therefore R is projectively normally flat along $\mathfrak{p}$, but R is not normally flat along $\mathfrak{p}$.

(23.3) Example for (IV). Let $R = k[[X,Y,Z,W]]/(W^2,WZ,WX,Z^2,XZ,YW) = k[[x,y,z,w]]$ and let $\mathfrak{p} = (x,z,w)$. Then we have for $n \geq 2$: $\mathfrak{p}^n/\mathfrak{p}^{n+1} = x^nR/x^{n+1}R \simeq R/xR + (0 : x^n) = R/\mathfrak{p}$, but $R/\mathfrak{p}^n$ is not Cohen-Macaulay for $n \geq 2$.

(23.4) Example for (V) and (VI). Let $R = k[[X,Y,Z,W]]/(W^2,WZ,XZ,YW) = k[[x,y,z,w]]$, and let $\mathfrak{p} = (x \cdot z,w)$. Then $\mathfrak{p}^2 = (x+z)\mathfrak{p}$, and $(x+z)$ is a non-zero-divisor of R. Hence we get: $ht(\mathfrak{p}) = s(\mathfrak{p}) = 1$ and $\mathfrak{p}^n/\mathfrak{p}^{n+1} \simeq \mathfrak{p}/\mathfrak{p}^2$. But $\mathfrak{p}/\mathfrak{p}^2$ is not Cohen-Macaulay, i.e. R is not projectively normally Cohen-Macaulay along $\mathfrak{p}$ (or equivalently R is not projectively normally flat along the regular ideal $\mathfrak{p}$).

The main point of the following part is to prove the implication (VI) under the essential assumption $\dim(R/I) + ht(I) = \dim(R)$. This is done by showing the inequality of Burch [9]:

$$s(I) \leq \dim R - \min_n\{\operatorname{depth}(R/I^n)\} .$$

This inequality was improved by Brodmann [8] as a consequence of the socalled "asymptotic property of ideals" (s. Proposition (23.6) and Corollary (23.7)), which we describe first.

(23.5) Lemma. Let $A = \bigoplus_{n\geq 0} A_n$ be a homogeneous noetherian graded ring over A_0, i.e. $A = A_0[A_1]$. Then there are integers $n_\circ$, $k_\circ \geq 0$ such that:

(1) for any $k \geq k_\circ$ there exists a weakly regular homogeneous element a of R with $\deg(a) = k$ and

(2) for any $n \geq n_\circ$ we have $(0 : a) \cap A_n = 0$, i.e. the element a is weakly regular.

Proof. Let $\{\mathfrak{p}_1,\ldots,\mathfrak{p}_r\}$ be the maximal members of $\operatorname{Ass}(A) \setminus V(A_+)$. For each $i \in \{1,\ldots,r\}$ we can choose a homogeneous element $a_i \in (A_+ \cap (\bigcap_{i \neq j} \mathfrak{p}_j)) \setminus \mathfrak{p}_i$. Let $k_\circ = \max\{\deg(a_i)\}$. For each i we choose a homogeneous element $x_i \in A_1 \setminus \mathfrak{p}_i$, and for each $k \geq k_\circ$

we put $a := \sum_{i=1}^{r} a_i x_i^{k-\deg(a_i)}$ so that $a \notin \bigcup_{i=1}^{r} \mathfrak{p}_i$ and $\deg(a) = k$. Let $(0) = Q_1 \cap \ldots \cap Q_s \cap Q_{s+1} \cap \ldots \cap Q_t$ be a primary decompositon of the ideal (0) such that $\sqrt{Q_i} \in \mathrm{Ass}(A) \setminus V(A_+)$ for $1 \leq i \leq s$ and $Q_i \supseteq (A_+)^{n_\circ}$ for $s+1 \leq i \leq t$ for some $n_\circ$. Then we have $(0 : a) \subseteq Q_1 \cap \ldots \cap Q_s$ and hence

$$(0 : a) \cap A_n \subseteq Q_1 \cap \ldots \cap Q_s \subseteq A_n \subseteq Q_1 \cap \ldots \cap Q_s \cap Q_{s+1} \cap \ldots \cap Q_t = (0)$$

for $n \geq n_\circ$.

<u>(23.6) Proposition</u>. Let $A = \bigoplus_{n \geq 0} A_n$ be a homogeneous noetherian graded ring over A_0. Then $\mathrm{Ass}_{A_0}(A_n)$ becomes constant for large n.

Proof. By Lemma (23.5) we find integers $k_\circ$, $n_\circ$ such that for all $k \geq k_\circ$ there is a non-zero-divisor a on $\bigoplus_{n \geq n_\circ} A_n$ with $\deg(a) = k$. Therefore $A_n \subseteq A_{n+k}$ for any $k \geq k_\circ$ and $n \geq n_\circ$, i.e. $\mathrm{Ass}_{A_0}(A_n) \subseteq \mathrm{Ass}_{A_0}(A_{n+k})$. But $\mathrm{Ass}_{A_0}(A)$ is a finite set, cf. [5] (7.G) and (9.A), and hence $\mathrm{Ass}_{A_0}(A_n)$ which is contained in $\mathrm{Ass}_{A_0}(A)$ becomes stable for large n.

<u>(23.7) Corollary.</u> Let $A = A_0[A_1]$ and assume in addition that A_0 is a local ring. Then $\mathrm{depth}_{A_0}(A_n)$ becomes stable for large n.

Proof. We proceed by induction on $\dim(A_0)$, the case $\dim(A_0) = 0$ being trivial. For $\dim A_0 > 0$, we consider two cases:

(1) If the maximal ideal $\mathfrak{m}_\circ \subset A_0$ belongs to $\mathrm{Ass}_{A_0}(A_n)$ for $n \gg 0$, then $\mathrm{depth}_{A_0}(A_n) = 0$ for $n \gg 0$.

(2) If $\mathfrak{m}_\circ \notin \mathrm{Ass}_{A_0}(A_n)$ for $n \gg 0$, we can choose an element $x \in \mathfrak{m}_\circ$ such that $\dim(A_0/xA_0) < \dim(A_0)$ and (by PRoposition (23.6) x is a non-zero-divisor on A_n for $n \gg 0$. Then $\mathrm{depth}_{A_0}(A_n/xA_n)$ is constant for large n by the inductive hypothesis. This completes the proof.

Applying these results to the associated graded ring $G(I,R) = \bigoplus_{n \geq 0} I^n/I^{n+1}$ of an ideal I in a local ring R, we get the following asymptotic properties.

<u>(23.8) Proposition</u>. For an ideal I of a local ring R we have:

(1) $\mathrm{Ass}_{R/I}(I^n/I^{n+1})$ is constant for large n.

(2) $\mathrm{Ass}_R(R/I^n)$ is constant for large n.

(3) $\mathrm{depth}(I^n/I^{n+1})$ is constant for large n.

(4) $\mathrm{depth}(R/I^n)$ is constant for large n.

Proof. We have already (1) and (3) by Proposition (23.6) and Corollary (23.7). To prove (2) consider the exact sequence $0 \to I^n/I^{n+1} \to R/I^{n+1} \to R/I^n \to 0$, which implies ([1], IV, § 1., Proposition 3): $\mathrm{Ass}(R/I^{n+1}) \subset \mathrm{Ass}(I^n/I^{n+1}) \cup \mathrm{Ass}(R/I^n)$. Moreover we know by Proposition (23.6), that $\mathrm{Ass}(I^n/I^{n+1}) = \mathrm{Ass}(I^{n-1}/I^n) \subset \mathrm{Ass}(R/I^n)$ for large n. Therefore $\mathrm{Ass}(R/I^{n+1}) \subset \mathrm{Ass}(R/I^n)$ for $n \gg 0$. Since $\mathrm{Ass}(R/I^n)$ is a finite set, it becomes stable for large n.

Property (4) follows from property (2) in the same way as we obtained Corollary (23.7) from Proposition (23.6), q.e.d.

Now we give a proof of Burch's inequality, which appears as a consequence of the following Lemma (23.9). Using the asymptotic properties of Proposition (23.6) and Corollary (23.7) this inequality will be improved in Theorem (23.11).

<u>(23.9) Lemma</u>. Let $(R,\mathfrak{m})$ be a local ring and I a proper ideal in R. Let $\mathfrak{a} := \mathfrak{m} \cdot G(I,R)$. Then we have:

$$\mathrm{depth}_{\mathfrak{a}}(G(I,R)) = \min_n\{\mathrm{depth}(I^n/I^{n+1})\} .$$

Proof. We put $t(I) = \min_n\{\mathrm{depth}(I^n/I^{n+1})\}$. We proceed by induction on $t(I)$. Assume that $t(I) = 0$. Then $\mathfrak{m} \in \mathrm{Ass}(I^n/I^{n+1})$ for some n, and for that n there exists a non-zero element $x^* \in I^n/I^{n+1}$ such that $\mathfrak{a}x^* = 0$. This shows that $\mathfrak{a}$ consists of zero-divisors of $G(I,R)$, i.e. $\mathrm{depth}_{\mathfrak{a}}(G(I,R)) = 0$. Now let $t(I) > 0$. Since $\bigcup_n \mathrm{Ass}_{R/I}(I^n/I^{n+1}) \subset \mathrm{Ass}_{R/I}(G(I,R))$ and $\mathrm{Ass}_{R/I}(G(I,R))$ is a finite set, cf. [5], one can choose an element $x \in \mathfrak{m}$ which is a non-zero-divisor on R/I^n for all $n > 0$. We have $xR \cap I^n = x \cdot I^n$ for all $n > 0$. Putting $\bar{R} = R/xR$ and $\bar{I} = I\bar{R}$, we get:

$$\overline{I}^n/\overline{I}^{n+1} \simeq I^n + xR/I^{n+1} + xR \simeq I^n/(xR \cap I^n) + I^{n+1}$$

$$= I^n/xI^n + I^{n+1} \simeq (I^n/I^{n+1})\big/x(I^n/I^{n+1}) \quad .$$

Moreover we know that $G(\overline{I},\overline{R}) = G(I,R)/xG(I,R)$, and that x is a non-zero-divisor on $G(I,R)$. Note that $t(\overline{I}) = t(I) - 1$. Then we get by induction hypothesis:

$$\operatorname{depth}_{\mathfrak{a}}(G(I,R)) - 1 = \operatorname{depth}_{\mathfrak{a}}(G(\overline{I},\overline{R})) = t(\overline{I}) = t(I) - 1 \quad .$$

<u>(23.10) Corollary</u>. Let $(R,\mathfrak{m})$ be a local ring and I a proper ideal in R . Then $s(I) \leqq \dim R - \min_n\{\operatorname{depth}(I^n/I^{n+1})\}$.

Proof. Since $\operatorname{depth}_{\mathfrak{a}}(G(I,R)) \leqq \operatorname{ht}(\mathfrak{a})$, $\mathfrak{a} = \mathfrak{m} \cdot G(I,R)$, we get by Lemma (23.9):

$$\begin{aligned} s(I) = \dim(G(I,R)/\mathfrak{a}) &\leqq \dim(G(I,R)) - \operatorname{ht}(\mathfrak{a}) \\ &\leqq \dim R - \operatorname{depth}_{\mathfrak{a}}(G(I,R)) \\ &= \dim R - \min_n\{\operatorname{depth}(I^n/I^{n+1})\} \quad , \text{ q.e.d.} \end{aligned}$$

To get a sharper bound for $s(I)$ in the sense of Brodmann [8] we define the following numbers, which make sense by Proposition (23.8):

$$a(I) := \operatorname{depth}(I^n/I^{n+1}) \quad \text{for large } n$$

$$b(I) := \operatorname{depth}(R/I^n) \quad \text{for large } n \quad .$$

<u>(23.11) Theorem</u>. Let $(R,\mathfrak{m})$ be a local ring and I a proper ideal of R . Then the analytic spread satisfies the following inequalities:

$$s(I) \leqq \dim R - a(I) \leqq \dim R - b(I) \quad .$$

Proof. By Proposition (23.8) we have $a(I) = a(I^n)$ and $b(I) = b(I^n)$ for $n \gg 0$. Moreover $s(I) = s(I^n)$ for all n . Therefore we may replace I by I^n , $n \gg 0$, to get $a(I) = \min_n\{\operatorname{depth}(I^n/I^{n+1})\}$ and $b(I) = \min_n\{\operatorname{depth}(R/I^n)\}$. Then the conclusion follows from Corollary (23.10) and from the fact that $a(I) \geqq b(I)$, q.e.d.

Now we turn to the announced hierarchy of equimultiplicity-conditions. First we prove the implications (V) and (VI).

(23.12) Theorem. Let I be a proper ideal of a local ring $(R,\mathfrak{m},k)$. Assume that one of the following conditions holds:

(1) $\operatorname{depth}(I^n/I^{n+1}) = \dim(R/I)$ for infinitely many n and $\dim R = \dim(R/I) + \operatorname{ht}(I)$;

(2) I^n/I^{n+1} is free over R/I for infinitely many n .

Then $\operatorname{ht}(I) = s(I)$.

Proof. Assume (1): Then we have by Proposition (23.8) and Theorem (23.11)

$$s(I) \leqq \dim(R) - a(I) = \dim(R) - \dim(R/I) = \operatorname{ht}(I) \quad ,$$

hence $s(I) = \operatorname{ht}(I)$.

Assume (2): Then we get for any $\mathfrak{p} \in \operatorname{Min}(I)$ and for infinitely many n the equality

$$(*) \quad \ell_{R_{\mathfrak{p}}}\left(I^nR_{\mathfrak{p}}/I^{n+1}R_{\mathfrak{p}}\right) = \ell_R(I^n/\mathfrak{m}I^n)\, \ell_{R_{\mathfrak{p}}}(R_{\mathfrak{p}}/IR_{\mathfrak{p}}) \quad .$$

Recall that $\ell_R(I^n/\mathfrak{m}I^n)$ is a polynomial function of degree $(s(I)-1)$ for large n. Also $\ell_{R_{\mathfrak{p}}}(I^nR_{\mathfrak{p}}/I^{n+1}R_{\mathfrak{p}})$ is a polynomial function of degree $(\operatorname{ht}(\mathfrak{p})-1)$ for large n . By assumption the equation (*) holds for infinitely many values of n , so comparing the degrees of the corresponding polynomials gives $\operatorname{ht}(\mathfrak{p}) = s(I)$ for all $\mathfrak{p} \in \operatorname{Min}(I)$, hence $\operatorname{ht}(I) = s(I)$.

(23.13) Remarks.

(i) If condition (1) of Theorem (23.12) holds, then I^n/I^{n+1} is Cohen-Macaulay for all large n .

(ii) Case (1) of Theorem (23.12) can also be proved by considering the multiplicity $e(\underline{x};I^n/I^{n+1})$ with respect to a suitable system of parameters $\underline{x}$ modulo I; s. [10], pp. 210.

Next we give a sufficient condition for the implication

$ht(I) = s(I) \Rightarrow R$ projectively normally flat along I. For that we need a lemma due to J. Lipman [13]:

(23.14) Lemma. Let R be a quasi-unmixed local ring, and let I, J be ideals of R with the property $ht(I+J) = s(I) + s(J)$. Then $I \cap J$ is integral over IJ.

Proof. We may assume that $R/\mathfrak{m}$ has infinitely many elements. If $s(I) = 0$ both $I \cap J$ and IJ are nilpotent, and the conclusion is clear. So we assume $s(I) > 0$ and $s(J) > 0$. Let $\mathfrak{p}$ be a minimal prime of R and put $R' = R/\mathfrak{p}$. Then we have by assumption:

$$\begin{aligned} ht(I+J) = s(I) + s(J) &\geqq s(IR') + s(JR') \\ &\geqq s(IR' + JR') \\ &\geqq ht(IR' + JR') \quad , \end{aligned}$$

and moreover (since R is quasi-unmixed): $ht(IR + JR') = ht(I + J + \mathfrak{p}) \geqq ht(I+J)$, hence we obtain equality at every place. So we may assume that R is a domain. Then we claim that: $Min(I) \cap Min(J) = \emptyset$. In fact, if $\mathfrak{p} \in Min(I) \cap Min(J)$, then we get: $s(I) + s(J) = ht(I+J) \leqq ht(\mathfrak{p}) = s(IR_{\mathfrak{p}}) \leqq s(I)$, hence $s(J) = 0$, so this is a contradiction. Using the results of Chapter III for the integral closure of ideals we know that $Ass(R/\overline{IJ}) = Min(IJ) = Min(I) \cup Min(J)$ being a disjoint union. Now take an element $x \in I \cap J$. We want to show: $x \in \overline{IJ}$. For any $\mathfrak{p} \in Ass(R/\overline{IJ})$, we have:

$$\frac{x}{1} \in \overline{(I \cap J)} R_{\mathfrak{p}} = \begin{cases} \overline{I}R_{\mathfrak{p}} & \text{if} \quad I \subset \mathfrak{p} \\ \overline{J}R_{\mathfrak{p}} & \text{if} \quad J \subset \mathfrak{p} \quad , \end{cases}$$

whence we get $\frac{x}{1} \in \overline{IJ}R_{\mathfrak{p}}$. Therefore

$$x \in \bigcap_{\mathfrak{p} \in Ass(R/\overline{IJ})} \overline{IJ}R_{\mathfrak{p}} \cap R = \overline{IJ} \quad ,$$

since $Ass(R/\overline{IJ}) = Min(IJ)$, q.e.d.

Now we are able to present a criterion for the condition $ht(I) = s(I)$.

<u>(23.15) Proposition</u>. Let R be a quasi-unmixed local ring and I a proper ideal of R. Then the following conditions are equivalent:

(1) $ht(I) = s(I)$.

(2) For any system of parameters $\underline{x}$ modulo I such that $\dim(R/\underline{x}R) = ht(I)$, the homomorphism $g(I,\underline{x}) : G(I,R)/\underline{x}G(I,R) \longrightarrow G(I+\underline{x}R/\underline{x}R, R/\underline{x}R)$ has a nilpotent kernel.

(3) $G(I,\underline{x})$ has a nilpotent kernel for some system of parameters of R/I .

Proof. (1) ⇒ (2): By the assumption (1) we get for a system of parameters $\underline{x}$ modulo I : $ht(\underline{x}R) = \dim(R/I)$ and $ht(I^n + \underline{x}R) = \dim R = ht(I^n) + \dim(R/I) = s(I^n) + s(\underline{x}R)$. Therefore since R is quasi-unmixed, we have $I^n \cap \underline{x}R$ is integral over $\underline{x}I^n$ for all $n > 0$ by Lemma (23.14). Let a^* be a homogeneous element of order n in $Ker(g(I,\underline{x}))$ and $a \in I^n$ a representative for a^* , i.e. $a \in (I^n \cap \underline{x}R) + I^{n+1}$. We find some $b \in I^{n+1}$ with: $a - b \in I^n \cap \underline{x}R$. Since $a - b$ is integral over $\underline{x}I^n$, there are elements $c_i \in (\underline{x}I^n)^i$, $1 \leq i \leq k$, such that $(a-b)^k + c_1(a-b)^{k-1} + \ldots + c_k = 0$ for some $k > 0$. Since $c_i(a-b)^{k-i} \subset \underline{x}I^{nk}$ and $b \in I^{n+1}$, we get: $a^k \in \underline{x}I^{nk} + I^{nk+1}$. This implies $(a^*)^k = 0$.

(2) ⇒ (1): Since $Ker(g(I,\underline{x}))$ is contained in the nilradical of $G(I,R)/\underline{x}G(I,R)$, we have

$$\begin{aligned}\dim(R/\underline{x}R) &= \dim(G(I+\underline{x}R/\underline{x}R, R/\underline{x}R)) \\ &= \dim(G(I,R)/\underline{x}G(I,R)) = \dim(G(I,R)/\mathfrak{m}G(I,R)) = s(I) .\end{aligned}$$

Since $\dim(R/\underline{x}R) = ht(I)$ by our choice of $\underline{x}$, we obtain $ht(I) = s(I)$.

(2) ⇒ (3): Putting $\overline{G} := G(I,R)/\underline{x}G(I,R)$, $H := G(I+\underline{x}R, R)$ and $\overline{H} := G(I+\underline{x}R/\underline{x}R, R/\underline{x}R)$, we get a commutative diagram:

$$\begin{array}{ccc} \overline{G}[\underline{X}] := \overline{G}[X_1,\ldots,X_r] & \xrightarrow{\varphi} & H \\ \downarrow{\scriptstyle \pi_1} & & \downarrow{\scriptstyle \pi_2} \\ \overline{G} & \xrightarrow{\varphi_0} & \overline{H} \end{array}$$

where $\varphi := G(I,\underline{x})$, $\varphi_\circ := g(I,\underline{x})$, and π_1 , π_2 are the canonical surjections with $\pi_1(X_i) = 0$ for each i . Since (2) implies (1) we obtain for $r = \dim R/I$ that: $\dim \overline{G}[\underline{X}] = \dim \overline{G} + r = s(I) + r = ht(I) + r = \dim R$. Since H is also quasi-unmixed by (18.24), for any minimal prime $\mathfrak{p}$ of H we have $\dim H/\mathfrak{p} = \dim R$. From these we conclude that every minimal prime of $\operatorname{Ker}\varphi$ is a minimal prime in $\overline{G}[\underline{X}]$. Let $\mathfrak{p}$ be any minimal prime of $\overline{G}[\underline{X}]$. Then there is a minimal prime $\mathfrak{p}'$ of $\overline{G}$ such that $\mathfrak{p} = \mathfrak{p}'\overline{G}[\underline{X}]$. Since $\varphi_\circ$ has a nilpotent kernel by assumption, we obtain $\operatorname{Ker}\varphi \subset \mathfrak{p}'\overline{G}[\underline{X}] + (\underline{X})\overline{G}[\underline{X}]$. Now take a minimal prime $\mathfrak{q}$ of $\operatorname{Ker}\varphi$ contained in $\mathfrak{p}'\overline{G}[\underline{X}] + (\underline{X})G[\underline{X}]$. Then $\mathfrak{q} \cap \overline{G} \subset (\mathfrak{p}'\overline{G}[\underline{X}] + (\underline{X})\overline{G}[\underline{X}]) \cap \overline{G} = \mathfrak{p}'$, hence $\mathfrak{q} = \mathfrak{p}$. This shows that $\operatorname{Ker}\varphi \subset \mathfrak{p}$ for any minimal prime $\mathfrak{p}$ of $\overline{G}[\underline{X}]$, i.e. $\operatorname{Ker}\varphi$ is nilpotent.

(3) ⇒ (1): If $G(I,\underline{x})$ has a nilpotnet kernel, then $\dim(G(I,\underline{x})/\underline{x}G(I,\underline{x})) + r = s(I) + r = \dim(G(I+\underline{x}R,R)) = \dim R$. This implies in the quasi-unmixed ring R : $s(I) = \dim(R) - \dim(R/I) = ht(I)$. This proves (23.15).

<u>(23.16) Corollary</u>. Let X be locally a noetherian Cohen-Macaulay scheme and Y a regular closed subscheme of X . We denote by $\pi' : X' \longrightarrow X$ the blowing up of X along Y , and by $\pi : E \longrightarrow Y$ the restriction of π' to the exceptional divisor. Then, if X is equimultiple along Y at $y \in Y$, and if the fibre E_y is reduced, π is flat at y .

Proof. We put $R = \mathcal{O}_{X,y}$. Let $\mathfrak{p}$ be the ideal of Y in R and $\underline{x}$ a regular system of parameters on $R/\mathfrak{p}$. By Proposition (23.15) $g(\mathfrak{p},\underline{x})$ has a nilpotent kernel. Since $E_y = \operatorname{Proj}(G_{\mathfrak{p}}(R))$ is reduced, we know by Lemma (22.17) that $\operatorname{Ker}(g(\mathfrak{p},\underline{x}))_n = 0$ for large n . Hence R is projectively normally flat along $\mathfrak{p}$ by Theorems (22.15) and (22.16), i.e. π is flat at y , q.e.d.

Finally we will describe in Theorem (23.21) a sufficient condition for the equivalence: $ht(I) = s(I) \Longleftrightarrow R$ is normally Cohen-Macaulay along I . We keep the following notations: (Q,M_0) denotes a local ring, $I_0 \subset M_0$ is an ideal of Q , $f_1,\dots,f_m \in I_0$, and $\mathfrak{a} = f_1Q + \dots + f_mQ$. Let $\underline{x} = \{x_1,\dots,x_r\} \subset Q$ be a system of parameters modulo I_0 and $V_0 = I_0 + \underline{x}Q$. We put: $R = Q/\mathfrak{a}$, $I = I_0/\mathfrak{a}$, $V = V_0/\mathfrak{a} = I + \underline{y}R$, where y_i is the image of x_i in R , so that $\underline{y} = \{y_1,\dots,y_r\}$ is a system of parameters modulo I .

(23.17) Lemma. Let $f \in I_0$ and $s = \mathrm{ord}(I_0)(f)$, the initial degree of f w.r.t. I_0. If the initial form $\mathrm{in}(I_0)(f)$ of f is weakly regular in $G(I_0,Q)$, and if f is a non-zero-divisor in Q, then

$$H^{(0)}[\underline{x},I_0/fQ,Q/fQ](n+s) = H^{(0)}[\underline{x},I_0,R](n+s) - H^{(0)}[\underline{x},I_0,R](n)$$

for large n, in particular: $e(\underline{x},I_0/fQ,Q/fQ) = s\, e(\underline{x},I_0,Q)$.

Proof. For large n we have the following exact sequence:

$$0 \longrightarrow I_0^n/I_0^{n+1} \xrightarrow{\beta} I_0^{n+s}/I_0^{n+s+1} \longrightarrow I_0^{n+s}/I_0^{n+s+1} + fI_0^{n+s} \longrightarrow 0 ,$$

where β is induced by the multiplication with f. Since $\underline{x}$ is a multiplicity system for all modules occurring in the above exact sequence, and since the multiplicity symbol $e(\underline{x},\ldots)$ is additive, we get immediately the assertion.

(23.18) Proposition. Let Q be a Cohen-Macaulay ring. Assume that

a) $e(V_0,Q) = e(\underline{x},I_0,Q)$,

b) $\mathrm{in}(I_0)(f_1),\ldots,\mathrm{in}(I_0)(f_m)$ is a weakly regular sequence in $G(I_0,Q)$, and

c) $\mathrm{in}(V_0)(f_1),\ldots,\mathrm{in}(V_0)(f_m)$ is a weakly regular sequence in $G(V_0,Q)$.

Then the following conditions are equivalent.

(1) $e(V,R) = e(\underline{y},I,R)$.

(2) $\mathrm{ord}(I_0)(f_i) = \mathrm{ord}(V_0)(f_i)$ for $i = 1,\ldots,m$.

Proof. First we note that $f_1,\ldots,f_m$ is a regular sequence in Q by c). Therefore we have for large n :

$$(f_1,\ldots,f_i) \cap I_0^n = \sum_{j=1}^{i} f_j\, I_0^{n-s_j} \quad \text{for} \quad i = 1,\ldots,m ,$$

where $s_j = \mathrm{ord}(I_0)(f_j)$. This allows to apply Lemma (23.17) inductively to get $e(\underline{y},I,R) = s_1 \cdots s_m e(\underline{x},I_0,Q)$ and

$e(V;R) = t_1 \cdots t_m e(V_0;Q)$, where $s_i = \mathrm{ord}(I_0)(f_i) \leq t_i = \mathrm{ord}(V_0)(f_i)$. Hence (1) and (2) are equivalent.

<u>(23.19) Theorem</u>. Assume that Q is a Cohen-Macaulay ring and that

$$H^{(0)}[\underline{x}, I_0, R](n) = (\Delta^r H^{(0)}[V_0, Q])(n) \quad , \quad \text{for} \quad n \gg 0 \quad ,$$

a) $\mathrm{in}(I_0)(f_1), \ldots, \mathrm{in}(I_0)(f_m)$ is a weakly regular sequence in $G(I_0, Q)$, and

b) $\mathrm{in}(V_0)(f_1), \ldots, \mathrm{in}(V_0)(f_m)$ is a weakly regular sequence in $G(V_0, Q)$.

Then the following conditions are equivalent:

(1) $e(V;R) = e(\underline{y}, I, R)$.

(2) $\mathrm{ord}(I_0)(f_i) = \mathrm{ord}(V_0)(f_i) = s_i$ for $i = 1, \ldots, m$

(3) $H^{(0)}[\underline{y}, I, R](n) = (\Delta^r H^{(0)}[V, R])(n)$ for $n \gg 0$.

Proof. The equivalence of (1) and (2) follows from (a) and (b) by Proposition (23.18). The implication (3) $\Rightarrow$ (1) is trivial. To prove (2) $\Rightarrow$ (3) , we note that $f_1, \ldots, f_m$ is a regular sequence. Moreover we obtain the standard-base-property for large n :

$$(f_1, \ldots, f_i) \cap I_0^n = \sum_{j=1}^{i} f_j I_0^{n-s_j} \quad , \quad n \gg 0 \quad ;$$

$$(f_1, \ldots, f_i) \cap V_0^n = \sum_{j=1}^{i} f_j V_0^{n-s_j} \quad , \quad n \gg 0 \quad .$$

Therefore we can make induction on m , and (3) follows in the same way as in Lemma (23.17).

<u>(23.20) Remark</u>. If we replace in a) and b) of Theorem (23.19) the property "weakly regular" by "regular", then the same conclusions as before can be made for all n . We formulate this in the following theorem:

(23.21) Theorem. Let Q be a Cohen-Macaulay ring such that $H^{(r)}[\underline{x},I_0,Q] = H^{(0)}[V_0,Q]$. Assume that

a) $in(I_0)(f_1),\dots,in(I_0)(f_m)$ is a regular sequence in $G(I_0,Q)$,

b) $in(V_0)(f_1),\dots,in(V_0)(f_m)$ is a regular sequence in $G(V_0,Q)$.

Then the following conditions are equivalent:

(1) $ht(I) = s(I)$; i.e. $e(V;R) = e(\underline{y},I,R)$.

(2) $ord(I_0)(f_i) = ord(V_0)(f_i)$ for $i = 1,\dots,m$.

(3) $H^{(r)}[\underline{y},I,R] = H^{(0)}[V,R]$.

(4) R is normally Cohen-Macaulay along I .

Proof. The equivalence (3)$\Longleftrightarrow$(4) was already proved in Theorem (22.23). The other equivalences follow from Remark (23.20).

(23.22) Corollary. Let Q be a Cohen-Macaulay ring and I_0 a permissible ideal in Q . If $in(I_0)(f_1),\dots,in(I_0)(f_m)$ form a regular sequence in $G(I_0,Q)$, i.e. the ideal $\mathfrak{a} = (f_1,\dots,f_m)Q$ is a so-called strict complete intersection with respect to I_0 , then the following conditions are equivalent:

(i) $e(R) = e(R_{I_0})$.

(ii) R is normally flat along I_0 .

(23.23) Remark. In Chapter IX we will show that for a proper ideal I in a quasi-unmixed local ring R with Cohen-Macaulay-Rees ring $B(I,R)$ the following implication is true: $ht(I) = s(I) \Rightarrow R$ is normally Cohen-Macaulay along I .

§ 24. Open conditions and transitivity properties

Let R be a ring, M an A-module and let **P** denote a property of modules over local rings. It is very important to know if the subset

$$\{\mathfrak{p} \in \mathrm{Spec}(R) \mid M_{\mathfrak{p}} \text{ satisfies } \mathbf{P}\} \subset \mathrm{Spec}(R)$$

is a (Zariski-) open set or not. A typical example is the following one. Let k be a field of $ch(k) = 0$ and let R be a reduced affine k-algebra. If $M := \Omega_{R/k}$ denotes the differential module of R over k, then

$$\{\mathfrak{p} \in \mathrm{Spec}(R) \mid R_{\mathfrak{p}} \text{ is regular}\} = \{\mathfrak{p} \in \mathrm{Spec}(R) \mid M_{\mathfrak{p}} \text{ is free over } R_{\mathfrak{p}}\}$$

is a non-empty open subset of $\mathrm{Spec}(R)$; see [4] , Theorem (7 . 2). We are going to refine the formulation. Let R be a noetherian ring and let $A = \bigoplus_{n\geq 0} A_n$ be a noetherian graded R-algebra generated by the elements of positive degree over $A_0 = R$. Let $M = \bigoplus_{n\in\mathbf{Z}} M_n$ be a finitely generated graded A-module. We put:

$$CM(M) = \{\mathfrak{p} \in \mathrm{Spec}(R) \mid \mathrm{depth}(M_n)_{\mathfrak{p}} = \dim R_{\mathfrak{p}} \text{ for all } M_n \neq 0\}$$

$$F(M) = \{\mathfrak{p} \in \mathrm{Spec}(R) \mid M_{\mathfrak{p}} \text{ is flat over } R_{\mathfrak{p}}\} .$$

We first prove the Krull-Seidenberg-Grothendieck Theorem, saying that $F(M)$ is an open subset of $\mathrm{Spec}(R)$. Applying this result to R/I and $A = M = G(I,R)$, where I is a proper ideal of R , we conclude that "normal flatness" is an open condition. Then we investigate the openness of $CM(G(I,R))$ and of the condition $ht(I) = s(I)$. It has to be noted that $CM(G(I,R))$ is not necessarily open in $\mathrm{Spec}(R/I)$: Hochster [12] constructed a noetherian ring S such that the Cohen-Macaulay locus $\{\mathfrak{p} \in \mathrm{Spec}(S) \mid S_{\mathfrak{p}}$ is Cohen-Macaulay$\}$ is not open. Now let $R = S[X]$ be a polynomial ring in one variable over this ring S and let $I = XR$. Since $I^n/I^{n+1} \simeq R/XR = S$, we see that $CM(G(I,R))$ is not open in $\mathrm{Spec}(R/I) = \mathrm{Spec}(S)$. Thus we need some restriction on R to get the openness of $CM(G(I,R))$. As an application of the Krull-Seidenberg-Grothendieck Theorem we will prove the openness of $CM(G(I,R))$ under the assumption that R is a homomorphic image of a regular ring.

For the proof of the Krull-Seidenberg-Grothendieck Theorem we need the following special case of the local criterion of flatness, for which we refer to [1], III, 5, no.2.

(24.1) Lemma. Let R be a noetherian ring, A a graded R-algebra of finite type and M a graded A-module of finite type. Assume that A is generated by homogeneous elements of positive degree as an R-algebra. Let I be an ideal of R contained in the Jacobson radical of R. Then the following conditions are equivalent:

(i) M is a flat R-module.

(ii) $M \otimes_R R/I$ is a flat R/I-module and $\mathrm{Tor}_1^R(M,R/I) = 0$.

(24.2) Lemma. Let R be a noehterian domain, A any R-algebra of finite type, and M any finitely generated A-module. Then there exists a non-zero element f of R such that M_f is a free R_f-module.

Proof. Consider the following filtration of M by A-submodules:

$$M = M_0 \supset M_1 \supset \ldots \supset M_n \supset M_{n+1} = (0) \quad ,$$

such that $M_i/M_{i+1} \simeq A/P_i$ for some $P_i \in \mathrm{Spec}(A)$. So we may assume that A is a domain and $M = A$. Moreover we may assume that R is a subring of A . Let K be the quotient field of R . Then $A \otimes K = AK$ is finitely generated as an algebra over K . We use induction on $n = \mathrm{tr.deg}_K AK < \infty$, the case $n = 0$ being trivial. By the noetherian normalization theorem ([5], (14.G)), the ring AK is integral over $K[y_1,\ldots,y_n]$, where $\underline{y} = \{y_1,\ldots,y_n\}$ are algebraically independent elements in A . Then, since A is finite over R , we find an element $0 \neq r \in R$ such that A_r is finite over the polynomial ring $T = R_r[\underline{y}]$. Assume that the maximal number of linearly independent elements over T in A_r is m . Consider the finitely generated torsion T-module $A' = A_r/T^m$. Since $\dim A'K < n$ we find by the induction hypothesis an element $0 \neq s \in R$ such that A'_s is R_f-free. Hence $f = rs$ is a suitable element in the sense of Lemma (24.2).

(24.3) Lemma. Let R be a noetherian ring, $A = \bigoplus_{n \geq 0} A_n$ a noetherian graded R-algebra, $A_0 = R$ and M a graded A-mdoule of finite type. Let $\mathfrak{p} \in \mathrm{Spec}(R)$ be a prime ideal of R such that $M_{\mathfrak{p}}$ is flat over R . Then there exists an element $f \in R \setminus \mathfrak{p}$ such that

(i) $(M/\mathfrak{p}M)_f$ is a flat $(R/\mathfrak{p})_f$-module,

(ii) $\mathrm{Tor}_1^R(M,R/\mathfrak{p})_f = 0$.

Proof. Since $M_{\mathfrak{p}}$ is $R_{\mathfrak{p}}$-flat by assumption we get by Lemma (24.1):

(I) $(M/\mathfrak{p}M)_{\mathfrak{p}}$ is $(R/P)_{\mathfrak{p}}$-flat (which is trivial),

(II) $\mathrm{Tor}_1^R(M,R/P)_{\mathfrak{p}} = 0$.

By Lemma (24.2) we even get (i). Moreover, since M is finitely generated over A , $\mathrm{Tor}_1^R(M,R/\mathfrak{p})$ is also finitely generated over A . Therefor (II) implies (ii).

<u>(24.4) Theorem (Krull-Seidenberg-Grothendieck)</u>. Let R be a noetherian ring and $A = \bigoplus_{n\geq 0} A_n$ a graded R-algebra generated by homogeneous elements of positive degree. Let M be a graded A-module of finite type. Then $F(M) = \{\mathfrak{p} \in \mathrm{Spec}(R) \mid M_{\mathfrak{p}}$ is $R_{\mathfrak{p}}$-flat$\}$ is an open subset of $\mathrm{Spec}(R)$.

Proof. We may assume that $F(M)$ is non-emty. Recall that $F(M)$ is open if and only if the following holds [5], (22.B):

(i) if $\mathfrak{p} \in F(M)$ and $\mathfrak{q}$ is a prime ideal contained in $\mathfrak{p}$, then $\mathfrak{q} \in F(M)$, and

(ii) if $\mathfrak{p} \in F(M)$, then there exists a non-empty open subset of $\mathrm{Spec}(R/\mathfrak{p})$ whose image in $\mathrm{Spec}(R)$ is contained in $F(M)$.

First of all, (i) is clear in our case. And (ii) follows from Lemma (24.1) and Lemma (24.3), q.e.d.

<u>(24.5) Corollary</u>. Let R be a noetherian ring and I a proper ideal of R . Then $F(G(I,R)) = \{\mathfrak{p} \in \mathrm{Spec}(R/I) \mid R_{\mathfrak{p}}$ is normally flat along $IR_{\mathfrak{p}}\}$ is an open set of $\mathrm{Spec}(R/I)$.

<u>(24.6) Corollary</u>. Let R and I be the same as in Corollary (24.5). Then $\{\mathfrak{p} \in \mathrm{Spec}(R/I) \mid R_{\mathfrak{p}}$ is projectively normally flat along $IR_{\mathfrak{p}}\}$ is an open subset of $\mathrm{Spec}(R/I)$.

Proof. Let $\mathfrak{p} \in \mathrm{Spec}(R/I)$ and $n_0 \in \mathbf{N}$ such that $I^nR_{\mathfrak{p}}/I^{n+1}R_{\mathfrak{p}}$ is flat over $(R/I)_{\mathfrak{p}}$ for $n \geq n_0$. Then apply Theorem (24.4) to the graded $G(I,R)$-module $M = \bigoplus_{n\geq n_0} I^n/I^{n+1}$.

<u>(24.7) Theorem</u>. Let R be a noetherian ring which is a homomorphic image of a regular ring S , and let I be a proper ideal of R . Then $CM(G(I,R))$ is an open subset of $\mathrm{Spec}(R/I)$.

Proof. Let $R = S/\mathfrak{a}$ for some ideal $\mathfrak{a}$ of S. Let J be the inverse image of I in S. Then $G(I,R)$ is a homomorphic image of a homogeneous graded polynomial ring $T = S[X_1,\dots,X_r]$ with $\deg(X_i) = 1$. Consider a free resolution of $G(I,R)$ as a graded T-module.

$$\dots \longrightarrow F_2 \xrightarrow{d_2} F_1 \xrightarrow{d_1} T \longrightarrow G(I,R) \longrightarrow 0 \quad .$$

We put $Z_i := \mathrm{Im}(d_i)$, and U denotes the set of those prime ideals $\mathfrak{p}$ in S such that either $J \not\subset \mathfrak{p}$ or $\mathrm{depth}(I^n/I^{n+1})_{\mathfrak{p}} = \dim(R/I)_{\mathfrak{p}}$ for $n \geq 0$ (as an S-module).

We show that U is an open set in $\mathrm{Spec}(S)$, which will imply that $CM(G(I,R))$ is an open set in $\mathrm{Spec}(R/I)$. For $\mathfrak{p} \in U$ we put $h = \mathrm{ht}(J_{\mathfrak{p}})$. Note that $(Z_h)_{\mathfrak{p}}$ is a free $S_{\mathfrak{p}}$-module, since the projective dimension of $(I^n/I^{n+1})_{\mathfrak{p}}$ is $\mathrm{pd}_{S_{\mathfrak{p}}}(I^n/I^{n+1})_{\mathfrak{p}} = h$ for all $n \geq 0$, and the homogeneous components of F_i are free S-modules for all i. Therefore using Lemma (24.3) we see that there is an element $a \in S \setminus \mathfrak{p}$ such that $(Z_h)_a$ is a flat S_a-module. Since $(S/J)_{\mathfrak{p}} = (R/I)_{\mathfrak{p}}$ is Cohen-Macaulay, every assocated prime $\mathfrak{p}'$ of J contained in $\mathfrak{p}$ has height h, i.e. we find an element

$$b \in \bigcap_{\substack{\mathfrak{p}' \in \mathrm{Min}(J) \\ \mathrm{ht}(\mathfrak{p}') \neq h}} \mathfrak{p}' \setminus \mathfrak{p} \quad .$$

We put $f = ab$. Then for any prime ideal $\mathfrak{q} \in D(f) \cap V(J)$, $(Z_h)_{\mathfrak{q}}$ is a free $S_{\mathfrak{q}}$-module and hence we get

$$\begin{aligned} \mathrm{depth}(I^n/I^{n+1})_{\mathfrak{q}} &= \mathrm{depth}\, S_{\mathfrak{q}} - \mathrm{pd}_{S_{\mathfrak{q}}}(I^n/I^{n+1})_{\mathfrak{q}} \\ &\geq \mathrm{depth}\, S_{\mathfrak{q}} - h = \dim S_{\mathfrak{q}} - \mathrm{ht}(J_{\mathfrak{q}}) \\ &= \dim\,(S/J)_{\mathfrak{q}} = \dim\,(R/I)_{\mathfrak{q}} \quad , \end{aligned}$$

i.e. $(I^n/I^{n+1})_{\mathfrak{q}}$ is Cohen-Macaulay with $\dim(R/I)_{\mathfrak{q}} = \mathrm{depth}(I^n/I^{n+1})_{\mathfrak{q}}$. Thus $CM(G(I,R))$ is an open set of $\mathrm{Spec}(R/I)$. q.e.d.

Using the same proof idea we get:

(24.8) Corollary. Let R and I be the same as in Theorem (24.7). Then the set $\{\mathfrak{p} \in \mathrm{Spec}(R/I) \mid R_{\mathfrak{p}}$ is projectively Cohen-Macaulay along $IR_{\mathfrak{p}}\}$ is open in $\mathrm{Spec}(R/I)$.

The openness of the equimultiplicity $\mathrm{ht}(I) = s(I)$ for some proper ideal I can be described and proved as follows. (For a geometric motivation see App. III, 1.4.9 and 1.4.10).

(24.9) Theorem. Let R be a noetherian ring and I a proper ideal of R. Let $U = \{\mathfrak{p} \in V(I) \mid \mathrm{ht}(IR_{\mathfrak{p}}) = s(IR_{\mathfrak{p}})\}$. Then U is a non-empty open subset in $V(I)$.

Proof. If R is local, then there are $s := s(I)$ elements $x_1,\ldots,x_s \in I$ such that

$$I^n = a_1 I^{n-n_1} + \ldots + a_s I^{n-n_s} \quad \text{for some } n, n_1, \ldots, n_s ,$$

see Chapter II, (10.11.1).

So, assuming $\mathrm{ht}(IR_{\mathfrak{p}}) = s(IR_{\mathfrak{p}}) = s$, we have

$$I^n R_{\mathfrak{p}} = a_1 I^{n-n_1} R_{\mathfrak{p}} + \ldots + a_r I^{n-n_s} R_{\mathfrak{p}}$$

for some n and suitable elements $a_i \in I$. Hence we can find some some $f \in R \setminus \mathfrak{p}$ such that

$$fI^n \subset a_1 I^{n-n_1} + \ldots + a_s I^{n-n_s} \quad \text{and} \quad \mathrm{ht}(IR_f) \geqq s .$$

Therefore for any $\mathfrak{q} \in D(f) \cap V(I)$ we have

$$I^n R_{\mathfrak{q}} = a_1 I^{n-n_1} R_{\mathfrak{q}} + \ldots + a_s I^{n-n_s} R_{\mathfrak{q}} ,$$

i.e. $s(IR_{\mathfrak{q}}) \leqq s$. Thus we get $s \geqq s(IR_{\mathfrak{q}}) \geqq \mathrm{ht}(IR_{\mathfrak{q}}) \geqq s$, which shows that $D(F) \cap V(I) \subset U$, q.e.d.

In the second part of this chapter we discuss the socalled transitivity of the condition "R is normally Cohen-Macaulay along I".

<u>(24.10) Lemma</u>. Let I be an ideal of a local ring R and $\underline{x} = \{x_1,\ldots,x_r\}$ a system of parameters modulo I. Put $J = (I,x_1,\ldots,x_s)$, $s \leq r$, and $\underline{x}' = \{x_{s+1},\ldots,x_r\}$. Then

(a) $H^{(s)}[\underline{x},I,R] \leq H^{(0)}[\underline{x}',J,R]$,

and the following conditions are equivalent:

(b) $H^{(s)}[\underline{x},I,R] = H^{(0)}[\underline{x}',J,R]$;

(c) $\operatorname{depth}(I^n/I^{n+1})_{\mathfrak{p}} = \dim(R/I)_{\mathfrak{p}}$ for all $\mathfrak{p} \in \operatorname{Assh}(R/J)$.

Proof. We put $\underline{x}'' = \{x_1,\ldots,x_s\}$, so that $\operatorname{Assh}(R/I^n+\underline{x}''R) = \operatorname{Assh}(R/J)$ for all $n > 0$. Then by the associativity-formula of the multiplicity-symbol we know that

$$(1) \qquad H^{(1)}[\underline{x},I,R](n) = e(\underline{x};R/I^{n+1})$$

$$= \sum_{\mathfrak{p}\in\operatorname{Assh}(R/J)} e(\underline{x}';R/\mathfrak{p})\ e(\underline{x}''R_{\mathfrak{p}};R_{\mathfrak{p}}/I^{n+1}R_{\mathfrak{p}}) .$$

This implies for all $i \geq 1$

$$(2) \qquad H^{(i)}[\underline{x},I,R] = \sum_{\mathfrak{p}\in\operatorname{Assh}(R/J)} e(\underline{x}';R/\mathfrak{p})\ H^{(i)}[\underline{x}''R_{\mathfrak{p}},IR_{\mathfrak{p}},R_{\mathfrak{p}}] .$$

Moreover by Chapter I, (3.8) we have

$$(3) \qquad H^{(s)}[\underline{x}''R_{\mathfrak{p}},IR_{\mathfrak{p}},R_{\mathfrak{p}}] \leq H^{(0)}[JR_{\mathfrak{p}},R_{\mathfrak{p}}] \quad \text{for} \quad \mathfrak{p} \in \operatorname{Assh}(R/J) .$$

Form (2) and (3) we get (a); namely

$$(4) \qquad H^{(s)}[\underline{x},I,R] \leq \sum_{\mathfrak{p}\in\operatorname{Assh}(R/J)} e(\underline{x}';R/\mathfrak{p})\ H^{(0)}[JR_{\mathfrak{p}},R_{\mathfrak{p}}]$$

$$= H^{(0)}[\underline{x}',J,R] .$$

To show the equivalence (b)$\Longleftrightarrow$(c), we consider the difference

(5) $$H^{(0)}[\underline{x}',J,R](n) - H^{(s)}[\underline{x},I,R](n) = \sum_{\mathfrak{p}\in \mathrm{Assh}(R/I)} e(\underline{x}';R/\mathfrak{p})\cdot\left\{H^{(0)}[J_{\mathfrak{p}},R_{\mathfrak{p}}](n) - H^{(s)}[\underline{x}''R_{\mathfrak{p}},IR_{\mathfrak{p}},R_{\mathfrak{p}}](n)\right\} .$$

By (3) the value in the bracket is not negative. Hence (b) holds if and only if $H^{(0)}[JR_{\mathfrak{p}},R_{\mathfrak{p}}] = H^{(s)}[\underline{x}''R_{\mathfrak{p}},IR_{\mathfrak{p}},R_{\mathfrak{p}}]$ for all $\mathfrak{p}\in \mathrm{Assh}(R/J)$. But this equality is equivalent to (c) by Theorem (22.23).

<u>(24.11) Theorem (Transitivity-property)</u>. Let $(R,\mathfrak{m})$ be a local ring and $I \subset J$ be ideals such that $\dim(R/I) = r$, $\dim(R/J) = s$. Assume that $\lambda_R(J/I + \mathfrak{m}J) = r-s$. Then the following conditions are equivalent:

(a) R is normally Cohen-Macaulay along I .

(b) R is normally Cohen-Macaulay along J and $R_{\mathfrak{p}}$ is normally Cohen-Macaulay along $IR_{\mathfrak{p}}$ for all $\mathfrak{p}\in \mathrm{Assh}(R/J)$.

Proof. By assumption we find elements $x_1,\ldots,x_{r-s}\in J$ such that $J = (I,x_1,\ldots,x_{r-s})$. Since $\dim(R/I) - \dim(R/J) = r-s$, those elements $x_1,\ldots,x_{r-s}$ form a part of a system of parameters modulo I . Now we choose a full system of parameters $\underline{x} = \{x_1,\ldots,x_{r-s},x_{r-s+1},\ldots,x_r\}$ modulo I , and we put $\underline{x}' = \{x_{r-s+1},\ldots,x_r\}$. By Lemma (24.10) we have $H^{(r)}[\underline{x},I,R] \leq H^{(s)}[\underline{x}',J,R] \leq H^{(0)}[V,R]$, where $V = (I,x_1,\ldots,x_r) = J + \underline{x}'R$. By Theorem (22.23), the condition (a) of the theorem is equivalent to $H^{(r)}[\underline{x},I,R] = H^{(0)}[V,R]$, and this holds if and only if

(*) $$H^{(r)}[\underline{x},I,R] = H^{(s)}[\underline{x}',J,R] = H^{(0)}[V,R] .$$

Now the first equality in (*) is equivalent to "$R_{\mathfrak{p}}$ is normally Cohen-Macaulay along $IR_{\mathfrak{p}}$ " by Lemma (24.10) and the second one is equivalent to "R is normally Cohen-Macaulay along J " by Theorem (22.23),

q.e.d.

As a corollary of Theroem (24.11) we obtain the following transitivity of normal flatness, which is due to Hironaka [11], Chapter II, Theorem 3.

<u>(24.12) Theorem</u>. Let R be a local ring and $\mathfrak{p},\mathfrak{q}$ prime ideals such that $\mathfrak{q}$ contains $\mathfrak{p}$ and $R/\mathfrak{p}$, $R/\mathfrak{q}$ are both regular. Then the following two conditions are equivalent:

(a) R is normally flat along $\mathfrak{p}$.

(b) R is normally flat along $\mathfrak{q}$ and $R_{\mathfrak{q}}$ is normally flat along $\mathfrak{p}R_{\mathfrak{q}}$.

In the next proposition we describe a corresponding transitivity property for the equimultiplicity condition $ht(I) = s(I)$.

<u>(24.13) Proposition</u>. Let $(R,\mathfrak{m})$ be a quasi-unmixed local ring and $I \subset J$ proper ideals in R such that $\dim(R/I) = r$ and $\dim(R/J) = s$. Assume that $\lambda_R(J/I + \mathfrak{m}J) = r-s$. Then the following conditions are equivalent:

(a) $ht(I) = s(I)$ and

$(b)_1$ $ht(J) = s(J)$

$(b)_2$ $ht(IR_{\mathfrak{p}}) = s(IR_{\mathfrak{p}})$ for all $\mathfrak{p} \in \mathrm{Assh}(R/J)$

$(b)_3$ $\dim(R_{\mathfrak{p}}/I_{\mathfrak{p}}) = r - s$ for all $\mathfrak{p} \in \mathrm{Assh}(R/J)$

We leave the proof to the reader as an easy application of (generalized) multiplicity theory. In the next chapters we only need the transitivity property of Theorem (24.11).

<u>(24.14) Remark</u>. In Chapter V we will show that this transitivity (24.11) is of some use to investigate Cohen-Macaulay properties under blowing up. To be more precise: we will compare the Rees ring of I and $J = I + \underline{x}R$, where $\underline{x}$ is a part of a system of parameters modulo I . For this situation we can prove a transitivity property for the Cohen-Macaulayness of both the Rees rings $B(I,R)$, $B(J,R)$ and the graded rings $G(I,R)$, $G(J,R)$, assuming that R itself is Cohen-Macaulay.

Of course the most important application of the transitivity in Theorem (24.12) was made by Hironaka in his resolution process in the characteristic zero case; see [11], also [3].

References - Chapter IV

Books

[1] N. Bourbaki, Algèbre Commutative, chapitres 1 à 4, Hermann, Paris, 1961 - 1965.

[2] A. Grothendieck and J. Dieudonné, Éléments de Géométrie Algébrique, Publ. Math. IHES Paris, No. 24 (1965).

[3] M. Herrmann, R. Schmidt and W. Vogel, Theorie der normalen Flachheit, Teubner-Texte zur Mathematik, Leipzig, 1977.

[4] E. Kunz, Kähler Differentials (Advanced Lectures in Mathematics), Vieweg & Sohn, Braunschweig, 1986.

[5] H. Matsumura, Commutative Algebra (2nd. Edition), W.A. Benjamin, 1980.

[6] J.P. Serre, Algèbre Locale-Multiplicités, Lecture notes in Mathematics, 11, Springer-Verlag, 1965.

[6*] O. Zariski and P. Samuel, Commutative Algebra II, Van Nostrand, Princeton, 1965.

Papers

[7] E. Böger, Einige Bemerkungen zur Theorie der ganz algebraischen Abhängigkeit von Idealen, Math. Ann., 185 (1970), 303 - 308.

[8] M. Brodmann, Asymptotic nature of analytic spreads, Math. Proc. Camb. Phil. Soc., 86 (1979), 35 - 39.

[9] L. Burch, Codimension and analytic spread, Math. Proc. Camb. Phil. Soc., 72 (1972), 369 - 373.

[10] M. Herrmann und U. Orbanz, Faserdimensionen von Aufblasungen lokaler Ringe und Äquimultiplizität, J. of Math. of Kyoto Univ., 20 (1980), 651 - 659.

[11] H. Hironaka, Resolution of singularities of an algebraic variety over a field of characteristic zero, I, II, Annals of Math., 79 (1964), 109 - 326.

[12] M. Hochster, Non-openness of loci in noetherian rings, Duke Math. J., 40 (1973), 215 - 219.

[13] J. Lipman, Equimultiplicity, reduction and blowing up, Commutative Algebra (Analytical Methods), Lecture Note in Pure and Applied Mathematics, 68, Marcel Dekker, New York and Basel, 1982, 111 - 147.

[14] U. Orbanz und L. Robbiano, Projective normal flatness and Hilbert functions, Trans. Amer. Math. Soc., 283 (1984), 33 - 47.

[15] L. Robbiano, On normal flatness and some related topics, Commutative Algebra, Proc. of the Trento Conference, Lecture Notes in Pure and Applied Mathematics, 84, Marcel Dekker, New York and Basel, 1983, 235 - 251.

Chapter V. EQUIMULTIPLICITY AND COHEN-MACAULAY PROPERTY OF BLOWING UP RINGS

The problem of describing the behaviour of a given variety X under blowing up a closed subvariety $Y \subset X$ should be phrased as follows: How does the blowing up morphism $X' \longrightarrow X$ depend on properties of Y ? Classically Y was chosen to be non-singular and equimultiple. For the non-hypersurface case equimultiplicity was refined to the notion of normal flatness by Hironaka, still assuming Y non-singular. But there are reasons to admit singular centers Y too. For example, in his theory of quasi-ordinary singularities, Zariski used generic projections of a surface to a plane, and blowing up a point in this plane induces the blowing up of a singular center in the original surface. In Chapter IV we gave three different algebraic generalizations of the classical equimultiplicity together with a numerical description of each condition. In Chapter VI we will indicate that the new conditions are useful in the study of the numerical behaviour of singularities under blowing up singular centers. In this Chapter V we want to show that these conditions are also of some use to investigate Cohen-Macaulay properties under blowing up, which are essential for the local and global study of algebraic varieties. Finally in Chapter IX we shall describe a general criterion of the Cohen-Macaulayness of Rees algebras in terms of local cohomology.

The starting point in this chapter is the following question: Given any local Cohen-Macaulay ring R and an equimultiple ideal I of R , when does it hold that

a) $B\ell(I,R)$ is Cohen-Macaulay

b) $G(I,R)$ is Cohen-Macaulay

c) $\mathcal{R}(I,R)$ is Cohen-Macaulay

d) $B(I,R)$ is Cohen-Macaulay ?

We will also treat the question how far it is necessary to assume R to be Cohen-Macaulay. Moreover we will describe a kind of transitivity of the Cohen-Macualyness of the Rees rings $B(I,R)$ and $B(J,R)$, where $J = I + \underline{x}R$ for a part $\underline{x}$ of a system of parameters mod I . Our main tools here are coming from multiplicity-theory.

§ 25. Graded Cohen-Macaulay rings

First we prove an auxiliary result which relates the Cohen-Macaulay properties of $G(I,R)$ and $G(I+\underline{x}R,R)$ via normal Cohen-Macaulayness of R along I, where $\underline{x}$ is a system of parameters $\bmod I$.

(25.1) Proposition. Let $(R,\mathfrak{m})$ be a local Cohen-Macaulay ring and I an equimultiple ideal of R. Let $\{x_1,\dots,x_r\}$ be a system of parameters $\bmod I$, $\underline{x}=(x_1,\dots,x_r)$ and $I(\underline{x}) = I+\underline{x}R$. Then the following conditions are equivalent:

(i) $G(I,R)$ is Cohen-Macaulay

(ii) $G(I(\underline{x}),R)$ is Cohen-Macaulay and R is normally Cohen-Macaulay along I.

(iii) $G(I(\underline{x})/\underline{x}R, R/\underline{x}R)$ is Cohen-Macaulay and R is normally Cohen-Macaulay along I.

Moreover, if $\underline{a}=(a_1,\dots,a_s)$, $s=\mathrm{ht}(I)$, generate a minimal reduction of I, then these conditions are equivalent to

(iv) $(\underline{x}R+\underline{a}R)\cap I^i = \underline{x}I^i + \underline{a}I^{i-1}$ for all $i \geq 1$.

Proof. (i) $\Rightarrow$ (ii). By Chapter II, Proposition (10.24) we know that $\mathrm{in}(I)(x_1),\dots,\mathrm{in}(I)(x_r)$ are part of a homogeneous system of parameters. Therefore the assumption implies that $x_1,\dots,x_r$ is a regular sequence on I^n/I^{n+1} for all n, i.e. R is normally Cohen-Macaulay along I. Since $x_1,\dots,x_r$ is also a regular sequence in R, we conclude from R normally Cohen-Macaulay along I that the canonical homomorphism

$$h : G(I,R)\otimes_R R/I(\underline{x})[X_1,\dots,X_r] \longrightarrow G(I(\underline{x}),R)$$

is an isomorphism (see Chapter IV, Theorem (21.9)). This proves (i) $\Rightarrow$ (ii), and also the equivalence (ii) $\Leftrightarrow$ (iii).

(iii) $\Rightarrow$ (i). $G(I,R)\otimes_R R/I(\underline{x})$ has a homogeneous system of parameters $\tilde{a}_1,\dots,\tilde{a}_s$, which is necessarily a regular sequence. If $\alpha_1,\dots,\alpha_s \in G(I,R)$ are homogeneous inverse images of $\tilde{a}_1,\dots,\tilde{a}_s$, then $\mathrm{in}(I)(x_1),\dots,\mathrm{in}(I)(x_r), \alpha_1,\dots,\alpha_s$ is a regular sequence in

$G(I,R)$ of length $r+s = \dim(G(I,R))$.

(i) $\Leftrightarrow$ (iv) . By Chapter II we know that the initial forms of $x_1,\dots,x_r$, $a_1,\dots,a_s$ in $G(I,R)$ are a homogeneous system of parameters, and therefore $G(I,R)$ is Cohen-Macaulay if and only if these initial forms are a regular sequence in $G(I,R)$. The assertion now follows from our standard arguments in Chapter II, § 10 in view of

$$\operatorname{ord}(I)(x_1) = 0 \ , \ 1 \leqq i \leqq r \quad , \quad \operatorname{ord}(I)(a_j) = 1 \ , \ 1 \leqq j \leqq s \quad .$$

(25.2) Remarks. a) From section 23 we know that R is normally Cohen-Macaulay along I if and only if R/I^{n+1} is Cohen-Macaulay for all $n \geqq 0$. Therefore, if R is normally Cohen-Macaulay along I , it is so along I^t for all $t \geqq 1$.

b) Assume R is normally Cohen-Macaulay along I . Then the isomorphism h above can be used to show that R is normally flat along I if and only if R is normally flat along $I(\underline{x})$, s. section 24.

(25.3) Proposition. Let $(R,\mathfrak{m})$ be a local Cohen-Macaulay ring and let I be an equimultiple ideal of R . If $G(I,R)$ is Cohen-Macaulay, then $G(I^t,R)$ is Cohen-Macaulay for all $t \geqq 1$.

Proof. Let $t \geqq 1$ be fixed. Since also I^t is equimultiple, by Remark (25.2) and Proposition (25.1), (iii) we are reduced to the case that I is $\mathfrak{m}$-primary. So let $a_1,\dots,a_d \in I$, $d = \dim R$, be elements, whose initial forms in $G(I,R)$ are a regular sequence. Let $s_i = \operatorname{ord}(I)(a_i)$, $1 \leqq i \leqq d$. Then $\operatorname{ord}(I)(a_i^t) = t\cdot s_i$, and $\operatorname{in}(I)(a_1^t),\dots,\operatorname{in}(I)(a_d^t)$ is a regular sequence in $G(I,R)$, so we have

$$\left(a_1^tR + \dots + a_d^tR\right) \cap I^n = \sum_{i=1}^{d} a_i^t I^{n-ts_i} \quad , \quad n \geqq 0$$

and in particular

$$\left(a_1^tR + \dots + a_d^tR\right) \cap I^{nt} = \sum_{i=1}^{d} a_i^t I^{t(n-s_i)} , \quad n \geqq 0 \quad .$$

Since $\operatorname{ord}(I^t)(a_i^t) = s_i$, we see that $\operatorname{in}(I^t)(a_1^t),\ldots,\operatorname{in}(I^t)(a_d^t)$ is a regular sequence in $G(I^t,R)$. A more general statement for any ideal I is given in Theorem (27.8). Now we are coming to the main theorem of this chapter.

<u>(25.4) Theorem.</u> Let $(R,\mathfrak{m})$ be a local ring and let I be an equimultiple ideal of R. Let $\underline{a} = (a_1,\ldots,a_s)$ generate a minimal reduction of I, where $s = \operatorname{ht}(I) = s(I) > 0$. Then the following conditions are equivalent:

(i) $B := B(I,R)$ is Cohen-Macaulay and R is Cohen-Macaulay

(ii) $G := G(I,R)$ is Cohen-Macaulay and $I^s \subseteq \underline{a}R$.

Proof. We may assume R to be Cohen-Macaulay from the beginning, since by Chapter II this is a consequence of $G(I,R)$ being Cohen-Macaulay. Let $\underline{b} = (b_1,\ldots,b_r)$, where $\{b_1,\ldots,b_r\}$ is a system of parameters mod I, and let

$$J = (a_1, a_1t-a_2,\ldots,a_st,b_1,\ldots,b_r)B$$

$$J' = (a_1,a_1t-a_2,\ldots,a_st)B, \quad B \cong R[It] \quad .$$

From (10.30) in Chapter II we know that J is a parameterideal of B. Therefore the Cohen-Macaulayness of B is characterized by the equality $e(J\,B) = \lambda_B(B/J)$. [With this we mean of course $e(JB_{\mathfrak{M}}) = \lambda_{B_{\mathfrak{M}}}(B_{\mathfrak{M}}/J_{\mathfrak{M}})$, where $\mathfrak{M}$ is the maximal homogeneous ideal of B.]

<u>Claim 1:</u> $e(J,B) = s\cdot e((\underline{a},\underline{b}),R)$.

Proof of claim 1. Using the fact that J' is a reduction of (I,It) we get:

$$e(J,B) = e(\underline{b},J',B) = e(\underline{b},(I,It),B),$$

moreover

$$e((\underline{a},\underline{b}),R) = e(\underline{b},\underline{a},R) = e(\underline{b},I,R) \quad .$$

Recall that $\frac{1}{s!}\cdot e(\underline{b},(I,It),B)$ is the highest coefficient of the Hilbert polynomial $(m \gg 0)$

$$H^{(0)}[\underline{b},(I,It),B](m) = e(\underline{b},(I,It)^m B/(I,It)^{m+1}B) =$$

$$e(\underline{b},\bigoplus_{i=0}^{m} I^m t^i/I^{m+1}t^i) = (m+1)\cdot e(\underline{b},I^m/I^{m+1}) = (m+1)\cdot H^{(0)}[\underline{b},I,R](m) \ .$$

[Note that the underlying ring for the computation of the generalized multiplicities is the ring $(B/(I,It)B)_{(m,It)B} = B/(I,It)B = R/I$] .

Since the highest coefficient of $H^{(0)}[\underline{b},I,R](m)$ is $\frac{1}{(s-1)!}e(\underline{b},I,R)$, if follows that

$$e(\underline{b},(I,It),B) = s\cdot e(\underline{b},I,R) \quad ,$$

which proves claim 1 .

<u>Claim 2</u>: $\lambda_B(B/J) = s\cdot\lambda_R(R/(\underline{a},\underline{b})) + \sum_{n\geq s}\lambda(I^n/\underline{a}I^{n-1}+\underline{b}I^n)$

$$+\sum_{n=1}^{s-1}\lambda(I^n\cap(\underline{a},\underline{b})/\underline{a}I^{n-1}+\underline{b}I^n) \quad .$$

Proof of claim 2. Since the residue fields $B/(m,It)$ and R/m are the same, we may compute the length as an R-module as well as an B-module. Thus

$$\lambda(B/J) = \lambda(B/(\underline{a},\underline{b},\underline{a}t)B)+\lambda((\underline{a},\underline{b},\underline{a}t)B/J)$$

$$= \sum_{n\geq 0}\lambda(I^n/\underline{a}I^{n-1}+\underline{b}I^n) + \sum_{n=2}^{s}\lambda(J+(a_n,\ldots,a_s)B/J+(a_{n+1},\ldots,a_s)B) \ .$$

In order to compute the terms of the second sum, we determine the kernel K of the following surjective map, given by multiplication with a_n where $n\in\{2,\ldots,s\}$:

$$B \xrightarrow{\cdot a_n} J+(a_n,\ldots,a_s)B/J+(a_{n+1},\ldots,a_s)B \quad .$$

Step 1: $(It,\underline{a},I^{n-1},\underline{b})B \subseteq K$.

This can be seen as follows:

a) $a_n\,(It) \subseteq I\, a_n t \subseteq \begin{cases} J & \text{for } n = s \\ (a_{n+1}) + J & \text{for } n < s \end{cases}$

b) For $i \in \{1,\dots,s\}$ we have:

$$a_n \cdot a_i \in \begin{cases} J & \text{for } i = 1 \\ a_n a_{i-1} t + J \subseteq (a_n + 1) + J & \text{for } i > 1 \end{cases}$$

c) $a_n I^{n-1} B \subseteq (a_{n-1} t) I^{n-1} B + J$

$$\subseteq a_{n-1}\, I^{n-2}\, B + J$$
$$\subseteq \dots$$
$$\subseteq a_1 I^0\, B + J \subseteq J$$

d) $a_n(\underline{b})B \subseteq J$.

Step 2: $K \subseteq (It, \underline{a},\, I^{n-1}, \underline{b})B$.

This can be checked as follows: Let $c \in K$. Since $(It)B$ is already contained in K , we may assume that $c \in R$. Now there exist elements $h_i, g_j, f_\ell \in B$ such that

$$ca_n = \sum_{i=1}^{s+1} h_i(a_{i-1}t - a_i) + \sum_{j=n+1}^{s} g_j a_j + \sum_{\ell=1}^{r} f_\ell b_\ell \quad ,$$

where $a_0 := a_{s+1} := 0$.

By comparing coefficients we find elements $r_k \in I^k$ with:

$$(c+r_0)a_n \in (a_2,\dots,a_{n-1},a_{n+1},\dots,a_s,\underline{b})R \quad ,$$
$$(r_0 - r_1)a_{n-1} \in (a_2,\dots,a_{n-2},a_n,\dots,a_s,\underline{b})R \quad ,$$
$$\dots\dots\dots\dots\dots\dots\dots\dots\dots\dots$$
$$(r_{n-2} - r_{n-1})a_1 \in (a_2,\dots,a_s,\underline{b})R \quad .$$

Since R is a Cohen-Macaulay-ring by assumption, $(\underline{a},\underline{b})$ is a regular sequence, hence c is contained in $(\underline{a},\underline{b}) + I^{n-1}$. This shows finally that

$$\lambda(B/J) = \lambda(R/(\underline{a},\underline{b})) + \sum_{n\geq 1} \lambda(I^n/\underline{a}I^{n-1} + \underline{b}I^n) + \sum_{n=2}^{s} \lambda(R/(\underline{a},I^{n-1},\underline{b})R)$$

$$= \sum_{n\geq 1} \lambda(I^n/\underline{a}I^{n-1} + \underline{b}I^n) + s\lambda(R/(\underline{a},b)) - \sum_{n=1}^{s-1} \lambda((\underline{a},\underline{b})+I^n/(\underline{a},\underline{b}))$$

$$= s\lambda(R/(\underline{a},b)) + \sum_{n\geq s} \lambda(I^n/\underline{a}I^{n-1} + \underline{b}I^n) + \sum_{n=1}^{s-1} \lambda(I^n \cap (\underline{a},\underline{b})/\underline{a}I^{n-1} + \underline{b}I^n) ,$$

which proves claim 2.

Now we are ready to finish the proof of Theorem (25.4). First remark that $e((\underline{a},\underline{b})R) = \lambda(R/(\underline{a},\underline{b}))$. Then claim 1 and claim 2 imply the following equivalences:

B is Cohen-Macaulay

$\overset{(1)}{\Longleftrightarrow}$ $e(JB) = \lambda(B/J)$

$\overset{(2)}{\Longleftrightarrow}$ $\begin{cases} I^s = \underline{a}I^{s-1} + \underline{b}I^s & \text{and} \\ (\underline{a},\underline{b}) \cap I^n = \underline{a}I^{n-1} + \underline{b}I^n & \text{for } 1 \leq n \leq s-1 \end{cases}$

$\overset{(3)}{\Longleftrightarrow}$ $\begin{cases} I^s = \underline{a}I^{s-1} & \text{and} \\ (\underline{a},\underline{b}) \cap I^n = \underline{a}I^{n-1} + \underline{b}I^n & \text{for all } n \in \mathbf{N} \end{cases}$

$\overset{(4)}{\Longleftrightarrow}$ $\begin{cases} I^s \subseteq (\underline{a}) & \text{and} \\ G \text{ is Cohen-Macaulay} \end{cases}$,

where (3) follows from Nakayama's lemma and (4) comes from Proposition (25.1). This proves Theorem (25.4).

<u>(25.5) Corollary</u>. Let $(R,\mathfrak{m})$ be a local ring and let I be an equimultiple ideal of height 1. Then the following conditions are equivalent:

(i) $B = B(I,R)$ is Cohen-Macaulay

(ii) I is principal and R is Cohen-Macaulay

(iii) I is principal and $G = G(I,R)$ is Cohen-Macaulay

Proof. By passing to the ring $R[X]_{\mathfrak{m}\cdot R[X]}$ we may assume that $R/\mathfrak{m}$ is infinite, since $IR[X]_{\mathfrak{m}R[X]}$ principal implies I is principal. Now let (a) be a minimal reduction of I .

(i) $\Rightarrow$ (ii): Since (a,at) is part of a regular sequence of B, we have $It \subseteq (at : a) = (at)$, hence $I \subseteq aR$. From this we get that $R = B/It\cdot B = B/(at)B$ is Cohen-Macaulay.

(ii) $\Rightarrow$ (iii): Because $B = R[at]$ is isomorphic to a polynomial ring over R and $G = B/(a)B$.

(iii) $\Rightarrow$ (i): This follows from Theorem (25.4).

(25.6) Corollary. Let $(R,\mathfrak{m})$ be a one-dimensional local ring. Then $B(\mathfrak{m},R)$ is Cohen-Macaulay if and only if R is regular.

The following Proposition (25.7) shows that the Cohen-Macaulayness of the Rees ring for an equimultiple ideal can be described by the Cohen-Macaulayness of the Rees ring of an $\mathfrak{m}$-primary ideal:

(25.7) Proposition. Let $(R,\mathfrak{m})$ be a local ring, I an equimultiple ideal of height s, $\{b_1,\ldots,b_r\}$ a part of a system of parameters mod I and $(\underline{b}) = (b_1,\ldots,b_r)$. Then the following conditions are equivalent:

(i) $B(I,R)$ is Cohen-Macaulay

(ii) $B(I+(\underline{b})/(\underline{b}), R/(\underline{b}))$ is Cohen-Macaulay and R is normally Cohen-Macaulay along I.

Proof. Since we may assume that $R/\mathfrak{m}$ is infinite, we find a minimal reduction $\underline{a} = (a_1,\ldots,a_s)$ of I with $s = ht(I)$.

(i) $\Rightarrow$ (ii) : First we show $(I^n : b_1) = I^n$: For every $c \in (I^n : b_1)$ we have

$$(cb_1)t^n a_1^n = c(a_1t)^n b_1 \in b_1\cdot B(I,R) .$$

Since (b_1,a_1) is part of a regular sequence of $B(I,R)$, it follows by comparing coefficients, that c is contained in I^n. Thus b_1 is a regular element in R/I^n for all $n \in \mathbf{N}$. This also shows that $(b_1) \cap I^n = b_1 I^n$, hence

$$B(I+(b_1)/(b_1), R/(b_1)) \simeq B(I,R)/(b_1) \quad \text{is Cohen-Macaulay.}$$

Now statement (ii) follows by induction.

(ii) $\Rightarrow$ (i): By passing to the ring $R/(b_2,\ldots,b_r)$ it is enough to consider the case $(\underline{b}) = (b_1)$: Since R is normally CM along I , we get

$$(0:b_1) \subseteq \bigcap_{n\geq 0}(I^n:b_1) = \bigcap_{n\geq 0} I^n = 0 \quad ,$$

hence b_1 is a regular element of R and therefore of $B(I,R)$ too. Moreover we get

$$B(I^n+(b_1)/(b_1) , R/(b_1)) \cong B(I,R)/(b_1) \quad ,$$

since $I^n \cap (b_1) = b_1 I^n$ for all n . Therefore $B(I,R)$ is Cohen-Macaulay.

Remark. Using this proposition one could easily prove Theorem (25.4), provided there would be an easy proof of this theorem for the $\mathfrak{m}$-primary case. But our proof of Theorem (25.4) cannot be simplified in this case. Moreover we remark that also another proof for the $\mathfrak{m}$-primary case, indicated by Goto-Shimoda in [8] , doesn't provide an easier approach. Thus our proof of Theorem (25.4) seems to be the shortest possible one for the moment. One of its high points is the observation, that the equimultiplicity of I lifts to the equimultiplicity of (I,It) in $B(I,R)$. Without the assumption on equimultiplicity, Theorem (25.4) is may be false. We will give a more general version of this theorem in Chapter IX.

§ 26. The case of hypersurfaces

We show that in the special case of hypersurfaces the criterion of Theorem (25.4) reduces to the condition $e(R) \leq \dim(R)$. Moreover the case $e(R) = 2$ is treated in detail. In particular we give an elementary proof for the fact that an excellent local ring of multiplicity 2, containing a field of characteristic zero and satisfying Serre's condition S_2 , is necessarily a hypersurface. This result will be extended in Chapter IX to any characteristic. The high point of proof is in both cases the proof of the direct summand conjecture of Hochster [2].

To avoid difficulties concerning the existence of reductions, we will assume throughout this section that R is a local ring having an infinite residue field. Of course by standard techniques all results can be extended to the general case.

(26.1) Definition. Let I be a proper ideal of a local ring R. The reduction exponent $r(I)$ of I is defined by

$$r(I) = \min \left\{ n \mid \text{there exists a minimal reduction } J \text{ of } I \text{ such that } I^n = JI^{n-1} \right\} .$$

(26.2) Definition. The local ring $(R,\mathfrak{m})$ is called a hypersurface if R is unmixed and $\operatorname{embdim}(R) \leq \dim(R) + 1$.

(26.3) Lemma. If R is a hypersurface then $G(\mathfrak{m},R)$ (and therefore R) is Cohen-Macaulay.

Proof. If $\operatorname{embdim}(R) = \dim(R)$ then R is regular, which clearly proves the claim.

If $\operatorname{embdim}(\hat{R}) = \operatorname{embdim}(R) = d + 1$, where $(\hat{R},\hat{\mathfrak{m}})$ is the completion of R, then $\hat{R} = S/\mathfrak{a}$ for a regular local ring S with $\dim S = d+1$. So we get $\operatorname{ht}(\mathfrak{a}) = 1$, hence $\mathfrak{a}$ is principal since R is unmixed. Therefore ($\hat{R}$ and) $G(\hat{\mathfrak{m}},\hat{R})$ are Cohen-Macaulay, q.e.d.

The next result will show that the above condition on $r(I)$ is closely related to a restriction of multiplicities.

(26.4) Proposition. Let $(R,\mathfrak{m})$ be a local ring and assume that $G(\mathfrak{m},R)$ is Cohen-Macaulay. Then we have:

(i) $r(\mathfrak{m}) \leq e(R)$

(ii) $r(\mathfrak{m}) = e(R)$ if and only if R is a hypersurface.

Proof. Let $r = r(\mathfrak{m})$ and let $\underline{x} = \{x_1,\ldots,x_d\}$ generate a minimal reduction of $\mathfrak{m}$ such that $\underline{x}\mathfrak{m}^{r-1} = \mathfrak{m}^r$. Since $\operatorname{in}(\mathfrak{m})(x_1),\ldots,\operatorname{in}(\mathfrak{m})(x_d)$ is a regular sequence in $G(\mathfrak{m},R)$ by assumption, by putting $\overline{R} = R/\underline{x}R$, $\overline{\mathfrak{m}} = \mathfrak{m}/\underline{x}R$, we see that

$$\overline{\mathfrak{m}}^i/\overline{\mathfrak{m}}^{i+1} \simeq \mathfrak{m}^i/\mathfrak{m}^{i+1} + \underline{x}\mathfrak{m}^{i-1}$$

(compare [9]). Therefore

$$e(R) = e(G(\mathfrak{m},\overline{R})) = I(G(\overline{\mathfrak{m}},\overline{R})) = \sum_{i=0}^{r-1} \dim_{R/\mathfrak{m}}(\overline{\mathfrak{m}}^i/\overline{\mathfrak{m}}^{i+1})$$

and furthermore

$$\dim_{R/\mathfrak{m}}(\overline{\mathfrak{m}}^i/\overline{\mathfrak{m}}^{i+1}) > 0 \quad \text{for} \quad 0 \leqq i < r$$

since r was minimal. This proves (i), and it shows that

$$r(\mathfrak{m}) = e(R) \Longleftrightarrow \dim_{R/\mathfrak{m}}(\overline{\mathfrak{m}}^i/\overline{\mathfrak{m}}^{i+1}) = 1 \quad \text{for} \quad 0 \leqq i < r$$

$$\Longleftrightarrow \overline{\mathfrak{m}} \text{ is principal },$$

proving (ii) too.[Note that "$G(\mathfrak{m},R)$ is CM" implies "R is CM" , hence R is unmixed] .

<u>(26.5) Corollary</u>. If $G(\mathfrak{m},R)$ is Cohen-Macaulay and $e(R) \leqq \dim R$, then $B(\mathfrak{m},R)$ is Cohen-Macaulay.

Proof. This follows immediately from Proposition (26.4) and Theorem (25.4).

<u>(26.6) Corollary</u>. If R is a hypersurface, then $B(\mathfrak{m},R)$ is Cohen-Macaulay if and only if $e(R) \leqq \dim R$.

Proof. Since our assumption implies that R and $G(\mathfrak{m},R)$ are Cohen-Macaulay, the corollary is a consequence of Proposition (26.4) and Theorem (25.4).

<u>(26.7) Remark</u>. Recall that for a local Cohen-Macaulay ring $(R,\mathfrak{m})$ we have the following inequality (see [5] and Appendix to Chapter V).

$$e(R) \geqq \operatorname{embdim}(R) - \dim(R) + 1 \quad .$$

Furthermore, equality holds if and only if $r(\mathfrak{m}) \leqq 2$, and in this case $G(\mathfrak{m},R)$ is Cohen-Macaulay ([5], p. 45). So for small values of $e(R)$ we can deduce the following list of cases in which $B(\mathfrak{m},R)$ is Cohen-

Macaulay:

<u>(26.8) Proposition</u>. Let $(R,\mathfrak{m})$ be a Cohen-Macaulay ring. Then

(i) $e(R) = 1$: $B(\mathfrak{m},R)$ is Cohen-Macaulay

(ii) $e(R) = 2$: R is a hypersurface and $B(\mathfrak{m},R)$ is Cohen-Macaulay if and only if $\dim R \geqq 2$.

(iii) $e(R) = 3$: 1) If $\dim(R) = 1$, $B(\mathfrak{m},R)$ is not Cohen-Macaulay.
2) If $\dim(R) = 2$, $B(\mathfrak{m},R)$ is Cohen-Macaulay if and only if R is <u>not</u> a hypersurface.
3) If $\dim R \geqq 3$, then $B(\mathfrak{m},R)$ is always Cohen-Macaulay.

Proof.(i): R is regular, hence $B(\mathfrak{m},R)$ is CM.

(ii): Since R is a hypersurface, $B(\mathfrak{m},R)$ is CM if and only if $\dim R \geqq e(R) = 2$ by Corollary (26.6).

(iii): Case 1 is clear. For 2) and 3) we remark [5] , that R is CM and $e(R) = 3$ imply $G(\mathfrak{m},R)$ is CM, hence $r(\mathfrak{m}) \leqq 3$ by Proposition (26.4). Therefore we get the following implications for case 2: $R \neq$ hypersurface $\Longleftrightarrow r(\mathfrak{m}) \leqq 2 = \dim(R) \Longleftrightarrow B(\mathfrak{m},R)$ is CM. Case 3 is now clear by Corollary (26.6).

In the following proposition we give sufficient conditions on I such that $G(I,R)$ becomes Cohen-Macaulay. For this we need two technical lemmas.

<u>(26.9) Lemma</u>. Let R be a local Cohen-Macaulay ring (with infinite residue field by our general assumption) of dimension d . Let I be a proper ideal of R and $\{x_1,\dots,x_r\}$ a system of parameters $\bmod I$. Put $(\underline{x}) = (x_1,\dots,x_r)$ and $I(\underline{x}) = I + \underline{x}R$, and let

$$G(I,\underline{x}) : G(I,R) \otimes_R R/I(\underline{x})[X_1,\dots,X_r] \longrightarrow G(I(\underline{x}),R)$$

denote the canonical homomorphism. Assume that $\operatorname{depth}(I/I^2) = r$. Then we have

(i) $\lambda(I(\underline{x})/I(\underline{x})^2) \leqq e(x;I/I^2) + r\lambda(R/I(\underline{x}))$, with equality if and only if $G(I,\underline{x})$ is an isomorphism in degree 1.

(ii) Assume in addition that $\operatorname{ht}(I) = s(I)$ and R/I is Cohen-Macaulay. Then

$$e(\underline{x};I/I^2) + r\lambda(R/I(\underline{x})) \leqq e(\underline{x},I,R) + (d-1)\lambda(R/I(\underline{x})) \quad ,$$

with equality if and only if $r(IR_{\mathfrak{p}}) \leqq 2$ for any $\mathfrak{p} \in \mathrm{Min}(I)$, the set of minimal primes of I .

Proof. (i) Since $\underline{x}$ is a regular sequence on I/I^2 , we have $e(\underline{x};I/I^2) = \lambda(I/I\cdot I(\underline{x}))$, and therefore the assertion follows by comparing terms of degree 1 for the surjective homomrphism $G(I,\underline{x})$.

(ii) Assume first that $\mathrm{ht}(I) > 0$. Then our assumptions imply $\mathrm{Assh}(R/I) = \mathrm{Min}(I) = \mathrm{Ass}(I/I^2)$. Therefore

$$e(\underline{x};I/I^2) = \sum_{\mathfrak{p}\in \mathrm{Min}(I)} e(\underline{x};R/\mathfrak{p})\lambda(IR_{\mathfrak{p}}/I^2R_{\mathfrak{p}}) \quad .$$

Putting $s = \mathrm{ht}(I) = d-r$, one can easily see that

$$(*) \qquad \lambda(IR_{\mathfrak{p}}/I^2R_{\mathfrak{p}}) \leqq e(IR_{\mathfrak{p}},R_{\mathfrak{p}}) + (s-1)\lambda(R_{\mathfrak{p}}/IR_{\mathfrak{p}})$$

by using the fact that $\lambda(IR_{\mathfrak{p}}/I^2R_{\mathfrak{p}}) = \lambda(R_{\mathfrak{p}}/I^2R_{\mathfrak{p}}) - \lambda(R_{\mathfrak{p}}/IR_{\mathfrak{p}})$ and $\lambda(R_{\mathfrak{p}}/I^2R_{\mathfrak{p}}) \leqq \lambda(R_{\mathfrak{p}}/\underline{y}R_{\mathfrak{p}}) + \lambda(\underline{y}R_{\mathfrak{p}}/\underline{y}IR_{\mathfrak{p}})$, where $\underline{y}R_{\mathfrak{p}}$ is a minimal reduction of $IR_{\mathfrak{p}}$. But $\underline{y}R_{\mathfrak{p}}/\underline{y}IR_{\mathfrak{p}} \cong (R_{\mathfrak{p}}/IR_{\mathfrak{p}})^3$ which proves (*) .

Equality holds in (*) if and only if for any minimal reduction $\underline{z}$ of $IR_{\mathfrak{p}}$ we have $\underline{z}IR_{\mathfrak{p}} = I^2R_{\mathfrak{p}}$, and this last condition is equivalent to $r(IR_{\mathfrak{p}}) \leqq 2$. Note that $\mathrm{ht}(I) = s(I)$ implies that any minimal reduction $\underline{z}$ of I remains a minimal reduction of $IR_{\mathfrak{p}}$ in $R_{\mathfrak{p}}$ for any prime $\mathfrak{p} \supseteq I$. Therefore, using the assumption that R/I is Cohen-Macaulay, (ii) follows from

$$e(\underline{x},I,R) = \sum_{\mathfrak{p}\in \mathrm{Min}(I)} e(\underline{x};R/\mathfrak{p})e(IR_{\mathfrak{p}},R_{\mathfrak{p}})$$

and

$$e(\underline{x};R/I) = \sum_{\mathfrak{p}\in \mathrm{Min}(I)} e(\underline{x};R/\mathfrak{p})\lambda(R_{\mathfrak{p}}/IR_{\mathfrak{p}}) = \lambda(R/I(\underline{x})) \quad .$$

Finally, if $\mathrm{ht}(I) = 0$ we have $\mathrm{Assh}(R/I) = \mathrm{Min}(I) \supseteq \mathrm{Ass}(I/I^2)$ and if $\mathfrak{p} \in \mathrm{Min}(I) \setminus \mathrm{Ass}(I/I^2)$, then $IR_{\mathfrak{p}} = 0$. Therefore we may conclude as above.

(26.10) Definition. Let R be a local ring and I a proper ideal of R. We say that I has generically reduction exponent t if $r(IR_{\mathfrak{p}}) \leqq t$ for all $\mathfrak{p} \in \mathrm{Min}(I)$.

(26.11) Proposition. Let R be a local Cohen-Macaulay ring and let I be a proper ideal of R. Assume that R is normally Cohen-Macaulay along I and that I has generically reduction exponent 2. Then $G(I;R)$ is Cohen-Macaulay.

Proof. Note that R Cohen-Macaulay implies $\dim(R) = \dim(R/I)+\mathrm{ht}(I)$ and therefore I is equimultiple. By Lemma (26.9) we know that for any system of parameters $\underline{x} = \{x_1,\ldots,x_r\} \bmod I$ we have (using the above notation with $I(x) = \underline{x}R + I$) :

$$\lambda(I(\underline{x})/I(\underline{x})^2) = e(\underline{x},I,R) + (d-1)\lambda(R/I(\underline{x}))$$

$$= e(I(\underline{x}),R) + (d-1)\lambda(R/I(\underline{x})) \quad ,$$

since $\mathrm{ht}(I) = s(I)$ implies $e(\underline{x},I,R) = e(I(\underline{x}),R)$. But this equality implies $r(I(\underline{x})) = 2$, hence

$$\underline{a}R \cap I(\underline{x})^i = \underline{a}I(\underline{x})^{i-1} \quad \text{for all } i \geqq 2 \quad (\text{for } i = 1 \text{ this}$$

is trivially true.) Therefore $G(I(\underline{x});R)$ is CM by Proposition (25.1), (iv) $\Rightarrow$ (i). From this we get from Proposition (25.1) (ii) $\Rightarrow$ (i), $G(I;R)$ is CM since R is normally CM along I by assumption, q.e.d.

As a direct consequence of Proposition (25.1) one can give a slightly more general version of Proposition (26.11).

(26.12) Proposition. Let $(R,\mathfrak{m})$ be a local Cohen-Macaulay ring and let I be an equimultiple ideal. Assume that R/I is Cohen-Macaulay and I has generically reduction exponent 2. Then $G(I,R)$ is Cohen-Macaulay.

Proof. Recall that $R/\mathfrak{m}$ is always infinite. Let $\underline{a} = (a_1,\ldots,a_s)$ be a minimal reduction of I. Consider the exact sequence

$$0 \longrightarrow \underline{a}/\underline{a}I \longrightarrow R/\underline{a}I \longrightarrow R/\underline{a} \longrightarrow 0 \quad ,$$

where $\underline{a}/\underline{a}I \simeq (R/I)^s$ is Cohen-Macaulay by assumption. Therefore $R/\underline{a}I$ is Cohen-Macaulay showing that $\underline{a}I$ is unmixed. This implies $I^2 = \underline{a}I$ since $I^2 R_{\mathfrak{p}} = \underline{a}IR_{\mathfrak{p}}$ for all $\mathfrak{p} \in \mathrm{Ass}(R/\underline{a}I) = \mathrm{Min}(R/\underline{a}I)$. (The equality comes from the fact that $\underline{a}I$ is unmixed.)

By Proposition (25.1) we have to show that

$$(*) \qquad I^n \cap (\underline{a},\underline{b}) = \underline{a}I^{n-1} + \underline{b}I^n \quad \text{for all} \quad n > 0 \quad ,$$

where $\underline{b}$ is a system of parameters $\bmod I$. But this is true for all $n \geq 2$ because of $I^2 = \underline{a}I$. Moreover for $n = 1$ we have

$$I \cap (\underline{a},\underline{b}) \subseteq \underline{a} + \underline{b} \cap I \quad ,$$

and $\underline{b} \cap I = \underline{b}I$ since R/I is CM. Hence $I \cap (\underline{a},\underline{b}) \subseteq \underline{a} + \underline{b}I$, which proves the assertion, q.e.d.

Now we discuss the case of multiplicity 2. First we prove a result which is a special case of Hochsters direct summand conjecture, see Hochster [2].

(26.13) Proposition. Let R be a local ring containing a field of charactersitic 0 and let S be a regular subring of R such that R is a finite module over S. Then S is a direct summand of R as S-module.

Proof. We may assume that R is complete (since $\hat{R}$ is faithfully flat over R). Now, to show that S is a direct summand of R, it is enough to construct an S-module homomorphism $\varphi : R \longrightarrow S$ such that $\varphi \circ \theta = 1$, where $S \xrightarrow{\theta} R$ is the canonical injection. Let $\mathfrak{P} \in \mathrm{Spec}(R)$ such that $\dim(R) = \dim(R/\mathfrak{P})$. Then $\mathfrak{P} \cap S = 0$, since $R/\mathfrak{P}$ is finite over $S/\mathfrak{P} \cap S$ and $\dim S = \dim R = \dim R/\mathfrak{P}$. Hence S can be thought as a subring of $R/\mathfrak{P}$. It is easy to see that if S is a direct summand of $R/\mathfrak{P}$, then S is a direct summand of R. Hence we may assume that R is a complete local domain which is finite over a regular local ring S. Let K and L be the quotient fields of S and R.

Let $Tr : L \longrightarrow K$ be the trace map, i.e. the K-linear map $Tr(\alpha) = \sum_{i=1}^{n} \sigma_i \alpha$, where $\sigma_1, \dots, \sigma_n$ are the distinct embeddings of L in an algebraic closure $\overline{K}$ of K, $\alpha \in L$ and $n = [L : K]$. Consider the map $\varphi := \frac{1}{n} Tr : R \longrightarrow K$. For $x \in R$ we see that $Tr(x)$ is integral over S. Since $\frac{1}{n} \in S$ by assumption of characteristic 0, we see that $\varphi(x) \in K$ is integral over S. Therefore $\varphi(x) \in S$, since S is integrally closed in K. Hence φ defines an S-module homomorphism $\varphi : R \longrightarrow S$ with the desired property.

<u>(26.14) Lemma</u>. Let $(R,\mathfrak{m})$ be a local Cohen-Macaulay ring and M a finitely generated R-module. Then M satisfies Serre's condition on (S_n), s. [4], if and only if any R-sequence $x_1, \dots, x_n$ is an M-sequence.

Proof. Suppose that M satisfies (S_n). Let $x_1, \dots, x_n$ be an R-sequence. Assume that $x_1, \dots, x_i$ is an M-sequence for $0 \leqq i < n$. Let $\mathfrak{P} \in Ass_R(M/(x_1, \dots, x_i)M)$. Then $\mathfrak{P} \supset (x_1, \dots, x_i)$, in particular $ht(\mathfrak{P}) \geqq i$. Hence by assumption we get $i = depth(M_{\mathfrak{P}}) \geqq \min\{n, ht(\mathfrak{P})\} \geqq i$. Therefore $ht(\mathfrak{P}) = i$ and $x_{i+1} \notin \mathfrak{P}$. This shows that x_{i+1} is $M/(x_1, \dots, x_i)M$ - regular. Hence $x_1, \dots, x_n$ is an M-sequence. To prove the converse, take any $\mathfrak{P} \in Spec(R)$ with $ht(\mathfrak{P}) = h$. Since R is Cohen-Macaulay, $\mathfrak{P}$ contains a regular sequence $x_1, \dots, x_h \in \mathfrak{P}$. Let $k = \min\{n, h\}$. Then $x_1, \dots, x_k$ is an M-sequence by assumption and $depth\, M_{\mathfrak{P}} \geqq k$. Hence M satisfies (S_n).

<u>(26.15) Lemma</u>. Let $(R,\mathfrak{m})$ be a quasi-unmixed local ring and S a subring of R such that R is finite over S and S is a Cohen-Macaulay local ring. Then a finitely generated R-module M satisfies (S_n) as R-module if and only if so does as S-module.

Proof. First suppose that M satisfies (S_n) as R-module. Let $x_1, \dots, x_n$ be an S-sequence. By Lemma (26.14) we show that $x_1, \dots, x_n$ is an M-sequence. Assume that $x_1, \dots, x_i$, $0 \leqq i < n$ is an M-sequence. Let $\mathfrak{P} \in Ass_R(M/(x_1, \dots, x_i)M)$. If $\mathfrak{P} \ni x_{i+1}$ we have $\mathfrak{P} \supset (x_1, \dots, x_{i+1})R$. Let $\mathfrak{p} = S \cap \mathfrak{P}$ and $\mathfrak{P}_0 \in Spec(R)$ be a minimal prime such that $\mathfrak{P}_0 \subset \mathfrak{P}$ and let $\mathfrak{p}_0 = S \cap \mathfrak{P}_0$.

Since R is quasi-unmixed and $R/\mathfrak{P}_0$ is finite over $S/\mathfrak{p}_0$ we see that $ht(\mathfrak{p}_0) = 0$.

By assumption S is Cohen-Macaulay, so $S/\mathfrak{p}_0$ is universally catenarian. Hence by the altitude formula we have

$$ht(\mathfrak{P}) \geq ht(\mathfrak{P}/\mathfrak{P}_0) = ht(\mathfrak{p}/\mathfrak{p}_0) = ht(\mathfrak{p}) = ht(x_1,\ldots,x_{i+1}) = i+1 \quad .$$

But this implies

$$i = \operatorname{depth} M_{\mathfrak{P}} \geq \min\{n, ht(\mathfrak{P})\} \geq i+1 \quad ,$$

a contradiction. Therefore we must have $x_{i+1} \notin \mathfrak{P}$ for any $\mathfrak{P} \in \operatorname{Ass}_R(M/(x_1,\ldots,x_i)M)$. We have shown that $x_1,\ldots,x_n$ is an M-sequence as required.
Conversely assume that M satisfies (S_n) as S-module. Let $\mathfrak{P} \in \operatorname{Spec}(R)$ and $k = \min\{n, ht(\mathfrak{P})\}$. If $\mathfrak{p} = \mathfrak{P} \cap S$ we have $ht(\mathfrak{P}) \leq ht(\mathfrak{p})$. Since S is Cohen-Macaulay one can find an S-sequence $x_1,\ldots,x_k \in \mathfrak{p}$ which forms an M-sequence by Lemma (26.14). Hence $x_1,\ldots,x_k$ is a $M_{\mathfrak{P}}$-regular sequence, i.e. $\operatorname{depth}(M_{\mathfrak{P}}) \geq k = \min\{n, ht(\mathfrak{P})\}$.

<u>(26.16) Proposition</u>. Let $(R,\mathfrak{m})$ be an excellent local ring containing a field of characteristic 0 . If $e(R) = 2$ and if R satisfies Serre's condition (S_2) , then R is a hypersurface.

<u>(26.17) Remark</u>. Note that the assumption " R is excellent" and " R satisfies (S_2) " imply, that R is quasi-unmixed.

Proof of (26.16). The completion $\hat{R}$ satisfies (S_2) , since R is excellent [1],[3]. Furthermore we have $e(\hat{R}) = 2$. Therefore we may assume that R is complete.
A) First we show that R is Cohen-Macaulay: Let k be the coefficient field of R (which is assumed to be infinite) and let $(x_1,\ldots,x_d)R$ be a minimal reduction of $\mathfrak{m}$. R is finite over its regular subring $S = k[[x_1,\ldots,x_d]]$, s. [6], hence the inclusion $S \subseteq R$ splits as S-module by Proposition (26.13), say

(1) $$R = S \oplus M$$

for some S-module M .
Recall that by definition $\operatorname{rank}_S N = \dim_K(K \otimes_S N)$ for a S-module N , where K is the fraction field of S .

We have that

$$e(R) = e(x_1,\dots,x_d;R) = e_S(R) = \mathrm{rank}_S(R) \quad ,$$

which gives in our case $\mathrm{rank}_S(M) = 1$.
Thus M satisfies (S_2) as an S-module by Lemma (26.15). So we may identify M with an ideal $\mathfrak{a} \neq 0$ of S . Observe that $\mathfrak{a}$ is a proper ideal by construction (otherwise R would be isomorphic to S). Let $\mathfrak{p} \in \mathrm{Ass}_S(S/\mathfrak{a})$. Assume that $\mathrm{ht}(\mathfrak{p}) \geqq s$. Applying $\mathrm{Ext}^i(k,-)$ on the exact sequence

$$0 \longrightarrow \mathfrak{a}S_{\mathfrak{p}} \longrightarrow S_{\mathfrak{p}} \longrightarrow S_{\mathfrak{p}}/\mathfrak{a}S_{\mathfrak{p}} \longrightarrow 0$$

with $\mathrm{depth}(S_{\mathfrak{p}}/\mathfrak{a}S_{\mathfrak{p}}) = 0$ and $\mathrm{depth}(S_{\mathfrak{p}}) \geqq 2$, we conclude from [3], p. 96. that $\mathrm{depth}(\mathfrak{a}S_{\mathfrak{p}}) = \mathrm{depth}(M_{\mathfrak{p}}) = 1$, which contradicts to (S_2) on M . Therefore $\mathrm{ht}(\mathfrak{p}) = 1$ for any $\mathfrak{p} \in \mathrm{Ass}(S/\mathfrak{a})$, hence $\mathfrak{a}$ is principal (since S is regular (see [4], Theorem 1.31 and Exercise 2).
Thus $M \cong \mathfrak{a} \cong S$ as S-module, i.e. M (and thus R) is a free S-module. It follows that R is Cohen-Macaulay.

B) To show that R is even a hypersurface,we take again a minimal reduction $\underline{x} = (x_1,\dots,x_d)$ of $\mathfrak{m}$. Thus

$$2 = e(R) = e(\underline{x};A) = \lambda(R/\underline{x}) \quad ,$$

since R is Cohen-Macaulay by step A . This implies $\lambda(\mathfrak{m}/\underline{x}) = 1$, i.e. $\mathfrak{m}/\underline{x}$ is generated $\mathrm{mod}\,\underline{x}$ by one element y . Take the map

$$\varphi : k\,[[X_1,\dots,X_d,Y]] \longrightarrow R$$

sending the indeterminates X_i to x_i and Y to y (φ is a map onto R , since $\mathrm{embdim}(R)=d+1$) . Since $\dim R = d$, the kernel of φ is a height one ideal of $k\,[[X_1,\dots,X_d,Y]]$, which corresponds to the zero-ideal of R . Since R is Cohen-Macaulay, $\ker\varphi$ is an unmixed ideal.Since R is factorial this ideal must be principal. This proves the assertion.

(26.18) Corollary. Let $(R,\mathfrak{m})$ be an excellent ring containing a field of characteristic 0 . If there exists a system $\underline{x} = \{x_1,\dots,x_d\}$ of parameters in R such that $\lambda(R/\underline{x}R) = 2$, then $B(\mathfrak{m},R)$ is Cohen-

Macaulay if and only if $e(R) \leqq \dim(R)$.

(26.19) Remark. The assumptions of Corollary (26.18) don't imply $e(R) = 2$. Take the example: $R = k[[X]]$ and $x_1 = X^2$. Then $\lambda(R/x_1R) = 2$, but $e(R) = 1$.

(26.20) Remark. Proposition (26.13) holds for any local ring R containing a field, by M. Hochster [2]. Proposition (26.16) can be generalized as follows: "Let R be a complete local ring containing a field and satisfying Serre's condition (S_n) for some $n \geqq 2$. If $e(R) \leqq n$, then R is Cohen-Macaulay", see [10]: As in the proof of Proposition (26.16) one shows that $R = S \oplus M$ and M satisfies (S_n) , where $\operatorname{rank}_S(M) \leqq n-1$. It comes out that M is an n-th syzygy . Then the next big step is to prove that a module M of finite projective dimension which is an n-th syzygy of $\operatorname{rank}_S(M) < n$, is free. This is a deep result of G. Evans and D. Griffith; see [7] for the long proof.

As an application of Proposition (26.16) we get the following result.

(26.21) Proposition. Let (R,m) be an excellent local ring of $\dim(R) \geqq 2$, containing a field of characteristic 0 . Let I be an equimultiple ideal of R with $\dim(R/I) = 1$. Assume that

(i) $B(I,R)$ is Cohen-Macaulay

(ii) $e(R) + e(R/I) \leqq \operatorname{embdim}(R/I) + 2$.

(iii) R is Cohen-Macaulay outside m .

Then R is a hypersurface.

Proof. (i) implies that R is normally CM along I and $\operatorname{depth}(R) \geqq \dim(R/I) + 1 = 2$, hence R satisfies (S_2) by (iii) (and R is quasi-unmixed).

Since in particular R/I is CM, we get $e(R/I) \geqq \operatorname{embdim}(R/I)$, which together with (ii) implies $e(R) \leqq 2$. This proves the assertion in view of Proposition (26.16).

(26.22) Remark. Using Huneke's generalization of Proposition (26.16), mentioned in Remark (26.20), one can prove Proposition (26.21) for any $\dim(R/I) \geqq 1$.

§ 27. Transitivity of Cohen-Macaulayness of Rees rings

We assume again that the given ring $(R,\mathfrak{m})$ is Cohen-Macaulay. Then we consider equimultiple ideals $J \subset I$ such that $I = J + \underline{x}R$, where $\underline{x}$ is part of a system of parameters $\bmod J$. For simplicity we are always working with an infinite residue field $R/\mathfrak{m}$.

(27.1) Theorem. (Transitivity of Cohen-Macaulay property.) Let $(R,\mathfrak{m})$ be a local Cohen-Macaulay ring with infinite residue field. Let J be an equimultiple ideal of R , let $\underline{x} = \{x_1,\dots,x_s\}$ be a part of a system of parameters $\bmod J$ and let $I = J + \underline{x}R$.

a) The following conditions are equivalent:

(i) $G(J,R)$ is Cohen-Macaulay.

(ii) $G(I,R)$ is Cohen-Macaulay; and $G(JR_{\mathfrak{p}},R_{\mathfrak{p}})$ is Cohen-Macaulay for all $\mathfrak{p} \in \mathrm{Min}(I)$.

b) If $\mathrm{ht}(J) > 0$, the following conditions are equivalent:

(i) $B(J,R)$ is Cohen-Macaulay.

(ii) $B(I,R)$ is Cohen-Macaulay, and $B(JR_{\mathfrak{p}},R_{\mathfrak{p}})$ is Cohen-Macaulay for all $\mathfrak{p} \in \mathrm{Min}(I)$.

Proof. a) Let $\underline{y}$ be a system of parameters $\bmod I$. Then $\underline{x} \cup \underline{y}$ is a system of parameters $\bmod J$.

(i) ⇒ (ii): Clearly $G(JR_{\mathfrak{p}},R_{\mathfrak{p}}) \simeq G(J,R) \otimes R_{\mathfrak{p}}$ is Cohen-Macaulay. By Proposition (25.1), $G(J,R)$ is Cohen-Macaulay if and only if $G(J+\underline{x}R+\underline{y}R,R)$ is Cohen-Macaulay and R is normally Cohen-Macaulay along J . This implies that R is normally Cohen-Macaulay along I , s. Chapter IV, Theorem (24.11).Using $G(J+\underline{x}R+\underline{y}R,R) = G(I+\underline{y}R,R)$, we see that $G(I,R)$ is Cohen-Macaulay, by Proposition (25.1) again.

(ii) ⇒ (i): By Chapter IV, Theorem (24.11) R is normally Cohen-

Macaulay along J, and $G(J+\underline{x}R+\underline{y}R,R)$ is Cohen-Macaulay, so $G(J,R)$ is Cohen-Macaulay.

b) By Theorem (25.4) we know that $B(J,R)$ is Cohen-Macaulay if and only if $G(J,R)$ is Cohen-Macaulay and $r(J) \leqq ht(J)$.

(i) $\Rightarrow$ (ii): Obviously we have $r(I) \leqq r(J) \leqq ht(J) \leqq ht(I)$, and also $r(JR_{\mathfrak{p}}) \leqq r(J) \leqq ht(J) = ht(JR_{\mathfrak{p}})$. Therefore the assertion follows from a), (i) $\Rightarrow$ (ii).

(ii) $\Rightarrow$ (i): By a) and Theorem (25.4) we have to show that $r(J) \leqq ht(J)$. Equivalently, taking any minimal reduction J' of J and putting $t = ht(J)$, we have to show that $J^t \subset J'$. Note that R/J' is Cohen-Macaulay, and therefore $Ass(R/J') = Min(J)$. So we are reduced to prove that $J^tR_{\mathfrak{Q}} \subset J'R_{\mathfrak{Q}}$ for all $\mathfrak{Q} \in Min(J)$. Now if $\mathfrak{Q} \in Min(J) = Min(J')$, we claim that $\mathfrak{Q} \subset \mathfrak{P}$ for some $\mathfrak{P} \in Min(I) = Ass(R/I)$ (note that R/I is CM by a)). Otherwise we would have $\mathfrak{Q} \not\subset \bigcup_{\mathfrak{P}\in Min(I)} \mathfrak{P}$, and therefore $\mathfrak{Q}$ would contain an element y which is a non-zero-divisor mod I. Since R/J is Cohen-Macaulay by a), any non-zero-divisor mod I is - as a parameter mod J - also a non-zero-divisor mod J, which gives a contradiction to $\mathfrak{Q} \in Min(J) = Ass(R/J)$. Now given $\mathfrak{P} \in Min(I)$ such that $\mathfrak{Q} \subset \mathfrak{P}$, we know from assumption (ii) that $J^tR_{\mathfrak{P}} \subset J'R_{\mathfrak{P}}$, and a forteriori $J^tR_{\mathfrak{Q}} \subset J'R_{\mathfrak{Q}}$, which completes the proof.

<u>(27.2) Remark</u>. The claim $\mathfrak{Q} \subset \mathfrak{p}$ for some $\mathfrak{p} \in Min(I)$ can also be seen as follows:

$$ht(I) \leqq ht(\mathfrak{Q}+\underline{x}R) = dim(R) - dim\,(R/\mathfrak{Q}+\underline{x}R)$$

$$= dim\,R - (dim(R/\mathfrak{Q}) - ht(\mathfrak{Q}+\underline{x}R/\mathfrak{Q})) \quad ,$$

since $R/\mathfrak{Q}$ is quasi-unmixed, see Chapter III. Hence we get

$$ht(I) \leqq ht(\mathfrak{Q}) + ht(\mathfrak{Q}+\underline{x}R/\mathfrak{Q}) \leqq ht(J) + s = ht(I) \quad ,$$

i.e. $ht(I) = ht(\mathfrak{Q}+\underline{x}R)$, thus $\mathfrak{Q}$ must be contained in some minimal prime ideal $\mathfrak{P} \in Min(I)$. Then we know from assumption (ii) that $J^tR_{\mathfrak{P}} \subset J'R_{\mathfrak{P}}$, and a forteriori $J^tR_{\mathfrak{Q}} \subset J'R_{\mathfrak{Q}}$, which completes the proof.

<u>(27.3) Corollary</u>. Let (R,m) be a Cohen-Macaulay ring and let $\mathfrak{P}$

be a permissible ideal in R (i.e. here $R/\mathfrak{P}$ is regular and $e(R) = e(R_{\mathfrak{P}})$). If $B(\mathfrak{P},R)$ is Cohen-Macaulay then $B(\mathfrak{Q}R_{\mathfrak{Q}},R_{\mathfrak{Q}})$ is Cohen-Macaulay for all prime ideals $\mathfrak{Q} \subset \mathfrak{P}$; in particular $B(\mathfrak{m},R)$ is Cohen-Macaulay.

(27.4) Example. $R = k[[X^2,XY,Y^2,XZ,YZ,Z]]$, X,Y,Z indeterminates, is Cohen-Macaulay. Consider the ideals

$$J = (X^2)R \subset I = (X^2,Y^2)R \subset H = (X^2,Y^2,Z)R \quad .$$

Since J,I,H are generated by regular sequences the corresponding Rees rings $B(J,R)$, $B(I,R)$ and $B(H,R)$ are Cohen-Macaulay. Since $(X^2,Y^2,Z)\,\mathfrak{m} = \mathfrak{m}^2$, we know that $G(\mathfrak{m},R)$ is CM, hence $B(\mathfrak{m},R)$ is Cohen-Macaulay for this ring. But the Rees ring $B(\mathfrak{P},R)$ for the ideal $\mathfrak{P} := (XZ,YZ,Z)$ is not Cohen-Macaulay, otherwise $\mathfrak{P}$ could be generated by one element. This suggests that generally the Cohen-Macaulay-property of $B(\mathfrak{P},R)$ for an ideal $\mathfrak{P} \neq \mathfrak{m}$ is in some sense "stronger" than the Cohen-Macaulay-property of $B(\mathfrak{m},R)$.

The next Theorem (27.5) shows that the assumption " R is Cohen-Macaulay " in the previous Theorem (27.1) is necessary. A second proof of this theorem is given in Chapter IX, where we use local cohomology.

(27.5) Theorem. Let $(R,\mathfrak{m})$ be a local ring, J an equimultiple ideal with $ht(J) = s$, $\underline{x} = \{x_1,\dots,x_k\}$ a part of a system of parameters $\bmod J$ and $I = J + \underline{x}R$. Assume that $k > 0$ and that $B(J,R)$, and $B(I,R)$ are Cohen-Macaulay. Then R is Cohen-Macaulay.

One essential in the "elementary" proof of this theorem is included in the following Lemma (27.6), which we verify first.

(27.6) Lemma. Let J be an equimultiple ideal in a local ring $(R,\mathfrak{m})$, let $\underline{z} = (z_1,\dots,z_r)$ be a minimal reduction of J . If the Rees ring $B(J,R)$ of J is Cohen-Macaulay, then

$$J^n \cap (z_1,\dots,z_i) = (z_1,\dots,z_i)J^{n-1}$$

for all $n \in \mathbf{N}$ and any fixed $i \in \{1,\dots,r\}$.

Proof. We know that $(z_r, z_r t - z_{r-1}, \dots, z_2 t - z_1, z_1 t)$ is a regular sequence on $B(J,R)$. Put $L_i := (z_i t - z_{i-1}, \dots, z_2 t - z_1, z_1 t)$. Let $c = \sum_{j=1}^{i} r_j z_j \in J^n \cap (z_1, \dots, z_i)$ with $r_j \in R$. Then we have:

$$\left(ct^n\right) \cdot z_r^n = \sum_{j=1}^{i} r_j z_j t^n z_r^n = \sum_{j=1}^{i} \tilde{r}_j z_j z_r^n \mod L_i$$

for suitable $\tilde{r}_j \in R$, since for $j = 1, \dots, i$, $m \in \mathbf{N}$ every term of the form $z_j t^m z_r^n$, $m \in \mathbf{N}$, is either 0 or congruent $z_{j-1} t^{m-1} z_r^n \mod L_i$. Since z_r is regular $\mod L_i$, we can cancel $z_r^n \mod L_i$, and we obtain by comparing coefficients

$$c \in (z_1, \dots, z_i) J^{n-1}, \text{ for } i \in \{1, \dots, r\} .$$

Proof of Theorem (27.5). We assume that $R/\mathfrak{m}$ is infinite. Take a minimal reduction $\underline{a} = (a_1, \dots, a_s)$ of J. Moreover we can assume that I is $\mathfrak{m}$-primary. This can be seen by using Proposition (25.1): R is normally Cohen-Macaulay along I, so every system $\underline{b}$ of parameters $\mod I$ is a regular sequence on R. Hence if $B(I,R)$ and $B(J,R)$ are CM, then $B(I + \underline{b}/\underline{b}, R/\underline{b})$ and $B(J + \underline{b}/\underline{b}, R/\underline{b})$ are CM and $R/\underline{b}R$ is CM iff R is CM.
First we show the following claim:

$$(*) \qquad J^n \cap ((\underline{a}, x_1, \dots, x_{k-1}) : x_k) \subseteq (\underline{a}, x_1, \dots, x_{k-1}) + J^{n+1}, \; n \in \mathbf{N} \cup \{0\} .$$

Proof of the claim. For every element c of the left side we have

$$\begin{aligned} c \cdot x_k \in (\underline{a}, x_1, \dots, x_{k-1}) \cap I^{n+1} &\overset{(27.6)}{=} (\underline{a}, x_1, \dots, x_{k-1}) I^n \\ &\subseteq (x_1, \dots, x_{k-1}) + (\underline{a}) \cdot (J, x_k)^n \\ &\subseteq (x_1, \dots, x_{k-1}) + J^{n+1} + (\underline{a}) \cdot x_k \quad . \end{aligned}$$

Since R is normally Cohen-Macaulay along J, x_k is a non-zero-divisor $\mod (x_1, \dots, x_{k-1}) + J^{n+1}$.

$$c \in (\underline{a}) + (x_1, \dots, x_{k-1}) + J^{n+1} \quad .$$

Now we get by (*) modulo $(\underline{a}, x_1, \dots, x_{k-1})$:

$$J^n \cap (0:x_k) \subseteq J^{n+1} \cap (0:x_k) \quad \text{for all} \quad n \in \mathbf{N} \cup \{0\} \quad .$$

This implies by the intersection theorem of Krull

$$(0:x_k) = 0 \bmod(\underline{a},x_1,\ldots,x_{k-1}) \quad ,$$

hence

$$(**) \qquad (\underline{a},x_1,\ldots,x_{k-1}) : x_k = (\underline{a},x_1,\ldots,x_{k-1}) \quad .$$

Let $\mathfrak{p} \neq \mathfrak{m}$ be a prime ideal of R . Then "$B(I,R)$ is Cohen-Macaulay" implies:

$$B(I,R) \otimes R_{\mathfrak{p}} = R_{\mathfrak{p}}[It] = R_{\mathfrak{p}}[t] \quad \text{is CM} \quad ,$$

hence $R_{\mathfrak{p}}$ is Cohen-Macaulay [note that we assume I is $\mathfrak{m}$-primary].

Using the associative formula for multiplicities we obtain:

$$e((\underline{a},\underline{x})R) = \sum_{\mathfrak{p}} e(a_k(R/\mathfrak{p}))\cdot e((\underline{a},x_1,\ldots,x_{k-1})R_{\mathfrak{p}})$$

where $\mathfrak{p} \in \operatorname{Assh}(R/(\underline{a},x_1,\ldots,x_{k-1}))$. Note that $\dim(R/\mathfrak{p}) = 1$ and $ht(\mathfrak{p}) = d-1$, since R is quasi-unmixed (s. Chapter III, 18.17). Therefore

$$\begin{aligned} e((\underline{a},\underline{x})R) &= \sum_{\mathfrak{p}} e(x_k(R/\mathfrak{p}))\cdot \lambda(R_{\mathfrak{p}}/(\underline{a},x_1,\ldots,x_{k-1})R_{\mathfrak{p}}) \\ &= e(x_k(R/(\underline{a},x_1,\ldots,x_{k-1}))) \\ &= \lambda(R/(\underline{a},\underline{x})) \quad \text{by } (**) \quad , \end{aligned}$$

showing that R is Cohen-Macaulay, q.e.d. (Theorem (27.5)).

Finally we show that the Cohen-Macaulay-property of $R(I,R)$, $G(I,R)$ and $B(I,R)$ is transfered to the corresponding rings of I^n . For that we first prove the following lemma.

<u>(27.7) Lemma</u>. Let S be a noetherian ring and let T be a noetherian subring of S , such that S is a finite module over T and T is a direct summand of S as T-module. Suppose that $\dim S = \dim S_{\mathfrak{n}}$ for any maximal ideal $\mathfrak{n}$ of S . If S is Cohen-Macaulay then T

is Cohen-Macaulay.

Proof. Let us write $S = T \oplus W$ for some T-module W. For any maximal ideal $\mathfrak{m}$ of T we have $S_{\mathfrak{m}} = T_{\mathfrak{m}} \oplus W_{\mathfrak{m}}$ and $S_{\mathfrak{m}}$ is Cohen-Macaulay. Hence we may assume that T is local. Let $\underline{x} = \{x_1, \dots, x_d\}$ be a system of parameters of T. Then $S/\underline{x}S$ has finite length as T-module, since S is finite over T. Since $S_{\mathfrak{n}}$ is Cohen-Macaulay of dimension $d = \dim T$ for any maximal ideal $\mathfrak{n}$ of S by assumption, one sees that $\underline{x}$ is a system of parameters of $S_{\mathfrak{n}}$. Since S is Cohen-Macaulay, $\underline{x}$ is an $S_{\mathfrak{n}}$-sequence for any maximal ideal $\mathfrak{n}$ of S, and it follows that $\underline{x}$ is an S-sequence. Hence we see that $\underline{x}$ is a T-sequence because $x_i \in T$ and T is a direct summand of S as T-module.

<u>(27.8) Theorem</u>. Let I be any ideal in the noetherian ring R with $\mathrm{ht}(I) > 0$. Then the following implications hold for any $n \in \mathbf{N}$:

(i) $\mathcal{R}(I,R)$ Cohen-Macaulay $\Rightarrow$ $\mathcal{R}(I^n,R)$ Cohen-Macaulay

(ii) $B(I,R)$ Cohen-Macaulay $\Rightarrow$ $B(I^n,R)$ Cohen-Macaulay

(iii) $G(I,R)$ Cohen-Macaulay $\Rightarrow$ $G(I^n,R)$ Cohen-Macaulay.

Proof. (i) Let $T = R[I^n t^n, u^n]$ and $W = \bigoplus_{i \in \mathbf{Z} - n\mathbf{Z}} I^i t^i$. Then we have $\mathcal{R}(I,R) = T \oplus_T W$, and Lemma (27.7) shows that the ring

$$\mathcal{R}(I^n,R) = R[I^n t,u] \cong R[(It)^n,u^n] = T$$

is Cohen-Macaulay.

(ii) Put $T = R[I^n t^n]$, $W = \bigoplus_{i \in \mathbf{N} - n\mathbf{N}} I^i t^i$ and $S = R[It] = B(I,R)$. Then apply again Lemma (27.7).

(iii) Follows from Proposition (25.3).

References - Chapter V

Books

[1] A. Grothendieck and J. Diendonné, Éléments de Géometrie Algébrique. Publ. Math. IHES Paris, No. 24 (1965).

[2] M. Hochster, Topics in the homological theory of modules over commutative rings, CBMS regional conference, Series in Math. 24, Amer. Math. Soc. 1975.

[3] H. Matsumura, Commutative Algebra, W.A. Benjamin 1980.

[4] M. Nagata, Local rings, Krieger Huntington, N.Y. 1975.

[5] J. Sally, Numbers of generators of ideals in local rings, New York, Dekker 1978.

[6] O. Zariski and P. Samuel, Commutative Algebra II, Van Nostrand, Princeton, 1960 - 1965.

Papers

[7] G. Evans and P. Griffith, The syzygy problem, Ann. of Math. (2) 114 (1981), 323 - 333.

[8] S. Goto and Y. Shimoda, On the Rees algebras of Cohen-Macaulay local rings, in Comm. Algebra: Analytic methods. Lecture Notes in Pure and Applied Math. 68, Dekker, N.Y. 1981.

[9] U. Grothe, M. Herrmann und U. Orbanz, Graded Cohen-Macaulay rings associated to equimultiple ideals, Math. Z. 186 (1984), 531 - 556.

[10] C. Huneke, A remark concerning multiplicities, Proc. Amer. Math. Soc. 85, (1982), 331 - 332.

Appendix: (K. Yamagishi and U. Orbanz)

Homogeneous domains of minimal multiplicity

Definition. A graded domain $A = \bigoplus_{n \geq 0} A_n$ will be called a homogeneous domain (over k) if

a) A is noetherian;

b) $A_0 = k$ is an algebraically closed field;

c) $A = k[A_1]$.

For such a homogeneous domain A we will denote by A_+ the unique maximal homogeneous ideal of A and we put

$$e(A) = e(A_+, A)$$
$$\operatorname{embdim}(A) = \dim_k A_1 \quad .$$

Below we will give in particular a simple proof of the well known fact (s. [1], § 12 and [2] for local Cohen-Macaulay rings) that

$$\operatorname{embdim}(A) - \dim A + 1 \leq e(A) \quad .$$

If equality holds above, we will say that A has minimal multiplicity (also called maximal embedding dimension by some authors). Homogeneous domains with minimal multiplicity, which correspond to projective varieties of minimal degree, are completely classified (geometrically by Bertini in 1924, for details of an algebraic approach see Eisenbud-Goto [3], § 4). The paper by Eisenbud-Goto is based on the theory of linear resolution and on graded local cohomology that will be described in this book in Chapter VII. In this appendix we will give an elementary proof of a structure theorem for homogeneous domains of minimal multiplicity, which has been split in Theorem (A.1) and (A.5) in this appendix.

(A.1) Theorem. If A is a homogeneous domain (over an algebraically closed field k) with minimal multiplicity , then

a) $\operatorname{embdim}(A) - \dim A + 1 \leq e(A)$;

b) if $\operatorname{embdim}(A) - \dim A + 1 = e(A)$ and $\dim A \geq 2$ then A is normal and Cohen-Macaulay.

Before giving the proof we need some preliminary results.

(A.2) Proposition. The Theorem is true if $\dim A = 2$.

Proof. Let $\bar{A}$ be the integral closure of A . Then $\bar{A}$ is graded and has a unique maximal homogeneous ideal $\mathfrak{m}$. Clearly the canonical map

$$(*) \qquad A_1 \longrightarrow \mathfrak{m}/\mathfrak{m}^2$$

is injective and therefore $\mathrm{embdim}(A) \leqq \dim_k \mathfrak{m}/\mathfrak{m}^2 = \mathrm{embdim}(B)$, when $B = \bar{A}_{\mathfrak{m}}$. Also, by the projection formula for multiplicities,

$$e(A_+,A) = e(A_1B,B) \geqq e(B_+,B) \quad .$$

Since $\dim B = 2$ and B is normal, it is Cohen-Macaulay and satisfies

$$e(B_+,B) \geqq \mathrm{embdim}(B) - \dim B + 1 \quad .$$

Therefore

$$(**) \qquad \mathrm{embdim}(A) - \dim A + 1 \leqq \mathrm{embdim}(B) - \dim B + 1 \leqq e(B_+,B) \leqq e(A_+,A)$$

which proves part a). If now equality holds in (**) then equality holds in (*) too, proving that $\mathfrak{m}$ is generated by A_1 and hence $\bar{A} = k[A_1] = A$.

Remark. An easy agrument shows that actually $\mathfrak{m}$ is integral over the ideal $A_1\bar{A}$, so $e(A_1\bar{A},\bar{A}) = e(B_+,B) = e(A_+,A)$ in general.

We are going to use the following well-known form of Bertini's theorem:

Theorem [5]. If A is a domain and $\dim A \geqq 3$ then there exists an $f \in A_1$, $f \neq 0$, and a prime ideal $\mathfrak{p} \subset A$ such that

$$f \cdot A = \mathfrak{p} \cap \mathfrak{q} \quad \text{with} \quad \sqrt{\mathfrak{q}} = A_+ \left(= \bigoplus_{n>0} A_n \right) \quad .$$

To make use of Bertini's theorem we need two lemmas.

(A.3) Lemma. Let R be a noetherian domain satisfying the altitude formula, let f be a nonzero element of R and assume

$$fR = \mathfrak{p} \cap \mathfrak{q} \quad \text{where } \mathfrak{p} \text{ is prime and } \operatorname{ht}\left(\sqrt{\mathfrak{q}}\right) > 1 \ .$$

Then $f\overline{R} \cap R = \mathfrak{p}$. (Here $\overline{R}$ denotes the integral closure of R).

Proof. Let

$$f\overline{R} = \mathfrak{q}_1 \cap \ldots \cap \mathfrak{q}_n$$

be the primary decomposition of f in the Krull domain $\overline{R}$. Then $\operatorname{ht}\left(\sqrt{\mathfrak{q}_i}\right) = 1$ for each i , and since R satisfies the altitude formula, we have

$$\operatorname{ht}\left(\sqrt{\mathfrak{q}_i} \cap R\right) = 1 \quad \text{for } i = 1,\ldots,n \ ,$$

so $\sqrt{\mathfrak{q}_i} \cap R = \mathfrak{p}$ for all i . Now we put $\mathfrak{q}' = \mathfrak{q}_1 \cap \ldots \cap \mathfrak{q}_n \cap R$. Then $\mathfrak{q}'$ is $\mathfrak{p}$-primary and

$$fR = \mathfrak{p} \cap \mathfrak{q} \cap \mathfrak{q}' \ .$$

By uniqueness of the isolated components of a primary decomposition we must have $\mathfrak{q}' = \mathfrak{p}$ and hence

$$f\overline{R} \cap R = \mathfrak{q}' = \mathfrak{p} \ .$$

(A.4) Lemma. Let R be a local noetherian domain such that the integral closure $\overline{R}$ is a finite R-module. Let f be a nonzero element of R and assume that

$$fR = \mathfrak{p} \cap \mathfrak{q} \ , \ \mathfrak{p} \text{ a prime and } \operatorname{ht}\left(\sqrt{\mathfrak{q}}\right) > 1 \ .$$

If $R/\mathfrak{p}$ is normal then R is normal.

Proof. First, if $\dim R = 1$ then $fR = \mathfrak{p}$ = maximal ideal, so R is regular. Now let $\dim R > 1$, and let $S = R \setminus \mathfrak{p}$. Then R_S satisfies the same hypothesis as R and hence, by the case $\dim R = 1$, R_S is normal. This means that

$$R_S = \overline{R}_S \quad , \quad \text{a local ring} ,$$

and hence $\overline{R}$ contains a unique prime ideal $\mathfrak{q}$ such that $\mathfrak{q} \cap R = \mathfrak{p}$. Moreover, $f\overline{R} \cap R = f\overline{R}_S \cap R = fR_S \cap R = \mathfrak{p}$ and $f\overline{R} = \mathfrak{q}$. So there is a natural injection

$$R/\mathfrak{p} \hookrightarrow \overline{R}/f\overline{R} \quad .$$

Now, denoting by $Q(-)$ the quotient field of a domain, we have

$$Q(R/\mathfrak{p}) = R_{\mathfrak{p}}/\mathfrak{p}R_{\mathfrak{p}} = \overline{R}_S/\mathfrak{q}\overline{R}_S = Q(\overline{R}/f\overline{R}) \quad .$$

Since $R/\mathfrak{p}$ was assumed to be normal, we have $R/\mathfrak{p} = \overline{R}/f\overline{R}$. This means that

$$\overline{R} = R + f\overline{R} \quad ,$$

and since $\overline{R}$ is a finite R-module, Nakayama's lemma implies $R = \overline{R}$.

Proof of Theorem (A.1). For $\dim A = 1$ we choose any nonzero element $f \in A_1$. Then $e(A) = \ell(A/fA) \geqq 1 + \dim_k (A_1/f\cdot k) = \dim_k A_1$, proving a) in this case. For $\dim A \geqq 2$ we use induction on the dimension, the starting point being the case $\dim A = 2$ which was treated in the Proposition. So assume now $\dim A \geqq 3$ and choose $f \in A_1$ as in Bertini's theorem. Let $A' = A/\mathfrak{p}$, where $\mathfrak{p} = \sqrt{fA}$. Then, by Lemma (A.3), we have $\mathfrak{p} = f\overline{A} \cap A$ and hence

$$\mathfrak{p} \cap A_1 = f\overline{A} \cap A_1 = f\cdot k \quad .$$

so

$$\operatorname{embdim}(A') = \operatorname{embdim}(A) - 1 \quad .$$

Clearly

$$\dim A' = \dim(A) - 1$$

and

$$e(A_+,A) = e(A_+/fA, A/fA) = e(A_+/\mathfrak{p}, A/\mathfrak{p})$$

(the first equality since f belongs to a minimal reduction of A_+, the second equality since the embedded component does not contribute to the multiplicity). So using the inductive assumption for A' we get the desired inequality for A.

Assume moreover that

$$\operatorname{embdim}(A) - \dim A + 1 = e(A_+, A) \quad .$$

Then, by our construction,

$$\operatorname{embdim}(A/\mathfrak{p}) = \dim(A/\mathfrak{p}) + 1 = e(A_+/\mathfrak{p}, A/\mathfrak{p}) \quad .$$

So our induction assumption implies that $A/\mathfrak{p}$ is normal and hence A is normal by Lemma (A.4). Finally, A normal implies

$$fA = \mathfrak{p} \quad ,$$

and so $A/\mathfrak{p} = A/fA$, which is Cohen-Macaulay by inductive assumption again. So A is Cohen-Macaulay.

Remark 1. The proof above gives the following additional information: If A is a domain of dimension $d \geqq 2$, satisfying

$$(***) \qquad \operatorname{embdim}(A) - \dim(A) + 1 = e(A_+, A)$$

then A_1 contains elements $f_1, \ldots, f_{d-2}$ such that

$$A/(f_1, \ldots, f_{d-2})A =: C$$

is a 2-dimensional normal domain satisfying (***) again. These C are known to be isomorphic to

$$K\left[t^n, t^{n-1}u, \ldots, u^{n-1}t, u^n\right] \quad , \quad n = e(C_+, C) \quad .$$

Using this information, one might hope to find an "easy" proof for the classification of projective (irreducible) varieties of minimal degree.

Remark 2. Lemmas 1 and 2 above may be generalized to reduced rings. Therefore Theorem (A.1) might be generalized as well if some assumption assures that

$$(\bar{A})_0 = \text{degree zero part of } \bar{A} = k = A_0 \quad .$$

In the next Theorem (A.5) we show when - under the same assumptions as in Theorem (A.1) - $\mathrm{Proj}(A)$ is regular. For the proof of this statement we refer to the following Proposition (*) on homomgeneous domains due to Abhyankar. For that let A be any homogeneous domain over a field k (not necessarily algebraically closed) and let L be a k-vector subspace of A_1 which has $\dim_k L = \dim_k(A_1) - 1$ $(=\mathrm{embdim}(A) - 1)$. We consider the homogeneous subdomain $C := k[L]$ of A and the homogeneous ideal LA of A generated by L. [$\mathfrak{m}$ denotes the maximal homogeneous ideal of A.]

Proposition (*) ([A_1], (12.1.6) and (12.3.4)). Suppose that $\sqrt{LA} \neq \mathfrak{m}$. Then we have the following statements:

(i) $LA \in \mathrm{Proj}(A)$;

(ii) if $\dim(C) = \dim(A)$, then

$$e(A) = e(C)\cdot[k(A) : k(C)] + e(A_{(LA)}) \quad ;$$

(iii) if $\dim(C) < \dim(A)$, then any element $Z \in A_1 \setminus L$ is algebraically independent over C and $A \cong C \otimes_k k[Z]$.

(A.5) Theorem. Let A be a homogeneous domain over an algebraically closed field k with minimal multiplicity. Then we find an homogeneous domain B contained in A and some elements $Z_1,\dots,Z_r \in A_1$, $r \geq 0$, algebraically independent over B, which satisfy the following properties:

(i) $A \cong B \otimes_k k[Z_1,\dots,Z_r]$

(ii) B has minimal multiplicty

(iii) if L is a k-vector subspace of B_1 with $\dim_k L = \dim_k(B_1) - 1$ and $\sqrt{LB} \neq B_+$, then $\dim(B) = \dim(k[L])$.

(iv) $\mathrm{Proj}(B)$ is regular.

Proof. a) We construct B and suitable elements Z_i by induction on $\dim(A)$. Note that any 1-dimensional normal homogeneous domain is a polynomial ring over k in one indeterminate, hence $\mathrm{embdim}(A) = \dim(A)$. Then A is a polynomial ring over k and we have nothing to prove. Therefore, we may assume that $\mathrm{embdim}(A) > \dim(A)$. Then we know that the normal ring A (see Theorem (A.1)) has $\dim(A) \geq 2$. It is easy to see that for $\dim(A) = 2$ there is no k-subspace

L of A_1 satisfying the following three conditions:

$$\dim_k L = \operatorname{embdim}(A) - 1 \ , \ \sqrt{LA} \neq \mathfrak{m} \ , \ \dim(k[L]) < \dim(A) \quad .$$

[Otherwise $k[L]$ would be an one-dimensional homogeneous domain with minimal multiplicty by (Proposition (*), (iii), and hence $k[L]$ had to be a one-dimensional normal homogeneous domain by Theorem (A.1). But this means that $k[L]$ would be a polynomial ring, which is a contradiction.] Therefore we may take $B = A$ and $r = 0$ in this case.

Assume that $\dim(A) \geqq 3$. We consider two cases.

<u>Case 1</u>: If there is no k-subspace L of A_1 such that $\dim_k L = \operatorname{embdim}(A) - 1$, $\sqrt{LA} \neq \mathfrak{m}$ and $\dim(k[L]) < \dim(A)$, then we take again $B = A$ and $r = 0$.

<u>Case 2</u>: If there exists a k-subspace L of A_1 with these three properties. Then any element $Z_1 \in A_1 \setminus L$ is algebraically independent over $k[L]$ by Proposition (*), (iii) and $A \cong k[L] \otimes_k k[Z_1]$. Applying the induction hypothesis to $k[L]$ we find a homogeneous subdomain B of $k[L]$ and algebraically independent elements $Z_2,\ldots,Z_r \in L$ over B, $r \geqq 1$, satisfying the following properties:

(i)' $k[L] \cong B \otimes_k k[Z_2,\ldots,Z_r]$,

(ii)' B has minimal multiplicity ,

(iii)' case (iii) in Proposition (*) doesn't occur for B .

Since $A \cong k[L] \otimes_k k[Z_1] \cong B \otimes_k k[Z_1,Z_2,\ldots,Z_r]$, we are done.

b) Finally we have to show that $\operatorname{Proj}(B)$ is regular. Since B is normal by Theorem (A.1) we may assume that $\dim(B) \geqq 3$. Let $P \in \operatorname{Proj}(B)$ so that $\dim(B/P) = 1$. Then $\dim_k P_1 = \dim_k B_1 - 1$. Let $C := k[P_1]$. Applying Proposition (*) to C we get from Condition (iii) of the theorem that

$$\dim(C) = \dim(B) \quad \text{and} \quad e(B) \geqq e(C) + e(B_{(P)}) \geqq e(C) + 1 \quad .$$

From these facts we conclude that

$$\begin{aligned} \operatorname{embdim}(B) - 1 &= \operatorname{embdim}(C) \\ &\leqq e(C) + \dim(C) - 1 \\ &\leqq (e(B) - 1) + \dim(B) - 1 \\ &= \operatorname{embdim}(B) - 1 \quad . \end{aligned}$$

This means $e(B) = e(C) + 1$, hence $e(B_{(P)}) = e(B_P) = 1$. Since B_P is Cohen-Macaulay we see that B_P is regular by Chapter I, (6.8). This proves Theorem (A.5).

Example 1. If k is not algebraically closed, our arguments do not work in general. In particular, the inequality $\operatorname{embdim}(A) \leqq e(A) + \dim(A) - 1$ must be changed even in the following special situation: Let A be a homogeneous domain over a field k and assume that the algebraic closure $\bar{k}$ of k is contained in the quotient field of A . Then we claim that

$$\operatorname{embdim}(A) \leqq e(A) + [\bar{k}:k]\cdot(\dim(A) - 1) \quad .$$

In fact, take $D := \bar{k}[A_1]$ in the quotient field of A . D is a homogeneous domain over the field $\bar{k}$, and it has the same quotient field as A . So we get

$$\operatorname{embdim}(D) \leqq e(D) + \dim(D) - 1$$

and (since A and D have the same quotient field)

$$e(A) = e(\mathfrak{m},D) \quad .$$

Since $\mathfrak{m}^n D = (D_+)^n$ for all $n > 0$, we get from the projection formula

$$e(\mathfrak{m},D) = [\bar{k} : k]\cdot e(D) \quad .$$

Therefore

$$\begin{aligned} \operatorname{embdim}(A) &\leqq [\bar{k}:k] \cdot \operatorname{embdim}(D) \\ &\leqq [\bar{k}:k] \cdot e(D) + [\bar{k}:k](\dim(D) - 1) \\ &\leqq e(A) + [\bar{k}:k] \cdot (\dim(A) - 1) \quad . \end{aligned}$$

For example take

$$A := \mathbf{R}[X,Y,iX,iY]$$

in $\mathbf{C}[X,Y]$, where $\mathbf{R}$ and $\mathbf{C}$ denote as usual the real number field and the complex number field, respectively and X,Y are indeterminates over $\mathbf{C}$. Then

$$\operatorname{embdim}(A) = 4 = e(A) + [\mathbf{C} : \mathbf{R}] \cdot (\dim(A) - 1) \quad .$$

Example 2. If A is not homogeneous, then again our arguments do not work in general. Recall that in the non-homogeneous cases

$$\operatorname{embdim}(A) = \dim_k \mathfrak{m}/\mathfrak{m}^2 \quad , \text{ where } \quad \mathfrak{m} = \bigoplus_{n>0} A_n \quad .$$

For example take

$$A := k[X^2,XY,Y] \quad ;$$
$$B := k[X^2,X^3,XY,Y]$$

in the polynomial ring $k[X,Y]$. Then

a) A is a Cohen-Macaulay ring which satisfies the equality $\operatorname{embdim}(A) = 3 = e(A) + \dim(A) - 1$, but A is not normal.

b) B does not satisfy even Abhyankar's inequality, since $\operatorname{embdim}(B) = 4 \nleq 3 = e(B) + \dim(B) - 1$.

Example 3. If A is not an integral domain our arguments do not hold in general. For example take

$$A := k[X,Y,Z]/(X^2) \cap (Y,Z) \quad ;$$

$$B := k[X,Y,Z,W]/(X,Y) \cap (Z,W) \quad ,$$

where X,Y,Z and W are indeterminates over $k = \bar{k}$. Then we get:

a) A satisfies the equality $\operatorname{embdim}(A) = 3 = e(A) + \dim(A) - 1$;

b) A is not Cohen-Macaulay;

c) B does not satisfy Abhyankar's inequality, since $\operatorname{embdim}(B) = 4 \nleq 3 = e(B) + \dim(B) - 1$.

On the other hand D. Eisenbud and S. Goto [3] have shown the Cohen-Macaulayness for homogeneous *reduced* rings A over an algebraically closed field satisfying the following inequality (which is contrary to Abhyankar's inequality)

$$e(A) \leqq 1 + \mathrm{embdim}(A) - \dim(A)$$

under the assumption of some topological condition (namely that A is connected in codimension 1). Without this assumption their arguments do not work. For example, the ring B above satisfies all conditions, but it is not Cohen-Macaulay. We remark that B is a so called Buchsbaum ring (see Chapter VIII, (41.14). For any Buchsbaum ring A we know that

$$\mathrm{embdim}(A) \leqq e(A) + \dim(A) - 1 + I(A) \quad ,$$

where $I(A)$ denotes the so called Buchsbaum invariant of A . In the mentioned example the invariant is $I(A) = 1$, so we get the equality

$$\mathrm{embdim}(B) = e(B) + \dim(B) - 1 + I(B) \quad .$$

We refer the reader to [4] for details on Buchsbaum rings which satisfy this equality.

References - Appendix - Chapter V

[1] S.S. Abhyankar, Resolution of singularities of embedded algebraic surfaces, Academic Press, New York and London, 1966.

[2] S.S. Abhyankar, Local rings of high embedding dimension, Amer. J. Math. 89, (1967), 1073 - 1077.

[3] D. Eisenbud and S. Goto, Linear free resolutions and minimal multiplicity, J. Alg. 88, (1984), 89 - 133.

[4] S. Goto, Buchsbaum rings of maximal embedding dimension, J. Alg. 76, (1982), 383 - 399.

[5] A. Seidenberg, The hyperplane sections of normal varieties, Trans. Amer. Math. Soc. 69, (1950), 357 - 386.

Chapter VI. CERTAIN INEQUALITIES AND EQUALITIES OF HILBERT FUNCTIONS AND MULTIPLICITIES

In this chapter we mainly study the behaviour of (generalized) Hilbert functions and (generalized) multiplicities of local rings R under blowing up an ideal $I \subset R$ such that R/I need not be regular. After some preliminaries in Section 28 we have to present in Section 29 a result of Singh on Hilbert functions under quadratic transformations. Using this result one can prove in Section 30 the semicontinuity of Hilbert functions by desingularizing curves. Finally for inequalities of Hilbert functions under blowing up other centers one has to apply this semicontinuity. The last Section 32 is related to equisingularity theory via flat families. As before (R,m,k) is again a noetherian local ring and I a proper ideal of R.

§ 28. Hyperplane sections

(28.1) Proposition. Let t a positive integer, let $f \in I^t$ and put

$$D_n = (I^{n+t} : fR)/I^n \quad \text{for all} \quad n \in \mathbf{Z}\ , \quad I^n = R \quad \text{for} \quad n < 0 \ .$$

Let $\overline{R} = R/fR$ and $\overline{I} = I/fR$. Then for any system of parameters $\underline{x} = (x_1,\dots,x_r)$ with respect to I we have for $n \in \mathbf{Z}$:

$$H^{(1)}[\underline{x},\overline{I},\overline{R}](n) = H^{(1)}[\underline{x},I,R](n) - H^{(1)}[\underline{x},I,R](n-t) + e(\underline{x};D_{n-t+1})$$

In particular we have

$$H^{(0)}[\underline{x},I,R] \leqq H^{(1)}[\underline{x},\overline{I},\overline{R}] \quad .$$

Proof. We may assume $n > 0$; then the claim of the proposition follows from the following exact sequence (see also [1], p. 162):

$$0 \longrightarrow D_{n-t+1} \longrightarrow R/I^{n-t+1} \xrightarrow{\cdot f} R/I^{n+1} \longrightarrow \overline{R}/\overline{I}^{n+1} \longrightarrow 0 \quad ,$$

using the additivity of $e(\underline{x};-)$.

(28.2) Corollary.

(i) If $in(I)(f)$ is a non-zero-divisor of degree t in $G(I;R)$, then

$$(*) \quad H^{(1)}[\underline{y},\bar{I},\bar{R}](n) = H^{(1)}[\underline{x},I,R](n) - H^{(1)}[\underline{x},I,R](n-t) \quad .$$

(ii) Conversely, if (*) holds for some system of parameters $\underline{x}$ of R/I and $y_i = x_i \bmod fR$ and if in addition $Ass(I^n/I^{n+1}) = Assh(R/I)$, $n \geq 0$, then $in(I)(f)$ is a non-zero-divisor of degree t in $G(I,R)$.

Proof of the Corollary.

(i) is clear by the proposition

(ii): If $r = \dim(R/I) = 0$ (in which case the assumption " $Ass(I^n/I^{n+1}) = Assh(R/I)$" is empty), we conclude from

$$e(\underline{x},D_n) = \lambda_R(D_n) = 0 \quad \text{for all} \quad n \quad ,$$

that $I^{n+t} : fR = I^n$, i.e. $in(I)(f)$ is a non-zero-divisor of degree t in $G(I,R)$. For general r we have

$$H^{(1)}\left[\underline{x},I,R\right] = \sum_{\mathfrak{p}\in Assh(R/I)} e(\underline{x};R/\mathfrak{p})\cdot H^{(1)}\left[IR_{\mathfrak{p}},R_{\mathfrak{p}}\right]$$

and

$$H^{(1)}\left[\underline{y},\bar{I},\bar{R}\right] = \sum_{\bar{\mathfrak{p}}\in Assh(\bar{R}/\bar{I})} e(\underline{y},\bar{R}/\bar{\mathfrak{p}})\cdot H^{(1)}\left[\bar{I}\bar{R}_{\bar{\mathfrak{p}}},\bar{R}_{\bar{\mathfrak{p}}}\right] \quad ,$$

where $\bar{\mathfrak{p}} = \mathfrak{p}/fR$. Therefore we get by assumption for all $\mathfrak{p} \in Assh(R/I)$

$$(**) \quad H^{(1)}\left[\bar{I}\bar{R}_{\bar{\mathfrak{p}}},\bar{R}_{\bar{\mathfrak{p}}}\right](n) = H^{(1)}\left[IR_{\mathfrak{p}},R_{\mathfrak{p}}\right](n) - H^{(1)}\left[IR_{\mathfrak{p}},R_{\mathfrak{p}}\right](n-t) \quad .$$

Now consider the homomorphism

$$I^n/I^{n+1} \xrightarrow{\cdot f} I^{n+t}/I^{n+t+1} \quad .$$

Then (**) and $r = 0$ imply that $f \otimes R_{\mathfrak{p}}$ is injective for all $\mathfrak{p} \in \mathrm{Assh}(R/I)$. Hence f is injective since $\mathrm{Assh}(R/I) = \mathrm{Ass}(I^n/I^{n+1})$.

Example. s. [1], p. 164: Let $K \subset \mathbf{P}^3_{\mathbf{C}}$ be the curve with the generic point $\{t_1^4, t_1^3 t_2, t_1 t_2^3, t_2^4\}$. Note that K has no singularities.

The affine cone X over K is defined in $\mathbf{A}^4_{\mathbf{C}}$ by the following prime ideal

$$P_X = \left(x_1x_4 - x_2x_3 \,,\, x_1^2x_3 - x_2^3 \,,\, x_1x_3^2 - x_2^2x_4 \,,\, x_2x_4^2 - x_3^3\right) \quad .$$

Let Y be the line on X, given by $x_1 = x_2 = x_3 = 0$. We denote by s the vertex of X. Let $(R,\mathfrak{m})$ be the local ring of X at s. Then we get by Proposition (28.1)

$$H^{(0)}[\mathfrak{m},R] \leqq H^{(3)}[\overline{\mathfrak{m}},\overline{R}] \quad ,$$

where $\overline{R} = R/(x_1,x_2,x_3)R$, $\overline{\mathfrak{m}} = \mathfrak{m}/(x_1,x_2,x_3)R$. Note that $\dim R - \dim \overline{R} = 1$. To get an inequality which allows a "geometric" interpretation by comparing the singularities of $(R,\mathfrak{m})$ and $(\overline{R},\overline{\mathfrak{m}})$, we would like to have on the right side the upper index 1 instead of 3. This can be reached as follows: Note that $\mathrm{rad}(x_1R) = (x_1,x_2,x_3)R$. Since $\mathrm{in}(\mathfrak{m})(x_1)$ is a non-zero-divisor of degree 1 in $G(\mathfrak{m},R)$, we get by Corollary (28.2):

$$H^{(0)}[\mathfrak{m},R] = H^{(1)}[\mathfrak{m}/x_1, R/x_1R] \quad .$$

Then we can say that the singularity $s \in X$ is of the "same complexity" as the point $x_1 = x_2 = x_3 = x_4 = 0$ on Y, now taken with a certain multiplicity.

§ 29. Quadratic transformations

The following theorem was proved by B. Singh [13] for the blowing up of any permissible ideal $\mathfrak{p}$ (i.e. R is normally flat along $\mathfrak{p}$ and $R/\mathfrak{p}$ is regular). We only need the case $\mathfrak{p} = \mathfrak{m}$ to show later a strong generalization of Singh's result.

(29.1) Theorem. Let $(R,\mathfrak{m},k)$ be a local ring and $(R',\mathfrak{m}',k')$ a quadratic transformation. Then we have

$$H^{(0)}[R] \geqq H^{(d)}[R'] \quad ,$$

where d is the transcendence degree of k' over k.

(29.2) Remark. Following Singh in full detail we first prove the theorem in the residually rational case, i.e. $k' \cong k$, then in case when k'/k is algebraic (necessarily finite) by induction on the field degree $[k' : k]$. Here the most difficult part is the purely inseparable case. (At this place the new ideas of Singh in [13] come in. All other cases were already contained in Hironaka's [7], [8] and Bennett's [4] work.) The proof of Theorem (29.1) will finally be completed by proving it in the case that k' is an arbitrary (finitely generated) extension of k. This will be done by induction on $d = \text{tr deg}_k k'$. Only in the residually rational case the proof for $\mathfrak{p} = \mathfrak{m}$ is shorter that this one for any permissible prime $\mathfrak{p}$.

Proof of Theorem (29.1).

Case 1. $k' \cong k$. We fix the following notation:

$$R_1 = R[\mathfrak{m}/t] \text{ , where } \mathfrak{m} = (t,x_1,\ldots,x_r) \text{ and}$$
$$\mathfrak{m}_1 = \mathfrak{m}R_1 + (x_1/t,\ldots,x_r/t)R_1 \text{ .}$$

$$R' = (R_1)_{\mathfrak{m}_1} \text{ , where } \mathfrak{m}R' = tR' \text{ .}$$

$$R'' = R'/\mathfrak{m}R' \text{ with maximal ideal } \mathfrak{m}'' \text{ .}$$

Note that $k'' = R''/\mathfrak{m}'' = k' = k$ in case 1 and $R'' = k[\bar{x}_1/t,\ldots,\bar{x}_r/\bar{t}]_{\mathfrak{m}_1/t}$, where $\bar{t}$, $\bar{x}_i$ are the respective initial forms in $G(\mathfrak{m};R)$.

We prove the following lemma:

(29.3) Lemma. a) $H^{(0)}[R] \geq H^{(1)}[R'']$ and

b) $H^{(1)}[R''] \geq H^{(0)}[R']$.

Proof of the Lemma. a) Since $H^{(0)}[R] = \dim_k(m^n/m^{n+1})$ and $H^{(1)}[R''] = \dim_k(R''/m''^{n+1})$, it is enough to show that there exists a surjective k-homomorphism

$$\varphi : m^n/m^{n+1} \longrightarrow R''/m''^{n+1} \quad .$$

For that we define a R-homomorphism

$$\psi : m^n \longrightarrow R''$$

by $\psi(f) = \eta\left(\frac{f}{t^n}\right)$, where η is the natural homomorphism $R' \longrightarrow R''$. Since $m \cdot m^n t^{-n} \subset mR_1$, we have $\psi(m^{n+1}) = \psi(m \cdot m^n) = 0$, i.e. ψ induces a k-vector space homomorphism $\overline{\psi} : m^n/m^{n+1} \longrightarrow R''$. Then we define φ to be the composite of $\overline{\psi}$ and the natural homomorphism $R'' \longrightarrow R''/m''^{n+1}$. Writing $x/t = (x_1/t,\dots,x_r/t)$ and $v = (\eta(x_1/t),\dots,\eta(x_r/t))$ we have

$$R'' = k[v]_{vk[v]} \quad \text{and} \quad R''/m''^{n+1} = k[v]/v^{n+1}k[v] \quad .$$

Therefore any element $g \in R''/m''^{n+1}$ is the image of an element $\sum_{|\alpha| \leq n} a_\alpha v^\alpha \in k[v]$, $a_\alpha \in k$. Taking a lift b_α of a_α , the element $f = \sum_{|\alpha| \leq n} b_\alpha t^{n-|\alpha|} x^\alpha \in m^n$ satisfies $\varphi(f \bmod m^{n+1}) = g$, hence φ is surjective.

b) Since $R'' = R'/tR'$, the second inequality follows from Proposition (28.1).

Case 2. k'/k is algebraic. We prove $H^{(0)}[R] \geq H^{(0)}[R']$ by induction on $[k':k]$. If $[k':k] = 1$ we have case 1. If $[k':k] > 1$, choose an element $\alpha \in k' \setminus k$ such that α is either separable or purely inseparable over k . Let $\overline{f}(X) \in k[X]$ be the irreducible monic polynomial of α over k and let $f(X) \in R[X]$ be a monic lift of $\overline{f}(X)$.

We consider the local ring $\tilde{R} = R[X]/(f(X))$ with the maximal ideal $\tilde{m} = m\tilde{R}$. We denote the natural homomorphism $R \longrightarrow \tilde{R}$ by η .

Claim. There exists a commutative diagram (i.e. there exists a suitable $\tilde{R}'$):

$$\begin{array}{ccc} R & \xrightarrow{\ h\ } & R' \\ {\scriptstyle\eta}\downarrow & & \downarrow{\scriptstyle\eta'} \\ \tilde{R} & \xrightarrow{\ \tilde{h}\ } & \tilde{R}' \end{array}$$

with a corresponding residue fields diagram

$$\begin{array}{ccc} k & \longrightarrow & k' \\ \downarrow & & \| \\ k(\alpha) & \hookrightarrow & k' \end{array}$$

satisfying the following properties. (Note that h is a blowing up of R with center m):

(i) $\tilde{h}$ is a quadratic blowing up of $\tilde{R}$

(ii) If α is separable, then η' is etale

(iii) If α is purely inseparable, then $\tilde{R}' = R'[X]/f(X) \cdot R'[X]$.

[With "η' is etale" we mean that four conditions are fulfilled:

1) $\tilde{R}'$ is flat over R'

2) $m'\tilde{R}' = \tilde{m}'$, where m' , $\tilde{m}'$ are the respective maximal ideals

3) $\tilde{R}'$ is a localization of an R'-algebra of finite type

4) $\tilde{k}'$ is a finite separable extension of the residue field of R'].

Proof of the claim. To define $\tilde{R}'$, η' and $\tilde{h}$, let a be a lift of α to R' . We put

$$n = m'[X] + (X-a)R'[X] \subset R'[X]$$

where m' is the maximal ideal of R' . Then we define

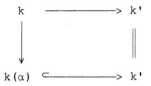

$$\tilde{R}' = (R'[X]/f(X)\cdot R'[X])_{\mathfrak{n}/f(X)} ,$$

and η' is the natural homomorphism $R' \longrightarrow \tilde{R}'$. Of course, $\tilde{h}$ is the homomorphism induced by h :

(i) Then by construction $\tilde{h}$ is a quadratic transform of $\tilde{R}$.

(ii) Moreover, since η is flat, the map η' is flat. The equality $\tilde{\mathfrak{m}}' = \mathfrak{m}'\tilde{R}'$ follows from the fact that $(f(X)R'[X] + \mathfrak{m}'[X])_{\mathfrak{n}} = ((X-a)R'[X] + \mathfrak{m}'[X])_{\mathfrak{n}}$ in case that α is separable. Since Properties 3) and 4) are trivially fulfilled for η', we see that η' is etale.

(iii) If α is purely inseparable, then $R'[X]/f(X)\cdot R'[X]$ is already local, since $\bar{f}(X) = (X-\alpha)^{p^r}\cdot k'[X]$. This proves the claim.

Now, since η is flat and $\tilde{\mathfrak{m}} = \mathfrak{m}\tilde{R}$, we have by Chapter I

$$G(\mathfrak{m},R) \otimes_{R/\mathfrak{m}} \tilde{R}/\tilde{\mathfrak{m}} \longrightarrow G(\tilde{\mathfrak{m}},\tilde{R}) ,$$

which means $H^{(0)}[R] = H^{(0)}[\tilde{R}]$. Since $[k' : k(\alpha)] \lneqq [k' : k]$, we know by induction hypothesis that

$$H^{(0)}[R] = H^{(0)}[\tilde{R}] \geqq H^{(0)}[\tilde{R}'] .$$

So it remains to be proved that $H^{(0)}[\tilde{R}'] \geqq H^{(0)}[R']$: If α is separable, then η' is etale, hence we have even equality.

The hard case is if α is purely inseparable. Then $\bar{f}(X) = X^q-\beta$ for some $\beta \in k$ and $q = p^r$, $p = \operatorname{char} k$; i.e..

$$\tilde{R}' = R'[X]/(X^q-b)R'[X] ,$$

where b is a lift of β to R. Note that $t := b-a^q \in \mathfrak{m}'$. Hence with $Y = X-a$ we get

$$\tilde{R}' = R'[Y]/(Y^q-t)R'[Y] .$$

The desired inequality $H^{(0)}[R'] \geqq H^{(0)}[R']$ now follows from Singh's main Lemma (29.4), which we prove now.

<u>(29.4) Lemma</u>. Let $(R,\mathfrak{m})$ be a (noetherian) local ring and let

$t \in \mathfrak{m}$. Then for $\tilde{R} := R[Y]/(Y^q-t)R[Y]$ with the maximal ideal, say $\tilde{\mathfrak{m}}$, we have:

$$H^{(0)}[\tilde{R}] \geq H^{(0)}[R] \quad .$$

Proof. We define the following sequence $\{\mathfrak{a}_n\}_{n\in Z}$ of ideals of R .

$$\mathfrak{a}_n = R \qquad \text{if} \quad n \leq 0$$

$$\mathfrak{a}_n = \mathfrak{m}^n + t\mathfrak{a}_{n-q} \qquad \text{if} \quad n > 0 \quad .$$

For all $n \in \mathbf{Z}$ one has

$$\mathfrak{m}\mathfrak{a}_n \subset \mathfrak{a}_{n+1} \subset \mathfrak{a}_n$$

$$t\mathfrak{a}_n \subset \mathfrak{a}_{n+q} \quad .$$

Putting $y := Y + (Y^q-t)\cdot R[Y]$ and indentifying R with its image in $\tilde{R}$ we have:

$$y^q = t \quad ,$$

$$\tilde{R} = \bigoplus_{i=0}^{q-1} y^i R \quad ,$$

$$\tilde{\mathfrak{m}} = \mathfrak{m} \oplus (\bigoplus_i y^i R) \quad .$$

To compute $\tilde{\mathfrak{m}}^n/\tilde{\mathfrak{m}}^{n+1}$ and $\dim_k(\tilde{\mathfrak{m}}^n/\tilde{\mathfrak{m}}^{n+1})$ we first consider $\tilde{\mathfrak{m}}^n$.

<u>Claim</u>. $\tilde{\mathfrak{m}}^n = \bigoplus_{i=0}^{q-1} \mathfrak{a}_{n-i} y^i$.

Proof (by induction on n). The case $n = 0$ is clear. For $n \geq 0$ one has to go to the following steps (where we omit some elementary details in the computation) by induction hypothesis:

$$\tilde{\mathfrak{m}}^{n+1} = \tilde{\mathfrak{m}}\cdot\tilde{\mathfrak{m}}^n = \left(\mathfrak{m} \oplus \left(\bigoplus_{i=1}^{q-1} y^i R\right)\right) \cdot \left(\bigoplus_{j=0}^{q-1} \mathfrak{a}_{n-j} y^i\right)$$

$$= \bigoplus_{j=0}^{q-1} \mathfrak{m}\mathfrak{a}_{n-j} y^i + \sum_{i+j\leq q-1} \mathfrak{a}_{n-j} y^{i+j} + \sum_{i+j\geq q} t\mathfrak{a}_{n-j} y^{i+j-q} \quad ,$$

where in the last two summations i and j vary over the sets $\{1,2,\dots,q-1\}$ and $\{0,1,\dots,q-1\}$ respectively. We denote the coefficient of y^h in this expression for $\tilde{\mathfrak{m}}^{n+1}$ by $\mathfrak{b}_h$, where $0 \leqq h \leqq q-1$. Now an easy computation shows that $\mathfrak{b}_h = \mathfrak{a}_{(n+1)-h}$. This proves the claim. Therefore we get

$$\tilde{\mathfrak{m}}^n/\tilde{\mathfrak{m}}^{n+1} \cong \bigoplus_{i=0}^{q-1} \mathfrak{a}_{n-i}/\mathfrak{a}_{n+1-i}$$

as R-modules, since the $\{y^i\}$ form a free R-base. Then considering that $k = R/\mathfrak{m} = \tilde{R}/\tilde{\mathfrak{m}}$, we obtain

$$H^{(0)}[\tilde{R}](n) = \lambda_R(\mathfrak{a}_{n+1-q}/\mathfrak{a}_{n+1}) \quad .$$

Now since $\mathfrak{m}^{n+1} \subset \mathfrak{a}_{n+1}$ we have the exact sequence

$$(1) \qquad 0 \longrightarrow \mathfrak{a}_{n+1}/\mathfrak{m}^{n+1} \longrightarrow \mathfrak{a}_{n+1-q}/\mathfrak{m}^{n+1} \longrightarrow \mathfrak{a}_{n+1-1}/\mathfrak{a}_{n+1} \longrightarrow 0 \quad .$$

Moreover we have the exact sequence

$$(2) \qquad 0 \longrightarrow \mathfrak{m}^n/\mathfrak{m}^{n+1} \longrightarrow \mathfrak{a}_{n+1-q}/\mathfrak{m}^{n+1} \longrightarrow \mathfrak{a}_{n+1-q}/\mathfrak{m}^n \longrightarrow 0 \quad .$$

Finally by multiplication by t we get the exact sequence

$$(3) \qquad \mathfrak{a}_{n+1-q}/\mathfrak{m}^n \xrightarrow{\cdot t} \mathfrak{a}_{n+1}/\mathfrak{m}^{n+1} \longrightarrow 0 \quad ;$$

for the exactness of (3) we use that $\mathfrak{a}_{n+1} = \mathfrak{m}^{n+1} + t\mathfrak{a}_{n+1-q}$. From these exact sequences we conclude that

$$H^{(0)}[\tilde{R}](n) \geqq H^{(0)}[R](n) \quad .$$

This proves the lemma and the Theorem (29.1) in the residually algebraic case.

<u>Case 3</u>. k' is an arbitrary extension of k (necessarily finitely generated). Let $d := \operatorname{tr} \deg_k k'$. We prove $H^{(0)}[R] \geqq H^{(d)}[R']$ by induction on d, since for $d = 0$ the inequality is already proved.

Assume $d \geqq 1$: The reduction procedure is now similar to that we had in Case 2 :

Let $\alpha \in k'$, transcendental over k, let a be a lift of α to R'. Define $R^* := R[X]_{\mathfrak{m}[X]}$ and $R^{*\prime} = R'[X]_{\mathfrak{n}}$, where $\mathfrak{n} := \mathfrak{m}'[X] + (X-a)R'[X]$. Then we have a commutative diagram

$$(*) \qquad \begin{array}{ccc} R & \xrightarrow{\ h\ } & R' \cong R^{*\prime}/(X-a)R^{*\prime} \\ {\scriptstyle \eta}\downarrow & & \downarrow{\scriptstyle \eta'} \\ R^* & \xrightarrow{\ h^*\ } & R^{*\prime} \end{array}$$

of local homomorphisms, where h^* is induced by h. Now it is clear by the flatness of η and by Proposition (28.1) that

$$H^{(0)}[R] = H^{(0)}[R^*] \quad \text{and} \quad H^{(1)}[R'] = H^{(0)}[R^{*\prime}] ,$$

hence $H^{(d-1)}[R^{*\prime}] = H^{(d)}[R']$. Then to prove $H^{(0)}[R] \geqq H^{(d)}[R']$ it is enough to show that $H^{(0)}[R^*] \geqq H^{(d-1)}[R^{*\prime}]$.

Clearly, h^* is a quadratic transformation (with center $\mathfrak{m}^* = \mathfrak{m}R^*$). Moreover the residue field diagram induced by $(*)$ is

$$\begin{array}{ccc} k & \longrightarrow & k' \\ \downarrow & & \\ k(X) & \longrightarrow & k' \end{array} \qquad \text{by} \quad X \longmapsto \alpha ,$$

hence $\operatorname{tr.deg}_{k(X)} k' = d-1$. Therefore by induction hypothesis we get for the quadratic transformation h^* the inequality

$$H^{(0)}[R^*] \geq H^{(d-1)}[R^{*\prime}] \quad .$$

This finishes the proof of Theorem (29.1).

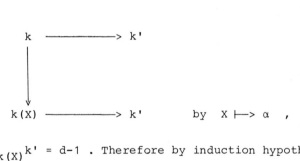

§ 30. Semicontinuity

We first prove a preliminary result (see [1], Lemma 2.28), page 56).

(30.1) Proposition. Let $(R,\mathfrak{m})$ be a local ring and $\mathfrak{p}$ a prime ideal in R with $\dim(R/\mathfrak{p}) = d \geq 1$. If $R/\mathfrak{p}$ is regular, then $H^{(0)}[R] \geq H^{(d)}[R_{\mathfrak{p}}]$.

Proof. a) Assume that $d = 1$. Then we have $\mathfrak{m} = fR + \mathfrak{p}$ for a suitable $f \in \mathfrak{m}$.
We may assume that for a fixed integer $n \geq 0$ the symbolic power $\mathfrak{p}^{(n+1)} = (0)$. Otherwise we consider the local ring $\overline{R} = R/\mathfrak{p}^{(n+1)}$ with the corresponding prime ideals $\overline{\mathfrak{m}}$ and $\overline{\mathfrak{p}}$. Clearly, $H^{(0)}[\overline{R}](n) \leq H^{(0)}[R](n)$. Moreover we have $H^{(1)}[\overline{R}_{\overline{\mathfrak{p}}}](n) = H^{(1)}[R_{\mathfrak{p}}](n)$ which can be easily seen. Now, since $\mathfrak{p}^{n+1} \subseteq \mathfrak{p}^{(n+1)} = (0)$, we have $\mathfrak{m}^{n+1} = f\cdot\mathfrak{m}^n$ and $((0) : f) = (0)$. [For the last relation we use that (0) is a $\mathfrak{p}$-primary ideal and $f \notin \mathfrak{p}$.]
This implies

$$
\begin{aligned}
H^{(0)}[R](n) &= \lambda(R/\mathfrak{m}^{n+1}) - \lambda(R/\mathfrak{m}^n) \\
&= \lambda(R/f\cdot\mathfrak{m}^n) - \lambda(fR/f\mathfrak{m}^n) = \lambda(R/fR) \quad .
\end{aligned}
$$

To compute $H^{(1)}[R_{\mathfrak{p}}](n)$, we note that

$$
\begin{aligned}
H^{(0)}[R_{\mathfrak{p}}](\nu) &= \lambda_{R_{\mathfrak{p}}}\left(\mathfrak{p}^{(\nu)}R_{\mathfrak{p}}/\mathfrak{p}^{(\nu+1)}R_{\mathfrak{p}}\right) \\
&\leq \lambda_R\left(\left(\mathfrak{p}^{(\nu)}/\mathfrak{p}^{(\nu+1)}\right)/\mathfrak{m}\left(\mathfrak{p}^{(\nu)}/\mathfrak{p}^{(\nu+1)}\right)\right) =: H(\nu) \quad .
\end{aligned}
$$

Since f is a non-zero-divisor in R and $\mathfrak{m} = (\mathfrak{p},f)$, the last term in this inequality is

$$
H(\nu) = \lambda_R\left(\mathfrak{p}^{(\nu)}/\mathfrak{p}^{(\nu+1)} + (fR \cap \mathfrak{p}^{(\nu)}\right) = \lambda_R\left(\mathfrak{p}^{(\nu)} + fR/\mathfrak{p}^{(\nu+1)} + fR\right) .
$$

From this we conclude

$$H^{(1)}[R_{\mathfrak{p}}](n) = \sum_{\nu=0}^{n} H^{(0)}[R_{\mathfrak{p}}](\nu) \leq \lambda(R/fR) = H^{(0)}[R](n) \quad .$$

b) For the general case we use induction on d : By assumption we have $\mathfrak{m} = (\mathfrak{p}, x_1, \ldots, x_d)$, where the images of $x_1, \ldots, x_d$ in $R/\mathfrak{p}$ form a regular system of parameters. Hence $(\mathfrak{p}, x_1)$ is a prime ideal in R of dimension $d-1$. By the induction hypothesis we know that

$$(1) \qquad H^{(0)}[R](n) \geqq H^{(d-1)}[R_{(\mathfrak{p},x_1)}](n) \quad .$$

By a) we have

$$(2) \qquad H^{(0)}[R_{(\mathfrak{p},x_1)}](n) \geqq H^{(1)}[R_{(\mathfrak{p},x_1)_{\mathfrak{p}}}](n) = H^{(1)}[R_{\mathfrak{p}}](n) \quad .$$

(1) and (2) imply

$$H^{(d)}[R_{\mathfrak{p}}](n) \leqq H^{(d-1)}[R_{(\mathfrak{p},x_1)}](n) \leqq H^{(0)}[R](n) \quad ,$$

as wanted.

The same inequality can be proven for any prime ideal $\mathfrak{p} \subset R$, if R is an excellent local ring by resolving the singularities of a curve, s. [1], Proposition 2.2.11, page 172 and also [4], page 77. It is then called Bennett's inequality:

<u>(30.2) Theorem (Semicontinuity)</u>. If $(R, \mathfrak{m})$ is an excellent local ring and $\mathfrak{p}$ a prime ideal with $\dim(R/\mathfrak{p}) = d$, then

$$H^{(0)}[R] \geqq H^{(d)}[R_{\mathfrak{p}}] \quad .$$

Note that every complete local ring is excellent. And in § 31 we may essentially assume that the given local ring is complete. Therefore the assumption of excellence is no real restriction for our purpose.

<u>(30.3) Corollary of Theorem (30.2)</u>. $H^{(0)}[R]$ is greater than or equal to $H^{(d)}[R_{\mathfrak{p}}]$ in the lexicographic order; i.e. if $H^{(0)}[R](j) = H^{(d)}[R_{\mathfrak{p}}](j)$ for $j < n$, then $H^{(0)}[R](n) > H^{(d)}[R_{\mathfrak{p}}](n)$.

(Note that lexicographic inequality imposes a linear ordering on the set of all sequences.)

Proof of Theorem (30.2). We may assume $d = 1$, since given any saturated chain

$$\mathfrak{p} = \mathfrak{p}_o \subset \mathfrak{p}_1 \subset \ldots \subset \mathfrak{p}_d = \mathfrak{m}$$

of prime in R, the inequalities

$$H^{(0)}\left[R_{\mathfrak{p}_j}\right] \geqq H^{(1)}\left[R_{\mathfrak{p}_{j-1}}\right]$$

imply that

$$H^{(0)}[R] \geqq H^{(d)}[R_{\mathfrak{p}}] \quad .$$

Now let

(1) $\quad R = R^{(0)} \longrightarrow R^{(1)} \longrightarrow \ldots \longrightarrow R^{(j)} \longrightarrow \ldots$

be any infinite sequence of (residually algebraic) quadratic transforms along $\mathfrak{p}$ (i.e. if $\mathfrak{p}^{(j+1)}$ denotes the strict transform of $\mathfrak{p}^{(j)}$ in $R^{(j+1)}$, then $\mathfrak{p}^{(j)} \neq R^{(j)}$ for any j.) We put $\overline{R}^{(j)} = R^{(j)}/\mathfrak{p}^{(j)}$ for all $j \geqq 0$. Then we get from (1) an infinite sequence of quadratic transforms of $R/\mathfrak{p}$:

(2) $\quad \overline{R} = \overline{R}^{(0)} \longrightarrow \overline{R}^{(1)} \longrightarrow \ldots \longrightarrow \overline{R}^{(j)} \longrightarrow \ldots$

Then we claim that $V := \bigcup_j \overline{R}^{(j)}$ is a discrete, rank 1, valuation ring of the quotient field K of $\overline{R}$ which dominates each $\overline{R}^{(j)}$, i.e. in our case the maximal ideal of V lies over the maximal $\overline{\mathfrak{m}}^{(j)}$ of $\overline{R}^{(j)}$, see [2], 35.3. For that take any valuation ring V' of the quotient field K of $\overline{R}$ which dominates V, i.e. in particular V' dominates $\overline{R} \subset K$. Hence by [2], 33.2 we know that V' is a discrete, rank 1, valuation ring. Denote the valuation of K associated to V' by v'. We want to show that $V = V'$. For that let $r/s \in V'$ with $r,s \in \overline{R}$. If s is a unit in $\overline{R}$, then $r/s \in V$. If $s \in \overline{\mathfrak{m}}$ (the maximal ideal of $\overline{R}$), then $r \in \overline{\mathfrak{m}}$ since $v'(r/s) \geqq 0$. Therefore, choosing $t \in \overline{\mathfrak{m}}$ such that $t \cdot \overline{R}^{(1)} = \overline{\mathfrak{m}}\,\overline{R}^{(1)}$, we have $r = r't$, $s = s't$, where

$r',s' \in \overline{R}^{(1)}$. So we get $r/s = r'/s'$ in $\overline{R}^{(1)}$ with $v'(s') < v'(s)$, since $t \in \overline{\mathfrak{m}}$. Then by induction we can express r/s as a fraction whose denominator is a unit in $\overline{R}^{(n)}$ for suitable large n , i.e. $r/s \in \overline{R}^{(n)} \subset V$. This proves the claim.

Let N be the nomalization of $\overline{R}$ and $\mathfrak{m}_V$ the maximal ideal of V . Then $\mathfrak{n} = N \cap \mathfrak{m}_V$ is a maximal ideal of N , and we have $\mathfrak{m}_V \cap N_{\mathfrak{n}} = \mathfrak{n}N_{\mathfrak{n}}$; since $\kappa \in N \setminus \mathfrak{n}$ implies κ is a unit of V . Moreover we have $N_{\mathfrak{n}} \subset V$, so that V dominates $N_{\mathfrak{n}}$. Since $N_{\mathfrak{n}}$ is a discrete, rank 1 valuation ring, it follows that $V = N_{\mathfrak{n}}$. Now we use the excellence of R . Then N is a finite $\overline{R}$-module, so that $N \subset \overline{R}^{(c)}$ for some c . If $\kappa \in N \setminus \mathfrak{n}$, then κ is a unit (in V and) in $\overline{R}^{(c)}$, i.e. $N_{\mathfrak{n}} \subset \overline{R}^{(c)}$. Therefore $\overline{R}^{(c)}$ dominates $N_{\mathfrak{n}}$. But then we have $\overline{R}^{(c)} = N_{\mathfrak{n}} = V$, see [2] , proof of 11.3 , hence $\overline{R}^{(c+i)} = \overline{R}^{(c)}$ for all $i \geqq 0$. For the regular ring $\overline{R}^{(c)}$ we may apply Proposition (30.1) concluding that

$$H^{(0)}\left[R^{(c)}\right] \geqq H^{(1)}\left[R^{(c)}_{\mathfrak{p}^{(c)}}\right] .$$

Note that by construction $\mathfrak{p}_o$ is outside of the center of the transform $R \longrightarrow R^{(c)}$, i.e. $R^{(c)}_{\mathfrak{p}^{(c)}} \cong R_{\mathfrak{p}}$. So, using Theorem (29.1) we get finally

$$H^{(0)}[R] \geqq H^{(1)}[R_{\mathfrak{p}}] .$$

§ 31. Permissibility and Blowing up of ideals

In this section we study the behaviour of Hilbert functions and multiplicities of a local ring $(R,\mathfrak{m})$ after blowing up an ideal I of R such that R/I need not be regular or even a domain. If I is a regular equimultiple prime ideal (and R quasi-unmixed), it is shown that the multiplicity cannot increase by blowing up I . Dade gave in [5] an unpublished proof of this fact and actually our proof which is due to Orbanz, is in part inspired by Dade's method. The other main ingredient is the use of generalized Hilbert functions.

First we recall several facts from Chapter II and fix some notation. Let $S = \bigoplus_{n \geqq 0} S_n$ be a graded ring which is generated by S_1 over S_0 . Recall that a homogeneous prime ideal Q in S is called relevant if $Q \cap S_1 \neq S_1$. By $S_{(Q)}$ we denote the homogeneous localization of S by Q , i.e. the subring of S_Q consisting of fractions $\frac{a}{b}$, where $a,b \in S$ are homogeneous of the same degree and $b \notin Q$. Let Q be a relevant prime ideal and $\mathfrak{n}$ the maximal ideal of $S_{(Q)}$. Then there exists by Chapter II, Corollary (12.18) an isomorphism

$$S_{(Q)}[T]_{\mathfrak{n}[T]} \longrightarrow S_Q$$

sending the indeterminate T to $t/1$, where $t \in S_1 \setminus Q$. From this we conclude that

$$H^{(i)}[S_{(Q)}] = H^{(i)}[S_Q] \quad \text{for all} \quad i \geq 0 \quad .$$

For the rest of this section let R' be either the completion of a local ring R or the local ring $R[T]_{\mathfrak{m}[T]}$. If R_1 is a local ring obtained by blowing up I in R which dominates R , then there is some R_1' obtained by blowing up IR' in R' which dominates R' and such that

$$(1) \qquad H^{(i)}[R_1'] = H^{(i)}[R_1] \quad \text{for all} \quad i \geq 0$$

and especially

$$(2) \qquad e(R_1') = e(R_1) \quad .$$

In the following we take $S = G(I;R)$. The unique maximal homogeneous ideal of S will be denoted by M^* and we put $I^* = \bigoplus_{i\geq 1} S_i$. If $\underline{x}$ is a system of parameters with respect to I , let $\underline{x}^*$ be the image of $\underline{x} \bmod I$ in S_{M^*} . Then $\underline{x}^*$ is a system of parameters with respect to $I^*S_{M^*}$, and since $G(I^*S_{M^*};S_{M^*}) \cong S$, we have

$$(3) \qquad H^{(i)}[\underline{x}^*,I^*S_{M^*},S_{M^*}] = H^{(i)}[\underline{x},I,R] \; , \quad i \geq 0 \quad ,$$

and especially

$$(4) \qquad e(\underline{x}^*,I^*S_{M^*},S_{M^*}) = e(\underline{x},I,R) \quad .$$

We also know that $s(I) = s(I^*S_{M^*})$ and $ht(I) = ht(I^*S_{M^*})$, hence we have

(5) $\quad ht(I) = s(I) \Longleftrightarrow ht(I^*S_{M^*}) = s(I^*S_{M^*})$.

If $R_1/IR_1 \cong S_{(Q)}$ for some homogeneous prime ideal Q of S, we put $t(R_1) = \dim S/Q - 1$. By the altitude formula (see Chapter III) we know that $t(R_1) = \dim R - \dim R_1$ if R is quasi-unmixed.

Now we come to the main-result of this chapter.

<u>(31.1) Theorem</u>. Let R_1 be a local ring of the blowing up of I in R which dominates R. Let $\underline{x}$ be a system of parameters with respect to I and let $r = \dim(R/I)$ and $t = t(R_1)$. Then the following holds:

(a) If R is normally Cohen-Macaulay with respect to I, then

$$H^{(t+1)}[R_1] \leqq H^{(r+1)}[\underline{x},I,R]$$

(b) If $ht(I) = s(I)$ and $t = \dim R - \dim R_1$, then

$$e(R_1) \leqq e(\underline{x},I,R) \quad .$$

Recall for the following corollary that an ideal is permissible in R, if R/I is regular and R is normally flat along I.

<u>(31.2) Corollary</u>. With the notations of the theorem we have

(c) If I is permissible in R, then

$$H^{(t)}[R_1] \leqq H^{(0)}[R]$$

(d) If $I = \mathfrak{p}$ is a prime ideal such that $R/\mathfrak{p}$ is Cohen-Macaulay and R normally flat along $\mathfrak{p}$, then

$$H^{(t+1)}[R_1] \leqq e(R/\mathfrak{p})\cdot H^{(r+1)}[R_\mathfrak{p}] \quad .$$

(e) If $I = \mathfrak{p}$ is a prime ideal such that $ht(P) = s(\mathfrak{p})$ and $t = \dim R - \dim R_1$ then

$$e(R_1) \leqq e(R/\mathfrak{p})e(R_\mathfrak{p}) \quad .$$

(f) If R is quasi-unmixed and $I = \mathfrak{p}$ is a prime ideal such that $e(R) = e(R_{\mathfrak{p}})$ and $R/\mathfrak{p}$ is regular, then

$$e(R_1) \leqq e(R) \quad .$$

(g) If I is $\mathfrak{m}$-primary and $t = \dim R - \dim R_1$ then

$$e(R_1) \leqq e(I,R) \quad .$$

Proof of Theorem (31.1). By (1) and (2) we may assume R to be complete. From Proposition (28.1) we know that

(5) $\quad H^{(0)}[R_1] \leqq H^{(1)}[R_1/IR_1] \quad , \quad IR_1 = tR_1 \quad$ for some t .

Now $R_1/IR_1 \cong S_{(Q)}$ for some homogeneous relevant prime ideal Q of S. Since S is excellent, we know that

(6) $\quad H^{(t+1)}[S_{(Q)}] = H^{(t+1)}[S_Q] \leqq H^{(0)}[S_{M^*}]$

by Bennett's inequality, and especially if $t = \dim R - \dim R_1$ then

(7) $\quad e(S_{(Q)}) \leqq e(S_{M^*}) \quad .$

If R is normally Cohen-Macaulay along I, then S_{M^*} is normally Cohen-Macaulay along $I^*S_{M^*}$ and therefore by Chapter IV, (22.23)

(8) $\quad H^{(0)}[\underline{x}^*S_{M^*} + I^*S_{M^*}, S_{M^*}] = H^{(r)}[\underline{x}^*, I^*S_{M^*}, S_{M^*}] = H^{(r)}[\underline{x}, I, R]$

by (3). Clearly $H^{(1)}[S_{M^*}] \leqq H^{(1)}[\underline{x}^*S_{M^*} + I^*S_{M^*}, S_{M^*}]$, so by (5), (6) and (8) we have

$$H^{(t+1)}[R_1] \leqq H^{(t+2)}[S_{(Q)}] \leqq H^{(1)}[S_{M^*}] \leqq H^{(r+1)}[\underline{x}, I, R] \quad .$$

We note that if R/I is regular and $I + \underline{x}R = \mathfrak{m}$, then we get by the same argument $H^{(t)}[R_1] \leqq H^{(r)}[\underline{x}, I, R]$. If now $ht(I) = s(I)$, then $ht(I^*S_{M^*}) = s(I^*S_{M^*})$ and therefore

$$e(\underline{x}^*S_{M^*} + I^*S_{M^*}, S_{M^*}) = e(\underline{x}^*, I^*S_{M^*}, S_{M^*}) = e(\underline{x}, I, R)$$

by Chapter IV, (20.5) and (4). Clearly $e(S_{M^*}) \leqq e(\underline{x}^* S_{M^*} + IS_{M^*}, S_{M^*})$, so from (5) and (7), where we assume $t = \dim R - \dim R_1$, we conclude

$$e(R_1) \leqq e(\underline{x}, I, R) \quad .$$

This completes the proof of the theorem. See end of App. III, § 3.

Proof of Corollary (31.2). If $R/\mathfrak{p}$ is regular, then R is normally flat along $\mathfrak{p}$ if and only if R is normally Cohen-Macaulay along $\mathfrak{p}$. Therefore (c) follows from the proof of (a) by taking a system $\underline{x}$ such that $\mathfrak{p} + \underline{x}R = \mathfrak{m}$ and observing that $H^{(r)}[\underline{x}, \mathfrak{p}, R] = H^{(0)}[\underline{x}R + \mathfrak{p}, R] = H^{(0)}[R]$. For (d) note first that $R/\mathfrak{p}$ Cohen-Macaulay and R normally flat along $\mathfrak{p}$. For (d) and (e) we may assume $R/\mathfrak{m}$ to be infinite, so we can choose a system $\underline{x}$ such that $e(\underline{x} \cdot R/\mathfrak{p}, R/\mathfrak{p}) = e(R/\mathfrak{p})$. Since $e(\underline{x}, R/\mathfrak{p}) = e(\underline{x} \cdot R/\mathfrak{p}, R/\mathfrak{p})$, (d) follows from (a). The same argument applies for (e) using (b). The assumptions of (f) imply that $\mathrm{ht}(\mathfrak{p}) = s(\mathfrak{p})$ by Chapter IV, (20.5). Taking $\underline{x}$ such that $\underline{x}R + \mathfrak{p} = \mathfrak{m}$, we know that

$$e(\underline{x}, \mathfrak{p}, R) = e(R) \quad ,$$

since $\mathrm{ht}(I) = s(I)$ implies $e(\underline{x}, I, R) = e(\underline{x}R + IR)$. Therefore we get $e(R_1) \leqq e(R)$ by (b). Finally (g) follows from (a), since the Cohen-Macaulay condition is void over a zero-dimensional ring.

(31.3) Remark. We cannot prove a corresponding result to statement (a) of Theorem (31.1) by replacing the condition " R is normally Cohen-Macaulay along I " by " R is projectively normally Cohen-Macaulay along I ". But we conjecture that the following might be true: Let R_1 be a local ring of the blowing up of an ideal I in R which dominates R. Let $\underline{x}$ be a system of parameters modulo I and let $r = \dim R/I$ and $t = t(R_1)$. Assume that $\mathrm{depth}(I^n/I^{n+1}) = \dim(R/I)$ for large n. Then

$$H^{(s+2)}[R_1/IR_1] \leqq H^{(1)}[I + \underline{x}R, R] \;,$$

which would imply that

$$H^{(s+1)}[R_1] \leqq H^{(1)}[I + \underline{x}R, R] \quad .$$

An idea of a proof was given in [6], but this proof is not complete.

§ 32. Transversal ideals and flat families

We first describe briefly two notions of transversality: Let X be an algebraic variety over a field k, z a point of X and $R = \mathcal{O}_{X,z}$. Any part of a system of parameters of R, say $\underline{t} = \{t_1,\ldots,t_s\}$, defines (near z) a projection $f : X \longrightarrow \mathbb{A}_k^s$ to an s-dimensional affine space over k, sending z to the origin $0 \in \mathbb{A}_k^s$. In general, the Hilbert function (resp. multiplicity) of $\mathcal{O}_{f^{-1}(0),z} = R/\underline{t}R$ will be worse than that of R. This suggests two notions of transversality, one for Hilbert functions (H-transversal) and a weaker one for multiplicities (e-transversal): The system $\underline{t}$ is called transversal, if the Hilbert function (resp. multiplicity) of R and $R/\underline{t}R$ coincide. For $s = \dim R$ we recover the notion of a transversal system of parameters introduced by Zariski for studying equisingularity problems. In the above set-up the numerical characters are defined with respect to the maximal ideal of R, but we are going to consider this problem for arbitrary ideals I using generalized Hilbert functions (resp. multiplicities). The result will be applied to derive some consequences for blowing ups with maximal Hilbert functions, resp. multiplicities. The study of e-transversal parameters was originally motivated by Lipman's characterization [9] of flat families with fibres of constant multiplicity. The last section of § 32 contains an analogous description of flat families with fibres of constant Hilbert function.

a) H-transversal parameters

Let (R,m) be a local ring, I a proper ideal of R and $\underline{x} = \{x_1,\ldots,x_r\} \subset R$ a system of parameters with respect to I.

(32.1) Definition. A subset $\underline{t} = \{t_1,\ldots,t_s\}$ of I is called H-transversal for $(\underline{x},I)$, if

$$H^{(0)}[\underline{x},I,R] = H^{(s)}[\underline{y},I/\underline{t}R,R/\underline{t}R] \quad ,$$

where $\underline{y}$ denotes the image of $\underline{x}$ in $R/\underline{t}R$. If $\underline{x} = \emptyset$, $\underline{t}$ will be called H-transversal for I.

(32.2 Proposition. Let R be a local ring, I a proper ideal of R, $\underline{x} = \{x_1,\ldots,x_r\}$ a system of parameters with respect to I and $\underline{t} = \{t_1,\ldots,t_s\}$ a subset of I. Consider the conditions:

(i) $\underline{t}$ is H-transversal for $(\underline{x},I)$

(ii) $in(I)(t_1),\ldots,in(I)(t_s)$ is a regular sequence of degree 1 in $G(I,R)$.

Then (ii) $\Rightarrow$ (i); and both conditions are equivalent, if $R/\underline{t}R$ is normally Cohen-Macaulay along $I/\underline{t}R$.

Proof. If $in(I)(t_1)$ is regular in $G(I,R)$, then

$$G(I/t_1R,R/t_1R) \cong G(I,R)/(in(I)(t_1))G(I,R) \quad .$$

Therefore (ii) $\Rightarrow$ (i) by Corollary (28.2) using induction.

For the converse, assume in addition that $R/\underline{t}R$ is normally Cohen-Macaulay along $I/\underline{t}R$. We will use induction on s, and we may assume $s > 0$, since the case $s = 0$ is trivial. Let

$$\overline{R} = R/t_1R + \ldots + t_{s-1}R \ , \quad I = I\overline{R}$$

and $\underline{z}$ the image of $\underline{x}$ in $\overline{R}$. Then (i) implies

(1) $$H^{(0)}[\underline{x},I,R] = H^{(s-1)}[\underline{z},\overline{I},\overline{R}] = H^{(s)}[\underline{y},I/\underline{t}R,R/\underline{t}R] \quad ,$$

i.e. $\{t_1,\ldots,t_{s-1}\}$ is H-transversal for $(\underline{x},I)$ and the image $\overline{t}_s$ of t_s in $\overline{R}$ is H-transversal for $(\underline{z},\overline{I})$. Let us show that $\overline{R}$ is normally Cohen-Macaulay along $\overline{I}$. For this we put $\overline{I}(\underline{z}) = \overline{I} + \underline{z}\overline{R}$. Then we have

$$H^{(r)}[\underline{x},I,R] = H^{(r+s-1)}[\underline{z},\overline{I},\overline{R}] \leq H^{(s-1)}[\overline{I}(\underline{z}),\overline{R}]$$
$$\leq H^{(s)}[\overline{I}(\underline{z})/\overline{t}_s\overline{R},\overline{R}/\overline{t}_s\overline{R}] = H^{(r+s)}[\underline{y},I/\underline{t}R,R/\underline{t}R] \quad ;$$

the last equality holds because $R/\underline{t}R$ is normally Cohen-Macaulay along $I/\underline{t}R$. We conclude that $H^{(r)}[\underline{z},\overline{I},\overline{R}] = H^{(0)}[\overline{I}(\underline{z}),\overline{R}]$, so that $\overline{R}$ is normally Cohen-Macaulay along $\overline{I}$. Now the inductive assumption implies that $in(I)(t_1),\ldots,in(I)(t_{s-1})$ is a regular sequence of degree 1 in $G(I,R)$, and in particular

$$G(\overline{I},\overline{R}) \cong G(I,R)/(in(I)(t_1),\ldots,in(I)(t_{s-1})) \quad .$$

Finally we know by Corollary (28.2) and (1) that $in(\overline{I})(\overline{t}_s)$ is a regular element of degree 1 in $G(\overline{I},\overline{R})$, which proves (ii) .

(32.3) Remarks.

(a) In the proof of Proposition (32.2) it was shown, without any extra assumption on R , that $R/\underline{t}R$ is normally Cohen-Macaulay along $I/\underline{t}R$ and $\underline{t}$ H-transversal for $(\underline{x},I)$ imply that R is normally Cohen-Macaulay along I .

(b) The second condition of (32.2) is independent of $\underline{x}$, and so is the property of being H-transversal for $(\underline{x},I)$, provided that $R/\underline{t}R$ is normally Cohen-Macaulay along $I/\underline{t}R$.

As a special case of Proposition (32.2) we obtain the following Corollary:

(32.4) Corollary. Let $(R,\mathfrak{m})$ be a local ring and I an $\mathfrak{m}$-primary ideal. Then a subset $\underline{t} = \{t_1,\ldots,t_s\}$ of I is H-transversal for I if and only if $in(I)(t_1),\ldots,in(I)(t_s)$ is a regular sequence in $G(I;R)$ of degree 1 .

(32.5) Corollary. If R is normally Cohen-Macaulay along I and $\underline{x}$ is any system of parameters with respect to I , then $\underline{x}$ is H-transversal for $I(\underline{x})$.

Proof. This follows immediately from Corollary (32.4) and Theorem (13.10) in Chapter II.

The title of this section 32 may need some explanation, since it is not clear from the Definition (32.1) that an H-transversal subset of I is part of a system of parameters of R , and in fact this may fail to hold in general. But in the situation of Proposition (32.2) it is clear that not only $\{t_1,\dots,t_s\}$ but even $\{x_1,\dots,x_r,t_1,\dots t_s\}$ is part of a system of parameters of R .

b) e-transveral parameters

Now $(R,\mathfrak{m})$ will denote a local ring with infinite residue field (to assure the existence of suitable minimal reductions). Furthermore, since we want to apply some fundamental results of multiplicity theory, we assume for this section that R is quasi-unmixed. From this we conclude that if I is an ideal of R with $\mathrm{ht}(I) = s(I)$, then $\mathrm{Assh}(R/I) = \mathrm{Min}(I)$. We also recall once more that if $\underline{x}$ is any system of parameters with respect to I , then the condition $\mathrm{ht}(I) = s(I)$ together with the formula

$$(2) \qquad H^{(0)}[\underline{x},I,R] = \sum_{\mathfrak{p}\in \mathrm{Assh}(R/I)} e(\underline{x};R/\mathfrak{p})H^{(0)}[IR_{\mathfrak{p}},R_{\mathfrak{p}}]$$

implies that

$$(3) \qquad e(\underline{x},I,R) = \sum_{\mathfrak{p}\in \mathrm{Assh}(R/I)} e(\underline{x};R/\mathfrak{p})e(IR_{\mathfrak{p}},R_{\mathfrak{p}}) \quad ,$$

where $e(\underline{x},I,R)$ is the multiplicity associated to the polynomial function $H^{(0)}[\underline{x},I,R]$.
In order to derive results for multiplicities parallel to those for Hilbert functions, we need the following

(32.6) Lemma. Let R be a quasi-unmixed local ring, I a proper ideal of R , $\underline{x}$ a system of parameters with respect to I and $\underline{t} = \{t_1,\dots,t_s\}$ a subset of I . Assume that $\mathrm{ht}(I) = s(I)$ and $\dim R/\underline{t}R = \dim R - s$. Then, putting $\overline{R} = R/\underline{t}R$ and $\overline{I} = I/\underline{t}R$, we have

$$(4) \qquad e(\underline{y},\overline{I},\overline{R}) = \sum_{\overline{\mathfrak{p}}\in \mathrm{Assh}(\overline{R}/\overline{I})} e(\underline{y};\overline{R}/\overline{\mathfrak{p}})e(\overline{I}\overline{R}_{\overline{\mathfrak{p}}},\overline{R}_{\overline{\mathfrak{p}}}) \quad ,$$

where $\underline{y}$ is the image of $\underline{x}$ in R as usual.

Proof. Comparing with (2) (applied to $\overline{R}$), it is clear that the assertion will follow from

(5) $\quad \mathrm{ht}(\bar{\mathfrak{p}}) = \mathrm{ht}(\bar{I}) \quad$ for all $\quad \mathfrak{p} \in \mathrm{Assh}(\bar{R}/\bar{I})$.

So let $\bar{\mathfrak{p}} \in \mathrm{Assh}(\bar{R}/\bar{I})$ and let $\mathfrak{p}$ be the inverse image of $\bar{\mathfrak{p}}$ in R . Let $\mathfrak{q}$ be any minimal prime of $\underline{t}R$ contained in $\mathfrak{p}$. Then $\mathrm{ht}(\mathfrak{q}) \leq s$, but $\dim R/\mathfrak{q} \leq \dim \bar{R} = \dim R - s$ by assumption, and $\dim R = \dim R/\mathfrak{q} + \mathrm{ht}(\mathfrak{q})$ since R is quasi-unmixed. It follows that $\mathrm{ht}(\mathfrak{q}) = s = \mathrm{ht}(\mathfrak{q}R_{\mathfrak{p}})$, hence by (18.13) $\dim R_{\mathfrak{p}}/\mathfrak{q}R_{\mathfrak{p}} = \dim R_{\mathfrak{p}} - s$ for all such $\mathfrak{q}$. We conclude that $\mathrm{ht}(\bar{\mathfrak{p}}) = \mathrm{ht}(\mathfrak{p}) - s$, which proves (5).

(32.7) Corollary. With the notations and assumptions as above, we have

(6) $\quad e(\underline{x},I,R) \leq e(\underline{y},I/\underline{t}R,R/\underline{t}R)$.

Proof. For $\underline{x} = \emptyset$ the assertion follows from Proposition (28.1) and the assumption $\dim R = \dim R/\underline{t}R + s$. From this the general case follows by comparing (3) and (4).

Now we want to characterize equality in (6), for which it seems convenient to treat the case $\underline{x} = \emptyset$ first. This case, in which I is $\mathfrak{m}$-primary, is of special interest, and it will simplify the argument for the general case.

(32.8) Definition. Let R be a quasi-unmixed local ring, I a proper ideal in R and $\underline{x}$ a system of parameters with respect to I . A subset $\underline{t} = \{t_1,\dots,t_s\}$ will be called e-transversal for $(\underline{x},I)$, if

(a) $\quad \dim R/\underline{t}R = \dim R - s$ and

(b) $\quad e(\underline{x},I,R) = e(\underline{y},I/\underline{t}R,R/\underline{t}R)$,

where $\underline{y}$ denotes the image of $\underline{x}$ in $R/\underline{t}R$. If $\underline{x} = \emptyset$, $\underline{t}$ will be called e-transversal for I .

(32.9) Proposition. Let $(R,\mathfrak{m})$ be a quasi-unmixed local ring (with infinite residue field), let I be an $\mathfrak{m}$-primary ideal and let $\underline{t} = \{t_1,\dots,t_s\}$ be a subset of I such that

$$\dim R/\underline{t}R = \dim R - s \quad .$$

Then the following conditions are equivalent:

(i) $\underline{t}$ is e-transversal for I .

(ii) For all $\mathfrak{p} \in \mathrm{Min}(\underline{t}R)$, $R_{\mathfrak{p}}$ is Cohen-Macaulay, and there are elements

$$u_{s+1},\dots,u_d \in I \quad (d = \dim R)$$

such that $t_1,\dots,t_s,u_{s+1},\dots,u_d$ generate a minimal reduction of I .

Proof. (i) $\Rightarrow$ (ii) . Choose $u_{s+1},\dots,u_d \in I$ such that their images in $I/\underline{t}R$ generate a minimal reduction of this ideal. Let $J = \underline{t}R + \underline{u}R$, where $\underline{u} = \{u_{s+1},\dots,u_d\}$. Then $J/\underline{t}R$ is a reduction of $I/\underline{t}R$, and therefore

$$e(I,R) \leq e(J,R) \leq e(J/\underline{t}R,R/\underline{t}R) = e(I/\underline{t}R,R/\underline{t}R) = e(I,R)$$

by Corollary (32.7) and (i). Therefore J is a reduction of I by Rees's Theorem. Now since $\mathrm{ht}(\underline{t}R) = s$, we have $\mathrm{Assh}(R/\underline{t}R) = \mathrm{Min}(\underline{t}R)$, and therefore

$$\begin{aligned} e(I,R) = e(J,R) &= \sum_{\mathfrak{p}\in \mathrm{Min}(\underline{t}R)} e(\underline{u};R/\mathfrak{p})e(\underline{t}R_{\mathfrak{p}},R_{\mathfrak{p}}) \\ &\leq \sum_{\mathfrak{p}\in \mathrm{Min}(\underline{t}R)} e(\underline{u};R/\mathfrak{p})\lambda(R_{\mathfrak{p}}/\underline{t}R_{\mathfrak{p}}) \\ &= e(J/\underline{t}R,R/\underline{t}R) = e(I/\underline{t}R,R/\underline{t}R) \quad . \end{aligned}$$

It follows that $e(\underline{t}R_{\mathfrak{p}},R_{\mathfrak{p}}) = \lambda(R_{\mathfrak{p}}/\underline{t}R_{\mathfrak{p}})$ for all $\mathfrak{p} \in \mathrm{Min}(\underline{t}R)$. (Note that $e(\underline{u};R/\mathfrak{p}) \neq 0$ for all $\mathfrak{p} \in \mathrm{Min}(\underline{t}R)$.) Since $\underline{t}$ is a system of parameters in $R_{\mathfrak{p}}$ for $\mathfrak{p} \in \mathrm{Min}(\underline{t}R)$, these $R_{\mathfrak{p}}$ are Cohen-Macaulay.

(ii) $\Rightarrow$ (i) . Let $\underline{u} = \{u_{s+1},\dots,u_d\}$ and let $\underline{v}$ be the image of $\underline{u}$ in $R/\underline{t}R$. Then by assumption

$$e(I,R) = e(\underline{t}R + \underline{u}R) = \sum_{\mathfrak{p}\in \mathrm{Assh}(R/\underline{t}R)} e(\underline{u};R/\mathfrak{p})e(\underline{t}R_{\mathfrak{p}},R_{\mathfrak{p}})$$

$$= \sum_{\mathfrak{p}\in \mathrm{Assh}(R/\underline{t}R)} e(\underline{u};R/\mathfrak{p})\lambda(R_{\mathfrak{p}},\underline{t}R_{\mathfrak{p}}) = e(\underline{v};R/\underline{t}R) = e(I/\underline{t}R,R/\underline{t}R) \quad .$$

(32.10) Theorem. Let (R,m) be a local ring (with infinite residue field), I a proper ideal of R, $\underline{x}$ a system of parameters with respect to I and $\underline{t} = \{t_1,\dots,t_s\}$ a subset of I. Assume that R is quasi-unmixed and $ht(I) = s(I)$. Then the following conditions are equivalent:

(i) $\underline{t}$ is e-transversal for $(\underline{x},I)$ and $ht(I/\underline{t}R) = s(I/\underline{t}R)$.

(ii) $R_{\mathfrak{p}}$ is Cohen-Macaulay for all $\mathfrak{p} \in Min(\underline{t}R)$, and $\underline{t}$ is part of a minimal set of generators of a minimal reduction of I.

Proof. (i) $\Rightarrow$ (ii). Note that $ht(I/\underline{t}R) = s(I/\underline{t}R)$ implies by Chapter II

$$e(\underline{x},I/\underline{t}R,R/\underline{t}R) = e(I(x)/\underline{t}R,R/\underline{t}R) \quad ,$$

where $I(x) = I + \underline{x}R$, and similarly

$$e(\underline{x},I,R) = e(I(x),R) \quad .$$

Therefore, by Proposition (32.9),

(*) $\quad R_{\mathfrak{p}}$ is Cohen-Macaulay for all $\mathfrak{p} \in Min(\underline{t}R)$.

Using $ht(I/\underline{t}R) = s(I/\underline{t}R)$ again we may choose $u_{s+1},\dots,u_h \in I$, $h = ht(I) = s + ht(I/\underline{t}R)$, such that the images in $R/\underline{t}R$ generate a minimal reduction of $I/\underline{t}R$. Let $\underline{u} = \{u_{s+1},\dots,u_h\}$ and $J = \underline{t}R + \underline{u}R$. Then $Min(J) = Min(I) = Assh(R/J)$, and by Proposition (32.9) we have (with the notation of (32.8))

$$\begin{aligned}\sum_{\mathfrak{p}\in Assh(R/I)} e(\underline{x};R/\mathfrak{p})e(IR_{\mathfrak{p}},R_{\mathfrak{p}}) &= e(\underline{x},I,R) = e(\underline{y},I/\underline{t}R,R/\underline{t}R)\\ &= e(\underline{y},J/\underline{t}R,R/\underline{t}R) = e(\underline{x},J,R) \quad \text{by } (*)\\ &= \sum_{\mathfrak{p}\in Assh(R/J)} e(\underline{x};R/\mathfrak{p})e(JR_{\mathfrak{p}},R_{\mathfrak{p}}) \quad .\end{aligned}$$

We conclude that $e(IR_{\mathfrak{p}},R_{\mathfrak{p}}) = e(JR_{\mathfrak{p}},R_{\mathfrak{p}})$ for all

$$\mathfrak{p} \in Min(I) = Assh(R/I) = Assh(R/J) = Min(J) \quad ,$$

and therefore J is a reduction of I by Böger's extension of Rees's Theorem (see Chapter III, (19.6)).

(ii) $\Rightarrow$ (i) . Clearly $ht(I/\underline{t}R) = s(I/\underline{t}R)$, and therefore (i) is a direct consequence of Proposition (32.9) and (3).

<u>(32.11) Corollary</u>. Let R be a quasi-unmixed local ring with infinite residue field, I a proper ideal of R and $\underline{x}$ a system of parameters with respect to I . Assume that $ht(I) = s(I)$ and that $\underline{x}$ is a regular sequence in R . Then $\underline{x}$ is e-transversal for $I(\underline{x})$.

Proof. If $\underline{u}$ generates a minimal reduction of I , then $\underline{u} \cup \underline{x}$ generates a minimal reduction of $I(\underline{x})$. Hence (ii) $\Rightarrow$ (i) of (32.10) applied to $I(\underline{x})$ and $\underline{x}$ proves the claim.

<u>(32.12) Remark</u>. We want to give some technical comments on the proof of Theorem (32.10). First we note that condition (ii) is independent of the system $\underline{x}$. Furthermore, for proving (ii) $\Rightarrow$ (i), it is enough to assume $R_{\mathfrak{p}}$ Cohen-Macaulay for those $\mathfrak{p} \in Min(\underline{t}R)$, which are contained in some minimal prime ideal of I . For the others it will follow automatically by (i) $\Rightarrow$ (ii) . Finally we point out that under the conditions of Theorem (32.10) any elements $u_{s+1},\ldots,u_h \in I$ generate, together with $\underline{t}$, a minimal reduction of I , provided that their images generate a minimal reduction of $I/\underline{t}R$.

<u>(32.13) Remark</u>. If R is quasi-unmixed, $ht(I) = s(I)$ and $\dim R/\underline{t}R = \dim R - s$, then $\underline{t}$ is H-transversal for $(\underline{x},I)$ trivially implies $\underline{t}$ e-transversal for $(\underline{x},I)$. If R is a hypersurface, i.e. $R = S/fS$ for some regular local ring S , then both notions are equivalent. This follows from the results in [12], where also more general cases are treated, in which H-transversal and e-transversal coincide, namely certain 'strict complete intersections'. If R is a hypersurface and $s = \dim R$, then $\underline{t}$ e-transversal for $\mathfrak{m}$ means that $\underline{t}$ is a transversal system of parameters in the sense of Zariski, and this is equivalent to saying that $\underline{t}$ generates a minimal reduction of $\mathfrak{m}$.

<u>(32.14) Proposition</u>. Let $(R,\mathfrak{m})$ be a quasi-unmixed local ring (with infinite residue field) and $\underline{t} = \{t_1,\ldots,t_s\}$ a subset of $\mathfrak{m}$. Assume that $\underline{t}$ is e-transversal for $\mathfrak{m}$ and that $R/\underline{t}R$ is Cohen-Macaulay.

Then R is Cohen-Macaulay.

Proof. By Theorem (32.10) we may choose $\underline{u} = \{u_{s+1},\ldots,u_d\}$, $d = \dim R$, such that $\underline{t} \cup \underline{u}$ generates a minimal reduction of $\mathfrak{m}$. Consequently

$$e(R) = e(\underline{t}R + \underline{u}R, R) = e(\underline{u}(R/\underline{t}R), R/\underline{t}R) = \lambda(R/\underline{t}R + \underline{u}R) \quad .$$

<u>(32.15) Remark</u>. The same argument as in (32.14) shows that, under the conditions of Theorem (32.10), if $\mathfrak{p}$ is any minimal prime ideal of I and $R_{\mathfrak{p}}/\underline{t}R_{\mathfrak{p}}$ is Cohen-Macaulay, then $R_{\mathfrak{p}}$ is Cohen-Macaulay.

Finally we want to apply the results obtained so far to derive some consequences for blowing-ups with "maximal" Hilbert functions resp. multiplicities in the sense of § 30. For this purpose we introduce some notations that will be kept fixed for the rest of this section.

<u>(32.16) Notation</u>. R is an excellent local ring, I is an ideal of R and $R \longrightarrow R_1$ is a local homomorphism obtained by blowing up R with center I. We put $A = G(I,R)$, so that $R_1/IR_1 \cong A_{(Q)}$ for a suitable homogeneous prime ideal Q of A. Let N be the unique homogeneous maximal ideal of A and $\mathfrak{m}(R_1)$ the maximal ideal of R_1. We put $s = \dim A/Q$ and we fix $t \in I$ such that $IR_1 = tR_1$. Finally, $\underline{x} = \{x_1,\ldots,x_r\}$ will denote a system of parameters with respect to I and $I(x) = I + \underline{x}R$.

Let R be normally Cohen-Macaulay along I. Then by the proof of Theorem (31.1) we have the following inequalities (see also [10]):

$$(1) \qquad H^{(s+1)}[R_1] \leq H^{(s+2)}[R_1/IR_1] = H^{(s+2)}[A_Q]$$

$$(2) \qquad \leq H^{(1)}[A_N]$$

$$(3) \qquad \leq H^{(r+1)}[\underline{x},I,R] \quad .$$

<u>(32.17) Proposition</u>. With the notations of (32.16), assume that R is normally Cohen-Macaulay along I. Then $H^{(s+1)}[R_1] = H^{(r+1)}[\underline{x},I,R]$ if and only if the following conditions are satisfied:

(a) $in(\mathfrak{M}(R_1))(t)$ is a regular element of degree 1 in $G(\mathfrak{m}(R_1),R_1)$.

(b) Q is a permissible ideal in A .

(c) R/I is regular and $\underline{x}$ is a regular system of parameters with respect to I .

Proof. By Corollary (28.2), (a) is equivalent to t being H-transversal for $\mathfrak{M}(R_1)$, which means equality in (1). (b) is equivalent to the equality in (2) by Chapter IV. For (c), let $\underline{y}$ be the image of $\underline{x}$ in $R/I \subset A$ and let

$$I^* = \bigoplus_{n>0} I^n/I^{n+1} \subset A \quad .$$

Then

$$H^{(r+1)}[\underline{x},I,R] = H^{(r+1)}[\underline{y},I^*A_N,A_N] \quad ,$$

and therefore equality in (3) means

(4) $$H^{(r+1)}[\underline{x},I^*A_N,A_N] = H^{(1)}[A_N] \quad .$$

Now obviously A_N is normally Cohen-Macaulay along I^*A_N , and therefore by Theorem (22.23) equality holds in (3) if and only if

$$H^{(1)}[\underline{y}A_N + I^*A_N,A_N] = H^{(1)}[A_N] \quad ,$$

i.e. if and only if (c) is satisfied.

Next we want to study the same question for multiplicities. We recall from the proof of 31.1 that, if R is excellent and quasi-unmixed, the condition $ht(I) = s(I)$ implies the following inequalities:

(5) $e(R_1) \leqq e(R_1/IR_1) = e(A_Q)$

(6) $\qquad \leqq e(A_N)$

(7) $\qquad \leqq e(\underline{x},I,R)$.

(32.18) Proposition. With the notations of (32.16), assume that R is quasi-unmixed and has an infinite residue field, and that $ht(I) = s(I)$. Then $e(R_1) = e(\underline{x},I,R)$ if and only if the following conditions are satisfied:

(a) t belongs to a minimal set of generators of a minimal reduction of $\mathfrak{m}(R_1)$.

(b) $e(A_Q) = e(A_N)$.

(c) $\underline{x}(R/I)$ is a minimal reduction of $\mathfrak{m}/I$, where $\mathfrak{m}$ denotes the maximal ideal of R.

Proof. It suffices to show that (c) is equivalent to the equality in (7). Using the notation of the preceeding proof, this equality is equivalent to

(8) $$e(\underline{y}A_N + I^*A_N, A_N) = e(A_N) .$$

Since A_N is quasi-unmixed, and since N and $\underline{y}A + I^*$ are homogeneous, this means that $\underline{y}A + I^*$ is a reduction of N (by the Theorem of Rees). Finally, taking into account the grading of $A = G(I;R)$, we see that (8) is equivalent to (c).

References - Chapter VI

Books

[1] M. Herrmann, R. Schmidt und W. Vogel, Theorie der normalen Flachheit, Teubner Texte zur Mathematik, Leipzig 1977.

[2] M. Nagata, Local rings, Huntington New York 1975.

[3] J.-P. Serre, Algèbre locale - Multiplicités, Lecture Notes in Math. 11, Springer-Verlag, Berlin - New York 1965.

Papers

[4] B.M. Bennett, On the characteristic function of a local ring, Ann. of Math. 91 (1970), 25 - 87.

[5] E.C. Dade, Multiplicity and monoidal transformations, Thesis Princeton 1960.

[6] M. Herrmann and U. Orbanz, On equimultiplicity, Math. Proc. Camb. Phil. Soc. 91 (1982), 207 - 213.

[7] H. Hironaka, Resolution of singularities of an algebraic variety over a field of characteristic zero I, Ann. of Math. 79 (1964), 169 - 236.

[8] H. Hironaka, Certain numerical characters of singularities, J. Math. Kyoto Univ. 10-1 (1970), 151 - 187.

[9] J. Lipman, Equimultiplicity, reduction and blowing up, In Comm. Algebra: Analytic methods, Lecture Notes in Pure and Appl. Math. 68, Marcel Dekker 1981.

[10] U. Orbanz, Multiplicites and Hilbert functions under blowing up, Man. Math. 36 (1981), 179 - 186.

[11] U. Orbanz and L. Robbiano, Projective normal flatness and Hilbert functions, Trans. Ann. Math. Soc. 283 (1984), 33 - 47.

[12] U. Orbanz, Transversal parameters and tangential flatness, Math. Proc. Camb. Phil. Soc. 98 (1985), 37 - 49.

[13] B. Singh, Effect of a permissible blow-up on the local Hilbert functions, Inv. Math. 26 (1974), 201 - 212.

[14] B. Singh, A numerical criterion for the permissibility of a blowing up, Comp. Math. 33 (1976), 15 - 28.

Chapter VII. LOCAL COHOMOLOGY AND DUALITY OF GRADED RINGS

In this chapter we give a summary of the theory of local cohomology and duality over graded rings, see [4],[6] and [13*]. To make the text as self-contained as possible we begin in § 33 with elementary properties of the category of graded modules over a graded ring $A = \bigoplus_{n \in \mathbf{Z}} A_n$. One should remark that most results in this chapter hold for any noetherian ring or any noetherian local ring R by regarding R as a graded ring with the trivial grading $R_0 = R$ and $R_n = 0$ for $n \neq 0$. On the other hand our theory of graded rings can be extended to any $\mathbf{Z}^n$-graded rings as Goto and Watanabe have done in [17].

However, it is important to recognize the difference between the category of *graded* A-modules and the category of A-modules. For example, there is an injective graded A-module which is not injective as an R-module (see Example (33.7)). In the first Sections 33 and 34 one can see that as far as finitely generated graded modules are concerned, the functor $\underline{\mathrm{Hom}}_A(-,-)$ of graded A-modules is the same as the usual functor $\mathrm{Hom}_A(-,-)$, where the grading is neglected. In general these two functors are different. For example, since in general injective graded modules are hardly finitely generated, one can not replace $\underline{\mathrm{Hom}}_A(\underline{E}_A,\underline{E}_A)$ by $\mathrm{Hom}_A(\underline{E}_A,\underline{E}_A)$ in the Matlis duality for graded rings in Theorem (34.8) of this chapter.

§ 33. Review on graded modules

The results of this section are more or less standard fact of homological algebra. A general theory of injective objects in abelian categories can be found in [21]. An extensive study of homological theory of graded rings was carried out in [16],[17]. Most of our results of this part were taken from these two papers. As general references for homological algebra we refer the reader to [18], [8] and [3].
Let $A = \bigoplus_{n \in \mathbf{Z}} A_n$ be a graded noetherian ring (except in Lemma (33.13)) and let $M = \bigoplus_{n \in \mathbf{Z}} M_n$ and $N = \bigoplus_{n \in \mathbf{Z}} N_n$ be graded A-modules. Let us denote the category of graded A-modules by $M^h(A)$. A homomorphism $f : M \longrightarrow N$ in $M^h(A)$ is an A-linear map such that $f(M_n) \subset N_n$ for all $n \in \mathbf{Z}$. We denote by $M(n)$ the graded A-module whose grading is defined by $[M(n)]_m = M_{n+m}$ for all $m \in \mathbf{Z}$. Let $\underline{\mathrm{Hom}}_A(M,N)_n$ be the abelian group of all homomorphisms in $M^h(A)$ from M into $N(n)$. Let $\underline{\mathrm{Hom}}_A(M,N) = \bigoplus_{n \in \mathbf{Z}} \underline{\mathrm{Hom}}_A(M,N)_n$. Then $\underline{\mathrm{Hom}}_A(M,N)$ is a graded A-module with grading given by $[\underline{\mathrm{Hom}}(M,N)]_n = \underline{\mathrm{Hom}}_A(M,N)_n$ for all $n \in \mathbf{Z}$.

(33.1) Lemma.

1) $\underline{\mathrm{Hom}}_A(M,N(n)) = \underline{\mathrm{Hom}}_A(M(-n),N) = \underline{\mathrm{Hom}}_A(M,N)(n)$.

2) If M is a finitely generated graded A-module then for any graded A-module N we have:
$\underline{\mathrm{Hom}}_A(M,N) = \mathrm{Hom}_A(M,N)$ as underlying A-modules.

Proof.

1) Straightforward.

2) Clearly $\underline{\mathrm{Hom}}_A(M,N) \subseteq \mathrm{Hom}_A(M,N)$.

To prove the opposite inclusion we first assume that M is a graded free A-module with homogeneous free basis $\{e_1,\ldots,e_n\}$. Let $\deg e_i = \nu_i$. Let $f \in \mathrm{Hom}_A(M,N)$ and $f(e_i) = \sum_\mu y_{i\mu}$, where $y_{i\mu} \in N_\mu$ and $y_{i\mu} = 0$ except for finitely many μ . Since M is free, one can define $f_\nu \in \underline{\mathrm{Hom}}_A(M,N)_\nu$ to be $f_\nu(e_i) = y_{i(\nu_i+\nu)}$ for $i = 1,\ldots,n$. Clearly $f = \sum_\nu f_\nu$, and $f_\nu = 0$ for all but finitely many ν .

If M is not free we consider an exact sequence of graded A-modules

$$G \longrightarrow F \longrightarrow M \longrightarrow 0$$

with F, G free and finitely generated.
This gives a commutative diagram with exact rows:

$$\begin{array}{ccccccc}
0 \longrightarrow & \underline{\mathrm{Hom}}_A(M,N) & \longrightarrow & \underline{\mathrm{Hom}}_A(F,N) & \longrightarrow & \underline{\mathrm{Hom}}_A(G,N) \\
 & \downarrow & & \| & & \| \\
0 \longrightarrow & \mathrm{Hom}_A(M,N) & \longrightarrow & \mathrm{Hom}_A(F,N) & \longrightarrow & \mathrm{Hom}_A(G,N)
\end{array}$$

Hence $\underline{\mathrm{Hom}}_A(M,N) = \mathrm{Hom}_A(M,N)$.

Recall that the tensor product $M \otimes_A N$ is a graded A-module whose homogeneous component $[M \otimes_A N]_n$ of degree n is the abelian group generated by the elements of the form $x \otimes y$ with $x \in M_i$, $y \in N_j$ and $i+j=n$.

(33.2) Lemma. Let $S = \bigoplus_{n \in \mathbf{Z}} S_n$ be a graded ring and $\varphi : S \longrightarrow A$ a homomorphism of graded rings (i.e. a ring homomorphism such that

$\varphi(S_n) \subset A_n$ for all $n \in \mathbf{Z}$) . Let L be a graded S-module. Regarding M and N as graded S-modules via φ , one gets a canonical isomorphism

$$\underline{\mathrm{Hom}}_A(M,\underline{\mathrm{Hom}}_S(N,L) \cong \underline{\mathrm{Hom}}_S(M \otimes_A N,L) \quad .$$

Proof. cf. [1].

Recall that a graded A-module E is injective in $M^h(A)$ (resp. projective in $M^h(A)$) if the functor $\underline{\mathrm{Hom}}_A(-,E)$ (resp. $\underline{\mathrm{Hom}}_A(E,-)$) from $M^h(A)$ into itself is an exact functor. Every graded free A-module is projective in $M^h(A)$ and every graded A-module is a factor module of a graded free A-module, i.e. $M^h(A)$ is an abelian category with enough projectives.

In order to show that $M^h(A)$ is an abelian category with enough injectives,we need the following lemma.

<u>(33.3) Lemma</u>. Let $\varphi : S \longrightarrow A$ be a homomorphism of graded rings and I an injective object in $M^h(S)$. Then for any flat graded A-module F the S-module $\underline{\mathrm{Hom}}_S(F,I)$ is injective in $M^h(A)$.

Proof. For any exact sequence of graded A-modules $0 \longrightarrow M \longrightarrow N$ we have by assumption on F an exact sequence $0 \longrightarrow M \otimes_A F \longrightarrow N \otimes_A F$. Since I is injective in $M^h(S)$, one has an exact sequence

$$\underline{\mathrm{Hom}}_S(N \otimes_A F,I) \longrightarrow \underline{\mathrm{Hom}}_S(M \otimes_A F,I) \longrightarrow 0 \quad .$$

By Lemma (33.2) we know that

$$\underline{\mathrm{Hom}}_S(M \otimes_A F,I) \cong \underline{\mathrm{Hom}}_A(M,\underline{\mathrm{Hom}}_S(F,I)) \quad \text{and}$$

$$\underline{\mathrm{Hom}}_S(N \otimes_A F,I) \cong \underline{\mathrm{Hom}}_A(N,\underline{\mathrm{Hom}}_S(F,I)) \quad .$$

Hence $\underline{\mathrm{Hom}}_S(F,I)$ is injective in $M^h(A)$.

<u>(33.4) Proposition</u>. $M^h(A)$ is an abelian category with enough injectives.

Proof. Let E be the direct product of the injective envelopes of the residue fields $A_0/\mathfrak{m}_0$ of A_0, where $\mathfrak{m}_0$ runs over all maximal ideals $\mathfrak{m}_0$ of A_0. Then E is an injective A_0-module. Let $\underline{E}$ be the graded A_0-module with $\underline{E}_0 = E$ and $\underline{E}_n = 0$ for $n \neq 0$. We set $I = \underline{\operatorname{Hom}}_{A_0}(A,\underline{E})$, where A is regarded as a graded A_0-module. I is an injective object in $M^h(A)$ by Lemma (33.3). To show that every graded A-module M is contained in an injective object of $M^h(A)$, we consider the graded A-module $M^* = \underline{\operatorname{Hom}}_A(M,I)$ and moreover a graded free A-module F such that $F \twoheadrightarrow M^*$.

Then we get an injective homomorphism

$$M^{**} := \underline{\operatorname{Hom}}_A(\underline{\operatorname{Hom}}_A(M,I),I) \longrightarrow \underline{\operatorname{Hom}}_A(F,I) \quad .$$

Since $\underline{\operatorname{Hom}}_A(F,I)$ is an injective object in $M^h(A)$, it is enough to show that the canonical homomorphism $\tau : M \longrightarrow M^{**}$ is an injection, in other words, for any homogeneous element $0 \neq x \in M$ there is a homomorphism $f \in \underline{\operatorname{Hom}}_A(M,I)$ such that $f(x) \neq 0$. For that let $n = \deg x$ and $\mathfrak{a} = \operatorname{ann}_A(x)$. Then $xA \cong A/\mathfrak{a}(-n)$. We will construct a non-zero homomorphism $g : A/\mathfrak{a}(-n) \longrightarrow I$. Since I is injective, g can be extended to a non-trivial homomorphism $f \in \underline{\operatorname{Hom}}_A(M,I)$. To construct a suitable $g \in \underline{\operatorname{Hom}}_A(A/\mathfrak{a}(-n),I)$ we first take a maximal homogeneous ideal $\mathfrak{m}$ of A containing $\mathfrak{a}$. Note that the o-th homogeneous component $[A/\mathfrak{m}]_0$ is a field $k = A_0/\mathfrak{m}_0$, where $\mathfrak{m}_0 = A_0 \cap \mathfrak{m}$. Then we have by Lemma (33.2) (applied to $A_0 \longrightarrow A$) :

$$\begin{aligned}\underline{\operatorname{Hom}}_A(A/\mathfrak{m}(-n),I) &\cong \underline{\operatorname{Hom}}_{A_0}(A/\mathfrak{m}(-n),\underline{E}) \\ &= \bigoplus_{\ell\in\mathbf{Z}} \underline{\operatorname{Hom}}_{A_0}(A/\mathfrak{m}(-n),\underline{E})_\ell \quad .\end{aligned}$$

By definition $\underline{\operatorname{Hom}}_{A_0}(A/\mathfrak{m}(-n),\underline{E})_{-n} = \underline{\operatorname{Hom}}_{A_0}(A/\mathfrak{m},\underline{E})_0$ is the set of A_0-linear maps φ such that $\varphi([A/\mathfrak{m}]_\ell \subset (\underline{E})_\ell$. But $\underline{E}_\ell$ is the zero-module for $\ell \neq 0$. Therefore, to get a non-zero map $h \in \underline{\operatorname{Hom}}_A(A/\mathfrak{m}(-n),I)$, it is enough to find a non-zero map $\alpha : [A/\mathfrak{m}]_0 \longrightarrow \underline{E}_0 = E$. But α exists by the construction of E. Now the canonical surjection $A/\mathfrak{a}(-n) \longrightarrow A/\mathfrak{m}(-n)$ induces an injection $j : \underline{\operatorname{Hom}}_A(A/\mathfrak{m}(-n),I) \hookrightarrow \underline{\operatorname{Hom}}_A(A/\mathfrak{a}(-n),I)$. Therefore we get a non-zero homomorphism $g = j(h) \in \underline{\operatorname{Hom}}_A(A/\mathfrak{a}(-n),I)$, q.e.d.

<u>(33.5) Definition</u>. An injective homomorphism $f : M \longrightarrow N$ in $M^h(A)$ is called essential if for any non-trivial graded A-submodule L of N we have $f(M) \cap L \neq (0)$.

Using Proposition (33.4) one shows as in the non-graded case, see [8], that a graded A-module M is injective in $M^h(A)$ if and only if there is no proper essential extension of M in $M^h(A)$.

<u>(33.6) Proposition and Definition</u>. For every graded A-module M there exists an injective module E in $M^h(A)$ and an essential morphism $M \hookrightarrow E$. E is uniquely determined up to isomorphisms in $M^h(A)$.

We denote E by $\underline{E}_A(M)$. $\underline{E}_A(M)$ is called the injective envelope of M in $M^h(A)$.

Proof. Take an embedding of M into an injective module I in $M^h(A)$ Then by Zorn's lemma there is a maximal graded A-submodule E of I containing M such that for any non-zero graded A-submodule L of E we have $M \cap L \neq (0)$. Clearly $M \hookrightarrow E$ is an essential extension of M in $M^h(A)$.

Let $\varphi : E \hookrightarrow L$ be an essential extension of E in $M^h(A)$. Since I is injective in $M^h(A)$, there is a morphism $\psi : L \longrightarrow I$ which makes the following diagram commute

$$\begin{array}{ccc} E & \overset{\varphi}{\hookrightarrow} & L \\ \downarrow & \swarrow{\scriptstyle \psi} & \\ I & & \end{array} ,$$

where $\varphi(E) \cap \ker \psi = (0)$. Since φ is an essential homomorphism we have $\ker \psi = (0)$.

Hence we may assume that L is a graded A-submodule of I . For any non-trivial graded A-submodule N of L we have $M \cap N \neq (0)$, since both $M \hookrightarrow E$ and $E \hookrightarrow L$ are essential in $M^h(A)$. By the maximality of E we get $E = L$. Hence E is injective.

To prove the uniqueness, take any injective module E' in $M^h(A)$ which satisfies the same property as E before. Then there is an injective morphism $\alpha : E \hookrightarrow E'$ which makes the following diagram commute:

$$\begin{array}{ccc} M & \hookrightarrow & E \\ \downarrow & \swarrow{\scriptstyle \alpha} & \\ E' & & \end{array} .$$

Since E is injective in $M^h(A)$, the image $\alpha(E)$ is a direct summand of E', say $E' = \alpha(E) \oplus F$ for some graded A-submodule F of E'. Regarding M as a submodule of E', we have $M \cap F = (0)$. But by the essentiality of $M \hookrightarrow E'$ we have $F = (0)$. Hence α is an isomorphism.

Note that in general the injective envelope of M as a graded module in $M^h(A)$ is not an injective envelope of M as an A-module.

<u>(33.7) Example</u>. Let k be a field and X an indeterminate over k. Let $A = k[X,X^{-1}]$ be a graded ring with $\deg X = 1$. Then A is the injective envelope of A in $M^h(A)$ (apply Lemma (33.3) to $k \hookrightarrow A$ and note that $A \cong \underline{\mathrm{Hom}}_k(A,k))$. But A is not an injective A-module since $\mathrm{Ext}^1_A(A/X-1),A) \cong A/(X-1) \neq (0)$.

<u>(33.8) Lemma</u>. Let S be a multiplicatively closed set of A consisting of homogeneous elements and $M \hookrightarrow N$ an essential homomorphism in $M^h(A)$. Then the induced homomorphism $S^{-1}M \hookrightarrow S^{-1}N$ is an essential homomorphism in $M^h(S^{-1}A)$.

Proof. We identify M with a submodule of N. We must show that for any non-zero homogeneous element $\frac{x}{s} \in S^{-1}N$ we have $\frac{x}{s} \cdot S^{-1}A \cap S^{-1}M \neq (0)$, where $s \in S$ and $x \in N$. We may replace $\frac{x}{s}$ by $\frac{x}{1}$. Let P be an associated prime of $\frac{x}{1} \cdot S^{-1}A$. Then there is an associated prime $\mathfrak{p}$ of N such that $P = \mathfrak{p}S^{-1}A$. Hence we find a homogeneous element $r \in A$ such that $\mathrm{ann}_A(rx) = \mathfrak{p}$ and $0 \neq \frac{rx}{1} \in S^{-1}N$. Therefore we may assume from the beginning that $\mathrm{ann}_A(x) = \mathfrak{p}$. Since $M \hookrightarrow N$ is essential, there is a homogeneous element $t \in A$ such that $0 \neq tx \in M$. Note that $t \notin \mathfrak{p}$. Now it is easy to see that $0 \neq \frac{tx}{1} \in S^{-1}M \cap \frac{x}{1}S^{-1}A$.

The rest of this section is devoted to the study of the structure of injective objects of $M^h(A)$.

<u>(33.9) Lemma</u>. Let $\mathfrak{p} \in \mathrm{Spec}(A)$ be a homogeneous prime ideal of A and let $x \notin \mathfrak{p}$ be a homogeneous element of A with $\deg x = n$. Then x induces an isomorphism $\underline{E}_A(A/\mathfrak{p})(-n) \xrightarrow[\sim]{\cdot x} \underline{E}_A(A/\mathfrak{p})$.

Proof. We put $\underline{E}(A/\mathfrak{p})$ instead of $\underline{E}_A(A/\mathfrak{p})$. First we observe that x is a non-zero-divisor on $\underline{E}(A/\mathfrak{p})$. For that let $e \in \underline{E}(A/\mathfrak{p})$ be an

element such that $xe = 0$. If $e \neq 0$ there is a homogeneous element
$r \in A$ such that $0 \neq re \in A/\mathfrak{p}$, since $\underline{E}(A/\mathfrak{p})$ is an essential extension
of $A/\mathfrak{p}$ in $M^h(A)$. But $x(re) = r(xe) = 0$ implies $re = 0$ since
$x \notin \mathfrak{p}$, a contradiction. Since x is a non-zero-divisor on $\underline{E}(A/\mathfrak{p})$,
we have an isomorphism $\underline{E}(A/\mathfrak{p})(-n) \xrightarrow{\sim} x\underline{E}(A/\mathfrak{p})$, i.e. $x\underline{E}(A/\mathfrak{p})$ is injective in $M^h(A)$. Therefore there is a graded A-submodule I of
$\underline{E}(A/\mathfrak{p})$ such that $\underline{E}(A/\mathfrak{p}) = x\underline{E}(A/\mathfrak{p}) \oplus I$. If $I \ni \alpha \neq 0$ we find a homogeneous element s of A such that $0 \neq s\alpha \in A/\mathfrak{p}$. But then
$0 \neq xs\alpha \in I \cap x\underline{E}(A/\mathfrak{p}) = (0)$, a contradiction. Thus we have $I = (0)$
and $\underline{E}(A/\mathfrak{p}) = x\underline{E}(A/\mathfrak{p})$, q.e.d.

(33.10) Corollary. $\underline{E}_A(A/\mathfrak{p})$ is a graded $A_{<\mathfrak{p}>}$-module, where $A_{<\mathfrak{p}>}$ is
the localization by the multiplicatively closed set $S = \{x \in A \mid x \notin \mathfrak{p},$
x homogeneous$\}$.

Proof. By Lemma (33.9) it is easy to see that the canonical homomor-
phism $\underline{E}_A(A/\mathfrak{p}) \longrightarrow \underline{E}_A(A/\mathfrak{p}) \otimes_A A_{<\mathfrak{p}>}$ is an isomorphism.

(33.11) Proposition. $\underline{E}_A(A/\mathfrak{p})$ is the injective envelope of $A_{<\mathfrak{p}>}/\mathfrak{p}A_{<\mathfrak{p}>}$
in $M^h(A_{<\mathfrak{p}>})$.

Proof. To prove that $\underline{E}_A(A/\mathfrak{p})$ is injective in $M^h(A_{<\mathfrak{p}>})$, take any
homogeneous ideal $\mathfrak{a}A_{<\mathfrak{p}>}$ and a homomorphism $\mathfrak{a}A_{<\mathfrak{p}>} \xrightarrow{\varphi} \underline{E}_A(A/\mathfrak{p})$ in
$M^h(A_{<\mathfrak{p}>})$. Since $\underline{E}_A(A/\mathfrak{p})$ is injective in $M^h(A)$, there is a homomor-
phism $\psi : A_{<\mathfrak{p}>} \longrightarrow \underline{E}_A(A/\mathfrak{p})$ in $M^h(A)$ which makes the following
diagram commute:

$$
\begin{array}{ccc}
& \underline{E}_A(A/\mathfrak{p}) & \\
& {}^{\varphi}\nearrow \quad \nwarrow{}^{\psi} & \\
0 \longrightarrow \mathfrak{a}A_{<\mathfrak{p}>} \longrightarrow & A_{<\mathfrak{p}>} &
\end{array}
$$

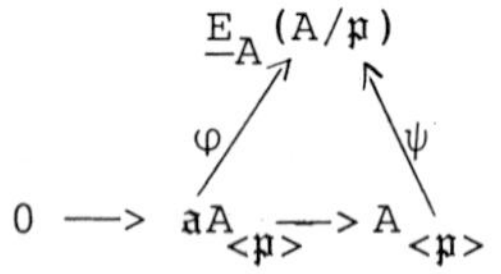

First we show that ψ is a homomorphism in $M^h(A_{<\mathfrak{p}>})$: Take any element
$\frac{x}{s} \in A_{<\mathfrak{p}>}$, $(s \notin \mathfrak{p}, x \in A)$. Then $s\psi(\frac{x}{s}) = \psi(x) = x\psi(1)$. Since the multiplication by s is an automorphism of $\underline{E}_A(A/\mathfrak{p})$ by Lemma (33.9) we
see that $\psi(\frac{x}{s}) = \frac{x}{s}\psi(1)$. Therefore ψ is a homomorphism in $M^h(A_{<\mathfrak{p}>})$.
Applying the functor $- \otimes_A A_{<\mathfrak{p}>}$ to the exact sequence

$$0 \longrightarrow A/\mathfrak{p} \longrightarrow \underline{E}_A(A/\mathfrak{p}) ,$$

we get an injection in $M^h(A_{<\mathfrak{p}>})$:

$$0 \longrightarrow A_{<\mathfrak{p}>}/\mathfrak{p}A_{<\mathfrak{p}>} \longrightarrow \underline{E}_A(A/\mathfrak{p}) \otimes_A A_{<\mathfrak{p}>} = \underline{E}_A(A/\mathfrak{p}) \quad ,$$

(see Corollary (33.10)), which is essential by Lemma (33.8), q.e.d.

Recall that a graded A-module M is said to be indecomposable if there are no non-trivial graded A-submodules M_1, M_2 of M such that $M = M_1 \oplus M_2$.
The importance of indecomposable injective modules in $M^h(A)$ can be seen in the following theorem which says that every injective object in $M^h(A)$ is a direct sum of indecomposable injective modules in $M^h(A)$.

(33.12) Theorem.

1) For any graded A-module M we have

$$\mathrm{Ass}_A(M) = \mathrm{Ass}_A(\underline{E}_A(M)) \quad .$$

2) An injective module I in $M^h(A)$ is indecomposable if and only if $I \cong \underline{E}_A(A/\mathfrak{p})(n)$ for some homogeneous prime ideal $\mathfrak{p}$ and some $n \in \mathbf{Z}$.

3) Every injective object in $M^h(A)$ is a direct sum of indecomposable injective modules in $M^h(A)$.

For the proof of Theorem (33.12) we need an auxiliary result, given by Lemma (33.13).

(33.13) Lemma. Let $A = \bigoplus_{n\in\mathbf{Z}} A_n$ be any graded ring. The the following conditions are equivalent:

1) A is a noetherian ring.

2) A satisfies the ascending chain condition for homogeneous ideals of A.

3) Every homogeneous ideal of A is finitely generated.

4) A_0 is a noetherian ring and A is of finite type over A_0.

5) Every direct sum of injective modules in $M^h(A)$ is injective in $M^h(A)$.

Proof. 1) ⇒ 2) ⇒ 3) and 4) ⇒ 1) are trivial.

<u>3) ⇒ 4):</u>

a) Let $\mathfrak{a}$ be an ideal of A_0. Then $\mathfrak{a}A$ is a homogeneous ideal of A and hence finitely generated. Let $a_1,\ldots,a_n \in \mathfrak{a}$ be generators of $\mathfrak{a}A$. Then $\mathfrak{a} = (a_1,\ldots,a_n)A_0$. Hence A_0 is a noetherian ring.

b) Let I be the ideal generated by $A_+ = \bigoplus_{n>0} A_n$ and $A_- = \bigoplus_{n<0} A_n$. By 3) one can choose an integer $n > 0$ so that $I = \sum_{i=1}^{n} A_iA + \sum_{i=1}^{n} A_{-i}A$ and $A_iA = \sum_{j=1}^{r_i} x_{ij}A$, $\deg x_{ij} = i$. Then

$$A = A_0\left[\left\{x_{ij} \mid i = -n,\ldots,-1,1,\ldots,n\ ;\ 1 \leq j \leq r_i\right\}\right].$$

<u>3) ⇒ 5):</u> Let $E = \bigoplus_{\lambda\in\Lambda} E_\lambda$ be a direct sum of injective modules in $M^h(A)$. Let I be a homogeneous ideal of A and $\varphi : I \longrightarrow E$ a homomorphism in $M^h(A)$.
Since I is finitely generated, the image $\varphi(I)$ is contained in a finite direct sum $E_{\lambda_1} \oplus\ldots\oplus E_{\lambda_n} \subset \bigoplus_{\lambda\in\Lambda} E_\lambda$. Since every finite direct sum of injective modules is injective, we can extend φ to a homomorphism $\Psi : A \longrightarrow \bigoplus_{i=1}^{n} E_{\lambda_i}$. Hence $E = \bigoplus_\lambda E_\lambda$ is injective in $M^h(A)$.

<u>5) ⇒ 2):</u> Let $\mathfrak{a}_0 \subseteq \mathfrak{a}_1 \subseteq \mathfrak{a}_2 \subseteq \ldots \subseteq \mathfrak{a}_n \subseteq \mathfrak{a}_{n+1}$ ... be an ascending chain of homogeneous ideals of A. Put $\mathfrak{a} = \bigcup_{i=0}^{\infty} \mathfrak{a}_i$. Let $E_i = \underline{E}_A(\mathfrak{a}/\mathfrak{a}_i)$ and let φ_i be the composition of $\mathfrak{a} \longrightarrow \mathfrak{a}/\mathfrak{a}_i$ and the canonical map $\mathfrak{a}/\mathfrak{a}_i \hookrightarrow \underline{E}_A(\mathfrak{a}/\mathfrak{a}_i)$. Then we define a map $\varphi : \mathfrak{a} \longrightarrow \bigoplus_{i=0}^{\infty} E_i$ by $\varphi(x) = (\varphi_i(x))_{0 \leq i < \infty} \in \bigoplus_{i=0}^{\infty} E_i$.

Note that this map is well defined, since for any $x \in \mathfrak{a}$ there is an ideal $\mathfrak{a}_i$ such that $x \in \mathfrak{a}_i$, hence $\varphi_j(x) = 0$ for $j \geq i$. By assumption $\bigoplus_i E_i$ is injective in $M^h(A)$, therefore φ can be extended to a homomorphism $\Psi : A \longrightarrow \bigoplus_i E_i$. Let $\Psi(1) = (e_i)_{0 \leq i < \infty}$ with $e_i \in E_i$. Then there is an integer $n \geq 0$ such that $e_i = 0$ for $i \geq n$, i.e. $\mathfrak{a} = \mathfrak{a}_n$.

Proof of Theorem (33.12).

<u>To 1):</u> It is clear that $\mathrm{Ass}_A(M) \subseteq \mathrm{Ass}_A(\underline{E}_A(M))$. So let $\mathfrak{p} \in \mathrm{Ass}_A(\underline{E}_A(M))$. We may assume that there is a homogeneous element $x \in \underline{E}_A(M)$ such that $\mathfrak{p} = \mathrm{ann}(x)$. Since $M \hookrightarrow \underline{E}_A(M)$ is essential, we have that $M \cap xA \neq (0)$ and that $\phi \neq \mathrm{Ass}_A(M \cap xA) \subset \mathrm{Ass}_A(xA) = \{\mathfrak{p}\}$. Hence $\{\mathfrak{p}\} = \mathrm{Ass}_A(M \cap xA) \subset \mathrm{Ass}_A(M)$.

To 2: Suppose that an injective module I in $M^h(A)$ is indecomposable. Let $\mathfrak{p} \in \mathrm{Ass}_A(I)$ and let x be a homogeneous element of I such that $\mathfrak{p} = \mathrm{ann}(x)$. Let $\underline{E}_A(xA)$ be the injective envelope of xA in $M^h(A)$. The inclusion $xA \hookrightarrow I$ can be extended to a homomorphism $\varphi : \underline{E}_A(xA) \longrightarrow I$. Since $xA \hookrightarrow \underline{E}_A(xA)$ is an essential extension, we see that φ is an injection. Hence $\underline{E}_A(xA)$ can be considered as a direct summand of I. Therefore $I \cong \underline{E}_A(xA)$, since I is indecomposable. If $\deg x = n$ we have: $\underline{E}_A(xA) \cong \underline{E}_A(A/\mathfrak{p}(-n)) \cong \underline{E}_A(A/\mathfrak{p})(-n)$. Suppose that $\underline{E}_A(A/\mathfrak{p})$ is not indecomposable. Then there are non-trivial graded A-submodules I_1, I_2 of $\underline{E}_A(A/\mathfrak{p})$ such that $\underline{E}_A(A/\mathfrak{p}) = I_1 \oplus I_2$. Let $x_1 \in I_1$ and $x_2 \in I_2$ be non-zero homogeneous elements. Since $A/\mathfrak{p} \hookrightarrow \underline{E}_A(A/\mathfrak{p})$ is essential, one can choose homogeneous elements $r_1, r_2 \in A$ so that $0 \neq r_1x_1 = r_2x_2 \in A/\mathfrak{p}$. But $r_1x_1 = r_2x_2 \in I_1 \cap I_2 = (0)$, a contradiction.

To 3: Let I be an injective object in $M^h(A)$. Let $\mathcal{F}$ be the set of graded A-submodules of I which can be written as a direct sum of indecomposable injective modules in $M^h(A)$. Clearly $\mathcal{F} \neq \phi$ if $I \neq (0)$. By Zorn's lemma one can find a maximal element E of $\mathcal{F}$ (i.e. maximal with respect to the inclusion).

By Lemma (33.13) we see that E is an injective module in $M^h(A)$. Hence $I = E \oplus J$ for some graded A-submodule J of I. We want to show that $J = (0)$: Suppose that $J \neq (0)$. Then we choose a homogeneous element $x \in J$, say $\deg x = n$, such that $\mathrm{ann}(x) = \mathfrak{p} \in \mathrm{Ass}(J)$. Since J is an injective module in $M^h(A)$, we may assume that the injective envelope $\underline{E}_A(xA)$ of xA is a submodule of J. Hence $E + \underline{E}_A(xA) = E \oplus \underline{E}_A(xA)$ and $\underline{E}_A(xA) \cong \underline{E}_A(A/\mathfrak{p})(-n)$. But this contradicts to the maximality of E. This completes the proof of Theorem (33.12).

(33.14) Remark. For the convenience of the reader we recall to Lemma (8.2) in Chapter II, saying that the following conditions are equivalent:

a) A is simple

b) $A = k$ or $k[X,X^{-1}]$, where k is a field and X an indeterminate over k with $\deg X = n$ for some $n > 0$.

c) Every graded A-module M is a free A-module.

Here the conclusions b) ⇒ a) and c) ⇒ a) were trivial, and for b) ⇒ c) we could assume $A = k[X,X^{-1}]$.

(33.15) Corollary. Let $\mathfrak{p}$ be a homogeneous prime ideal of A . Then:

$$\underline{\mathrm{Hom}}_{A_{<\mathfrak{p}>}}(A_{<\mathfrak{p}>}/\mathfrak{p}A_{<\mathfrak{p}>},\underline{E}_A(A/\mathfrak{p})) \cong A_{<\mathfrak{p}>}/\mathfrak{p}A_{<\mathfrak{p}>} .$$

Proof. By definition $A_{<\mathfrak{p}>}/\mathfrak{p}A_{<\mathfrak{p}>}$ is simple. Hence $A_{<\mathfrak{p}>}/\mathfrak{p}A_{<\mathfrak{p}>} \cong k$ or $k[X,X^{-1}]$ by Remark (33.14). Assume that $A_{<\mathfrak{p}>}/\mathfrak{p}A_{<\mathfrak{p}>} \cong k[X,X^{-1}]$. Let f be the canonical injection

$$A_{<\mathfrak{p}>}/\mathfrak{p}A_{<\mathfrak{p}>} \hookrightarrow \underline{E}_{A_{<\mathfrak{p}>}}(A_{<\mathfrak{p}>}/\mathfrak{p}A_{<\mathfrak{p}>}) \cong \underline{E}_A(A/\mathfrak{p})$$

and $g \in \underline{\mathrm{Hom}}_{A_{<\mathfrak{p}>}}(A_{<\mathfrak{p}>}/\mathfrak{p}A_{<\mathfrak{p}>},\underline{E}_A(A/\mathfrak{p}))$ a homogeneous homomorphism. Note that $\underline{\mathrm{Hom}}_{A_{<\mathfrak{p}>}}(A_{<\mathfrak{p}>}/\mathfrak{p}A_{<\mathfrak{p}>},\underline{E}_A(A/\mathfrak{p}))$ is an $A_{<\mathfrak{p}>}/\mathfrak{p}A_{<\mathfrak{p}>}$-module. Hence there is an integer $m \in \mathbf{Z}$ such that $(X^{-m}g)(1) \in [A_{<\mathfrak{p}>}/\mathfrak{p}A_{<\mathfrak{p}>}]_0 = k$. Let $(X^{-m}g)(1) = \alpha \in k$. Then $g = \alpha X^m f$. Since $\underline{\mathrm{Hom}}_{A_{<\mathfrak{p}>}}(A_{<\mathfrak{p}>}/\mathfrak{p}A_{<\mathfrak{p}>},\underline{E}(A/\mathfrak{p}))$ is a free $A_{<\mathfrak{p}>}/\mathfrak{p}A_{<\mathfrak{p}>}$-module by Remark (33.14) and $\deg f = 0$, we get the assertion in the case $A_{<\mathfrak{p}>}/\mathfrak{p}A_{<\mathfrak{p}>} \cong k[X,X^{-1}]$. The other case can be shown similarly.

In the following definition we recall to injective resolutions and give a brief description of derived functors. The aim is to define Bass numbers in a proper way.

(33.16) Definition. 1) Let M be a graded A-module. An injective resolution of M in $M^h(A)$ is an exact sequence

$$0 \longrightarrow M \xrightarrow{\partial^{-1}} I^0 \xrightarrow{\partial^0} I^1 \xrightarrow{\partial^1} I^2 \longrightarrow \cdots \xrightarrow{\partial^{n-1}} I^n \xrightarrow{\partial^n} I^{n+1} \longrightarrow \cdots$$

in $M^h(A)$ such that I^n is an injective module in $M^h(A)$ for all $n \geqq 0$.

2) An injective resolution of M in $M^h(A)$ is called minimal if $I^0 \cong \underline{E}_A(M)$ and $I^n \cong \underline{E}_A(\ker \partial^n) \cong \underline{E}_A(\mathrm{coker}\,\partial^{n-2})$.

(33.17) Lemma. For any graded A-module M there exists a minimal injective resolution of M in $M^h(A)$.

Proof. Let I^0 be the injective envelope $\underline{E}_A(M)$ of M in $M^h(A)$ and $\partial^{-1} : M \longrightarrow I^0$ be the canonical injection. Suppose that we have already constructed a sequence

$$0 \xrightarrow{\partial^{-2}} M \xrightarrow{\partial^{-1}} I^0 \xrightarrow{\partial^0} I^1 \longrightarrow \dots \xrightarrow{\partial^{n-1}} I^n$$

so that $I^n \cong \underline{E}_A(\operatorname{coker} \partial^{n-2})$ for $n \geqq 0$, $(\partial^{-2} = 0)$. Then we have an exact sequence

$$I^{n-1} \xrightarrow{\partial^{n-1}} I^n \xrightarrow[\alpha]{} \operatorname{coker} \partial^{n-1} \longrightarrow 0 \quad .$$

Let $I^{n+1} := \underline{E}_A(\operatorname{coker} \partial^{n-1})$ and let $\beta : \operatorname{coker} \partial^{n-1} \longrightarrow I^{n+1}$ be the canonical injection. We set $\partial^n = \beta \circ \alpha$. Then we get an exact sequence

$$0 \longrightarrow M \xrightarrow{\partial^{-1}} I^0 \xrightarrow{\partial^0} I^1 \longrightarrow \dots \xrightarrow{\partial^{n-1}} I^n \xrightarrow{\partial^n} I^{n+1}$$

such that $I^{n+1} \cong \underline{E}_A(\operatorname{coker} \partial^{n-1})$. Thus one can construct inductively a minimal injective resolution of M in $M^h(A)$.

<u>(33.18) Lemma</u>. Let M be a graded A-module and

$$0 \longrightarrow M \longrightarrow I^0 \xrightarrow{\partial^0} I^1 \dots \xrightarrow{\partial^{n-1}} I^n \xrightarrow{\partial^n} I^{n+1} \longrightarrow \dots$$

an injective resolution of M in $M^h(A)$.
Then the injective resolution $0 \longrightarrow M \longrightarrow I^\bullet$ is minimal if and only if for any homogeneous prime ideal $\mathfrak{p}$ of A the induced homomorphism

$$\bar{\partial}^n : \underline{\operatorname{Hom}}_{A_{<\mathfrak{p}>}}\left(A_{<\mathfrak{p}>}/\mathfrak{p}A_{<\mathfrak{p}>}, I^n_{<\mathfrak{p}>}\right) \longrightarrow \underline{\operatorname{Hom}}_{A_{<\mathfrak{p}>}}\left(A_{<\mathfrak{p}>}/\mathfrak{p}A_{<\mathfrak{p}>}, I^{n+1}_{<\mathfrak{p}>}\right)$$

is trivial for $n \geqq 0$.

Proof. Suppose that the injective resolution $0 \longrightarrow M \longrightarrow I^\bullet$ is minimal. Let

$$f \in \underline{\operatorname{Hom}}_{A_{<\mathfrak{p}>}}(A_{<\mathfrak{p}>}/\mathfrak{p}A_{<\mathfrak{p}>}, I^n_{<\mathfrak{p}>})$$

be a homogeneous homomorphism. Since $I^n_{<\mathfrak{p}>} = \underline{E}_A(\operatorname{coker} \partial^{n-2})_{<\mathfrak{p}>}$ by Lemma (33.8), we find a homogeneous element $x \in A_{<\mathfrak{p}>} - \mathfrak{p}A_{<\mathfrak{p}>}$ such that $0 \neq xf(1) \in \operatorname{im}\partial^{n-1} \otimes 1_{A_{<\mathfrak{p}>}} \cong (\operatorname{coker} \partial^{n-2})_{<\mathfrak{p}>}$, where $\partial^{n-1} \otimes 1_{A_{<\mathfrak{p}>}} : I^{n-1}_{<\mathfrak{p}>} \longrightarrow I^n_{<\mathfrak{p}>}$ is the induced homomorphism. Hence we have

$\bar{\partial}^n(xf) = x\bar{\partial}^n(f) = 0$. This implies $\bar{\partial}^n(f) = 0$, since every graded $A_{<\mathfrak{p}>}/\mathfrak{p}A_{<\mathfrak{p}>}$-module is free. Conversely assume that the injective resolution is not minimal. Then there is an $n \geq 0$ such that the induced homomorphism

$$0 \longrightarrow \operatorname{im}\partial^{n-1} \longrightarrow I^n$$

is not essential. By the proof of Proposition (33.6) we may assume that $\underline{E}_A(\operatorname{Im}\partial^{n-1})$ is contained in I^n . Hence $I^n \cong \underline{E}_A(\operatorname{im}\partial^{n-1}) \oplus J^n$ for some graded A-submodule J^n of I^n . Since $\underline{E}_A(\operatorname{im}\partial^{n-1}) \supset \operatorname{im}\partial^{n-1}$, we see that J^n is isomorphically mapped into I^{n+1} by ∂^n . Therefore $I^{n+1} \cong E \oplus J^n$ for some graded A-submodule E of I^{n+1} , and the restriction of ∂^n to J^n is an isomorphism $\partial^n \mid J^n \xrightarrow{\sim} J^n$. Hence

$$\bar{\partial}_n : \underline{\operatorname{Hom}}_{A_{<\mathfrak{p}>}}\left(A_{<\mathfrak{p}>}/\mathfrak{p}A_{<\mathfrak{p}>}, I^n_{<\mathfrak{p}>}\right) \longrightarrow \underline{\operatorname{Hom}}_{A_{<\mathfrak{p}>}}\left(A_{<\mathfrak{p}>}/\mathfrak{p}A_{<\mathfrak{p}>}, I^{n+1}_{<\mathfrak{p}>}\right)$$

is not trivial for $\mathfrak{p} \in \operatorname{Ass}_A J^n$.

<u>(33.19) Remark</u>. The proof of Lemma (33.18) shows that a minimal injective resolution can be obtained by deleting superfluous direct summands from a given injective resolution.

Now we mention the notion of a derived functor. Let $\mathcal{A}$ be an abelian category with enough injectives and F an additive functor from $\mathcal{A}$ to another abelian category $\mathcal{L}$. If F is a left exact covariant functor the i-th derived functor $R^iF(-)$ of F is defined as follows: For any object A of $\mathcal{A}$ we find an injective resolution

$$0 \longrightarrow A \xrightarrow{\partial^{-1}} I^0 \xrightarrow{\partial^0} I^1 \xrightarrow{\partial^1} \cdots \xrightarrow{\partial^{n-1}} I^n \xrightarrow{\partial^n} I^{n+1} \longrightarrow \cdots$$

of A . $R^iF(A)$ is defined to be the i-th cohomology of the following complex in $\mathcal{L}$:

$$0 \longrightarrow F(I^0) \xrightarrow{F(\partial^0)} F(I^1) \xrightarrow{F(\partial^1)} \cdots \xrightarrow{F(\partial^{n-1})} F(I^n) \xrightarrow{F(\partial^n)} F(I^{n+1})$$

i.e. $R^iF(A) := \ker F(\partial^i)/\operatorname{im}F(\partial^{i-1})$.

The functors $R^iF : \mathcal{A} \longrightarrow \mathcal{B}$, $i = 0,1,2,\ldots$ have the following properties:

1) For any object A of $\mathcal{A}$, $R^iF(A)$ does not depend on the injective resolution of A

2) $R^0F(A) = F(A)$

3) For any injective object I we have $R^iF(I) = 0$ for $i > 0$

4) For any short exact sequence $0 \longrightarrow A' \longrightarrow A \longrightarrow A'' \longrightarrow 0$ in $\mathcal{A}$ we have a long exact sequence

$$0 \longrightarrow R^0F(A') \longrightarrow R^0F(A) \longrightarrow R^0F(A'') \longrightarrow R^1F(A') \longrightarrow R^1F(A) \longrightarrow$$
$$R^1F(A'') \ldots \longrightarrow R^{n-1}F(A'') \longrightarrow R^nF(A') \longrightarrow R^nF(A) \longrightarrow R^nF(A'')$$
$$\longrightarrow R^{n+1}F(A') \longrightarrow \ldots \quad \text{in } \mathcal{C} \ .$$

5) If a family of additve functors $\{T^i : T^i : \mathcal{A} \text{ — } \mathcal{C}\ ,\ i = 0,1,2,\ldots\}$ satisfies the properties 2), 3) and 4) then T^i is isomorphic to R^iF for all $i \geq 0$.

Similarly one defines the derived functors of a right exact contravariant functor by using projective resolutions. For the general theory of derived functors we refer the reader to [18], [3] and [8]. In particular, we may apply the theory of derived functors to the category $M^h(A)$ of graded A-modules.

<u>(33.20) Definition</u>. Let M be a graded A-module. For $i \geq 0$ the functor $\underline{\mathrm{Ext}}^i_A(M,-)$ is defined to be the i-th derived functor of $\underline{\mathrm{Hom}}_A(M,-)$. Note that, as in the non-graded case, $\underline{\mathrm{Ext}}^i_A(M,N)$ can be computed by a projective resolution of M for any graded A-module N . By Lemma (33.1) we know that if M is a finitely generated graded A-module, then $\underline{\mathrm{Hom}}_A(M,N) = \mathrm{Hom}_A(M,N)$ as underlying A-modules. Since A is noetherian, we find for any finitely generated graded A-module M a projective resolution of M by finitely generated projective A-modules. Hence we get the following result:

<u>(33.21) Lemma</u>. Let M be a finitely generated graded A-module. Then, for any graded A-module N, we have $\underline{\mathrm{Ext}}^i_A(M,N) \cong \mathrm{Ext}^i_A(M,N)$ for all $i > 0$.

Now we are ready to introduce the notion of Bass number, which was defined in [15] to study Gorenstein rings.

(33.22) Definition. For any noetherian ring R and for any R-module M we define

$$\mu^i(\mathfrak{p},M) := \dim_{\kappa(\mathfrak{p})} \operatorname{Ext}^i_{R_{\mathfrak{p}}}(\kappa(\mathfrak{p}),M_{\mathfrak{p}}) \quad ,$$

where $\mathfrak{p} \in \operatorname{Spec}(A)$ and $\kappa(\mathfrak{p}) = R_{\mathfrak{p}}/\mathfrak{p}R_{\mathfrak{p}}$ and call it the i-th Bass number of M at $\mathfrak{p}$.

Recall that a noetherian local ring $(R,\mathfrak{m},k)$ is Cohen-Macaulay if and only if $\operatorname{Ext}^i_R(k,R) = (0)$ for $i < \dim R$ or equivalently $\mu^i(\mathfrak{m},R) = (0)$, $i < \dim R$.

(33.23) Definition. 1) A noetherian local ring $(R,\mathfrak{m},k)$ is called Gorenstein if

$$\mu^d(\mathfrak{m},R) = \begin{cases} 1 & \text{if } d = \dim R \\ 0 & \text{otherwise} \end{cases}$$

2) A noetherian ring R is Gorenstein if for any prime ideal $\mathfrak{p} \in \operatorname{Spec}(R)$ the local ring $R_{\mathfrak{p}}$ is Gorenstein.

(33.24) Remark. 1) If a local ring $(R,\mathfrak{m},k)$ is Gorenstein then $R_{\mathfrak{p}}$ is Gorenstein for all $\mathfrak{p} \in \operatorname{Spec}(R)$. For completeness we will prove this well-known fact in the appendix, see Corollary (A3), where we also show that a local ring R is Gorenstein if and only if it has finite injective dimension as an R-module.

2) P. Roberts sketched a proof in [23] that the condition $\mu^d(\mathfrak{m},R) = 1$ is in itself sufficient to imply that R is Gorenstein, which was a conjecture of W. Vasconcelos. Roberts used for his proof the technique of dualizing complexes which is not within the frame of this book.

The following result describes the main property of Bass numbers:

(33.25) Theorem. Let A be a (noetherian) graded ring and M a graded A-module. Let

$$0 \longrightarrow M \longrightarrow I^0 \xrightarrow{\partial^0} I^1 \xrightarrow{\partial^1} \ldots \longrightarrow I^n \xrightarrow{\partial^n} I^{n+1} \longrightarrow \ldots$$

be a minimal injective resolution of M in $\underline{M}^h(A)$. Then for any homogeneous prime ideal $\mathfrak{p} \in \operatorname{Spec}(A)$ the Bass number $\mu^i(\mathfrak{p},M)$ is equal to the number of the modules of the form $\underline{E}_A(A/\mathfrak{p})(n)$, $n \in \mathbf{Z}$, which appear in I^i as direct summands.

Proof. Consider the complex

$$0 \longrightarrow \underline{\mathrm{Hom}}_{A_{<\mathfrak{p}>}}\left(A_{<\mathfrak{p}>}/\mathfrak{p}A_{<\mathfrak{p}>}, I^0_{<\mathfrak{p}>}\right) \longrightarrow \ldots \longrightarrow \underline{\mathrm{Hom}}_{A_{<\mathfrak{p}>}}\left(A_{<\mathfrak{p}>}/\mathfrak{p}A_{<\mathfrak{p}>}, I^{i-1}_{<\mathfrak{p}>}\right) \longrightarrow$$

$$\longrightarrow \underline{\mathrm{Hom}}_{A_{<\mathfrak{p}>}}\left(A_{<\mathfrak{p}>}/\mathfrak{p}A_{<\mathfrak{p}>}, I^{i}_{<\mathfrak{p}>}\right) \longrightarrow \underline{\mathrm{Hom}}_{A_{<\mathfrak{p}>}}\left(A_{<\mathfrak{p}>}/\mathfrak{p}A_{<\mathfrak{p}>}, I^{i+1}_{<\mathfrak{p}>}\right) \longrightarrow \ldots \quad .$$

$\underline{\mathrm{Ext}}^i_{A_{<\mathfrak{p}>}}(A_{<\mathfrak{p}>}/\mathfrak{p}A_{<\mathfrak{p}>}, M_{<\mathfrak{p}>})$ is isomorphic to the i-th cohomology of this complex. Hence, by Lemma (33.18), we have:

$$\underline{\mathrm{Ext}}^i_{A_{<\mathfrak{p}>}}\left(A_{<\mathfrak{p}>}/\mathfrak{p}A_{<\mathfrak{p}>}, M_{<\mathfrak{p}>}\right) \cong \underline{\mathrm{Hom}}_{A_{<\mathfrak{p}>}}\left(A_{<\mathfrak{p}>}/\mathfrak{p}A_{<\mathfrak{p}>}, I^i_{<\mathfrak{p}>}\right) \quad .$$

By Theorem (33.12) we know that I^i is a direct sum of terms $\underline{E}_A(A/\mathfrak{q})(n)$, where $n \in \mathbf{Z}$ and $\mathfrak{q}$ is a homogeneous prime ideal of A . By Lemma (33.9) we see that

$$\underline{\mathrm{Hom}}_{A_{<\mathfrak{p}>}}\left(A_{<\mathfrak{p}>}/\mathfrak{p}A_{<\mathfrak{p}>}, \underline{E}_A(A/\mathfrak{q})_{<\mathfrak{p}>}\right) \neq (0)$$

if and only if $\mathfrak{p} = \mathfrak{q}$. Hence $\underline{\mathrm{Ext}}^i_{A_{<\mathfrak{p}>}}(A_{<\mathfrak{p}>}/\mathfrak{p}A_{<\mathfrak{p}>}, M_{<\mathfrak{p}>})$ is isomorphic to a direct sum of modules of the form

$$\underline{\mathrm{Hom}}_{A_{<\mathfrak{p}>}}\left(A_{<\mathfrak{p}>}/\mathfrak{p}A_{<\mathfrak{p}>}, \underline{E}_A(A/\mathfrak{p})\right)(n) \cong A_{<\mathfrak{p}>}/\mathfrak{p}A_{<\mathfrak{p}>}(n) \quad ,$$

by Corollary (33.15). Since $A_{\mathfrak{p}}$ is a localization of $A_{<\mathfrak{p}>}$ we have the required assertion.

Let $\mathfrak{p}$ be a non-homogeneous prime ideal of a graded ring A and let $\mathfrak{p}^*$ be in the sequel the homogeneous ideal $H(\mathfrak{p})$ generated by the homogeneous elements contained in $\mathfrak{p}$, which was introduced in Chapter II. The following result shows that the Bass numbers of a graded A-module at $\mathfrak{p}$ can be computed from those of $\mathfrak{p}^*$. (Note that $\mathfrak{p}^*$ is a prime ideal.) Furthermore one can prove that a finitely generated graded A-module has finite injective dimension as an underlying module if and only if it has finite injective dimension in $M^h(A)$.

(33.26) Corollary. Let $A, \mathfrak{p}$, and $\mathfrak{p}^*$ be as above. Then we have $\mu^i(\mathfrak{p}^*,M) = \mu^{i+1}(\mathfrak{p},M)$ for any graded A-module M .

Proof. Since $\mathfrak{p}^* \neq \mathfrak{p}$ we know that $A_{<\mathfrak{p}^*>}/\mathfrak{p}^*A_{<\mathfrak{p}^*>} \cong k[X,X^{-1}]$ for some field k and an indeterminate X over k .

Hence we see that $\mathfrak{p}A_{<\mathfrak{p}^*>}/\mathfrak{p}^*A_{<\mathfrak{p}^*>}$ is a principal ideal and hence there is an element $f \in \mathfrak{p} - \mathfrak{p}^*$ such that $\mathfrak{p}A_{<\mathfrak{p}^*>} = (\mathfrak{p}^*,f)A_{<\mathfrak{p}^*>}$. From the exact sequence

$$0 \longrightarrow A/\mathfrak{p}^* \xrightarrow{\cdot f} A/\mathfrak{p}^* \longrightarrow A/(\mathfrak{p}^*,f) \longrightarrow 0$$

we obtain the long exact sequence

$$\begin{array}{l}
\underline{\mathrm{Ext}}^i_{A_{<\mathfrak{p}^*>}}\left(A_{<\mathfrak{p}^*>}/\mathfrak{p}^*A_{<\mathfrak{p}^*>}, M_{<\mathfrak{p}^*>}\right) \xrightarrow{f} \\
\underline{\mathrm{Ext}}^i_{A_{<\mathfrak{p}^*>}}\left(A_{<\mathfrak{p}^*>}/\mathfrak{p}^*A_{<\mathfrak{p}^*>}, M_{<\mathfrak{p}^*>}\right) \longrightarrow \\
\mathrm{Ext}^{i+1}_{A_{<\mathfrak{p}^*>}}\left(A_{<\mathfrak{p}^*>}/(\mathfrak{p}^*,f)A_{<\mathfrak{p}^*>}, M_{<\mathfrak{p}^*>}\right) \longrightarrow \\
\underline{\mathrm{Ext}}^{i+1}_{A_{<\mathfrak{p}^*>}}\left(A_{<\mathfrak{p}^*>}/\mathfrak{p}^*A_{<\mathfrak{p}^*>}, M_{<\mathfrak{p}^*>}\right) \xrightarrow{f} \\
\underline{\mathrm{Ext}}^{i+1}_{A_{<\mathfrak{p}^*>}}\left(A_{<\mathfrak{p}^*>}/\mathfrak{p}^*A_{<\mathfrak{p}^*>}, M_{<\mathfrak{p}^*>}\right) \longrightarrow \cdots \, .
\end{array}$$

Since $\underline{\mathrm{Ext}}^i_{A_{<\mathfrak{p}^*>}}(A_{<\mathfrak{p}^*>}/\mathfrak{p}^*A_{<\mathfrak{p}^*>}, M_{<\mathfrak{p}^*>})$ is a free $A_{<\mathfrak{p}^*>}/\mathfrak{p}^*A_{<\mathfrak{p}^*>}$-module for all $i \geqq 0$ by Lemma (33.14) we see that

$$(*) \qquad \mathrm{Ext}^{i+1}_{A_{<\mathfrak{p}^*>}}\left(A_{<\mathfrak{p}^*>}/\mathfrak{p}A_{<\mathfrak{p}^*>}, M_{<\mathfrak{p}^*>}\right) \cong \frac{\underline{\mathrm{Ext}}^i_{A_{<\mathfrak{p}^*>}}\left(A_{<\mathfrak{p}^*>}/\mathfrak{p}^*A_{<\mathfrak{p}^*>}, M_{<\mathfrak{p}^*>}\right)}{f\,\underline{\mathrm{Ext}}^i_{A_{<\mathfrak{p}^*>}}\left(A_{<\mathfrak{p}^*>}/\mathfrak{p}^*A_{<\mathfrak{p}^*>}, M_{<\mathfrak{p}^*>}\right)}$$

is a free $A_{<\mathfrak{p}^*>}/\mathfrak{p}A_{<\mathfrak{p}^*>}$-module.

Noting that $\kappa(\mathfrak{p})$ and $\kappa(\mathfrak{p}^*)$ are localizations of $A_{<\mathfrak{p}^*>}/\mathfrak{p}A_{<\mathfrak{p}^*>}$ and

$A_{<\mathfrak{p}^*>}/\mathfrak{p}^*A_{<\mathfrak{p}^*>}$ respectively, we have:

$$\begin{aligned}\mu^i(\mathfrak{p}^*,M) &:= \dim_{\kappa(\mathfrak{p}^*)} \operatorname{Ext}^i_{A_{\mathfrak{p}^*}}(\kappa(\mathfrak{p}^*),M_{\mathfrak{p}^*})\\ &= \operatorname{rank}_{A_{<\mathfrak{p}^*>}/\mathfrak{p}^*A_{<\mathfrak{p}^*>}} \underline{\operatorname{Ext}}^i_{A_{<\mathfrak{p}^*>}}(A_{<\mathfrak{p}^*>}/\mathfrak{p}^*A_{<\mathfrak{p}^*>},M_{<\mathfrak{p}^*>})\\ &= \operatorname{rank}_{A_{<\mathfrak{p}^*>}/\mathfrak{p}A_{<\mathfrak{p}^*>}} \operatorname{Ext}^{i+1}_{A_{<\mathfrak{p}^*>}}(A_{<\mathfrak{p}^*>}/\mathfrak{p}A_{<\mathfrak{p}^*>},M_{<\mathfrak{p}^*>})\\ &= \dim_{\kappa(\mathfrak{p})} \operatorname{Ext}^{i+1}_{A_{\mathfrak{p}}}(\kappa(\mathfrak{p}),M_{\mathfrak{p}})\\ &=: \mu^{i+1}(\mathfrak{p},M) \quad .\end{aligned}$$

<u>(33.27) Corollary</u>. Let A be a graded ring. Then A is Cohen-Macaulay (resp. Gorenstein) if and only if $A_{\mathfrak{p}}$ is Cohen-Macaulay (resp. Gorenstein) for any homogeneous prime ideal $\mathfrak{p}$ of A.

Proof. Note that for any non-homogeneous prime ideal $\mathfrak{p}$ of A we have $\operatorname{Hom}_A(A/\mathfrak{p},A)_{\mathfrak{p}} = 0$, since every associated prime of A is homogeneous. Hence $\mu^\circ(\mathfrak{p},A) = 0$ (which also follows from (33.26)). Therefore we get:

a) $A_{\mathfrak{p}}$ is Cohen-Macaulay $\iff \mu^i(\mathfrak{p},A) = 0$ for $i < \operatorname{ht}\mathfrak{p}$

$\iff \mu^i(\mathfrak{p}^*,A) = 0$ for $i < \operatorname{ht}\mathfrak{p}^*$

$\iff A_{\mathfrak{p}^*}$ is Cohen-Macaulay,

where $\operatorname{ht}(\mathfrak{p}) - 1 = \operatorname{ht}(\mathfrak{p}^*)$.

b) Since $A_{\mathfrak{p}}$ is Gorenstein if and only if

$$\mu^{t-1}(\mathfrak{p}^*,A) = \mu^t(\mathfrak{p},A) = \begin{cases} 1 & (t = \operatorname{ht}\mathfrak{p}) \\ 0 & (t \neq \operatorname{ht}\mathfrak{p}) \end{cases},$$

the assertion for the Gorenstein property follows from Step a).

<u>(33.28) Definition</u>. Let M be a graded A-module. We say that M has finite injective dimension in $M^h(A)$ if there is an injective resolution

$$0 \longrightarrow M \longrightarrow I^0 \longrightarrow I^1 \longrightarrow \dots \longrightarrow I^n \longrightarrow 0 \quad .$$

of M in $M^h(A)$ for some $n \geqq 0$.

If M has finite injective dimension in $M^h(A)$, then there exists an integer $n \geqq 0$ such that $\underline{\mathrm{Ext}}_A^i(A/\mathfrak{p},M) = (0)$, for all homogeneous prime ideals $\mathfrak{p}$ of A , if $i > n$. Also the converse is true, as we show in the following lemma which is a corollary of the Theorems (33.12) and (33.25).

<u>(33.29) Lemma</u>. A graded A-module M has finite injective dimension if and only if there is an integer $n \geqq 0$ such that $\underline{\mathrm{Ext}}_A^i(A/\mathfrak{p},M) = (0)$ for all $i > n$ and for all homogeneous prime ideals $\mathfrak{p}$ of A .

Proof. If $\underline{\mathrm{Ext}}_A^i(A/\mathfrak{p},M) = (0)$ for all homogeneous prime ideals $\mathfrak{p}$, then the i-th module I^i of a minimal injective resolution of M has no direct summand of the form $\underline{E}_A(A/\mathfrak{p})(m), m \in \mathbf{Z}$, by Theorem (33.25). By Theorem (33.12) we know that $I^i = 0$.

<u>(33.30) Definition</u>. 1) For a graded A-module M we define

$$\underline{\mathrm{id}}_A M = \min\left\{ n \geqq 0 \mid \underline{\mathrm{Ext}}_A^i(-,M) = 0 \quad \text{for} \quad n < i \right\}$$

and call it the injective dimension of M in $M^h(A)$, where $\underline{\mathrm{Ext}}_A^i(-,M) = 0$ means that $\underline{\mathrm{Ext}}_A^i(N,M) = (0)$ for any graded A-module N .

2) $\mathrm{id}_A M$ denotes the injective dimension of M in the category of A-modules.

<u>(33.31) Lemma</u>. A graded A-module M has finite injective dimension in $M^h(A)$ if and only if M has finite injective dimension as underlying A-module. In this case we have $\underline{\mathrm{id}}_A M + 1 \geqq \mathrm{id}_A M$.

Proof. This follows from Corollary (33.26).

§ 34. MATLIS DUALITY

Part I: Local case

In this section we recall the Matlis duality for local rings (cf. [20]) which will be needed for the graded case. Throughout this section $(R,\mathfrak{m},k)$ denotes a noetherian local ring and E_R denotes the injective envelope of k. The completion of R is denoted by $\hat{R}$.

Since the inclusion $k \hookrightarrow E_R$ is essential, we have:

$$(*) \qquad \mathrm{Hom}_R(k,E_R) \cong k \quad .$$

(34.1) Lemma. For any R-module M of finite length the canonical homomorphism $M \longrightarrow \mathrm{Hom}_R(\mathrm{Hom}_R(M,E_R),E_R)$ is an isomorphism.

Proof. Using (*), we get the isomorphism $k \xrightarrow{\sim} \mathrm{Hom}_R(\mathrm{Hom}_R(k,E_R),E_R)$. Since $\mathrm{Hom}_R(\mathrm{Hom}_R(-,E_R),E_R)$ is an exact functor, we get the claim by using induction on the length $\lambda(M)$ of M.

Note that we have the following statements:

1) $R/\mathfrak{m}^n \cong \mathrm{Hom}_R(\mathrm{Hom}_R(R/\mathfrak{m}^n,E_R),E_R)$ for $n > 0$.

2) $\mathrm{Ass}_R(E_R) = \{\mathfrak{m}\}$.

Hence for any element $0 \neq x \in E_R$ we have $\phi \neq \mathrm{Ass}_R(Rx) \subset \mathrm{Ass}_R(E_R) = \{\mathfrak{m}\}$. From this we see that $\mathrm{ann}_R(Rx)$ is an $\mathfrak{m}$-primary ideal. Hence we have $E_R = \bigcup_{n>0} (0 :_{E_R} \mathfrak{m}^n) = \varinjlim_n \mathrm{Hom}_R(R/\mathfrak{m}^n,E_R)$.

3) For any $n > 0$ we get a commutative diagram by Lemma (34.1)

$$\begin{array}{ccc} R/\mathfrak{m}^n & \xrightarrow{\sim} & \mathrm{Hom}_R(\mathrm{Hom}_R(R/\mathfrak{m}^n,E_R),E_R) \\ \uparrow \varphi_n & & \uparrow \Psi_n \\ R/\mathfrak{m}^{n+1} & \xrightarrow{\sim} & \mathrm{Hom}_R(\mathrm{Hom}_R(R/\mathfrak{m}^{n+1},E_R),E_R) \end{array} ,$$

where φ_n is the canonical surjection and Ψ_n is obtained from the injection $(0 :_{E_R} \mathfrak{m}^n) \hookrightarrow (0 :_{E_R} \mathfrak{m}^{n+1})$. From these properties we get:

(34.2) Proposition. $\hat{R} \xrightarrow{\sim} \mathrm{Hom}_R(E_R, E_R)$.

Proof. By [1], Chapter II, § 6, n°6, Proposition 11, we get

$$\begin{aligned} \mathrm{Hom}_R(E_R,E_R) &\cong \mathrm{Hom}_R(\varinjlim_n \mathrm{Hom}_R(R/m^n,E_R),E_R) \\ &\cong \varprojlim_n \mathrm{Hom}_R(\mathrm{Hom}_R(R/m^n,E_R),E_R) \\ &\cong \varprojlim_n R/m^n \cong \hat{R} . \end{aligned}$$

Note that if R is complete, we have

$$(**) \qquad \mathrm{Hom}_R(E_R,E_R) \cong R \quad .$$

As we will see in the sequel, the isomorphism (**) is the heart of the duality theory of local (graded) rings. One may say that all of the results in the rest of this section are consequences of the isomorphism (**).

For the proof of Matlis duality (Theorem (34.4)) we need an auxiliary result.

(34.3) Lemma. Let R be a complete local ring and M an R-module. Then M is an artinian R-module if and only if M is a submodule of a finite direct sum of copies of E_R .

Proof. 1) Suppose that $M \subset E_R \oplus \ldots \oplus E_R$. It is enough to show that E_R is artinian. If E_R is not artinian there is an infinite descending chain of submodules

$$E_R \supsetneq E_1 \supsetneq E_2 \supsetneq \ldots \quad .$$

From the surjection $E_R \longrightarrow E_R/E_i$ we get an injection

$$\mathrm{Hom}_R(E_R/E_i,E_R) \hookrightarrow \mathrm{Hom}_R(E_R,E_R) \cong R$$

by Proposition (34.2). Hence we have an infinite ascending chain

$$(E_R/E_1)^* \subsetneq (E_R/E_2)^* \subsetneq \ldots \subsetneq R \quad , \quad \text{(check)}$$

where $(-)^* = \mathrm{Hom}_R(-,E_R)$, a contradiction.

2) Suppose that M is artinian. We will show that even $E_R(M)$ is a finite direct sum of copies of E_R : Since $\mathrm{Ass}_R(E_R(M)) = \mathrm{Ass}_R(M) = \{\mathfrak{m}\}$, $E_R(M)$ must be a direct sum of E_R (cf. Theorem (33.12)).
Suppose $E_R(M) = \bigoplus_{\lambda\in\Lambda} E_\lambda$, $E_\lambda \cong E_R$ and $\#\Lambda = \infty$. Then there is an infinite descending chain $\Lambda \supsetneqq \Lambda_1 \supsetneqq \Lambda_2 \supsetneqq \dots$ of subsets of Λ. Hence we get a chain $E_R(M) \supsetneqq \bigoplus_{\lambda\in\Lambda_1} E_\lambda \supsetneqq \bigoplus_{\lambda\in\Lambda_2} E_\lambda \supsetneqq \dots$ of submodules of $E_R(M)$.
Since $E_R(M)$ is an essential extension of M, we have an infinite chain

$$M \supsetneqq M \cap \left(\bigoplus_{\lambda\in\Lambda_1} E_\lambda\right) \supsetneqq M \cap \left(\bigoplus_{\lambda\in\Lambda_2} E_\lambda\right) \supsetneqq \dots \quad .$$

This contradicts to the fact that M is artinian.

<u>(34.4) Theorem</u>. (*Matlis duality*) Let $(R,\mathfrak{m},k)$ be a complete local ring and E_R be the injective envelope of k. Then we have:

a) $\mathrm{Hom}_R(E_R,E_R) \cong R$.

b) Let M be a noetherian (resp. artinian) R-module. Then $\mathrm{Hom}_R(M,E_R)$ is an artinian (resp. noetherian) R-module. In other words, the functor $\mathrm{Hom}_R(-,E_R)$ gives an equivalence between the category of artinian and noetherian R-modules.

c) If M is a notherian or an artinian R-module then the canonical homomorphism $M \longrightarrow \mathrm{Hom}_R(\mathrm{Hom}_R(M,E_R),E_R)$ is an isomophism.

Proof. a) was already proved in Proposition (34.2).

b) If M is a noetherian R-module, i.e. a finitely generated R-module, there is a finitely generated free R-module F and a surjection $F \twoheadrightarrow M$. Hence $\mathrm{Hom}_R(M,E_R)$ can be identified with a submodule of $\mathrm{Hom}_R(F,E_R)$ which is isomorphic to a finite direct sum of copies of E_R. Therefore $\mathrm{Hom}_R(M,E_R)$ is artinian by Lemma (34.3).

If M is an artinian R-module, then M is a submodule of a finite direct sum of copies of E_R. Hence $\mathrm{Hom}_R(M,E_R)$ is a homomorphic image of a finitely generated free R-module by Proposition (34.2). Therefore $\mathrm{Hom}_R(M,E_R)$ is noetherian.

c) We first assume that M is a noetherian R-module. If $M = R$ we have by Proposition (34.2):

$$R \cong \mathrm{Hom}_R(E_R,E_R) \cong \mathrm{Hom}_R(\mathrm{Hom}_R(R,E_R),E_R) \quad .$$

Hence c) is proved for any finitely generated free R-module. If M is not free we consider an exact sequence $G \longrightarrow F \longrightarrow M \longrightarrow 0$, where F and G are finitely generated free R-modules. We get a commutative diagram with exact rows

$$\begin{array}{ccccccc} G^{**} & \longrightarrow & F^{**} & \longrightarrow & M^{**} & \longrightarrow & 0 \\ \uparrow \wr & & \uparrow \wr & & \uparrow & & \\ G & \longrightarrow & F & \longrightarrow & M & \longrightarrow & 0 \end{array} \quad ,$$

where $(-)^{**} = \mathrm{Hom}_R(\mathrm{Hom}_R(-,E_R),E_R)$. Hence we have an isomorphism $M \xrightarrow{\sim} M^{**}$.

Next we assume that M is an artinian R-module. Since $\mathrm{Hom}_R(M,E_R)$ is a noetherian R-module by b), one can apply the last result to $\mathrm{Hom}_R(M,E_R)$ in order to get an isomorphism

$$\mathrm{Hom}_R(M,E_R) \xrightarrow{\sim} \mathrm{Hom}_R(\mathrm{Hom}_R(\mathrm{Hom}_R(M,E_R),E_R),E_R) \, .$$

This implies that

$$M \longrightarrow \mathrm{Hom}_R(\mathrm{Hom}_R(M,E_R),E_R)$$

is an isomorphism.

For arbitrary local rings we have the following result:

<u>(34.5) Proposition</u>. Let R be a local ring and M a finitely generated R-module. Then

$$\mathrm{Hom}_R(\mathrm{Hom}_R(M,E_R),E_R) \cong M \otimes_R \hat{R} \quad .$$

Proof. By Proposition (34.2) we have

$$\mathrm{Hom}_R(\mathrm{Hom}_R(R,E_R),E_R) \cong \mathrm{Hom}_R(E_R,E_R) \cong \hat{R} \quad .$$

Using the same argument as in the proof of Theorem (34.4), b), we get the assertion.

Part II: Graded case

Now we come to Matlis duality for graded rings, which is the main topic in this section.

(34.6) Definition. 1) Let $A = \bigoplus_{n\geq 0} A_n$ be a non-negatively graded noetherian ring. We say that A is defined over a local ring if A_0 is a local ring. If A is defined over a local ring $(A_0, \mathfrak{m}_0)$, A has the unique maximal homogeneous ideal

$$\mathfrak{M} = \mathfrak{m}_0 \oplus \left(\bigoplus_{n>0} A_n\right) .$$

2) Let $\hat{A}_0$ be the completion of A_0. We denote the graded ring $A \otimes_{A_0} \hat{A}_0$ by $\hat{A}$.

3) Let E_0 be the injective envelope of $A_0/\mathfrak{m}_0$ as A_0-module and $\underline{E}_0$ be the graded A_0-module whose grading is given by $[\underline{E}_0]_0 = E_0$ and $[\underline{E}_0]_n = (0)$ for $n \neq 0$.

4) We put $\underline{E}_A := \underline{\mathrm{Hom}}_{A_0}(A, \underline{E}_0)$, where A is considered as a graded A_0-module.

(34.7) Lemma. Let A be a noetherian graded ring defined over a local ring $(A_0, \mathfrak{m}_0)$. Then we have:

a) $\underline{\mathrm{Hom}}_A(\underline{E}_A, \underline{E}_A) \cong \hat{A}$

b) $\underline{E}_A$ is the injective envelope of $A/\mathfrak{M}$ in the category $M^h(A)$.

Proof. a) By Lemma (33.2) and Proposition (34.5) we get:

$$\begin{aligned}
\underline{\mathrm{Hom}}_A(\underline{E}_A, \underline{E}_A) &\cong \underline{\mathrm{Hom}}_A(\underline{\mathrm{Hom}}_{A_0}(A, \underline{E}_0), \underline{\mathrm{Hom}}_{A_0}(A, \underline{E}_0)) \\
&\cong \underline{\mathrm{Hom}}_{A_0}(\underline{\mathrm{Hom}}_{A_0}(A, \underline{E}_0), \underline{E}_0) \\
&\cong \bigoplus_{n\in\mathbf{Z}} \underline{\mathrm{Hom}}_{A_0}(\underline{\mathrm{Hom}}_{A_0}(A, \underline{E}_0), \underline{E}_0)_n \\
&\cong \bigoplus_{n\in\mathbf{Z}} \underline{\mathrm{Hom}}_{A_0}(\underline{\mathrm{Hom}}_{A_0}(A, \underline{E}_0(-n)), \underline{E}_0)_0 \\
&\cong \bigoplus_{n\in\mathbf{Z}} \mathrm{Hom}_{A_0}(\mathrm{Hom}_{A_0}(A_n, E_0), E_0) \\
&\cong \bigoplus_{n\in\mathbf{Z}} (A_n \otimes \hat{A}_o) \cong \hat{A} \quad .
\end{aligned}$$

b) By Lemma (33.3) we know that $\underline{E}_A = \underline{\mathrm{Hom}}_{A_0}(A, \underline{E}_0)$ is an injective A-module in $M^h(A)$. Therefore by Theorem (33.12) and Corollary (33.15) it is enough to show that

1) $\mathrm{Supp}(\underline{E}_A) = \{\mathfrak{m}\}$.

2) $\underline{\mathrm{Hom}}_A(A/\mathfrak{m}, \underline{E}_A) \cong A/\mathfrak{m}$.

For 1) take any homogeneous element $f \in \underline{E}_A$ with $\deg(f) = -n$. By definition f is an A_0-homomorphism $A \longrightarrow \underline{E}_0$ such that $f(A_k) \subset [\underline{E}_0]_{k-n}$ for all k, i.e. $f(A_n) \subset \underline{E}_0$ and $f(A_k) = (0)$ for $n \neq k$. Hence for any homogeneous element $x \in A$ of $\deg(x) > n$, we have $(xf)(r) = f(xr) = 0$ for any $r \in A$, because A is non-negatively graded. Therefore $xf = 0$. Since A_n is a finitely generated A_0-module and E_0 is the injective envelope of $A_0/\mathfrak{m}_0$, we have $\mathfrak{m}_0^k f(A_n) = (0)$ for large $k>>0$. Hence $\mathrm{Supp}(\underline{E}_A) = \{\mathfrak{m}\}$. For 2) let $A/\mathfrak{m} = A_0/\mathfrak{m}_0 = k$. Then we get by Lemma (33.2):

$$\underline{\mathrm{Hom}}_A(A/\mathfrak{m}, \underline{E}_A) \cong \underline{\mathrm{Hom}}_A(k, \underline{\mathrm{Hom}}_{A_0}(A, \underline{E}_0) \cong \underline{\mathrm{Hom}}_{A_0}(k, \underline{E}_0) \cong k \quad , \qquad \text{q.e.d.}$$

We have seen that for a graded ring A the injective envelope $\underline{E}_A$ of $A/\mathfrak{m}$ in the category $M^h(A)$ has the similar property as the injective envelope of the residue field of a local ring. More precisely we can obtain Matlis duality for graded rings defined over a local ring as follows:

(34.8) Theorem. (Matlis duality for *graded* rings) Let $A = \bigoplus_{n \geq 0} A_n$ be a noetherian graded ring defined over a complete local ring $(A_0, \mathfrak{m}_0)$. Then one has:

a) $\underline{\mathrm{Hom}}_A(\underline{E}_A, \underline{E}_A) \cong A$.

b) Let M be a noetherian (resp. artinian) graded A-module. Then $\underline{\mathrm{Hom}}_A(M, \underline{E}_A)$ is an artinian (resp. noetherian) graded A-module.

c) If M is a noetherian or an artinian graded A-module, then $\underline{\mathrm{Hom}}_A(\underline{\mathrm{Hom}}_A(M, \underline{E}_A), \underline{E}_A) \cong M$ in $M^h(A)$.

Proof. The proof follows as that of Theorem (34.4) using the fact that $\underline{E}(A)$ is artinian which follows from Remark (34.9).

(34.9) Remark. $\underline{E}_A$ is still injective as an underlying A-module, in fact $\underline{E}_A \cong E_A(A/\mathfrak{m})$. Hence $\mathrm{Hom}_A(\underline{E}_A,\underline{E}_A) \cong (A_{\mathfrak{m}})^{\wedge}$. (Note that now we are in the underlying category of $A_{\mathfrak{m}}$-modules.) To see this, it is enough to show that $\mu^1(\mathfrak{p},\underline{E}_A) = 0$ for all $\mathfrak{p} \in \mathrm{Spec}(A)$, cf. Theorem (33.25): If $\mathfrak{p}$ is homogeneous clearly $\mu^1(\mathfrak{p},\underline{E}_A) = 0$. If $\mathfrak{p}$ is not homogeneous we consider the maximum homogeneous prime ideal $\mathfrak{p}^*$ contained in $\mathfrak{p}$. Then by Corollary (33.26) we get $\mu^\circ(\mathfrak{p}^*,\underline{E}_A) = \mu^1(\mathfrak{p},\underline{E}_A)$. But $\mathfrak{p}^* \neq \mathfrak{m}$ implies $\mathrm{Hom}_{A_{\mathfrak{p}^*}}(\kappa(\mathfrak{p}^*),(\underline{E}_A)_{\mathfrak{p}^*}) = (0)$, since $\mathrm{Supp}(\underline{E}_A) = \{\mathfrak{m}\}$. Hence $\mu^1(\mathfrak{p},\underline{E}_A) = 0$. Note that $\mathrm{Hom}_A(\underline{E}_A,\underline{E}_A) \cong \mathrm{Hom}_{A_{\mathfrak{m}}}(\underline{E}_A,\underline{E}_A) \cong (A_{\mathfrak{m}})^{\wedge}$, since $\underline{E}_A$ is the injective envelope of $A/\mathfrak{m}$ as $A_{\mathfrak{m}}$-module. This shows that $\mathrm{Hom}_A(\underline{E}_A,\underline{E}_A)$ is vastly bigger than $\underline{\mathrm{Hom}}_A(\underline{E}_A,\underline{E}_A) \cong A$.

§ 35. Local cohomology

Let $A = \bigoplus_{n\in\mathbf{Z}} A_n$ be a noetherian graded ring, $\mathfrak{a}$ a homogeneous ideal of A and M a graded A-module.

(35.1) Definition. We define for $i \geq 0$

$$\underline{H}^i_{\mathfrak{a}}(M) = \underset{n}{\underrightarrow{\lim}}\ \underline{\mathrm{Ext}}^i_A(A/\mathfrak{a}^n,M)$$

and call this the i-th local cohomology module of M with respect to $\mathfrak{a}$.

From the definition we see that $\underline{H}_{\mathfrak{a}}(-)$ is an additive functor from $M^h(A)$ to itself. The local cohomology provides a powerful tool for the study of homological properties of local rings and graded rings (cf. Grothendieck [4], Herzog-Kunz [6] and Goto-Watanabe [16],[17]). In this section we give basic properties of local cohomomology.

Note that, since $A/\mathfrak{a}^n$ is a finitely generated A-module, we have by (33.1): $\underline{\mathrm{Hom}}_A(A/\mathfrak{a}^n,M) = \mathrm{Hom}_A(A/\mathfrak{a}^n,M)$ for all $n > 0$. Hence we can deduce that $\underline{\mathrm{Ext}}^i_A(A/\mathfrak{a}^n,M) = \mathrm{Ext}^i_A(A/\mathfrak{a}^n,M)$ for all $i \geq 0$.

In the sequel for any noetherian ring A and for any ideal $\mathfrak{a}$ of A we denote the functor $\underset{n}{\underrightarrow{\lim}}\ \mathrm{Ext}^i_A(A/\mathfrak{a}^n,-)$ by $H^i_{\mathfrak{a}}(-)$. By this remark we get the following result.

(35.2) Lemma. Let A be a graded ring and $\mathfrak{a}$ a homogeneous ideal of A. Then for any graded A-module M we have $\underline{H}^i_{\mathfrak{a}}(M) = H^i_{\mathfrak{a}}(M)$ for all $i \geq 0$ as underlying A-modules.

(35.3) Proposition. The functor $\underline{H}^i_{\mathfrak{a}}(-)$ is the i-th derived functor of $\underrightarrow{\lim}_n \underline{\mathrm{Hom}}_A(A/\mathfrak{a}^n,-) = \underline{H}^0_{\mathfrak{a}}(-)$.

Proof. We must show the following facts:

(1) If I is an injective module in $M^h(A)$, then $H^i_{\mathfrak{a}}(I) = (0)$ for $i > 0$.

(2) From a short exact sequence $0 \to M' \to M \to M'' \to 0$ in $M^h(A)$ we have a long exact sequence

$$0 \to \underline{H}^0_{\mathfrak{a}}(M') \to \underline{H}^0_{\mathfrak{a}}(M) \to \underline{H}^0_{\mathfrak{a}}(M'') \to \underline{H}^1_{\mathfrak{a}}(M') \to \ldots$$

$$\underline{H}^i_{\mathfrak{a}}(M') \to \underline{H}^i_{\mathfrak{a}}(M) \to \underline{H}^i_{\mathfrak{a}}(M'') \to H^{i+1}_{\mathfrak{a}}(M') \to \ldots \quad .$$

The statement (1) follows from $\underline{\mathrm{Ext}}^i_A(A/\mathfrak{a}^n,I) = (0)$ for $i > 0$ for any injective module I in $M^h(A)$.

For (2): For any integer $n > 0$ we get from the short exact sequence a long exact sequence of $\underline{\mathrm{Ext}}$'s:

$$\to \underline{\mathrm{Ext}}^i_A(A/\mathfrak{a}^n,M') \to \underline{\mathrm{Ext}}^i_A(A/\mathfrak{a}^n,M) \to \underline{\mathrm{Ext}}^i_A(A/\mathfrak{a}^n,M'')$$

$$\to \underline{\mathrm{Ext}}^{i+1}_A(A/\mathfrak{a}^n,M') \to \underline{\mathrm{Ext}}^{i+1}_A(A/\mathfrak{a}^n,M) \to \ldots \quad .$$

Since the direct limit of a direct system of exact sequences is exact, (cf. [1]) we get the required long exact sequence.

Note that for any graded A-module M

$$\underrightarrow{\lim}_n \underline{\mathrm{Hom}}_A(A/\mathfrak{a}^n,M) = \bigcup_{n>0}(0 : \mathfrak{a}^n)_M$$
$$= \left\{x \in M \mid \mathfrak{a}^n x = 0 \text{ for some } n > 0\right\} \quad .$$

From this fact we get the following auxiliary results.

(35.4) Lemma. $\underline{H}^i_{\mathfrak{a}}(-)$ depends only on the radical $\sqrt{\mathfrak{a}}$.

Proof. Let $\mathfrak{b}$ be an ideal of A such that $\sqrt{\mathfrak{a}} = \sqrt{\mathfrak{b}}$ and let M be a graded A-module . Then

$$\underline{H}^0_{\mathfrak{a}}(M) = \bigcup_{n>0}(0 : \mathfrak{a}^n)_M = \bigcup_{n>0}(0 : \mathfrak{b}^n)_M = \underline{H}^0_{\mathfrak{b}}(M) .$$

(35.5) Lemma. $\mathrm{Supp}_A(\underline{H}^i_{\mathfrak{a}}(M)) \subset V(\mathfrak{a})$ for any graded A-module M , i.e. for any element $x \in \underline{H}^i_{\mathfrak{a}}(M)$ there is an integer $n > 0$ such that $\mathfrak{a}^n x = 0$.

Proof. This follows from Proposition (35.3).

(35.6) Definition. Let R be a noetherian ring, $\mathfrak{a}$ an ideal of R and M a finitely generated R-module such that $M \neq \mathfrak{a}M$. We denote the maximal length of M-sequences in $\mathfrak{a}$ by $\mathrm{depth}_{\mathfrak{a}}M$, see [9*].

(35.7) Proposition. For any noetherian ring R and for any finitely generated R-module M we have $\inf\left\{i \mid H^i_{\mathfrak{a}}(M) \neq (0)\right\} = \mathrm{depth}_{\mathfrak{a}}M$.

Proof. By induction on $t = \mathrm{depth}_{\mathfrak{a}}M$. If $t = 0$ every element of $\mathfrak{a}$ is a zero-divisor on M . Hence $\mathfrak{a} \subset \mathfrak{p}$ for some associated prime $\mathfrak{p}$ of M . There is an element $0 \neq x \in M$ such that $\mathfrak{p}x = 0$. Therefore $0 \neq x \in (0 : \mathfrak{a})_M \subset \underline{H}^0_{\mathfrak{a}}(M)$.

Let $t > 0$ and assume that the assertion is true for any finitely generated R-module N such that $\mathrm{depth}_{\mathfrak{a}} N < t$. Let $a_1, \ldots, a_t \in \mathfrak{a}$ be a maximal M-sequence in $\mathfrak{a}$. Then $\mathrm{depth}_{\mathfrak{a}}(M/a_1M) = t - 1$. By induction hypothesis $H^i_{\mathfrak{a}}(M/a_1M) = (0)$ for $i < t - 1$ and $H^{t-1}_{\mathfrak{a}}(M/a_1M) \neq (0)$. From the exact sequence $0 \longrightarrow M \xrightarrow{.a_1} M \longrightarrow M/a_1M \longrightarrow 0$ we obtain an exact sequence $(0) = H^{i-1}_{\mathfrak{a}}(M/a_1M) \longrightarrow H^i_{\mathfrak{a}}(M) \xrightarrow{.a_1} H^i_{\mathfrak{a}}(M)$ for $i < t$. This shows that a_1 is a non-zero-divisor on $H^i_{\mathfrak{a}}(M)$ for $i < t$. Take $x \in H^i_{\mathfrak{a}}(M)$. By Lemma (35.5) $a_1^n x = 0$ for some $n > 0$. Hence $x = 0$ and

$H^i_{\mathfrak{a}}(M) = (0)$ for $i < t$. From the following exact sequence, where $H^{t-1}_{\mathfrak{a}}(M/a_1M) \neq 0$,

$$0 \longrightarrow H^{t-1}_{\mathfrak{a}}(M/a_1M) \longrightarrow H^t_{\mathfrak{a}}(M)$$

we see that $\inf\left\{i \mid H^i_{\mathfrak{a}}(M) \neq 0\right\} = \mathrm{depth}(M)$.

<u>(35.8) Remark</u>. By Proposition (35.7) we see that $\inf\left\{i \mid H^i_{\mathfrak{a}}(M) \neq (0)\right\}$ does not stand for the maximal length of M-sequences consisting of homogeneous elements in $\mathfrak{a}$, since for any graded A-module M $\underline{H}^i_{\mathfrak{a}}(M) = H^i_{\mathfrak{a}}(M)$. For example, let $A = k[X,Y,XT,YT] \subset k[X,Y,T]$, where k is a field and X,Y,T are indetermiantes over k with $\deg X = \deg Y = 0$ and $\deg T = 1$. Then A is a Cohen-Macaulay graded ring of dimension 3. But A has no regular sequence of length 3 consisting of homogeneous elements and $\underline{H}^i_{\mathfrak{a}}(A) = 0$ for $i < 3$, where $\mathfrak{a} = (X,Y,XT,YT)$.

Our next purpose is to show that the local cohomology is a direct limit of a direct system of Koszul homology. First we recall once more some properties of Koszul complexes for any commutative ring R (see Chapter II, (11.6), [10] and [12]). Let $a_1,\ldots,a_r$ be elements of R . Let e_i be a free base of $K_1(a_i;R) \cong R$. Then $K_p(a_1,\ldots,a_r;R)$ is isomorphic to the free R-module with the free basis

$$e_{i_1} \otimes e_{i_2} \otimes \ldots \otimes e_{i_p}, \quad 1 \leqq i_1 < i_2 < \ldots < i_p \leqq r ,$$

where the boundary operator $\partial^b : K_p(\underline{a};R) \longrightarrow K_{p-1}(\underline{a};R)$ is given by

$$\partial^p\left(e_{i_1} \otimes \ldots \otimes e_{i_p}\right) = \sum_{k=1}^{p} (-1)^{k+1} a_{i_k} e_{i_1} \otimes \ldots \otimes \check{e}_{i_k} \otimes \ldots \otimes e_{i_p} ,$$

If M is an R-module then $K.(a_1,\ldots,a_r;M) \cong K.(a_1,\ldots,a_r;R) \otimes_R M$. By $H_i(a_1,\ldots,a_r;M)$ we denote its i-th homology. Recall that the dual complex of $K.(a_1,\ldots,a_r;R)$ with respect to M is the complex

$$0 \longrightarrow \mathrm{Hom}_R(K_0(a_1,\ldots,a_r;R),M) \xrightarrow{d_1^*}$$
$$\mathrm{Hom}_R(K_1(a_1,\ldots,a_r;R),M) \longrightarrow \ldots \xrightarrow{d_{r-1}^*}$$
$$\mathrm{Hom}_R(K_{r-1}(a_1,\ldots,a_r;R),M) \xrightarrow{d_r}$$
$$\mathrm{Hom}_R(K_r(a_1,\ldots,a_r;R),M) \longrightarrow 0 \quad ,$$

where $d_i^* = \mathrm{Hom}_R(d_i,M)$. We denote this complex by $K^{\bullet}(a_1,\ldots,a_r;M)$ and its cohomology by $H^i(a_1,\ldots,a_r;M)$. Note that for $F = K_1(a_1,\ldots,a_r;R)$ we have (if ΛF denotes the exterior algebra of F)

$$\mathrm{Hom}_R(K_i(a_1,\ldots,a_r;R),M) \cong \mathrm{Hom}_R(\overset{i}{\Lambda} F,M)$$
$$\cong \mathrm{Hom}_R(\overset{i}{\Lambda} F,R) \otimes_R M \cong \overset{r-i}{\Lambda} F \otimes_R M \quad ,$$

where the last isomorphism comes from the canonical isomorphism $\mathrm{Hom}_R(\overset{i}{\Lambda} F,R) \cong \overset{r-i}{\Lambda} F$, cf. [1]. It is not hard to see that

$$H^i(a_1,\ldots,a_r;M) \cong H_{r-i}(a_1,\ldots,a_r;M) \quad .$$

In the following propositions we list some properties of the Koszul homology which are needed in the following. These properties of the Koszul homology are well known. So we omit the proof and we refer the reader to suitable references.

(35.9) Proposition. Let R , $a_1,\ldots,a_r$, and M be the same as above. Then we have:

1) $H_0(a_1,\ldots,a_r;M) = M/(a_1,\ldots,a_r)M$

2) $H_r(a_1,\ldots,a_r;M) = (0 : (a_1,\ldots,a_r))_M = \{x \in M \mid a_i x = 0 \text{ for } 1 \leq i \leq r\}$

3) $H_i(a_1,\ldots,a_r;M) = (0)$ for $i > r$ and $i < 0$.

For the proof see [12].

(35.10) Proposition. Under the same assumptions as in Proposition (35.9) we have $(a_1,\ldots,a_r) \cdot H_i(a_1,\ldots,a_r;M) = (0)$ for $0 \leq i \leq r$.

Proof. See [12], IV-7, Proposition 4, and [10], 8.3, Theorem 3.

(35.11) Proposition. From a short exact sequence of R-modules $0 \longrightarrow M' \longrightarrow M \longrightarrow M'' \longrightarrow 0$ we get an exact sequence of complexes

$$0 \longrightarrow K.(a_1,\ldots,a_r;M') \longrightarrow K.(a_1,\ldots,a_r;M)$$

$$\longrightarrow K.(a_1,\ldots,a_r;M'') \longrightarrow 0 \quad .$$

Consequently we have a long exact sequence

$$\longrightarrow H_i(a_1,\ldots,a_r;M') \longrightarrow H_i(a_1,\ldots,a_r;M)$$

$$\longrightarrow H_i(a_1,\ldots,a_r;M'') \longrightarrow H_{i-1}(a_1,\ldots,a_r;M')$$

$$\longrightarrow H_{i-1}(a_1,\ldots,a_r;M) \longrightarrow \ldots \quad .$$

This follows from the fact that the Koszul complex is a complex of free R-modules, cf. [10], 8, Theorem 2 .

(35.12) Proposition. Under the same assumption as above there is an exact sequence

$$0 \longrightarrow H_r(a_1,\ldots,a_r;M) \longrightarrow H_{r-1}(a_1,\ldots,a_{r-1};M)$$

$$\xrightarrow{\partial_{r-1}} H_{r-1}(a_1,\ldots,a_{r-1};M) \xrightarrow{\sigma_{r-1}} H_{r-1}(a_1,\ldots,a_r;M)$$

$$\xrightarrow{\tau_{r-1}} H_{r-2}(a_1,\ldots,a_{r-1};M) \xrightarrow{\partial_{r-2}} H_{r-2}(a_1,\ldots,a_{r-1};M)$$

$$\ldots \xrightarrow{\partial_1} H_i(a_1,\ldots,a_{r-1};M) \xrightarrow{\sigma_i} H_i(a_1,\ldots,a_r;M)$$

$$\xrightarrow{\tau_i} H_{i-1}(a_1,\ldots,a_{r-1};M) \xrightarrow{\partial_{i-1}} H_{i-1}(a_1,\ldots,a_{r-1};M) \longrightarrow \ldots \quad ,$$

where ∂_i is the multiplication by a_r , and σ_i , τ_i are canonical.

For the proof see [10], 8, Proposition 2, p. 365.

In Proposition (35.9) and (35.12) we did not assume that R is noetherian. But for the following two propositions we must assume that R is noetherian.

<u>(35.13) Proposition</u>. Let R be a noetherian ring and $\mathfrak{a} = (a_1,\dots,a_r)$ an ideal of R contained in the Jacobson radical of R. Then for any finitely generated R-module M we get the following statements:

(1) $\operatorname{depth}_{\mathfrak{a}}(M) + \max\{n \mid H_n(a_1,\dots,a_r;M) \neq 0\} = r$.

(2) If $t = \operatorname{depth}_{\mathfrak{a}}(M)$ and if $b_1,\dots,b_t$ is a M-sequence in $\mathfrak{a}$, then $H_{r-t}(a_1,\dots,a_r;M) \cong [(b_1,\dots,b_t)M \underset{M}{:} \mathfrak{a}]/(b_1,\dots,b_t)M$.

See for the proof [10], 8.5, Theorem 6.

<u>(35.14) Proposition</u>. Let R , $\mathfrak{a} = (a_1,\dots,a_r)$, and M be the same as in Proposition (35.13). If $H_p(a_1,\dots,a_r;M) = 0$ for some $p \geq 0$ then $H_i(a_1,\dots,a_r;M) = 0$ for $i \geq p$.

See for the proof [10], proof of Theorem 6, in 8.5.

Here we introduce the Čech complex which is the direct limit of a certain direct system of Koszul complexes. Our aim is to show that the local cohomology is obtained as the cohomology of a Čech complex and hence it is a direct limit of Koszul homologies.

In the rest of this section $A = \bigoplus_{n\in\mathbb{Z}} A_n$ denotes a noetherian graded ring. Although all the results in the rest of this section are formulated only for graded rings and graded modules, the corresponding results hold for any noetherian ring R by regarding R as a graded ring with a trivial grading, i.e. $A_0 = R$ and $A_n = 0$ for $n \neq 0$.

Let a be a homogeneous element of A, $\deg a = \nu$ and $m > n > 0$ be non-negative integers. Consider the commutative diagram of graded A-modules

$$\begin{array}{lccccccc}
K^{\bullet}(a^n;A) : & 0 & \longrightarrow & A & \xrightarrow{\;.a^n\;} & A(n\nu) & \longrightarrow & \\
 & & & \| & & \downarrow{\scriptstyle a^{m-n}} & & \\
K^{\bullet}(a^m;A) : & 0 & \longrightarrow & A & \xrightarrow[\;.a^m\;]{} & A(m\nu) & \longrightarrow & 0
\end{array}$$

Then one can form a direct system of Koszul complexes $\{K^{\bullet}(a^n;A\}_{n\geq 0}$.

The limit complex can be identified with $0 \longrightarrow A \longrightarrow A_a \longrightarrow 0$, graded in the natural way, where A_a is the localization of A by a, cf. [5], N°4.

We denote the limit complex by $K^\bullet(a^\infty;A)$. The canonical map $\varphi_n^\bullet : K^\bullet(a^n;A) \longrightarrow K^\bullet(a^\infty;A)$ is given in the following way

$$\begin{array}{ccccccc} 0 & \longrightarrow & A & \xrightarrow{\cdot a^n} & A(n\nu) & \longrightarrow & 0 \\ & & \Big\| \varphi_n^\circ & & \Big\downarrow \varphi_n^1 & & \\ 0 & \longrightarrow & A & \longrightarrow & A_a & \longrightarrow & 0 \end{array} ,$$

where φ_n° is the identity and $\varphi_n^1(1) = \frac{1}{a^n}$, $\frac{1}{a}$ being the inverse of a in A_a.

Let $a_1,\dots,a_r$ be homogeneous elements of A with $\deg a_i = \nu_i$ and $m > n$ non-negative integers. From the maps $K^\bullet(a_i^n;A) \to K^\bullet(a_i^m;A)$, $1 \leq i \leq r$, one gets a complex homomorphism

$$\psi_{m,n} : K^\bullet(a_1^n,\dots,a_r^n;A) = \bigotimes_{i=1}^{r} K^\bullet(a_i^n;A) \longrightarrow \bigotimes_{i=1}^{r} K^\bullet(a_i^m;A) = K^\bullet(a_1^m,\dots,a_r^m;A) .$$

Then for any graded A-module M the map $\psi_{m,n}$ induces an homomorphism $\varphi_{m,n} : H^i(a_1^n,\dots,a_r^n;M) \longrightarrow H^i(a_1^m,\dots,a_r^m;M)$. Using these homomorphisms, we get two direct systems $\{K^\bullet(a_1^n,\dots,a_r^n;M\}_{n\geq 0}$ and $\{H^i(a_1^n,\dots,a_r^n;M)\}_{n\geq 0}$. Since the direct limit commutes with the tensor product (cf. [1]) we get

$$\begin{aligned} & \varinjlim_n K^\bullet(a_1^n,\dots,a_r^n;M) \\ = & \varinjlim_n (K^\bullet(a_1^n;A) \otimes \dots \otimes K(a_r^n;A) \otimes M) \\ = & \varinjlim_n K^\bullet(a_1^n;A) \otimes \dots \otimes \varinjlim_n K^\bullet(a_r^n;A) \otimes M \\ = & K^\bullet(a_1^\infty;A) \otimes \dots \otimes K^\bullet(a_r^\infty;A) \otimes M , \end{aligned}$$

where all the tensor products are taken over A.

(35.15) Definition. We denote the complex $K^\bullet(a_1^\infty;A) \otimes \ldots \otimes K^\bullet(a_r^\infty;A)$ by $C^\bullet(a_1,\ldots,a_r;A)$ and call it the Čech complex with respect to $a_1,\ldots,a_r$. For any graded A-module M we define

$$C^\bullet(a_1,\ldots,a_r;M) := C^\bullet(a_1,\ldots,a_r;A) \otimes_A M \quad .$$

Note that the direct limit commutes with the homology. Hence we have

(35.16) Lemma.

$$\varinjlim_n H^i(a_1^n,\ldots,a_r^n;M) = H^i\left(\varinjlim_n K^\bullet(a_1^n,\ldots,a_r^n;M)\right) = H^i(C^\bullet(a_1,\ldots,a_r;M)) .$$

By definition the Čech complex $C^\bullet(a_i;A)$ with respect to one element a_i is the complex of flat A-modules such that $C^0(a_i;A) = A$, $C^1(a_i;A) = A_{a_i}$, and $C^p(a_i;A) = 0$ for $p \neq 0,1$, and the boundary operator $\partial : C^0(a_i;A) \longrightarrow C^1(a_i;A)$ is given by the canonical map of localization. From this we see that the Čech complex

$$C^\bullet(a_1,\ldots,a_r;A) = \bigotimes_{i=1}^{r} (0 \longrightarrow A_i \longrightarrow A_{a_i} \longrightarrow 0)$$

with respect to $a_1,\ldots,a_r$ is the complex of flat A-modules whose p-th module is given by

$$C^p(a_1,\ldots,a_r;A) = \bigoplus_{1\leq i_1<\ldots<i_p\leq r} A_{a_{i_1}\cdots a_{i_p}}$$

for $0 \leq p \leq r$ and $C^p(a_1,\ldots,a_r;A) = 0$ for $p \neq 0,1,\ldots,r$. For an element $\sigma \in C^p(a_1,\ldots,a_r;A)$ we denote the $A_{a_{i_1}\cdots a_{i_p}}$-coordinate component of σ by $\sigma_{i_1\cdots i_p}$.

Then $\partial^p : C^p(a_1,\ldots,a_r;A) \longrightarrow C^{p+1}(a_1,\ldots,a_r;A)$ is given as follows:

For $x \in A_{a_{i_1}\cdots a_{i_p}}$ and $0 \leq p \leq r$ we have

$$(\mathrm{I}) \qquad \partial^p(x)_{j_1\cdots j_{p+1}} = \begin{cases} (-1)^k \frac{x}{1} & \text{if } \{j_1,\dots,j_{p+1}\} - \{i_1,\dots,i_p\} = \{\ell\} \\ & \text{and } i_1<\dots<i_k<\ell<i_{k+1}<\dots<i_p , \\ 0 & \text{otherwise} , \end{cases}$$

where $\frac{x}{1}$ is the canonical image of $x \in A_{a_{i_1}\cdots a_{i_p}}$ in $A_{a_{j_1}\cdots a_{j_{p+1}}}$.

In particular if $p = r-1$, we see that $\partial^{r-1} : C^{r-1}(a_1,\dots,a_r;A) \longrightarrow$ $\longrightarrow C^r(a_1,\dots,a_r;A)$ is given by

$$(\mathrm{II}) \qquad \partial^{r-1}((x_1,\dots,x_r)) = \sum_{k=1}^{r} (-1)^{k+1} \frac{x_k}{1} ,$$

where $x_i \in A_{a_1\cdots \check{a}_i \cdots a_r}$.

Next we observe that the Čech-complex can be constructed as mapping cone. Recall the definition of mapping cone. Let $\varphi' : \{C^\cdot,\partial\} \to \{D^\cdot,\delta\}$ be a homomorphism of complexes of A-modules, i.e. we are given a commutative diagram

$$\begin{array}{ccccccccc} \cdots \longrightarrow & C^{p-1} & \xrightarrow{\partial^{p-1}} & C^p & \xrightarrow{\partial^p} & C^{p+1} & \longrightarrow \cdots \\ & \downarrow{\scriptstyle \varphi^{p-1}} & & \downarrow{\scriptstyle \varphi^p} & & \downarrow{\scriptstyle \varphi^{p+1}} & \\ \cdots \longrightarrow & D^{p-1} & \xrightarrow[\delta^{p-1}]{} & D^p & \xrightarrow[\delta^p]{} & D^{p+1} & \longrightarrow \cdots \end{array} .$$

Then the mapping cone $M^\cdot(\varphi^\cdot)$ is the complex such that $M^p(\varphi^\cdot) = C^p \oplus D^{p-1}$ and $d^p : M^p(\varphi^\cdot) \longrightarrow M^{p+1}(\varphi^\cdot)$ is given by $d((\sigma,\tau)) = (\partial^p(\sigma) , \varphi^p(\sigma) - \delta^{p-1}(\tau))$ for any $\sigma \in C^p$, $\tau \in D^{p-1}$.

By definition we see that there is an exact sequence of complexes $0 \longrightarrow D^\cdot[-1] \longrightarrow M^\cdot(\varphi^\cdot) \longrightarrow C^\cdot \longrightarrow 0$, where $D^\cdot[-1]$ is the complex whose p-th module is D^{p-1} and whose boundary operator is $-\delta^{p-1}$.

By a diagram chase we know that there is an exact sequence

$$\longrightarrow H^{i-1}(D^\bullet) \longrightarrow H^i(M^\bullet(\varphi)) \longrightarrow H^i(C^\bullet)$$

$$\xrightarrow{\psi^i} H^i(D^\bullet) \longrightarrow H^{i+1}(M^\bullet(\varphi^\bullet)) \longrightarrow \dots \quad ,$$

such that ψ^i is the canonical map induced from φ^i .

Let us return to the Čech-complex. Consider the complex $C^\bullet(a_1;A)_{a_2}$, the localization of $C^\bullet(a_1;A)$ by a_2 , and the canonical map $\eta : C^\bullet(a_1;A) \longrightarrow C^\bullet(a_1;A)_{a_2}$, i.e. the commutative diagram

$$\begin{array}{ccccccc} 0 & \longrightarrow & A & \xrightarrow{\partial} & A_{a_1} & \longrightarrow & 0 \\ & & \downarrow{\scriptstyle \eta^\circ} & & \downarrow{\scriptstyle \eta^1} & & \\ 0 & \longrightarrow & A_{a_2} & \xrightarrow[\partial']{} & A_{a_1a_2} & \longrightarrow & 0 \end{array} \quad ,$$

where η° , η^1 are the localizations and ∂' is the homomorphism induced from ∂ . Then it is easy to see that $C^\bullet(a_1,a_2;A)$ is isomorphic to the mapping cone $M^\bullet(\eta^\bullet)$. Let $2 \leqq s < r$, and consider the canonical homomorphism $\eta^\bullet : C^\bullet(a_1,\dots,a_s;A) \longrightarrow C^\bullet(a_1,\dots,a_s;A)_{a_{s+1}}$ of localization of the Čech-complex $C^\bullet(a_1,\dots,a_s;A)$. Since

$$C^{p-1}(a_1,\dots,a_s;A)_{a_{s+1}} = \Big(\bigoplus_{1\leqq i<\dots<i_{p-1}\leqq s} A_{a_{i_1}\cdots a_{i_{p-1}}}\Big)_{a_{s+1}} \quad ,$$

$$= \bigoplus_{1\leqq i<\dots<i_{p-1}\leqq s} A_{a_{i_1}\cdots a_{i_{p-1}}a_{s+1}} \quad ,$$

we have: $M^p(\eta^\bullet) = C^p(a_1,\dots,a_s;A) \oplus C^{p-1}(a_1,\dots,a_s;A)_{a_{s+1}}$

$$= \bigoplus_{1\leqq i_1<\dots<i_p\leqq s+1} A_{a_{i_1}\cdots a_{i_p}} \quad .$$

By induction on s , we see that $d^p : M^p(\eta^\bullet) \longrightarrow M^{p+1}(\eta^\bullet)$ coincides with $\partial^p : C^p(a_1,\dots,a_{s+1};A) \longrightarrow C^{p+1}(a_1,\dots,a_{s+1};A)$. Hence we have an isomorphism $C^\bullet(a_1,\dots,a_{s+1};A) \xrightarrow{\sim} M^\bullet(\eta^\bullet)$. Thus we have shown the following fact.

(35.17) Lemma. There is an exact sequence

$$(*)\qquad 0 \longrightarrow C^{\bullet}(a_1,\dots,a_s;A)_{a_{s+1}}[-1] \longrightarrow C^{\bullet}(a_1,\dots,a_{s+1};A) \longrightarrow C^{\bullet}(a_1,\dots,a_s;A) \longrightarrow 0 \quad .$$

Thus we can construct the Čech-complex $C^{\bullet}(a_1,\dots,a_r;A)$ by succession of mapping cone-construction starting from $C^{\bullet}(a_1;A)$.

Now we come to the main result of this section.

(35.18) Theorem. Let A be a noetherian graded ring and M a graded A-module. Then for any homogeneous ideal $\mathfrak{a}$ and for any homogeneous elements $a_1,\dots,a_r$ such that $\sqrt{\mathfrak{a}} = \sqrt{(a_1,\dots,a_r)}$, we have

$$\underline{H}^i_{\mathfrak{a}}(M) \cong \varinjlim_n H^i(a_1^n,\dots,a_r^n;M) \cong H^i(C^{\bullet}(a_1,\dots,a_r;M) \quad .$$

Proof. The second isomorphism has been already proved in Lemma (35.16). So it is enough to show that $\underline{H}^i_{\mathfrak{a}}(M) \cong H^i(C^{\bullet}(a_1,\dots,a_r;M))$: By definition we see that the functor $H^i(C^{\bullet}(a_1,\dots,a_r;-))$ is an additive functor. We will prove that $H^i(C^{\bullet}(a_1,\dots,a_r;-))$ is the i-th derived functor of $\underline{H}^0_{\mathfrak{a}}(-)$.

(1) $\underline{H}^0_{\mathfrak{a}}(M) = H^0(C^{\bullet}(a_1,\dots,a_r;M))$:

By definition

$$\partial^0 : C^0(a_1,\dots,a_r;M) = M \longrightarrow C^1(a_1,\dots,a_r;M) = \bigoplus_{i=1}^{r} M_{a_i}$$

is given by $\partial^0(x) = \left(\frac{x}{1},\dots,\frac{x}{1}\right) \in \bigoplus_{i=1}^{r} M_{a_i}$ for $x \in M$.

Since $\sqrt{\mathfrak{a}} = \sqrt{(a_1,\dots,a_r)}$ we have:

$x \in \ker \partial^0 = H^0(C^{\bullet}(a_1,\dots,a_r;M) \Longleftrightarrow$ the image of x in M_{a_i} is zero for all $1 \leq i \leq r \Longleftrightarrow \mathfrak{a}^n x = 0$ for some $n > 0 \Longleftrightarrow x \in \underline{H}^0_{\mathfrak{a}}(M)$.

(2) A short exact sequence induces a long exact sequence: Let $0 \longrightarrow M' \longrightarrow M \longrightarrow M'' \longrightarrow 0$ be a short exact sequence of graded

A-modules. Then, since $C^\bullet(a_1,\dots,a_r;A)$ is a complex of flat A-modules, we have an exact sequence $0 \longrightarrow C^\bullet(a_1,\dots,a_r;M') \longrightarrow$ $\longrightarrow C^\bullet(a_1,\dots,a_r;M) \longrightarrow C^\bullet(a_1,\dots,a_r;M'') \longrightarrow 0$. This induces the required long exact sequence of cohomology modules.

(3) For any injective module I in $M^h(A)$ we have $H^i(C^\bullet(a_1,\dots,a_r;I)) = 0$ for $i > 0$: To see this we use induction on r . Let $r = 1$. By Theorem (33.12) we may assume that $I = \underline{E}_A(A/\mathfrak{p})$ for some homogeneous prime ideal $\mathfrak{p}$. If $a_1 \in \mathfrak{p}$ then we have $\underline{E}_A(A/\mathfrak{p}) \otimes_A A_{a_1} = 0$ by Theorem (33.12),1). Hence $H^1(C^\bullet(a_1;\underline{E}_A(A/\mathfrak{p}))) = 0$. If $a_1 \notin \mathfrak{p}$ holds then $\underline{E}_A(A/\mathfrak{p}) \otimes_A A_{a_1} = \underline{E}_A(A/\mathfrak{p})$ by Lemma (33.9). Hence $\partial^0 : C^0(a_1;\underline{E}_A(A/\mathfrak{p})) \longrightarrow C^1(a_1;\underline{E}(A/\mathfrak{p}))$ is an isomorphism and hence $H^1(C^\bullet(a_1;\underline{E}_A(A/\mathfrak{p}))) = 0$.

Let $r \geq 2$. By Lemma (35.17) we have an exact sequence

$$
\begin{array}{ll}
 & 0 \longrightarrow H^0(C^\bullet(a_1,\dots,a_r;I)) \longrightarrow H^0(C^\bullet(a_1,\dots,a_{r-1};I)) \\
 & \xrightarrow{\psi^\circ} H^0(C^\bullet(a_1,\dots,a_{r-1};I))_{a_r} \longrightarrow H^1(C^\bullet(a_1,\dots,a_r;I)) \\
(\#) & \longrightarrow H^1(C^\bullet(a_1,\dots,a_{r-1};I)) \xrightarrow{\psi^1} H^1(C^\bullet(a_1,\dots,a_{r-1};I))_{a_r} \\
 & \longrightarrow \dots \longrightarrow H^{i-1}(C^\bullet(a_1,\dots,a_{r-1};I))_{a_r} \longrightarrow \\
 & H^i(C^\bullet(a_1,\dots,a_r;I)) \longrightarrow H^i(C^\bullet(a_1,\dots,a_{r-1};I)) \xrightarrow{\psi^i} \quad ,
\end{array}
$$

where ψ° is the canonical map of localization by a_r . By inductive hypothesis $H^i(C^\bullet(a_1,\dots,a_{r-1};I) = 0$ for $i > 0$, hence $H^i(C^\bullet(a_1,\dots,a_r;I) = 0$ for $i \geq 2$ by (#) . It remains to show that $H^1(C^\bullet(a_1,\dots,a_r;I) = 0$. Again we may assume that $I = \underline{E}_A(A/\mathfrak{p})$ for some homogeneous prime ideal $\mathfrak{p}$. If $a_r \in \mathfrak{p}$ we have $H^0(C^\bullet(a_1,\dots,a_{r-1};\underline{E}_A(A/\mathfrak{p})))_{a_r} = 0$. From (#) we conclude that $H^1(C^\bullet(a_1,\dots,a_r;\underline{E}_A(A/\mathfrak{p}))) = 0$. If $a_r \notin \mathfrak{p}$ we see that ψ° is an isomorphism by Lemma (33.9). By (#) we have $H^1(C^\bullet(a_1,\dots,a_r;\underline{E}_A(A/\mathfrak{p}))) = 0$. This completes the proof of Theorem (35.18).

<u>(35.19) Corollary</u>. Let M be a graded A-module such that $\mathrm{Supp}(M) \subset V(\mathfrak{a})$ for some homogeneous ideal $\mathfrak{a}$ of A . Then $H^i_{\mathfrak{a}}(M) = 0$ for $i > 0$.

Proof. From the assumption it follows that $M_a = 0$ for any $0 \neq a \in \mathfrak{a}$. Let $\mathfrak{a} = (a_1,\dots,a_r)$. Then $C^i(a_1,\dots,a_r;M) = 0$ for $i > 0$ and $\underline{H}^i_{\mathfrak{a}}(M) = 0$ for $i > 0$.

<u>(35.20) Corollary</u>. Let $\varphi : A \longrightarrow B$ be a ring homomorphism of noetherian graded rings, $\mathfrak{a}$ a homogeneous ideal of A and M a B-module. Then $\underline{H}^i_{\mathfrak{a}}(M_\varphi) \cong \underline{H}^i_{\mathfrak{a}B}(M)$ for all i, where $M_\varphi = M$, regarded as an A-module via φ.

Proof. Let $a_1,\dots,a_r$ be homogeneous generators of $\mathfrak{a}$. Then for any $n > 0$ we have $H^i(a_1^n,\dots,a_r^n;M_\varphi) = H^i(a_1^n,\dots,a_r^n;M)$ for all i. Hence the result follows.

The next result is particularly useful in many applications of local cohomology.

<u>(35.21) Corollary</u>. Let $a_1,\dots,a_r$ be homogeneous elements of A with $\deg a_i = \nu_i$ and M a graded A-module. Consider the direct system

$$M/(a_1^n,\dots,a_r^n)M(n\nu) \xrightarrow{(a_1 \cdots a_r)^{m-n}} M/(a_1^m,\dots,a_r^m)M(m\nu) \quad ,$$

where $m > n \geqq 0$ and $\nu = \sum_{i=1}^{r} \nu_i$. Let $\mathfrak{a} = (a_1,\dots,a_r)$. Then we have

$$\varinjlim_{n} M/(a_1^n,\dots,a_r^n)M(n\nu) \cong \underline{H}^r_{\mathfrak{a}}(M) \quad .$$

Proof. By construction $K^r(a_1^n,\dots,a_r^n;A) \cong A(n\nu)$, $K^r(a_1^m,\dots,a_r^m;A) \cong A(m\nu)$ and $A(n\nu) \longrightarrow A(m\nu)$ is given by the multiplication by $(a_1 \cdots a_r)^{m-n}$. Hence the induced map

$$\begin{array}{ccc} H^r(a_1^n,\dots,a_r^n;M) & \longrightarrow & H^r(a_1^m,\dots,a_r^m;M) \\ \Vert\wr & & \Vert\wr \\ M/(a_1^n,\dots,a_r^n)M(n\nu) & \longrightarrow & M/(a_1^m,\dots,a_r^m)M(m\nu) \end{array}$$

is given by

$(x \bmod (a_1^n,\dots,a_r^n)M) \longmapsto ((a_1 \cdots a_r)^{m-n} x \bmod (a_1^m,\dots,a_r^m)M)$, $x \in M$. This proves Corollary (35.21).

(35.22) Corollary. Let $\mathfrak{a}$ and M be as in Corollary (35.21). Then we have:

(1) There is an exact sequence

$$\bigoplus_{i=1}^{r} M_{a_1\ldots\check{a}_i\ldots a_r} \xrightarrow{\partial^{r-1}} M_{a_1\ldots a_r} \longrightarrow \underline{H}^r_{\mathfrak{a}}(M) \longrightarrow 0 \text{ , with } \partial^{r-1}((x_1,\ldots,x_r)) =$$

$\sum_{i=1}^{r}(-1)^{i+1}\frac{x_i}{1}$ for $x_i \in M_{a_1\ldots\check{a}_i\ldots a_r}$, and $\frac{x_i}{1}$ is the image of x_i in $M_{a_1\ldots a_r}$.

(2) The canonical map $\varphi_n : M/(a_1^n,\ldots,a_r^n)M(n\,) \longrightarrow \underline{H}^r_{\mathfrak{a}}(M)$ is given by $\varphi_n(x \bmod (a_1^n,\ldots,a_r^n)M = \left[\frac{x}{(a_1\ldots a_r)^n}\right]$, where $\left[\frac{x}{(a_1\ldots a_r)^n}\right]$ denotes the residue class of $\frac{x}{(a_1\ldots a_r)^n}$ in $\underline{H}^i_{\mathfrak{a}}(M)$.

Proof. This follows from the construction.

(35.23) Corollary. Let $\mathfrak{a}$ and M be as in Corollary (35.21). Then there is a canonical isomorphism $\underline{H}^r_{\mathfrak{a}}(M) \cong M \otimes_A \underline{H}^r_{\mathfrak{a}}(A)$.

Proof. Since the direct limit commutes with the tensor product we have

$$\begin{aligned}\underline{H}^r_{\mathfrak{a}}(M) \cong \varinjlim_n (M/(a_1^n,\ldots,a_r^n)M(n\nu)) &\cong \varinjlim_n (M \otimes_A A/(a_1^n,\ldots,a_r^n)(n\nu)) \\ &\cong M \otimes_A \varinjlim_n A/(a_1^n,\ldots,a_r^n)(n\nu)) \\ &\cong M \otimes_A \underline{H}^r_{\mathfrak{a}}(A) \quad .\end{aligned}$$

We will close this section with the geometric meaning of the Čech-complex and the local cohomology. Let $A = \bigoplus_{n\geq 0} A_n$ be a non-negatively graded noetherian ring, $A_0 = k$, M a finitely generated graded A-module and $f_0,\ldots,f_d$ homogeneous elements such that $\sqrt{A_+} = \sqrt{(f_0,\ldots,f_d)}$. Let $X = \mathrm{Proj}(A)$ and $M^\sim$ the corresponding coherent sheaf on X. If we set $X_{f_i} = D_X(f_i) := \{\mathfrak{p} \in \mathrm{Proj}\,A \mid \mathfrak{p} \not\ni f_i\}$, then X_{f_i} form an affine covering $\mathfrak{A} \subset X$. Hence we can define the Čech-complex $\check{C}^\bullet(\mathfrak{A};M(n)^\sim)$ whose cohomology is the Grothendieck cohomology $H^\bullet(X,M(n)^\sim)$, cf.[5].

Recall that $X_{f_i} \cong \mathrm{Spec}((A_{f_i})_0)$ and $\Gamma(X_{f_i}, M(n)^{\sim}) = (M_{f_i}(n))_0$. Let us denote the n-th homogeneous piece of the Čech-complex $C^{\cdot}(f_o,\dots,f_d;M)$ by $C_n^{\cdot}(f_o,\dots,f_d;M)$. By definition of $\mathfrak{C}^{\cdot}(\mathfrak{A};M(n)^{\sim})$, we see that there is an isomorphism of complexes $\{\eta^i\}_{i \geq 0}$:

$$\eta^i : \mathfrak{C}^i(\mathfrak{A};M(n)^{\sim}) \cong C_n^{i+1}(f_o,\dots,f_d;M) \quad .$$

Let $\mathfrak{a} = (f_o,\dots,f_d)$. Then, by Theorem (35.18) we get isomorphisms $\underline{H}^0_{\mathfrak{a}}(M) \cong \ker(M \longrightarrow \bigoplus_{n\in\mathbf{Z}} \Gamma(X,M(n)^{\sim}))$, $\underline{H}^1_{\mathfrak{a}}(M) \cong \mathrm{coker}(M \longrightarrow \bigoplus_{n\in\mathbf{Z}} \Gamma(X,M(n)^{\sim}))$ and $\underline{H}^i_{\mathfrak{a}}(M) = \bigoplus_{n\in\mathbf{Z}} H^{i-1}(X,M(n)^{\sim})$ for $i \geq 2$.

§ 36. Local duality for graded rings

Throughout this section we use the following notation.

1) A denotes a non-negatively graded noetherian ring ($A_n = 0$ for $n < 0$) such that A_0 is a local ring with maximal ideal $\mathfrak{m}_o$. In this case we say that A is defined over a local ring A_0.

2) For such a ring A and the completion $\hat{A}_0$ of A_0 we denote $A \otimes_{A_0} \hat{A}_0$ by $\hat{A}$.

3) The maximal homogeneous ideal $\mathfrak{m}_o \oplus A_1 \oplus A_2 \oplus \dots$ of A is denoted by $\mathfrak{M}$ and the maximal homogeneous ideal of $\hat{A}$ is denoted by $\hat{\mathfrak{M}}$.

The purpose of this section is to present the local duality for graded rings defined over local rings. The duality theorem which we are going to prove is a generalization of Serre's projective duality theorem. The statement of Serre's duality theorem is as follows:

Let $\mathcal{F}$ be a coherent sheaf on projective r-space $X = \mathbf{P}^r$ over a field k. Then there is a perfect pairing $H^i(X,\mathcal{F}) \times \mathrm{Ext}^{r-i}_{\mathcal{O}_X}(\mathcal{F},\Omega) \longrightarrow k$, $i \geq 0$, where Ω is the sheaf of differential r-forms on $\mathbf{P}^r$ (cf. E.G.A. [5]). Let A be a Cohen-Macaulay graded ring defined over a complete local ring. Our duality theorem may be stated as follows: For any finitely generated graded A-module M there is a perfect pairing of graded A-modules $\underline{H}^i_{\mathfrak{M}}(M) \times \underline{\mathrm{Ext}}^{d-i}_A(M,\underline{K}_A) \longrightarrow \underline{E}_A$ for $i \geq 0$, where $d = \dim A$ and $\underline{K}_A$ is the canonical module of A which is defined to be $\underline{\mathrm{Hom}}_A(\underline{H}^d_{\mathfrak{m}}(A),\underline{E}_A)$. If, in particular, A is the homogeneous polynomial ring $k[X_0,\ldots,X_r]$ over a field k with $\deg X_i = 1$ the assertions of the above duality theorems are essentially the same(cf. Serre [22], Grothendieck [4] and Goto-Watanabe [16]). Our definition of the canonical module $\underline{K}_A$ is as that given in Goto-Watanabe [16], if A_0 is a field. To describe the theory of canonical modules, we follow Goto-Watanabe [16], Herzog-Kunz [6] and Grothendieck [4]. We begin with the elementary properties of the functor $\underline{H}^i_{\mathfrak{m}}(-)$.

(36.1) Lemma. For any finitely generated graded A-module M , $\underline{H}^n_{\mathfrak{m}}(M)$ is an artinian A-module for $n \geqq 0$.

Proof. Let $0 \longrightarrow M \longrightarrow I^0 \longrightarrow I^1 \longrightarrow \ldots$ be a minimal injective resolution of M in $M^h(A)$. By Proposition (35.3) we know that $\underline{H}^i_{\mathfrak{m}}(M)$ is the i-th cohomology of the complex

$$0 \longrightarrow \underline{H}^0_{\mathfrak{m}}(I^0) \longrightarrow \underline{H}^0_{\mathfrak{m}}(I^1) \longrightarrow \ldots \longrightarrow \underline{H}^0_{\mathfrak{m}}(I^n) \longrightarrow \underline{H}^0_{\mathfrak{m}}(I^{n+1}) \longrightarrow \ldots .$$

By Theorem (33.12) and Lemma (33.9) we see that $\underline{H}^0_{\mathfrak{m}}(I^n) = \{x \in I^n \mid \mathfrak{m}^i x = 0$ for some $i > 0\}$ is a finite direct sum of the modules of the form $\underline{E}_A(k)$. Since we may assume that A_0 is complete, we see by Theorem (34.8) that $\underline{E}_A$ is an artinian A-module. Hence $\underline{H}^n_{\mathfrak{m}}(M)$ is an artinian A-module.

(36.2) Lemma. Let M be a finitely generated graded A-module, and $\dim M = d$. Then $\underline{H}^i_{\mathfrak{m}}(M) = 0$ for $i > d$.

Proof. Let $\mathfrak{a} = \mathrm{ann}_A(M)$ and $\bar{A} = A/\mathfrak{a}$. Then we have $\underline{H}^i_{\mathfrak{m}}(M) = \underline{H}^i_{\bar{\mathfrak{m}}}(M)$ for $i \geqq 0$, where $\bar{\mathfrak{m}}$ is the maximal homogeneous ideal of $\bar{A}$, by Corollary (35.20). Hence we may assume that $\dim A = \dim M = d$. Note that

$$\begin{aligned}\underline{H}^i_{\mathfrak{m}}(M) &= \varinjlim_n \underline{\mathrm{Ext}}^i_A(A/\mathfrak{m}^n, M) \cong \varinjlim_n \mathrm{Ext}^i_A(A/\mathfrak{m}^n, M) \otimes_A A_{\mathfrak{m}} \\ &\cong \varinjlim_n \mathrm{Ext}^i_{\mathfrak{m}}(A_{\mathfrak{m}}/\mathfrak{m}^n A_{\mathfrak{m}}, M_{\mathfrak{m}}) \cong H^i_{\mathfrak{m}A_{\mathfrak{m}}}(M_{\mathfrak{m}}) .\end{aligned}$$

Hence we may assume that A is a local ring with maximal ideal $\mathfrak{m}$ and M is a finitely generated A-module with $\dim M = \dim A = d$. Let

$a_1,\dots,a_d$ be a system of parameters of A . Then, by Theorem (35.18), we have $H^i_{\mathfrak{m}}(M) = \varinjlim_n H^i(a_1^n,\dots,a_d^n;M)$. Hence $H^i_{\mathfrak{m}}(M) = 0$ for $i > d$.

<u>(36.3) Lemma</u>. Let M be as in Lemma (36.2). Then $\operatorname{depth}_{\mathfrak{m}} M \geq t$ if and only if $\underline{H}^i_{\mathfrak{m}}(M) = 0$ for $i < t$. In particular M is Cohen-Macaulay if and only if $\underline{H}^i_{\mathfrak{m}}(M) = 0$ for $i < d = \dim M$.

Proof. This follows from Proposition (35.7) and from the fact that $\underline{H}^i_{\mathfrak{m}}(M) = H^i_{\mathfrak{m}}(M)$ as underlying A-modules.

In what follows $\operatorname{depth}_{\mathfrak{m}} M$ is denoted by $\operatorname{depth} M$.

Now we come to the definition of canonical modules.

<u>(36.4) Definition</u>.

(1) If A_0 is <u>complete</u> we define $\underline{K}_A = \underline{\operatorname{Hom}}_A(\underline{H}^d_{\mathfrak{m}}(A), \underline{E}_A)$, where $d = \dim A$, and call $\underline{K}_A$ the canonical module of A .

(2) If A_0 is not complete then a graded A-module $\underline{K}_A$ is a canonical module of A if $\underline{K}_A \otimes_A \hat{A} \cong \underline{K}_{\hat{A}}$.

<u>(36.5) Lemma</u>. (Nakayama's lemma). Let M be a finitely generated A-module and N a graded A-submodule of M . If $M = N + \mathfrak{m}M$, then $M = N$.

Proof. From the assumption we get $M_{\mathfrak{m}} = N_{\mathfrak{m}}$ which implies $M = N$.

<u>(36.6) Proposition</u>. If A has a canonical module $\underline{K}_A$ then $\underline{K}_A$ is a finitely generated A-module and is unique up to isomorphisms.

Proof. Since $\underline{H}^d_{\hat{\mathfrak{m}}}(\hat{A})$ is an artinian $\hat{A}$-module by Lemma (36.1), $\underline{\operatorname{Hom}}_{\hat{A}}(\underline{H}^d_{\hat{\mathfrak{m}}}(\hat{A}), \underline{E}_{\hat{A}}) = \underline{K}_{\hat{A}}$ is a finitely generated A-module by Theorem (34.8).

Since $\hat{A}$ is faithfully flat over A, $\underline{K}_A \otimes_A \hat{A} \cong \underline{K}_{\hat{A}}$ is finitely generated over $\hat{A}$ if and only if so is $\underline{K}_A$ over A . To prove the uniqueness we take another canonical module $\underline{K}'_A$ of A . Then, by definition $\underline{K}_A \otimes_A \hat{A} \cong \underline{K}'_A \otimes_A \hat{A} \cong \underline{K}_{\hat{A}}$. The uniqueness follows from the following lemma.

<u>(36.7) Lemma</u>. Let L and N be finitely generated graded A-modules. If $L \otimes_A \hat{A} \cong N \otimes_A \hat{A}$ then $L \cong N$.

Proof. Let $f \in \underline{\mathrm{Hom}}_{\hat{A}}(\hat{L},\hat{N})_0 \cong \underline{\mathrm{Hom}}_A(L,N)_0 \otimes_{A_0} \hat{A}_0$ be an isomorphism of $\deg f = 0$, where $\hat{L} = L \otimes_A \hat{A}$ and $\hat{N} = N \otimes_A \hat{A}$. We have to show that there is an isomorphism $f' : L \longrightarrow N$.

Since L and N are finitely generated A-modules we know that $\underline{\mathrm{Hom}}_A(L,N)_0$ is a finitely generated A_0-module. Hence $\underline{\mathrm{Hom}}_A(L,N)_0 \otimes_{A_0} \hat{A}_0$ is the completion of $\underline{\mathrm{Hom}}_A(L,N)_0$, denoted by $\underline{\mathrm{Hom}}_A(L,N)_0^{\wedge}$. By definition of the completion, for any $n > 0$ there is a homomorphism $f_n \in \underline{\mathrm{Hom}}_A(L,N)_0$ such that $f - f_n \in \mathfrak{m}_\circ^n \underline{\mathrm{Hom}}_A(L,N)_0^{\wedge}$.
Note that f and f_n induce the same isomorphism $L/\mathfrak{m}_\circ^n L \xrightarrow{\sim} N/\mathfrak{m}_\circ^n N$, that means in particular, by tensoring with $A/\mathfrak{m}$, we get $L/\mathfrak{m}L \cong N/\mathfrak{m}N$. By Lemma (36.5) there exist finitely generated graded free A-modules F and G such that $\mathrm{rank}_A F = \mathrm{rank}_A G = \dim_{A/\mathfrak{m}} L/\mathfrak{m}L$ and such that there are surjections $\varphi : F \twoheadrightarrow L$ and $\psi : G \twoheadrightarrow N$. Let $S = \ker \varphi$ and $T = \ker \psi$. Then we have a commutative diagram

$$\begin{array}{ccccccccc} 0 & \longrightarrow & S & \longrightarrow & F & \longrightarrow & L & \longrightarrow & 0 \\ & & \downarrow h_n & & \downarrow g_n & & \downarrow f_n & & \\ 0 & \longrightarrow & T & \longrightarrow & G & \longrightarrow & N & \longrightarrow & 0 \end{array}$$

with exact rows, where g_n and h_n are homomorphisms induced from f_n. Since $F/\mathfrak{m}F \cong L/\mathfrak{m}L \cong N/\mathfrak{m}N \cong G/\mathfrak{m}G$, f_n and g_n are surjective. Hence we know that g_n is an isomorphism since F and G are free A-modules of the same rank. Since f_n induces an isomorphism $L/\mathfrak{m}_\circ^n L \cong N/\mathfrak{m}_\circ^n N$ for all $n > 0$, we have $h_n(S) + \mathfrak{m}_\circ^n G = T + \mathfrak{m}_\circ^n G$. Hence $T \subset h_n(S) + \mathfrak{m}_\circ^n G \cap T$. By Artin-Rees lemma there is an $r > 0$ such that $\mathfrak{m}_\circ^n G \cap T = \mathfrak{m}_\circ^{n-r}(\mathfrak{m}_\circ^r G \cap T)$ for all $n > r$. For $n > r$, we have: $T = h_n(S) + \mathfrak{m}_\circ^{n-r}(\mathfrak{m}_\circ^r G \cap T) =$ $= h_n(S) + \mathfrak{m}_\circ^{n-r} T$. By Nakayama's lemma we get $T = h_n(S)$. Hence h_n is an isomorphism. Consequently f_n is an isomorphism, q.e.d.

The following theorem is called the local duality theorem, which will play the central role in the rest of this book.

<u>(36.8) Theorem</u>. (*Local duality*). Let A be a graded ring defined over a complete local ring with $d = \dim A$. Then A is Cohen-Macaulay if

and only if there is an isomorphism ∂ of functors

$$\underline{\mathrm{Hom}}_A(\underline{H}^i_{\mathfrak{m}}(-),\underline{E}_A) \cong \underline{\mathrm{Ext}}^{d-i}_A(-,\underline{K}_A)$$

on the category of finitely generated graded A-modules for all i, i.e. for any finitely generated graded A-module M:
$\underline{\mathrm{Hom}}_A(\underline{H}^i_{\mathfrak{m}}(M),\underline{E}_A) \cong \underline{\mathrm{Ext}}^{d-i}_A(M,\underline{K}_A)$ for all i, and the isomorphism is functorial in M.

(36.9) Remark. The functoriality of this isomorphism σ means in particular that for $M = A/I$, where I is a proper ideal in A, σ is also an A/I-linear map.

Proof.

(1): Suppose that A is Cohen-Macaulay. Let $\widetilde{M}^h(A)$ be the category of finitely generated graded A-modules. Let $T^i(-)$ be the functor $\underline{\mathrm{Hom}}_A(\underline{H}^{d-i}_{\mathfrak{m}}(-),\underline{E}_A)$. Since $\underline{H}^{d-i}_{\mathfrak{m}}(M)$ is an artinian A-module for any finitely generated A-module M by Lemma (36.1), $T^i(-)$ defines an additive functor from $\widetilde{M}^h(A)$ to itself. We will show that $T^i(-)$ is the i-th left derived functor of the functor $\underline{\mathrm{Hom}}_A(-,\underline{K}_A)$ on the category $\widetilde{M}^h(A)$. It is enough to show that $T^i(-)$ satisfies the following conditions (cf. [3] or [8]):

1) For any finitely generated graded projective A-module P we have $T^i(P) = 0$ for $i > 0$.
2) $T^0(M) \cong \underline{\mathrm{Hom}}_A(M,\underline{K}_A)$ for any finitely generated graded A-module M.
3) From a short exact sequence $0 \to M' \to M \to M'' \to 0$ in $M^h(A)$, we have a long exact sequence

$$0 \to T^0(M'') \to T^0(M) \to T^0(M') \to T^1(M'')$$
$$\to T^1(M) \to T^1(M') \to \dots \to T^{n-1}(M') \to$$
$$T^n(M'') \to T^n(M) \to T^n(M') \to T^{n+1}(M'') \to \dots \quad .$$

Since the functor $\underline{\mathrm{Hom}}_A(-,\underline{E}_A)$ is exact, 3) follows from the long exact sequence of local cohomology modules obtained from the given short exact sequence.

For 2), recall that $\underline{H}^d_{\mathfrak{m}}(M) \cong \underline{H}^d_{\mathfrak{m}}(A) \otimes_A M$ (cf. Corollary (35.23).). Then it follows that

$$\underline{\mathrm{Hom}}_A(\underline{H}^d_{\mathfrak{m}}(M),\underline{E}_A) \cong \underline{\mathrm{Hom}}_A(\underline{H}^d_{\mathfrak{m}}(A) \otimes_A M , \underline{E}_A)$$

$$\cong \underline{\mathrm{Hom}}_A(M,\underline{\mathrm{Hom}}_A(\underline{H}^d_{\mathfrak{m}}(A),\underline{E}_A))$$

$$\cong \underline{\mathrm{Hom}}_A(M,\underline{K}_A) \quad .$$

It remains to show 1): By Lemma (36.5) we see that every finitely generated graded projective A-module is free. Hence it is enough to show that $T^i(A) = \underline{\mathrm{Hom}}_A(H^{d-i}_{\mathfrak{m}}(A),\underline{E}_A) = 0$ for $i > 0$. But this is clear since A is Cohen-Macaulay (cf. Lemma (36.3)).

(2): To prove the "if"-part, take $M = A$. Then by assumption we have $\underline{\mathrm{Hom}}_A(\underline{H}^{d-i}_{\mathfrak{m}}(A),\underline{E}_A) \cong \underline{\mathrm{Ext}}^i_A(A,\underline{K}_A) = 0$ for $i > 0$. But $\underline{\mathrm{Hom}}_A(\underline{H}^{d-i}_{\mathfrak{m}}(A),\underline{E}_A) = 0$ if and only if $\underline{H}^{d-i}_{\mathfrak{m}}(A) = 0$ by (34.7). Hence, by Lemma (36.3), A is Cohen-Macaulay.

<u>(36.11) Corollary</u>. Let A be a graded ring defined over a local ring having a canonical module $\underline{K}_A$. Suppose that A is Cohen-Macaulay. Then $\underline{K}_A$ is a Cohen-Macaulay graded A-module with $\operatorname{depth} \underline{K}_A = \dim A$ and $\underline{K}_A$ has finite injective dimension in $M^h(A)$.

Proof.

1) Since $\hat{A}$ is faithfully flat over A, $\underline{K}_A$ is Cohen-Macaulay if and only if $\underline{K}_{\hat{A}} \cong \underline{K}_A \otimes_A \hat{A}$ is Cohen-Macaulay. Hence we may assume that $A = \hat{A}$. By Theorem (36.8) we have $\underline{\mathrm{Hom}}_A(\underline{H}^{d-i}(A/\mathfrak{m}),\underline{E}_A) \cong \underline{\mathrm{Ext}}^i_A(A/\mathfrak{m},\underline{K}_A)$ for all $i \in \mathbf{Z}$, where $d = \dim A$.

Since $\dim A/\mathfrak{m} = 0$, we have $\underline{H}^i_{\mathfrak{m}}(A/\mathfrak{m}) = 0$ for $i \neq 0$ by Lemma (36.2). Hence $\underline{\mathrm{Ext}}^i_A(A/\mathfrak{m},\underline{K}_A) = 0$ for $i < d$. This shows that $\underline{K}_A$ is Cohen-Macaulay with $\operatorname{depth} \underline{K}_A = d$.

2) For any finitely generated graded A-module N we have $\underline{\mathrm{Hom}}_{\hat{A}}(\underline{H}^{d-i}_{\hat{\mathfrak{m}}}(N \otimes_A \hat{A}),\underline{E}_{\hat{A}}) \cong \underline{\mathrm{Ext}}^i_{\hat{A}}(N \otimes_A \hat{A},\underline{K}_{\hat{A}}) = 0$ for $i > d$.

Since $\underline{\mathrm{Ext}}^i_A(N,\underline{K}_A) \otimes_A \hat{A} \cong \underline{\mathrm{Ext}}^i_{\hat{A}}(N \otimes_A \hat{A},\underline{K}_{\hat{A}})$, we see that $\underline{\mathrm{Ext}}^i_A(N,\underline{K}_A) = 0$ for $i > d$.
By Theorem (33.25) we know that $\underline{K}_A$ has finite injective dimension in $M^h(A)$.

(36.12) Corollary. A is Gorenstein if and only if A is Cohen-Macaulay and $\underline{K}_A \cong A(n)$ for some $n \in \mathbf{Z}$.

Proof. Recall that by Corollary (33.27) and Remark (33.24) A is Gorenstein if and only if for suitable $n \in \mathbf{Z}$ we have

$$\underline{\mathrm{Ext}}^i_A(A/\mathfrak{m},A) = \begin{cases} A/\mathfrak{m}(-n) & \text{for } i = \dim A = d \,. \\ 0 & \text{for } i \neq \dim A \,. \end{cases}$$

1) If A is Gorenstein then $\hat{A}$ is Gorenstein and we have $\underline{H}^d_{\hat{\mathfrak{m}}}(\hat{A}) = \underline{E}_{\hat{A}}(-n)$ for some $n \in \mathbf{Z}$ by Theorem (33.25). Hence

$$\underline{K}_{\hat{A}} = \underline{\mathrm{Hom}}_{\hat{A}}(\underline{H}^d_{\hat{\mathfrak{m}}}(\hat{A}),\underline{E}_{\hat{A}}) \cong \underline{\mathrm{Hom}}_{\hat{A}}(\underline{E}_{\hat{A}}(-n),\underline{E}_{\hat{A}}) \cong \hat{A}(n) \qquad \text{by Lemma (34.7).}$$

By the uniqueness of canonical modules we have $\underline{K}_A \cong A(n)$.

2) Now assume that A is Cohen-Macaulay and $\underline{K}_A \cong A(n)$. By Theorem (36.8) we get:

$$\underline{\mathrm{Ext}}^i_{\hat{A}}(\hat{A}/\hat{\mathfrak{m}},\hat{A}(n)) \cong \underline{\mathrm{Hom}}_{\hat{A}}(\underline{H}^{d-i}_{\hat{\mathfrak{m}}}(\hat{A}/\hat{\mathfrak{m}}),\underline{E}_{\hat{A}}) \qquad \text{for all} \quad i \geq 0 \quad .$$

Since $\underline{H}^{d-i}_{\hat{\mathfrak{m}}}(\hat{A}/\hat{\mathfrak{m}}) = 0$ for $i \neq d$ and $\underline{H}^0_{\hat{\mathfrak{m}}}(\hat{A}/\hat{\mathfrak{m}}) = \hat{A}/\hat{\mathfrak{m}}$, we have:

$$\underline{\mathrm{Ext}}^i_A(A/\mathfrak{m},A(n)) \cong \underline{\mathrm{Ext}}^i_A(A/\mathfrak{m},A(n)) \otimes_A \hat{A} \cong \underline{\mathrm{Ext}}^i_{\hat{A}}(\hat{A}/\hat{\mathfrak{m}},\hat{A}(n))$$

$$= \begin{cases} \hat{A}/\hat{\mathfrak{m}} = A/\mathfrak{m} & \text{for } i = d \\ 0 & \text{for } i \neq d \,. \end{cases}$$

Hence A is Gorenstein.

(36.13) Remark. Let $a = \max\{n \mid \underline{H}^d_{\mathfrak{m}}(A)_n \neq 0\}$. If A is Gorenstein we have $\underline{K}_A \cong A(a)$. In the sequel we denote this number by $a(A)$ and call it the a-invariant of A .

(36.14) Corollary. Let $A \longrightarrow S$ be a finite homomorphism of graded rings defined over local rings. Suppose that A is Cohen-Macaulay and A has a canonical module $\underline{K}_A$. Then $\underline{K}_S = \underline{\mathrm{Ext}}_A^r(S,\underline{K}_A)$, where $r = \dim A - \dim S$.

Proof. Let $\hat{S}_0$ be the completion of S_0. Since S_0 is finite over A_0 we have $\hat{S}_0 = S_0 \otimes_{A_0} \hat{A}_0$. Let N be the maximal homogeneous ideal of S. Then $\hat{S} = S \otimes_{A_0} \hat{A}_0$ and $\hat{N} = N \otimes_{A_0} \hat{A}_0$.

We first claim that $\underline{\mathrm{Hom}}_{\hat{A}}(\hat{S},\underline{E}_{\hat{A}})$ is the injective envelope of $\hat{S}/\hat{N}$ in $M^h(\hat{S})$: In fact, by Lemma (33.3) $\underline{\mathrm{Hom}}_{\hat{A}}(\hat{S},\underline{E}_{\hat{A}})$ is an injective module in $M^h(\hat{S})$, and $\mathrm{Supp}_{\hat{S}}(\underline{\mathrm{Hom}}_{\hat{A}}(\hat{S},\underline{E}_{\hat{A}}) = \{\hat{N}\}$. To prove the claim it is enough to show that

$$\underline{\mathrm{Hom}}_{\hat{S}}(\hat{S}/\hat{N},\underline{\mathrm{Hom}}_{\hat{A}}(\hat{S},\underline{E}_{\hat{A}})) \cong \hat{S}/\hat{N} \quad .$$

But, by Lemma (33.2) we have:

$$\begin{aligned}
&\underline{\mathrm{Hom}}_{\hat{S}}(\hat{S}/\hat{N},\underline{\mathrm{Hom}}_{\hat{A}}(\hat{S},\underline{E}_{\hat{A}}))\\
\cong\ &\underline{\mathrm{Hom}}_{\hat{A}}(\hat{S}/\hat{N},\underline{E}_{\hat{A}})\\
\cong\ &\underline{\mathrm{Hom}}_{\hat{A}}(\hat{S}/\hat{N} \otimes_{\hat{A}} \hat{A}/\hat{\mathfrak{m}},\underline{E}_{\hat{A}})\\
\cong\ &\underline{\mathrm{Hom}}_{\hat{A}}(\hat{S}/\hat{N},\underline{\mathrm{Hom}}_{\hat{A}}(\hat{A}/\hat{\mathfrak{m}},\underline{E}_{\hat{A}}))\\
\cong\ &\underline{\mathrm{Hom}}_{\hat{A}}(\hat{S}/\hat{N},\hat{A}/\hat{\mathfrak{m}})\\
\cong\ &\hat{S}/\hat{N} \quad .
\end{aligned}$$

Now we have, since $\sqrt{\hat{\mathfrak{m}}\hat{S}} = \hat{N}$,

$$\begin{aligned}
\underline{K}_{\hat{S}} &\cong \underline{\mathrm{Hom}}_{\hat{S}}(\underline{H}^s_{\hat{N}}(\hat{S}),\underline{E}_{\hat{S}}) \qquad (s = \dim S)\\
&\cong \underline{\mathrm{Hom}}_{\hat{S}}(\underline{H}^s_{\hat{\mathfrak{m}}}(\hat{S}),\underline{\mathrm{Hom}}_{\hat{A}}(\hat{S},\underline{E}_{\hat{A}}))\\
&\cong \underline{\mathrm{Hom}}_{\hat{A}}(\underline{H}^s_{\hat{\mathfrak{m}}}(\hat{S}),\underline{E}_{\hat{A}})\\
&\cong \underline{\mathrm{Ext}}^r_{\hat{A}}(\hat{S},\underline{K}_{\hat{A}}) \ , \qquad \text{by local duality}\\
&\cong \underline{\mathrm{Ext}}^r_A(S,\underline{K}_A) \otimes_A \hat{A}\\
&\cong \underline{\mathrm{Ext}}^r_A(S,\underline{K}_A) \otimes_S (S \otimes_A \hat{A})\\
&\cong \underline{\mathrm{Ext}}^r_A(S,\underline{K}_A) \otimes_S \hat{S} \quad .
\end{aligned}$$

By the uniqueness of $\underline{K}_S$ we have $\underline{K}_S \cong \underline{\mathrm{Ext}}_A^r(S,\underline{K}_A)$, q.e.d.

(36.15) Corollary. If moreover A is Gorenstein in Corollary (36.14) then $\underline{K}_S = \underline{\mathrm{Ext}}_A^r(S,A)(n)$ for some $n \in \mathbf{Z}$.

Proof. This follows from Corollary (36.12) and (36.14).

(36.16) Corollary. Suppose that A is Gorenstein. Then for a finitely generated graded A-module M of dimension n the following are equivalent:

1) M is a Cohen-Macaulay A-module

2) $\underline{\mathrm{Ext}}_A^i(M,A) = 0$ for $i \neq \dim A - n$.

Proof. This follows from Theorem (36.8), Corollary (36.12) and Lemma (36.3).

By Proposition (35.7) and Lemma (36.3) we know that if A is a Cohen-Macaulay ring with $\dim A = d$ then $\underline{H}_{\mathfrak{m}}^d(A) \neq 0$. The following result shows that the same is true for any graded ring defined over a local ring.

(36.17) Theorem. Let A be a graded ring defined over a local ring of $\dim A = d$. Then $\underline{H}_{\mathfrak{m}}^d(A) \neq 0$.

Proof. Since $\underline{H}_{\hat{\mathfrak{m}}}^d(\hat{A}) \cong \underline{H}_{\mathfrak{m}}^d(A) \otimes_A \hat{A}$ and $d = \dim \hat{A}$ we may assume that A_0 is complete. Then there is a complete regular local ring B such that A_0 is a homomorphic image of B . Hence A is a homomorphic image of a graded polynomial ring $S = B[X_1,\ldots,X_n]$. Since S is Gorenstein we have by the local duality: $\underline{\mathrm{Hom}}_S(\underline{H}_{\mathfrak{m}}^d(A),\underline{E}_S) \cong \underline{\mathrm{Ext}}_S^r(A,S(\ell))$ for some $\ell \in \mathbf{Z}$, where $r = \dim S - \dim A$.

Let $\mathfrak{N}$ be the maximal homogeneous ideal of S . Then it is sufficient to show that

$$\mathrm{Ext}_S^r(A,S) \otimes_S S_{\mathfrak{N}} \cong \mathrm{Ext}_{S_{\mathfrak{N}}}^r(A_{\mathfrak{N}},S_{\mathfrak{N}}) \neq 0 \quad .$$

Suppose that $A = S/\mathfrak{a}$. Then $\mathrm{ht}\,\mathfrak{a} = \mathrm{ht}\,\mathfrak{a}S_{\mathfrak{N}} = r$. Since $S_{\mathfrak{N}}$ is a regular local ring we know that $\mathrm{depth}_{\mathfrak{a}S_{\mathfrak{N}}} S_{\mathfrak{N}} = r$. Now the theorem follows from the following lemma.

(36.18) Lemma. Let R be a noetherian local ring and $\mathfrak{a}$ an ideal of R . Then for any finitely generated R-module M we have $\mathrm{depth}_{\mathfrak{a}} M = \inf\{i \mid \mathrm{Ext}^i_R(R/\mathfrak{a},M) \neq 0\}$.

Proof. See Matsumura [9].

As an immediate consequence of Theorem (36.17) we have:

(36.19) Corollary. Let M be a finitely generated graded A-module. Then we have $\dim M = \max\{i \mid \underline{H}^i_{\mathfrak{m}}(M) \neq 0\}$.

Proof. Let $\mathfrak{p}$ be a minimal prime ideal of M such that $\dim(A/\mathfrak{p}) = \dim M$. Note that $\mathfrak{p}$ is homogeneous. From the exact sequence $0 \to \mathfrak{p}M \to M \to M/\mathfrak{p}M \to 0$ we get an exact sequence $\underline{H}^d_{\mathfrak{m}}(M) \to \underline{H}^d_{\mathfrak{m}}(M/\mathfrak{p}M) \to \underline{H}^{d+1}_{\mathfrak{m}}(\mathfrak{p}M)$, where $d = \dim M$. Since $\dim \mathfrak{p}M \leq \dim M$ we have $\underline{H}^{d+1}_{\mathfrak{m}}(\mathfrak{p}M) = 0$ by Lemma (36.2). Hence we may assume that A is a domain and $\dim A = \dim M$, since $\dim M/\mathfrak{p}M = \dim M$ and since $\underline{H}^d_{\mathfrak{m}}(M/\mathfrak{p}M) = \underline{H}^d_{\mathfrak{m}/\mathfrak{p}}(M/\mathfrak{p}M)$ by Corollary (35.20).

Let $T = \{x \in M \mid rx = 0 \text{ for some } 0 \neq r \in A\}$. Then $\dim T < \dim M$. From the exact sequence $0 \to T \to M \to M/T \to 0$ we know that $\underline{H}^d_{\mathfrak{m}}(M) \cong \underline{H}^d_{\mathfrak{m}}(M/T)$ by Lemma (36.2). Hence we may assume M is torsion free. In this case there is a finitely generated free A-module F containing M such that $\dim F/M < \dim M$. Now we have an exact sequence $\underline{H}^d_{\mathfrak{m}}(M) \to \underline{H}^d_{\mathfrak{m}}(F) \to 0$ obtained from the exact sequence $0 \to M \to F \to F/M \to 0$. Therefore $\underline{H}^d_{\mathfrak{m}}(M) \neq 0$ by Theorem (36.17).

(36.20) Corollary. If A has a canonical module $\underline{K}_A$, then

a) $\underline{K}_A \neq 0$

b) $\dim \underline{K}_A = \dim A$

c) $\underline{K}_A$ satisfies Serre's (S_2)-condition.

Proof. $\underline{K}_A \otimes_A \hat{A} \cong \underline{K}_{\hat{A}} \cong \underline{\mathrm{Hom}}_{\hat{A}}(H^d_{\hat{\mathfrak{m}}}(\hat{A}), \underline{E}_{\hat{A}}) \neq 0$ by Theorem (36.17), where $d = \dim A$. Hence $\underline{K}_A \neq 0$. To prove the second assertion b), we consider a Gorenstein graded ring S defined over a complete local ring such that $\hat{A}$ is a homomorphic image of S and $\dim \hat{A} = \dim S$. Then, by Corollary (36.15) we get $\underline{K}_{\hat{A}} \cong \underline{\mathrm{Hom}}_S(\hat{A}, S)(n)$ for some $n \in \mathbb{Z}$. Let $\hat{A} = S/\mathfrak{a}$. Then we have: $\underline{\mathrm{Hom}}_S(\hat{A}, S) \cong (0 : \mathfrak{a})_S$. We have to show that $\dim(0 :_S \mathfrak{a}) = \dim \hat{A}$. But this is clear, since $\mathrm{Ass}_S(0 :_S \mathfrak{a}) \subset \mathrm{Ass}_S S$ and since $\dim S/\mathfrak{p} = \dim S$ for $\mathfrak{p} \in \mathrm{Ass}_S S$. For the last assertion it is enough to prove that $\underline{K}_{\hat{A}}$ satisfies (c) (cf. [9],[9*]). Hence if $\dim A \geq 2$, it is enough to show that if $a, b \in S$ is any S-sequence then a, b is a $\underline{K}_{\hat{A}}$-sequence. Since $\underline{K}_{\hat{A}} \cong (0 : \mathfrak{a})_S \subset S$, a must be $\underline{K}_{\hat{A}}$-regular. From the exact sequence

$$0 \longrightarrow S \xrightarrow{\cdot a} S \longrightarrow S/aS \longrightarrow 0$$

we get an exact sequence

$$\begin{array}{ccccc} 0 \longrightarrow \underline{\mathrm{Hom}}_S(\hat{A}, S) & \xrightarrow{\cdot a} & \underline{\mathrm{Hom}}_S(\hat{A}, S) & \longrightarrow & \mathrm{Hom}_S(\hat{A}, S/aS) \\ \Big\downarrow \wr & & \Big\downarrow \wr & & \\ \underline{K}_{\hat{A}} & & \underline{K}_{\hat{A}} & & \end{array}$$

Therefore $\underline{K}_{\hat{A}}/a\underline{K}_{\hat{A}} \hookrightarrow \underline{\mathrm{Hom}}_S(\hat{A}, S/aS) \hookrightarrow S/aS$. Since a, b is an S-sequence, b is a non-zero-divisor on S/aS and hence on $\underline{K}_{\hat{A}}/a\underline{K}_{\hat{A}}$. Therefore a, b is a $\underline{K}_{\hat{A}}$-sequence as wanted.

Appendix.

Characterization of local Gorenstein-rings by its injective dimension.

We characterize in Theorem (A2) Gorenstein rings R by the property that $(R, \mathfrak{m})$ has finite injective dimension as an R-module. The injective dimension of a module M is denoted by $\mathrm{id}_R(M)$, its projective dimension by $\mathrm{pd}_R M$. The same proof works for graded rings A defined over a complete local ring and finitely generated graded A-modules.

First we need the following lemma.

(A1) Lemma. Let M be a non-trivial finitely generated module over a local ring $(R,\mathfrak{m},k)$ with $\mathrm{id}_R M < \infty$. Then $\mathrm{id}_R M = \mathrm{depth}\, R$.

Proof. Let $t = \mathrm{depth}\, R$ and $n = \mathrm{id}_R M$. Take a maximal R-sequence $\underline{a} := a_1,\dots,a_t \in \mathfrak{m}$. Then we have (see [7], p. 127):

$$\mathrm{pd}_R(R/(\underline{a})R) = t \quad \text{and} \quad \mathrm{Ext}_R^t(R/\underline{a}R,M) \cong M/\underline{a}M \quad .$$

This shows that $n \geqq t$.

In order to prove the other inequality we claim that $\mathrm{Ext}_R^n(k,M) \neq 0$. For this we will show that $\mathrm{Ext}_R^n(k,M) = 0$ would imply $\mathrm{id}_R M < n$, i.e. $\mathrm{Ext}_R^n(N,M) = 0$ for any finitely generated R-module N.

Assuming that $\mathrm{Ext}_R^n(k,M) = 0$, one can easily see by induction on $\lambda(N)$ that $\mathrm{Ext}_R^n(N,M) = 0$ for any R-module N with $\lambda(N) < \infty$. If $\dim N > 0$ consider a filtration, $N_0 = N \supset N_1 \supset \dots \supset N_r \supset N_{r+1} = 0$ such that $N_i/N_{i+1} \cong R/\mathfrak{p}_i$ for some $\mathfrak{p}_i \in \mathrm{Spec}(R)$ for $0 \leqq i \leqq r$. So we may assume $N \cong R/\mathfrak{p}$ for some $\mathfrak{p} \in \mathrm{Spec}(R)$. Since $\dim N > 0$, we can choose $x \in \mathfrak{m}-\mathfrak{p}$. From the exact sequence

$$0 \longrightarrow R/\mathfrak{p} \xrightarrow{\cdot x} R/\mathfrak{p} \longrightarrow R/(\mathfrak{p},x) \longrightarrow 0$$

we get an exact sequence

$$\mathrm{Ext}_R^n(R(\mathfrak{p},x),M) \longrightarrow \mathrm{Ext}_R^n(R/\mathfrak{p},M) \xrightarrow{\cdot x} \mathrm{Ext}_R^n(R/\mathfrak{p},M) \longrightarrow 0 \quad ,$$

since $\mathrm{id}_R M = n$. Applying Nakayama's lemma we have $\mathrm{Ext}_R^n(R/\mathfrak{p},M) = 0$. And this is in contradiction with $\mathrm{id}_R M = n$. Hence we have $\mathrm{Ext}_R^n(k,M) \neq 0$.

Since $\mathrm{depth}\, R/\underline{a}R = 0$, we have an exact sequence

$$0 \longrightarrow k \longrightarrow R/\underline{a}R \quad ,$$

which - together with $\mathrm{id}_R M = n$ - induces an exact sequence

$$\mathrm{Ext}_R^n(R/\underline{a}R,M) \longrightarrow \mathrm{Ext}_R^n(k,M) \longrightarrow 0 \quad .$$

Hence $n \leqq t$, since $\mathrm{Ext}_R^n(R/\underline{a}R,M) \neq 0$ by the first part of the proof.

(A2) Theorem. Let $(R,\mathfrak{m},k)$ be a noetherian local ring. Then the following conditions are equivalent:

(1) R is Gorenstein (in the sense of (33.23),1))

(2) $\mathrm{id}_R R < \infty$.

Proof. (1) ⇒ (2) : Let

$$0 \longrightarrow R \longrightarrow I^0 \xrightarrow{f^\circ} I^1 \longrightarrow \dots \longrightarrow I^n \xrightarrow{f^n} I^{n+1} \longrightarrow \dots$$

be an injective resolution of R and let $J = \ker f^d$, where $d = \dim R$.

We want to show that J is an injective R-module. To see this it is enough to prove that

$$\mathrm{Ext}_R^n(N,J) \overset{\sigma}{=} \mathrm{Ext}_R^{n+d}(N,R) = 0 \quad \text{for all } n > 0 ,$$

for any finitely generated R-module N . [Note that via this isomorphism σ we know that $\mathrm{Ext}_R^n(N,J)$ is finitely generated] : By assumption we have $\mathrm{Ext}_R^n(k,R) = 0$ for $n > d$. Now we proceed by induction on $\dim N$: If $\dim N = 0$ we see, by induction on $\lambda(N)$, that $\mathrm{Ext}_R^n(N,R) = 0$ for $n > d$. If $\dim N > 0$ we take a filtration

$$N_0 = N \supset N_1 \supset \dots \supset N_r \supset N_{r+1} = 0 ,$$

such that $N_i/N_{i+1} = R/\mathfrak{p}_i$ for some $\mathfrak{p}_i \in \mathrm{Spec}(R)$ for $0 \leq i \leq r$. We note that $\dim R/\mathfrak{p}_i \leq \dim N$. So we assume $N \cong R/\mathfrak{p}$ for some $\mathfrak{p} \in \mathrm{Spec}(R)$. Since $\dim N = \dim R/\mathfrak{p} > 0$, we get an exact sequence

$$0 \longrightarrow R/\mathfrak{p} \xrightarrow{\cdot x} R/\mathfrak{p} \longrightarrow R/(\mathfrak{p},x) \longrightarrow 0$$

for some $x \in \mathfrak{m} - \mathfrak{p}$. This yields an exact sequence

$$(*) \qquad \mathrm{Ext}_R^n(R/\mathfrak{p},x),R) \longrightarrow \mathrm{Ext}_R^n(R/\mathfrak{p},R) \xrightarrow{\cdot x} \mathrm{Ext}_R^n(R/\mathfrak{p},R)$$

$$\longrightarrow \mathrm{Ext}_R^{n+1}(R/(\mathfrak{p},x),R) \longrightarrow \dots .$$

Since $\dim R/(\mathfrak{p},x) < \dim N$ we get by inductive hypothesis that $\mathrm{Ext}_R^n(R/(\mathfrak{p},x),R) = 0$ for $n > d$.

Therefore by using Nakayama's lemma we obtain from sequence (*) that

$$\mathrm{Ext}_R^n(R/\mathfrak{p},R) = 0 \quad \text{for} \quad n > d \quad .$$

Hence we have $\mathrm{Ext}_R^n(N,R) = 0$ for $n > d$.

(2) ⇒ (1): Let $n = \mathrm{id}_R R$ and let

$$(**) \qquad 0 \longrightarrow R \longrightarrow I^0 \longrightarrow I^1 \longrightarrow \ldots \longrightarrow I^{n-1} \longrightarrow I^n \longrightarrow 0$$

be a minimal injective resolution of R . This implies $H_{\mathfrak{m}}^i(R) = 0$ for all $i > n$. Therefore we have $n \geq \dim R$ by (36.17), hence R is Cohen-Macaulay by Lemma (A1) . Now using Lemma (33.9) and Theorem (33.25), one can show that

$$H_{\mathfrak{m}}^0(I^i) = 0 \quad \text{for} \quad i < d \quad ,$$

since $\mathrm{Ext}_R^i(k,R) = 0$ for $i < d$. Therefore we have $H_{\mathfrak{m}}^d(R) \cong H_{\mathfrak{m}}^0(I^d) \cong E_R(k)^r$, where $r = \dim_k \mathrm{Ext}_R^d(k,R)$.
So it remains to prove that $r = 1$. For that let $\underline{a} = \{a_1,\ldots,a_d\}$, be an R-sequence. Then we consider the exact sequence

$$0 \longrightarrow H_{\mathfrak{m}}^{d-1}(R/a_1R) \longrightarrow H_{\mathfrak{m}}^d(R) \xrightarrow{\cdot a_1} H_{\mathfrak{m}}^d(R) \longrightarrow 0 \ .$$

Hence we get

$$H_{\mathfrak{m}}^{d-1}(R/a_1R) \cong ((0 : a_1)_{E_R(k)})^r \cong (E_{R/a_1R}(k))^r \quad ,$$

since $E_{R/a_1R}(k) \cong \mathrm{Hom}_R(R/a_1R,E_R(k)) \cong (0 : a_1)E_R(k)$. Inductively we finally obtain $R/\underline{a}R \cong H_{\mathfrak{m}}^0(R/\underline{a}R) = (E_{R/\underline{a}R}(k))^r$. Hence $r = 1$.

<u>(A3) Corollary</u>. Let $(R,\mathfrak{m},k)$ be a Gorenstein local ring in the sense of (33.23),1). Then $R_{\mathfrak{p}}$ is Gorenstein for all $\mathfrak{p} \in \mathrm{Spec}(R)$.

Proof. Since R is Gorenstein, we get an injective resolution $(d = \dim R)$:

$$0 \longrightarrow R \longrightarrow I^0 \longrightarrow I^1 \longrightarrow \ldots \longrightarrow I^d \longrightarrow 0 \quad ,$$

by the theorem. Hence we have an exact sequence

$$0 \longrightarrow R_{\mathfrak{p}} \longrightarrow I^0_{\mathfrak{p}} \longrightarrow I^1_{\mathfrak{p}} \longrightarrow \ldots \longrightarrow I^d_{\mathfrak{p}} \longrightarrow 0 \quad .$$

Therefore it is enough to show that $I^i_{\mathfrak{p}}$ is an injective $R_{\mathfrak{p}}$-module. for any $\mathfrak{p} \in \mathrm{Spec}(R)$. But this follows from Lemma (33.13) and Proposition (33.11).

References - Chapter VII

Books

[1] N. Bourbaki, Algébre, Ch. I - III. Herman Paris 1970.

[2] N. Bourbaki, Algébre commutative, Ch. I - IV. Herman Paris 1961.

[3] H. Cartan - Eilenberg, Homological Algebra. Princeton, N.J.: Princeton University Press 1956.

[4] A. Grothendieck, Local Cohomology. Lecture Notes in Math. 41, Springer Verlag 1967.

[5] A. Grothendieck, J. Dieudonné, Elements de Géométrie Algébrique IV,No. 4 - No. 24. I.H.E.S. Paris 1965.

[6] J. Herzog, E. Kunz, Der kanonische Modul eines Cohen-Macaulay-Rings. Lecture Notes in Math. 238, Springer Verlag 1971.

[7] I. Kaplansky, Commutative rings. The University of Chicago Press, Chicago 1970/74.

[8] S. MacLane, Homology. Berlin-Göttingen-Heidelberg: Springer Verlag 1963.

[9] H. Matsumura, Commutative algebra. Benjamin New York 1970.

[10] D.G. Northcott, Lessons on rings, modules and multiplicities, Cambridge Univ. Press 1968.

[11] D.G. Northcott, Finite free resolutions, Cambridge Tracts No. 71, 1976.

[12] J.P. Serre, Algébre Locala: Multiplicités. Lecture Notes in Math. 11, Springer Verlag 1965.

[13] D.W. Sharpe, P. Vámos, Injective modules, Cambridge Tract No. 62, 1971.

[13*] W. Vogel, J. Stückrad, Buchsbaum rings and applications. Springer Verlag 1987.

Papers

[14] M. Auslander, D.A. Buchsbaum, Codimension and multiplicity, Ann. of Math. 68 (1958), 625 - 657.

[15] H. Bass, On the ubiquity of Gorenstein rings, Math. Z. 82 (1963), 8 - 28.

[16] S. Goto, K. Watanabe, On graded rings I, J. Math. Soc. Japan 30 (1978), 179 - 213.

[17] S. Goto, K. Watanabe, On graded rings II ($\mathbf{Z}^n$-graded rings),

[18] A. Grothendieck, Sur quelque points d'algèbre homologique, Tohoku Math. J., vol. IX (1957), 119 - 221.

[19] S. Ikeda, The Cohen-Macaulayness of the Rees algebras of local rings, Nagoya Math. J. 89 (1983), 47 - 63.

[20] E. Matlis, Injective modules over noetherian rings, Pacific J. Math. 8 (1958), 511 - 528.

[21] P. Gabriel, Objets injectifs dans les catégories abéliennes, Sém. Dubreil-Pisot Fas. 12, Exp. 17 (1958/59).

[22] J.P. Serre, Algèbre locale-multiplicités, Lecture Notes in Math., No. 11, Springer 1965.

[23] P. Roberts, Rings of type 1 are Gorenstein, Bull. London Math. Soc. 15 (1983), 48 - 50.

[9*] H. Matsumura, Commutative ring theory. Cambridge University Press 1986.

Chapter VIII. GENERALIZED COHEN-MACAULAY RINGS AND BLOWING UP

In this chapter we investigate the properties of local rings $(A,\mathfrak{m},k)$ such that $\lambda_A(H^i_\mathfrak{m}(A)) < \infty$ for $i < \dim A$. Rings of this type appear in algebraic geometry frequently. For example, if $X \subseteq \mathbf{P}^n_k$ is an irreducible, non-singular projective variety over a field k, then the local ring at the vertex of the affine cone over X satisfies this property (cf. Hartshorne [1]; see also the remark at the end of § 35 in Chapter VII). The purpose of this chapter is to present the results on "generalized Cohen-Macaulay rings" in a unified manner. We develop the theory according to S. Goto [7] and N.V. Trung [17]. Throughout the next two chapters $(A,\mathfrak{m},k)$ denotes a noetherian local ring with $\dim A = d$. The reason for this deviation from our principle to denote local rings by R and graded rings by A is the fact that we want to use R for "Rees rings" in the sequel.

§ 37. Finiteness of local cohomology

We first recall the notion of a reducing system of parameters.

(37.1) Definition. A system of parameters $a_1,\ldots,a_d$ of a local ring $(A,\mathfrak{m})$ is said to be a reducing system if

$$e_A(a_1,\ldots,a_d;A) = \lambda_A(A/(a_1,\ldots,a_d)A) - \lambda_A\left(\frac{(a_1,\ldots,a_{d-1}):a_d}{(a_1,\ldots,a_{d-1})}\right) .$$

(37.2) Lemma. Let $(A,\mathfrak{m},k)$ be a local ring such that $A_\mathfrak{p}$ is Cohen-Macaulay and $\mathrm{ht}(\mathfrak{p}) + \dim A/\mathfrak{p} = \dim A$ for all $\mathfrak{p} \in \mathrm{Spec}(A) - \{\mathfrak{m}\}$. Then every system of parameters of A is a reducing system.

Proof. If $d = \dim A = 1$ every system of parameters of A is reducing by definition. We proceed by induction on d. Let $d \geq 2$ and $a_1,\ldots,a_d$ a system of parameters of A.

First we note that $\mathrm{ht}(a_1A_\mathfrak{p}) = 1$ for any $\mathfrak{p} \in V(a_1A)$: Assume that there is a minimal prime $\mathfrak{p}$ of A containing a_1; then we get by assumption:

$$d = \mathrm{ht}(\mathfrak{p}) + \dim A/\mathfrak{p} \leqq \dim A/a_1A = d-1 ,$$

which is a contradiction.

Hence a_1 is part of a system of parameters of $A_{\mathfrak{p}}$ for any $\mathfrak{p} \in V(a_1A)$. If $\mathfrak{p} \neq \mathfrak{m}$, then $(A/a_1A)_{\mathfrak{p}}$ is a Cohen-Macaulay ring and

$$\mathrm{ht}(\mathfrak{p}/a_1A) + \dim A/\mathfrak{p} = \mathrm{ht}(\mathfrak{p}) - 1 + \dim A/\mathfrak{p} = d-1$$

by assumption.

Now one can apply the inductive hypothesis to $\bar{A} = A/a_1A$, i.e.

$$(*) \qquad e(\bar{a}_2,\dots,\bar{a}_d;\bar{A}) = \lambda_{\bar{A}}(\bar{A}/(\bar{a}_2,\dots,\bar{a}_d)\bar{A}) - \lambda_{\bar{A}}\left(\frac{(\bar{a}_2,\dots,\bar{a}_{d-1}):\bar{a}_d}{(\bar{a}_2,\dots,\bar{a}_{d-1})}\right) .$$

But

$$e(\bar{a}_2,\dots,\bar{a}_d;\bar{A}) = e(a_1,\dots,a_d;A) + e(\bar{a}_2,\dots,\bar{a}_d;(0 \underset{A}{:} a_1)) .$$

Moreover we have $(0 : a_1)_{\mathfrak{p}} = 0$ for $\mathfrak{p} \in \mathrm{Spec}(A) - \{\mathfrak{m}\}$, and therefore $\lambda_A((0 : a_1)) < \infty$. Hence $e(\bar{a}_2,\dots,\bar{a}_d;(0 : a_1)) = 0$. And the result follows from (*).

<u>(37.3) Lemma</u>. Let M be a finitely generated module over a local ring $(A,\mathfrak{m})$ with $\dim M = n$. Then the following conditions are equivalent:

1) $\lambda_A(H^i_{\mathfrak{m}}(M)) < \infty$ for $i < n$.

2) $H^i_{\mathfrak{m}}(M)$ is a finitely generated A-module for $i < n$.

3) There is an $\mathfrak{m}$-primary ideal I such that $IH^i_{\mathfrak{m}}(M) = 0$ for $i < n$.

Proof. Since $H^i_{\mathfrak{m}}(M)$ is an artinian A-module by Chapter VII, Lemma (36.1) the implications 1)$\Longleftrightarrow$2) and 2)$\Rightarrow$3) are clear.

3)$\Rightarrow$1): We apply induction on $n = \dim M$. For $n \leqq 1$ there is nothing to prove. Let $n \geqq 2$. We put $\bar{M} = M/H^0_{\mathfrak{m}}(M)$. Then from the exact sequence

$$0 \longrightarrow H^0_{\mathfrak{m}}(M) \longrightarrow M \longrightarrow \overline{M} \longrightarrow 0$$

we get an exact sequence

$$(\#)\quad \begin{array}{l} 0 \longrightarrow H^0_{\mathfrak{m}}(H^0_{\mathfrak{m}}(M)) \longrightarrow H^0_{\mathfrak{m}}(M) \longrightarrow H^0_{\mathfrak{m}}(\overline{M}) \longrightarrow H^1_{\mathfrak{m}}(H^0_{\mathfrak{m}}(M)) \longrightarrow \ldots \\ \ldots \longrightarrow H^i_{\mathfrak{m}}(H^0_{\mathfrak{m}}(M)) \longrightarrow H^i_{\mathfrak{m}}(M) \longrightarrow H^i_{\mathfrak{m}}(\overline{M}) \longrightarrow H^{i+1}_{\mathfrak{m}}(H^0_{\mathfrak{m}}(M)) \longrightarrow \ldots \end{array} .$$

By Chapter VII, Lemma (36.2) we have $H^i_{\mathfrak{m}}(H^0_{\mathfrak{m}}(M)) = 0$ for $i > 0$, since $\dim H^0_{\mathfrak{m}}(M) = 0$. Hence (#) yields certain isomorphisms $H^i_{\mathfrak{m}}(M) \cong H^i_{\mathfrak{m}}(\overline{M})$ for $i > 0$ and $H^0_{\mathfrak{m}}(\overline{M}) = 0$. So we may assume $\operatorname{depth} M > 0$.

Let $a \in I$ be a non-zero divisor on M. From the exact sequence:

$$0 \longrightarrow M \xrightarrow{\cdot a} M \longrightarrow M/aM \longrightarrow 0$$

we obtain an exact sequence

$$\ldots \longrightarrow H^i_{\mathfrak{m}}(M) \longrightarrow H^i_{\mathfrak{m}}(M/aM) \xrightarrow{\varphi} H^{i+1}_{\mathfrak{m}}(M) \xrightarrow{\cdot a} H^{i+1}_{\mathfrak{m}}(M) \longrightarrow \ldots$$

for $i < n-1$. Hence $I^2 H^i_{\mathfrak{m}}(M/aM) = 0$ for $i < n-1$. By inductive hypothesis we know that $H^i_{\mathfrak{m}}(M/aM)$ is of finite length. Since $aH^i_{\mathfrak{m}}(M) = 0$ for $i < n$ by assumption, φ is surjective. Hence $H^{i+1}_{\mathfrak{m}}(M)$ has finite length for $i < n-1$, q.e.d.

The following result is a direct application of local duality.

(37.4) Theorem. For a local ring $(A, \mathfrak{m}, k)$ the following conditions are equivalent.

1) $\lambda_A(H^i_{\mathfrak{m}}(A)) < \infty$ for $i < d$.

2) For $\mathfrak{p} \in \operatorname{Spec}(\hat{A}) - \{\mathfrak{m}\hat{A}\}$ we have $\dim \hat{A}_{\mathfrak{p}} + \dim \hat{A}/\mathfrak{p} = d$, and $\hat{A}_{\mathfrak{p}}$ is Cohen-Macaulay, where $\hat{A}$ is the completion of A.

Proof. 1) ⇒ 2): Since $H^i_{\mathfrak{m}\hat{A}}(A) \cong H^i_{\mathfrak{m}}(A) \otimes_A \hat{A} \cong \widehat{H^i_{\mathfrak{m}}(A)}$ we may assume that A is complete. By Cohen's structure theorem A is a homomorphic image of a complete regular local ring $(S, \mathfrak{n}, k)$ with $\dim S = n$.

By the local duality-theorem we have

$$\mathrm{Hom}_R(H^i_{\mathfrak{m}}(A), E_S(k)) \cong \mathrm{Ext}^{n-i}_S(A,S)$$

for all i. Hence $\mathrm{Ext}^{n-i}_S(A,S)$ as an S-module has finite length for $i \neq d$. Therefore we get $\mathrm{Ext}^{n-i}_{S_{\mathfrak{p}}}(A_{\mathfrak{p}},S_{\mathfrak{p}}) = 0$ for $i \neq d$ and $\mathfrak{p} \in \mathrm{Spec}(S) - \{\mathfrak{n}\}$. This shows that $A_{\mathfrak{p}}$ is a Cohen-Macaulay ring with $\dim A_{\mathfrak{p}} = \dim S_{\mathfrak{p}} - (n-d) = d - (n - \dim S_{\mathfrak{p}}) = d - \dim S/\mathfrak{p}$, provided that $A_{\mathfrak{p}} \neq 0$.

<u>2) ⇒ 1)</u>: We may assume that A is complete. Let S be the same as in the proof of 1) ⇒ 2). For $\mathfrak{p} \in \mathrm{Supp}_S(A) - \{\mathfrak{n}\}$ we have

$$\begin{aligned}\dim S_{\mathfrak{p}} - \dim A_{\mathfrak{p}} &= \dim S_{\mathfrak{p}} - (d - \dim S/\mathfrak{p}) \\ &= \dim S_{\mathfrak{p}} + \dim S/\mathfrak{p} - d = n-d\end{aligned}$$

by assumption. Moreover $A_{\mathfrak{p}}$ is Cohen-Macaulay for $\mathfrak{p} \in \mathrm{Supp}_S(A) - \{\mathfrak{n}\}$. Hence, by Corollary (36.15) we have

$$\mathrm{Ext}^i_{S_{\mathfrak{p}}}(A_{\mathfrak{p}},S_{\mathfrak{p}}) = 0$$

for $i \neq n-d$ and for any $\mathfrak{p} \in \mathrm{Spec}(S) - \{\mathfrak{n}\}$. Therefore $\mathrm{Ext}^i_S(A,S)$ has finite length for $i \neq n-d$. And by the local duality we get:

$$\begin{aligned}H^i_{\mathfrak{m}}(A) &\cong \mathrm{Hom}_S(\mathrm{Hom}_S(H^i_{\mathfrak{m}}(A),E_S(k)),E_S(k)) \\ &\cong \mathrm{Hom}_S(\mathrm{Ext}^{n-i}_S(A,S),E_S(k)) \quad ,\end{aligned}$$

hence we know that $H^i_{\mathfrak{m}}(A)$ has finite length for $i < d$.

<u>(37.5) Definition</u>. A noetherian local ring $(A,\mathfrak{m},k)$ is called generalized Cohen-Macaulay if $\lambda_A(H^i_{\mathfrak{m}}(A)) < \infty$ for $i < \dim A$.

Note that a Cohen-Macaulay local ring is a generalized Cohen-Macaulay ring, and every local ring of dimension 1 is a generalized Cohen-Macaulay ring.

<u>(37.6) Corollary</u>. If A is a generalized Cohen-Macaulay ring then for any $\mathfrak{p} \in \mathrm{Spec}(A) - \{\mathfrak{m}\}$ the localization $A_{\mathfrak{p}}$ is Cohen-Macaulay and $\dim A_{\mathfrak{p}} + \dim A/\mathfrak{p} = \dim A$.

Proof. Let $\hat{A}$ be the completion of A and $\mathfrak{P} \in \operatorname{Spec}(\hat{A})$ be a minimal prime ideal of $\mathfrak{p}\hat{A}$ with $\operatorname{ht}(\mathfrak{P}) = \operatorname{ht}(\mathfrak{p}\hat{A})$. Then

$$\dim \hat{A}_{\mathfrak{P}} = \dim A_{\mathfrak{p}} + \dim \hat{A}_{\mathfrak{P}}/\mathfrak{p}\hat{A}_{\mathfrak{P}}$$
$$\operatorname{depth} \hat{A}_{\mathfrak{P}} = \operatorname{depth} A_{\mathfrak{p}} + \operatorname{depth} \hat{A}_{\mathfrak{P}}/\mathfrak{p}\hat{A}_{\mathfrak{P}} .$$

By Theorem (37.4) we have $\dim \hat{A}_{\mathfrak{P}} = \operatorname{depth} \hat{A}_{\mathfrak{P}}$. Therefore $\dim A_{\mathfrak{p}} = \operatorname{depth} A_{\mathfrak{p}}$, i.e. $A_{\mathfrak{p}}$ is Cohen-Macaulay, since $\dim \hat{A}_{\mathfrak{P}}/\mathfrak{p}\hat{A}_{\mathfrak{P}} = \operatorname{depth} \hat{A}_{\mathfrak{P}}/\mathfrak{p}\hat{A}_{\mathfrak{P}} = 0$. Moreover

$$d = \dim \hat{A}_{\mathfrak{P}} + \dim \hat{A}/\mathfrak{P} \leq \dim A_{\mathfrak{p}} + \dim \hat{A}/\mathfrak{p}\hat{A}$$
$$= \dim A_{\mathfrak{p}} + \dim A/\mathfrak{p} \leq d .$$

Hence

$$\dim A = \dim A_{\mathfrak{p}} + \dim A/\mathfrak{p} .$$

Note that the converse of Corollary (37.6) does not hold in general. But if A is a homomorphic image of a local Cohen-Macaulay ring or if every formal fibre of A is Cohen-Macaulay, then the converse of (37.6) is true, see [5].

(37.7) Corollary. Let A be a generalized Cohen-Macaulay ring and $a \in \mathfrak{m}$ a part of a system of parameters of A. Then $(0 : a)$ has finite length, and there is an exact sequence

$$\cdots \longrightarrow H^i_{\mathfrak{m}}(A) \xrightarrow{.a} H^i_{\mathfrak{m}}(A) \longrightarrow H^i_{\mathfrak{m}}(A/aA) \longrightarrow H^{i+1}_{\mathfrak{m}}(A) \xrightarrow{.a} H^{i+1}_{\mathfrak{m}}(A) \longrightarrow \cdots .$$

Proof. For the first assertion we discuss the cases $\dim A = 1$ and $\dim A = 2$ separately.

$d = \dim A = 1$: Since $H^0_{\mathfrak{m}}(A)$ has finite length, it is sufficient to show that $(0 : a) \subseteq H^0_{\mathfrak{m}}(A)$. For that let $x \in (0 : a)$ and let $0 = \mathfrak{q}_1 \cap \mathfrak{q}_2 \cap \ldots \cap \mathfrak{q}_s \cap \mathfrak{q}$ be a primary decomposition with primary ideals $\mathfrak{q}_i$ belonging to minimal primes $\mathfrak{p}_i$ for $1 \leq i \leq s$, and an $\mathfrak{m}$-primary ideal $\mathfrak{q}$.

Since a is a system of parameters of A, it is not contained in any minimal prime ideal. Therefore $x \in \mathfrak{q}_1 \cap \ldots \cap \mathfrak{q}_s$. Moreover there exists an integer $n > 0$ such that $x \cdot \mathfrak{m}^n \subseteq \mathfrak{m}^n \subseteq \mathfrak{q}$. So we know that $x \cdot \mathfrak{m}^n \subseteq \mathfrak{q}_1 \cap \ldots \cap \mathfrak{q}_s \cap \mathfrak{q} = 0$ and $x \in (0 : \mathfrak{m}^n) \subseteq H^0_{\mathfrak{m}}(A)$.

<u>$d \geq 2$</u>: Since $(0 : a)$ is a finitely generated A-module, it is enough to show that $\mathrm{Supp}_A(0 : a) = \{\mathfrak{m}\}$ for the case $(0 : a) \neq 0$. Let $\mathfrak{p} \in \mathrm{Spec}(A) - \{\mathfrak{m}\}$. We have to discuss two cases:

i) If $a \notin \mathfrak{p}$, then $(0 : a)_{\mathfrak{p}} = 0$.

ii) If $a \in \mathfrak{p}$, then $\mathrm{ht}(aA_{\mathfrak{p}}) > 0$. Assume that there is a minimal prime $\mathfrak{p}$ of A containing a. Then by assumption:

$$d = \mathrm{ht}(\mathfrak{p}) + \dim(A/\mathfrak{p}) \leq \dim(A/aA) = d - 1$$

which is a contradiction. Hence a is a part of a system of parameters of $A_{\mathfrak{p}}$. But by assumption $A_{\mathfrak{p}}$ is Cohen-Macaulay and therefore a is $A_{\mathfrak{p}}$-regular, hence we know $(0 : a)_{\mathfrak{p}} = (0 : aA_{\mathfrak{p}}) = 0$.

For the second assertion consider the following diagram with exact row and column:

$$\begin{array}{ccccccccc}
 & & 0 & & & & & & \\
 & & \uparrow & & & & & & \\
0 & \longrightarrow & aA & \longrightarrow & A & \longrightarrow & A/aA & \longrightarrow & 0 \\
 & & \uparrow & \nearrow_{\cdot a} & & & & & \\
 & & A & & & & & & \\
 & & \uparrow & & & & & & \\
 & & (0 : a) & & & & & & \\
 & & \uparrow & & & & & & \\
 & & 0 & & & & & &
\end{array}$$

Since $\lambda_A((0 : a)) < \infty$, we have $H^i_{\mathfrak{m}}((0 : a)) = 0$ for $i > 0$, by Lemma (36.2). Therefore we obtain a commutative diagram with exact rows and columns for $i \geq 0$ as follows:

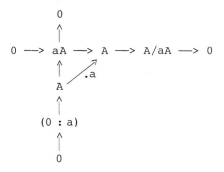

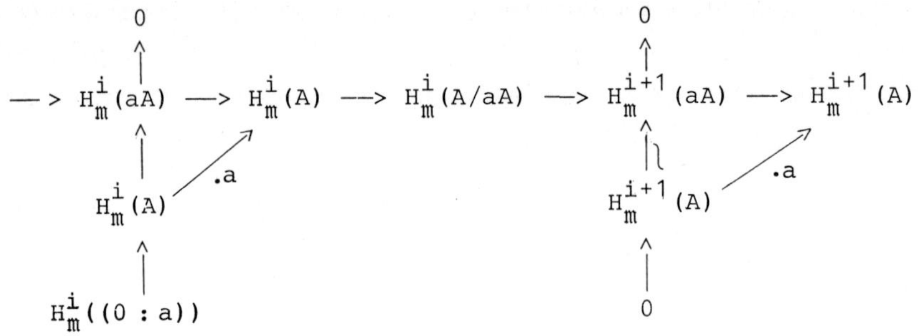

And this proves the second assertion.

(37.8) Corollary. Let A and a be as in Corollary (37.7). Then

$$\lambda_A(H^i_{\mathfrak{m}}(A/aA)) \leqq \lambda_A(H^i_{\mathfrak{m}}(A)) + \lambda_A(H^{i+1}_{\mathfrak{m}}(A)) \quad \text{for} \quad i < d-1 \quad .$$

The equality holds for $0 \leqq i < d-1$ if and only if

$$aH^i_{\mathfrak{m}}(A) = 0 \quad \text{for} \quad 0 \leqq i < d \quad .$$

In particular if a is a part of a system of parameters of a generalized Cohen-Macaulay ring A, then A/aA is generalized Cohen-Macaulay

Proof. This follows from Corollary (37.7).

(37.9) Corollary. Let A be a generalized Cohen-Macaulay ring and $a_1,\ldots,a_d$ be any system of parameters of A. Then

$$\lambda_A(A/\underline{a}A) - e(\underline{a}A;A) \leqq \sum_{i=0}^{d-1} \binom{d-1}{i} \lambda_A(H^i_{\mathfrak{m}}(A)) \quad ,$$

where $\underline{a}A$ is the ideal generated by $a_1,\ldots,a_d$.

Proof. By Lemma (37.2) we have

$$\lambda_A(A/\underline{a}A) - e(\underline{a}A;A) = \lambda_A\left(\frac{(a_1,\ldots,a_{d-1}):a_d}{(a_1,\ldots,a_{d-1})}\right) \quad .$$

Since $\dfrac{(a_1,\ldots,a_{d-1}):a_d}{(a_1,\ldots,a_{d-1})} \subseteq H^0_{\mathfrak{m}}(A/(a_1,\ldots,a_{d-1}))$, it is enough to show tha

$$\lambda_A(H^0_{\mathfrak{m}}(A/(a_1,\ldots,a_{d-1}))) \leqq \sum_{i=0}^{d-1} \binom{d-1}{i} \lambda_A(H^i_{\mathfrak{m}}(A)) \quad .$$

By Corollary (37.8) we have

$$\lambda_A(H^i_{\mathfrak{m}}(A/(a_1,\ldots,a_j))) \leqq \lambda_A(H^i_{\mathfrak{m}}(A/(a_1,\ldots,a_{j-1}))) + \lambda_A(H^{i+1}_{\mathfrak{m}}(A/(a_1,\ldots,a_{j-1})))$$

for $i+j<d$. Hence by induction on $\dim A$ we get the required inequality.

The following theorem shows that also the converse of Corollary (37.9) is true.

<u>(37.10) Theorem</u>. For a noetherian local ring A with $\dim A = d > 0$ the following conditions are equivalent:

1) A is a generalized Cohen-Macaulay ring.

2) There is an integer $c \geqq 0$ such that

$$\lambda_A(A/\mathfrak{q}) - e(\mathfrak{q};A) \leqq c$$

for any parameter ideal $\mathfrak{q}$ of A .

3) There is an $\mathfrak{m}$-primary ideal I such that

$$(a_1,\dots,a_{d-1}) : a_d = (a_1,\dots,a_{d-1}) : I$$

for any system of parameters $a_1,\dots,a_d$ contained in I .

Proof. 1) ⇒ 2): by Corollary (37.9).

2) ⇒ 3): First we claim that every system of parameters $a_1,\dots,a_d$ of A is a reducing system. For this we observe that for any $n > 0$ we have

$$\lambda_A(A/(a_1,\dots,a_{d-1},a_d^n)) - e(a_1,\dots,a_{d-1},a_d^n;A)$$

(*)

$$= \lambda_A\left(\frac{(a_1,\dots,a_{d-1}):a_d^n}{(a_1,\dots,a_{d-1}}\right) + \sum_{i=1}^{d-1} e(\bar{\mathfrak{q}}/\mathfrak{q}_i;(\mathfrak{q}_{i-1}:a_i)/\mathfrak{q}_{i-1}) \leqq c \quad ,$$

where $\bar{\mathfrak{q}} = (a_1,\dots,a_{d-1},a_d^n)$ and $\mathfrak{q}_i = (a_1,\dots,a_i)$ for $0 \leqq i < d$. Note that:

$$(**) \qquad e(\bar{\mathfrak{q}}/\mathfrak{q}_i;(\mathfrak{q}_{i-1}:a_i)/\mathfrak{q}_{i-1}) = n\cdot e(\mathfrak{q}/\mathfrak{q}_i;(\mathfrak{q}_{i-1}:a_i)/\mathfrak{q}_{i-1}) \quad ,$$

where $\mathfrak{q} = (a_1,\dots,a_d)$.

Since (*) is true for all $n > 0$, we conclude from (**) that

$$e(\mathfrak{q}/\mathfrak{q}_i;(\mathfrak{q}_{i-1}:a_i)/\mathfrak{q}_{i-1}) = 0 ; \quad 1 \leq i \leq d-1 \quad .$$

Hence $a_1,\ldots,a_d$ is a reducing system.

Let $M := (a_1,\ldots,a_{d-1}):a_d/(a_1,\ldots,a_{d-1})$. Then

$$c \geqq \lambda_A(M) = \sum_{i \geqq 0} \lambda_A(\mathfrak{m}^i M/\mathfrak{m}^{i+1} M) \quad .$$

Therefore we must have $\mathfrak{m}^c M = 0$, i.e.

$$(a_1,\ldots,a_{d-1}):a_d \subseteq (a_1,\ldots,a_{d-1}):\mathfrak{m}^c \quad .$$

So it is enough to define $I = \mathfrak{m}^c$.

3) $\Rightarrow$ 1): By Lemma (37.3) it is enough to show that $IH^i_{\mathfrak{m}}(A) = 0$ for $i < d$. We apply induction on d : Let $d = 1$, and let $a \in I$ be a system of parameters of A . Then we have:

$$H^0_{\mathfrak{m}}(A) = (0:a^n) \quad \text{for} \quad n >> 0 \quad .$$

By assumption we know that $(0:a) = (0:a^n)$. Hence $IH^0_{\mathfrak{m}}(A) = 0$.

Let $d \geqq 2$ and let $a \in I^2$ be a part of a system of parameters of A . We claim that $(0:a)$ has finite length. To see this we complete a to a full system of parameters $a_1,\ldots,a_{d-1},a_d := a$ of A , contained in I . So by assumption we have:

$$(0:a) = (\bigcap_{n>0}(a_1^n,\ldots,a_{d-1}^n):a) = \bigcap_{n>0}((a_1^n,\ldots,a_{d-1}^n):a)$$

$$= \bigcap_{n>0}(a_1^n,\ldots,a_{d-1}^n):I = (0:I) \quad .$$

This proves the above claim. To continue with the proof of 3) $\Rightarrow$ 1) note that by the proof of Corollary (37.7) we have an exact sequence

$$(\#) \quad H^i_{\mathfrak{m}}((0:a)) \longrightarrow H^i_{\mathfrak{m}}(A) \xrightarrow{\cdot a} H^i_{\mathfrak{m}}(A) \longrightarrow H^i_{\mathfrak{m}}(A/aA) \longrightarrow H^{i+1}_{\mathfrak{m}}(A) \xrightarrow{\cdot a} H^{i+1}_{\mathfrak{m}}(A)$$

But $I\cdot A/aA$ has the corresponding property of I . Hence by inductive hypothesis we know that

$$IH^i_{\mathfrak{m}}(A/aA) = 0 \quad \text{for} \quad i < d-1 \quad .$$

From the exact sequence (#) we get for $i < d$:

$$(0:a)_{H^i_{\mathfrak{m}}(A)} \subseteq (0:I)_{H^i_{\mathfrak{m}}(A)} \subseteq (0:I^2)_{H^i_{\mathfrak{m}}(A)} \subseteq (0:a)_{H^i_{\mathfrak{m}}(A)} \quad ,$$

since $a \in I^2$. Now it is easy to see that

$$(0:I)_{H^i_{\mathfrak{m}}(A)} = (0:I^n)_{H^i_{\mathfrak{m}}(A)}$$

for all $n > 0$, by induction on n .
Therefore we have

$$H^i_{\mathfrak{m}}(A) = \bigcup_{n>0} (0:I^n)_{H^i_{\mathfrak{m}}(A)} = (0:I)_{H^i_{\mathfrak{m}}(A)}$$

for $i < d$. In other words, we get $IH^i_{\mathfrak{m}}(A) = 0$ for $i < d$.

§ 38. Standard system of parameters

Throughout this section $(A,\mathfrak{m},k)$ denotes a generalized Cohen-Macaulay local ring. We put

$$I(A) := \sum_{i=0}^{d-1} \binom{d-1}{i} h^i(A) \quad ,$$

where $h^i(A) = \lambda_A(H^i_{\mathfrak{m}}(A))$.

(38.1) Definition. A system of parameters $a_1,\ldots,a_d$ of a generalized Cohen-Macaulay ring A is called standard if

$$\lambda_A(A/(a_1,\ldots,a_d)) - e(a_1,\ldots,a_d;A) = I(A) \quad .$$

An $\mathfrak{m}$-primary ideal I of A is called standard if every system of parameters $a_1,\ldots,a_d$ in I is standard. We put

$$\lambda_A(A/(a_1,\ldots,a_d)) - e(a_1,\ldots,a_d;A) = I(\underline{a};A) \quad .$$

<u>(38.2) Lemma</u>. Let A and $a_1,\ldots,a_d$ be as in Definition (38.1). We put $\bar{A} := A/a_1A$. Then we have:

1) $I(a_1,\ldots,a_d;A) = I(\bar{a}_2,\ldots,\bar{a}_d;\bar{A})$ if $d \geqq 2$.

2) Let $0 < n_i \leqq m_i$ $(1 \leqq i \leqq d)$ be integers. Then:

$$I\left(a_1^{n_1},\ldots,a_d^{n_d};A\right) \leqq I\left(a_1^{m_1},\ldots,a_d^{m_d};A\right) \quad .$$

3) $I(A)$ is the smallest upper bound for the $I(a_1,\ldots,a_d;A)$.

Proof. 1) By Chapter I we have:

$$e(\bar{a}_2,\ldots,\bar{a}_d;\bar{A}) = e(a_1,\ldots,a_d;A) + e(\bar{a}_2,\ldots,\bar{a}_d;(0 : a_1)) \quad .$$

Since $\lambda_A((0 : a_1)) < \infty$ we know that $e(\bar{a}_2,\ldots,\bar{a}_d;(0:a_1)) = 0$ if $d \geqq 2$. Hence $e(a_1,\ldots,a_d;A) = e(\bar{a}_2,\ldots,\bar{a}_d;\bar{A})$. Since $\lambda_A(A/(a_1,\ldots,a_d)) =$ $= \lambda_{\bar{A}}(\bar{A}/(\bar{a}_2,\ldots,\bar{a}_d))$, then 1) follows.

2) Recall that $a_1,\ldots,a_d$ is a reducing system by Lemma (37.2) and Corollary (37.6), and the same is true for any permutation $a_{\nu_1},\ldots,a_{\nu_d}$ of $a_1,\ldots,a_d$. Therefore it will be enough to show that

$$I\left(a_1^{n_1},\ldots,a_{d-1}^{n_{d-1}},a_d^{n_d} ; A\right) \leqq I\left(a_1^{n_1},\ldots,a_{d-1}^{n_{d-1}},a_d^{n_d+1} ; A\right) \quad .$$

Since every system of parameters of A is reducing we have

$$I\left(a_1^{n_1},\ldots,a_{d-1}^{n_{d-1}},a_d^{n_d};A\right) = \lambda_A\left((a_1^{n_1},\ldots,a_d^{n_{d-1}}) : a_d^{n_d}/(a_d^{n_1},\ldots,a_{d-1}^{n_{d-1}})\right)$$

for all $n_i > 0$, that means

$$I\left(a_1^{n_1},\ldots,a_{d-1}^{n_{d-1}},a_d^{n_d} ; A\right) \leqq I\left(a_1^{n_1},\ldots,a_{d-1}^{n_{d-1}},a_d^{n_d+1} ; A\right) \quad .$$

3) By Corollary (37.9) $I(A)$ is an upper bound for the term $I(a_1,\ldots,a_d;A)$. If I is a $\mathfrak{m}$-primary ideal satisfying condition 3)

of Theorem (37.10), we have $I H_{\mathfrak{m}}^{i}(A/(a_1,\ldots,a_j)) = 0$ for $i+j<d$ by the proof 3) ⇒ 1) of Theorem (37.10), where $a_1,\ldots,a_d \in I$. Hence, by Corollary (37.8) and by induction on $\dim A$:

$$\lambda_A(H_{\mathfrak{m}}^0(A/(a_1,\ldots,a_{d-1})) = \sum_{i=0}^{d-1}\binom{d-1}{i} h^i(A) \quad .$$

Finally

$$I(a_1,\ldots,a_d;A) = \lambda_A\left(\frac{a_1,\ldots,a_{d-1}):a_d}{(a_1,\ldots,a_{d-1})}\right) = \lambda(H_{\mathfrak{m}}^0(A/(a_1,\ldots,a_{d-1}))) \quad .$$

(38.3) Proposition. Let $a_1,\ldots,a_d$ be a system of parameters of a generalized Cohen-Macaulay ring A. Then the following conditions are equivalent:

1) $a_1,\ldots,a_d$ is standard.

2) $\mathfrak{q} H_{\mathfrak{m}}^{i}(A/(a_1,\ldots,a_j)) = 0$ for $i+j<d$, where $\mathfrak{q} = (a_1,\ldots,a_d)$.

3) $I(a_1^2,\ldots,a_d^2;A) = I(a_1,\ldots,a_d;A)$.

Proof. 1) ⇒ 2): By Lemma (37.2) and the proof of Corollary (37.9), we have

$$I(a_1,\ldots,a_d;A) = \lambda_A\left(\frac{(a_1,\ldots,a_{d-1}):a_d}{(a_1,\ldots,a_{d-1})}\right) \leq \lambda_A(H_{\mathfrak{m}}^0(A/(a_1,\ldots,a_{d-1}))) \leq I(A).$$

Since $\underline{a} = \{a_1,\ldots,a_d\}$ is standard we have

$$\lambda_A(H_{\mathfrak{m}}^0(A/(a_1,\ldots,a_{d-1})) = \sum_{i=0}^{d-1}\binom{d-1}{i} h^i(A) \quad .$$

By Corollary (37.8) we see that

$$a_{j+1} H_{\mathfrak{m}}^{i}(A/(a_1,\ldots,a_j)) = 0 \quad \text{for} \quad i+j<d \ .$$

Hence 2) implies 1).

2) ⇒ 3): This follows from Corollary (37.8) and from the fact that $I(A)$ is the upper bound for $I(\underline{a};A)$.

3) $\Rightarrow$ 1): By induction on d : If $d = 1$ we have $I(a_1;A) = \lambda_A((0:a_1))$ and $I(a_1^2;A) = \lambda_A((0:a_1^2))$. By assumption we get $(0:a_1) = (0:a_1^2)$ and consequently $(0:a_1) = (0:a_1^n)$ for all $n > 0$. Hence $(0:a_1) = H_m^0(A)$, i.e. a_1 is standard.

Let $d \geq 2$. Since every system of parameters of A is a reducing system, by Lemma (38.2) we get:

$$\lambda_A\left(\frac{(a_1^2,\ldots,a_{d-1}^2):a_d}{(a_1^2,\ldots,a_{d-1}^2)}\right) \geq \lambda_A\left(\frac{(a_1,\ldots,a_{d-1}):a_d}{(a_1,\ldots,a_{d-1})}\right) =$$

$$\lambda_A\left(\frac{(a_1^2,\ldots,a_{d-1}^2):a_d^2}{(a_1^2,\ldots,a_{d-1}^2)}\right) \geq \lambda_A\left(\frac{(a_1^2,\ldots,a_{d-1}^2):a_d}{(a_1^2,\ldots,a_{d-1}^2)}\right) .$$

Hence we know that

$$(a_1^2,\ldots,a_{d-1}^2) : a_d = (a_1^2,\ldots,a_{d-1}^2) : a_d^2 .$$

Consequently

$$(a_1^2,\ldots,a_{d-1}^2) : a_d = (a_1^2,\ldots,a_{d-1}^2) : a_d^n$$

for all $n > 0$. This implies by Lemma (38.2)

$$\begin{aligned} I(a_1,\ldots,a_d;A) &= I(a_1^2,\ldots,a_d^2;A) = I(a_1^2,\ldots,a_{d-1}^2,a_d^n;A) \\ &\geq I(a_1,\ldots,a_{d-1},a_d^n;A) \geq I(a_1,\ldots,a_d;A) , \end{aligned}$$

i.e. we have $I(a_1^2,\ldots,a_{d-1}^2,a_d^n;A) = I(a_1,\ldots,a_{d-1},a_d^n;A)$. Passing to the ring $\bar{A} := A/a_d^nA$, for any n we have

$$I(\bar{a}_1,\ldots,\bar{a}_{d-1};\bar{A}) = I(\bar{a}_1^2,\ldots,\bar{a}_{d-1}^2;\bar{A})$$

by Lemma (38.2). If n is sufficiently large then $I(\bar{A}) = I(A)$ by Corollary (37.8).

By inductive hypothesis we have for $n >> 0$

$$I(A) = I(\bar{A}) = I(\bar{a}_1,\dots,\bar{a}_{d-1};\bar{A}) = I(a_1,\dots,a_d;A)$$

by Lemma (38.2),1), q.e.d.

Condition 3) in Proposition (38.3) is particularly important because it does not involve information about the local cohomology.

Recall that $I(A)$ is the smallest upper bound for $I(\underline{a};A)$. Hence if $\underline{a} = \{a_1,\dots,a_d\}$ is a standard system of parameters of A then $a_1^{n_1},\dots,a_d^{n_d}$ is standard by Lemma (38.2). Therefore we have proved:

<u>(38.4) Lemma</u>. Let $a_1,\dots,a_d$ be a standard system of parameters of A. Then for any $n_i > 0$, $i = 1,\dots,d$ the system of parameters $a_1^{n_1},\dots,a_d^{n_d}$ of A is standard. -

Here we recall the definition of a d-sequence which was introduced in Huneke [10]. The notion of d-sequences will play a crucial role in the study of standard systems of parameters.

<u>(38.5) Definition</u>. A sequence $a_1,\dots,a_n$ in a noetherian ring A is called a d-sequence if

1) $a_1,\dots,a_n$ form a minimal base of the ideal $(a_1,\dots,a_n)$.

2) $(a_1,\dots,a_i) : a_{i+1}a_k = (a_1,\dots,a_i) : a_k$ for all $i = 0,\dots,n-1$ and $k \geq i+1$.

<u>(38.6) Lemma</u>.

a) If $a_1,\dots,a_n$ is a d-sequence in a noetherian ring A we have $(a_1,\dots,a_i) : a_{i+1} = (a_1,\dots,a_i) : \mathfrak{q}$ for $i = 0,\dots,n-1$, where $\mathfrak{q} = (a_1,\dots,a_n)$.

b) Condition 2) of definition (38.5) is equivalent to 2)' : $((a_1,\dots,a_i) : a_{i+1}) \cap (a_1,\dots,a_n) = (a_1,\dots,a_i)$ for $i = 0,\dots,n-1$.

Proof. a) By 2) of Definition (38.5) we get

$$(a_1,\dots,a_i) : a_{i+1} \subseteq (a_1,\dots,a_i) : a_{i+1}a_k = (a_1,\dots,a_i) : a_k$$

for $k \geqq i+1$. Hence $(a_1,\dots,a_i) : a_{i+1} = (a_1,\dots,a_i) : \mathfrak{q}$.

b) <u>2) ⇒ 2)':</u> Let $x \in ((a_1,\dots,a_i) : a_{i+1}) \cap (a_1,\dots,a_n)$. Then write

$$x = \sum_{j=1}^{k} a_j y_j , \quad y_j \in A , \quad 1 \leqq k \leqq n ,$$

and apply induction on k. If $k \leqq i$ there is nothing to prove. Suppose $k \geqq i+1$. By a) we have $a_k x \in (a_1,\dots,a_i)$. Hence

$$a_k x = a_1 a_k y_1 + \dots + a_{k-1} a_k y_{k-1} + a_k^2 y_k \in (a_1,\dots,a_i) .$$

Therefore, by condition 2), we have

$$y_k \in ((a_1,\dots,a_{k-1}) : a_k^2 = (a_1,\dots,a_{k-1}) : a_k .$$

This implies $x \in (a_1,\dots,a_{k-1})$. By inductive hypothesis we get $x \in (a_1,\dots,a_i)$.

<u>2)' ⇒ 2):</u> Let $x \in (a_1,\dots,a_i) : a_k a_{i+1}$, $k \geqq i+1$. Then by condition 2)' :

$$a_k x \in (a_1,\dots,a_i) : a_{i+1}) \cap (a_1,\dots,a_n) = (a_1,\dots,a_i) .$$

Therefore

$$(a_1,\dots,a_i) : a_k a_{i+1} \subseteq (a_1,\dots,a_i) : a_k .$$

The other inclusion is clear.

<u>(38.7) Definition</u>. A d-sequence $a_1,\dots,a_n$ is called permutable if for any permutation $\nu_1,\dots,\nu_n$ of $1,\dots,n$ the sequence $a_{\nu_1},\dots,a_{\nu_n}$ is again a d-sequence.

<u>(38.8) Example</u>. Let $A = k[[X,Y]]$, where k is a field and X,Y

are indeterminates over k . Then the sequence XY, X^2 is a d-sequence, but X^2, XY is not a d-sequence. Hence not every d-sequence is permutable. Clearly every regular sequence in a local ring is a permutable d-sequence.

(38.9) Proposition. Every standard system of parameters of a generalized Cohen-Macaulay ring is a permutable d-sequence.

Proof. Let $a_1,\dots,a_d$ be a standard system of parameters of a generalized Cohen-Macaulay ring $(A,\mathfrak{m})$. By induction on $d = \dim A$ we will show the condition 2)' in Lemma (38.6):

1) Let $d = 1$. Let $x \in (0 : a_1) \cap (a_1)$. Then $x = a_1 y$ for some $y \in A$ and we have $a_1^2 y = 0$. Hence $y \in (0 : a_1^2) \subseteq H^0_{\mathfrak{m}}(A)$. By Proposition (38.3) we have $a_1 H^0_{\mathfrak{m}}(A) = 0$. Therefore $y \in (0 : a_1)$, and consequently $x = a_1 y = 0$.

2) Let $d \geq 2$. By Proposition (38.3) and Corollary (37.8) we know that the images $\bar{a}_2,\dots,\bar{a}_d$ of $a_2,\dots,a_d$ in $\bar{A} := A/a_1A$ remain standard. By inductive hypothesis we get

$$((\bar{a}_2,\dots,\bar{a}_i) : \bar{a}_{i+1}) \cap (\bar{a}_2,\dots,\bar{a}_d) = (\bar{a}_2,\dots,\bar{a}_i)$$

for $i = 1,\dots,d-1$. This yields

$$((a_1,\dots,a_i) : a_{i+1}) \cap (a_1,\dots,a_d) = (a_1,\dots,a_i)$$

for $1 \leq i \leq d-1$.

3) We must still show that (see Lemma (38.6))

$$(0 : a_1) \cap (a_1,\dots,a_d) = 0 \quad .$$

To prove this, by induction on k , we show the following statement:

$$(0 : a_1) \cap (a_1,\dots,a_k) = 0 \quad \text{for} \quad 1 \leq k \leq d \quad .$$

Let $k = 1$: It is enough to show that $(0 : a_1) = (0 : a_1^2)$. By Corollary (37.7) we have $(0 : a_1^2) \subseteq H^0_{\mathfrak{m}}(A)$. On the other hand, by

Proposition (38.3) we know that $a_1 H^0_{\mathfrak{m}}(A) = 0$. Hence $(0 : a_1) = (0 : a_1^2)$.

Let $k \geqq 2$. Take $x \in (0 : a_1) \cap (a_1,\ldots,a_k)$. If $x = \sum_{i=1}^{k} a_i x_i$ we have $a_k^2 x_k \in (a_1,\ldots,a_{k-1})$, since $(0 : a_1) \subset H^0_{\mathfrak{m}}(A)$ and since $(a_1,\ldots,a_d)\cdot H^0_{\mathfrak{m}}(A) = 0$ by Proposition (38.3). By inductive hypothesis (on we know that $\bar{a}_2,\ldots,\bar{a}_d$ is a d-sequence in $\bar{A}$. Hence $a_k x_k \in (a_1,\ldots,a_{k-1})$. And by inductive hypothesis on k we have

$$x \in (0 : a_1) \cap (a_1,\ldots,a_{k-1}) = 0 \quad .$$

Since the above argument does not depend on the order of the elements $a_1,\ldots,a_d$ we have completed the proof.

<u>(38.10) Definition</u>. Let R be a noetherian ring and $\mathfrak{a}$ an ideal of R such that $\dim R/\mathfrak{a} < \infty$. Let $\mathfrak{a} = \mathfrak{q}_1 \cap \ldots \cap \mathfrak{q}_n$ be a primary decomposition of $\mathfrak{a}$. We define

$$U(\mathfrak{a}) := \bigcap_{\dim R/\mathfrak{q}_i = \dim R/\mathfrak{a}} \mathfrak{q}_i$$

and call $U(\mathfrak{a})$ the unmixed part of $\mathfrak{a}$. Note that $U(\mathfrak{a})$ is independent of the primary decomposition. If $\mathfrak{a} = (a_1,\ldots,a_d)$ we denote $U(\mathfrak{a})$ by $U(a_1,\ldots,a_d)$.

<u>(38.11) Lemma</u>. Let $a_1,\ldots,a_d$ be a standard system of parameters of a generalized Cohen-Macaulay ring $(A,\mathfrak{m},k)$ and let $\mathfrak{q} = (a_1,\ldots,a_d)$. Then we have

$$U(a_1,\ldots,a_i) = (a_1,\ldots,a_i) : \mathfrak{q} = (a_1,\ldots,a_i) : a_{i+1} \quad \text{for} \quad 0 \leqq i < d \quad .$$

Proof. If $(a_1,\ldots,a_i) = U(a_1,\ldots,a_i)$ the assertion is clear because a_{i+1} is not in any $\mathfrak{p} \in \mathrm{Assh}_A(A/(a_1,\ldots,a_i))$.

Suppose that $(a_1,\ldots,a_i) \neq U(a_1,\ldots,a_i)$ and let $(a_1,\ldots,a_i) = U(a_1,\ldots,a_i) \cap \mathfrak{q}_1 \cap \ldots \cap \mathfrak{q}_r$ be a primary decomposition of $(a_1,\ldots,a_i)$, such that $\dim A/\mathfrak{q}_j < d-i$ for $j = 1,\ldots,r$. Hence there is an element

$$b \in \mathfrak{q}_1 \cap \ldots \cap \mathfrak{q}_r - \bigcup_{\mathfrak{p}\in \mathrm{Assh}(A/(a_1,\ldots,a_i))} \mathfrak{p}$$

such that $a_1,\ldots,a_i, b$ is a part of a system of parameters of A. Then we have

$$(a_1,\ldots,a_i) : b = U(a_1,\ldots,a_i) : b = U(a_1,\ldots,a_i) \quad .$$

By Corollary (37.7) we know that

$$\lambda_A\left(\frac{(a_1,\ldots,a_i):b}{(a_1,\ldots,a_i)}\right) < \infty \quad .$$

Hence there is a natural number n such that

$$H^0_{\mathfrak{m}}(A/(a_1,\ldots,a_i)) = \frac{(a_1,\ldots,a_i):\mathfrak{m}^n}{(a_1,\ldots,a_i)}$$

and such that

$$(a_1,\ldots,a_i) : b \subset (a_1,\ldots,a_i) : \mathfrak{m}^n \subset (a_1,\ldots,a_i) : b^n = U(a_1,\ldots,a_i) \,,$$

i.e. $$\frac{U(a_1,\ldots,a_i)}{(a_1,\ldots,a_i)} = H^0_{\mathfrak{m}}(A/(a_1,\ldots,a_i)) \quad .$$

Therefore by Proposition (38.3) we have

$$\mathfrak{q}\cdot U(a_1,\ldots,a_i) \subset (a_1,\ldots,a_i) \quad .$$

We may choose n so that $\mathfrak{m}^n \subset \mathfrak{q}$. Hence we get

$$U(a_1,\ldots,a_i) \subseteq (a_1,\ldots,a_i) : \mathfrak{q} \subseteq (a_1,\ldots,a_i) : \mathfrak{m}^n \subseteq U(a_1,\ldots,a_i) \,,$$

i.e. $U(a_1,\ldots,a_i) = (a_1,\ldots,a_i) : \mathfrak{q}$.

The second equality follows from Lemma (38.6), since $a_1,\ldots,a_d$ is a d-sequence.

(38.12) Definition. Let $a_1,\dots,a_n$ be elements of a ring A. For a non-empty subset Γ of $\{1,\dots,n\}$ we define

$$a_\Gamma = \prod_{i\in\Gamma} a_i \quad \text{and} \quad \mathfrak{q}_\Gamma = \sum_{i\in\Gamma} a_i A \quad ,$$

the ideal generated by the elements a_i, $i \in \Gamma$. For $\Gamma = \phi$ we define $a_\phi = 1$ and $\mathfrak{q}_\phi = \langle 0\rangle$.

The following result is fundamental in what follows.

(38.13) Theorem. Let $a_1,\dots,a_d$ be a standard system of parameters of a generalized Cohen-Macaulay ring A with $d = \dim A$. Then we have:

1) For any $0 \leq k < d$ and $n > 0$

$$U(a_1^n,\dots,a_k^n) = \sum_{\Gamma\subseteq\{1,\dots,k\}} a_\Gamma^{n-1} U(\mathfrak{q}_\Gamma)$$

2) For any $2 \leq k \leq d$ and $n > 0$ we have:

$$(a_1^{n+1},\dots,a_k^{n+1}) : a_1 \cdots a_k = \sum_{\Gamma\subsetneqq\{1,\dots,k\}} a_\Gamma^{n-1} U(\mathfrak{q}_\Gamma) \quad .$$

3) For any $2 \leq k < d$ we have:

$$U(a_1,\dots,a_k) \cap \left(\sum_{i=1}^{k} U(a_1,\dots,\check{a}_i,\dots,a_{k+1})\right) = \sum_{i=1}^{k} U(a_1,\dots,\check{a}_i,\dots,a_k) \quad .$$

3') For $k = 1$ we have:

$$U(a_1) \cap U(a_2) = (a_1 \cap a_2) + U(0) \quad .$$

Proof. 1) By induction on k: Let $k = 1$ and $x \in U(a_1^n)$. Then, by Lemma (38.11) we have $a_2 x \in (a_1^n)$. Writing $a_2 x = a_1^n y$ for some $y \in A$, we have $y \in (a_2) : a_1^n$.

Since a_2, a_1 is a d-sequence by Proposition (38.9), we get

$y \in (a_2) : a_1$. Hence $a_1 y = a_2 z$ for some $z \in (a_1) : a_2$. Therefore $a_2 x = a_1^{n-1} a_2 z$ which implies

$$x \in a_1^{n-1} U(a_1) + U(0) \quad ,$$

since $z \in (a_1) : a_2 = U(a_1)$ by (38.11).

The other inclusion is clear by Lemma (38.11). Let $k \geq 2$. For any $n > 0$ we know that $a_1, \ldots, a_{k-1}, a_k^n, a_{k+1}, \ldots, a_d$ is a standard system of parameters of A by Lemma (38.4). Hence the images $\overline{a}_1, \ldots, \overline{a}_{k-1}$ of $a_1, \ldots, a_{k-1}$ in $\overline{A} := A/a_k^n A$ form a part of a standard system of parameters of $\overline{A}$. By inductive hypothesis and Lemma (38.11) we have

$$U(a_1^n, \ldots, a_k^n)/(a_k^n) \cong U(\overline{a}_1^n, \ldots, \overline{a}_{k-1}^n) = \sum_{\Gamma \subseteq \{1, \ldots, k-1\}} \overline{a}_\Gamma^{n-1} U(\overline{\mathfrak{q}}_\Gamma)$$

where $\overline{\mathfrak{q}}_\Gamma = \mathfrak{q}_\Gamma + (a_k^n)/(a_k^n)$.

From this we see that

$$(*) \qquad U(a_1^n, \ldots, a_k^n) = \sum_{\Gamma \subseteq \{1, \ldots, k-1\}} a_\Gamma^{n-1} U(\mathfrak{q}_\Gamma, a_k^n) \quad .$$

From the case $k = 1$ we know that

$$U(\mathfrak{q}_\Gamma, a_k^n) = a_k^{n-1} U(\mathfrak{q}_\Gamma, a_k) + U(\mathfrak{q}_\Gamma) \quad .$$

And by (*) we have

$$U(a_1^n, \ldots, a_k^n) = \sum_{\Gamma \subseteq \{1, \ldots, k-1\}} a_\Gamma^{n-1} (a_k^{n-1} U(\mathfrak{q}_\Gamma, a_k) + U(\mathfrak{q}_\Gamma))$$

$$= \sum_{\Gamma \subseteq \{1, \ldots, k\}} a_\Gamma^n U(\mathfrak{q}_\Gamma) \quad .$$

For the proof of 2) we need the following lemma.

<u>(38.14) Lemma</u>. With the same assumption as in Theorem (38.13) we get for $1 \leq k \leq d$ and $n, \ell > 0$:

$$(a_1^{n+\ell}, \ldots, a_k^{n+\ell}) : (a_1 \ldots a_k)^\ell = (a_1^{n+1}, \ldots, a_k^{n+1}) : (a_1 \ldots a_k) \quad .$$

Proof. By induction on k : If $k = 1$ the assertion is clear. Let $k \geq 2$ and take $x \in (a_1^{n+\ell},\ldots,a_k^{n+\ell}) : (a_1\cdot\ldots\cdot a_k)^{\ell}$. Then

$$(a_1\cdot\ldots\cdot a_k)^{\ell}x = \sum_{i=1}^{k} a_i^{n+\ell}y_i \text{ , for some } y_i \in A \quad .$$

Since $a_1^{n+\ell},\ldots,a_{k-1}^{n+\ell}$, a_k^{ℓ} form a part of a standard system of parameters of A by Lemma (38.4) , we get

$$\begin{aligned}(a_1\cdot\ldots\cdot a_{k-1})^{\ell}x - a_k^n y_k &\in (a_1^{n+\ell},\ldots,a_{k-1}^{n+\ell}) : a_k^{\ell} \\ &= U(a_1^{n+\ell},\ldots,a_{k-1}^{n+\ell})\end{aligned}$$

by Lemma (38.11).

Again from Lemma (38.11) we have furthermore

$$a_k((a_1\cdot\ldots\cdot a_{k-1})^{\ell}x - a_k^n y_k) \in (a_1^{n+\ell},\ldots,a_{k-1}^{n+\ell}) \quad .$$

Passing to the ring $\bar{A} := A/a_k^{n+1}A$ we obtain

$$(\bar{a}_1\cdot\ldots\cdot\bar{a}_{k-1})^{\ell}\bar{a}_k\bar{x} \in (\bar{a}_1^{n+\ell},\ldots,\bar{a}_{k-1}^{n+\ell}) \quad .$$

By inductive hypothesis we know that

$$\bar{a}_k\bar{x} \in (\bar{a}_1^{n+1},\ldots,\bar{a}_{k-1}^{n+1}) : (\bar{a}_1\ldots\bar{a}_{k-1}) \quad .$$

Hence

$$a_k x \in (a_1^{n+1},\ldots,a_{k-1}^{n+1}, a_k^{n+1}) : (a_1\ldots a_{k-1}) ,$$

so that

$$x \in (a_1^{n+1},\ldots,a_k^{n+1}) : a_1\ldots a_k \text{ , proving Lemma (38.14)} \quad .$$

Now let us turn to the proof of (38.13). It is enough to show that

$$(a_1^{n+1},\ldots,a_k^{n+1}) : (a_1\ldots a_k) \subseteq \sum_{\Gamma \subsetneq \{1,\ldots,k\}} a_{\Gamma}^{n-1}U(\mathfrak{q}_{\Gamma}) \quad ,$$

because the other inclusion is clear by Lemma (38.11). We apply induction on k : Let $\underline{k = 2}$ and $x \in (a_1^{n+1}, a_2^{n+1}) : a_1 a_2$. Then

$$a_1 a_2 x - a_2^{n+1} y \in (a_1^{n+1})$$

for some $y \in A$. Hence

$$a_1 x - a_2^n y \in U(a_1^{n+1}) = a_1^n U(a_1) + U(0)$$

by 1). For a suitable $z \in U(a_1)$:

$$a_1 x - a_2^n y - a_1^n z \in U(0) \cap (a_1,\dots,a_d) = 0$$

by Lemma (38.11), Proposition (38.9) and Lemma (38.6). Hence

$$x - a_1^{n-1} z \in (a_2^n) : a_1 = U(a_2^n) = a_2^{n-1} U(a_2) + U(0) \quad ,$$

by 1). Since $z \in U(a_1)$, then:

$$x \in a_1^{n-1} U(a_1) + a_2^{n-1} U(a_2) + U(0) \quad ,$$

which completes the proof for $k = 2$.

Let $\underline{k \geq 3}$. Take any $x \in (a_1^{n+1},\dots,a_k^{n+1}) : (a_1 \dots a_k)$. Then for a suitable $y \in A$ we have by 1):

$$(a_1 \dots a_{k-1}) x - a_k^n y \in (a_1^{n+1},\dots,a_{k-1}^{n+1}) : a_k = U(a_1^{n+1},\dots,a_{k-1}^{n+1})$$

$$= \sum_{\Gamma \subseteq \{1,\dots,k-1\}} a_\Gamma^n U(\mathfrak{q}_\Gamma)$$

$$= (a_1 \cdots a_{k-1})^n U(a_1,\dots,a_{k-1}) + \sum_{\Gamma \subsetneqq \{1,\dots,k-1\}} a_\Gamma^n U(\mathfrak{q}_\Gamma)$$

But by inductive hypothesis we know that

$$\sum_{\Gamma \subsetneqq \{1,\dots,k-1\}} a_\Gamma^n U(\mathfrak{q}_\Gamma) = (a_1^{n+2},\dots,a_{k-1}^{n+2}) : (a_1 \dots a_{k-1}) \quad .$$

Hence, for some $z \in U(a_1,\dots,a_{k-1})$ we have:

$$(a_1 \dots a_{k-1}) x - a_k^n y - (a_1 \dots a_{k-1})^n z \in (a_1^{n+2},\dots,a_{k-1}^{n+2}) : (a_1 \dots a_{k-1}) \quad .$$

Passing to the ring $\overline{A} := A/a_k^n A$ we get

$$(\bar{a}_1\dots\bar{a}_{k-1})^2(\bar{x}-(\bar{a}_1\dots\bar{a}_{k-1})^{n-1}\bar{z}) \in (\bar{a}_1^{n+2},\dots,\bar{a}_{k-1}^{n+2}) \quad .$$

By Lemma (38.14) and inductive hypothesis one can see

$$\begin{aligned}(I) : \bar{x}-(\bar{a}_1\dots\bar{a}_{k-1})^{n-1}\bar{z} &\in (\bar{a}_1^{n+2},\dots,\bar{a}_{k-1}^{n+2}) : (\bar{a}_1\dots\bar{a}_{k-1})^2 \\ &= (\bar{a}_1^{n+1},\dots,\bar{a}_{k-1}^{n+1}) : (\bar{a}_1\dots\bar{a}_k) \\ &= \sum_{\Gamma\subsetneqq\{1,\dots,k-1\}} \bar{a}_\Gamma^{n-1}U(\bar{\mathfrak{q}}_\Gamma) \quad .\end{aligned}$$

Since we know by 1) that

$$U(\bar{\mathfrak{q}}_\Gamma) \cong U(\mathfrak{q}_\Gamma,a_k^n)/(a_k^n) = \frac{a_k^{n-1}U(\mathfrak{q}_\Gamma,a_k)+U(\mathfrak{q}_\Gamma)}{(a_k^n)}$$

we see from (I) that

$$\begin{aligned}x-(a_1\dots a_{k-1})^{n-1}z &\in \sum_{\Gamma\subsetneqq\{1,\dots,k-1\}} (a_k^{n-1}a_\Gamma^{n-1}U(\mathfrak{q}_\Gamma,a_k)+a_\Gamma^{n-1}U(\mathfrak{q}_\Gamma)) \\ &= \sum_{\Gamma\subsetneqq\{1,\dots,k\}} a_\Gamma^{n-1}U(\mathfrak{q}_\Gamma) \quad .\end{aligned}$$

Since $z\in U(a_1,\dots,a_{k-1})$, then

$$x \in \sum_{\Gamma\subsetneqq\{1,\dots,k\}} a_\Gamma^{n-1}U(\mathfrak{q}_\Gamma) \quad .$$

This completes the proof of statement 2 of (38.13).

3): Let $\bar{A} := A/a_{k+1}A$. Then we know by 2)

$$\sum_{i=1}^{k} U(a_1,\dots,\check{a}_i,\dots,a_{k+1})/(a_{k+1}) \cong (\bar{a}_1^2,\dots,\bar{a}_k^2) : (\bar{a}_1\dots\bar{a}_k)$$

Hence for any

$$x\in U(a_1,\dots,a_k) \cap (\sum_{i=1}^{k} U(a_1,\dots,\check{a}_i,\dots,a_{k+1}))$$

we have

$$a_1 \ldots a_k x \in (a_1^2,\ldots,a_k^2) + (a_{k+1}) \quad .$$

Since $x \in U(a_1,\ldots,a_k)$, then $a_{k+1}x \in (a_1,\ldots,a_k)$ by Lemma (38.11). Therefore

$$\text{(II):} \quad a_1 \ldots a_{k+1} x \in a_1 \ldots a_k (a_1,\ldots,a_k) \cap (a_{k+1}(a_1^2,\ldots,a_k^2) + (a_{k+1}^2))$$

$$\subseteq (a_1^2,\ldots,a_k^2) \cap (a_{k+1}(a_1^2,\ldots,a_k^2) + (a_{k+1}^2))$$

$$= a_{k+1}(a_1^2,\ldots,a_k^2) + (a_1^2,\ldots,a_k^2) \cap (a_{k+1}^2) \quad .$$

But, by Lemma (38.4) and Lemma (38.11) we have

$$(a_1^2,\ldots,a_k^2) \cap (a_{k+1}^2) = a_{k+1}^2((a_1^2,\ldots,a_k^2) : a_{k+1}^2)$$

$$= a_{k+1}^2 U(a_1^2,\ldots,a_k^2) \subseteq a_{k+1}(a_1^2,\ldots,a_k^2) \quad .$$

So from (II) we conclude that

$$a_1 \ldots a_k a_{k+1} x \in a_{k+1}(a_1^2,\ldots,a_k^2) \quad .$$

Therefore

$$x \in (a_1^2,\ldots,a_k^2) : a_1 \ldots a_k = \sum_{i=1}^{k} U(a_1,\ldots,\check{a}_i,\ldots,a_k)$$

by 2).

3') $(k = 1)$: We only have to show "$\subseteq$" : Let $x \in U(a_1) \cap U(a_2)$. Then there are $y,z \in A$ with $xa_2 = ya_1$ and $xa_1 = za_2$, i.e. $ya_1^2 = za_2^2$. Hence $y \in (a_2^2 : a_1^2) = (a_2^2 : a_1)$, since a_1, a_2^2 is a d-sequence. From this we can check

$$xa_2 = ya_1 \in (a_2^2) \cap (a_1) = a_2^2(a_1 : a_2^2) = a_2^2(a_1 : a_2) = a_2(a_1 \cap a_2) \quad ,$$

i.e. $x \in (a_1 \cap a_2) + U(0)$, q.e.d. (38.13).

§ 39. The computation of local cohomology of generalized Cohen-Macaulay rings

This section is mainly an application of Theorem (38.13). Throughout this section we fix the following notations:

1) $(A,\mathfrak{m},k)$ denotes a generalized Cohen-Macaulay local ring and $d = \dim A$;

2) $a_1,\ldots,a_d$ is a standard system of parameters of A ;

3) $\mathfrak{q}_i = (a_1,\ldots,a_i)$, $0 \leq i < d$, $\mathfrak{q} = (a_1,\ldots,a_d)$;

4) $C^{\bullet}(a_1,\ldots,a_i;A)$ is the Čech complex with respect to $a_1,\ldots,a_i$ (cf. Chapter VII, Definition (35.15)) .

Recall that for all $0 \leq i \leq j \leq d$ we have by VII, Theorem (35.18)

$$H^i(C^{\bullet}(a_1,\ldots,a_j;A)) \cong \varinjlim_n H^i(a_1^n,\ldots,a_j^n;A) \cong H^i_{\mathfrak{q}_j}(A) \quad .$$

We denote the complex $C^{\bullet}(a_1,\ldots,a_i;A) \otimes_A A_{a_{i+1}}$ by $C^{\bullet}(a_1,\ldots,a_i;A)_{a_{i+1}}$.

Then, by Chapter VII, Lemma (35.17), there is an exact sequence

$$(*) \quad \longrightarrow H^{i-1}_{\mathfrak{q}_j}(A) \xrightarrow{\psi^{i-1}} H^{i-1}_{\mathfrak{q}_j}(A)_{a_{j+1}} \longrightarrow H^i_{\mathfrak{q}_{j+1}}(A) \longrightarrow H^i_{\mathfrak{q}_j}(A) \xrightarrow{\psi^i} H^i_{\mathfrak{q}_j}(A)_{a_{j+1}} \longrightarrow$$

where ψ^i is the canonical map of localization.

We are going to compute $H^i_{\mathfrak{m}}(A)$ for $i < d$ in terms of $a_1,\ldots,a_d$.

<u>(39.1) Theorem</u>. Let $a_1,\ldots,a_d$ be a standard system of parameters in a generalized Cohen-Macaulay ring $(A,\mathfrak{m},k)$ with $d = \dim A > 0$. Then we have

$$H^0_{\mathfrak{m}}(A) = (0 : \mathfrak{q}) \quad \text{and} \quad H^i_{\mathfrak{m}}(A) \cong U(a_1,\ldots,a_i)/(a_1^2,\ldots,a_i^2) : (a_1\ldots a_i)$$

for $0 < i < d$.

Proof. By induction on j we will show that

$$H^0_{\mathfrak{q}_j}(A) \cong (0 : \mathfrak{q}) \quad \text{and}$$

$$H^i_{\mathfrak{q}_j}(A) \cong U(a_1,\dots,a_i)/(a_1^2,\dots,a_i^2) : (a_1 \dots a_i)$$

for $0 < i < j$.

Since $H^i_{\mathfrak{q}}(A) \cong H^i_{\mathfrak{m}}(A)$, this will prove the theorem.

Let $\underline{j = 1}$. Then $H^0_{a_1A}(A) = (0 : a_1^n)$ for some $n > 0$. Since a_1 is a part of a standard system of parameters of A , then $(0 : a_1^n) = (0 : a_1) = (0 : \mathfrak{q})$, by Lemma (38.6) and the fact that $a_1,\dots,a_d$ is a d-sequence. Hence $H^0_{a_1A}(A) = (0 : \mathfrak{q})$.

Let $\underline{j > 1}$ and assume that $H^0_{\mathfrak{q}_{j-1}}(A) = (0 : \mathfrak{q})$ and

$$H^i_{\mathfrak{q}_{j-1}}(A) = U(a_1,\dots,a_i)/(a_1^2,\dots,a_i^2) : (a_1 \dots a_i) \quad \text{for } i < j-1 .$$

Since $\mathfrak{q}_i \subset (a_1^2,\dots,a_i^2) : (a_1 \dots a_i)$ and since $\lambda_A(U(a_1,\dots,a_i)/\mathfrak{q}_i) < \infty$ for $i < d$ we know from the assumption that $\lambda_A(H^i_{\mathfrak{q}_{j-1}}(A)) < \infty$ for $i < j-1$.

By the exact sequence (*), just before Theorem (39.1), we have $H^i_{\mathfrak{q}_j}(A) \cong H^i_{\mathfrak{q}_{j-1}}(A)$ for $i < j-1$. - Hence it remains to be shown that

$$H^{j-1}_{\mathfrak{q}_j}(A) \cong U(a_1,\dots,a_{j-1})/(a_1^2,\dots,a_{j-1}^2) : (a_1 \dots a_{j-1}) .$$

By (*) we have an exact sequence

$$0 \longrightarrow H^{j-1}_{\mathfrak{q}_j}(A) \longrightarrow H^{j-1}_{\mathfrak{q}_{j-1}}(A) \xrightarrow{\psi^{j-1}} H^{j-1}_{\mathfrak{q}_{j-1}}(A)_{a_j} ,$$

where ψ^{j-1} is the localization by a_j . Hence $H^{j-1}_{\mathfrak{q}_j}(A) \cong \ker \psi^{j-1}$.

Recall that $\varinjlim_n (A/(a_1^n,\dots,a_{j-1}^n)) \cong H^{j-1}_{\mathfrak{q}_{j-1}}(A)$ by Chapter VII, Corollary (35.21). Let $\varphi_n : A/(a_1^n,\dots,a_{j-1}^n) \longrightarrow H^{j-1}_{\mathfrak{q}_{j-1}}(A)$ be the canonical map. We claim that

$$\ker \varphi_n = \frac{(a_1^{n+1},\dots,a_{j-1}^{n+1}) : (a_1 \dots a_{j-1})}{(a_1^n,\dots,a_{j-1}^n)} \quad \text{for all } n > 0 .$$

In fact take any element

$$\bar{x} = x \bmod (a_1^n,\ldots,a_{j-1}^n) \in A/(a_1^n,\ldots,a_{j-1}^n)$$

such that $\varphi_n(\bar{x}) = 0$. Then by the properties of the direct system $\{A/(a_1^n,\ldots,a_{j-1}^n)\}_{n\geqq 0}$ we have

$$(a_1\ldots a_{j-1})^{m-n}x \in (a_1^m,\ldots,a_{j-1}^m)$$

for some $m > n$, cf. Chapter VII, Corollary (35.21). By Lemma (38.14) we know that

$$x \in (a_1^{n+1},\ldots,a_{j-1}^{n+1}) : (a_1\ldots a_{j-1}) \quad , \text{ hence}$$

$$\ker \varphi_n \subseteq \frac{(a_1^{n+1},\ldots,a_{j-1}^{n+1}) : (a_1\ldots a_{j-1})}{(a_1^n,\ldots,a_{j-1}^n)} \quad .$$

The other inclusion is clear. This proves the claim. Now let $\alpha \in H^{j-1}_{\mathfrak{q}_{j-1}}(A)$. Suppose that α is represented by

$$\bar{z} = z \bmod (a_1^n,\ldots,a_{j-1}^n) \in A/(a_1^n,\ldots,a_{j-1}^n) \quad \text{for some} \quad n > 0 \quad .$$

Consider the commutative diagram for $n < m$:

$$\begin{array}{ccc} A/(a_1^n,\ldots,a_{j-1}^n) & \xrightarrow{\eta_n} & (A/(a_1^n,\ldots,a_{j-1}^n))_{a_j} \\ \downarrow{\scriptstyle \varphi_{m,n}} & & \downarrow{\scriptstyle \bar{\varphi}_{m,n}} \\ A/(a_1^m,\ldots,a_{j-1}^m) & \xrightarrow{\eta_m} & (A/(a_1^m,\ldots,a_{j-1}^m))_{a_j} \end{array} \quad ,$$

where η_m , η_n are the localizations, and $\varphi_{m,n} , \bar{\varphi}_{m,n}$ are the same as in Chapter VII, Corollary (35.21).

Then $\alpha \in \operatorname{Ker} \psi^{j-1}$ if and only if $\eta_m \cdot \varphi_{m,n}(\bar{z}) = 0$ for some $m > n$. If $\eta_m \cdot \varphi_{m,n}(\bar{z}) = 0$ there is an integer $\ell > 0$ such that

$$a_j^{\ell}(a_1\ldots a_{j-1})^{m-n}z \in (a_1^m,\ldots,a_{j-1}^m) \quad .$$

Then by Theorem (38.13) we have

$$(a_1\ldots a_{j-1})^{m-n}z \in (a_1^m,\ldots,a_{j-1}^m) : a_j^{\ell} = U(a_1^m,\ldots,a_{j-1}^m)$$

$$= (a_1\ldots a_{j-1})^{m-1}U(a_1,\ldots,a_{j-1}) + ((a_1^{m+1},\ldots,a_{j-1}^{m+1}) : a_1\ldots a_{j-1}) \quad .$$

Hence one can choose an element $y \in U(a_1,\ldots,a_{j-1})$ so that

$$(a_1\ldots a_{j-1})^{m-n}(z-(a_1\ldots a_{j-1})^{n-1}y) \in (a_1^{m+1},\ldots,a_{j-1}^{m+1}) : a_1\ldots a_{j-1} \quad .$$

By Lemma (38.14) we get

$$(**) \quad \begin{aligned} z-(a_1\ldots a_{j-1})^{n-1}y &\in (a_1^{m+1},\ldots,a_{j-1}^{m+1}) : (a_1\ldots a_{j-1})^{m-n+1} \\ &= (a_1^{n+1},\ldots,a_{j-1}^{n+1}) : (a_1\ldots a_{j-1}) \quad . \end{aligned}$$

Since $\ker\varphi_n = (a_1^{n+1},\ldots,a_{j-1}^{n+1}) : (a_1\ldots a_{j-1})/(a_1^n,\ldots,a_{j-1}^n)$, we know that $\alpha \in \ker\psi^{j-1}$ is represented by some $y \bmod (a_1,\ldots,a_{j-1}) \in A/(a_1,\ldots,a_{j-1})$ such that $y \in U(a_1,\ldots,a_{j-1})$. We have shown that

$$\ker\psi^{j-1} \subseteq \varphi_1(U(a_1,\ldots,a_{j-1})/(a_1,\ldots,a_{j-1})) \quad .$$

The other inclusion is trivially true. Hence

$$\ker\psi^{j-1} = \varphi_1(U(a_1,\ldots,a_{j-1})/(a_1,\ldots,a_{j-1}))$$

$$\cong \frac{U(a_1,\ldots,a_{j-1})}{(a_1^2,\ldots,a_{j-1}^2) : (a_1\ldots a_{j-1})} \quad , \qquad \text{q.e.d.}$$

§ 40. Blowing up of a standard system of parameters

In this section we keep the notations of § 39. We ask first of all for the Cohen-Macaulay property of $Bl(\mathfrak{q},A)$ and we want to compute the local cohomology of the associated graded ring $G = \bigoplus_{n\geq 0}\mathfrak{q}^n/\mathfrak{q}^{n+1}$ with respect to an ideal generated by a standard system of parameters of A .

We start with an auxiliary lemma, (see also (18.23) and (12.6)).

(40.1) Lemma. Let $a_1,\dots,a_d$ be a system of parameters of a local ring $(A,\mathfrak{m})$ with $d = \dim A > 0$ and $\mathfrak{q} = (a_1,\dots,a_d)$. Then for any closed point $x \in X := Bl(\mathfrak{q},A)$, $\dim \mathcal{O}_{X,x} = d$.

Proof. We assume that $x \in X = Bl(\mathfrak{q},A)$ corresponds to a maximal ideal $\mathfrak{n}$ of $B := A\left[\frac{\mathfrak{q}}{a_i}\right]$ for some i. From the analytic independence of a system of parameters we conclude that

$$(B/a_iB)_{red} = (B/\mathfrak{q}B)_{red} \cong A/\mathfrak{m}[T_1,\dots,T_{i-1},T_{i+1},\dots,T_d]$$

where the T_j are indeterminates over $A/\mathfrak{m}$.

Hence every maximal ideal of B/a_iB has a height $d-1$. Since $X \longrightarrow \mathrm{Spec}(A)$ is proper, we see that $\mathfrak{n} \cap A = \mathfrak{m}$ and $\mathfrak{n} \supset a_iB$. But a_i is a non-zero-divisor of B, so

$$\dim \mathcal{O}_{X,x} = \dim B_{\mathfrak{n}} = \dim B_{\mathfrak{n}}/a_iB_{\mathfrak{n}} + 1 = d \quad .$$

(40.2) Lemma. Let $a_1,\dots,a_d$ be a standard system of parameters of a generalized Cohen-Macaulay ring. Then for $1 \leqq i \leqq d$ and for all $n > 0$ and $\mathfrak{q} = (a_1,\dots,a_d)$ we get

$$(a_1,\dots,a_i) \cap \mathfrak{q}^n = (a_1,\dots,a_i)\mathfrak{q}^{n-1} \quad .$$

Proof. We apply descending induction on i and ascending induction on n. If $i = d$ or if $n = 1$ the conclusion is clear. Let $i < d$. It is enough to show that

$$(a_1,\dots,a_i) \cap \mathfrak{q}^n \subset (a_1,\dots,a_i)\mathfrak{q}^{n-1} \quad .$$

Let $x \in (a_1,\dots,a_i) \cap \mathfrak{q}^n$. Then by inductive hypothesis on i we have

$$x \in (a_1,\dots,a_{i+1}) \cap \mathfrak{q}^n = (a_1,\dots,a_{i+1})\mathfrak{q}^{n-1} \quad .$$

Hence one can write $x = \sum_{j=1}^{i+1} a_j y_j$ with $y_j \in \mathfrak{q}^{n-1}$. Since we may assume $n > 1$, we obtain by inductive hypothesis on n (using Lemma (38.6) and Proposition (38.9)):

$$y_{i+1} \in ((a_1,\dots,a_i) : a_{i+1}) \cap \mathfrak{q}^{n-1} = (a_1,\dots,a_i) \cap \mathfrak{q}^{n-1}$$

$$= (a_1,\dots,a_i)\mathfrak{q}^{n-2} \quad .$$

And therefore

$$x \in (a_1,\dots,a_i)\mathfrak{q}^{n-1} + a_{i+1}(a_1,\dots,a_i)\mathfrak{q}^{n-2} \subseteq (a_1,\dots,a_i)\mathfrak{q}^{n-1} \quad .$$

Now we can prove that the blowing up $Bl(\mathfrak{q},A)$ of a generalized Cohen-Macaulay ring with respect to a standard system of parameters of A is always Cohen-Macaulay.

(40.3) Proposition. Let $(A,\mathfrak{m},k)$ be a generalized Cohen-Macaulay ring, $a_1,\dots,a_d$ a standard system of parameters of A and $\mathfrak{q} = (a_1,\dots,a_d)$. Then $Bl(\mathfrak{q},A)$ is Cohen-Macaulay.

Proof. By [20], Chapter 0, (10.3.1) we know that there is a flat extension $\bar{A}$ of A such that $\bar{A}/\mathfrak{m}\bar{A} = \bar{k}$, the algebraic closure of k , and such that $\bar{A}$ is a noetherian local ring. Since

$$\ell_{\bar{A}}(\bar{A}/\mathfrak{q}\bar{A}) - e(\mathfrak{q}\bar{A};\bar{A}) = \ell_A(A/\mathfrak{q}) - e(\mathfrak{q};A)$$

and since

$$H^i_{\mathfrak{m}\bar{A}}(\bar{A}) \cong H^i_{\mathfrak{m}}(A) \otimes_A \bar{A}$$

clearly $I(A) = I(\bar{A})$. Therefore $a_1,\dots,a_d$ form a standard system of parameters of $\bar{A}$.

Moreover the induced morphism $Bl(\mathfrak{q}\bar{A},\bar{A}) \longrightarrow Bl(\mathfrak{q},A)$ is flat. Hence it is enough to show that $Bl(\mathfrak{q}\bar{A},\bar{A})$ is Cohen-Macaulay. We may assume therefore that k is algebraically closed.

Let $P \in Bl(\mathfrak{q},A)$ be a closed point and suppose that P corresponds to a maximal ideal $\mathfrak{n}$ of $B = A\left[\frac{a_1}{a_d},\dots,\frac{a_{d-1}}{a_d}\right]$. Since $Bl(\mathfrak{q},A) \longrightarrow Spec(A)$ is proper we have $\mathfrak{n} \cap A = \mathfrak{m}$, hence in particular $\mathfrak{n} \supset a_dB$. Moreover we know that the maximal ideal $\mathfrak{n}/\sqrt{a_dB}$ of

$B/\sqrt{a_dB} = (B/a_dB)_{red} \cong k\left[\frac{\bar a_1}{a_d},\dots,\frac{\bar a_{d-1}}{a_d}\right]$ can be generated by the elements of the form $\frac{\bar a_1}{a_d} - \alpha_1,\dots,\frac{\bar a_{d-1}}{a_d} - \alpha_{k-1}$, where $\frac{\bar a_i}{a_d}$ is the image of $\frac{a_i}{a_d}$ in $B/\sqrt{a_dB}$ and $\alpha_i \in k = \bar k$.
Therefore $\frac{a_1}{a_d} - r_1,\dots,\frac{a_{d-1}}{a_d} - r_{d-1}$, a_d form a system of parameters of $B_{\mathfrak{n}}$ for some $r_i \in A$ since $\dim B_{\mathfrak{n}} = d$ by Lemma (40.1). Note that $a_1 - r_1a_d,\dots,a_{d-1} - r_{d-1}a_d, a_d$ is a standard system of parameters of A and that

$$A\left[\frac{a_1}{a_d},\dots,\frac{a_{d-1}}{a_d}\right] = A\left[\frac{a_1 - r_1a_d}{a_d},\dots,\frac{a_{d-1} - r_{d-1}a_d}{a_d}\right] .$$

Hence it is enough to show that for any standard system $a_1,\dots,a_d$ the elements $\frac{a_1}{a_d},\dots,\frac{a_{d-1}}{a_d}$, a_d form a regular sequence of $A\left[\frac{a_1}{a_d},\dots,\frac{a_{d-1}}{a_d}\right]$:

Let $x \in \left(\frac{a_1}{a_d},\dots,\frac{a_i}{a_d}\right)B : \frac{a_{i+1}}{a_d}$ for $0 \leqq i < d-1$. Then x can be written in the form $x = \frac{y}{a_d^n}$ for some $n > 0$ and $y \in \mathfrak{q}^n$.

We can choose an integer $m \geqq n$ so that

$$\frac{a_{i+1}}{a_d}\cdot\frac{y}{a_d^n} = \sum_{j=1}^{i}\frac{a_j}{a_d}\cdot\frac{y_j}{a_d^m}\ ; \qquad y_j \in \mathfrak{q}^m .$$

This relation holds at the localization A_{a_d} too, i.e. there is an $\ell > 0$ such that

$$a_d^{\ell}(a_{i+1}a_d^{m-n}y - \sum_{j=1}^{i}a_jy_j) = 0 .$$

This implies

$$a_{i+1}a_d^{m-n}y - \sum_{j=1}^{i}a_jy_j \in (0 : a_d^{\ell}) \cap \mathfrak{q} = 0 .$$

Then, by Lemma (38.6) and Lemma (38.11):

$$y \in [(a_1,\dots,a_i) : a_d^{m-n} a_{i+1}] \cap \mathfrak{q}^n \subseteq U(a_1,\dots,a_i) \cap \mathfrak{q}^n$$

$$= (a_1,\dots,a_i) \cap \mathfrak{q}^n \quad ,$$

and from Lemma (40.2) we conclude that $y \in (a_1,\dots,a_i)\mathfrak{q}^{n-1}$. This shows that $x = \frac{y}{a_d^n} \in \left(\frac{a_1}{a_d},\dots,\frac{a_i}{a_d}\right) B$.

Now one checks with similar methods as before that

$$\left(\frac{a_1}{a_d},\dots,\frac{a_{d-1}}{a_d}\right) B : a_d = \left(\frac{a_1}{a_d},\dots,\frac{a_{d-1}}{a_d}\right) B \quad .$$

<u>(40.4) Proposition</u>. Let $G = \bigoplus_{n \geq 0} \mathfrak{q}^n/\mathfrak{q}^{n+1}$ be the associated graded ring of a generalized Cohen-Macaulay ring $(A,\mathfrak{m},k)$ with respect to a parameter ideal $\mathfrak{q} = (a_1,\dots,a_d)$, generated by a standard system of parameters of A . Then $\lambda_G(\underline{H}^i_{\mathfrak{M}}(G)) < \infty$ for $i < d$, where $\mathfrak{M}$ is the maximal homogeneous ideal of G .

Proof. Note that G is a graded ring defined over an artinian local ring. Since $A/\mathfrak{q}$ is complete, G is a homomorphic image of a regular graded ring S defined over a regular local ring. Then, by Theorem (36.8) and Corollary (36.11) we have

$$\underline{\mathrm{Hom}}_S(\underline{H}^i_{\mathfrak{M}}(G), \underline{E}_S) \cong \underline{\mathrm{Ext}}_S^{s-i}(G,S)(m)$$

for some $m \in \mathbf{Z}$, where $s = \dim S$. Hence it is enough to show that $\underline{\mathrm{Ext}}_S^{s-i}(G,S)$ has finite length for $i < d$. By Corollary (36.16) we have to prove the following statements:

1) For any homogeneous prime $\mathfrak{p} \in \mathrm{Spec}(G) - \{\mathfrak{M}\}$ we have $\dim(G) = \dim(G/\mathfrak{p}) + \mathrm{ht}(\mathfrak{p})$.

2) $G_{\mathfrak{p}}$ is Cohen-Macaulay for any homogeneous $\mathfrak{p} \in \mathrm{Spec}(G) - \{\mathfrak{M}\}$.

Statement 1) is true because A is quasi-unmixed and so we get G is locally quasi-unmixed by (18.24).

For 2). By Proposition (40.3) we know $B\ell(\mathfrak{q},A)$ is Cohen-Macaulay, so that $G_{\mathfrak{p}}$ is Cohen-Macaulay for all $\mathfrak{p} \in \mathrm{Proj}(G) = \{\mathfrak{p} \in \mathrm{Spec}(G) :$ $\mathfrak{p}$ is homogeneous and $\mathfrak{p} \neq \mathfrak{M}\}$.

<u>(40.5) Remark</u>. In general it is not true that the Cohen-Macaulayness of the blowing up $\mathrm{Bl}(I,A)$ of an $\mathfrak{m}$-primary ideal I in a local ring $(A,\mathfrak{m})$ implies the finiteness of the local cohomology of the associated graded ring $G(I,A) = \bigoplus_{n\geq 0} I^n/I^{n+1}$. For example, let k be a field and let X,Y,Z be indeterminates over k and let $A = k[[X,Y,Z]]/((X) \cap (y,Z))$. Then $G(\mathfrak{m};A) \cong k[X,Y,Z]/((X) \cap (Y,Z)) =$ $= k[x,y,z]$. It is easy to verify that the rings $G(\mathfrak{m};A)_x$, $G(\mathfrak{m};A)_y$ and $G(\mathfrak{m};A)_z$ are Cohen-Macaulay, in particular $\mathrm{Bl}(\mathfrak{m};A)$ is Cohen-Macaulay. Let $G = G(\mathfrak{m};A)$ and $\mathfrak{M}$ the maximal homogeneous ideal of $k[X,Y,Z]$, i.e. $\mathfrak{M} = (X,Y,Z)$. From the exact sequence

$$0 \longrightarrow G \longrightarrow k[X,Y,Z]/(X) \oplus k[X,Y,Z]/(Y,Z) \longrightarrow k \longrightarrow 0$$

we get an exact sequence

$$0 \longrightarrow k \longrightarrow \underline{H}^i_{\mathfrak{M}}(G) \longrightarrow \underline{H}^i_{\mathfrak{M}}(k[X,Y,Z]/(Y,Z)) \longrightarrow 0$$
$$\underline{H}^i_{\mathfrak{M}}(k[X,Y,Z]/(Y,Z)) \cong \underline{H}^1_{Xk[X]}(k[X]) \quad .$$

Since the Čech-complex $C^{\bullet}(X;k[X])$ is given by

$$0 \longrightarrow k[X] \longrightarrow k[X,X^{-1}] \longrightarrow 0 \quad ,$$

the cohomology $\underline{H}^1_{Xk[X]}(k[X])$ is isomorphic to $X^{-1}k[X^{-1}]$, which is not of finite dimension as a k-vector space. Hence $\underline{H}^1_{\mathfrak{M}}(G)$ is not of finite length.

To compute explicitly the local cohomology of the Rees algebra and the associated graded ring of a standard parameter ideal in a generalized Cohen-Macaulay ring we need the following four technical lemmas.

(40.6) Lemma. Let $A, a_1, \ldots, a_d$, $\mathfrak{q}$ and G be the same as in Proposition (40.4). Then for $0 \leqq i < d$ we have for the initial froms a_i^* :

$$\left[(a_1^*, \ldots, a_i^*) : a_{i+1}^* / (a_1^*, \ldots, a_i^*)\right]_n = \begin{cases} U(a_1, \ldots, a_i) + \mathfrak{q}/\mathfrak{q} & \text{for } n = 0 \\ 0 & \text{otherwise.} \end{cases}$$

Proof. This is an immediate consequence of Lemma (40.2). So we omit the proof.

(40.7) Lemma. For any standard system of parameters $a_1, \ldots, a_d$ of A, we have

$$U(a_1, \ldots, a_i) \cap U(a_d) \subset (a_d) + U(0) \quad \text{for} \quad 1 \leqq i \leqq d \quad .$$

Proof. This is an easy application of Theorem (38.13).

(40.8) Lemma. Let $A, a_1, \ldots, a_d$ and $\mathfrak{q}$ be as in Proposition (40.4). Then, for $n \geqq 2$, we have

$$\left(\sum_{\substack{\Gamma \subsetneqq \{1, \ldots, d\} \\ |\Gamma| < n}} a_\Gamma U(\mathfrak{q}_\Gamma)\right) \cap \mathfrak{q}^n \subset (a_1^2, \ldots, a_d^2) \quad .$$

(See (38.12) for notation).

Proof. By induction on n. Let $n = 2$. Then we have to show that

$$\left(\sum_{i=1}^{d} a_i U(a_i) + U(0)\right) \cap \mathfrak{q}^2 \subset (a_1^2, \ldots, a_d^2) \quad .$$

Take any $x \in \left(\sum_{i=1}^{d} a_i U(a_i) + U(0)\right) \cap \mathfrak{q}^2$ and write $x = \sum_{i=1}^{d} a_i u_i + u$ with

$u_i \in U(a_i)$ and $u \in U(0)$. Since $u \in U(0) \cap \mathfrak{q} = 0$, we have

$$x = \sum_{i=1}^{d} a_i u_i \in (a_1,\ldots,a_{d-1})\mathfrak{q} + (a_d^2) = \mathfrak{q}^2 \quad .$$

This implies that for some $y \in A$

$$u_d - a_d y \in ((a_1,\ldots,a_{d-1}) : a_d) \cap U(a_d) = U(a_1,\ldots,a_{d-1}) \cap U(a_d)$$

$$\subset (a_d) + U(0) \quad , \text{ by Lemma (40.7).}$$

Therefore $a_d u_d \in (a_d^2)$. Similarly one can prove that $a_i u_i \in (a_i^2)$ for all $1 \leqq i \leqq d$.

Let $n \geqq 3$. Assume that the assertion is true for $n-1$. Take

$$x \in \left(\sum_{\substack{\Gamma \subsetneqq \{1,\ldots,d\} \\ |\Gamma| < n}} a_\Gamma U(\mathfrak{q}_\Gamma) \right) \cap \mathfrak{q}^n \quad .$$

Then we can write

$$x = \sum_{\substack{\Gamma \subsetneqq \{1,\ldots,d\} \\ |\Gamma| = n-1}} a_\Gamma u_\Gamma + w \quad ,$$

where $u_\Gamma \in U(\mathfrak{q}_\Gamma)$ and $w \in \sum_{\substack{\Gamma \subsetneqq \{1,\ldots,d\} \\ |\Gamma| < n-1}} a_\Gamma U(\mathfrak{q}_\Gamma)$.

By inductive hypothesis we get

$$w \in \left(\sum_{\substack{\Gamma \subsetneqq \{1,\ldots,d\} \\ |\Gamma| < n-1}} a_\Gamma U(\mathfrak{q}_\Gamma) \right) \cap \mathfrak{q}^{n-1}$$

$$\subseteq (a_1^2,\ldots,a_d^2) \quad , \text{ since } n-1 \geqq 2 \quad .$$

Note that if $n > d$ then $\mathfrak{q}^n \subset (a_1^2,\ldots,a_d^2)$, which allows us to assume $n \leqq d$.

Now we fix $\Gamma \subset \{1,\ldots,d\}$ such that $|\Gamma| = n-1$. After a renumbering we may assume $\Gamma = \{1,\ldots,n-1\}$. Then

$$a_\Gamma u_\Gamma \in (a_1^2,\dots,a_{n-1}^2,a_n,\dots,a_d)$$

and

$$a_n a_\Gamma u_\Gamma \in a_\Gamma(a_1,\dots,a_{n-1}) \subset (a_1^2,\dots,a_{n-1}^2) \quad .$$

Therefore by Lemmata (38.4), (38.6) and Proposition (38.9)

$$a_\Gamma u_\Gamma \in (a_1^2,\dots,a_{n-1}^2,a_n,\dots,a_d) \cap U(a_1^2,\dots,a_{n-1}^2) = (a_1^2,\dots,a_{n-1}^2) \quad .$$

<u>(40.9) Lemma</u>. Let A, $a_1,\dots,a_d$ and $\mathfrak{q}$ be as in Proposition (40.4) for any $n > 1$ we have

$$(a_1^2,\dots,a_d^2) \cap \mathfrak{q}^n = (a_1^2,\dots,a_d^2)\mathfrak{q}^{n-2} \quad .$$

Proof. By induction on $d = \dim A$. Clearly we may assume $n \geqq 3$.

If $d = 1$, then $(a_1^2) \cap \mathfrak{q}^n \subset (a_1) \cap \mathfrak{q}^n = a_1\mathfrak{q}^{n-1}$ by Lemma (40.2). Hence $x \in (a_1^2) \cap \mathfrak{q}^n$ can be written in the form $x = a_1 y$, $y \in \mathfrak{q}^{n-1}$. So $a_1 y \in (a_1^2)$, and

$$y \in ((a_1) + U(0)) \cap \mathfrak{q}^{n-1} = (a_1) \cap \mathfrak{q}^{n-1} = a_1\mathfrak{q}^{n-2}$$

by Lemma (40.2), i.e. $x = a_1 y \in a_1^2\mathfrak{q}^{n-2}$

Let $d \geqq 2$. Define $\bar{A} := A/a_dA$. Then

$$(\bar{a}_1^2,\dots,\bar{a}_{d-1}^2) \cap \bar{\mathfrak{q}}^n = (\bar{a}_1^2,\dots,\bar{a}_{d-1}^2)\bar{\mathfrak{q}}^{n-2}$$

by inductive hypothesis.

Hence we have

$$(a_1^2,\dots,a_d^2) \cap \mathfrak{q}^n \subset (a_1^2,\dots,a_{d-1}^2)\mathfrak{q}^{n-2} + (a_d) \cap \mathfrak{q}^n$$

$$= (a_1^2,\dots,a_{d-1}^2)\mathfrak{q}^{n-2} + a_d\mathfrak{q}^{n-1}$$

by Lemma (40.2).

Therefore $x \in (a_1^2,\dots,a_d^2) \cap \mathfrak{q}^n$ can be expressed as

$$x = \sum_{i=1}^{d-1} a_i^2 y_i + a_d y$$

with $y_i \in \mathfrak{q}^{n-2}$ and $y \in \mathfrak{q}^{n-1}$. So $a_d y \in (a_1^2,\dots,a_d^2) \cap \mathfrak{q}^n$. We must show that $a_d y \in (a_1^2,\dots,a_d^2) \cap \mathfrak{q}^{n-2}$: but now

$$\begin{aligned} y &\in ((a_1^2,\dots,a_d^2) : a_d) \cap \mathfrak{q}^{n-1} \\ &= \left(\sum_{\Gamma\subseteq\{1,\dots,d-1\}} a_\Gamma U(\mathfrak{q}_\Gamma) + (a_d) \right) \cap \mathfrak{q}^{n-1} \\ &= \sum_{\substack{\Gamma\subseteq\{1,\dots,d-1\} \\ |\Gamma| \geq n-1}} a_\Gamma U(\mathfrak{q}_\Gamma) + \left(\sum_{\substack{\Gamma\subseteq\{1,\dots,d-1\} \\ |\Gamma| < n-1}} a_\Gamma U(\mathfrak{q}_\Gamma) + (a_d) \right) \cap \mathfrak{q}^{n-1} \end{aligned}$$

by Theorem (38.13).

If $|\Gamma| \geq n-1$, we have $a_d a_\Gamma U(\mathfrak{q}_\Gamma) \subset a_\Gamma(\{a_i \mid a_i \in \Gamma\}) \subset \sum_{i\in\Gamma} a_i^2 \mathfrak{q}^{n-2}$.

Therefore we may assume

$$y \in \left(\sum_{\substack{|\Gamma| < n-1 \\ \Gamma\subseteq\{1,\dots,d-1\}}} a_\Gamma U(\mathfrak{q}_\Gamma) + (a_d) \right) \cap \mathfrak{q}^{n-1} .$$

Then the image $\bar{y}$ of y in $\bar{A} = A/a_d A$ belongs to

$$\left(\sum_{\substack{\Gamma\subseteq\{1,\dots,d-1\} \\ |\Gamma| < n-1}} \bar{a}_\Gamma U(\bar{\mathfrak{q}}_\Gamma) \right) \cap \bar{\mathfrak{q}}^{n-1} \subset (\bar{a}_1^2,\dots,\bar{a}_{d-1}^2) .$$

by Lemma (40.8). (Note that $n-1 \geq 2$). Hence, by induction :

$$y \in (a_1^2,\dots,a_{d-1}^2,a_d) \cap \mathfrak{q}^n \subseteq (a_1^2,\dots,a_{d-1}^2)\mathfrak{q}^{n-3} + a_d\mathfrak{q}^{n-2} ,$$

therefore

$$a_d y \in (a_1^2,\dots,a_{d-1}^2) a_d \mathfrak{q}^{n-3} + a_d^2 \mathfrak{q}^{n-2} \subset (a_1^2,\dots,a_d^2)\mathfrak{q}^{n-2} , \quad \text{q.e.d.}$$

Now we come to the main result of this section.

<u>(40.10) Theorem</u>. Let $a_1,\ldots,a_d$ be a standard system of parameters of a generalized Cohen-Macaulay ring $(A,\mathfrak{m})$, $\mathfrak{q} = (a_1,\ldots,a_d)$, $G = \bigoplus_{n\geq 0}\mathfrak{q}^n/\mathfrak{q}^{n+1}$ and $\mathfrak{M}$ the maximal homogeneous ideal of G. Then for $i < d$:

$$\underline{H}^i_{\mathfrak{M}}(G)_n = \begin{cases} H^i_{\mathfrak{m}}(A) & \text{if } n = -i \\ 0 & \text{otherwise} \end{cases}$$

and

$$\underline{H}^d_{\mathfrak{M}}(G)_n = (0) \quad \text{for } n > -d \quad .$$

Proof. Let a_i^* be the initial form of a_i with respect to $\mathfrak{q}$. We will show that $a_1^*,\ldots,a_d^*$ is a standard system of parameters of $G_{\mathfrak{M}}$, which is generalized Cohen-Macaulay by Proposition (40.4).

By Proposition (38.3) it suffices to show that

$$I(a_1^*,\ldots,a_d^*;G_{\mathfrak{M}}) = I(a_1^{*2},\ldots,a_1^{*2};G_{\mathfrak{M}})$$

Now

$$I(a_1^*,\ldots,a_d^*;G_{\mathfrak{M}}) = \lambda(G_{\mathfrak{M}}/(a_1^*,\ldots,a_d^*)) - e(a_1^*,\ldots,a_d^*;G_{\mathfrak{M}})$$

$$= \lambda(A/\mathfrak{q}) - e(a_1,\ldots,a_d;A) = I(A) \quad .$$

On the other hand, we know from Lemma (40.9)

$$I(a_1^{*2},\ldots,a_d^{*2};G_{\mathfrak{M}}) = \lambda(G_{\mathfrak{M}}/(a_1^{*2},\ldots,a_d^{*2})) - e(a_1^{*2},\ldots,a_d^{*2};G_{\mathfrak{M}})$$

$$= \sum_{n\geq 0}\lambda(\mathfrak{q}^n/(a_1^2,\ldots,a_d^2)\mathfrak{q}^{n-2} + \mathfrak{q}^{n+1}) - e(a_1^2,\ldots,a_d^2;A)$$

$$= \sum_{n\geq 0}\lambda(\mathfrak{q}^n/(a_1^2,\ldots,a_d^2) \cap \mathfrak{q}^n + \mathfrak{q}^{n+1}) - e(a_1^2,\ldots,a_d^2;A)$$

$$= \lambda(A/(a_1^2,\ldots,a_d^2)) - e(a_1^2,\ldots,a_d^2;A) = I(A) \quad .$$

Hence $a_1^*,\ldots,a_d^*$ is standard in $G_{\mathfrak{M}}$.

By Theorem (39.1) and its proof we see that for $0 \leq i < d$:

$$\underline{H}^i_{\mathfrak{m}}(G) \cong \underline{H}^i_{\mathfrak{q}^*_{i+1}}(G) \cong \operatorname{Ker}\left(\underline{H}^i_{\mathfrak{q}^*_i}(G) \longrightarrow \underline{H}^i_{\mathfrak{q}^*_i}(G)_{a^*_{i+1}}\right)$$

$$\cong \left[U(a^*_1,\ldots,a^*_i)/(a^{*2}_1,\ldots,a^{*2}_i) : (a^*_1\ldots a^*_i)\right](i) \quad ,$$

where $\mathfrak{q}^*_i = (a^*_1,\ldots,a^*_i)$.
$\left(\text{Note that } \varinjlim_n \left[G/(a^{*n}_1,\ldots,a^{*n}_i)\right](n\cdot i) \cong \underline{H}^i_{\mathfrak{q}^*_i}(G)\right)$.
By Lemma (40.6) we have for $0 \leq i < d$

$$\cong \left[\frac{U(a^*_1,\ldots,a^*_i)}{(a^{*2}_1,\ldots,a^{*2}_i) : a^*_1\ldots a^*_i}\right]_0 \cong \frac{U(a_1,\ldots,a_i) + \mathfrak{q}}{\sum_{j=1}^{i} U(a_1,\ldots,\check{a}_j,\ldots,a_i) + \mathfrak{q}}$$

$$\cong \frac{U(a_1,\ldots,a_i)}{\mathfrak{q} \cap U(a_1,\ldots,a_i) + ((a^2_1,\ldots,a^2_i) : a_1\ldots a_i)}$$

$$\cong \frac{U(a_1,\ldots,a_i)}{(a^2_1,\ldots,a^2_i) : a_1\ldots a_i} \cong H^i_{\mathfrak{m}}(A) \quad , \text{ and}$$

$$\left[\frac{U(a^*_1,\ldots,a^*_i)}{(a^{*2}_1,\ldots,a^{*2}_i) : a^*_1\ldots a^*_i}\right]_n = 0 \quad \text{for} \quad n \neq 0 \quad .$$

Thus we have for $0 \leq i < d$

$$\underline{H}^i_{\mathfrak{m}}(G)_n = \begin{cases} H^i_{\mathfrak{m}}(A) & \text{for} \quad n = -i \\ 0 & \text{otherwise} \end{cases} \quad .$$

Let $\alpha \in \underline{H}^d_{\mathfrak{m}}(G)$ be a homogeneous element of $\nu = \deg\alpha > -d$ and let α be represented by an element of

$$\left[G/(a^{*n}_1,\ldots,a^{*n}_d)(nd)\right]_\nu = \left[G/(a^{*n}_1,\ldots,a^{*n}_d)\right]_{nd+\nu} \qquad \text{for some} \quad n > 0 \ .$$

But $nd + \nu \geq nd - d + 1 = d(n-1) + 1$ implies that

$$(a^*_1,\ldots,a^*_d)^{d(n-1)+1} \subset (a^{*n}_1,\ldots,a^{*n}_d) \quad .$$

Therefore $\left[G/(a^{*n}_1,\ldots,a^{*n}_d)\right]_{dn+\nu} = 0$, i.e.

$$\underline{H}^d_{\mathfrak{m}}(G)_n = 0 \quad \text{for} \quad n > -d \quad , \text{ q.e.d. (48.10).}$$

As an application of this result we compute the local cohomology of the Rees algebra $R = \bigoplus_{n \geq 0} \mathfrak{q}^n \cong A[\mathfrak{q}T]$.

<u>(40.11) Corollary</u>. Let $A, a_1, \ldots, a_d$ and $\mathfrak{q}$ be the same as in Theorem (40.10) and let $R = \bigoplus_{n \geq 0} \mathfrak{q}^n \cong A[\mathfrak{q}T]$ be the Rees algebra with respect to $\mathfrak{q}$, and $\mathfrak{N}$ the maximal homogeneous ideal of R . Then we have:

1) If $\dim A = 1$, then $\underline{H}^0_{\mathfrak{N}}(R) \cong \underline{H^0_{\mathfrak{m}}(A)}$ and $\underline{H}^1_{\mathfrak{N}}(R) = 0$, where $\underline{H^0_{\mathfrak{m}}(A)}$ is the graded R-module $[\underline{H^0_{\mathfrak{m}}(A)}]_0 = H^0_{\mathfrak{m}}(A)$ and $[\underline{H^0_{\mathfrak{m}}(A)}]_n = 0$ for $n \neq 0$.

2) If $\dim A = 2$, then $\underline{H}^0_{\mathfrak{N}}(R) \cong \underline{H^0_{\mathfrak{m}}(A)}$; $\underline{H}^1_{\mathfrak{N}}(R) = 0$ and $\underline{H}^2_{\mathfrak{N}}(R) = 0$.

3) If $\dim A = d \geq 3$, then

$$\underline{H}^0_{\mathfrak{N}}(R) \cong \underline{H^0_{\mathfrak{m}}(A)} \ , \ \underline{H}^1_{\mathfrak{N}}(R) = \underline{H}^2_{\mathfrak{N}}(R) = 0 \ ,$$

and for $3 \leq i \leq d$

$$\underline{H}^i_{\mathfrak{N}}(R)_n = \begin{cases} H^{i-1}_{\mathfrak{m}}(A) & \text{for } -i+2 \leq n \leq -1 \\ 0 & \text{otherwise} \end{cases} .$$

Proof. We first observe that $R_{\mathfrak{N}}$ is a generalized Cohen-Macaulay ring, i.e. $\lambda(\underline{H}^i_{\mathfrak{N}}(R)) < \infty$ for $i \leq d$: We may assume that A is complete. By Theorem (37.4) we know that A is quasi-unmixed and hence R is so by (18.23) . Since R is a graded ring defined over a complete local ring, by duality it is enough to show that $R_{\mathfrak{P}}$ is Cohen-Macaulay for homogeneous $\mathfrak{P} \in \operatorname{Spec}(R) - \{\mathfrak{N}\}$, see proof of (40.4).
But by Proposition (40.3) we know that $\operatorname{Bl}(\mathfrak{q},A)$ is Cohen-Macaulay and hence $R_{\mathfrak{P}}$ is Cohen-Macaulay for $\mathfrak{P} \in \operatorname{Spec}(R) - V(R_+)$, where $R_+ = \bigoplus_{n>0} \mathfrak{q}^n$, which we will denote by I in this proof.

If $\mathfrak{P} \in \operatorname{Spec}(R) - \{\mathfrak{N}\}$ contains R_+ we have $\mathfrak{P} \cap A = \mathfrak{p} \subsetneqq \mathfrak{m}$, since $\mathfrak{P} \neq \mathfrak{N}$. Therefore $R_{\mathfrak{p}} \cong A_{\mathfrak{p}}[X]$, where X is an indeterminate over $A_{\mathfrak{p}}$. Since $R_{\mathfrak{P}}$ is a localization of $R_{\mathfrak{p}}$ we know that $R_{\mathfrak{P}}$ is Cohen-Macaulay.

Next we compute $\underline{H}^0_{\mathfrak{N}}(R)$. Since $R_{\mathfrak{N}}$ is a generalized Cohen-Macaulay ring and a_1 is a part of a system of parameters of R_N we have for large $n >> 0$ (by Lemma (38.6))

$$\underline{H}_N^0(R) = (0 : a_1^n)_R = \bigoplus_{i=0}^{\infty} (0 : a_1^n) \cap q^i = \underline{(0 : a_1^n)} = \underline{H_m^0(A)} \quad .$$

Now we compute $\underline{H}_N^1(R)$. Consider the two exact sequences

$$(*) \qquad \begin{array}{l} 0 \longrightarrow I \longrightarrow R \longrightarrow A \longrightarrow 0 \\ 0 \longrightarrow I(1) \longrightarrow R \longrightarrow G \longrightarrow 0 \end{array}$$

and

$$(**) \qquad \begin{array}{l} \underline{H_m^0(A)} \longrightarrow \underline{H}_N^1(I) \xrightarrow{\varphi^1} \underline{H}_N^1(R) \longrightarrow \underline{H_m^1(A)} \longrightarrow \ldots \\ \underline{H}_N^0(G) \longrightarrow \underline{H}_N^1(I)(1) \xrightarrow{\psi^1} \underline{H}_N^1(R) \longrightarrow \underline{H}_N^1(G) \longrightarrow \ldots \end{array} \quad .$$

Since $\underline{H_m^0(A)}$ and $\underline{H_m^1(A)}$ are concentraded in degree 0 , φ^1 induces isomorphisms $\varphi_n^1 : \underline{H}_N^1(I)_n \longrightarrow \underline{H}_N^1(R)_n$ for $n \neq 0$. Now $\underline{H}_N^i(G)_n = 0$ for $n \neq -i$ and $i < d$ by Theorem (40.10), so one has from (**) isomorphisms

$$\psi_n^1 : \underline{H}_N^1(I)_{n+1} \longrightarrow \underline{H}_N^1(R)_n$$

for $n > 0$, a surjection $\psi_0^1 : \underline{H}_N^1(I)_1 \longrightarrow \underline{H}_N^1(R)_0$ and injections $\psi_n^1 : \underline{H}_N^1(I)_{n+1} \longrightarrow \underline{H}_N^1(R)_n$ for $n < 0$. Since $\lambda(\underline{H}_N^1(R)) < \infty$, we have $\underline{H}_N^1(R)_n = 0$ for $n << 0$ and $n >> 0$. By a diagram chase it also follows that $\underline{H}_N^1(R) = 0$. - Once can prove similarly that $\underline{H}_N^2(R) = 0$.

Let $d \geqq 3$. The exact sequence (*) induces the exact sequence

$$(***) \qquad \begin{array}{l} \underline{H}_N^{i-1}(R) \longrightarrow \underline{H_m^{i-1}(A)} \longrightarrow \underline{H}_N^i(I) \xrightarrow{\varphi^i} \underline{H}_N^i(R) \longrightarrow \underline{H_m^i(A)} \longrightarrow \ldots \\ \underline{H}_N^{i-1}(R) \longrightarrow \underline{H}_N^{i-1}(G) \longrightarrow \underline{H}_N^i(I)(1) \xrightarrow{\psi^i} \underline{H}_N^i(R) \longrightarrow \underline{H}_N^i(G) \longrightarrow \end{array}$$

for $3 \leqq i \leqq d$.

Let $\varphi_n^i : \underline{H}_N^i(I)_n \longrightarrow \underline{H}_N^i(R)_n$ and $\psi_n^i : \underline{H}_N^i(I)_{n+1} \longrightarrow \underline{H}_N^i(R)_n$ be induced homomorphisms. Then we get the following statements by Theorem (40.10

1) φ_n^i is an isomorphism for $n \neq 0$

2) ψ_n^i is an isomorphism for $n \geqq -i + 2$

3) ψ_{-i+1}^i is a surjection

4) ψ_n^i is an injection for $n \leqq -i$.

Since $\lambda(\underline{H}_N^i(R)) < \infty$, it follows by a diagram chase that

$$\begin{aligned} H_{\mathfrak{m}}^{i-1}(A) &\cong \underline{H}_N^i(I)_0 \\ &\cong \underline{H}_N^i(R)_{-1} \\ &\cong \underline{H}_N^i(R)_{-i+2} \end{aligned}$$

and

$$\underline{H}_N^i(R)_n = 0 \quad \text{for} \quad n \leqq -i + 1 \quad .$$

This completes the proof of (40.11).

§ 41. Standard ideals and Buchsbaum rings

In the last three sections we have studied the properties of standard systems of parameters of generalized Cohen-Macaulay rings. The purpose of this section is to characterize $\mathfrak{m}$-primary ideals I in a generalized Cohen-Macaulay ring $(A,\mathfrak{m},k)$ such that every system of parameters contained in I is standard. Recall (see (38.1)):

(41.1) Definition. Let $(A,\mathfrak{m},k)$ be a generalized Cohen-Macaulay ring and I an $\mathfrak{m}$-primary ideal of A . I is called standard if every system of parameters contained in I is a standard system of parameters.

(41.2) Remark. We first note that for a given generalized Cohen-Macaulay ring $(A,\mathfrak{m},k)$ there exists a standard ideal I . In fact, since A is a generalized Cohen-Macaulay ring there exists an $\mathfrak{m}$-primary ideal I such that $(a_1,\ldots,a_{d-1}) : a_d = (a_1,\ldots,a_{d-1}) : I$ for any system of parameters $a_1,\ldots,a_d$ contained in I by Theorem (37.10).

Then, by Lemma (38.2),3), we see that $a_1,\dots,a_d \in I$ is standard.

For the following we need a technical lemma.

<u>(41.3) Lemma</u>. Let $(A,\mathfrak{m})$ be a noetherian local ring, let $I = (y_1,\dots,y_r)$ an $\mathfrak{m}$-primary ideal with generators $y_1,\dots,y_r$, $r \geq d := \dim A$, and $a_1,\dots,a_d$ a system of parameters of A. Then there exists an element $x \in I$, such that

i) $a_1,\dots,a_{d-1},x$ is a system of parameters of A

ii) $(x,y_2,\dots,y_r) = I$.

Proof. First note, that

a) $a_1,\dots,a_{d-1},x$ is a system of parameters of A if and only if x is not contained in any $\mathfrak{p} \in \mathrm{Assh}_A(A/(a_1,\dots,a_{d-1}))$, and

b) $\mathrm{Assh}_A(A/(a_1,\dots,a_{d-1}))$ is a finite set.

Let $\mathrm{Assh}_A(A/(a_1,\dots,a_{d-1})) = \{\mathfrak{p}_1,\dots,\mathfrak{p}_n\}$ and assume $y_1 \notin \mathfrak{p}_1,\dots,\mathfrak{p}_t$ but $y_1 \in \mathfrak{p}_{t+1},\dots,\mathfrak{p}_n$ for $0 \leq t \leq n$. If $t = n$, put $x := y_1$. If $t < n$, then $\mathfrak{a} := (y_2,\dots,y_r) \cap \mathfrak{p}_1 \cap \dots \cap \mathfrak{p}_t$ is an ideal contained in I and $\mathfrak{a} \not\subseteq \mathfrak{p}_j$ for $t+1 \leq j \leq n$. Therefore there exists an element $x' \in \mathfrak{a} \setminus \bigcup_{j=t+1}^{n} \mathfrak{p}_j$. Let then $x = x' + y_1$. Then $x \notin \bigcup_{j=1}^{n} \mathfrak{p}_j$, and i) and ii) are satisfied.

<u>(41.4) Corollary</u>. Let $(A,\mathfrak{m})$, $I = (y_1,\dots,y_r)$ and $a_1,\dots,a_d$ be as in (41.2).

a) Then there are generators $x_1,\dots,x_r$ of I such that $a_1,\dots,a_{d-1},x_i$ is a system of parameters for $1 \leq i \leq r$.

b) There exists a system of parameters $x_1,\dots,x_d$ of A such that $I = (x_1,\dots,x_d,y_{d+1},\dots,y_r)$.

c) There are generators $x_1,\dots,x_r$ of I such that $x_{i_1},\dots,x_{i_d}$ is a system of parameters for all $1 \leq i_1 < \dots < i_d \leq r$.

Proof. a) and b) are an immediate consequence of Lemma (41.3).
c): Let $I = (y_1,\dots,y_r)$ and $d \leq k \leq r$. We show by induction on k

that there are $x_1,\dots,x_k \in A$ such that

$$(*)\quad \begin{aligned} & I = (x_1,\dots,x_k,y_{k+1},\dots,y_r) \quad \text{and} \\ & x_{i_1},\dots,x_{i_d} \ \text{is a system of parameters for all} \ 1 \leqq i_1<\dots<i_d \leqq k . \end{aligned}$$

The case $k = d$ follows from b).

Now let $d < k \leqq r$. By inductive hypothesis there exist $x_1,\dots x_{k-1}$ with the same properties as in (*).

$$\text{Let } \mathfrak{N} := \left\{ \mathfrak{p} \in \operatorname{Spec}(A) \;\middle|\; \begin{array}{l} \exists 1 \leqq i_1 < \dots < i_{d-1} \leqq k-1 \ \text{and} \\ \mathfrak{p} \in \operatorname{Assh}_A(A/(x_{i_1},\dots,x_{i_{d-1}})) \end{array} \right\} .$$

In the same way as in the proof of Lemma (41.3) we can find $x_k \notin \bigcup_{\mathfrak{p}\in\mathfrak{N}} \mathfrak{p}$ and $I = (x_1,\dots,x_k,y_{k+1},\dots,y_r)$. Now let $1 \leqq i_1<\dots<i_k \leqq k$. If $i_d \leqq k-1$, then $x_{i_1},\dots,x_{i_d}$ is a system of parameters of A by induction. If $i_d = k$, then x_k is not contained in any $\mathfrak{p} \in \operatorname{Assh}_A(A/(x_{i_1},\dots,x_{i_{d-1}}))$ and therefore $x_{i_1},\dots,x_{i_d}$ form a system of parameters of A.

Our first characterization of standard ideals can be stated as follows:

<u>(41.5) Theorem</u>. Let $(A,\mathfrak{m},k)$ be a noetherian local ring with $d = \dim A > 0$ and I an $\mathfrak{m}$-primary ideal of A. Then the following conditions are equivalent:

1) A is a generalized Cohen-Macaulay ring and I is a standard ideal.

2) For any system of parameters $a_1,\dots,a_d$ of A contained in I, we have:

$$(a_1,\dots,a_{d-1}) : a_d = (a_1,\dots,a_{d-1}) : I .$$

3) For any system of parameters $a_1,\dots,a_d$ of A contained in I, we have:

$$(a_1,\dots,a_{d-1}) : a_d = (a_1,\dots,a_{d-1}) : a_d^2 .$$

Proof. 1) ⇒ 3). Let $a_1,\dots,a_d \in I$ be a system of parameteres of A. By Proposition (38.9) $a_1,\dots,a_d$ is a d-sequence. From the Definition (38.5) of d-sequences it follows that

$$(a_1,\dots,a_{d-1}) : a_d = (a_1,\dots,a_{d-1}) : a_d^2 \quad .$$

3) ⇒ 2). Let $a_1,\dots,a_d \in I$ be a system of parameters of A. Then, by assumption 3)

$$(a_1,\dots,a_{d-1}) : a_d = (a_1,\dots,a_{d-1}) : a_d^n$$

for all $n > 0$. Consequently, $(a_1,\dots,a_{d-1}) : a_d = U(a_1,\dots,a_{d-1})$, cf. Definition (38.10). By Corollary (41.4) we conclude that there are generators $x_1,\dots x_r$ of I such that $a_1,\dots,a_{d-1}, x_i$ is a system of parameters of A for $1 \leqq i \leqq r$. Hence $U(a_1,\dots,a_{d-1}) =$
$= (a_1,\dots,a_{d-1}) : x_i$ for $1 \leqq i \leqq r$. Therefore
$(a_1,\dots,a_{d-1}) : a_d = U(a_1,\dots,a_{d-1}) = (a_1,\dots,a_{d-1}) : I$.

2) ⇒ 1). This has been shown in Remark (41.2).

(41.5*) Remark. Under the assumptions of Theorem (41.5) one can show that the conditions (2) and (3) are equivalent to the following ones:

2') For any system of parameters $a_1,\dots,a_d$ of A contained in I we have:

$$(a_1,\dots,a_i) : a_{i+1} = (a_1,\dots,a_i) : I, \quad 0 \leqq i \leqq d \quad .$$

3') Any system of parameters of A contained in I is a d-sequence.

Now we are going to characterize standard ideals by the Koszul-(co)-homology. First we fix some notations: Let $a_1,\dots,a_r$ be elements of a noetherian ring A and let $K.(a_1,\dots,a_r;A)$ be the Koszul complex with respect to $a_1,\dots,a_r$. We identify $K.(a_1,\dots,a_r;A)$ with the exterior algebra ΛF, where F is a free A-module of rank r with free basis $e_1,\dots,e_r$, cf. Chapter VII, § 35.
For $p \geqq 0$ elements of $K_p(a_1,\dots,a_r;A) = \Lambda^p F$ will be denoted by greek letters $\sigma,\tau\dots$. If $\sigma \in K_p(a_1,\dots,a_r;A)$ belongs to the p-th cycle $Z_p(a_1,\dots,a_r;A)$ of $K.(a_1,\dots,a_r;A)$, the homology class of σ will be denoted by $[\sigma] \in H_p(a_1,\dots,a_r;A)$. Let ∂ be the boundary operator of $K.(a_1,\dots,a_r;A)$. Then $\partial(\sigma \wedge \tau) = \partial(\sigma) \wedge \tau + (-1)^p \sigma \wedge \partial(\tau)$

for $\sigma \in K_p(a_1,\ldots,a_r;A)$, $\tau \in K_q(a_1,\ldots,a_r;A)$.

(41.6) Lemma. Let $a_1,\ldots,a_r$ be elements in a noetherian ring A . We consider the Koszul complex $K.(a_1,\ldots,a_r,0;A) = \Lambda F$, where F is a free A-module of rank $r+1$ with free basis $e_1,\ldots,e_{r+1}$, and the boundary operator ∂ is given by $\partial(e_i) = a_i$ for $1 \leqq i \leqq r$ and $\partial(e_{r+1}) = 0$.
Then there is an isomorphism

$$H_p(a_1,\ldots,a_r,0;A) \cong H_p(a_1,\ldots,a_r;A) \oplus H_{p-1}(a_1,\ldots,a_r;A)$$

for all $p \geqq 0$. Identifying $H_0(a_1,\ldots,a_r;A)$ with $A/(a_1,\ldots,a_r)$, for $p = 1$ the isomorphism is given by

$$\left[\sum_{i=1}^{r+1} x_i e_i\right] \longmapsto \left(\left[\sum_{i=1}^{r} x_i e_i\right], x_{r+1} \bmod (a_1,\ldots,a_r)\right)$$

for any element $\sum_{i=1}^{r+1} x_i e_i$ such that $\sum_{i=1}^{r} a_i x_i = 0$.

Proof. By Proposition (35.12) we have an exact sequence

$$0 \longrightarrow H_p(a_1,\ldots,a_r;A) \longrightarrow H_p(a_1,\ldots,a_r,0;A) \xrightarrow{\alpha_p} H_{p-1}(a_1,\ldots,a_r;A) \longrightarrow 0$$

for all $p \geqq 0$. We can describe α_p as follows: Let E be the free A-submodule generated by $e_1,\ldots,e_r$ and let $K. = \wedge E$ be the Koszul complex with respect to $a_1,\ldots,a_r$. Then $K_p(a_1,\ldots,a_r,0;A) = K_p \oplus K_{p-1} \wedge e_{r+1}$ for $p \geqq 0$. If $\sigma \in K_p(a_1,\ldots,a_r,0;A)$ is a cycle, i.e. $\partial(\sigma) = 0$, where $\sigma = \tau + \theta \wedge e_{r+1}$ with $\tau \in K_p$, $\theta \in K_{p-1}$, we get $\partial(\theta) = 0$. Then $\alpha_p(\sigma) = [\theta] \in H_{p-1}(a_1,\ldots,a_r;A)$.
Consider the A-linear map $\beta : \Lambda^{p-1}E \longrightarrow \Lambda^p F$, given by $\beta(\sigma) = \sigma \wedge e_{r+1}$. If σ is a cycle we have:

$$\partial(\sigma \wedge e_{r+1}) = \partial(\sigma) \wedge e_{r+1} + (-1)^{p-1} \sigma \wedge \partial(e_{r+1}) = 0 ,$$

i.e. $\beta(\sigma)$ is a cycle in $K_p(a_1,\ldots,a_r,0;A)$. If moreover σ is a boundary, say $\partial(\tau) = \sigma$ for some $\tau \in K_p(a_1,\ldots,a_r;A)$, then $\beta(\sigma) = \partial(\tau \wedge e_{r+1})$. From this we see that β induces an homomorphism

$\overline{\beta} : H_{p-1}(a_1,\ldots,a_r ; A) \longrightarrow H_p(a_1,\ldots,a_r,0;A)$, such that $\alpha_p \cdot \overline{\beta} = id$. Hence we obtain the required decomposition. The second assertion is obvious.

Now we can give a characterization of standard ideals by means of Koszul homology.

(41.7) Proposition. Let $(A,\mathfrak{m},k)$ be a noetherian local ring with $d = \dim A > 0$ and I an $\mathfrak{m}$-primary ideal. Then the following conditions are equivalent:

1) A is a generalized Cohen-Macaulay ring and I is standard.

2) For any system of parameters $a_1,\ldots,a_d$ of A contained in I, we have: $I \cdot H_p(a_1,\ldots,a_d;A) = 0$ for $p > 0$.

3) For any system of parameters $a_1,\ldots,a_d$ of A contained in I, we have

$$I \cdot H_1(a_1,\ldots,a_d;A) = 0 \quad .$$

Proof. 1) ⇒ 2). By induction on d . If $d = 1$, we have $H_1(a_1;A) = (0 : a_1) = (0 : I)$.

Let $d \geq 2$ and $a_1,\ldots,a_d$ be a system of parameters contained in I . Since $a_1,\ldots,a_d$ is standard by assumption 1) we have $(0 : a_d) = (0 : I) = H^0_{\mathfrak{m}}(A)$.

Put $A' = A/H^0_{\mathfrak{m}}(A)$. Then we have an exact sequence

$$\begin{array}{c} 0 \longrightarrow a_dA \longrightarrow A \longrightarrow A/a_dA \longrightarrow 0 \\ \| \\ A' \end{array}$$

Taking $\overline{A} := A/a_dA$, we obtain an exact sequence

$$H_p(a_1,\ldots,a_d;A') \longrightarrow H_p(a_1,\ldots,a_d;A) \xrightarrow{\varphi_p} H_p(a_1,\ldots,a_d;\overline{A})$$

for $p > 0$. We claim that φ_p is an injection. To see this, it suffices to show that if a cycle $\sigma \in K_p(a_1,\ldots,a_d;A)$ satisfies $\sigma \in a_dK_p(a_1,\ldots,a_d;A) + B_p(a_1,\ldots,a_d;A)$, then $\sigma \in B_p(a_1,\ldots,a_d;A)$, where $B_p(a_1,\ldots,a_d;A)$ is the p-th boundary of $K.(a_1,\ldots,a_d;A)$. Let $\sigma = a_d\tau + \partial(\theta)$ with $\tau \in K_p(a_1,\ldots,a_d;A)$, $\theta \in K_{p+1}(a_1,\ldots,a_d;A)$. Then

$0 = \partial(\sigma) = a_d \cdot \partial(\tau)$ implies that $\partial(\tau) \in (0 : a_d)K_{p-1}(a_1,\dots,a_d;A)$.
Since $(0 : a_d) \cap (a_1,\dots,a_d) = 0$, we get:

$$\partial(\tau) \in (0 : a_d)K_{p-1}(a_1,\dots,a_d;A) \cap (a_1,\dots,a_d)K_{p-1}(a_1,\dots,a_d;A) = 0 .$$

Hence τ is a cycle. But $(a_1,\dots,a_d)H_p(a_1,\dots,a_d;A) = 0$ by Proposition (35.10). Hence $a_d\tau$ is a boundary. Therefore the claim is proved.

If $p \geqq 2$, $H_p(a_1,\dots,a_d;A)$ is a submodule of

$$H_p(a_1,\dots,a_d;\bar{A}) = H_p(\bar{a}_1,\dots,\bar{a}_{d-1},\bar{0};\bar{A})$$

$$= H_p(\bar{a}_1,\dots,\bar{a}_{d-1};\bar{A}) \oplus H_{p-1}(\bar{a}_1,\dots,\bar{a}_{d-1};\bar{A})$$

by Lemma (41.6). By inductive hypothesis $I\cdot H_p(\bar{a}_1,\dots,\bar{a}_{d-1};\bar{A}) = 0$ for $p > 0$. Hence $I\cdot H_p(a_1,\dots,a_d;A) = 0$ for $p \geqq 2$.
Let $p = 1$. Then by Lemma (41.6) we have

$$H_1(a_1,\dots,a_d;A) \longrightarrow H_1(\bar{a}_1,\dots,\bar{a}_{d-1};\bar{A}) \oplus A/(a_1,\dots,a_d) .$$

By the second assertion of Lemma (41.6) we get

$$H_1(a_1,\dots,a_d;A) \hookrightarrow H_1(\bar{a}_1,\dots,\bar{a}_{d-1};\bar{A}) \oplus \frac{(a_1,\dots,a_{d-1}) : a_d + (a_d)}{(a_1,\dots,a_d)}$$

Since $(a_1,\dots,a_{d-1}) : a_d = (a_1,\dots,a_{d-1}) : I$ by Theorem (41.5), we have $I\cdot H_1(a_1,\dots,a_d;A) = 0$, which proves 2).

2) $\Rightarrow$ 3). Trivial.

3) $\Rightarrow$ 1). Let $a_1,\dots,a_d \in I$ be a system of parameters of A . By Proposition (35.12) we get an exact sequence

$$\begin{array}{ccccc} H_1(a_1,\dots,a_d;A) & \longrightarrow & H_0(a_1,\dots,a_{d-1};A) & \xrightarrow{a_d} & H_0(a_1,\dots,a_{d-1};A) \\ & & \| & & \| \\ & & A/(a_1,\dots,a_{d-1}) & & A/(a_1,\dots,a_{d-1}) \end{array} .$$

Hence there is a surjection

$$H_1(a_1,\dots,a_d;A) \twoheadrightarrow \frac{((a_1,\dots,a_{d-1}) : a_d)}{(a_1,\dots,a_d)} .$$

Since $I\cdot H_1(a_1,\dots,a_d;A) = 0$ by assumption, we know that

$$(a_1,\dots,a_{d-1}) : a_d = (a_1,\dots,a_{d-1}) : I \quad .$$

Then by Theorem (41.5) we get 1). This proves Proposition (41.7).

To prove the main result of this section we need some more auxiliary results.

(41.8) Lemma. (Y. Yoshino): Let $(A,\mathfrak{m})$ be a noetherian local ring and let

$$C. := 0 \longrightarrow C_n \xrightarrow{\partial_n} C_{n-1} \xrightarrow{\partial_{n-1}} \dots \longrightarrow C_1 \xrightarrow{\partial_1} C_0 \longrightarrow 0$$

be a complex of finitely generated A-modules such that $H_0(C.) \neq 0$ and $\lambda(H_i(C.)) < \infty$ for all i. Then there is a subcomplex $D.$ of $C.$ such that

(1) $H_i(D.) = 0$ for $i > 0$ and $\lambda(H_0(D.)) < \infty$.

(2) $D_0 = C_0$, and there is a non-negative integer ℓ such that $\mathfrak{m}^{\ell} C_i \subseteq D_i$ for $1 \leq i \leq n$.

Proof. Let $Z.$ and $B.$ be the cycles and boundaries of $C.$ respectively. By assumption there is an integer $s \geq 0$ such that $\mathfrak{m}^s Z_i \subset B_i$ for all i. By the Artin-Rees lemma there exist integers $s_0,\dots,s_n$ such that

$$\mathfrak{m}^r C_i \cap Z_i \subseteq \mathfrak{m}^{r-s_i}(\mathfrak{m}^{s_i} C_i \cap Z_i) \subseteq \mathfrak{m}^{r-s_i} Z_i$$

for all $r \geq s_i$. Hence if r is large enough, $\mathfrak{m}^r C_i \cap Z_i \subset \mathfrak{m}^{r-s_i-s} B_i$.
Now let $r_i = s_i - s \; (0 \leq i \leq n)$ and let ℓ be an integer such that for all $i \geq 0$:

$$\mathfrak{m}^{\ell-r_0-r_1-\dots-r_{i-1}} C_i \cap Z_i \subset \mathfrak{m}^{\ell-r_0-\dots-r_i} B_i \quad .$$

Then define

$$D_0 := C_0 \quad \text{and} \quad D_i := \mathfrak{m}^{\ell-r_0-\dots-r_i} B_i + \mathfrak{m}^{\ell-r_0-\dots-r_{i-1}} C_i$$

for $1 \leq i \leq n$. By the constructions we get

$$\partial_i(D_i) = \mathfrak{m}^{\ell-r_0-\ldots-r_{i-1}} B_{i-1} \subset D_{i-1}$$

for $i > 0$. Hence $D.$ is a subcomplex of $C.$. For $i > 0$ the i-th cycle of $D.$ is given by

$$D_i \cap Z_i = \left(\mathfrak{m}^{\ell-r_0-\ldots-r_i} B_i + \mathfrak{m}^{\ell-r_0-\ldots-r_{i-1}} C_i\right) \cap Z_i =$$

$$= \mathfrak{m}^{\ell-r_0-\ldots-r_i} B_i = \partial_{i+1}(D_{i+1}) \quad .$$

Hence $H_i(D.) = 0$ for $i > 0$.

(2) is clear by construction.

This result in particular applies to the Koszul complex $K.(x_1,\ldots,x_r;A)$ with respect to generators $x_1,\ldots,x_r$ of an $\mathfrak{m}$-primary ideal of a local ring $(A,\mathfrak{m})$.

<u>(41.9) Corollary</u>. Let $I = (x_1,\ldots,x_r)$ and $J = (y_1,\ldots,y_s)$ be $\mathfrak{m}$-primary ideals in a local ring $(A,\mathfrak{m})$ and let ℓ be an integer such that $I \subset \mathfrak{m}^\ell J$. Let $\varphi., \psi. : K.(x_1,\ldots,x_r;A) \longrightarrow K.(y_1,\ldots,y_s;A)$ be complex homomorphisms such that

$$\varphi_0 = \psi_0 = \mathrm{id} \quad \text{and} \quad \varphi_i \otimes_A A/\mathfrak{m}^\ell = \psi_i \otimes_A A/\mathfrak{m}^\ell = 0 \quad \text{for} \quad i > 0 \quad .$$

Then, if ℓ is large enough, $\varphi.$ and $\psi.$ are homotopic.

Proof. Let $D.$ be a subcomplex of $K.(\underline{y};A)$ obtained by Lemma (41.8). If ℓ is large enough both $\varphi.$ and $\psi.$ factor through $D.$ Since $\varphi_0 = \psi_0 = \mathrm{id}$ and since $H_i(D.) = 0$ for $i > 0$ we see that $\varphi.$ and $\psi.$ are homotopic , q.e.d.

Let $I = (x_1,\ldots,x_r)$ be an $\mathfrak{m}$-primary ideal of a noetherian local ring $(A,\mathfrak{m})$. By Chapter VII, Theorem (35.18), we have an isomorphism $\varinjlim_n H^i(x_1^n,\ldots,x_r^n;A) \cong H^i_{\mathfrak{m}}(A)$ and hence there is a canonical homomorphism

$$\varphi^i_n : H^i(x_1^n,\ldots,x_r^n;A) \longrightarrow H^i_{\mathfrak{m}}(A) \quad .$$

Let $J = (y_1,\dots,y_s)$ be an $\mathfrak{m}$-primary ideal contained in I and let $y_i = \sum_{j=1}^{r} a_{ij}x_j$, $1 \leq i \leq s$, for some $a_{ij} \in A$. Then we get a homomorphism of complexes

$$\alpha. : K.(y_1,\dots,y_s;A) \longrightarrow K.(x_1,\dots,x_r;A)$$

such that $\alpha_1 : K_1(y_1,\dots,y_s;A) \cong A^s \longrightarrow K_1(x_1,\dots,x_r;A) \cong A^r$ is given by multiplication with the matrix

$$\begin{pmatrix} a_{11},\dots\dots,a_{1r} \\ a_{21},\dots\dots,a_{2r} \\ \dots\dots\dots\dots \\ a_{s1},\dots\dots,a_{sr} \end{pmatrix}$$

and $\alpha_p = {}^p\Lambda_1$ for all $p \geqq 1$, and $\alpha_0 = \mathrm{id}_A$. Applying the functor $\mathrm{Hom}_A(-,A)$ we get a complex map

$$\alpha^{\cdot} : K.(x_1,\dots,x_r;A) \longrightarrow K^{\cdot}(y_1,\dots,y_s;A)$$

which induces $\bar{\alpha}^i : H^i(x_1,\dots,x_r;A) \longrightarrow H^i(y_1,\dots,y_s;A)$ for all $i \geqq 0$.

For any integer $n > 0$ one can find an integer $m > 0$ such that $(y_1^n,\dots,y_s^n) \supset (x_1^m,\dots,x_s^m)$, and $k > n$, $\ell > m$ such that $(x_1^m,\dots,x_s^m) \supset (y_1^k,\dots,y_s^k) \supset (x_1^\ell,\dots,x_r^\ell)$.

Then we can construct homomorphisms

$$\beta^i_{n,m} : H^i(y_1^n,\dots,y_s^n;A) \longrightarrow H^i(x_1^m,\dots,x_r^m;A)$$

$$\alpha^i_{m,k} : H^i(x_1^m,\dots,x_s^m;A) \longrightarrow H^i(y_1^k,\dots,y_s^k;A) \ ,\dots,\ \text{etc.}$$

which make the following diagram commute for a suitable choice of k,ℓ,m,n by Corollary (41.9):

$$\begin{array}{ccc}
 & & H^i(y_1^n,\dots,y_s^n;A) \\
 & \swarrow{\scriptstyle \beta^i_{n,m}} & \downarrow{\scriptstyle \psi_{n,k}} \\
H^i(x_1^m,\dots,x_r^m;A) & \xrightarrow{\alpha^i_{m,k}} & H^i(y_1^k,\dots,y_s^k;A) \\
\downarrow{\scriptstyle \varphi_{m,\ell}} & \swarrow{\scriptstyle \beta^i_{k,\ell}} & \\
H^i(x_1^\ell,\dots,x_r^\ell;A) & &
\end{array}$$

Taking the limit, we obtain automorphisms

$$\tilde{\alpha}^i, \tilde{\beta}^i : H^i_{\mathfrak{m}}(A) \xrightarrow{\sim} H^i_{\mathfrak{m}}(A) \quad \text{such that} \quad \tilde{\alpha}^i \circ \tilde{\beta}^i = \tilde{\beta}^i \circ \tilde{\alpha}^i = \mathrm{id} \quad ,$$

and a commutative diagram

$$\begin{array}{ccc}
H^i(x_1,\dots,x_r;A) & \xrightarrow{\bar{\alpha}^i} & H^i(y_1,\dots,y_s;A) \\
\downarrow{\scriptstyle \varphi^i_1} & & \downarrow{\scriptstyle \psi^i_1} \\
H^i_{\mathfrak{m}}(A) & \xrightarrow[\tilde{\alpha}^i]{\sim} & H^i_{\mathfrak{m}}(A)
\end{array}$$

Thus we conclude the following lemma:

<u>(41.10) Lemma</u>. Let $I = (x_1,\dots,x_r)$ and $J = (y_1,\dots,y_s)$ be $\mathfrak{m}$-primary ideals in a noetherian local ring $(A,\mathfrak{m})$ such that $I \supset J$ and let $\varphi^i_1 : H^i(x_1,\dots,x_r;A) \longrightarrow H^i_{\mathfrak{m}}(A)$ and $\psi^i_1 : H^i(y_1,\dots,y_s;A) \longrightarrow H^i_{\mathfrak{m}}(A)$ be canonical homomorphisms. Then there is an automorphism $\tilde{\alpha}^i : H^i_{\mathfrak{m}}(A) \longrightarrow H^i_{\mathfrak{m}}(A)$ which makes the following diagram commute

$$\begin{array}{ccc}
H^i(x_1,\dots,x_r;A) & \xrightarrow{\bar{\alpha}^i} & H^i(y_1,\dots,y_s;A) \\
\downarrow{\scriptstyle \varphi^i_1} & & \downarrow{\scriptstyle \psi^i_1} \\
H^i_{\mathfrak{m}}(A) & \xrightarrow[\tilde{\alpha}^i]{\sim} & H^i_{\mathfrak{m}}(A)
\end{array} \quad .$$

After that we need one more auxiliary result to prove the main result of this section.

(41.11) Lemma. Let I be a standard $\mathfrak{m}$-primary ideal in a generalized Cohen-Macaulay ring $(A,\mathfrak{m})$ and $U = H^0_{\mathfrak{m}}(A)$. Then there exists a system $x_1,\ldots,x_r$ of generators of I such that

$$\eta^i : H^i(x_1,\ldots,x_r;U) \longrightarrow H^i(x_1,\ldots,x_r;A)$$

is an injection for $0 \leq i \leq d = \dim A$.

Proof. By a prime avoidance argument (see proof of Corollary (41.4)) we can choose the generators $x_1,\ldots,x_r$ so that any $d(=\dim A)$ elements of $\{x_1,\ldots,x_r\}$ form a system of parameters of A.
By the definition of the Koszul complex, η^i is just the map induced from $K_{r-i}(x_1,\ldots,x_r;U) \longrightarrow K_{r-i}(x_1,\ldots,x_r;A)$, cf. Chapter VII, § 35.
Let $e_1,\ldots,e_r$ be the free basis of $K_1(x_1,\ldots,x_r;A) \cong A^r$ and ∂ the differential map such that $\partial(e_i) = x_i$. Let

$\sigma = \sum_{1 \leq i_1 < \ldots < j_{r-i}} a_{j_1 \ldots j_{r-i}} e_{j_1} \wedge \ldots \wedge e_{j_{r-i}} \in K_{r-i}(x_1,\ldots,x_r;U)$ be a cycle such that $\sigma \in \partial(K_{r-i+1}(x_1,\ldots,x_r;A))$. We have to show that $\sigma = 0$. By the definition of ∂ we see that

$$a_{j_1 \ldots j_{r-i}} \in \left(x_{k_1},\ldots,x_{k_i}\right) ,$$

where $\{j_1,\ldots,j_{r-i}\} \cup \{k_1,\ldots,k_i\} = \{1,\ldots,r\}$. By the choice of $x_1,\ldots,x_r$ we know that $x_{k_1},\ldots,x_{k_i}$ form a part of a standard system of parameters, since I is standard. Hence

$$a_{j_1 \ldots j_{r-i}} \in U \cap \left(x_{k_1},\ldots,x_{k_i}\right) = 0 .$$

Therefore $\sigma = 0$.

The following is the main result of this section.

(41.12) Theorem. Let I be an $\mathfrak{m}$-primary ideal in a noetherian ring A with $d = \dim A > 0$. Then the following conditions are equivalent:

1) I is a standard ideal in a generalized Cohen-Macaulay ring A.

2) There is a system of generators $x_1,\dots,x_r$ of I such that the canonical homomorphism $\varphi^i : H^i(x_1,\dots,x_r;A) \longrightarrow H^i_{\mathfrak{m}}(A)$ is surjective for $i<d$.

Proof. 1) ⇒ 2). We choose a system $x_1,\dots,x_r$ of generators of I as in Lemma (41.11). We proceed by induction on d.

Let $d=1$ and let $a\in I$ be a system of parameters of A: Then $H^0(x_1,\dots,x_r;A) \cong H_r(x_1,\dots,x_r;A) = (0:I) = (0:a)$ by Theorem (41.5). Since a is standard we know that $(0:a) = H^0_{\mathfrak{m}}(A)$. Hence φ^0 is an isomorphism. Let $d \geq 2$:

Case 1: $\operatorname{depth} A > 0$. Let $a\in I$ be a non-zero-divisor of A and let $\bar{A} = A/aA$. From the exact sequence

$$0 \longrightarrow A \xrightarrow{\cdot a} A \longrightarrow \bar{A} \longrightarrow 0$$

we obtain a commutative diagram with exact rows

$$(*)\qquad \begin{array}{ccccccccc}
H^i(\underline{x};A) & \xrightarrow{\cdot a} & H^i(\underline{x};A) & \longrightarrow & H^i(\underline{x};\bar{A}) & \longrightarrow & H^{i+1}(\underline{x};A) & \xrightarrow{a} & \dots \\
\downarrow{\scriptstyle \varphi^i} & & \downarrow{\scriptstyle \varphi^i} & & \downarrow{\scriptstyle \bar{\varphi}^i} & & \downarrow{\scriptstyle \varphi^{i+1}} & & \\
H^i_{\mathfrak{m}}(A) & \xrightarrow[\cdot a]{} & H^i_{\mathfrak{m}}(A) & \longrightarrow & H^i_{\mathfrak{m}}(\bar{A}) & \longrightarrow & H^{i+1}_{\mathfrak{m}}(A) & \xrightarrow{a} & \dots
\end{array}$$

where $\bar{\varphi}^i$ is the canonical homomorphism.

By Proposition (38.3) $aH^{i+1}_{\mathfrak{m}}(A)=0$ for $i<d-1$. Since the image $\bar{I}$ of I in $\bar{A}$ is standard, there is a system of generators $\bar{y}_1,\dots,\bar{y}_s$ of $\bar{I}$ such that the canonical map

$$\psi^i : H^i(\bar{y}_1,\dots,\bar{y}_s;\bar{A}) \longrightarrow H^i_{\mathfrak{m}}(\bar{A})$$

is surjective. Since $(\bar{x}_1,\dots,\bar{x}_r) = (\bar{y}_1,\dots,\bar{y}_s)$ and since $H^i(x_1,\dots,x_r;\bar{A}) = H^i(\bar{x}_1,\dots,\bar{x}_r;\bar{A})$, by Lemma (41.10) we get a commutative diagram

$$\begin{array}{ccc}
H^i_{\mathfrak{m}}(\bar{x}_1,\dots,\bar{x}_r;\bar{A}) & \longleftarrow & H^i_{\mathfrak{m}}(\bar{y}_1,\dots,\bar{y}_s;\bar{A}) \\
{\scriptstyle \bar{\varphi}^i}\downarrow & & \downarrow{\scriptstyle \psi^i} \\
H^i_{\mathfrak{m}}(\bar{A}) & \xleftarrow[\beta^i]{\sim} & H^i_{\mathfrak{m}}(\bar{A})
\end{array},$$

where β is an automorphism of $H^i_{\mathfrak{m}}(\bar{A})$. Since ψ^i is surjective, $\bar{\varphi}^i$ is surjective too. From (*) we know that φ^{i+1} is surjective for $0 < i < d-1$ and $H^0_{\mathfrak{m}}(A) = 0$;hence φ^0 and so φ_i , $1 \leq i \leq d$ are surjective.

Case 2: depth $A = 0$. Let $U = H^0_{\mathfrak{m}}(A)$ and $A' := A/U$. Then depth $A' > 0$. It is not hard to see that the image I' in A' is standard. Hence, by Lemma (41.10) the canonical map

$$'\varphi^i : H^i(x_1,\ldots,x_r;A') \longrightarrow H^i_{\mathfrak{m}}(A')$$

is surjective for $i < d$. From the exact sequence

$$0 \longrightarrow U \longrightarrow A \longrightarrow A' \longrightarrow 0$$

we get a commutative diagram with exact rows

$$\begin{array}{ccccccccc}
\ldots \to H^i(x_1,\ldots,x_r;U) & \to & H^i(x_1,\ldots,x_r;A) & \xrightarrow{\alpha^i} & H^i(x_1,\ldots,x_r;A') & \to & H^{i+1}(x_1,\ldots,x_r;U) & \to \\
\downarrow & & \downarrow \varphi^i & & \downarrow\!\downarrow {'\varphi^i} & & \downarrow & \\
\ldots \longrightarrow H^i_{\mathfrak{m}}(U) & \longrightarrow & H^i_{\mathfrak{m}}(A) & \longrightarrow & H^i_{\mathfrak{m}}(A') & \longrightarrow & H^{i+1}_{\mathfrak{m}}(U) & \longrightarrow .
\end{array} \tag{\#}$$

For $0 < i \leq d$ we have $H^i_{\mathfrak{m}}(U) = 0$ and α^i is surjective by Lemma (41.11). Since $'\varphi^i$ is surjective, we know that φ^i is surjective for $0 < i < d$. If $i = 0$, we have $H^0(x_1,\ldots,x_r;A) = (0 : I) = H^0_{\mathfrak{m}}(A)$ since I is standard. Hence φ^0 is an isomorphism. This completes the proof of 1) $\Rightarrow$ 2).

2) $\Rightarrow$ 1). Since $H^i(x_1,\ldots,x_r;A)$ is finitely generated and φ^i is surjective by assumption, the given ring A is generalized Cohen-Macaulay. To show that I is a standard ideal we use induction on d:
Let $d = 1$. Then, by assumption 2) and Chapter VII, Proposition (35.10), we get $IH^0_{\mathfrak{m}}(A) = 0$. Hence by Proposition (38.3), we see that every system of parameters of A contained in I is standard.
Let $d \geq 2$. We first claim that for any system of parameters $a_1,\ldots,a_d$ contained in I and for any integer $n \geq 2$ the system $a_1,\ldots,a_{d-1}, a_d^n$ is a standard system of parameters of A . For this it is enough to show that the canonical map

$$\bar{\varphi}^i : H^i(x_1,\ldots,x_r;A/a_d^nA) \longrightarrow H^i_{\mathfrak{m}}(A/a_d^nA)$$

is surjective for $i < d-1$. Because, if this is true, then by inductive hypothesis the images of $a_1,\dots,a_{d-1}$ in A/a_d^nA form a standard system of parameters of A/a_d^nA and $I(A) = I(A/a_d^nA)$ by Corollary (37.8). Put $\bar{A} = A/a_d^nA$. Then from the exact sequence

$$0 \longrightarrow a_d^nA \longrightarrow A \longrightarrow \bar{A} \longrightarrow 0$$

we get a commutative diagram with exact rows

$$\begin{array}{ccccccccc}
\to H^i(x_1,\dots,x_r;a_d^nA) & \to & H^i(x_1,\dots,x_r;A) & \xrightarrow{\beta^i} & H^i(x_1,\dots,x_r;\bar{A}) & \to & H^{i+1}(x_1,\dots,x_r;a_d^nA) & \to \\
\downarrow \tau^i & & \downarrow \varphi^i & & \downarrow \bar{\varphi}^i & & \downarrow \tau^{i+1} & \\
\dots \to H_{\mathfrak{m}}^i(a_d^nA) & \xrightarrow{\alpha^i} & H_{\mathfrak{m}}^i(A) & \longrightarrow & H_{\mathfrak{m}}^i(\bar{A}) & \longrightarrow & H_{\mathfrak{m}}^{i+1}(a_d^nA) & .
\end{array}$$

Since $\lambda_A((0 : a_d^n)) < \infty$, τ^i is surjective for $i < d$. Therefore, if β^i is injective for $0 \leqq i \leqq d$, the map $\bar{\varphi}^i$ is surjective for $i < d-1$. Hence it is enough to prove that β^i is injective: For that we show the following implication: If a cycle $\sigma \in K^i(x_1,\dots,x_r;A)$ satisfies $\sigma \in a_d^nK^i(x_1,\dots,x_r;A) + B^i(x_1,\dots,x_r;A)$, then $\sigma \in B^i(x_1,\dots,x_r;A)$: Let $\sigma = a_d^n\tau + \partial(\theta)$, with $\tau \in K^i(x_1,\dots,x_r;A)$, $\theta \in K^{i-1}(x_1,\dots,x_r;A)$. Then $0 = \partial(\sigma) = a_d^n\partial(\tau)$. Hence $\partial(\tau) \in (0 : a_d^n)K^{i+1}(x_1,\dots,x_r;A)$. But by assumption we see that $H_{\mathfrak{m}}^0(A) = (0 : a_d) = (0 : a_d^n)$. So we have $a_d\tau \in Z^i(x_1,\dots,x_r;A)$. By Proposition (35.10), $a_d^n\tau \in B^i(x_1,\dots,x_r;A)$ for $n \geqq 2$. This proves the claim. — By induction $I \cdot A/a_d^nA$ is standard for $n \geqq 2$. So we get, by Theorem (41.5):

$$(a_1,\dots,a_{d-2},a_d^n) : a_{d-1} = (a_1,\dots,a_{d-2},a_d^n) : I$$

for all $n \geqq 2$. This implies

(**) $$(a_1,\dots,a_{d-2}) : a_{d-1} = (a_1,\dots,a_{d-2}) : I \quad .$$

By Proposition (35.12), we have a surjection

$$H_1(a_1,\dots,a_{d-1},a_d^2;A) \twoheadrightarrow (a_1,\dots,a_{d-1}) : a_d^2/(a_1,\dots,a_{d-1}) \quad .$$

If one can show that $IH_1(a_1,\dots,a_{d-1},a_d^2;A) = 0$, then it follows that

$$(a_1,\dots,a_{d-1}) : a_d^2 \subseteq (a_1,\dots,a_{d-1}) : I \subseteq (a_1,\dots,a_{d-1}) : a_d \quad ,$$

i.e. $(a_1,\dots,a_{d-1}) : a_d^2 = (a_1,\dots,a_{d-1}) : a_d$. Using the same argument for the injectivity of β^i in (#) we see that there is an injection

$$\begin{array}{ccc} H_1(a_1,\dots,a_{d-1},a_d^2;A) & \hookrightarrow & H_1(a_1,\dots,a_{d-1},a_d^2;A/a_d^2A) \\ & & \| \wr \\ & & H_1(a_1,\dots,a_{d-1};A/a_d^2) \oplus A/(a_1,\dots,a_{d-1},a_d^2) \end{array} \quad ,$$

cf. Lemma (41.6).

Let $\sigma = y_1e_1 + \dots + y_de_d \in K_1(a_1,\dots,a_{d-1},a_d^2;A)$ be a cycle, i.e. $a_1y_1 + \dots + a_{d-1}y_{d-1} + a_d^2y_d = 0$. Since $IH_1(a_1,\dots,a_{d-1};A/a_d^2A) = 0$ by inductive hypothesis and Proposition (41.7), by (41.6) it is enough to show that $Iy_d \subset (a_1,\dots,a_{d-1})$.

From Lemma (41.10) we have a commutative diagram

$$\begin{array}{ccccc} H^{d-1}(x_1,\dots,x_r;A) & \xrightarrow{\gamma^{d-1}} & H^{d-1}(a_1,\dots,a_{d-1},a_d^2;A) & = & H_1(a_1,\dots,a_{d-1},a_d^2;A) \\ & \searrow^{\varphi^{d-1}} & & \swarrow_{\psi^{d-1}} & \\ & & H^{d-1}_{\mathfrak{m}}(A) & & \end{array} \quad .$$

Hence ψ^{d-1} is surjective since φ^{d-1} is surjective. This shows that

$$H_1(a_1,\dots,a_{d-1},a_d^2;A) = \operatorname{Im}\gamma^{d-1} + \ker\psi^{d-1} \quad .$$

Since I annihilates $\operatorname{Im}\gamma^{d-1}$ we may assume that $[\sigma] \in \ker\psi^{d-1}$. By construction of direct limit maps we see that

$$\begin{aligned} y_d &\in \left(a_1^{k+1},\dots,a_{d-1}^{k+1}\right) : (a_1 \dots a_{d-1})^k \\ &= \left(a_1^2,\dots,a_{d-1}^2\right) : (a_1 \dots a_{d-1}) \quad \text{by Lemma (38.14)} \quad . \end{aligned}$$

Since $a_1,\dots,a_{d-1},a_d^2$ is standard, we get for

<u>d = 2:</u> $y_d \in (a_1^2) : a_1 = (a_1) + U(0)$. By (**) we have $U(0) = (0 : a_1) = (0 : I)$ and therefore $Iy_d \subseteq (a_1)$.

<u>d ≧ 3:</u> $y_d \in (a_1^2,\dots,a_{d-1}^2) : (a_1 \dots a_{d-1})$

$$= \sum_{\Gamma \subsetneq \{1,\dots,d-1\}} U(\mathfrak{q}_\Gamma)$$

$$= \sum_{i=1}^{d-1} (a_1,\dots,\check{a}_i,\dots,a_{d-1}) : a_i$$

$$= \sum_{i=1}^{d-1} (a_1,\dots,\check{a}_i,\dots,a_{d-1}) : I$$

by Theorem (38.13)b, Lemma (38.11) and (**). Therefore we get $Iy_d \subseteq (a_1,\dots,a_{d-1})$.

This finishes the proof of Theorem (41.12).

Using the implication (1) ⇒ (2) in the proof of Theorem (41.12) and Lemma (41.9), we get the following proposition.

<u>(41.13) Proposition</u>. Let I be a standard ideal of a generalized Cohen-Macaulay ring $(A,\mathfrak{m})$. Then for any generators $x_1,\dots,x_r$ of I the canonical map

$$H^i(x_1,\dots,x_r;A) \longrightarrow H^i_{\mathfrak{m}}(A)$$

is surjective for $i < d$.

Now we introduce the notion of Buchsbaum rings.

<u>(41.14) Definition</u>. 1) A noetherian ring $(A,\mathfrak{m})$ is called Buchsbaum if A is a generalized Cohen-Macaulay ring and if the maximal ideal $\mathfrak{m}$ is a standard ideal of A .

2) A sequence $a_1,\dots,a_n$ in a noetherian local ring $(A,\mathfrak{m})$ is called a weak sequence if

$$(a_1,\dots,a_i) : a_{i+1} = (a_1,\dots,a_i) : \mathfrak{m} \quad \text{for } 0 \leq i < n .$$

The notion of Buchsbaum rings was first introduced in Stückrad-Vogel [14] and [0] in which they gave an answer to the following question of Buchsbaum [4]: Is it true that for any parameter ideal $\mathfrak{q}$ in a noetherian local ring the difference $\lambda_A(A/\mathfrak{q}) - e(\mathfrak{q};A)$ is an invariant of A not depending on $\mathfrak{q}$? Stückrad and Vogel gave counter examples to Buchsbaum's question and they could show that a local ring for which Buchsbaum's question is true has the remarkable property that every system of parameters is a weak sequence. They called such a local ring Buchsbaum. The next result shows that our definition of Buchsbaum rings is equivalent to Stückrad-Vogel's definition.

<u>(41.15), Theorem</u>. Let $(A,\mathfrak{m})$ be a noetherian local ring with $d = \dim A > 0$. Then the following are equivalent:

1) A is a Buchsbaum ring.

2) Every system of parameters of A is a weak sequence.

3) There is an integer $c \geqq 0$ such that

$$\lambda_A(A/\mathfrak{q}) - e(\mathfrak{q};A) = c$$

for any parameter ideal $\mathfrak{q}$ of A .

4) There is a system of generators $x_1,\ldots,x_r$ of the maximal ideal $\mathfrak{m}$ such that the canonical map

$$H^i(x_1,\ldots,x_r;A) \longrightarrow H^i_{\mathfrak{m}}(A)$$

is surjective for $i < d$.

Proof. This is an immediate consequence of Theorem (41.5) and Theorem (41.12).

<u>(41.16) Remark</u>. If A is a Buchsbaum ring and if $(a_1,\ldots,a_r)$ form a part of a system of parameters in A , then $A/(a_1,\ldots,a_r)$ is a Buchsbaum ring too. The converse is not true in general: Take a Buchsbaum ring, which is not Cohen-Macaulay; let X be an indeterminate over A . Then $A[X]$ is not Buchsbaum but $A[X]/X \cong A$ is so, see also [19]).

The following result of Stückrad has played a dominant role in the theory of Buchsbaum rings.

(41.17) Corollary. Let $(A,\mathfrak{m},k)$ be a noetherian local ring with $d = \dim A > 0$. Suppose that the canonical map

$$\psi^i : \operatorname{Ext}^i_A(k,A) \longrightarrow H^i_{\mathfrak{m}}(A)$$

is surjective for all $i < d$. Then A is Buchsbaum.

Proof. Let $x_1,\ldots,x_r$ be generators of the maximal ideal $\mathfrak{m}$ and let $\ldots \to F_n \to F_{n-1} \to \ldots \to F_1 \to F_0 \to k \to 0$ be a minimal free resolution of k. Then there is a map $\alpha.$ of complexes:

$$\begin{array}{ccccccccccccc}
0 \to \Lambda^r A^r & \to & \Lambda^{r-1}A^r & \to & \ldots & \to & \Lambda^2 A^r & \to & \Lambda^1 A^r & \to & A & \to & 0 \\
\downarrow \alpha_r & & \downarrow \alpha_{r-1} & & & & \downarrow \alpha_2 & & \downarrow \alpha_1 & & \downarrow \alpha_{0=\mathrm{id}} & & \\
\ldots \to F_r & \to & F_{r-1} & \to & \ldots & \to & F_2 & \to & F_1 & \to & F_0 & \to & 0
\end{array}$$

This yields an homomorphism $\alpha^i : \operatorname{Ext}^i_A(k,A) \longrightarrow H^i(x_1,\ldots,x_r;A)$ which makes the following diagram commute

$$\begin{array}{ccc}
\operatorname{Ext}^i_A(k,A) & \xrightarrow{\alpha^i} & H^i(x_1,\ldots,x_r;A) \\
& \psi^i \searrow \quad \swarrow \varphi^i & \\
& H^i_{\mathfrak{m}}(A) & ,
\end{array}$$

where φ^i is the canonical map, cf. Corollary (41.9).

By assumption ψ^i is surjective and hence φ^i is surjective for $i < d$. By Theorem (41.15) A is Buchsbaum, q.e.d.

Now applying Theorem (41.15), we know that if A is a Buchsbaum ring, $H^i_{\mathfrak{m}}(A)$ is a finite dimensional k-vectorspace. The converse of this is not true in general.

The last topic of this section is an estimation of the multiplicity of Buchsbaum rings. First we need an auxiliary result.

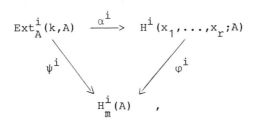

<u>(41.18) Lemma.</u> Let $a_1,\dots,a_d$ be a standard system of parameters of a generalized Cohen-Macaulay ring $(A,\mathfrak{m})$ with $d = \dim A > 0$. Then we have for $1 \leq k \leq d$:

$$\lambda_A\left(\frac{(a_1^2,\dots,a_k^2) : (a_1\dots a_k)}{(a_1,\dots,a_k)}\right) = \sum_{i=0}^{k-1}\binom{k}{i} h^i(A) \quad ,$$

where $h^i(A) = \lambda_A(H^i_{\mathfrak{m}}(A))$.

Proof. By induction on k : Let $k = 1$. Then $(a_1^2 : a_1) = (a_1) + U(0)$. Hence

$$\lambda_A\left(\frac{(a_1^2 : a_1)}{(a_1)}\right) = \lambda_A(U(0)/(a_1) \cap U(0)) = \lambda_A(U(0)) = h^0(A) \quad .$$

Let $k \geq 2$. Recall that, by Theorem (38.13), we have:

$$(a_1^2,\dots,a_k^2) : (a_1\dots a_k) = \sum_{i=1}^{k} U(a_1,\dots,\check{a}_i,\dots,a_k) \quad .$$

Let $\mathfrak{q}_i = (a_1,\dots,a_i)$ for $1 \leq i \leq d$.

Consider the exact sequence

$$(*)\qquad 0 \longrightarrow \frac{\mathfrak{q}_k + \sum_{i=1}^{k-1} U(a_1,\dots,\check{a}_i,\dots,a_k)}{\mathfrak{q}_k} \longrightarrow \frac{(a_1^2,\dots,a_k^2) : a_1\dots a_k}{\mathfrak{q}_k} \longrightarrow \frac{\sum_{i=1}^{k} U(a_1,\dots,\check{a}_i,\dots,a_k)}{\mathfrak{q}_k + \sum_{i=1}^{k-1} U(a_1,\dots,\check{a}_i,\dots,a_k)} \longrightarrow 0 \quad .$$

The first term of the exact sequence (*) is isomorphic to $\dfrac{(\bar{a}_1^2,\dots,\bar{a}_{k-1}^2) : (\bar{a}_1\dots\bar{a}_{k-1})}{(\bar{a}_1,\dots,\bar{a}_{k-1})}$, where $^-$ means modulo reduction by a_kA . Hence by induction

$$\lambda_A\left(\frac{\mathfrak{q}_k + \sum_{i=1}^{k-1} U(a_1,\dots,\check{a}_i,\dots,a_k)}{\mathfrak{q}_k}\right) = \sum_{i=0}^{k-2}\binom{k-1}{i} h^i(\bar{A})$$

$$(**) \qquad = \sum_{i=0}^{k-2}\binom{k-1}{i}\left(h^i(A) + h^{i+1}(A)\right)$$

$$= (k-1)h^{k-1}(A) + \sum_{i=0}^{k-2}\binom{k}{i} h^i(A) \quad .$$

The last term in (*) is isomorphic to

$$\frac{U(a_1,\dots,a_{k-1})}{\mathfrak{q}_{k-1} + \sum_{i=1}^{k-1} U(a_1,\dots,\check{a}_i,\dots,a_{k-1})} = \frac{U(a_1,\dots,a_{k-1})}{(a_1^2,\dots,a_{k-1}^2):a_1\dots a_{k-1}} \cong H_{\mathfrak{m}}^{k-1}(A)$$

by Theorem (38.13) and Theorem (39.1).

From (*) and (**) we get the required formula.

(41.19) Theorem. Let $(A,\mathfrak{m})$ be a Buchsbaum ring with $d = \dim A > 0$. Then

$$e(A) \geqq 1 + \sum_{i=1}^{d-1}\binom{d-1}{i-1} h^i(A) \quad .$$

where $h^i(A) = \lambda_A(H_{\mathfrak{m}}^i(A))$.

Proof. If necessary, passing to the local ring $A[X]_{\mathfrak{m}A[X]}$, we may assume that A has an infinite residue field, cf. Theorem (41.15). Then there is a minimal reduction $\mathfrak{q} = (a_1,\dots,a_d)$ of $\mathfrak{m}$. Since $e(A) = e(\mathfrak{q};A)$ we have by Lemma (41.18)

$$\begin{aligned} e(A) &= \lambda_A(A/\mathfrak{q}) - I(A) \\ &= \lambda_A(A/\mathfrak{m}) + \lambda_A(\mathfrak{m}/(a_1^2,\dots,a_d^2) : (a_1\dots a_d)) + \lambda_A\left(\frac{(a_1^2,\dots,a_d^2):(a_1\dots a_d)}{\mathfrak{q}}\right) \end{aligned}$$

$$- \sum_{i=0}^{d-1} \binom{d-1}{i} h^i(A)$$

$$\geq 1 + \sum_{i=0}^{d-1} \binom{d}{i} h^i(A) - \sum_{i=0}^{d-1} \binom{d-1}{i} h^i(A)$$

$$= 1 + \sum_{i=1}^{d-1} \binom{d-1}{i-1} h^i(A) \quad , \text{q.e.d.}$$

(41.20) Remark. In Theorem (41.19) the equality holds if and only if $\mathfrak{m} = (a_1^2,\ldots,a_d^2) : (a_1\ldots a_d)$. In this case we have $\mathfrak{m}^2 = (a_1,\ldots,a_d)\mathfrak{m}$, see [7].

(41.21) Proposition. Let $(A,\mathfrak{m})$ be a Buchsbaum ring. Then $Bl(\mathfrak{q},A)$ is Cohen-Macaulay for all parameter ideals $\mathfrak{q} \subset A$.

Proof. Use Proposition (40.3) and the fact that in Buchsbaum rings every system of parameters is standard.

(41.22) Remark. There is a more general result [6], saying that the following statements are equivalent:

(i) $Bl(\mathfrak{q},A)$ is Cohen-Macaulay for all parameter ideals

(ii) $A/H_{\mathfrak{m}}^0(A)$ is Buchsbaum.

(41.23) Remark. At the end of this chapter we have to mention that Theorem (40.10) and Corollary (40.11) can be proved under more general assumptions. This has been discussed by M. Brodmann and - in a more extensive context - by S. Goto and K. Yamagishi. We will give here a glimpse of these developments:

(1) In [3] M.Brodmann has introduced a socalled "permutable standard sequence", and he studied the blowing-up of rings by ideals generated by those sequences. To indicate this, we define for a noetherian local ring $(A,\mathfrak{m})$ an integer

$$t := \max\left\{r \mid \lambda(H_{\mathfrak{m}}^i(A)) < \infty \quad \text{for all} \quad i < r\right\} .$$

Then a sequence $a_1, a_2, \ldots, a_t \in \mathfrak{m}$ is called a permutable standard sequence if the following two conditions hold in any order:

(i) There is some integer $n > 0$ such that

$$(a_1, \ldots, a_{k-1}) : a_k \subset (a_1, \ldots, a_{k-1}) : \mathfrak{m}^n$$

for all $k = 1, \ldots, t$;

(ii) For all integers i, j $(0 \leq i + j < t)$,

$$\mathfrak{q} \cdot H^i_{\mathfrak{m}}(A/(a_1, \ldots, a_j)) = (0) \quad ,$$

where $\mathfrak{q} = (a_1, \ldots, a_t)$.

Then one can prove the same statements as in Theorem (40.10) and Corollary (40.11) provided we replace the notion of a "standard system of parameters" by the notion of a "permutable standard sequence" and the integer $d = \dim A$ by the integer t . It is obvious that Brodmann's work has coverd a wider class of local rings than the class of generalized Cohen-Macaulay rings.

(2) S. Goto and K. Yamagishi has introduced in [9] the new notion of "an unconditioned strong d-sequence". In this work a sequence $a_1, a_2, \ldots, a_s$ of elements in A is called an "unconditioned strong d-sequence" if every power $a_1^{n_1}, a_2^{n_2}, \ldots, a_s^{n_s}$ (where $n_i > 0$) form a d-sequence in any order (actually it is enough to assume it only for $n_i = 1$ or 2 .) One key-point of their work is to emphasize "sequence-properties". Recall that the notions of a "regular sequence" or a "weak sequence" were useful to characterize the class of Cohen-Macaulay rings or Buchsbaum rings. Therefore Goto and Yamagishi asked for a good "sequence-property" to characterize the class of generalized Cohen-Macaulay rings too. This is the main motivation for their new notion of an unconditioned strong d-sequence. It comes out that if there exists a system of parameters in a noetherian local ring A forming an unconditioned strong d-sequence, then the given A is a generalized Cohen-Macaulay ring. On the other hand, for any system $a_1, a_2, \ldots, a_d$ of parameters in a generalized Cohen-Macaulay ring there exists an integer $n > 0$ such that $a_1^n, a_2^n, \ldots, a_d^n$ is an unconditioned strong d-sequence. Using this notion Goto and Yamagishi have developed a theory which unifies the whole facts on (sub-) systems of parameters

for Buchsbaum rings, generalized Cohen-Macaulay rings, and for the wider class of local rings in the sense of Brodmann. The local cohomology functors in their work are the direct limits of the Koszul cohomology functors w.r.t. an unconditioned strong d-sequence. [These functors coincide with the original local cohomology functors in case that the ring A is a noetherian local ring]. This is one of the reasons that their arguments don't need any assumptions on finiteness conditons for the used rings (and modules).

Another interesting result concerning unconditioned strong d-sequences was recently given by N. Suzuki [16]:

Let E be any A-module and let I be an injective A-module. Assume that $a_1,\dots,a_d$ is an unconditioned strong d-sequence on E, then it is also an unconditioned strong d-sequence on $\mathrm{Hom}_A(H^d_{\mathfrak{q}}(E),I)$, where $\mathfrak{q}$ is the ideal in A generated by $a_1,\dots,a_d$.

§ 42. Examples.

First we prove the following useful result, which gives a sufficient condition for the Buchsbaum property of a local ring A.

(42.1) Proposition. Let $(A,\mathfrak{m},k)$ be a local ring with $t := \operatorname{depth} A < d := \dim A$. If $H^i_{\mathfrak{m}}(A) = 0$ for $i \neq t, d$ and $\mathfrak{m}H^t_{\mathfrak{m}}(A) = 0$, then A is a Buchsbaum ring.

Proof. By Corollary (41.17) it is enough to show that the canoncical map $\psi^t : \mathrm{Ext}^t_A(k,A) \longrightarrow H^t_{\mathfrak{m}}(A)$ is surjective. For that let

$$0 \longrightarrow A \longrightarrow I^0 \xrightarrow{\partial^0} I^1 \xrightarrow{\partial^1} \dots \longrightarrow I^d \xrightarrow{\partial^d} \dots$$

be a minimal injective resolution of A.

Since $\operatorname{depth} A = t$ we have $\mathrm{Ext}^i_A(k,A) = 0$ for $i < t$; therefore $H^0_{\mathfrak{m}}(I^i) = 0$ for $i < t$ by Theorem (33.25). From this we obtain the following commutative diagram with exact rows:

$$\begin{array}{ccccccc}
0 \longrightarrow & \mathrm{Ext}_A^t(k,A) & \longrightarrow & \mathrm{Hom}_A(k,I^t) & \xrightarrow{\partial^t} & \mathrm{Hom}_A(k,I^{t+1}) \\
 & \cap \downarrow \psi^t & & \cap \downarrow & & \cap \downarrow \\
0 \longrightarrow & H_{\mathfrak{m}}^t(A) & \longrightarrow & H_{\mathfrak{m}}^0(I^t) & \xrightarrow{\partial^t} & H_{\mathfrak{m}}^0(I^{t+1})
\end{array}$$

Since $I^{\cdot}$ is minimal, ∂^t is the zero-map; see Chapter VII, Lemma (33.18). This implies

$$\begin{aligned}
\mathrm{Ext}_A^t(k,A) &= \mathrm{Hom}_A(k,I^t) \\
&= \mathrm{Hom}_A(k,H_{\mathfrak{m}}^0(I^t)) \\
&= (0 : \mathfrak{m})_{H_{\mathfrak{m}}^0(I^t)}
\end{aligned}$$

Since $\mathfrak{m}H_{\mathfrak{m}}^t(A) = 0$ by the assumption, we get

$$H_{\mathfrak{m}}^t(A) \subset (0 : \mathfrak{m})_{H_{\mathfrak{m}}^0(I^t)} = \mathrm{Ext}_A^t(k,A) \subset H_{\mathfrak{m}}^t(A) \quad .$$

Hence $\mathrm{Ext}_A^t(k,A) = H_{\mathfrak{m}}^t(A)$, as submodules of I^t, and this shows that ψ^t is an isomorphism.

As an application of this result we present several examples. We always assume in the sequel that k is a field and $X,Y,Z,\ldots$ are indeterminates. In the Examples (42.2) and (42.3) we consider rings A which are not domains. It seems to be much harder to find interesting examples where A is a domain, see for that Examples (42.4) and (42.5).

(42.2) Example. Let $A = k[[X,Y,Z,W]]/(X,Y) \cap (Z,W) = k[[x,y,z,w]]$. From the exact sequence

$$0 \longrightarrow A \longrightarrow A/(x,y) \oplus A/(z,w) \longrightarrow A/(x,y,z,w) = k \longrightarrow 0$$

we get: $H_{\mathfrak{m}}^1(A) \simeq k$ and $H_{\mathfrak{m}}^0(A) = 0$, which shows that A is Buchsbaum (since $\dim A = 2$).

(42.3) Example. Let $A = k[[X,Y,Z,W]]/(X,Y)\cap(Z,W)\cap(X^2,Y,Z^2,W) = k[[x,y,z,w]]$.

Claim: $\mathfrak{m}H^0_{\mathfrak{m}}(A) = \mathfrak{m}H^1_{\mathfrak{m}}(A) = 0$, but A is not Buchsbaum. [In this case A is said to be quasi-Buchsbaum.]

Proof. It is easy to see that $H^0_{\mathfrak{m}}(A) = (x,y)\cap(z,w)$ and $\mathfrak{m}H^0_{\mathfrak{m}}(A) = 0$. Then $A/H^0_{\mathfrak{m}}(A)$ is isomorphic to the local ring considered in Example (42.2 Hence $H^1_{\mathfrak{m}}(A) = H^1_{\mathfrak{m}}(A/H^0_{\mathfrak{m}}(A)) = k$. Assume that A is Buchsbaum. Then $A/(x+z)A$ has to be Buchsbaum by Theorem (41.15) (2), since $(x+z)$ is a part of a system of parameters. But

$$\bar{A} := A/(x+z)A \simeq k[[X,Y,W]]/(X^3,XW,XY,YW) = k[[\bar{x},\bar{y},\bar{w}]]$$

is not Buchsbaum, because $\bar{x}\in H^0_{\mathfrak{m}}(\bar{A})$ and $0\neq \bar{x}^2 \in \mathfrak{m}\cdot H^0_{\mathfrak{m}}(\bar{A})$.

(42.4) Example. Let $A = k[[X^4,X^3Y,XY^3,Y^4]] \subset B = k[[X^4,X^3Y,X^2Y^2,XY^3,Y^4]]$.

Claim: A is a Buchsbaum ring.

Proof. Note that $B = k[[(X,Y)^4]] \subset k[[X,Y]]$ is a Cohen-Macaulay ring since X^4, Y^4 form a regular sequence in B (but not in A). Now for any $f\in B$ (considered as an A-module) we have for suitable power series $f_i(X^4,Y^4)$ in X^4 and Y^4

$$f = f_0(X^4,Y^4) + f_1(X^4,Y^4)X^3Y + f_2(X^4,Y^4)X^2Y^2 + f_3(X^4,Y^4)XY^3$$

with

$$f_2(X^4,Y^4)\cdot X^2Y^2 = \alpha_0 X^2Y^2 + \sum_{\substack{i>0 \\ \text{or} \\ j>0}} \alpha_{i,j} X^{4i}Y^{4j}\cdot X^2Y^2 ,$$

where $\alpha_0,\alpha_{ij}\in k$. Since the second term is in A , we conclude that B/A is a k-vector space of dimension 1, i.e. we have an exact sequence of A-modules

$$0 \longrightarrow A \longrightarrow B \longrightarrow k \longrightarrow 0 .$$

This implies $H^1_{\mathfrak{m}}(A) = k$; moreover we have $H^0_{\mathfrak{m}}(A) = 0$. Therefore A is Buchsbaum by Proposition (42.1).

(42.5) Example. Let $A = k[[X^5, X^4Y, X^3Y^2, XY^4, Y^5]]$, $B = k[[X^5, X^4Y, XY^4, Y^5]]$ and $C = k[[X^5, X^4Y, X^3Y^2, X^2Y^3, XY^4, Y^5]]$.

Claim: A is Buchsbaum; and B is generalized Cohen-Macaulay, but not Buchsbaum.

Proof. As in Example (42.4) we obtain the Buchsbaum-property of A from the exact sequence of A-modules

$$0 \longrightarrow A \longrightarrow C \longrightarrow k \longrightarrow 0 ,$$

where C is Cohen-Macaulay. Now every two-dimensional complete local domain is a generalized Cohen-Macaulay ring (by Theorem (37.4)), so B is a generalized Cohen-Macaulay ring. We denote by $\mathfrak{n}$ the maximal ideal of B. Then we conclude from the exact sequence of B-modules

$$0 \longrightarrow B \longrightarrow C \longrightarrow C/B \longrightarrow 0$$

that $H^0_{\mathfrak{n}}(C/B) \simeq H^1_{\mathfrak{n}}(B)$.

Note that $X^5 \cdot (X^3Y^2) = (X^4Y)^2 \in B$ and $(Y^5)^2(X^3Y^2) = (XY^4)^3 \in B$. Hence the image of X^3Y^2 in C/B belongs to $H^0_{\mathfrak{n}}(C/B)$. But $(X^4Y)\cdot(X^3Y^2) = X^7Y^3 \notin B$. Thus $\mathfrak{n}H^1_{\mathfrak{n}}(B) \neq 0$, hence B is not Buchsbaum.

Finally we mention two easy examples where the ring A has $\operatorname{depth} A = 0$.

(42.6) Example. Let $A = k[[X,Y,Z]]/(X^2, XY, XZ)k[[X,Y,Z]] = k[[x,y,z]]$.

Claim: A is a Buchsbaum ring.

Proof. Let $\mathfrak{a} = (X^2, XY, XZ) = (X) \cap (X^2, Y, Z)$. Then the exact sequence

$$0 \longrightarrow \frac{k[[X,Y,Z]]}{\mathfrak{a}} \longrightarrow \frac{k[[X,Y,Z]]}{(X)} \oplus \frac{k[[X,Y,Z]]}{(X^2,Y,Z)} \longrightarrow k \longrightarrow 0$$

implies $H^1_{\mathfrak{m}}(A) = 0$, where $\mathfrak{m}$ denotes the maximal ideal of A. Moreover $H^0_{\mathfrak{m}}(A) = (x)$ and $\mathfrak{m}H^0_{\mathfrak{m}}(A) = 0$. Therefore A is Buchsbaum by Proposition (42.1).

(42.7) Example. Let $A = k[[X,Y,Z]]/(X^3,XY,XZ)k[[X,Y,Z]] = k[[x,y,z]]$.

Claim: A is generalized Cohen-Macaulay, but not quasi-Buchsbaum (hence not Buchsbaum).

Proof. The corresponding exact sequence of (42.6) for $\mathfrak{a} = (X) \cap (X^3,Y,Z)$ implies again that $H^1_{\mathfrak{m}}(A) = 0$. Moreover $H^0_{\mathfrak{m}}(A) = (x)$, $\mathfrak{m}^2 H^0_{\mathfrak{m}}(A) = 0$, but $\mathfrak{m}H^0_{\mathfrak{m}}(A) \neq 0$.

(42.8) Example (Goto). $B := k[[s^2,s^3,st,t]]$ is a non-Cohen-Macaulay ring with $\operatorname{depth}(B) = 1$ and $\dim(B) = 2$. Now it is easy to show, that B is Buchsbaum: Let x_1,x_2 be a system of parameters of B . From the exact sequence

$$0 \longrightarrow B \longrightarrow k[[s,t]] \longrightarrow k \longrightarrow 0$$

we get an exact sequence for the Koszul homology:

$$H_1(x_1,x_2;B) \longrightarrow H_1(x_1,x_2;k[[s,t]]) \longrightarrow H_1(x_1,x_2;k) \longrightarrow$$

$$B/(x_1,x_2)B \longrightarrow k[[s,t]]/(x_1,x_2) \longrightarrow k \longrightarrow 0 \quad .$$

Since $k[[s,t]]$ is Cohen-Macaulay and $H_1(x_1,x_2;k) \simeq k^2$ it follows that

$$\lambda(B/(x_1,x_2)B) - e(x_1,x_2;B) = \lambda(k[[s,t]]/(x_1,x_2)) - e(x_1,x_2;B) + 1$$

$$= \lambda(k[[s,t]]/(x_1,x_2)) - e(x_1,x_2;k[[s,t]]) + 1 = 1 \quad .$$

Hence B is Buchsbaum, by Theorem (41.15).

The ring $A := k[[s^2,s^5,st,t]]$ is generalized Cohen-Macaulay but not Buchsbaum, as one can see as follows: From the exact sequence $0 \longrightarrow A \longrightarrow k[[s,t]] \longrightarrow k[[s,t]]/A \longrightarrow 0$ we get:

$$H^0_{\mathfrak{m}}(A) = 0 \quad \text{and}$$
$$H^1_{\mathfrak{m}}(A) = H^0_{\mathfrak{m}}(k[[s,t]]/A) \cong k[[s,t]]/A \cong k^2 \quad ,$$

where $\mathfrak{m}$ is the maximal ideal in A .

It is easy to check, that the local cohomology $H^i_m(A)$ is annulated by the m-primary ideal (s^4,s^5,st,t) in A for all $i < \dim(A) = 2$. Therefore A must be generalized Cohen-Macaulay, and every system of parameters which is contained in (s^4,s^5,st,t) is a standard system. But the system $\{s^4,t\}$ is not a weak sequence since

$$(s^4 : t) = (s^4,s^5) \neq (s^4,s^7) = (s^4 : m) \quad .$$

Hence A is not Buchsbaum by (41.15). Moreover we obtain for $A = k[[s^2,s^5,st,t]]$ the following relations:

$$I(A) = 2$$
$$e(s^2,t;A) = 2$$
$$\lambda(A/(s^2,t)) = 3 \quad ,$$

hence

$$2 = I(A) \underset{\neq}{>} \lambda(A/(s^2,t)) - e(s^2,t;A) = 1 \quad ,$$

i.e. (s^2,t) is not a standard system of parameters in A. Note that: (t,s^2) do not form a d-sequence since $(t : s^4) \not\ni st$, but $st \in (t : s^2)$.

References - Chapter VIII

Books

[0] J. Stückrad and W. Vogel, Buchsbaum rings and applications, Springer-Verlag 1987.

[1] R. Hartshorne, Residues and Duality, Springer-Verlag 1966.

Papers

[2] Auslander-Buchsbaum, Codimension and multiplicity, Annals Math., 68 (1958), 625 - 657.

[3] M. Brodmann, Local cohomology of certain Rees- and form-rings, I, J. Alg., 81 (1983), 29 - 57.

[4] D.A. Buchsbaum, Complexes in local ring theory, in "Some Aspects of Ring Theory", 223 - 228, C.I.M.E. Roma 1965.

[5] N.T. Cuong, N.V. Trung and P. Schenzel, Verallgemeinerte Cohen-Macaulay-Moduln, M. Nachr., 85 (1978), 57 - 73.

[6] S. Goto, Blowing-up of Buchsbaum rings, Comm. Alg.: Durham 1981, London M.S.LN 72, 140 - 162.

[7] S. Goto, On the associated graded rings of parameter ideals in Buchsbaum rings, J. Alg., 85 (1983), 490 - 534.

[8] S. Goto and Y. Shimoda, On the Rees algebras over Buchsbaum rings, J. M. Kyoto U., 20 (1980), 699 - 708.

[9] S. Goto and K. Yamagishi, The theory of unconditioned strong d-sequences and modules of finite local cohomology, Preprint 1985

[10] C. Huneke, The theory of d-sequences and powers of ideals, Ad. Math., 46 (1982), 249 - 279.

[11] P. Schenzel, Standard systems of parameters and their blowing-up rings, J. reine und angew. M., 344 (1983), 201 - 220.

[12] P. Schenzel, Applications of dualizing complexes to Buchsbaum rings, Ad. Math., 44 (1982), 61 - 77.

[13] J. Stückrad, Über die kohomologische Charakterisierung von Buchsbaum-Moduln, M. Nachr., 95 (1980), 265 - 272.

[14] J. Stückrad and W. Vogel, Eine Verallgemeinerung der Cohen-Macaulay-Ringe und Anwendungen auf ein Problem der Multiplizitätstheorie, J. M. Kyoto U., 13 (1973), 513 - 528.

[15] J. Stückrad and W. Vogel, Toward a theory of Buchsbaum singularities, AJM., 100 (1978), 727 - 746.

[16] N. Suzuki, Canonical duality for unconditioned strong d-sequences to appear in J. M. Kyoto U. 1986/87.

[17] N.V. Trung, Toward a theory of generalized Cohen-Macaulay modules Nagoya Math. J. 102 (1986), 1 - 49.

[18] W. Vogel, Über eine Vermutung von D.A. Buchsbaum, J. Alg., 25 (1973), 106 - 112.

[19] W. Vogel, A nonzero-divisor characterization of Buchsbaum modules, Michigan J. 28 (1981), 147 - 152.

[20] A. Grothendieck and J. Dieudonné, Elements de Géométrie Algébrique III, No. 11, I.H.E.S. Paris 1961.

Chapter IX. APPLICATIONS OF LOCAL COHOMOLOGY TO THE COHEN-MACAULAY-BEHAVIOUR OF BLOWING UP RINGS

Here the results of Chapter V are partially extended and rephrased in a different context. One motivation for this chapter is the fact, that the Cohen-Macaulay (CM) properties of an algebraic variety X and its blowing up X' with center $Y \subset X$ are totally unrelated, unless we have suitable properties for Y and e.g. the local cohomology modules of the affine vertex over X have finite length in all orders $\leq \dim X$. Hence, replacing again X by a local ring $(A,\mathfrak{m})$ and Y by an ideal $I \subset A$ we want to relate the CM-property of the Rees ring $B(I,A) = \bigoplus_{n\geq 0} I^n$ to the CM-properties of A, $B(\mathfrak{m},A)$ and $G(I,A) \otimes A/\mathfrak{m}$ under suitable cohomological conditions. Our first aim is to give a general criterion of the Cohen-Macaulayness of Rees algebras in terms of local cohomology, see main Theorem (44.1). Then we ask this question for Rees rings of equimultiple ideals I, in particular of $\mathfrak{m}$-primary ideals and of ideals $\mathfrak{q}$ and $\mathfrak{q}^{\nu}$, where $\mathfrak{q}$ is generated by a system of parameters.

Finally we give special applications in §48 to rings $(A,\mathfrak{m})$ with low multiplicity. Here we follow the idea that those rings are "better" than rings of high multiplicities. Corresponding to standard geometric situations we will restrict ourselves to equimultiple ideals I, being different from $\mathfrak{m}$-primary ideals. Several examples will demonstrate our idea. There are similar results for Rees rings which have the Gorenstein property. This can be found in [16].

§ 43. Generalized Cohen-Macaulay-rings with respect to an ideal

Generalized CM-rings $(A,\mathfrak{m})$ has been characterized in Chapter VIII, Lemma (37.3) by the condition that there exists an $\mathfrak{m}$-primary ideal I such that $IH^i_{\mathfrak{m}}(A) = 0$ for $i < \dim A$, i.e. all $H^i_{\mathfrak{m}}(A)$ with $i < \dim A$ are killed by a power of $\mathfrak{m}$. Algebraically we may consider the corresponding property for an arbitrary ideal $\mathfrak{a} \subset A$. Geometrically this property is motivated by the following fact (see in particular (47.9)). If $(A,\mathfrak{m})$ is quasi-unmixed and if $B\ell(\mathfrak{a},R)$ is Cohen-Macaulay then there is an integer $n > 0$ such that $\mathfrak{a}^n H^i_{\mathfrak{m}}(A) = 0$ for $i < \dim A$. Therefore we make the following definition.

(43.1) Definition. A noetherian local ring $(A,\mathfrak{m})$ is called generalized Cohen-Macaulay with respect to an ideal $\mathfrak{a}$, if there exists an integer $n > 0$ such that $\mathfrak{a}^n H^i_{\mathfrak{m}}(A) = 0$ for $i < \dim A$.

As mentioned above, for $\mathfrak{a} = \mathfrak{m}$ we get Definition (37.5) in Chapter VIII for a generalized Cohen-Macaulay ring.

(43.2) Lemma. If $(A,\mathfrak{m})$ is the homomorphic image of a Gorenstein ring and if $\mathfrak{a}$ is an ideal in A then the following statements are equivalent:

(1) A is generalized Cohen-Macaulay with respect to $\mathfrak{a}$.

(2) $A_{\mathfrak{p}}$ is Cohen-Macaulay and $\dim A/\mathfrak{p} + \dim A_{\mathfrak{p}} = \dim A$ for $\mathfrak{p} \in \operatorname{Spec} A - V(\mathfrak{a})$.

Proof. Let $(B,\mathfrak{n})$ be a Gorenstein local ring such that $A = B/I$ for some ideal I of B. Let $d = \dim A$ and $n = \dim B$.

$(1) \Rightarrow (2)$: We may assume that A and B are complete (see Corollary (37.6) in Chapter VIII). Then by local duality we have for $i \geq 0$ the following A-linear isomorphisms

$$\operatorname{Hom}_B(H^i_{\mathfrak{m}}(A), E_B) \cong \operatorname{Ext}^{n-i}_B(A,B) \quad ,$$

where E_B is the injecitve envelope of $B/\mathfrak{n}$. By assumption (1) we have $\mathfrak{a}^k H^i_{\mathfrak{m}}(A) = 0$ for $i < d$. Since $\operatorname{Hom}_B(-,E_B)$ is an exact functor, we get $\mathfrak{a}^k \operatorname{Ext}^{n-i}_B(A,B) = 0$ for $i < d$. By Chapter VII, Corollary (36.15), we know that $A_{\mathfrak{p}}$ is a Cohen-Macaulay-ring such that $\dim B_{\mathfrak{P}} - \dim A_{\mathfrak{p}} = n - d$ for all $\mathfrak{p} \in \operatorname{Spec} A \setminus V(\mathfrak{a})$ and for the inverse image $\mathfrak{P}$ of $\mathfrak{p}$ in B. Hence $d = \dim B - \dim B_{\mathfrak{P}} + \dim A = \dim A/\mathfrak{p} + \dim A$

$(2) \Rightarrow (1)$: Let $\mathfrak{p} \in \operatorname{Spec}(A) - V(\mathfrak{a})$ and let $\mathfrak{P}$ be the inverse image of $\mathfrak{p}$ in B. Then $\operatorname{Ext}^i_{B_{\mathfrak{P}}}(A_{\mathfrak{p}}, B_{\mathfrak{P}}) = 0$ for $i \neq \dim B_{\mathfrak{P}} - \dim A_{\mathfrak{p}} = n - d$ by Corollary (36.15). Hence $\operatorname{Supp}_A\left(\operatorname{Ext}^i_B(A,B)\right) \subset V(\mathfrak{a})$ for $i \neq n - d$. Therefore there is an integer $k > 0$ such that $\mathfrak{a}^k \operatorname{Ext}^i_B(A,B) = 0$ for $i > n - d$. Hence by the local duality we have:

$$\mathfrak{a}^k H^i_{\mathfrak{m}}(A) = 0 \quad \text{for} \quad i < d \;, \qquad \text{q.e.d.}$$

The graded version of Lemma (43.2) is as follows.

(43.3) Lemma. Let $R = \bigoplus_{n\geq 0} R_n$ be a noetherian graded ring defined over a local ring which is a homomorphic image of a Gorenstein ring S, and let $\mathfrak{m}$ be the maximal homogeneous ideal of R. Let $\mathfrak{a} = R_+ = \bigoplus_{n>0} R_n$. Then the following statements are equivalent:

(1) $R_{\mathfrak{m}}$ is generalized Cohen-Macaulay with respect to $\mathfrak{a}R_{\mathfrak{m}}$.

(2) $R_{\mathfrak{p}}$ is Cohen-Macaulay and $\dim R/\mathfrak{p} + \dim R_{\mathfrak{p}} = \dim R$ for $\mathfrak{p} \in \mathrm{Proj}(R)$.

(3) There is an integer $n > 0$ such that $\underline{H}^i_{\mathfrak{m}}(R)_\nu = 0$ for $\nu < -n$ and $i < d = \dim R$.

In this case $\mathfrak{a}^k \underline{H}^i_{\mathfrak{m}}(R) = 0$ for $i < d$ and some $k \geq 0$.

Proof. To (1)⟺(2): By Corollary (33.27), we know that $R_{\mathfrak{p}^*}$ is Cohen-Macaulay if and only if $R_{\mathfrak{p}}$ is Cohen-Macaulay, where $\mathfrak{p}^* = H(\mathfrak{p})$. Moreover R is universally catenarian by [3], 14.B, 16.E (Theorem 31). Then the equivalence of (1) and (2) follows from Lemma (43.2).

(3) ⇒ (1): Note that $\underline{H}^i_{\mathfrak{m}}(R) = H^i_{\mathfrak{m}}(R) = H^i_{\mathfrak{m}R_{\mathfrak{m}}}(R_{\mathfrak{m}})$ for all $i \geq 0$ (see Chapter VII, proof of Lemma (36.2)) and that $\underline{H}^i_{\mathfrak{m}}(R)$ is an artinian R-module, i.e. in particular: $\underline{H}^i_{\mathfrak{m}}(R)_\nu = 0$ for ν large. Hence we easily see that (3) implies (1).

(1) ⇒ (3): We may assume that R_0 and S_0 are complete. Then by the local duality of graded rings defined over complete local rings R_0 and S_0 we have:

$$\underline{\mathrm{Hom}}_S(H^i_{\mathfrak{m}}(R), \underline{E}_S) \cong \underline{\mathrm{Ext}}_S^{n-i}(R,S)(a)$$

where $a = a(S)$ is the a-invariant of S (Remark (36.12), Chapter VII), and $n = \dim S$. By assumption we get an integer $k \geq 0$ such that

$$\mathfrak{a}^k \underline{\mathrm{Ext}}_S^{n-i}(R,S) = 0 \quad \text{for} \quad i < d = \dim R \quad .$$

By Nakayama's lemma for graded rings which will be proved later as Lemma (43.4), $\underline{\mathrm{Ext}}_S^{n-i}(R,S)_\nu = 0$ for large $\nu >> 0$. Then again by local duality and Matlis duality we get $\underline{H}^i_{\mathfrak{m}}(R)_\nu = 0$ for $i < d$ and $\nu << 0$. This completes the proof of Lemma (43.3).

<u>(43.4) Lemma</u>. Let $R = \bigoplus_{n\geq 0} R_n$ be a graded noetherian ring defined over a local ring R_0 and let $\mathfrak{a} = \bigoplus_{n>0} R_n$. Let M be a finitely generated graded R-module such that $\mathfrak{a}^k M = 0$ for some $k \geq 0$. Then $M_n = 0$ for all large $n >> 0$.

Proof. Let $\mathfrak{m}_\circ$ be the maximal ideal of the local ring R_0. We put $\overline{R} = R/\mathfrak{m}_\circ R$ and $\overline{M} = M/\mathfrak{m}_\circ M$. Then by assumption $\overline{M}$ is annihilated by a power of the maximal homogeneous ideal of $\overline{R}$. Therefore $\overline{M}$ has finite length as $\overline{R}$-module and hence $\overline{M}_n = 0$ for $n >> 0$. Since $\overline{M}_n \cong M_n/\mathfrak{m}_\circ M_n$ and since M_n is a finitely generated R_0-module we have $M_n = 0$ for $n >> 0$, by Nakayama's lemma.

<u>(43.5) Remark</u>. By Matlis duality the statement of Lemma (43.4) is equivalent to the following assertion on artinian graded R-modules: Let M be a graded artinian module over R such that $\mathfrak{a}^k M = 0$ for some $k \geq 0$. Then $M_n = 0$ for $n << 0$.

Proof. To see this we may assume that R_0 is complete. In this case $\underline{\mathrm{Hom}}_R(M,\underline{E}_R)$ is a finitely generated graded R-module which satisfies the condition of Lemma (43.4). Hence by Lemma (43.4) we have $\left[\underline{\mathrm{Hom}}_R(M,\underline{E}_R)\right]_n = 0$ for $n >> 0$. By Matlis duality we get $M_n = 0$ for $n << 0$. The converse can be shown similarly.

§ 44. The Cohen-Macaulay property of Rees algebras

Now we are ready to prove the main theorem of this chapter, s. [19].

<u>(44.1) Theorem</u>. Let $(A,\mathfrak{m})$ be a noetherian local ring and I an ideal of A with $\mathrm{ht}(I) > 0$. We put $R := B(I,A)$ and $G := G(I,A)$. Let $\mathfrak{M}$ and $\mathfrak{N}$ be the maximal homogeneous ideal of R and G. Then the following conditions are equivalent:

(1) R is Cohen-Macaulay.

(2) For $i < d = \dim A$ we have:

$$\left[\underline{H}^i_{\underline{N}}(G)\right]_n = \begin{cases} H^i_{\mathfrak{m}}(A) & \text{for } n=-1 \\ 0 & \text{for } n\neq -1 \end{cases} \quad \text{and} \quad a(G)<0 \quad .$$

Proof. (1) ⇒ (2): Let $\mathfrak{a} = R_+$ and $\mathfrak{a}(1)$ be the corresponding module with the degree shifted by 1. Then we have exact sequences

$$(*)\qquad \begin{array}{l} 0 \longrightarrow \mathfrak{a} \longrightarrow R \longrightarrow A \longrightarrow 0 \\ 0 \longrightarrow \mathfrak{a}(1) \longrightarrow R \longrightarrow G \longrightarrow 0 \quad . \end{array}$$

Applying the local cohomology functor we get exact sequences:

$$(\#)\qquad \begin{array}{l} \underline{H}^i_{\underline{\mathfrak{m}}}(R) \longrightarrow H^i_{\mathfrak{m}}(A) \longrightarrow \underline{H}^{i+1}_{\underline{\mathfrak{m}}}(\mathfrak{a}) \longrightarrow \underline{H}^{i+1}_{\underline{\mathfrak{m}}}(R) \\ \underline{H}^i_{\underline{\mathfrak{m}}}(R) \longrightarrow \underline{H}^i_{\underline{N}}(G) \longrightarrow \underline{H}^{i+1}_{\underline{\mathfrak{m}}}(\mathfrak{a})(1) \longrightarrow \underline{H}^{i+1}_{\underline{\mathfrak{m}}}(R) \quad . \end{array}$$

Since R is Cohen-Macaulay and $ht(I) > 0$, we know that

$$\underline{H}^i_{\underline{\mathfrak{m}}}(R) = 0 \quad \text{for} \quad i \leqq d \quad .$$

From (#) it follows for $i < d$:

$$\underline{H}^{i+1}_{\underline{\mathfrak{m}}}(\mathfrak{a}) \cong \underline{H}^i_{\underline{N}}(G)(-1) \cong H^i_{\mathfrak{m}}(A) \quad .$$

Therefore we have for $i < d$

$$\left[\underline{H}^i_{\underline{N}}(G)\right]_n = \begin{cases} H^i_{\mathfrak{m}}(A) & \text{for } n=-1 \\ 0 & \text{for } n\neq -1 \end{cases} \quad .$$

To prove $a(G) < 0$ we consider the following exact sequences (recall that $\dim R = d+1$ and $\underline{H}^d_{\underline{\mathfrak{m}}}(R) = 0$ by assumption (1)):

$$(**)\qquad \begin{array}{l} 0 \longrightarrow H^d_{\mathfrak{m}}(A) \longrightarrow \underline{H}^{d+1}_{\underline{\mathfrak{m}}}(\mathfrak{a}) \xrightarrow{\varphi} \underline{H}^{d+1}_{\underline{\mathfrak{m}}}(R) \longrightarrow 0 \\ 0 \longrightarrow \underline{H}^d_{\underline{N}}(G) \longrightarrow \underline{H}^{d+1}_{\underline{\mathfrak{m}}}(\mathfrak{a})(1) \xrightarrow{\Psi} \underline{H}^{d+1}_{\underline{\mathfrak{m}}}(R) \longrightarrow 0 \quad . \end{array}$$

By the second sequence it is enough to show that $\left[H_{\mathfrak{m}}^{d+1}(\mathfrak{a})\right]_n = 0$ for $n > 0$. Now φ induces isomorphisms of A-modules

$$\varphi_n : \left[H_{\mathfrak{m}}^{d+1}(\mathfrak{a})\right]_n \xrightarrow{\sim} \left[H_{\mathfrak{m}}^{d+1}(R)\right]_n \quad \text{for} \quad n \neq 0 \ ,$$

whereas ψ induces surjections

$$\psi_n : \left[H_{\mathfrak{m}}^{d+1}(\mathfrak{a})\right]_{n+1} \twoheadrightarrow \left[H_{\mathfrak{m}}^{d+1}(R)\right]_n \quad \text{for all} \quad n \ .$$

Since $H_{\mathfrak{m}}^{d+1}(R)$ and $H_{\mathfrak{m}}^{d+1}(\mathfrak{a})$ are artinian R-modules we have: $\left[H_{\mathfrak{m}}^{d+1}(R)\right]_\nu = \left[H_{\mathfrak{m}}^{d+1}(\mathfrak{a})\right]_\nu = 0$ for all large $\nu >> 0$. Then by a diagram chase we get $\left[H_{\mathfrak{m}}^{d+1}(\mathfrak{a})\right]_n = 0$ for $n > 0$, as wanted.

(2) $\Rightarrow$ (1): We may assume that A is complete. First we claim that R is generalized Cohen-Macaulay with respect to $\mathfrak{a} = R_+$ (i.e. $R_{\mathfrak{m}}$ is so with respect to $\mathfrak{a}R_{\mathfrak{m}}$). For that we have to prove by Lemma (43.3) that $R_{\mathfrak{p}}$ is Cohen-Macaulay and that $\dim R/\mathfrak{p} + \dim R_{\mathfrak{p}} = \dim R$ for all $\mathfrak{p} \in \mathrm{Proj}(R)$. Note that the complete local ring A is the homomorphic image of a Gorenstein local ring. First we will indicate that we may assume $\mathfrak{p} \supset IR$: Suppose that $\mathfrak{p} \not\supset IR$. If every minimal prime $\mathfrak{Q}$ of $\mathfrak{p} + IR$ contains R_+ then $\mathfrak{p} + IR \supset R_+^n$ for some $n > 0$. This means that $\mathfrak{p}_n + I^{n+1} \supset I^n$, where $\mathfrak{p}_n$ is the n-th homogeneous component of $\mathfrak{p}$. Hence by Nakayama's lemma $\mathfrak{p}_n = I^n$, i.e. $\mathfrak{a} = R_+ \subset \mathfrak{p}$, a contradiction. Hence if $\mathfrak{p} \not\supset IR$ there must be a minimal prime $\mathfrak{Q}$ of $\mathfrak{p} + IR$ such that $\mathfrak{Q} \not\supset \mathfrak{a}$. Therefore it is enough to show that $R_{\mathfrak{Q}}$ is Cohen-Macaulay for all $\mathfrak{Q} \supset IR$. For that let us identify R with $A[IX]$. Choose an element $a \in I$ such that $aX \notin \mathfrak{Q}$. Then $G_{\mathfrak{Q}} \cong (R/IR)_{\mathfrak{Q}} \cong R_{\mathfrak{Q}}/aR_{\mathfrak{Q}}$. Since by Lemma (43.3), (3) $\Rightarrow$ (2), $G_{\mathfrak{Q}}$ is Cohen-Macaulay and a is a non-zero-divisor of $R_{\mathfrak{Q}}$, we see that $R_{\mathfrak{Q}}$ is Cohen-Macaulay. To prove the dimension condition we may assume that $\mathfrak{p} \in \mathrm{Proj}(R)$ is a minimal prime of R since A as a complete ring is universally catenary and hence so is R , see [3], 14.B. As before one can choose a minimal prime $\mathfrak{Q}$ of $\mathfrak{p} + IR$ such that $\mathfrak{Q} \not\supset \mathfrak{a}$. Then for a suitable $a \in I$ such that $aX \notin \mathfrak{Q}$ we have $G_{\mathfrak{Q}} \cong R_{\mathfrak{Q}}/aR_{\mathfrak{Q}}$. Then we get by Lemma (43.3):

$$\dim R/\mathfrak{p} + \mathrm{ht}\,\mathfrak{p} = \dim R/\mathfrak{p} = \dim R/\mathbb{Q} + \mathrm{ht}\,\mathbb{Q}/P$$

$= \dim G - \mathrm{ht}\,\mathbb{Q}/IR + \mathrm{ht}\,\mathbb{Q}/\mathfrak{p}$, since $G_{\mathfrak{N}}$ is generalized Cohen-Macaulay with respect to G_+

$= d - (\mathrm{ht}\,\mathbb{Q} - 1) + \mathrm{ht}\,\mathbb{Q}/\mathfrak{p}$, since $IR_{\mathbb{Q}} = \mathfrak{a} R_{\mathbb{Q}}$

$= d + 1$, since $R_{\mathbb{Q}}$ is Cohen-Macaulay .

This completes the proof of the first claim.
To continue the proof of (2) ⇒ (1), consider the exact sequences

$$(*)\qquad \begin{array}{l} 0 \longrightarrow \mathfrak{a} \longrightarrow R \longrightarrow A \longrightarrow 0 \\ 0 \longrightarrow \mathfrak{a}(1) \longrightarrow R \longrightarrow G \longrightarrow 0 \end{array} ,$$

implying for $i \leqq d$ the exact sequences:

$$(**)\qquad \begin{array}{l} H^{i-1}_{\mathfrak{m}}(A) \longrightarrow \underline{H}^{i}_{\mathfrak{m}}(\mathfrak{a}) \xrightarrow{\varphi^i} \underline{H}^{i}_{\mathfrak{m}}(R) \longrightarrow H^{i}_{\mathfrak{m}}(A) \\ \underline{H}^{i-1}_{\mathfrak{m}}(G) \longrightarrow \underline{H}^{i}_{\mathfrak{m}}(\mathfrak{a})(1) \xrightarrow{\psi^i} \underline{H}^{i}_{\mathfrak{m}}(R) \longrightarrow \underline{H}^{i}_{\mathfrak{m}}(G) \end{array} .$$

Since $H^{i}_{\mathfrak{m}}(A)$ is concentrated in degree 0 we have isomorphisms of A-modules

$$\varphi^i_n : \left[\underline{H}^{i}_{\mathfrak{m}}(\mathfrak{a})\right]_n \xrightarrow{\sim} \left[\underline{H}^{i}_{\mathfrak{m}}(R)\right]_n \quad \text{for } n \neq 0 ,$$

and from the assumption (2) of the theorem we have surjections

$$\psi^i_n : \left[\underline{H}^{i}_{\mathfrak{m}}(\mathfrak{a})\right]_{n+1} \twoheadrightarrow \left[\underline{H}^{i}_{\mathfrak{m}}(R)\right]_n \quad \text{for } n \geqq 0 \text{, and injections}$$

$$\psi^i_n : \left[\underline{H}^{i}_{\mathfrak{m}}(\mathfrak{a})\right]_{n+1} \hookrightarrow \left[\underline{H}^{i}_{\mathfrak{m}}(R)\right]_n \quad \text{for } n \leqq -2 .$$

Now the first claim garantees by Lemma (43.3) that $\left[\underline{H}^{i}_{\mathfrak{m}}(R)\right]_n = 0$ for all $n \ll 0$ and moreover, since $\underline{H}^{i}_{\mathfrak{m}}(R)$ is an artinian R-module we have $\left[\underline{H}^{i}_{\mathfrak{m}}(R)\right]_n = 0$ for $n \gg 0$. By a diagram chase we see that $\underline{H}^{i}_{\mathfrak{m}}(R) = 0$ for $i \leqq d$. This completes the proof of Theorem (44.1).

(44.2) Corollary. Let A and I be as in Theorem (44.1). If $R := B(I,A)$ is Cohen-Macaulay then A is generalized Cohen-Macaulay with respect to I. In particular if I is $\mathfrak{m}$-primary, A is a generalized Cohen-Macaulay ring.

Proof. By Theorem (44.1) we have for $i < d$

$$\left[\underline{H}_{N}^{i}(G)\right]_{-1} \cong H_{\mathfrak{m}}^{i}(A) \quad .$$

Hence $IH_{\mathfrak{m}}^{i}(A) = 0$ for $i < d$.

§ 45. Rees algebras of $\mathfrak{m}$-primary ideals

Now we are going to apply Theorem (44.1) to $\mathfrak{m}$-primary ideals I in a local ring $(A,\mathfrak{m},k)$. The first application is an estimation of the reduction exponent of I. Recall the definition of the reduction exponent of a minimal reduction $\mathfrak{q}$ of I (see (26.1)):

$$r_I(\mathfrak{q}) = \min\left\{ r \mid I^r = \mathfrak{q}I^{r-1}\right\} \quad .$$

We put

$$r(I) = \min\left\{ r_I(\mathfrak{q}) \mid \mathfrak{q} \text{ is a minimal reduction of } I \right\}$$

and call it the reduction exponent of I.

(45.1) Lemma. Let R be a noetherian graded ring defined over an artinian local ring R_0 such that $R = R_0[R_1]$, and let $a_1,\dots,a_d$ be a homogeneous system of parameters of R with $\deg a_i = 1$. $\mathfrak{M}$ denotes the maximal homogeneous ideal of R. Then

$$a(R) + \dim R \le \max\left\{ n \mid [R/(a_1,\dots,a_d)]_n \neq 0\right\} \le \max_{0\le i\le d} \{\alpha_i + i\} \quad ,$$

where $\alpha_i = \max\{n \mid \left[\underline{H}_{\mathfrak{M}}^{i}(R)\right]_n \neq 0\}$.

Proof. We may assume that the residue field $R_0/\mathfrak{m}_0$ of R_0 is infinite. Since $\max\{n \mid [R/(a_1,\dots,a_d)]_n \neq 0\}$ does not depend on the particular choice of generators of the ideal $(a_1,\dots,a_d)$, we may assume that a_1 does not belong to any associated prime ideal $\mathfrak{p}$ of R such that $\mathfrak{p} \neq \mathfrak{m}$. This implies that $(0 : a_1)$ has finite length. By the same argument as in the proof of Corollary (37.7) we get an exact sequence

$$\underline{H}^i_{\mathfrak{m}}(R)(-1) \xrightarrow{a_1} \underline{H}^i_{\mathfrak{m}}(R) \longrightarrow \underline{H}^i_{\mathfrak{m}}(R/a_1R) \longrightarrow \underline{H}^{i+1}_{\mathfrak{m}}(R)(-1) .$$

Hence we have

$$\alpha_i' := \max\left\{n \mid \left[\underline{H}^i_{\mathfrak{m}}(R/a_1R)\right]_n \neq 0\right\} \leqq \max\{\alpha_i, \alpha_{i+1}+1\} .$$

Since the second inequality is trivial if $d = \dim R = 0$, by induction on $\dim R$ we get

$$\max\{n \mid [R/(a_1,\dots,a_d)]_n \neq 0\} \leqq \max_{0\leqq i\leqq d-1} \{\alpha_i' + i\}$$

$$\leqq \max_{0\leqq i\leqq d} \{\alpha_i + i\} .$$

To prove the first inequality, we observe that for $i > \max\{n \mid [R/(a_1,\dots,a_d)]_n \neq 0\}$, we have $\mathfrak{a}^i = (a_1,\dots,a_d)\mathfrak{a}^{i-1}$, where $\mathfrak{a} = R_+$. So it is enough to show that if $\mathfrak{a}^i = (a_1,\dots,a_d)\mathfrak{a}^{i-1}$, then $\left[\underline{H}^d_{\mathfrak{m}}(R)\right]_n = 0$ for $i - d \leqq n$: Let $x \in \left[\underline{H}^d_{\mathfrak{m}}(R)\right]_n$ be a homogeneous element of $\deg x \geqq i - d$. Then by Chapter VII, Corollary (35.22), x can be represented by a homogeneous element $\dfrac{f}{(a_1\cdots a_d)^k} \in R_{a_1\cdots a_d}$ such that $\deg f = \deg x + kd$ and that $k > 0$. Note that $\deg f \geqq i - d + kd = i + d(k-1)$ and hence $f \in (a_1^k,\dots,a_d^k)\mathfrak{a}^{\deg(f)-k}$. This shows that $x = 0$ by Corollary (35.22), as desired.

As an immediate consequence of this lemma we get:

<u>(45.2) Proposition</u>. Let I be an $\mathfrak{m}$-primary ideal of a noetherian local ring $(A,\mathfrak{m})$ with infinite residue field and let $d = \dim A > 0$. Suppose that the Rees algebra $B(I,A)$ is Cohen-Macaulay. Then $I^d = (a_1,\dots,a_d)I^{d-1}$ for any minimal reduction $(a_1,\dots,a_d)$ of I.

Proof. Let $\mathfrak{q} = (a_1,\dots,a_d)$ be a minimal reduction of I and $a_1^*,\dots,a_d^*$ be the initial forms of $a_1,\dots,a_d$ in $G(I,A) = \bigoplus_{n\geq 0} I^n/I^{n+1}$. By definition we get

$$r_I(\mathfrak{q}) - 1 = \max\{n \mid [G(I)/(a_1^*,\dots,a_d^*)]_n \neq 0\} \quad ,$$

and by Theorem (44.1) we have:

$$\left[\underline{H}_{\mathfrak{m}}^i(G(I))\right]_n = 0 \quad \text{for} \quad i < d \quad \text{and} \quad n \neq -1 \quad \text{and}$$

$$\left[\underline{H}_{\mathfrak{m}}^d(G(I))\right]_n = 0 \quad \text{for} \quad n \geq 0 \quad .$$

Therefore Lemma (45.1) (second inequaltiy) implies $r_I(\mathfrak{q}) \leq d$.

As a corollary we get again the previous result (25.4) of Chapter V:

<u>(45.3) Corollary</u>. Let I be an $\mathfrak{m}$-primary ideal of a Cohen-Macaulay local ring $(A,\mathfrak{m},k)$ with $d = \dim A > 0$ such that $|k| = \infty$. Then $B(I,A)$ is Cohen-Macaulay if and only if $G(I,A)$ is Cohen-Macaulay and there is a minimal reduction $\mathfrak{q}$ of I such that $I^d = \mathfrak{q}I^{d-1}$.

Proof. Obvious from Theorem (44.1) and Lemma (45.1) applied to $G(I,A)$.

Without any assumption on the ring A we get the following result for equimultiple ideals I .

<u>(45.4) Proposition</u>. Let I be an equimultiple ideal of a noetherian local ring $(A,\mathfrak{m})$ with infinite residue field and $s = \mathrm{ht}(I) > 0$. If $R = B(I,A) = A[It]$ is Cohen-Macaulay then the following is true:

(i) $I^s = (a_1,\dots,a_s)I^{s-1}$ for any minimal reduction $(a_1,\dots,a_s)$ of I .

(ii) $\operatorname{depth} A \geq \dim A/I + 1$

(iii) A is normally Cohen-Macaulay along I .

Proof. to (i): Since $A[It]$ is CM , we know that $a_1, a_2 - a_1t, \dots, a_st$ is an $R_{\mathfrak{m}}$-sequence , by (10.30). Then for any $a \in I^s$ we have the following congruences $\mathrm{mod}(a_2 - a_1t, a_3 - a_2t, \ldots, a_st)$:

$$a_1at^s \equiv a_2at^{s-1} \equiv \ldots \equiv a_sat \equiv 0 \quad ,$$

hence $at^s \in (a_2 - a_1t, \ldots, a_st)R_{\mathfrak{m}}$. So we can find an equation in R of the form

$$rat^s = (a_2 - a_1t)f_1 + \ldots + a_stf_s \quad , \quad \text{where}$$

$r, f_1, \ldots, f_s \in R \quad , \quad r \notin \mathfrak{m}$.

Comparing the coefficients of t^s in this equation we obtain (i), since the constant term of r must be a unit in A . For (ii) and (iii) we first remark that for any minimal reduction $(z_1, \ldots, z_s)$ of I and for any system $\{b_1, \ldots, b_r\}$ of parameters with respect to I the sequence $\{z_1, z_2 - z_1t, \ldots, z_st, b_1, \ldots, b_r\}$ is an $R_{\mathfrak{m}}$-sequence. We consider the exact sequence

$$0 \longrightarrow \frac{(z_1, z_1t)R}{z_1R} \longrightarrow \frac{R}{z_1R} \longrightarrow \frac{R}{(z_1, z_1t)R} \longrightarrow 0 \quad ,$$

where

$$\frac{(z_1, z_1t)R}{z_1R} \simeq \frac{R}{(z_1R : z_1t)} (-1) \quad .$$

Since z_1 is a non-zero-divisor on R we have:

$$(z_1R : z_1t) = IR \quad , \quad \text{by comparison the degrees in } A[t] \quad .$$

Hence we have the exact sequence

$$(1) \qquad 0 \longrightarrow G(I,A)(-1) \longrightarrow R/z_1R \longrightarrow R/(z_1, z_1t)R \longrightarrow 0 \quad .$$

To prove (ii) and (iii) we use induction of $r = \dim A/I$. If $r = 0$ then (iii) is clear and $\operatorname{depth} A \geq 1$ (z_1 is a non-zero-divisor in A). If $r > 0$ then $\{z_1, b_1\}$ is an $R_{\mathfrak{m}}$-sequence. By the exact sequence (1) b_1 is a non-zero-divisor on $G(I,A)$. Therefore $b_1A \cap I^n = b_1I^n$ for $n \geq 0$. Hence

$$B(I + b_1A/b_1A, A/b_1A) \cong B(I,A)/b_1B(I,A) = R/b_1R$$

is CM since b_1 is a non-zero-divisor on R. Note that for $\overline{A} = A/b_1A$ and $\overline{I} = I\overline{A}$ we have again equimultiplicity $ht(\overline{I}) = s(\overline{I})$. Therefore by induction hypothesis

$$\operatorname{depth} A/b_1A \geq \dim A/(I,b_1) + 1 = \dim A/I ,$$

hence $\operatorname{depth} A \geq \dim A/I + 1$. Since

$$(I^n + b_1A)/(I^{n+1} + b_1A) \simeq I^n/(I^{n+1} + b_1I^n) \simeq (I^n/I^{n+1})/b_1(I^n/I^{n+1})$$

and since b_1 is a non-zero-divisor on I^n/I^{n+1}, we have

$$\operatorname{depth} I^n/I^{n+1} = \operatorname{depth}((I^n + b_1A)/(I^{n+1} + b_1A)) + 1 = \dim(A/(I,b_1)) + 1$$

by induction hypothesis. Hence we obtain $\operatorname{depth} I^n/I^{n+1} = \dim A/I$ for $n \geq 0$ as required.

<u>(45.5) Theorem</u>. Let $(A,\mathfrak{m})$ be a d-dimensional local ring, J an equimultiple ideal of A, $\underline{x} = \{x_1,\ldots,x_s\}$ part of a system of parameters mod J and $I = J + \underline{x}A$. Assume that $s > 0$ and that $B(J,A)$ and $B(I,A)$ are Cohen-Macaulay. Then A is Cohen-Macaulay.

Proof. We may assume that A has infinite residue field and $h(J) > 0$. Since $ht(J) = s(J)$ and $B(J,A)$ is Cohen-Macaulay we know by Proposition (45.4) that A is normally Cohen-Macaulay along J. Therefore $\underline{x}$ is a regular sequence on J^i/J^{i+1} for all $i \geq 0$. But for any finitely generated A-module M and any submodule N of M such that $\underline{x}$ is a M/N-regular sequence we get an exact sequence

$$0 \longrightarrow N/\underline{x}N \longrightarrow M/\underline{x}M \longrightarrow M/\underline{x}M + N \longrightarrow 0$$

by using the exact sequence for the homology modules of the Koszul-complexes of N,M and M/N with respect to $\underline{x}$; (see Chapter II, (11.9) and [5], 8.5, Theorem 7). Hence we have $\underline{x}M \cap N = \underline{x}N$. This means exactly

$$\underline{x}A \cap J^i = (\underline{x})J^i$$

in our case, implying $\underline{x}A \cap I^i = \underline{x}I^{i-1}$ for $i \geqq 1$. Then it follows by Chapter II, § 13 that

$$(x_1,\ldots,x_j)A \cap I^i = (x_1,\ldots,x_j)I^{i-1}$$

for $i \geqq 1$ and $1 \leqq j \leqq s$, and the initial forms $x_1^*,\ldots,x_s^*$ with respect to I form a regular sequence in $G(I,A)$.

Note that $\deg_J(x) = 0$, but $\deg_I(x) = 1$ since $\underline{x}$ is part of a minimal reduction of I (see Chapter II). We put:

$$G_I = G(I,A) \quad , \quad G_J = G(J,A)$$

$$G_I^{(0)} = G_I \quad , \quad G_I^{(j)} = G_I \,/\, (x_1^*,\ldots,x_j^*) \quad , \quad 1 \leqq j \leqq s \quad ,$$

Then we consider the exact sequence

$$(1) \qquad 0 \longrightarrow G_I^{(j)}(-1) \xrightarrow{\cdot x_{j+1}^*} G_I^{(j)} \longrightarrow G_I^{(j+1)} \longrightarrow 0 \quad .$$

Now set $G_1 := G_I^{(s)}$ and $G_2 := G_J/\underline{x}G_J$. Denote by M_J and M_I the unique maximal homogeneous ideals of $B(J,A)$ and $B(I,A)$ respectively. Then we get from (1) the long exact sequence for the local cohomology

$$(2) \qquad \ldots \longrightarrow \underline{H}_{M_I}^{i-1}(G_1) \longrightarrow \underline{H}_{M_I}^{i}(G_I^{(s-1)})(-1) \xrightarrow{\delta} \underline{H}_{M_I}^{i}(G_I^{(s-1)}) \longrightarrow \ldots \quad ,$$

where δ is defined by multiplying with x_s^* . Now $G_1 \cong G_2$ over $S := B(J,A)/\underline{x}B(J,A) \cong B(I,A)/(\underline{x},\underline{x}t) \cong A[It]/\bigoplus_{n\geqq 0}(\underline{x}A \cap I^n)t^n$. Since $\underline{x}B(J,A)$ is a regular sequence on $B(J,A)$, S is Cohen-Macaulay. Hence we get by Theorem (44.1):

$$(3) \qquad \left[\underline{H}_{M_I}^{i-1}(G_1)\right]_n \simeq \left[\underline{H}_{M_J}^{i-1}(G_2)\right]_n = 0 \quad \text{for} \quad n \geqq 0 \; , \; i \leqq d-s \quad .$$

This implies that

$$\left[\underline{H}_{M_I}^{i}(G_I^{(s-1)})\right]_{n-1} \xrightarrow{\delta_n} \left[\underline{H}_{M_I}^{i}(G_I^{(s-1)})\right]_n \quad \text{for} \quad n \geqq 0 \; , \; i \leqq d-s$$

is injective. For any $\alpha \in \underline{H}_{M_I}^{i}(G_I^{(s-1)})$ we find a positive integer m

such that $x_s^{*m}\cdot\alpha = 0$, (see Lemma (35.5)). Since the multiplication with x_s^* is injective, we get $\bigoplus_{n\geq -1}\left[\underline{H}^i_{M_I}(G_I^{(s-1)})\right]_n = 0$.

By induction on j we see that

$$\left[\underline{H}^i_{M_I}(G_I^{(s-j)})\right]_n = 0 \quad \text{for} \quad n \geq -j \ , \ i \lneq d-s+j \ , \ 0 \leq j \leq s \quad .$$

For $j = s$ and $i \leq d-1$ this implies in particular:
$0 = \underline{H}^i_{M_I}(G_I)_{-1} = H^i_{\mathfrak{m}}(A)$, where the second equality follows from Theorem (44.1). This completes the proof.

Now we are going to relate the notion of standard ideals to the Cohen-Macaulayness of a Rees algebra of an $\mathfrak{m}$-primary ideal. For this we first need a technical lemma.

<u>(45.6) Lemma</u>. Let I be an $\mathfrak{m}$-primary ideal in a local ring $(A,\mathfrak{m},k)$ with $\dim A = d \geq 2$ such that $B(I,A)$ is Cohen-Macauly. Then, for any minimal reduction $(a_1,\ldots,a_d)$ of I and for any integers $n_1,\ldots,n_d > 0$ we have

$$\left(a_1^{n_1},\ldots,a_{d-1}^{n_{d-1}}\right) : a_d = \left(a_1^{n_1},\ldots,a_{d-1}^{n_{d-1}}\right) : a_d^{n_d} \quad .$$

Proof. It is enough to show that

$$\left(a_1^{n_1},\ldots,a_{d-1}^{n_{d-1}}\right) : a_d = \left(a_1^{n_1},\ldots,a_{d-1}^{n_{d-1}}\right) : a_d^2 \quad .$$

We identify $B(I,A)$ with the subring $A[It]$ of the polynomial ring $A[t]$. Consider the ideal

$$Q = \left(a_1^{n_1}, a_2^{n_2} - a_1t, \ldots, a_{d-1}^{n_{d-1}} - a_{d-2}t, a_d - a_{d-1}t, a_dt\right)$$

of $B(I,A)$. It is not hard to see that $\sqrt{Q} = \mathfrak{m}B(I,A) + B(I,A)_+ = \mathfrak{M}$, the maximal homogeneous ideal of $B(I,A)$. Let
$x \in \left(a_1^{n_1},\ldots,a_{d-1}^{n_{d-1}}\right) : a_d^2$ and

$$J := \left(a_1^{n_1}, a_2^{n_2} - a_1t, \ldots, a_{d-1}^{n_{d-1}} - a_{d-2}t, a_dt\right) \quad .$$

Writing $xa_d^2 = \sum_{i=1}^{d-1} a_i^{n_i}y_i$, $y_i \in A$ we get the following congruences modulo J :

$$a_d x(a_d - a_{d-1}t)^d \equiv a_d^{d+1} x$$

$$\equiv a_d^{d-1}\left(\sum_{i=1}^{d-1} a_i^{n_i} y_i\right)$$

$$\equiv a_d^{d-1}\left(\sum_{i=2}^{d-1} (a_{i-1}t) y_i\right)$$

$$\equiv 0 \quad .$$

Since $B(I,A)$ is Cohen-Macaulay the given generators of Q form a regular sequence of $B(I,A)_{\mathfrak{M}}$. Hence we have

$$a_d x \in \left(a_1^{n_1}, a_2^{n_2} - a_1 t, \ldots, a_{d-1}^{n_{d-1}} - a_{d-2}t, a_d t\right) B(I,A)_{\mathfrak{M}} \quad .$$

Therefore there is an element $f \in B(I,A) - \mathfrak{M}$ such that

$$a_d x f \in \left(a_1^{n_1}, a_2^{n_2} - a_1 t, \ldots, a_{d-1}^{n_{d-1}} - a_{d-2}t, a_d t\right) \quad .$$

Noting that the homogeneous component of degree 0 of f is a unit of A, we get $a_d x \in \left(a_1^{n_1}, \ldots, a_{d-1}^{n_{d-1}}\right)$, q.e.d.

The following theorem, due to Grothe [11], IV (3.4), is essential in the study of Cohen-Macaulayness of Rees algebras of parameter ideals and maximal ideals in local rings.

<u>(45.7) Theorem</u>. Let $(A,\mathfrak{m},k)$ be a noetherian local ring with $d = \dim A > 0$ and I an $\mathfrak{m}$-primary ideal of A. If $B(I,A)$ is Cohen-Macaulay then I is standard.

Proof. We may assume that k is infinite. In this case every minimal reduction of I is generated by d elements. Let $q = (a_1, \ldots, a_d)$ be a minimal reduction of I. Suppose $d = 1$. Then by Proposition (45.2) and (45.4) we know that A is Cohen-Macaulay. So we may assume $d \geq 2$. We claim that $\{a_1, \ldots, a_d\}$ is a standard system of parameters of A. To see this, it is sufficient to show that

$$I(a_1^2,\dots,a_d^2;A) = I(a_1,\dots,a_d;A) = \lambda_A(A/\underline{a}A) - e(\underline{a};A)$$

by Chapter VIII, Proposition (38.3). Since A is generalized Cohen-Macaulay by Corollary (44.2) we know that $a_1^2,\dots,a_d^2$ is reducing by Chapter VIII, Lemma (37.2). Therefore we get by Lemma (38.2) and Lemma (45.6)

$$\begin{aligned} I(a_1^2,\dots,a_d^2;A) &= \lambda\left(\frac{(a_1^2,\dots,a_{d-1}^2):a_d^2}{(a_1^2,\dots,a_{d-1}^2)}\right) \\ &= \lambda\left(\frac{(a_1^2,\dots,a_{d-1}^2):a_d}{(a_1^2,\dots,a_{d-1}^2)}\right) \\ &= I\left(a_1^2,\dots,a_{d-1}^2,a_d;A\right) \\ &= \lambda\left(\frac{(a_1^2,\dots,a_{d-2}^2,a_d):a_{d-1}^2}{(a_1^2,\dots,a_{d-2}^2,a_d)}\right) \\ &= \lambda\left(\frac{(a_1^2,\dots,a_{d-2}^2,a_d):a_{d-1}}{(a_1^2,\dots,a_{d-2}^2,a_d)}\right) \\ &= I\left(a_1^2,\dots,a_{d-2}^2,a_{d-1},a_d;A\right) \quad . \end{aligned}$$

This yields finally $I(a_1^2,\dots,a_d^2;A) = I(a_1,\dots,a_d;A)$. Thus we have shown that every system of minimal generators of a minimal reduction of I is a standard system of parameters. To complete the proof of Theorem (45.7) we must prove that *every* system of parameters $\{x_1,\dots,x_d\}$ of A contained in I is standard. By induction on i we will prove the following statement: "If $x_1,\dots,x_{d-i}$ form a part of a system of minimal generators of a minimal reduction of I, then $x_1,\dots,x_d$ is standard". The statement is true for $i = 0$ by the above claim. Let $i > 0$. Recall that for $s := s(I)$ the elements $z_1,\dots,z_s$ give a minimal system of generators of a minimal reduction of I, if and only if their initial forms $z_1^*,\dots,z_s^*$ form a homogeneous system of parameters in $\overline{G} := G(I,A) \otimes A/\mathfrak{m}$, where z_j^* has degree 1 in $\overline{G}$ for $1 \leq j \leq s$. Since A is noetherian and since $W_{\mathfrak{p}} := ((\mathfrak{p} \cap I) + \mathfrak{m}I)/\mathfrak{m}I$ and $V_{\mathfrak{P}} := [\mathfrak{P}]_1$ are proper k-vector-spaces in $[\overline{G}]_1 = I/mI$ for any $\mathfrak{p} \in \mathrm{Assh}_A(A/(x_1,\ldots,x_{d-1}))$ and $\mathfrak{P} \in \mathrm{Assh}_{\overline{G}}(\overline{G}/(x_1^*,\ldots,x_{d-i}^*))$, one can find an $r \in \mathbf{N}$ and successively suitable generators $y_1,\ldots,y_r$ of I such that their initial forms

$$y_j^* \notin \bigcup_{\mathfrak{p}\in \mathrm{Assh}_A(A_{(d-1)})} W_{\mathfrak{p}} \cup \bigcup_{\mathfrak{P}\in \mathrm{Assh}_{\overline{G}}(\overline{G}_{(d-i)})} V_{\mathfrak{P}} \cup k < y_1^*,\ldots,y_{j-1}^* >$$

for all $1 \leqq j \leqq r$, where $A_{(d-1)} := A/(x_1,\ldots,x_{d-1})$ and $\overline{G}_{(d-i)} := \overline{G}/(x_1^*,\ldots,x_{d-i}^*)$, and $k < y_1^*,\ldots,y_{j-1}^* >$ is the k-vector-space generated by $y_1^*,\ldots,y_{j-1}^*$. In this way we can choose generators $y_1,\ldots,y_r$ of I such that:

(1) $x_1,\ldots,x_{d-i},y_j$ form a part of a system of minimal generators of a minimal reduction of I for any j with $1 \leqq j \leqq r$ and

(2) $x_1,\ldots,x_{d-i},\ldots,x_{d-1},y_j$ is a system of parameters of A for all j .

We can find an integer $n \geqq 2$ such that

$$(x_1,\ldots,x_{d-1}) : y_j \subset (x_1,\ldots,x_{d-1}) : x_d^n \quad \text{for all} \quad j \ .$$

By induction hypothesis and Chapter VIII, Lemma (37.2) we get:

$$\begin{aligned} I(A) &= I(x_1,\ldots,x_{d-i},y_j,x_{d-i+1},\ldots,x_{d-1};A) \\ &\leqq I(x_1,\ldots,x_{d-1},x_d^n;A) \leqq I(A) \quad . \end{aligned}$$

Form this we know that

$$(x_1,\ldots,x_{d-1}) : y_j = (x_1,\ldots,x_{d-1}) : x_d^n \quad \text{for all} \quad j$$

and consequently we get:

$$\begin{aligned} &(x_1,\ldots,x_{d-1}) : x_d^n = (x_1,\ldots,x_{d-1}) : I = \\ &\subseteq (x_1,\ldots,x_{d-1}) : x_d \subseteq (x_1,\ldots,x_{d-1}) : x_d^n \quad . \end{aligned}$$

This implies

$$\begin{aligned} I(x_1,\dots,x_d;A) &= I(x_1,\dots,x_{d-1},x_d^n;A) \\ &= I(x_1,\dots,x_{d-1},y_j;A) \\ &= I(A) \quad , \end{aligned}$$

as required.

Now we give several applications of Theorem (45.7).

<u>(45.8) Corollary</u>. Let $(A,\mathfrak{m},k)$ be a noetherian local ring. If $B(\mathfrak{m},A)$ is Cohen-Macaulay then A and $G(\mathfrak{m},A)$ are Buchsbaum.

Proof. By Theorem (45.7) we know that $\mathfrak{m}$ is standard and hence A is Buchsbaum by definition. Let $\mathfrak{M} = (f_1,\dots,f_n)$ be the maximal homogeneous ideal of $B(\mathfrak{m},A)$ and let $H^i(M) = H^i(f_1,\dots,f_n;M)$ denote the i-th Koszul cohomology of a $B(\mathfrak{m},A)$-module M with respect to $f_1,\dots,f_n$. By Chapter VIII, Theorem (41.12), we have to show that the canonical maps $H^i(G(\mathfrak{m},A)) \longrightarrow \underline{H}^i_{\mathfrak{M}}(G(\mathfrak{m},A))$ are surjective for $i<d$: Let $\mathfrak{a} = \bigoplus_{n>0}\mathfrak{m}^n$. From the exact sequences

$$0 \longrightarrow \mathfrak{a} \longrightarrow B(\mathfrak{m},A) \longrightarrow A \longrightarrow 0$$

$$0 \longrightarrow \mathfrak{a}(1) \longrightarrow B(\mathfrak{m},A) \longrightarrow G(\mathfrak{m},A) \longrightarrow 0$$

we get commutative diagrams

$$\begin{array}{ccc} H^i(A) & \xrightarrow{\ \sim\ } & \underline{H}^{i+1}(\mathfrak{a}) \\ {\scriptstyle\varphi_i}\downarrow & & \downarrow{\scriptstyle\alpha} \\ H^i_{\mathfrak{M}}(A) & \xrightarrow{\ \sim\ } & \underline{H}^{i+1}_{\mathfrak{M}}(\mathfrak{a}) \end{array} \qquad \text{and} \qquad \begin{array}{ccc} \underline{H}^i(G(\mathfrak{m},A)) & \xrightarrow{\ \sim\ } & \underline{H}^{i+1}(\mathfrak{a})(1) \\ {\scriptstyle\Psi_i}\downarrow & & \downarrow{\scriptstyle\alpha} \\ \underline{H}^i_{\mathfrak{M}}(G(\mathfrak{m},A)) & \xrightarrow{\ \sim\ } & \underline{H}^{i+1}_{\mathfrak{M}}(\mathfrak{a})(1) \end{array}$$

for $i<d$ because $B(\mathfrak{m},A)$ is Cohen-Macaulay. By Chapter VIII, Proposition (41.13), φ_i is surjective and hence so is α. From the second commutative diagram we see that Ψ_i is surjective. Hence $G(\mathfrak{m},A)$ is Buchsbaum.

This result was first proved by Ikeda [15], using a result of Schenzel [21a] obtained by his dualizing-complex-criterion of Buchsbaum rings.

§ 46. The Rees algebra of parameter ideals

Now we want to characterize the Cohen-Macaulayness of the Rees algebra of a parameter ideal. Blowing up parameter ideals is a well known procedure in the classical resolution process for singularities: Zariski and Jung used for the desingularization of surfaces generic projections and embedded resolution of the discriminant locus. Blowing up a point on the discriminant induces blowing up of a "thick" point on the given surface. Algebraically this situation can be described as follows: $(A,\mathfrak{m})$ is a two dimensional local domain which is the quotient of a regular local ring. In this case $A_{\mathfrak{p}}$ is Cohen-Macaulay and $\dim A/\mathfrak{p} + \dim A_{\mathfrak{p}} = \dim A$ for $\mathfrak{p} \in \operatorname{Spec} A \setminus \{\mathfrak{m}\}$. Therefore $(A,\mathfrak{m})$ is generalized Cohen-Macaulay by Lemma (43.2). Algebraically this is one of the essentials of the procedure of Zariski and Jung.

(46.1) Theorem. Let $(A,\mathfrak{m},k)$ be a local ring with $d = \dim A > 1$ and $\mathfrak{q} = (a_1,\ldots,a_d)$ be a parameter ideal of A. Then the following are equivalent:

1) $B(\mathfrak{q},A)$ is Cohen-Macaulay

2) $H^i_{\mathfrak{m}}(A) = 0$ for $i \neq 1,d$ and $\mathfrak{q}H^1_{\mathfrak{m}}(A) = 0$.

3) There is a finite ring extension B of A such that B is Cohen-Macaulay and $\mathfrak{q}B \subset A$ and $\operatorname{depth} A > 0$.

Proof. 1) $\Rightarrow$ 2): By Theorem (45.7) we see that $\mathfrak{q}$ is standard and hence we see by Theorem (41.12) and Corollary (44.2), that $H^i_{\mathfrak{m}}(A) = 0$ for $i \neq 1,d$ and $\mathfrak{q}H^1_{\mathfrak{m}}(A) = 0$.

2) $\Rightarrow$ 3): Note first that A is generalized Cohen-Macaulay and a_i is a non-zero-divisor of A. Consider the Čech-complex $C^{\bullet}(a_1,\ldots,a_d;A)$

$$0 \longrightarrow A \xrightarrow{\delta_0} \bigoplus_{i=1}^{d} A_{a_i} \xrightarrow{\delta_1} \bigoplus_{1\le i\le j\le d} A_{a_ia_j} \longrightarrow \ldots$$

It is not hard to see that $\ker\delta_1 = \bigcap_{i=1}^{d} A_{a_i} \subset A_{a_1\cdots a_d}$. We put $B = \bigcap_{i=1}^{d} A_{a_i}$. Then we have the exact sequence

$$(*) \qquad 0 \longrightarrow A \longrightarrow B \longrightarrow H^1_{\mathfrak{m}}(A) \longrightarrow 0 \quad .$$

Since $\mathfrak{q}H^1_{\mathfrak{m}}(A) = 0$, we can conclude that B is finite over A and $\mathfrak{q}B \subset A$.

It remains to show that B is Cohen-Macaulay: From (*) we get an exact sequence

$$(**) \qquad 0 \longrightarrow H^0_{\mathfrak{m}}(B) \longrightarrow H^1_{\mathfrak{m}}(A) \longrightarrow H^1_{\mathfrak{m}}(A) \longrightarrow H^1_{\mathfrak{m}}(B) \longrightarrow 0 \quad .$$

Note that B is a subring of the total quotient ring of A. Therefore $\operatorname{depth} B > 0$ and hence from (**) we know that $H^1_{\mathfrak{m}}(B) = 0$. Since $H^i_{\mathfrak{m}}(A) = 0$ for $2 \leqq i < d$, we get by (*) that $H^i_{\mathfrak{m}}(B) \xrightarrow{\sim} H^i_{\mathfrak{m}}(A) = 0$ for $2 \leqq i < d$. Thus B is Cohen-Macaulay.

3) ⇒ 2): From the exact sequence $0 \to A \to B \to B/A \to 0$ we get isomorphisms $H^i_{\mathfrak{m}}(A) \xrightarrow{\sim} H^i_{\mathfrak{m}}(B) = 0$ for $i \geqq 2$. Since $\mathfrak{q}B \subset A$ we have $H^0_{\mathfrak{m}}(B/A) = B/A$, in particular $\mathfrak{q}H^i_{\mathfrak{m}}(A) = 0$.

2) ⇒ 1): Suppose for the moment that $\mathfrak{q}$ is a standard ideal. Then, by Chapter VIII, Theorem (40.10), we see that $B(\mathfrak{q},A)$ is Cohen-Macaulay. Hence it is sufficient to show that $\mathfrak{q}$ is standard. For that it is enough to show by Theorem (41.12) that the canonical map

$$\varphi^i : H^i(a_1,\ldots,a_d;A) \longrightarrow H^i_{\mathfrak{m}}(A)$$

is surjectiv for $i < d$. Since $H^i_{\mathfrak{m}}(A) = 0$ for $i \neq 1,d$, we have only to show that φ^1 is surjective. Note that a_1 is a non-zero-divisor on A [because A is generalized Cohen-Macaulay and a_1 is a parameter element of A, i.e. $\lambda(0 : a_1) < \infty$ by (37.7) and therefore $(0 : a_1) \subseteqq (0 : \mathfrak{m}^n) \subseteqq H^0_{\mathfrak{m}}(A) = 0$]. Hence we have an exact sequence

$$0 \longrightarrow A \xrightarrow{\cdot a_1} A \longrightarrow A/a_1A \longrightarrow 0 \quad ,$$

and a commutative diagram

$$\begin{array}{ccccccccc}
H^0(\underline{a};A) & \longrightarrow & H^0(\underline{a};A/a_1) & \xrightarrow{\partial_\circ} & H^1(\underline{a};A) & \xrightarrow{\cdot a_1} & H^1(\underline{a};A) & \longrightarrow & H^1(\underline{a};A/a_1) \longrightarrow \\
\downarrow & & \downarrow \tilde{\varphi}^0 & & \downarrow \varphi^1 & & \downarrow \varphi^1 & & \downarrow \tilde{\varphi}^1 \\
0 & \longrightarrow & H^0_{\mathfrak{m}}(A/a_1) & \xrightarrow{\delta_\circ} & H^1_{\mathfrak{m}}(A) & \xrightarrow{\cdot a_1} & H^1_{\mathfrak{m}}(A) & \longrightarrow & H^1_{\mathfrak{m}}(A/a_1) \longrightarrow ,
\end{array}$$

where $\tilde{\varphi}^i$ are the corresponding canonical maps with respect to the ring A/a_1. From $\mathfrak{q}H^1_{\mathfrak{m}}(A) = 0$ we conclude that $\delta_\circ$ is an isomorphism. By [5], Chapter VIII, § 5, we have

$$H^0(\underline{a};A/a_1) \cong H_d(\underline{a};A/a_1) = (0 :_{A/(a_1)} \mathfrak{q}) .$$

Moreover we know already that $\mathfrak{q}\cdot H^0_{\mathfrak{m}}(A/a_1) = 0$. Therefore

$$\begin{array}{ccccc}
(0 :_{A/(a_1)} \mathfrak{q}) & \subseteqq & \bigcup_{n\in\mathbf{N}} (0 :_{A/(a_1)} \mathfrak{q}^n) & \subseteqq & (0 :_{A/(a_1)} \mathfrak{q}) , \\
\| & & \| & & \\
H^0(\underline{a},A) & & H^0_{\mathfrak{m}}(A/a_1) & &
\end{array}$$

i.e. $\tilde{\varphi}^\circ$ is an isomorphism, hence φ^1 is an isomorphism.

As an immediate consequence of this theorem we get the following result due to S. Goto and Y. Shimoda, [9].

(46.2) Corollary. Let $(A,\mathfrak{m},k)$ be a noetherian local ring with $\dim A = d > 0$. Then the following are equivalent:

1) A is a Buchsbaum ring such that $H^i_{\mathfrak{m}}(A) = 0$ for $1 \neq 1,d$.

2) $B(\mathfrak{q},A)$ is Cohen-Macaulay for any parameter ideal $\mathfrak{q}$ of A.

Proof. 1) $\Rightarrow$ 2). Since A is Buchsbaum we have $\mathfrak{m}H^1_{\mathfrak{m}}(A) = 0$ by Chapter VIII, Theorem (41.12). Hence $\mathfrak{q}H^1_{\mathfrak{m}}(A) = 0$ for any parameter ideal of A, and by Theorem (46.1) we see that $B(\mathfrak{q},A)$ is Cohen-Macaulay.

2) $\Rightarrow$ 1). Since $B(\mathfrak{q},A)$ is Cohen-Macaulay for any parameter ideal $\mathfrak{q}$ of A, by Theorem (45.7) and Theorem (46.1), we see that A is a Buchsbaum ring with $H^i_{\mathfrak{m}}(A) = 0$ for $i \neq 1,d$.

§ 47. The Rees algebra of powers of parameter ideals

At the beginning of § 46 we have indicated that it is quite natural to blow up "thick" points on a surface. Then one idea is that the more complicated the singularity is (e.g. the more it differs from being Cohen-Macaulay), the more one has to choose the blowing up center as a "very thick" point to "simplify" the given singularity (cf. Corollary (47.7)).

Let $\mathfrak{q} = (a_1,\ldots,a_d)$ be a parameter ideal in a local ring $(A,\mathfrak{m},k)$. We want to discuss the Cohen-Macaulayness of the Rees algebra $B(\mathfrak{q}^n,A)$. To this end, we need an auxiliary lemma from the homological algebra, cf. [1], Chapter IV, § 11, Exercises (11.5).

(47.1) Lemma. Let A and B be abelian categories. Suppose that A has enough injectives. Let $F : A \longrightarrow B$ a left exact covariant additive functior and M an object of A. Assume that there is an exact sequence in A

$$0 \longrightarrow M \longrightarrow J^0 \xrightarrow{\delta^0} J^1 \xrightarrow{\delta^1} J^2 \longrightarrow \ldots \longrightarrow J^n \xrightarrow{\delta^n} J^{n+1} \longrightarrow$$

such that $R^iF(J^n) = 0$ for all $i > 0$ and $n > 0$. Then $R^iF(M)$ is isomorphic to the i-th cohomology of the complex

$$0 \longrightarrow F(J^0) \xrightarrow{F(\delta^0)} F(J^1) \longrightarrow \ldots \longrightarrow F(J^n) \xrightarrow{F(\delta^n)} F(J^{n+1}) \longrightarrow \ldots .$$

Let $R = \bigoplus_{n\geq 0} R_n$ be a noetherian graded ring defined over a local ring, $\mathfrak{m}$ the maximal homogeneous ideal of R and M a graded R-module.

(47.2) Definition. 1) $R^{(n)} = \bigoplus_{k\geq 0} R_{nk}$. $R^{(n)}$ is called the n-th Veronesian subring of R.

2) $M^{(n)} = \bigoplus_{k\geq 0} M_{nk}$.

Clearly $M^{(n)}$ is an $R^{(n)}$-module and $M \longrightarrow M^{(n)}$ defines a functor $(n) : M^h(R) \longrightarrow M^h(R^{(n)})$.

Now recall from Chapter VII that every injective module in the category $M^h(R)$ is a direct sum of modules of the form $\underline{E}_R(R/\mathfrak{p})(n)$, where $n \in \mathbf{Z}$ and $\mathfrak{p} \in \mathrm{Spec}(R)$ is homogeneous.

<u>(47.3) Definition</u>. We say that a graded R-module M has property (A) if there is a homogeneous element $x \in \mathfrak{m}$ such that the multiplication by x is an isomorphism $M(-\nu) \xrightarrow[\sim]{x} M$ or $\operatorname{Supp}(M) = \{\mathfrak{m}\}$.

By Chapter VII, Lemma (33.9) we see that every indecomposable injective module in $M^h(R)$ has property (A). [$\mathfrak{m}$ = maximal homogeneous ideal in R.]

<u>(47.4) Lemma</u>. Let R and M be as above. If M has property (A) then $\underline{H}^i_{\mathfrak{m}}(M) = 0$ for $i > 0$.

Proof. Suppose first that there is a homogeneous element $x \in \mathfrak{m}$ of degree n such that $M(-n) \xrightarrow[\sim]{\cdot x} M$. Then $\underline{H}^i_{\mathfrak{m}}(M)(-n) \xrightarrow[\sim]{\cdot x} \underline{H}^i_{\mathfrak{m}}(M)$. For any $\alpha \in \underline{H}^i_{\mathfrak{m}}(M)$ there is an integer $k > 0$ such that $\alpha x^k = 0$. Hence $\alpha = 0$, i.e. $\underline{H}^i_{\mathfrak{m}}(M) = 0$ for any $i > 0$. Now suppose that $\operatorname{Supp}(M) = \{\mathfrak{m}\}$. Then $\underline{H}^i_{\mathfrak{m}}(M) = 0$ by Chapter VII.

Let I be an injective module in $M^h(R)$. Then I is a direct sum of modules with property (A), cf. Chapter VII. It is easy to see that the property of I, being a direct sum of modules with property (A) is preserved by the functor $(n) : M^h(R) \longrightarrow M^h(R^{(n)})$. Thus we get the following result.

<u>(47.5) Proposition</u>. Let R and M be as above and $\mathfrak{N}$ be the maximal homogeneous ideal of $R^{(n)}$. Then $\underline{H}^i_{\mathfrak{N}}(M^{(n)}) = (\underline{H}^i_{\mathfrak{m}}(M))^{(n)}$ for all $i \geq 0$.

Proof. Let $0 \to M \to I^0 \to I^1 \to I^2 \to \ldots$ be an injective resolution in $M^h(R)$. Then $(I^k)^{(n)}$ is a direct sum of graded $R^{(n)}$-modules with property (A). Hence $\underline{H}^i_{\mathfrak{N}}((I^k)^{(n)}) = 0$ for all $i > 0$. Moreover $0 \to M^{(n)} \to (I^0)^{(n)} \to \ldots$ is exact in $M^h(R^{(n)})$. By Lemma (47.1) applied to the functor $\underline{H}^0_{\mathfrak{N}}(-)$ we see that $\underline{H}^i_{\mathfrak{N}}(M^{(n)}) \cong (\underline{H}^i_{\mathfrak{m}}(M))^{(n)}$.

<u>(47.6) Corollary</u>. If R is Cohen-Macaulay then $R^{(n)}$ is Cohen-Macaulay for all $n > 0$.

Proof. $\underline{H}^i_{\mathfrak{N}}(R^{(n)}) = (\underline{H}^i_{\mathfrak{m}}(R))^{(n)} = 0$ for $i < \dim R$.

(47.7) Corollary. Let $(A,\mathfrak{m},k)$ be a generalized Cohen-Macaulay ring with $\operatorname{depth} A > 0$ and $\mathfrak{q}$ a standard parameter ideal of A. Then $B(\mathfrak{q}^{d-1},A)$ is Cohen-Macaulay.

Proof. Since $B(\mathfrak{q}^{d-1},A) = B(\mathfrak{q},A)^{(d-1)}$, the claim follows from Proposition (47.5) and Chapter VIII, Corollary (40.11), 3).

(47.8) Proposition. Let $(A,\mathfrak{m},k)$ be a generalized Cohen-Macaulay ring of $\dim A \geqq 3$. If $H^i_{\mathfrak{m}}(A) = 0$ for $t < i < d$ and for $i = 0$, $t \in \mathbf{N}$ fixed, then there exists an $\mathfrak{m}$-primary ideal $I \subset A$ such that $B(\mathfrak{q}^t,A)$ is Cohen-Macaulay for all parameter ideals $\mathfrak{q} \subset I$.

Proof. This follows from Chapter VIII, Corollary (40.11) and Propsition (47.5).

(47.9) Corollary. Let $(A,\mathfrak{m},k)$ be a noetherian ring with $\operatorname{depth} A > 0$. Then the following conditions are equivalent:

1) A is generalized Cohen-Macaulay

2) There is a parameter ideal $\mathfrak{q}$ such that $B\ell(\mathfrak{q},A)$ is Cohen-Macaulay.

3) There is an $\mathfrak{m}$-primary ideal I such that $B(I,A)$ is Cohen-Macaulay.

Proof. 1) ⇒ 2): If $\mathfrak{q}$ is a standard parameter ideal then $B\ell(\mathfrak{q},A)$ is Cohen-Macaulay by Chapter VIII, Proposition (40.3).

2) ⇒ 1): We may assume that A is complete. By Chapter VIII, Lemma (40.1), we know that A is equidimensional. Since $B\ell(\mathfrak{q},A)$ is Cohen-Macaulay, we see that $A_{\mathfrak{p}}$ is Cohen-Macaulay for $\mathfrak{p} \in \operatorname{Spec}(A) - \{\mathfrak{m}\}$. Now we can conclude that A is generalized Cohen-Macaulay by Chapter VIII, Theorem (37.4).

3) ⇒ 1): This follows from Corollary (44.2).

1) ⇒ 3): Let $\mathfrak{q}$ be a standard parameter ideal of A and $I := \mathfrak{q}^{d-1}$. Then $B(I,A)$ is Cohen-Macaulay by Corollary (47.7).

§ 48. Applications to rings of low multiplicity. Examples

Part I: The Rees ring of the maximal ideal.

Restrictions to the multiplicity $e(A)$ occur necessarily for one- or two-dimensional rings $(A,\mathfrak{m})$ with Cohen-Macaulay Rees-rings.

We know already that a one-dimensional ring A has a Cohen-Macaulay Rees ring $B(\mathfrak{m},A)$ if and only if A is regular. For two-dimensional rings $(A,\mathfrak{m})$ we have the following result due to Goto and Shimoda [10].

(48.1) Proposition. Let $(A,\mathfrak{m})$ be a local ring of $\dim A = 2$, $|A| = \infty$. Then the following conditions are equivalent:

(i) $B(\mathfrak{m},A)$ is Cohen-Macaulay.

(ii) A is Cohen-Macaulay and $e(A) = \text{embdim}(A) - 1$.

Proof. (i) $\Rightarrow$ (ii): Condition (i) implies $\operatorname{depth} A \geqq 2$, see Chapter V. Hence A is CM and $e(A) \geqq \text{embdim}(A) - 1$ by App. to Chap. V. Moreover by Chapter IX, Proposition (45.2), we know that $\mathfrak{m}^2 = (a_1,a_2)\mathfrak{m}$ for any minimal reduction (a_1,a_d) of $\mathfrak{m}$. That means $e(A) = \text{emb}(A) - 1$.

(ii) $\Rightarrow$ (i): Condition (ii) implies [6] that $G(\mathfrak{m},A)$ is CM and $r(\mathfrak{m}) \leqq 2$. Then $B(\mathfrak{m},A)$ is CM by Chapter V, Theorem (25.4). Moreover we know by (26.6) that a hypersurface $(A,\mathfrak{m})$ has a Cohen-Macaulay Rees ring $B(\mathfrak{m},A)$ if and only if $e(A) \leqq \dim A$. The situation is more complicated if we consider non-Cohen-Macaulay rings A. The following proposition shows that the multiplicity $e(A)$ cannot become arbitrarily small, if $B(\mathfrak{m},A)$ is Cohen-Macaulay.

(48.2) Proposition. If $(A,\mathfrak{m})$ is a non-Cohen-Macaulay ring with a Cohen-Macaulay Rees-ring $B(\mathfrak{m},A)$, then $e(A) \geqq \dim(A)$.

Proof. By Chapter IX, Corollary (45.8), we know that A is a Buchsbaum ring. Hence by Chapter VIII, Theorem (41.19), we can use Goto's inequality:

$$e(A) \geqq 1 + \sum_{i=1}^{d-1} \binom{d-1}{i-1} h^i(A) \quad .$$

Since $\operatorname{depth} A \geq 2$ we get $h^{\circ}(A) = h^1(A) = 0$. Now, by assumption there exists at least one i with $2 \leq i \leq d-i$ such that $h^i(A) \neq 0$. This proves the claim.

<u>(48.3) Corollary</u>. For a local ring $(A,\mathfrak{m})$ with $e(A) \lneq \dim(A)$ the following conditions are equivalent:

(i) $B(\mathfrak{m},A)$ is Cohen-Macaulay

(ii) $G(\mathfrak{m},A)$ is Cohen-Macaulay.

Proof. This follows from Theorem (25.4) and Corollary (26.5) in Chapter V and Proposition (48.2).

<u>(48.4) Example</u>. $(e(A) = \dim A)$:

$$A = k[[X_1,X_2,X_3,Y_1,Y_2,Y_3]] \,/\, (X_1Y_1 + X_2Y_2 + X_3Y_3, (Y_1,Y_2,Y_3)^2) \quad ,$$

where k is a field and X_i, Y_i are indeterminates. We will show that A is a non-Cohen-Macaulay Buchsbaum ring with $e(A) = \dim A = 3$ and that $B(\mathfrak{m},A)$ is Cohen-Macaulay:

Proof. Let $G = G(\mathfrak{m},A)$. Then it is clear that

$$\begin{aligned} G &= k[X_1,X_2,X_3,Y_1,Y_2,Y_3]/(X_1Y_1 + X_2Y_2 + X_3Y_3, (Y_1,Y_2,Y_3)^2) \\ &= k[x_1,x_2,x_3,y_1,y_2,y_3] \quad . \end{aligned}$$

Let S be the subring of G generated by x_1,x_2,x_3. Since x_1,x_2,x_3 is a system of parameters of G, we see that G is finite over S and S is isomorphic to a polynominal ring in three variables. Let

$$0 \longrightarrow S(-3) \xrightarrow{f_3} S^3(-2) \xrightarrow{f_2} S^3(-1) \xrightarrow{f_1} S \longrightarrow 0$$

be the Koszul complex $K.(x_1,x_2,x_3;S)$ and let $E = (\operatorname{Im} f_2)(1)$. Consider the symmetric algebra

$$S(E) = \bigoplus_{n=0}^{\infty} \operatorname{Sym}_n(E) \cong S[Y_1,Y_2,Y_3]/(X_1Y_1 + X_2Y_2 + X_3Y_3) \quad .$$

Now we have

$$S \oplus E \cong S(E) / \bigoplus_{n\geq 2} \operatorname{Sym}_n(E) \cong \frac{k[X_1,X_2,X_3,Y_1,Y_2,Y_3]}{(X_1Y_1+X_2Y_2+X_3Y_3,(Y_1,Y_2,Y_3)^2)} .$$

Let $\mathfrak{m}$ be the maximal homogeneous ideal of G . From the exact sequences

$$0 \longrightarrow E \longrightarrow S^3 \longrightarrow S(1) \longrightarrow \underline{k}(1) \longrightarrow 0$$

$$0 \longrightarrow S(-2) \longrightarrow S^3(-1) \longrightarrow E \longrightarrow 0$$

we get

$$\underline{H}^2_{\mathfrak{m}}(G) = \underline{H}^2_{\underline{x}S}(S \oplus E) = \underline{H}^2_{\underline{x}S}(E) \cong \underline{k}(1) \ , \text{ i.e. } \ \underline{H}^2_{\mathfrak{m}}(G)_n = 0 \quad \text{for} \quad n \neq -1 \ ,$$

$$\underline{H}^0_{\mathfrak{m}}(G) = \underline{H}^1_{\mathfrak{m}}(G) = 0 \qquad \text{and}$$

$$\underline{H}^3_{\mathfrak{m}}(G) = \underline{H}^3_{\underline{x}S}(S \oplus E)_n = 0 \quad \text{for} \quad n \geq -1 \ ,$$

since $\underline{H}^3_{\underline{x}S}(S)_n = 0$ for $n \geq -2$.
Since A is isomorphic to the completion of G we get

$$H^0_{\mathfrak{m}}(A) = H^1_{\mathfrak{m}}(A) = 0 \qquad \text{and}$$

$$H^2_{\mathfrak{m}}(A) \cong k \quad .$$

By Theorem (44.1) we can conclude that $B(\mathfrak{m},A)$ is Cohen-Macaulay. In particular A is Buchsbaum by Corollary (45.8). In order to compute the multiplicity $e(A) = e(G)$ we note that for any finitely generated graded G-module M we have $e(M) = rk_S M$. Hence we have

$$e(G) = rk_S(S \oplus E) = 1 + 2 = 3 \quad .$$

<u>(48.5) Example</u>. $(e(A) < \dim A)$:

$A = k[[X]]/I_2(X)$, where $X = (X_{ij})$ is the 2×3 matrix of indeterminates X_{ij} over a field k and $I_2(X)$ is the ideal generated by the 2×2 minors of X . Then A is Cohen-Macaulay with the properties:

$$e(A) = 3 < \dim A = 4$$

$$\operatorname{embdim}(A) = 6 = e(A) + d - 1 \quad .$$

Therefore $G(\mathfrak{m},A)$ is CM. Hence $B(\mathfrak{m},A)$ is CM by Corollary (48.3).

Part II: Weakly permissible ideals

In this Part II we call a prime ideal $\mathfrak{p}$ also permissible (instead of weakly permissible) if $A/\mathfrak{p}$ is regular and $e(A) = e(A_{\mathfrak{p}})$. We know that such a permissible ideal is equimultiple.
In part II we moot this question: If the multiplicity $e(A)$ is small with respect to the height of a given permissible ideal $\mathfrak{p} \neq \mathfrak{m}$ which has a Cohen-Macaulay Rees ring $B(\mathfrak{p},A)$, what can we say about the behaviour of A and $B(\mathfrak{m},A)$?
The case $\operatorname{ht}(\mathfrak{p}) = 1$ is clear: If $B(I,A)$ is CM for any equimultiple ideal I of height 1 , then I is principal and A is CM; see Chapter V, Theorem (25.4). Moreover if this ideal $I = \mathfrak{p}$ is also permissible, i.e. if $\mathfrak{m} = \mathfrak{p} + xR$ for a suitable parameter $\bmod \mathfrak{p}$, then we know by the transitivity of the CM-property for Rees rings that $B(\mathfrak{m},A)$ must be CM.

Permissible prime ideals $\mathfrak{p}$ of height 1 with a Cohen-Macaulay Rees ring are too special as we can see in the following lemma.

(48.6) Lemma. Let $\mathfrak{p} \neq \mathfrak{m}$ be a permissible ideal in A of height 1. Then $B(\mathfrak{p},A)$ is Cohen-Macaulay if and only if A is regular.

Geometric interpretation of Lemma (48.6): Let X be an algebraic variety, $x \in X$. Assume that there is an equimultiple subvariety Z of codimension 1 through x which is non-singular at x . If the vertex of the affine cone over the blowing up of $\operatorname{Spec} \mathcal{O}_{X,x}$ along $\operatorname{Spec} \mathcal{O}_{Z,x}$ is a Cohen-Macaulay singularity then X is non-singular at x .

Proof of Lemma (48.6). Let $B(\mathfrak{p},A)$ be CM. Since $\operatorname{ht}(\mathfrak{p}) = s(\mathfrak{p}) = 1$, $\mathfrak{p}$ is generated by one element f . By assumption we have $\mathfrak{m} = fA + (x_1,\ldots,x_{d-1})A$ where $x_1,\ldots,x_{d-1}$ form a regular system of parameters $\bmod \mathfrak{p}$. Hence A is regular. The converse is obvious since A and $A/\mathfrak{p}$ regular imply that $\mathfrak{p}$ is generated by a regular sequence.

For arbitrary height we get the following restriction on the reduction exponent $r(\mathfrak{m})$ for permissible ideals $\mathfrak{p}$ with Cohen-Macaulay Rees ring $B(\mathfrak{p},A)$.

(48.7) Lemma. If there exists a permissible ideal $\mathfrak{p} \neq \mathfrak{m}$ in $(A,\mathfrak{m})$ with $B(\mathfrak{p},A)$ Cohen-Macaulay then

$$r(\mathfrak{m}) \leqq ht(\mathfrak{p}) < \dim(A) \quad .$$

Proof. By assumption $\mathfrak{p}^t = \underline{a}\mathfrak{p}^{t-1}$, where $t = ht(\mathfrak{p})$ and $\underline{a}$ is a minimal reduction of $\mathfrak{p}$, see Proposition (45.2). Since $\mathfrak{m} = \mathfrak{p} + \underline{x}A$ for a suitable system of parameters $\underline{x} \bmod \mathfrak{p}$, we obtain:

$$\mathfrak{m}^t = \mathfrak{p}^t + \underline{x}\mathfrak{p}^{t-1} = \underline{a}\mathfrak{p}^{t-1} + \underline{x}\mathfrak{p}^{t-1} \quad , \quad \text{hence}$$

$$\mathfrak{m}^t \subseteq \underline{a}\mathfrak{m}^{t-1} + \underline{x}\mathfrak{m}^{t-1} = (\underline{a}A + \underline{x}A)\mathfrak{m}^{t-1} \quad , \quad \text{i.e.}$$

$$\mathfrak{m}^t = (\underline{a}A + \underline{x}A)\mathfrak{m}^{t-1} \quad .$$

(48.8) Remark. If $t = 2$ then either $r(\mathfrak{m}) = 1$ or $r(\mathfrak{m}) = 2$. In the first case A is regular. In the second case we have $\mathfrak{m}^2 = (\underline{a}A + \underline{x}A)\mathfrak{m}$. The next Proposition (48.9) shows that then A is CM, hence $G(\mathfrak{m},A)$ is CM, i.e. $B(\mathfrak{m},A)$ is CM by Chapter V, (25.4).

(48.9) Proposition. Let $\mathfrak{p} \neq \mathfrak{m}$ be an ideal in a local ring $(A,\mathfrak{m})$ such that

(i) $B(\mathfrak{p},A)$ is Cohen-Macaulay

(ii) $ht(\mathfrak{p}) = s(\mathfrak{p}) = 2$

(iii) $A/\mathfrak{p}$ regular .

Then A and $B(\mathfrak{m},A)$ are Cohen-Macaulay.

Proof. By assumption $\mathfrak{m} = \mathfrak{p} + \underline{x}A$, where $\underline{x} = \{x_1,\ldots,x_r\}$ form a regular system of parameters $\bmod \mathfrak{p}$, $r = \dim A/\mathfrak{p}$. $R(\mathfrak{p})$ Cohen-Macaulay and $ht(\mathfrak{p}) = s(\mathfrak{p})$ imply that A is normally Cohen-Macaulay along $\mathfrak{p}$, see Proposition (45.2). Therefore $\underline{x}$ is a regular sequence on $\mathfrak{p}^n/\mathfrak{p}^{n+1}$ for $n \geq 0$, hence on A too. Furthermore $\underline{x}$ is part of a system of parameters on $B(\mathfrak{p},A)$ i.e. $B(\mathfrak{p},A)/\underline{x}B(\mathfrak{p},A) \cong B(\overline{\mathfrak{m}},\overline{A})$ is Cohen-Macaulay, where $\overline{\mathfrak{m}} = \mathfrak{m}/\underline{x}A$ is the maximal ideal of $\overline{A} = A/\underline{x}A$. Hence we know that $\operatorname{depth} A \geq 2$. But $\dim \overline{A} = \operatorname{ht}(\mathfrak{p}) = 2$, hence $\overline{A}$ and A must be Cohen-Macaulay. Therefore $B(\mathfrak{m},A)$ is Cohen-Macaulay by the transitivity property (27.1).

(48.10) Remark. If $\mathfrak{p}$ be a prime ideal in A such that

(i) $B(\mathfrak{p},A)$ is Cohen-Macaulay

(ii) $\operatorname{ht}(\mathfrak{p}) = s(\mathfrak{p}) = 2$,

then possibly A has to be always Cohen-Macaulay. For ideals which are not prime we have the following counterexample:

(48.11) Example. $A = k[[s^2,s^3,st,t]]$ with s,t indeterminates over a field k. Then $\dim A = 2$, $\operatorname{depth} A = 1$ and $x_1 = s^2$, $x_2 = t$ is a system of parameters in A. Since A is a Buchsbaum local domain of dimension 2 one can deduce from Theorem (46.1) that for $I = x_1A + x_2A$ the Rees-ring $B(I,A)$ is Cohen-Macaulay.

For $\operatorname{ht}(\mathfrak{p}) = s(\mathfrak{p}) = 3$, $\mathfrak{p}$ a prime ideal, we also know a counterexample:

(48.12) Example. Let

$$A = k[[X_1,X_2,X_3,Y_1,Y_2,Y_3]]/(X_1Y_2 + X_2Y_2 + X_3Y_3,(X_1,Y_2,Y_3)^2) \quad .$$

Take $R := A[[T_1,\dots,T_n]]$ with indeterminates $T_1,\dots,T_n$. Let $\mathfrak{p} = \mathfrak{n}R$, where $\mathfrak{n}$ denotes the maximal ideal of A. Then $\mathfrak{p}$ is an equimultiple prime ideal of $\operatorname{ht}(\mathfrak{p}) = 3$. Since $B(\mathfrak{n},A)$ is CM (see Example (48.4) to Proposition (48.2)) and R is faithfully flat over A, $B(\mathfrak{p},R)$ is CM, but R does not even satisfy S_3, as we know. Note that $B(\mathfrak{m},R)$ is not CM for the maximal ideal $\mathfrak{m}$ of R, otherwise R has to be a Buchsbaum ring by Corollary (45.6).

(48.13) Remark. Note that for any prime ideal $\mathfrak{p}$, " $B(\mathfrak{p},A)$ is

Cohen-Macaulay" implies $B(\mathfrak{p}A_{\mathfrak{p}}, A_{\mathfrak{p}})$ is CM. This means for $ht(\mathfrak{p}) = s(\mathfrak{p}) = 1$ that $A_{\mathfrak{p}}$ is regular. If $ht(\mathfrak{p}) = s(\mathfrak{p}) = 2$, we deduce from Corollary (45.8) that $A_{\mathfrak{p}}$ is Cohen-Macaulay, which is not true in general for $ht(\mathfrak{p}) = s(\mathfrak{p}) \geq 3$. We also get for $ht(\mathfrak{p}) = 2$ by Chapter IX, Proposition (45.2) and Chapter V, Theorem (25.4), that $\operatorname{depth} A \geq \dim A - 1$ and $r(\mathfrak{p}) \leq 2$. These informations may justify our question whether A is already CM under the conditions (i) and (ii) in (48.10). We have seen in Lemma (48.7) that a permissible ideal $\mathfrak{p} \neq \mathfrak{m}$ with $B(\mathfrak{p}, A)$ Cohen-Macaulay has to satisfy the condition $r(\mathfrak{m}) \leq ht(\mathfrak{p}) < \dim A$. If A is a hypersurface, this means $e(A) \leq ht(\mathfrak{p}) < \dim A$, hence $B(\mathfrak{m}, A)$ is Cohen-Macaulay by (26.6) in that case. This motivates the restriction on $e(A)$ in the next theorem and the following propositions.

<u>(48.14) Theorem</u>. Let $\mathfrak{p} \neq \mathfrak{m}$ be a permissible ideal in a local ring $(A, \mathfrak{m})$ with $t = ht(\mathfrak{p}) \geq 2$. Assume that

(i) $B(\mathfrak{p}, A)$ is Cohen-Macaulay

(ii) $e(A) < ht(\mathfrak{p})$.

Then A and $B(\mathfrak{m}, A)$ are Cohen-Macaulay.

Proof. By assumption $\mathfrak{m} = \mathfrak{p} + \underline{x}A$, where $\underline{x} = \{x_1, \ldots, x_r\}$ form a regular system of parameters $\operatorname{mod} \mathfrak{p}$, $r = \dim A/\mathfrak{p}$. By Chapter IV, §20 we may assume that $\{x_1, \ldots, x_r\}$ is a sequence of superficial elements with $e(A/\underline{x}A) = e(A)$.
We put $\overline{A} = A/\underline{x}A$ and $\overline{\mathfrak{m}} = \mathfrak{m}/\underline{x}A$. As in the proof of Proposition (48.9) we deduce from condition (i) and Corollary (45.8) that $\overline{R}$ is a Buchsbaum ring of $\operatorname{depth} \overline{A} \geq 2$, i.e. $h^{\circ}(\overline{A}) = h^1(\overline{A}) = 0$. The case $t = 2$ is clear by Proposition (48.9). Let $t \geq 3$ and assume that $\overline{A}$ is a non-Cohen-Macaulay ring. Now we use explicitly that $\overline{A}$ is Buchsbaum: There is at least one i, $2 \leq i \leq t-1$, with $h^i(\overline{A}) \neq 0$, hence by Goto's inequality for multiplicities in Buchsbaum rings (see Theorem (41.19))

$$e(\overline{A}) \geq 1 + \sum_{i=1}^{t-1} \binom{t-1}{i-1} h^i(\overline{A})$$

we get $e(A) = e(\overline{A}) \geq t$, in contradiction to (ii). Therefore $\overline{A}, A$ and $B(\mathfrak{m}, A)$ are Cohen-Macaulay-rings.

We give a counterexample for $e(A) = ht(\mathfrak{p})$:

(48.15) Example.

We take the ring $R = A[[T_1,\dots,T_n]]$ of a previous example, where $A = k[[X_1,X_2,X_3,Y_1,Y_2,Y_3]]/(X_1Y_1+X_2Y_2+X_3Y_3,(Y_1,Y_2,Y_3)^2)$. The ideal $\mathfrak{p} = \mathfrak{n}R$ (where $\mathfrak{n}$ is the maximal ideal in A) is permissible in R . Now we have $e(R) = ht(\mathfrak{p}) = 3$. But we know already that R and $B(\mathfrak{m},R)$ are not CM.
This example shows that we cannot replace condition (ii) in Theorem (48.14) by $e(R) \leqq ht(\mathfrak{p})$.

(48.16) Remark. Ikeda [16] gave the following counterexample for $e(R) > ht(\mathfrak{p})$, this needs a lot of computations, which we omit here.

Let k be a field of $char(k) = 2$.
Let $A = k[[X_1,X_2,X_3,Y_1,Y_2,Y_3,Y_4]]/I$, where

$$I = (X_1Y_1+X_2Y_2+X_3Y_3;Y_1^2,Y_2^2,Y_3^2,Y_4^2,Y_1Y_4,Y_2Y_4,Y_3Y_4,$$
$$Y_1Y_2-X_3Y_4,Y_2Y_3-X_1Y_4,Y_1Y_3-X_2Y_4) .$$

$\mathfrak{n}$ denotes the maximal ideal of A . By Ikeda $(A,\mathfrak{n})$ is a Buchsbaum ring with $e(A) = 4$ and $\dim A = 3$. Moreover $B(\mathfrak{n},A)$ is Gorenstein. Now let $R = A[[T_1,\dots,T_s]]$, $\mathfrak{p} = \mathfrak{n}R$. Then $\mathfrak{p}$ is permissible and $B(\mathfrak{p},R)$ is Cohen-Macaulay. But R is not Cohen-Macaulay, hence $B(\mathfrak{m},R)$ is not Cohen-Macaulay by Theorem (45.5). Note that now $e(R) = 4 > ht(\mathfrak{p}) = 3$.

The next proposition tells us that for rings of multiplicity 2 we don't need the strong condition (ii) of Theorem (48.14).

(48.17) Proposition. Let $\mathfrak{p} \neq \mathfrak{m}$ be a permissible ideal in $(A,\mathfrak{m})$ with $ht(\mathfrak{p}) \geqq 2$. Assume that

(i) $B(\mathfrak{p},A)$ is Cohen-Macaulay

(ii) $e(A) = 2$.

Then A and $B(\mathfrak{m},A)$ are Cohen-Macaulay.

Proof. The claim follows from Proposition (48.9) if $ht(\mathfrak{p}) = 2$ and from Theorem (48.14) if $ht(\mathfrak{p}) > 2$.

For permissible ideals in excellent rings containing a field, we mention another bound for multiplicities in the following two propositions using a result of Huneke [14], based on the direct summand conjecture. This result we cannot prove in the book. The reader is referred to [2], [14]. One may take these two propsitions as a remark.

(48.18) Proposition. Let $(A,\mathfrak{m})$ be an excellent ring containing a field. Let $\mathfrak{p}$ be a permissible ideal in A , $ht(\mathfrak{p}) \geqq 2$. Assume that

(i) A is Cohen-Macaulay outside of $\mathfrak{m}$

(ii) $B(\mathfrak{p},A)$ is Cohen-Macaualy

(iii) $e(A) \leqq \dim(A/\mathfrak{p}) + 2$.

Then A and $B(\mathfrak{m},A)$ are Cohen-Macaulay rings.

Proof. By the same arguments as in the proofs of Proposition (48.9) and Theorem (48.2) we find a regular system $\underline{x}$ of parameters mod $\mathfrak{p}$ such that the ring $\overline{A} = A/\underline{x}A$ has $\operatorname{depth} \overline{A} \geqq 2$. Since $\underline{x}$ is a regular sequence on A , we get that $\operatorname{depth} A \geqq r + 2$ with $r = \dim A/\mathfrak{p}$, hence A satisfies S_{r+2} by (i). Therefore A is Cohen-Macaulay by [14], i.e. $B(\mathfrak{m},A)$ is Cohen-Macaulay.

(48.19) Remark. For $ht(\mathfrak{p}) = 2$ condition (iii) means $e(A) \leqq \dim A$. But for $ht(\mathfrak{p}) = 2$ we know already by Proposition (48.9) that $B(\mathfrak{m},A)$ is CM - without assuming (i) and (iii).

A corresponding statement to Proposition (48.18) is true for the geometric blowing up. For that let $A \to A_1$ be a local homomorphism obtained by blowing up $\mathfrak{p}$ in A (see Chapter II and VI).

(48.20) Proposition. Let $(A,\mathfrak{m})$ be an excellent local ring containing a field. Let $\mathfrak{p} \neq \mathfrak{m}$ be a regular prime ideal in A . Assume that

(i) $\operatorname{Proj}(G(\mathfrak{p},A)) \to \operatorname{Spec}(A/\mathfrak{p})$ is a flat morphism.

(ii) A_1 is Cohen-Macaulay outside the maximal ideal $\mathfrak{m}_1 \subset A_1$.

(iii) $e(A) \leqq \dim(A/\mathfrak{p}) + 1$.

Then A_1 is Cohen-Macaulay.

Proof. $\mathfrak{m} = \mathfrak{p} + \underline{x}A$, where $\underline{x} = \{x_1,\ldots,x_r\}$ is a system of parameters mod $\mathfrak{p}$ and $r = \dim A/\mathfrak{p}$. Condition (i) implies by Chapter IV that $\mathfrak{p}^n/\mathfrak{p}^{n+1}$ is Cohen-Macaulay over the (regular) ring $A/\mathfrak{p}$ for $n >> 0$. Furthermore $\underline{x}$ is a regular sequence on $A_1/\mathfrak{p}A_1$, i.e. $\operatorname{depth} A_1 \geqq r+1$. Hence A_1 satisfies S_{r+1} by (ii). Since $\mathfrak{p}$ is equimultiple, we have $e(A_1) \leqq e(A) \leqq r+1$ by the above condition (iii).

A_1 is the localization of a ring of finite type over an excellent ring, hence it is an excellent ring too, containing a field. Then the claim of the proposition follows again from Huneke's Theorem [14].

Finally we ask for the Cohen-Macaulay property of the coordinate ring $F_{\mathfrak{m}} = G(\mathfrak{p},A) \otimes_A A/\mathfrak{m}$ of the fibre of the blowing up morphism $\operatorname{Proj} B(\mathfrak{p},A) \longrightarrow \operatorname{Spec} A$ over the closed point $\mathfrak{m}$, if $B(\mathfrak{p},A)$ is CM.

(48.21) Theorem. Let $\mathfrak{p} \neq \mathfrak{m}$ be a permissible ideal in $(A,\mathfrak{m})$ of $\operatorname{ht}(\mathfrak{p}) \geqq 3$. Assume that $B(\mathfrak{p},A)$ is Cohen-Macaulay. Then the following conditions are equivalent:

(i) $F_{\mathfrak{m}}$ is Cohen-Macaulay

(ii) $B(\mathfrak{m},A)$ is Cohen-Macaulay

(iii) A is Cohen-Macaulay.

Proof. As in the proof of Theorem (48.14) we may assume that $\mathfrak{m} = \mathfrak{p} + \underline{x}A$, where $\underline{x} = \{x_1,\ldots,x_r\}$ with $r = \dim A/\mathfrak{p}$ is a sequence of superficial elements such that $e(A/\underline{x}A) = e(A)$. Furthermore $\underline{x}$ is a regular sequence of homogeneous elements on $G(\mathfrak{p},A)$ (since A is normally Cohen-Macaulay along $\mathfrak{p}$ by the assumption on $B(\mathfrak{p},A)$), and $B(\overline{\mathfrak{m}},\overline{A})$ is CM (see proof of Proposition (48.9)). Putting $\overline{A} = A/\underline{x}A$ and $\overline{\mathfrak{m}} = \mathfrak{m}/\underline{x}A$, we have:

$$F_{\mathfrak{m}} = G(\mathfrak{p},A)/\underline{x}G(\mathfrak{p},A) \cong G(\overline{\mathfrak{m}},\overline{A}) \quad .$$

Since $B(\overline{\mathfrak{m}},\overline{A})$ is CM, $F_{\mathfrak{m}} \cong G(\overline{\mathfrak{m}},\overline{A})$ is CM if and only if $\overline{A}$ (and therefore A) is CM, see Chapter V, Theorem (25.4). Since $B(\mathfrak{p},A)$ is CM by assumption, A is CM if and only if $B(\mathfrak{m},A)$ is CM by Theorem (45.5). This proves the claim.

(48.22) Remark. For non-Cohen-Macaulay rings $(A,\mathfrak{m})$ one can characterize the fibres $F_{\mathfrak{m}}$ completely if $e(A) = 3$, using results of Ikeda [17]: Let $\mathfrak{p} \neq \mathfrak{m}$ be a permissible ideal of height $t \geq 3$ in a non-Cohen-Macaulay ring $(A,\mathfrak{m})$. Assume that

(i) $B(\mathfrak{p},A)$ is Cohen-Macaulay

(ii) $e(A) = 3$.

Then $F_{\mathfrak{m}} \cong \dfrac{k[X_1,X_2,X_3,Y_1,Y_2,Y_3]}{(X_1Y_1+X_2Y_2+X_3Y_3,(Y_1,Y_2,Y_3)^2)}$, i.e. $F_{\mathfrak{m}}$ is uniquely determined up to isomorphisms. The proof is basically the same as the procedure in the proofs of Theorem (48.21) and Theorem (48.14). Using the same notations as in the proofs of these theorems one sees that $(\overline{A},\overline{\mathfrak{m}})$ is a non-Cohen-Macaulay Buchsbaum ring with $e(\overline{A}) = \dim \overline{A} = 3$, satisfying S_2 (since $\operatorname{depth} \overline{A} \geq 2$). Then Ikeda could recently show that $G(\overline{\mathfrak{m}},\overline{A})$ is uniquely determined (up to isomorphisms) by

$$G(\overline{\mathfrak{m}},\overline{A}) \cong \frac{k[X_1,X_2,X_3,Y_1,Y_2,Y_3]}{(X_1Y_1+X_2Y_2+X_3Y_3,(Y_1,Y_2,Y_3)^2)} .$$

Part III: Arbitrary equimultiple ideals

Besides using equimultiple "testideals" I as before in Proposition (48.9) and Theorem (48.14) (where I was permissible) we now assume above all restrictions on the multiplicity of A and A/I to get informations about the structure of $B(\mathfrak{m},A)$. Recall that the typical properties of a "testideal" I ($B(I,A)$ is CM and $\operatorname{ht}(I) = s(I) =: t$) imply that A is normally CM along I, and as a consequence we get $\operatorname{depth} A \geq d-t+1$. The permissibility of I in Part II will now be partly "replaced" by the Buchsbaum property of A.

First we start with an auxiliary result which characterizes the Cohen-Macaulay-property of $B(\mathfrak{m},A)$ by a numerical condition.

(48.23) Proposition. Let $(A,\mathfrak{m})$ be a three dimensional local-ring with an infinite residue field k. Let $\mathfrak{q} = (a,b,c)$ be a minimal reduction of $\mathfrak{m}$. We put

$$I = ((a,b) : c) + ((b,c) : a) + ((a,c) : b) + \mathfrak{m}^2 .$$

Then the following conditions are equivalent:

(1) $B(\mathfrak{m},A)$ is Cohen-Macaulay

(2) $\lambda(I/\mathfrak{m}^2) = 3(\lambda(A/\mathfrak{q}) - e(A)) + 3$.

Proof. We identify $B(\mathfrak{m},A)$ with the subring $A[\mathfrak{m}t]$ of the polynominal ring $A[t]$. For simplicity we put $S = B(\mathfrak{m},A)$, and $\mathfrak{M}$ denotes the maximal homogeneous ideal of S . As we have seen before, the sequence $\{a, b-at, c-bt, ct\}$ generates a minimal reduction of $\mathfrak{M}S_{\mathfrak{M}}$. Let $\mathfrak{Q} = (a, b - at, c - bt, ct)S$. Then S is Cohen-Macaulay if and only if $\lambda_S(S/\mathfrak{Q}) = e(S_{\mathfrak{M}})$. By the proof of (25.4) we know that $e(S_{\mathfrak{M}}) = 3\, e(A)$. To prove the proposition we have to determine the length $\lambda_S(S/\mathfrak{Q})$. For the computation we observe that $\mathfrak{P} := (a,b,c,at,bt,ct)S = \mathfrak{Q} + (b,c)A$.

Hence we get

$$(*) \quad \lambda_S(S/\mathfrak{Q}) = \lambda_S(S/\mathfrak{P}) + \lambda_S(\mathfrak{P}/\mathfrak{Q}) = \sum_{n\geq 0} \lambda(\mathfrak{m}^n/\mathfrak{q}\mathfrak{m}^{n-1}) + \lambda((b,c)A \cap \mathfrak{Q}) ,$$

where $\mathfrak{m}^{-1}$ is supposed to be A . We claim that

$$(**) \qquad (b,c)A \cap \mathfrak{Q} = (b\mathfrak{m}, cI) .$$

To prove the claim, we show that $(b,c)A \cap \mathfrak{Q} \supset (b\mathfrak{m}, cI)$:

a) For any $x \in \mathfrak{m}$ we have

$$bx \equiv (at)x \equiv a(xt) \equiv 0 \mod \mathfrak{Q} , \text{ i.e. } b\mathfrak{m} \subset (b,c)A \cap \mathfrak{Q}$$

b) Let $y \in (b,c) : a$. Then one can find elements $z, w \in \mathfrak{m}$ such that $ay = bz + cw$, since a,b,c are analytically independent. Therefore we have:

$$cy \equiv (bt)y \equiv b(yt) \equiv ayt^2 \equiv (bz + cw)t^2 \equiv c(zt) \equiv 0 \mod \mathfrak{Q} .$$

Similarly one can show that $c((a,b) : c)$, $c((a,c) : b)$ and $c\mathfrak{m}^2$ are contained in $(b,c)A \cap \mathfrak{Q}$, i.e. $cI \subset (b,c)A \cap \mathfrak{Q}$.
To prove the other inclusion take $f \in (b,c)A \cap \mathfrak{Q}$ and write

$$f = af_1 + (b-at)f_2 + (c-bt)f_3 + ctf_4$$

with $f_i \in S$. Let $f_i = \sum_{j\geq 0} f_{ij}t^j$, $i = 1,\ldots,4$, $f_{ij} \in \mathfrak{m}^j$. Then we get:

$$f = af_{10} + bf_{20} + cf_{30}$$

$$(\#) \qquad 0 = af_{11} + bf_{21} - af_{20} + cf_{31} - bf_{30} + cf_{40}$$

$$0 = af_{12} + bf_{22} - af_{21} + cf_{32} - bf_{31} + cf_{41} \quad .$$

Since $f \in (b,c)A$ we get $f_{10} \in ((b,c) : a)$. Hence we can write $af_{10} = bx + cy$ with $x \in \mathfrak{m}$ and $y \in ((a,b) : c)$. Therefore we have $af_{10} \in (b\mathfrak{m}, cI)$. Note that $f_{20} \in \mathfrak{m}$. From (#) we conclude:

$$f_{21} - f_{30} \in ((a,c) : b)$$

$$f_{12} - f_{21} \in ((b,c) : a) \quad .$$

Since $f_{12} \in \mathfrak{m}^2$, this yields:

$$f_{30} \in ((a,c) : b) + ((b,c) : a) + \mathfrak{m}^2 \quad .$$

Summing up we obtain

$$f = af_{10} + bf_{20} + cf_{30} \in (b\mathfrak{m}, cI) \quad .$$

This proves the above claim.

Using (*) and (**) we see that

$$(***) \qquad \begin{aligned} \lambda(S/\mathfrak{Q}) &= \sum_{n\geq 0} \lambda(\mathfrak{m}^n/\mathfrak{q}\mathfrak{m}^{n-1}) + \lambda((b,c)A/(b\mathfrak{m}, cI)) \\ &= \sum_{n\geq 0} \lambda(\mathfrak{m}^n/\mathfrak{q}\mathfrak{m}^{n-1}) + 1 + \lambda(A/I) \quad . \end{aligned}$$

Now the proof of Proposition (48.23) runs as follows:

(1) ⇒ (2) : Since $S = B(\mathfrak{m}, A)$ is Cohen-Macaulay and $\dim A = 3$ we have $\mathfrak{m}^3 = \mathfrak{q}\mathfrak{m}^2$ by Proposition (45.2). Hence with (***) and $\lambda(S/\mathfrak{Q}) = 3e(A)$ we obtain:

$$3e(A) = \lambda(A/\mathfrak{q}) + \lambda(\mathfrak{m}/\mathfrak{q}) + \lambda(\mathfrak{m}^2/\mathfrak{q}\mathfrak{m}) + 1 + \lambda(A/I)$$

$$= 3\lambda(A/\mathfrak{q}) + 3 - \lambda(I/\mathfrak{m}^2) \quad .$$

(2) ⇒ (1): By (***) we have $\lambda(S/\mathfrak{Q}) = 3e(A) = e(S_{\mathfrak{M}})$. Hence S is Cohen-Macaulay.

(48.24) Proposition. Let $(A,\mathfrak{m})$ be a Buchsbaum ring of dimension $d \geq 3$. Let I be a proper ideal in A of height $t < d$. Assume that

(i) $B(I,A)$ is Cohen-Macaulay

(ii) $ht(I) = s(I) \geq 2$.

Then the following statements hold:

a) If $e(A) < d$, A is Cohen-Macaulay.

b) If $e(A) = d$ and $r(\mathfrak{m}) \geq 3$, A is Cohen-Macaulay.

c) If $e(A) \leq 3$, $B(\mathfrak{m},A)$ is Cohen-Macaulay.

Proof. a) Conditions (i) and (ii) imply by Proposition (45.4) that $\operatorname{depth} A \geq d - t + 1$, i.e. $h^i(A) = 0$ for $i \leq d - t$ in our case. Therefore we get (see Chapter VIII and [23])

$$(1) \qquad e(A) = 1 + \binom{d-1}{d-t} h^{d-t+1} + \ldots + \binom{d-1}{d-3} h^{d-2} + (d-1)h^{d-1} + \lambda(\mathfrak{m}/J) \ ,$$

where $J = \sum_{1}^{d} (y_1,\ldots,\check{y}_i,\ldots,y_d) : y_i$ and $(y_1,\ldots,y_d)$ is a minimal reduction of $\mathfrak{m}$. Since $e(A) < d$, $h^i(A) = 0$ for $i \neq d$, i.e. A is CM.

b) Let $e(A) = d$. Assume that A is not Cohen-Macaulay. Then equation (1) implies $h^{d-t+1} = \ldots = h^{d-2} = 0$, $h^{d-1} = 1$ and $\lambda(\mathfrak{m}/J) = 0$. The last equality yields $r(\mathfrak{m}) \leq 2$. But that is a contradiction to $r(\mathfrak{m}) \leq 3$ in b) .

c) We test the multiplicities $e(A) = 1,2$ and 3 : If $e(A) = 1$ then A is regular, hence $B(\mathfrak{m},A)$ is Cohen-Macaulay. If $e(A) = 2$, A is Cohen-Macaulay by statement a). Then we know that $\operatorname{embdim}(A) \leq e(A) + d - 1 = d + 1$, i.e. A is hypersurface with $e(A) \leq d$, hence $B(\mathfrak{m},A)$ is Cohen-Macaulay, see Chapter V, (26.6). If $e(A) = 3$, we consider two cases:

Case 1: A is Cohen-Macaulay. Then $\mathrm{embdim}(A) \leqq d+2$: If $\mathrm{embdim}(A) \leqq d+1$ then we have again the hypersurface case with $e(A) \leqq d$, i.e. $B(\mathfrak{m},A)$ is Cohen-Macaulay. If $\mathrm{embdim}(A) = d+2$, i.e. $\mathrm{embdim}(A) = e(A) + d - 1$, then we have $\mathfrak{m}^2 = \underline{a}\mathfrak{m}$ for a minimal reduction $\underline{a} = (a,b,c)$ of $\mathfrak{m}$ by [6] , hence $G(\mathfrak{m},A)$ and $B(\mathfrak{m},A)$ are Cohen-Macaulay, see Chapter V, Theorem (25.4).

Case 2: A is a non-Cohen-Macaulay-ring. Then relation (1) tells us that

1) $d = \dim A = 3$

2) $h^2(A) = 1$, $h^i(A) = 0$ for $i \neq 2,3$

3) $\lambda(\mathfrak{m}/J) = 0$.

Therefore we have $\mathfrak{m}^2 = \underline{a}\mathfrak{m}$.
As in Proposition (48.23) we consider the ideal $I := J + \mathfrak{m}^2$. Then one has

$$\lambda(I/\mathfrak{m}^2) = \lambda(\mathfrak{m}/\mathfrak{m}^2) + \lambda(J + \mathfrak{m}^2/J) ,$$

hence $\lambda(I/\mathfrak{m}^2) = \mathrm{embdim}(A)$.
By definition of the invariant $I(A)$ of a Buchsbaum-ring we know that

$$\begin{aligned} e(A) &= \lambda(A/\underline{a}A) - I(A) \\ &= \lambda(A/\mathfrak{m}) + \lambda(\mathfrak{m}/\underline{a}A + \mathfrak{m}^2) + \lambda(\underline{a}A + \mathfrak{m}^2/\underline{a}A + \mathfrak{m}^3) + \ldots - I(A) \\ &= 1 + \lambda(\mathfrak{m}/\underline{a}A + \mathfrak{m}^2) - 1 , \end{aligned}$$

because $\mathfrak{m}^2 = \underline{a}\mathfrak{m}$ and $I(A) = 1$ in our case. Since

$$\lambda(\mathfrak{m}/\underline{a}A + \mathfrak{m}^2) = \mathrm{embdim}(A) - 3 \quad \text{and} \quad e(A) = 3 ,$$

we get finally $\lambda(I/\mathfrak{m}^2) = 6$.

So we have $3(\lambda(A/\underline{a}A) - e(A)) + 3 = 6$, hence $B(\mathfrak{m},A)$ is Cohen-Macaulay by Proposition (48.23).

Question 1: Is Proposition (48.24) true without assuming A is Buchsbaum? Note that this property implies together with the conditions (i) and (ii) that A satisfies S_2 . And only that fact has to be used in the proof of a), b) and c).

<u>Question 2</u>: Assume that

(i) $(A,\mathfrak{m})$ is Buchsbaum with $\dim A \geqq 3$.

(ii) $B(I,A)$ is Cohen-Macaulay.

(iii) $ht(I) = s(I) = 2$.

How far is A from being Cohen-Macaulay?

The next Proposition (48.25) gives a partial answer to question 2. Recall that $\mathfrak{p}^* := G(\mathfrak{m},\mathfrak{p}\subset A)$ is the ideal of the initial forms of $\mathfrak{p}$ with respect to $\mathfrak{m}$.

<u>(48.25) Proposition</u>. Let $(A,\mathfrak{m})$ be a Buchsbaum ring of dimension $d \geqq 3$ with an algebraically closed residue field $A/\mathfrak{m}$. Let $\mathfrak{p}$ be a prime ideal such that $\mathfrak{p}^*$ is prime. Assume that

(i) $B(\mathfrak{p},A)$ is Cohen-Macaulay

(ii) $ht(\mathfrak{p}) = s(\mathfrak{p}) = 2$.

If $e(A) = 3$, then A and $B(\mathfrak{m},A)$ are Cohen-Macaulay.

Proof. a) $B(\mathfrak{m},A)$ is Cohen-Macaulay by Proposition (48.24), c).
b) Assume A is not Cohen-Macaulay. Then we get by the same arguments (and with the same notations) as in the proof of Proposition (48.24):

$$d = \dim A = 3 \ , \ h^2(A) = 1 \ \text{ and } \ \lambda(\mathfrak{m}/J) = 0 \ .$$

Note that $\mathfrak{p}^* \subset G = G(\mathfrak{m},A)$ is an ideal of height two. Since $G/\mathfrak{p}^*$ is a k-algebra-domain of dimension one, which is generated by homogeneous elements of degree one over an algebraically closed field, we have:

$$G/\mathfrak{p}^* \cong k[X_1,\ldots,x_n]/\mathfrak{P} \ , \quad \text{for some } n \ ,$$

where $\mathfrak{P} = (f_1,\ldots,f_{n-1})$ is a homogeneous prime ideal with height $ht(\mathfrak{P})=n-1$ and $\deg f_i = 1$. From this we conclude by a suitable coordinate change that $G/\mathfrak{p}^* \cong k[Z]$, where Z is an indeterminate over k . This means that $A/\mathfrak{p}$ is regular since $G/\mathfrak{p}^* \cong G(\mathfrak{m}/\mathfrak{p},A/\mathfrak{p})$.

But this property cannot occur together with $B(\mathfrak{p},A)$ is Cohen-Macaulay and $ht(\mathfrak{p}) = s(\mathfrak{p}) = 2$ for a non-Cohen-Macaulay ring A by Proposition (48.9). Therefore A must be Cohen-Macaulay.

(48.26) Proposition. Let $(A,\mathfrak{m})$ be a local ring and let I be a complete intersection in A. Assume that

(i) $B(I,A)$ is Cohen-Macaulay,

(ii) $e(A/I) = e(A)$.

Then A is Cohen-Macaulay.

Proof. If $r = \dim A/I = 0$, A is CM. In the general case we may assume that A has an infinite residue field. Let $x_1,\ldots,x_r$ be a system of parameters mod I , such that $\overline{x}_1,\ldots,\overline{x}_r$ form a minimal reduction of $\mathfrak{m}/I$ in A/I . Since $B(I,A)$ is CM and $ht(I) = s(I)$, A is normally CM along I and $\underline{x} = (x_1,\ldots,x_r)$ is a regular sequence on I^n/I^{n+1} , hence on A too. From A/I is CM and $\dim(A/I+\underline{x}A) = 0$ we conclude:

$$e(A) = e(A/I) = e((\overline{x}_1,\ldots,\overline{x}_r))$$

$$= e(A/I+\underline{x}A) \geqq e(A/\underline{x}A) \geqq e(A) \quad .$$

Putting $\overline{A} = A/\underline{x}A$, we see that $e(\overline{A}) = e(\overline{A}/I\overline{A})$, where $I\overline{A}$ is generated by $\dim\overline{A} = d-r$ parameter elements, i.e. $\overline{A}$ is CM by step 1. Therefore A is CM.

In Proposition (48.24) we have excluded rings with $d = \dim A = 2$. For this case we mention an easy consequence of Theorem (25.4) in Chap. V.

(48.27) Lemma. Let $(A,\mathfrak{m})$ be a local ring with $e(A) = \dim A = 2$. Then $B(\mathfrak{m},A)$ is Cohen-Macaulay if and only if A is Cohen-Macaulay.

Proof. A is CM implies $\operatorname{embdim}(A) \geqq 3 = d+1$, i.e. A is a hypersurface with $e(A) = \dim A$, hence $B(\mathfrak{m},A)$ is CM. The converse is clear.

(48.28) Example.

$$A = k[[X_1,X_2,X_3,X_4]] / (X_3\cdot X_4 - X_2\cdot X_1 ; X_2^2 - X_3^2 + X_1\cdot X_3^2 ; \\ X_4^2 + X_1^3 - X_1^2 ; X_1\cdot X_3 - X_2\cdot X_4 - X_1^2\cdot X_3)$$

$$\simeq k[[x_1,x_2,x_3,x_4]] \ .$$

A defines a rational surface C in $\mathbf{A}_k^4$.
Consider the complete intersection $V \subset \mathbf{A}_k^4$, defined by the equations

$$X_4X_3 - X_2X_1 = 0 \ , \quad X_1X_3 - X_2X_4 - X_1^2X_3 = 0 \quad .$$

Furthermore let L be the union of planes in $\mathbf{A}_k^4$:

$$X_1 = X_4 = 0 \qquad \text{and} \qquad X_2 = X_3 = 0 \quad ;$$

L is a Buchsbaum-surface.
Since $V = C \cup L$, i.e. C and L are linked by the complete intersection V , the ring A is a non-Cohen-Macaulay Buchsbaum ring with $I(A) = 1$, (see [21b] for this last remark). Furthermore $\mathfrak{m}^2 = (X_1,X_2)\mathfrak{m}$, where $\mathfrak{m}$ is the maximal ideal of A , hence $e(A) = \operatorname{embdim}(A) - d - I(A) + 1 = 2$. So $B(\mathfrak{m},A)$ is not CM by the lemma.

(48.29) Example. $A = k[[s^2,s^3,st,t]]$, $e(A) = \dim A = 2$. Since A is not CM, $B(\mathfrak{m},A)$ is not CM.

Part IV: Equimultiple ideals in rings containing a field.

(48.30) Lemma. If a catenary local ring A satisfies (S_2) then $\dim A = \dim A/\mathfrak{p}$ for all $\mathfrak{p} \in \operatorname{Ass}(A)$.

Proof. Suppose that there is an associated prime $\mathfrak{p} \in \operatorname{Ass}(A)$ such that $\dim A/\mathfrak{p} < \dim A$. Let

$$(0) = \mathfrak{q}_1 \cap \ldots \cap \mathfrak{q}_r \cap \mathfrak{q}_{r+1} \cap \ldots \cap \mathfrak{q}_n$$

be a primary decomposition such that $\dim A/\mathfrak{p}_i = \dim A$ for $1 \leq i \leq r$ and $\dim A/\mathfrak{p}_i < \dim A$ for $r+1 \leq i \leq n$ where $\mathfrak{p}_i = \sqrt{\mathfrak{q}_i}$. Put $I := \bigcap_{i=1}^{r} \mathfrak{q}_i$ and $J := \bigcap_{i=r+1}^{r} \mathfrak{q}_i$. A has no embedded prime because of (S_2) , and we get $\dim A/I + J \leq \dim(A) - 2$. Let $\mathfrak{Q} \in \mathrm{Ass}(A/I + J)$. Then $\mathfrak{Q}$ contains one of $\mathfrak{q}_i$, $1 \leq i \leq r$, say $\mathfrak{Q} \supset \sqrt{\mathfrak{q}_1} = \mathfrak{p}_1$. Since A is catenary we have

$$\mathrm{ht}(\mathfrak{Q}) \geq \mathrm{ht}(\mathfrak{Q}/\mathfrak{p}_1) = \dim A/\mathfrak{p}_1 - \dim A/\mathfrak{Q} \geq 2 \quad .$$

From the exact sequence

$$0 \longrightarrow A_{\mathfrak{Q}} \longrightarrow (A/I)_{\mathfrak{Q}} \oplus (A/J)_{\mathfrak{Q}} \longrightarrow (A/I + J)_{\mathfrak{Q}} \longrightarrow 0$$

we get $\operatorname{depth} A_{\mathfrak{Q}} = 1$, contradicting (S_2) .

Recall Proposition (26.21) in Chapter V, where we considered equimultiple ideals I in an excellent local ring $(A,\mathfrak{m})$ containing a field of characteristic 0 , such that $\dim A/I = 1$. Assuming that A is CM outside $\mathfrak{m}$ and moreover that $e(A) + e(A/I) \leq \mathrm{embdim}(A/I) + 2$ we saw that the Cohen-Macaulayness of $B(I,A)$ implies the Cohen-Macaulayness of A . In this final part of Chapter IX we extend this result to excellent rings $(A,\mathfrak{m})$ containing a field of <u>any</u> characteristic. Furthermore we prove a similar statement for the geometric blowing up.

<u>(48.31) Definition</u>. Let $(A,\mathfrak{m})$ be a local ring with $d = \dim A$ and let $a_1,\dots,a_d$ be a system of parameters of A . We say that $a_1,\dots,a_d$ has the monomial property if

$$(a_1 \dots a_d)^n \notin (a_1^{n+1},\dots,a_d^{n+1}) \quad \text{for all} \quad n \geq 0 \quad .$$

Recall that there is a canonical map

$$\varphi_n : A/(a_1^n,\dots,a_d^n) \longrightarrow H^d_{\mathfrak{m}}(A) \cong \varinjlim_n A/(a_1^n,\dots,a_d^n) \quad ,$$

and

$$\mathrm{Ker}(\varphi_n) = \bigcup_{k>0} \frac{(a_1^{n+k},\dots,a_d^{n+k}) : (a_1 \dots a_d)^k}{(a_1^n,\dots,a_d^n)} \quad ,$$

(cf. Chapter VII, Corollary (35.21)).
Hence $a_1,\dots,a_d$ has monomial property if and only if $\varphi_1(1) \neq 0$. By Chapter VII, Theorem (36.19) we have $H^d_{\mathfrak{m}}(A) \neq 0$. Therefore $\varphi_n(1) \neq 0$ for all $n \gg 0$ and hence $a_1^n,\dots,a_d^n$ has monomial property for $n \gg 0$.

Now we are proving the equivalence of the "monomial property" and the "direct summand property".

(48.32) Proposition. Let $(A,\mathfrak{m},k)$ be a local ring containing a field. Then the following statements are equivalent:

(1) Every system of parameters of A has monomial property.

(2) Every subring S of A such that S is a regular local ring and A is finite over S is a direct summand of A as S-module.

Proof. (1) ⇒ (2): Let $a_1,\ldots,a_d$ be a regular system of parameters of S and let $\mathfrak{n} = (a_1,\ldots,a_d)$ be the maximal ideal of S. Then $\underline{a} = \{a_1,\ldots,a_d\}$ is a system of parameters of A. Look at the following commutative diagram

$$\begin{array}{ccc} S/\mathfrak{n} = S/\underline{a}S & \hookrightarrow & A/\underline{a}A \\ \Psi_1 \downarrow & & \downarrow \varphi_1 \\ H^d_{\mathfrak{n}}(S) & \xrightarrow{f^*} & H^d_{\mathfrak{m}}(A) \end{array} ,$$

where Ψ_1 and φ_1 are canonical maps, and f^* is the map induced from the inclusion $f : S \hookrightarrow A$. By assumption (1) we know that $\varphi_1(1) \neq 0$. Hence we have $0 \neq \varphi_1(1) \otimes_A 1 \in H^d_{\mathfrak{m}}(A) \otimes_A \hat{A} = H_{\mathfrak{m}\hat{A}}(\hat{A})$. Moreover S is a direct summand of A if and only if $\hat{S}$ is a direct summand of $\hat{A}$. So we may assume that A and S are complete. Since $\underline{a}$ is an S-sequence, Ψ_1 is an injection. We know that $H^d_{\mathfrak{n}}(S)$ is the injective envelope of $S/\mathfrak{n}$ as S-module. Therefore, since $\varphi_1(1) \neq 0$, the map f^* is injective, hence $H^d_{\mathfrak{n}}(S)$ as an injective submodule is a direct summand of $H^d_{\mathfrak{m}}(A)$ as S-module. Thus we have an S-homomorphism $g : H^d_{\mathfrak{m}}(A) \longrightarrow H^d_{\mathfrak{n}}(S)$ such that $g \circ f^* = id$. Noting that $H^d_{\mathfrak{m}}(A) \cong H^d_{\mathfrak{n}}(S) \otimes_S A$, we get:

$$\begin{aligned} g \in & \operatorname{Hom}_S(H^d_{\mathfrak{m}}(A), H^d_{\mathfrak{n}}(S)) \\ & \cong \operatorname{Hom}_S(H^d_{\mathfrak{n}}(S) \otimes_S A, H^d_{\mathfrak{n}}(S)) \\ & \cong \operatorname{Hom}_S(A, \operatorname{Hom}_S(H^d_{\mathfrak{n}}(S), H^d_{\mathfrak{n}}(S))) \\ & \cong \operatorname{Hom}_S(A,S) \end{aligned}$$

by Matlis duality and by the fact that $H^d_{\mathfrak{n}}(S) \cong E_S(S/\mathfrak{n})$. Regarding now g as an homomorphism $A \longrightarrow S$, we see that $g \circ f = id_S$. Hence S is a direct summand of A .

$(2) \Rightarrow (1)$: We may assume that A is complete, since $(a_1 \dots a_d)^n \notin (a_1^{n+1},\dots,a_d^{n+1})A \Longleftrightarrow (a_1 \dots a_d)^n \notin (a_1^{n+1},\dots,a_d^{n+1})\hat{A}$ for any elements $a_1,\dots,a_d \in A$. Let $a_1,\dots,a_d$ be a system of parameters of A and let $S = k[[a_1,\dots,a_d]]$. Then S satisfies the properties of (2). Hence by assumption S is a direct summand of A as S-module, say $A = S \oplus M$, for some S-module M . Putting $\mathfrak{n} = (a_1,\dots,a_d)S$, we get a commutative diagram

$$\begin{array}{ccc} S/\mathfrak{n} & \hookrightarrow & A/\underline{a}A \\ \Psi_1 \downarrow & & \downarrow \varphi_1 \\ H^d_{\mathfrak{n}}(S) & \hookrightarrow & H^d_{\mathfrak{m}}(A) \cong H^d_{\mathfrak{n}}(S) \oplus H^d_{\mathfrak{n}}(M) \end{array} ,$$

where Ψ_1 and φ_1 are canonical as before. From this we see that $\varphi_1(1) \neq 0$. Hence $a_1,\dots,a_d$ has monomial property.

<u>(48.33) Remark</u>. Note that we didn't use the assumption "A contains a field" for the implication $(1) \Rightarrow (2)$. Also the implication $(2) \Rightarrow (1)$ is true without this assumption by Hochster [2], but the proof of this fact is much more difficult then our proof given here.

<u>(48.34) Proposition</u>. Let $(A,\mathfrak{m})$ be a local ring containing a field k of $char(k) = p > 0$. Then every system of parameters of A has monomial property.

Proof. Let $a_1,\dots,a_d$ be a system of parameters of A . Suppose that there is an $n > 0$ such that

$$(a_1 \dots a_d)^n = \sum_{i=1}^{d} r_i a_i^{n+1} \quad , \text{ for some } \quad r_i \in A \quad .$$

By taking the p^e-th powers we get:

$$\left(a_1^{p^e} \dots a_d^{p^e}\right)^n = \sum_{i=1}^{d} r_i^{p^e} a_i^{p^e(n+1)}$$

for any $e > 0$. But this contradicts to the fact that $a_1^{p^e}, \dots, a_d^{p^e}$ has monomial property for sufficiently large e , as we have pointed out before.

(48.35) Corollary. Let A be a local ring containing a field of any characteristic. Let S be a regular subring of A such that A is finite over S . Then S is a direct summand of A as S-submodule.

Proof. This follows from Chapter V, Proposition (26.13) and from the Proposition (48.32) and (48.34).

(48.36) Remark. The direct summand conjecture would be always true if there would exist a so-called "big Cohen-Macaulay module" for any noetherian local ring $(A,\mathfrak{m})$. The converse question is open. Here we mean by a big Cohen-Macaulay module a module M with $M \neq \mathfrak{m}M$ such that there exists a system of parameters of A which is an M-sequence.

(48.37) Corollary. Let A be an excellent local ring containing a field. Suppose A satisfies Serre's condition (S_2) , and the multiplicity $e(A) \leq 2$. Then A is a hypersurface, see [14].

(48.38) Proposition. Let $(A,\mathfrak{m})$ be an excellent local ring containing a field. Let I be an ideal of the principal class of A defining a curve (i.e. $\dim(A/I) = 1$) and let $A \longrightarrow A_1$ be a local homomorphism obtained by blowing up I in A . Assume that

a) $\operatorname{depth}(I^n/I^{n+1}) = \dim(A/I)$ for large n

b) $e(A/I) \leq 2$

c) A_1 is Cohen-Macaulay outside the maximal ideal $\mathfrak{m}_1$ of A_1 .

Then A_1 is a hypersurface.

Proof. Let x be a parameter $\operatorname{mod} I$. Then a) implies that x is a non-zero-divisor on A_1/IA_1 , (see Chapter IV), hence $\operatorname{depth} A_1 \geq 2$, i.e. Serre condition (S_2) is satisfied by A_1 . By Chapter VI, Theorem (31.1) we know that $e(A_1) \leq e(x,I,A)$ for any choice of x . Since I is an ideal of the principal class, we have

$$e(x,I,A) = \sum_{\mathfrak{p}\in \mathrm{Assh}(A/I)} e(x,A/\mathfrak{p})\cdot e(IA_{\mathfrak{p}},A_{\mathfrak{p}})$$

$$\leq \sum e(x,A/\mathfrak{p})\cdot \lambda(A_{\mathfrak{p}}/IA_{\mathfrak{p}}) = e(x,A/I) \quad .$$

Since $A/\mathfrak{m}$ was assumed to be infinite, we get for a suitable choice of x that $e(x;A/I) = e(A/I)$. This shows that $e(A_1) \leq e(A/I) \leq 2$, proving the assertion in view of Corollary (48.37).

<u>(48.39) Proposition</u>. Let $(A,\mathfrak{m})$ be an excellent local ring of $\dim A \geq 2$ containing a field. Let I be an equimultiple ideal of A with $\dim A/I = 1$. Assume that

(a) $B(I,A)$ is Cohen-Macaulay,

(b) $e(A) + e(A/I) \leq \mathrm{embdim}(A/I) + 2$,

(c) A is Cohen-Macaulay outside $\mathfrak{m}$.

Then A is a hypersurface.

Finally we remark that most probably Corollary (48.37) (as well as Propositions (48.38) and (48.39) are valid without the assumption that $(A,\mathfrak{m})$ contains a field. If A is not a domain Goto and Ikeda found independently elementary proofs for this conjecture, which cannot be transferred to the domain case. We reproduce this proof in the sequel. (see also [18]).

<u>(48.40) Proposition</u>. Let $(A,\mathfrak{m})$ be a complete local ring which is not a domain. If A satisfies (S_2) , and $e(A) = 2$, then A is Cohen-Macaulay.

Proof. By the associativity formula

$$2 = e(A) = \sum_{\mathfrak{p}\in \mathrm{Assh}(A)} \lambda(A_{\mathfrak{p}})\cdot e(A_{\mathfrak{p}})$$

and by the fact (see Lemma (48.30)) that $\mathrm{Ass}(A) = \mathrm{Assh}(A)$, we see that $\mathrm{Ass}(A)$ has at most two elements.

Case 1: $\mathrm{Ass}(A) = \{\mathfrak{p}_1,\mathfrak{p}_2\}$.

In this case we have $e(A/\mathfrak{p}_1) = e(A/\mathfrak{p}_2) = 1$, hence $A/\mathfrak{p}_i$ is regular

for $i = 1,2$ by [4].
For $\mathfrak{P} \in \mathrm{Ass}(A/\mathfrak{p}_1 + \mathfrak{p}_2)$ we consider the exact sequence

$$0 \longrightarrow A_{\mathfrak{P}} \longrightarrow (A/\mathfrak{p}_1)_{\mathfrak{P}} \oplus (A/\mathfrak{p}_2)_{\mathfrak{P}} \longrightarrow (A/\mathfrak{p}_1 + \mathfrak{p}_2)_{\mathfrak{P}} \longrightarrow 0 \quad .$$

Assuming $\mathrm{ht}(\mathfrak{P}) \geqq 2$ we get from Serre's condition S_2

$$\operatorname{depth} A_{\mathfrak{P}} \geqq \min\{2, \mathrm{ht}(\mathfrak{P})\} = 2 \quad ;$$

but $\operatorname{depth} A_{\mathfrak{P}} = 1$, which gives a contradiction. Hence we have

$$\mathrm{ht}(\mathfrak{P}) = 1 \quad \text{for all} \quad \mathfrak{P} \in \mathrm{Ass}(A/\mathfrak{p}_1 + \mathfrak{p}_2) \quad .$$

Therefore $\mathfrak{p}_1 + \mathfrak{p}_2/\mathfrak{p}_1$ is a principal ideal of the regular local ring $A/\mathfrak{p}_1$, in particular

$$\operatorname{depth}(A/\mathfrak{p}_1 + \mathfrak{p}_2) = \operatorname{depth}(A/\mathfrak{p}_1)/(\mathfrak{p}_1 + \mathfrak{p}_2/\mathfrak{p}_1) = d - 1 \quad .$$

Now, using the exact sequence

$$0 \longrightarrow A \longrightarrow A/\mathfrak{p}_1 \oplus A/\mathfrak{p}_2 \longrightarrow A/\mathfrak{p}_1 + \mathfrak{p}_2 \longrightarrow 0 \quad ,$$

we get $\operatorname{depth} A = \dim A$, i.e. A is Cohen-Macaulay.

Case 2: $\mathrm{Ass}(A) = \{\mathfrak{p}\}$, where $\mathfrak{p} \neq (0)$.
Then $A \supset M \cong A/\mathfrak{p}$ and $\mathfrak{p}^n = 0$. Since A is not a domain we get $\lambda(A_{\mathfrak{p}}) = 2$, hence $A/\mathfrak{p}$ is regular. It follows that $\mathfrak{p}^2 A_{\mathfrak{p}} = 0$ and therefore $\mathfrak{p}^2 = 0$. So we can think of $\mathfrak{p}$ as an $A/\mathfrak{p}$-module. From the exact sequence (of A-modules)

$$0 \longrightarrow \mathfrak{p} \longrightarrow A \longrightarrow A/\mathfrak{p} \longrightarrow 0$$

we conclude that $\mathfrak{p}$ satisfies (S_2) as an A-module and as an $A/\mathfrak{p}$-module. Moreover we have (see Chapter I, (1.3))

$$2 = e(A) = e(\mathfrak{p}) + e(A/\mathfrak{p}) = \mathrm{rk}_{A/\mathfrak{p}}(\mathfrak{p}) + \mathrm{rk}_{A/\mathfrak{p}}(A/\mathfrak{p}) \quad ,$$

i.e. $\mathrm{rk}_{A/\mathfrak{p}}(\mathfrak{p}) = 1$. Since $\mathfrak{p}$ is torsionfree as an $A/\mathfrak{p}$-module, it may be regarded as a submodule of the quotient field $Q(A/\mathfrak{p})$, say

$\mathfrak{p} = \sum_i A/\mathfrak{p}\cdot\frac{\overline{z}_i}{\overline{x}}$ with $\overline{z}_i, \overline{x} \in A/\mathfrak{p}$. Thus $\mathfrak{p}$ is isomorphic to an ideal $\mathfrak{a} = \overline{x}\cdot\mathfrak{p}$ of $A/\mathfrak{p}$.

Denoting $A/\mathfrak{p}$ by $\overline{A}$, we consider the exact sequences:

$$0 \longrightarrow \mathfrak{a}_{\mathfrak{P}} \longrightarrow \overline{A}_{\mathfrak{P}} \longrightarrow \overline{A}_{\mathfrak{P}}/\mathfrak{a}_{\mathfrak{P}} \longrightarrow 0$$

for $\mathfrak{P} \in \operatorname{Ass}(\overline{A}/\mathfrak{a})$. Since $\mathfrak{a}$ satisfies (S_2) , we see by similar arguments as in Case 1 that $\operatorname{ht}(\mathfrak{P}) = 1$ for all $\mathfrak{P} \in \operatorname{Ass}(\overline{A}/\mathfrak{a})$. Therefore $\mathfrak{a}$ must be a principal ideal, i.e. $\mathfrak{p}$ is a free $A/\mathfrak{p}$-module. This implies that $\mathfrak{p}$ (as well as $A/\mathfrak{p}$ itself) are Cohen-Macaulay A-modules, hence A is Cohen-Macaulay, q.e.d.

References - Chapter IX

Books

[1] P.J. Hilton and U. Stammbach, A Course in Homological Algebra, Graduate Text in Math. 4, Springer-Verlag, 1971.

[2] M. Hochster, Topics in the homological theory of modules over commutative rings, the CBMS Regional Conference Series in Mathematics 24, Amer. Math. Soc., 1975.

[3] H. Matsumura, Commutative algebra, W.A. Benjamin 1980.

[4] M. Nagata, Local rings, Krieger, Huntington, New York 1975.

[5] D.G. Northcott, Lessons on rings and modules, Cambridge Tracts no. 71, 1976.

[6] J. Sally, Numbers of generators of ideals in local rings, New York: Dekker 1978.

Papers

[7] R. Fossum, H.-B. Foxby, P. Griffith and I. Reiten, Minimal injective resolutions with applications to dualizing modules and Gorenstein modules, Publ. Math. I.H.E.S., 45 (1975), 193 - 215.

[8] S. Goto, On the Cohen-Macaulay-fication of certain Buchsbaum rings, Nagoya Math. J., 80 (1980), 107 - 116.

[9] S. Goto and Y. Shimoda, On the Rees algebras over Buchsbaum rings, J. of Math. of Kyoto Univ., 20 (1980), 691 - 708.

[10] S. Goto and Y. Shimoda, On the Rees algebras of Cohen-Macaulay local rings, Commutative Algebra (Analytical Methods), Lecture Notes in Pure and Applied Mathematics 68 (1982), 201 - 231.

[11] U. Grothe, Zur Cohen-Macaulay-Struktur von Aufblasungsringen, Inaugural-Dissertation, Köln Univ., 1985.

[12] M. Herrmann and S. Ikeda, Remarks on liftings of Cohen-Macaulay property, Nagoya Math. J. 92 (1983), 121 - 132.

[13] M. Herrmann and U. Orbanz, Between equimultiplicity and normal flatness, Algebraic Geometry, Proceedings La Rabida 1981 (ed. Aroca-Buchweiz-Giusti-Merle), Lecture Notes in Mathematics 961, Springer, Berlin and New York (1982), 200 - 232.

[14] C. Huneke, A remark concerning multiplicity, Proc. Amer. Math. Soc., 85 (1982), 331 - 332.

[15] S. Ikeda, The Cohen-Macaulayness of the Rees algebras of local rings, Nagoya Math. J., 89 (1983), 47 - 63.

[16] S. Ikeda, On the Gorensteinness of Rees algebras over local rings Nagoya Math. J., 102 (1986), 135 - 154.

[17] S. Ikeda, Remarks on Rees algebras and graded rings with multiplicity 3, Preprint.

[18] S. Ikeda, Conductor ideals of Gorenstein domains and local rings with multiplicity 2, Preprint.

[19] S. Ikeda and N.V. Trung, When is the Rees algebra Cohen-Macaulay? Preprint.

[20] J.D. Sally, On the associated graded rings of a local Cohen-Macaulay ring, J. of Math. of Kyoto Univ., 17 (1977), 19 - 21.

[21a] P. Schenzel, Applications of dualizing complexes to Buchsbaum rings, Ad. in Math., 44 (1982), 61 - 77.

[21b] P. Schenzel, Notes on liason and duality, J. Math. Kyoto Univ., 22 (1982), 485 - 498.

[22] G. Valla, Certain graded algebras are always Cohen-Macaulay, J. of Alg., 42 (1976), 537 - 548.

[23] S. Goto, Buchsbaum rings of maximal embedding dimension, J. of Alg. 76 (1982), 383 - 399.

APPENDIX

GEOMETRIC EQUIMULTIPLICITY

INTRODUCTION

The idea of a complex space emerged slowly over the decades as a natural generalization of the idea of a Riemann surface and its higher dimensional analogues, the complex manifolds. As in the classical theory of holomorphic functions of one variable, complex spaces arise in the attempt to understand holomorphic functions of several variables by constructing their natural home, "das analytische Gebilde", i.e. the maximal natural domain of definition. The nonuniformizable points, nowadays called singularities, caused great conceptual difficulties, so that a satisfactory definition had to wait until the 50's of this century when it was given by Behnke and Stein and, somewhat later in some greater generality, by Cartan and Serre. Subsequently it became clear that if one wants to gain a deeper understanding of complex manifolds, even of curves, complex spaces with nilpotents in their structure sheaf inevitably show up, be it in inductive proofs, or be it in the construction of such important geometric objects as moduli spaces of various, sometimes very classical, structures. This step was taken by Grauert and Grothendieck in the early 60's, who introduced the now generally accepted definition of, possibly nonreduced, complex spaces.

Aside from their intricate and important global properties, complex spaces possess a very rich and interesting local geometry, due to the presence of singularities. The algebraization of this local geometry was initiated by Weierstraß, who formulated his famous Preparation Theorem. Rückert, in a fundamental paper of 1931, was the first to use systematically algebraic tools in the local theory, and the consequent use of local algebra was further systematized in the Cartan Seminer of 1960/61, and Abhyankar's book of 1964 on local analytic geometry. It then became clear that the local geometry of complex spaces and the algebraic structure of the corresponding local rings are completely equivalent. In this way, then, algebraic statements within the category of local analytic algebras (i.e. quotients of convergent power series algebras) have an equivalent geometric interpretation which can be systematically exploited. Conversely, geometric considerations may provide particular insights and suggest natural algebraic statements which possibly would not have shown up easily within a pure algebraic context. It is this interplay between algebra and geometry which makes local analytic algebras a particularly intersting category, and a "testing ground" for conjectures and concepts in local algebra.

This Appendix sets out to give an introduction to Local Complex Analytic Geometry, to give the geometric interpretation of some fundamental algebraic concepts as dimension, system of parameters, multiplicity, and finally to explore to some extent the geometric meaning of the equimultiplicity results of Chapter IV. Thus, it is concerned with the material of the first four Chapters of this book.

I now give a quick overview over the contents and intentions of the three parts;more details are provided in the introductory remarks of the various parts and their paragraphs.

In Part I, my intention was to give a rapid introduction to the local theory of complex spaces, but at the same time to maintain the contradictory principle of giving all main lines of thought, in order not to discourage the nonspecialist by refering constantly to a labyrinthic and sometimes extreme technical literature. The main results are the Equivalence Theorem 3.3.3, which establishes the equivalence of the algebraic and geometric viewpoint; and the Local Representation Theorem 6.3.1.This local description of a complex space as a branched cover, which was, in principle,known to Weierstraß,lies at the heart of algebraization of the analytic theory, expressing the fact that any complex spacegerm gives rise to a "relative algebraic situation" over a smooth germ. This geometric situation is the local analogue of the Noether Normalization and contains the notions of dimension, system of parameters, and multiplicity, in its geometry. Technically, I have tried to emphasize two points. Firstly, I have made constant use of the General Division Theorem of Grauert-Hironaka from the beginning. From my point of view, it is a natural and systematic procedure which classifies many technical points. Moreover, it is basic for Hironaka's resolution of complex space singularities (see III, 1.3.5) and its effective algorithmic character may someday point the way to an explicit resolution procedure. (Presently, at least, it provides an effective algorithm for computing standard bases, and so Hilbert functions and tangent cones, see I, 2.4.4) Secondly, following Grothendieck's treatment in [64], I have postponed the introduction of coherence to the point where it really becomes indispensible; since, in the complex analytic case, coherence is a deep and not at all obvious property, it should be used only for the proof of those results which depend crucially on it (in our case, the property that openness of a finite map at a point implies the map being open near that point). Large parts of the exposition are taken from [28], and I refer to it and [40],[64] for complete details.

In Part II, I expose the geometric theory of local multiplicity as a local mapping degree; for more historical and geometrical background I refer to the introductory remarks to that Part. The main technical concept, introduced in § 1, is that of a compact Stein neighbourhood. This concept allows to relate properties of nearby analytic local rings of a complex space to one unifying algebraic object, the coördinate ring of a compact Stein neighbourhood. This gives a systematic way of deducing local properties of complex spaces from results of local algebra, and vice versa. Here, coherence enters in a fundamental way, and it is via coherence and the Equivalence Theorem I 3.3.3 that local, not only punctual, properties of complex spaces can be deduced by doing local algebra. This technique seems to have originated in [33], and has been exploited by various authors to deduce results in Complex Analytic Geometry from corresponding results in Algebraic Geometry, starting with [4]; see [5], [29], [38], and [63]. Here, I have simplified the treatment by dropping the requirement of semianalyticity for the compact Stein neighbourhoods, thus avoiding the highly nontrivial stratification theory of semianalytic sets.

Part III, finally, deals with the geometric theory of equimultiplicity, and forms the central part of the Appendix. It also gives various instances of the method of compact Stein neighbourhoods. In § 1, we deduce properties of normal flatness in the complex analytic case from the algebraic case; in § 2 we give a geometric proof of the equivalence of the conditions $e(R) = e(R_{\mathfrak{p}})$ and $ht(\mathfrak{p}) = s(\mathfrak{p})$ of Chapter IV, Theorem (20.9); and in § 3, finally, we turn this principle around and establish the geometric contents of equimultiplicity via Theorem (20.5) of Chapter IV. Further, bearing in mind the title of a well-known paper by Lipman [49] I have made comments on the connections with, and the geometric significance of, the algebraic notions of reduction and integral dependence. The underlying fundamental geometric principle, which unifies equimultiplicity, reduction, and integral dependence, is the notion of transversality (this is a basic principle in the work of Teissier [69]); this becomes particularly clear from the geometric description of multiplicity as the mapping degree of a projection (see the introductory remarks to III, III § 2, and III § 3 below).

On one hand, this Appendix was intended to give an overview of the geometric significance of equimultiplicity and not to be a full detailed treatment. On the other hand, I felt that it would have been of little value just to state the results without providing some insights into the machinery producing them, especially as there seems to be some interest on the side of algebraists to become more acquainted with complex-analytic methods. In connection with the confinements of space, time, and perseverance of the author, there results that the prsentation oscillates between rigour and loose writing, a dilemma I have been unable to solve. I can only offer my apologies and hope that those who approve of the one and disapprove of the other will appreciate seeing their approvals met instead of complaining about seeing their disapprovals aroused.

Concerning the notation, local rings are usually denoted R etc. instead of $(R,\mathfrak{m})$. The maximal ideal of R etc. is then denoted by $\mathfrak{m}_R$, and its nilradical by $\mathfrak{n}_R$. The notation $\mathfrak{m}_n$, $n \in \mathbb{N}$, refers specially to the maximal ideal of $\mathbb{k}\{z_1,\ldots,z_n\}$. If $(X,\mathcal{O}_X)$ is a complex space, $\mathfrak{m}_{X,x}$ or $\mathfrak{m}_x$, denotes the maximal ideal of $\mathcal{O}_{X,x}$, and $N_{X,x}$, or N_x , its nilradical. References within this Appendix usually are by full address; II 5.2.1 for instance refers to 5.2.1 of Part II. When they are made within one Part, the corresponding numbers I, II, III are suppressed. Numbers in brackets refer to formulas; I (2.3.1) for instance means the formula numbered (2.3.1) in Part I.

I wish to take the opportunity to express my profound indebtness to Professor Manfred Herrmann for the suggestion to include this work as a part of the book. I thank him, and O. Villamayor, for the interest they took in this work and for numerous hours of discussion, which saved me from error more than once. It goes without saying that all the remaining errors and misconceptions are entirely within the author's responsibility. I further express may gratitude to the Max-Planck-Institut für Mathematik and its director, Professor F. Hirzebruch, to be able to work in a stimulating atmosphere, and for financial support. Finally, I thank Mrs. Pearce for her skilful typing and for the patience with which she bore many hours of extra work and the everlasting threat of possible changes.

I. LOCAL COMPLEX ANALYTIC GEOMETRY

In this chapter I give an overview over the basic facts of the local theory of complex analytic spaces. The main references are the Cartan seminar [64], especially the exposés 9 - 11, 13 - 14 of Grothendieck and 18 - 21 of Houzel, and the excellent book [28]. For further information, one can also consult the book [40].

The main results are the Equivalence Theorem 3.3.3, which establishes the equivalence of the category of local analytic algebras and the category of complex space germs, the Integrality Theorem 4.4.1., which characterizes finiteness geometrically and algebraically, and, finally, the Local Representation Theorem 6.3.1., which is a local analogue of Noether normalization. It allows to represent a complex space germ locally as a branched cover of an affine space, and this gives the geometric interpretation of the dimension and of a system of parameters of the corresponding local ring. Moreover, this setup will be fundamental for the description of the multiplicity of this local ring in the next chapter.

§ 1. Local analytic algebras

In this section, I describe the category $\underline{\underline{la}}$ of local analytic algebras, which will be basic to all what follows. Its objects, the local analytic algebras, are the algebraic counterparts to the geometric objects formed by the germs of analytic spaces, or singularities, which will be introduced in § 3.

In what follows, $\mathbb{k}$ denotes any complete valued field. Proofs are mostly sketched, or omitted. For details I refer to [26], Kapitel 1, § 0 - 1; [40], and § 21.

1.1. Formal power series

I assume known the notion of a formal power series in n indeterminates $X_1,\ldots,X_n$. They form a ring denoted $\mathbb{k}[[X_1,\ldots,X_n]]$, or $\mathbb{k}[[X]]$ if n is understood. I use the multiindex notation; a monomial $X_1^{A^1}\ldots X_n^{A^n}$ will be denoted X^A with $A = (A^1,\ldots,A^n) \in \mathbb{N}^n$. Let $M(n) \subseteq \mathbb{k}[[X]]$ be the space of monomials; then

$$\text{(1.1.1)} \qquad \log : M(n) \longrightarrow \mathbb{N}^n$$
$$X^A \longrightarrow A$$

induces an isomorphism $(M(n),\cdot,X^0) \longrightarrow (\mathbb{N}^n,+,0)$ of monoids which I will freely use; in this way, one may view monomials as lattice points in $\mathbb{R}^n$, and divisibility properties of monomials turn into combinatorial properties of lattice points. This interplay between algebra and combinatorics will be quite crucial in establishing in § 2 fundamental properties of power series rings such as the Division Theorem, the noetherian property, or the Krull Intersection Theorem.

In the multiindex notation, $|A| := \sum_{j=1}^{n} A^j$, so that $|X^A| := |A|$ is the usual degree. Formal power series will be written as $f = \sum_{M \in M(n)} f_M M = \sum_{A \in \mathbb{N}^n} f_A X^A$, with $f_M, f_A \in \mathbb{k}$. We define

$$\text{(1.1.2)} \qquad \operatorname{supp}(f) := \{M \in M(n) \mid f_M \neq 0\} \quad ,$$

the <u>support of</u> f, and

$$\text{(1.1.3)} \qquad \nu(f) := \min\{|M| \mid M \in \operatorname{supp}(f)\}$$

the <u>order</u> or <u>subdegree</u> of f. We will make use of the following properties of $\mathbb{k}[[X_1,\dots,X_n]]$:

<u>Proposition 1.1.1.</u>

(i) $\mathbb{k}[[X_1,\dots,X_n]]$ <u>is a commutative ring with unit, and in fact a</u> $\mathbb{k}$<u>-algebra.</u>

(ii) $f \in \mathbb{k}[[X]]$ <u>is a unit if and only if</u> $f_0 \neq 0$.

(iii) $\mathbb{k}[[X_1,\dots,X_n]]$ <u>is a local ring with maximal ideal</u>
$\hat{\mathfrak{m}}_n := \{f \mid \nu(f) \geq 1\} = (X_1,\dots,X_n)\cdot \mathbb{k}[[X_1,\dots,X_n]]$.

(iv) $\forall k \in \mathbb{N} : \hat{\mathfrak{m}}_n^k = \{f \mid \nu(f) \geq k\}$; <u>especially</u>

$$\bigcap_{k=0}^{\infty} \hat{\mathfrak{m}}_n^k = \{0\} \quad .$$

These properties are elementary. (i) is clear. For (ii), note that, when $f := 1 - u$ with $\nu(u) \geqq 1$, $\sum_{j=0}^{\infty} u^j$ exists in $\mathbb{k}[[X]]$. Finally, if f has $\nu(f) \geqq k$, it can be written as

$$(1.1.4) \qquad f = \sum_{\substack{M \in \mathcal{M}(n) \\ |M| = k}} M \cdot f^{(M)} \quad , \qquad f^{(M)} \in \mathbb{k}[[X]] \quad ,$$

with the $\operatorname{supp}(M \cdot f^{(M)})$ pairwise disjoint (this will be systematized later on in the Division Algorithm 2.3.1.). This shows that $\{f \mid \nu(f) \geqq k\} \subseteq (X_1, \dots, X_n)^k$, which implies, together with (ii), the statements (iii) and (iv).

1.2. Convergent power series

Let $\mathbb{A}^n$ be the affine n-space over $\mathbb{k}$. A polyradius ρ is an element $\rho = (\rho^1, \dots, \rho^n) \in (\mathbb{R}_{>0})^n$, and if $z_0 = (z_0^1, \dots, z_0^n)$ is a point in $\mathbb{A}^n$, the set

$$(1.2.1) \qquad P(z_0;\rho) := \left\{ z \in \mathbb{A}^n \mid \forall\, 1 \leqq i \leqq n : |z^i - z_0^i| < \rho^i \right\}$$

is called the polycylinder around z_0 of (poly-)radius ρ.

Proposition 1.2.1. For a formal power series $f \in \mathbb{k}[[X]]$, the following properties are equivalent:

(i) $\exists$ a polyradius $\rho \in (\mathbb{R}_{>0})^n$ such that the family $(f_A z^A)_{A \in \mathbb{N}^n}$ is summable in $\mathbb{k}$ for $z \in P(0;\rho)$.

(ii) $\exists$ a polyradius $\rho \in (\mathbb{R}_{>0})^n$ such that

$$(1.2.2) \qquad \|f\|_\rho := \sum_{A \in \mathbb{N}^n} |f_A| \rho^A < \infty$$

(iii) $\exists$ constants $C, N \in \mathbb{R}_{>0}$ such that

$$(1.2.3) \qquad |f_A| \leqq C \cdot N^{|A|}$$

for all $A \in \mathbb{N}^n$.

Moreover, in these cases there is the "Cauchy estimate"

(1.2.4) $$|f_A| \leq \|f\|_\rho \cdot \rho^{-A} \quad \text{for all} \quad A \in \mathbb{N}^n \quad .$$

Definition 1.2.2. A formal power series $f \in \mathbb{k}[[X]]$ satisfying one of the properties of Proposition 1.2.1. is called a convergent power series. The convergent power series form a commutative, unitary ring and a $\mathbb{k}$-algebra, denoted $\mathbb{k}\{X_1,\ldots,X_n\}$, or $\mathbb{k}\{X\}$ for short.

The "norm" $\|\ \|_\rho$ defined in (1.2.2) is the main technical tool in manipulating convergent power series. Introduce the following subalgebras, for $\rho \in (\mathbb{R}_{>0})^n$, of $\mathbb{k}\{X\}$:

$$\mathbb{k}\{X\}_\rho := \{f \in \mathbb{k}[[X]] \mid \|f\|_\rho < \infty\}$$

That these are in fact subalgebras follows from

Proposition 1.2.4. $\mathbb{k}\{X\}_\rho$ is a $\mathbb{k}$-Banach-algebra with norm $\|\ \|_\rho$, and has no zerodivisors.

The proof uses the Cauchy estimate (1.2.4).

We now find the units of $\mathbb{k}\{X\}$:

Lemma 1.2.5. For $f \in \mathbb{k}\{X\}$, $\lim_{\rho\to 0} \|f\|_\rho = |f_0|$.

Proof. Write, as in (1.1.4), $f = f_0 + \sum_{j=1}^{n} X_j f_j$ with the $\mathrm{supp}(X_j f_j)$ disjoint, then $\|f\|_\rho = |f_0| + \sum_{j=1}^{n} \rho^j \|f_j\|_\rho$, whence the claim.

Hence, if $f = 1 - u$ with $\nu(u) \geq 1$, $\|u\|_\rho < 1$ for suitable ρ, and so $\sum_{j=0}^{\infty} u^j$ in fact exists not only in $\mathbb{k}[[X]]$ but in $\mathbb{k}\{X\}$ because of Proposition 1.2.4. This proves

Proposition 1.2.6. $f \in \mathbb{k}\{X\}$ is a unit if and only if $f_0 \neq 0$.

Corollary 1.2.7. $\mathbb{k}\{X\}$ is a local ring with maximal ideal $\mathfrak{m}_n = (X_1,\ldots,X_n)\cdot \mathbb{k}\{X\}$.

Proof. $\mathbb{k}\{X\}$ is local by Proposition 1.2.6, with maximal ideal $m_n := \hat{m}_n \cap \mathbb{k}\{X\}$. By the proof of Lemma 1.2.5 we may write $f \in m_n$ as $f = \sum_{j=1}^{n} f_j X_j$ with $\|f\|_\rho = \sum_{j=1}^{n} \rho^j \|f_j\|_\rho$ which shows the f_j are in $\mathbb{k}\{X\}$.

Finally, reasonings analoguous to those above show the following lemma

Lemma 1.2.8. $m_n^k = \{f \in \mathbb{k}\{X\} \mid \nu(f) \geq k\} = (X_1,\dots,X_n)^k \cdot \mathbb{k}\{X\}$.

Corollary 1.2.9. $\bigcap_{k=0}^{\infty} m_n^k = \{0\}$.

1.3. Local analytic $\mathbb{K}$-algebras

We are now in a position to describe the category la/$\mathbb{k}$ of local analytic $\mathbb{k}$-algebras. The proofs are sketched, for more details see [26] or [40]; they are more or less straigthforward with the notations and results of 1.1. and 1.2.

The following definition makes sense because of Corollary 1.2.7.

Definition 1.3.1. Let R be a $\mathbb{k}$-algebra. R is called a local analytic $\mathbb{k}$-algebra if and only if R is isomorphic to a quotient algebra $\mathbb{k}\{X_1,\dots,X_n\}/I$, where $I \subseteq \mathbb{k}\{X_1,\dots,X_n\}$ is a finitely generated ideal.

The assumption on I being finitely generated is in fact superfluous due to the following famous theorem.

Theorem 1.3.2 (Rückert Basissatz). A local analytic $\mathbb{k}$-algebra is noetherian.

This is a nontrivial result. I will give a proof in 2.4. which makes it clear that this property comes from a combinatorial property of the monomials which puts the noetherianness of $\mathbb{k}[X]$, $\mathbb{k}[[X]]$ and $\mathbb{k}\{X\}$ on an equal footing. ("Dickson's Lemma"; see Proposition 2.2.1).

Here, we assume Theorem 1.3.2.

The local analytic $\mathbb{k}$-algebras with the local $\mathbb{k}$-algebra homomorphims form a category which I will call la/$\mathbb{k}$. The following remark is sometimes useful:

Remark 1.3.3 (Serre). Any $\mathbb{k}$-algebra homomorphism of local $\mathbb{k}$-algebras is local.

The proof is simple and left to the reader.

The following theorem is the main result of this section; it characterizes the convergent power series in $\underline{la/\mathbb{k}}$.

Theorem 1.3.4. The algebras $\mathbb{k}\{X_1,\dots,X_n\}$ are free objects in $\underline{la/\mathbb{k}}$.

In other words, given a local analytic $\mathbb{k}$-algebra R and n elements $f_1,\dots,f_n \in m_R$, there is a unique $\mathbb{k}$-algebra homomorphism $\varphi : \mathbb{k}\{X_1,\dots,X_n\} \longrightarrow R$ with $\varphi(X_j) = f_j$ for $j = 1,\dots,n$.
This property will be an essential step in the proof of the Equivalence Theorem 3.3.3; see Proposition 3.3.1.

Sketch of proof of 1.3.4.

For existence we may assume $R = \mathbb{k}\{U_1,\dots,U_m\}$ is a convergent power series ring. Let $f_1,\dots,f_n \in m_R$ be given. Write $g \in \mathbb{k}\{X_1,\dots,X_n\}$ as

$$g = \sum_{k=0}^{\infty} g_k \quad , \tag{1.3.1}$$

where the g_k are homogeneous polynomials of degree k. Then $g_k(f_1,\dots,f_n)$ is a formal power series with $\nu(g_k(f_1,\dots,f_n)) \geqq k$, and so $g(f_1,\dots,f_n) := \sum_{k=0}^{\infty} g_k(f_1,\dots,f_n)$ is a well-defined formal power series. If then $\sigma \in (\mathbb{R}_{>0})^n$ is such that $\|g\|_\sigma < \infty$, there is a $\rho \in (\mathbb{R}_{>0})^m$ with $\|g(f_1,\dots,f_n)\|_\rho \leqq \|g\|_\sigma$; this follows from Lemma 1.2.5. So $g(f_1,\dots,f_n) \in \mathbb{k}\{U\}_\rho \subseteq \mathbb{k}\{U\}$, and we put $\varphi(g) := g(f_1,\dots,f_n)$.

For uniqueness assume $\varphi,\psi : \mathbb{k}\{X\} \longrightarrow R$ are such that $\varphi(X_j) = \psi(X_j)$, $1 \leqq j \leqq n$. Then, with the notation (1.3.1), $(\varphi-\psi)(g) = (\varphi-\psi)\left(\sum_{k=p}^{\infty} g_k\right)$ for all $g \in \mathbb{k}\{X\}$ and $p \in \mathbb{N}$. By Lemma 1.2.8., $\sum_{k=p}^{\infty} g_k \in m_n^p$, so $(\varphi-\psi)(g) \in \bigcap_{p=0}^{\infty} m_R^p$, but $\bigcap_{p=0}^{\infty} m_R^p = \{0\}$ because of Theorem 1.3.2. and the Krull Intersection Theorem (see Theorem 2.4.5, or [1], 10.19).

§ 2. Local Weierstrass Theory I: The Division Theorem

The classical Weierstrass Preparation and Division Theorem lie at the foundation of local analytic geometry and are the most basic and important results of the theory. In their classical appearance, their use in proofs requires always induction on the dimension, which makes sometimes these proofs appear not very transparent. A more natural statement of the Division Theorem has been found independently by Grauert [23] and Hironaka [35], the main point being to divide a power series not by a single other one, but by several others at the same time. This is also related to the construction of standard bases, i.e. computing equations of tangent cones (for which by now an effective algorithm exists), and seems also to be of crucial importance in Hironaka's desingularization theory, since it allows to put generators of an ideal of power series into a canonical form. I will sketch a proof here which I think is the most simple one and clearly exhibits that it is based on a manifest division algorithm suggested by the usual euclidean algorithm for polynomials in one variable, the sole difference being that one divides with respect to ascending monomial degree instead of descending degree. See also [8], [18]-[21], and [62].

In this section, $\mathbb{k}\langle X\rangle$ will stand either for $\mathbb{k}[[X]]$ or $\mathbb{k}\{X\}$.

2.1. Ordering the monomials

Usually, in order to prove noetherianness of power series rings, or the Weierstrass theorems, one uses the valuation on power series given by the subdegree $\nu \in \mathbb{N}$ (1.1.3). The crucial idea of getting a refined division theorem is to manipulate power series by using the finer valuation given by the monomial degree $\log(M) = A \in \mathbb{N}^n$ for $M = X^A$. For this, one has to choose an ordering on the monomials, or, equivalently (because of (1.1.1)) on the monoid $\mathbb{N}^n$. The idea of putting an order on the monomials appears for the first time in a famous paper of Macaulay ([52], p. 533). We require that this order is compatible with the monoid structure. Nevertheless, there are quite a lot of orders fulfilling these requirements; they have been classified by Robbiano [58] and, in fact, there are infinitely many. We will temporarily work with the following one.

Definition 2.1.1. The lexicographic degree order on $\mathcal{M}(n)$ is defined as follows:

(2.1.0) $X^A < X^B$ <u>if and only if</u>
<u>either</u> $|A| < |B|$,
<u>or</u> $|A| = |B|$, <u>and the last nonzero coordinate</u> of
<u>of</u> $A - B$ <u>is negative</u>.

It has the properties

(2.1.1) (i) $1 < M$ for all $M \in (n)$;

(ii) $M < M' \Rightarrow MN < M'N$ for all N ;

(iii) $<$ is a well-ordering.

<u>Definition 2.1.2.</u> <u>For</u> $f \in \mathbb{k}[[X]]$,

$$LM(f) := \min(\operatorname{supp}(f)) \in \mathcal{M}(n) \cup \{\infty\}$$

<u>is called the leitmonomial of</u> f , <u>with the convention</u> $LM(0) = \infty$.

Recall that $f \in \mathbb{k}[[X]]$ has a unique decomposition $f = \sum_{k=\nu(f)}^{\infty} f_k$, with $f_k \in \mathbb{k}[[X]]$ homogeneous of degree k ; $f_{\nu(f)} =: L(f) =: in(f)$ is called the <u>leitform</u> or <u>initial form</u> of f . The following properties are immediate from the definitions:

(2.1.2) (i) $LM(f) = LM(L(f))$, <u>and so</u> $|LM(f)| = \nu(f)$;

(ii) $LM(f+g) \geq \operatorname{Min}(LM(f),LM(g))$, <u>with equality holding</u> <u>when</u> $LM(f) \neq LM(g)$;

(iii) $LM(f \cdot g) = LM(f) \cdot LM(g)$.

<u>2.2. Monomial ideals and leitideals</u>

A monomial ideal $I \subseteq \mathbb{k}<X>$ is defined to be an ideal generated by monomials. The following lies at the heart of the noetherian property of $\mathbb{k}[X]$, $\mathbb{k}[[X]]$, and $\mathbb{k}\{X\}$:

<u>Proposition 2.2.1</u> ("Dickson's Lemma"). <u>A monomial ideal is finitely generated. A canonical basis consists of those monomials which are minimal with respect to the divisibility relation</u>.

For this, introduce the "<u>stairs of</u> I", $E(I)$, for a monomial ideal I:

$$E(I) := \left\{ A \in \mathbb{N}^n \;\middle|\; X^A \in I \right\} \quad \text{(see Fig. 1)} \quad .$$

$E(I)$ is translation invariant: $E(I) + \mathbb{N}^n \subseteq E(I)$. In 1913, Dickson studied numbers with only finitely many given prime factors and proved ([11]):

<u>2.2.2.</u> $E \subseteq \mathbb{N}^n$ <u>is translation invariant if and only if</u> E <u>can be written as</u>

$$E = \bigcup_{j=1}^{k} \left(A_k + \mathbb{N}^n \right)$$

<u>for some</u> $A_j \in \mathbb{N}^n$, $j = 1,\dots,k$. <u>These</u> A_j <u>are unique up to permutation when they are taken as the minimal elements of</u> E <u>with respect to the partial orders</u> $A < B :\Longleftrightarrow \forall 1 \leq j \leq n : A^j \leq B^j$ <u>on</u> $\mathbb{N}^n$.

Looking at Figure 1, this is intuitively clear, since when approaching the coordinate hyperplanes the steps of the stairs decrease by integral amounts in the coordinate directions, which can happen only finitely many times. The precise proof is left to the reader. This result proves Proposition 2.2.1.

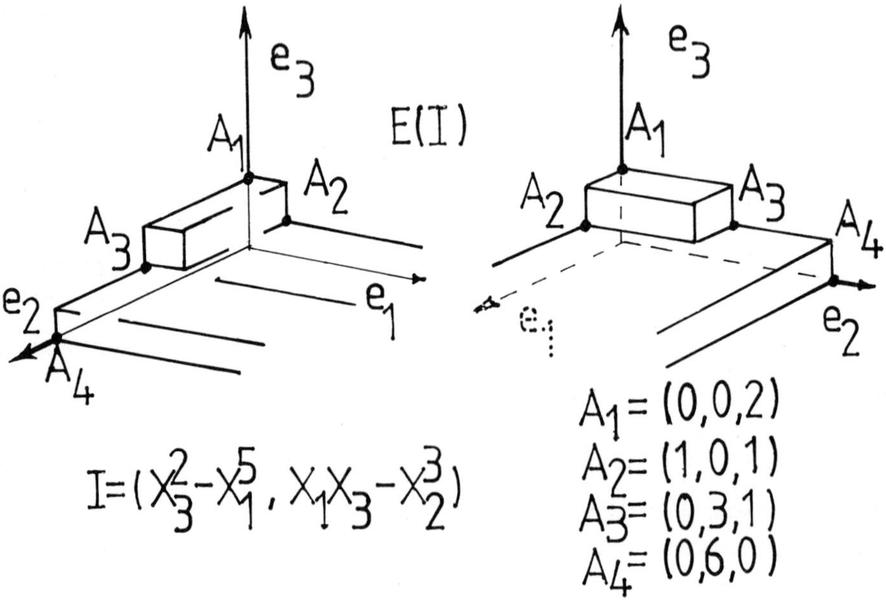

Fig. 1

If I is monomial, $f \in \mathbb{k}<X>$ belongs to I if and only if all $M \in \mathrm{supp}(f)$ belong to I ; this is analogous to the fact that a polynomial belongs to a homogeneous ideal if and only if all its homogeneous components belong to it. The crucial property of monomial ideals is now that membership of a monomial is effectively decidable if generators are known, since a monomial belongs to it if and only if it is divisible by the generators. But testing the divisibility of monomials is a simple effective operation; this operation will be put to work in the Division Algorithm 2.3.1. below. One therefore associates to any ideal I a monomial ideal $LM(I)$:

Definition 2.2.3. Let $I \subseteq \mathbb{k}<X>$ be an ideal. The monomial ideal

$$LM(I) := \text{ideal generated by the } LM(f) \text{ for } f \in I$$

is called the leitideal of I .

$LM(I)$ reflects many properties of I . For instance, a famous result of [ ] is that, if I is homogeneous, the Hilbert function $H(I,t)$ of I equals $H(M(I),t)$, and we will see in Section 2.4 that a base of an ideal I whose leitmonomials generate $M(I)$ has special pleasant properties and allows to deduce in an elegant way various facts about ideals in the rings $\mathbb{k}<X>$; see 2.4.3, 2.4.4. , and 2.5.2.

2.3. The Division Theorem

In order to give some motivation for the Division Theorem, consider the problem of finding a finite basis for an ideal I . The idea of how to obtain a finite basis is as follows: By Dickson's Lemma there are finitely many $f^1,\dots,f^k \in I$ such that the $LM(f^1),\dots,LM(f^k)$ generate $LM(I)$. Given $f \in I$, we then may write

$$LM(f) = g_1^{(0)} LM(f^1) + \dots + g_k^{(0)} LM(f^k) \tag{2.3.1}$$

for some $g_1^{(0)},\dots,g_k^{(0)}$; note this step is constructive. We regard this as the 0-th approximation to a wanted equation

$$f = g_1 f^1 + \dots + g_k f^k \quad . \tag{2.3.2}$$

For the first approximation, we form

$$(2.3.3) \qquad f^{(1)} := f - \left(g_1^{(0)} f^1 + \ldots + g_k^{(0)} f^k\right)$$

and iterate the step (2.3.1) with f replaced by $f^{(1)}$. Continuing this way, we get formal solutions $g_j = \sum_{p=0}^{\infty} g_j^{(p)}$ to (2.3.2) (which actually converge). This process is constructive when the f^j are given, and so it is feasible to call it an algorithm. The development of this idea leads to the Division Algorithm 2.3.1, which technically proceeds a little differently. Of course, this is only one aspect of the Division Algorithm, and its full power can only be seen from the consequences to which it will lead.

We begin with elements $f^1,\ldots,f^k \in \mathbb{k}\langle X\rangle$. Let $LM(f^j) = X^{A_j}$, $j = 1,\ldots,k$, and fix the ordering $(A_1,A_2,\ldots,A_k)$ of the A_j.

Definition 2.3.1. Let $f^1,\ldots,f^h \in \mathbb{k}\langle X\rangle$. The Division Algorithm with respect to $(f^1,\ldots,f^k)$ is defined by the following recursion scheme:

Start: For $f \in \mathbb{k}\langle X\rangle$ put

$$(2.3.4) \quad \text{(i)} \quad \forall 1 \leq j \leq k : g_j^{(-1)} := 0 \;, \quad h^{(-1)} := 0 \;,$$

$$\text{(ii)} \quad f^{(0)} := f \;.$$

Recursion: Let $g_1^{(0-1)},\ldots,g_k^{(p-1)}$, $h^{(p-1)}, f^{(p)}$ be defined for $0 \leq p \leq q$ Then put

$$(2.3.5) \qquad f^{(q)} =: g_1^{(q)}\cdot X^{A_1} + \ldots + g_k^{(q)}\cdot X^{A_k} + h^{(q)} \;,$$

where the $g_j^{(q)}$, $j = 1,\ldots,k$ and $h^{(q)}$ are defined uniquely by the requirements

(2.3.6) (i) $\mathrm{supp}(g_j^{(q)}\cdot X^{A_j})$ and $\mathrm{supp}(h^{(q)})$ are pairwise disjoint for $1 \leq j \leq n$;

(ii) if $X^B \in \mathrm{supp}(g_j^{(q)} X^{A_j})$, B is in no $A_i + \mathbb{N}^n$ where A_i precedes A_j in the given order. (In other words, one first collects all $M \in \mathrm{supp}(f^{(q)})$ divisible by X^{A_1} to obtain $g_1^{(q)}$, then those divisible by X^{A_2} to obtain $g_2^{(q)}$, and so on.)

Finally, put

$$(2.3.7)\qquad f^{(q+1)} := f^{(q)} - g_1^{(q)} f^1 - \dots - g_k^{(q)} f^k - h^{(q)}$$
$$= g_1^{(q)} (X^{A_1} - f^1) + \dots + g_1^{(q)} (X^{A_k} - f^k) \quad .$$

From (2.3.7), it is easy to see that, because of (2.1.2), one gets a strictly increasing sequence

$$LM(f^{(0)}) < LM(f^{(1)}) < LM(f^{(2)}) < \dots \quad ,$$

which implies that, for $1 \leqq j \leqq n$,

$$(2.3.8)\qquad g_j := \sum_{q=0}^{\infty} g_j^{(q)}$$

and

$$(2.3.9)\qquad h := \sum_{q=0}^{\infty} h^{(q)}$$

exist in $\mathbb{k}[[X]]$, and so

$$(2.3.10)\qquad f = g_1 f^1 + \dots + g_k f^k + h$$

holds in $\mathbb{k}[[X]]$, with the g_j and h uniquely determined by (2.3.6).

The miracle which now happens is that, if $f \in \mathbb{k}\{X\}$, the g_j and h are also in $\mathbb{k}\{X\}$, and (2.3.10) holds in $\mathbb{k}\{X\}$. I will just collect together the necessary estimates and leave the details, which are elementary after all, to the reader.

<u>(i)</u>. The conditions (2.3.6) guarantee , because of (2.3.5):

$$(2.3.11)\qquad \|g_1^{(q)}\|_\rho \rho^{A_1} + \dots + \|g_k^{(q)}\|_\rho \rho^{A_k} + \|h^{(q)}\|_\rho = \|f^{(q)}\|_\rho$$

for all q and ρ , and so, fixing some ρ :

$$(2.3.12)\qquad \forall 1 \leqq j \leqq k : \ \|g_j^{(q)}\|_\rho \leqq \|f^{(q)}\|_\rho \rho^{-A_k}$$

$$\text{and}\qquad \|h^{(q)}\|_\rho \leqq \|f^{(q)}\|_\rho$$

for all q .

<u>(ii)</u>. Because of (2.3.7), (2.3.12) implies

$$(2.3.13)\quad \|f^{(q+1)}\|_\rho \leqq \left(\|X^{A_1}-f^1\|_\rho \rho^{-A_1}+\ldots+\|X^{A_k}-f^k\|_\rho \rho^{-A_k}\right)\|f^{(q)}\|_\rho$$

for all q .

<u>The crucial point is now that the expression in brackets can be made smaller then a given</u> ε <u>for</u> $\rho = \lambda\rho_0$, <u>where</u> $\rho_0 \in (\mathbb{R}_{>0})^n$ <u>is suitable, and</u> $\lambda \in (0,1)$ <u>arbitrary</u>. This is a tedious, but elementary point which the reader should try to convince himself of; trouble is caused by those monomials of f^j which are different from X^{A_j} but have the same degree. It depends on the fact that all monomials in $\mathrm{supp}(X^{A_j}-f^j)$ are strictly greater then X^{A_j} , and on the lexicographic order (see [23], Satz 2.). Hence, choosing $\varepsilon \in (0,1)$ and $\rho_0 \in (\mathbb{R}_{>0})^n$ suitably we get from (2.3.11), by summing over q and using (2.3.13):

$$(2.3.14)\qquad \|g_1\|_\rho \rho^{A_1} + \ldots + \|g_k\|_\rho \rho^{A_k} + \|h\|_\rho \leqq \frac{1}{1-\varepsilon}\|f\|_\rho$$

for all $\rho = \lambda\rho_0$, $\lambda \in (0,1) \subseteq \mathbb{R}$, which gives the desired estimates on the $\|g_j\|_\rho$ and $\|h\|_\rho$ to ensure $g_j, h \in \mathbb{k}\{X\}$, $1 \leqq j \leqq n$.

This establishes the Division Algorithm with respect to the lexicographic degree order. There are, however, further important orders which arise, more generally, from a strictly positive linear form

$$(2.3.15)\qquad \begin{aligned} \Lambda : \mathbb{R}^n &\longrightarrow \mathbb{R} \\ (x^1,\ldots,x^n) &\longrightarrow \sum_{i=1}^n \lambda_i x^j \end{aligned} ,$$

with $\lambda_1,\ldots,\lambda_n \in \mathbb{R}_{>0}$, by defining

$X^A <_\Lambda X^B$ <u>if and only if either</u> $\Lambda(A) < \Lambda(B)$ <u>or</u> $\Lambda(A) = \Lambda(B)$ <u>and the last nonzero coordinate of</u> $A - B$ <u>is negative</u>.

Call such an order a <u>linear order</u>. It again defines, for $f \in \mathbb{k}\langle X\rangle$, a leitmonomial which I denote by $LM_\Lambda(f)$, or $LM(f)$ if Λ is understood. With these leitmonomials, one can again set up the Division Algorithm 2.3.1. To arrive at the estimates (2.3.12) and (2.3.13), one changes the definition of $\|\ \|_\rho$ to

$$\|f\|_{\rho,\Lambda} := \sum_{A\in\mathbb{N}^n} |f_A|\cdot\rho^{\Lambda(A)}$$

with $\rho^{\Lambda(A)} := (\rho^1)^{\lambda_1 A^1} \ldots (\rho^n)^{\lambda_n A^n}$; this norm clearly is equivalent to the former norm $\|\ \|_\rho$ defined by (1.2.2). One gets again the estimates (2.3.12) and (2.3.13), with $\rho^{\pm A_j}$ replaced by $\rho^{\pm\Lambda(A_j)}$, and the conclusion that the bracket in (2.3.13) can be made arbitrarily small still holds.

Finally, one may even allow positive linear forms, i.e. with $\lambda_i \in \mathbb{R}_{\geqslant 0}$, since a generic small perturbation of the λ_i defines a strictly positive linear form with the <u>same</u> division algorithm. Summing up one gets

<u>Theorem 2.3.2</u> (The Division Theorem, or Weierstrass préparé à la Grauert-Hironaka"). <u>Let</u> $\Lambda : \mathbb{R}^n \longrightarrow \mathbb{R}$ <u>be a positive linear form. Let</u> $f^1,\ldots,f^k \in \mathbb{k}\langle X\rangle$, <u>and</u> $LM_\Lambda(f^j) = X^{A_j}$, $1 \leq j \leq k$, <u>be the leitmonomials with respect to the linear order on</u> $M(n)$ <u>induced by</u> Λ. <u>Fix the order</u> $(f^1,\ldots,f^k)$ <u>of the</u> f^j, <u>and put recursively</u>

$$\Delta_0 := \emptyset \quad ,$$

$$\Delta_j := (A_j + \mathbb{N}^n) - \coprod_{i<j} \Delta_j \ , \quad j = 1,\ldots,k \quad ,$$

$$\bar{\Delta} := \mathbb{N}^n - \coprod_{j\leq k} \Delta_j \quad .$$

<u>Finally, let</u> $f \in \mathbb{k}\langle X\rangle$. <u>Then the following statements hold:</u>

(i) <u>The Division Algorithm 2.3.1. gives a unique representation</u>

$$f = g_1 f^1 + \ldots + g_k f^k + h$$

<u>with</u> $g_j = \sum_{A\in\Delta_j} g_{jA} X^{A-A_j}$, $1 \leq j \leq k$, <u>and</u> $h = \sum_{A\in\bar{\Delta}} h_A X^A$ <u>power series</u> in $\mathbb{k}[[X]]$.

(ii) <u>If</u> $f \in \mathbb{k}\{X\}$, <u>then for any</u> ε <u>with</u> $0 < \varepsilon < 1$ <u>there exists a neighbourhood basis</u> $\mathcal{B}$ <u>of</u> $0 \in \mathbb{A}^n$ <u>consisting of polycylinders</u> $P(0;\rho)$ such that for any $P(0;\rho) \in \mathcal{B}$ the estimate

$$\|g_1\|_{\rho,\Lambda}\,\rho^{\Lambda(A_1)} + \ldots + \|g_k\|_{\rho,\Lambda}\,\rho^{\Lambda(A_k)} + \|h\|_{\rho,\Lambda} \leq \frac{1}{1-\varepsilon}\,\|f\|_{\rho,\Lambda}$$

holds. In particular, the g_j *and* h *are in* $\mathbb{k}\{X\}$.

2.4. Division with respect to an ideal; standard bases

We are now in a position to carry out the suggestions at the beginning of 2.3. and prove the Rückert Basissatz, Theorem 1.3.2. We also give a proof of the Krull Intersection Theorem.

Let $I \subseteq \mathbb{k}\langle X\rangle$ be an ideal. Fix some linear order and choose $f^1,\dots,f^k \in I$ such that the leitmonomials $LM(f^1),\dots,LM(f^k)$ generate the leitideal $LM(I)$, which is possible by Dickson's Lemma 2.2.1.

Proposition 2.4.1 (Division with respect to I). *Let* $f \in \mathbb{k}\langle X\rangle$, *and let* I *and* $f^1,\dots,f^k \in I$ *be as above*.

(i) *In the representation*

$$f = g_1 f^1 + \dots + g_k f^k + h$$

given by the Division Theorem 2.3.2, h *does not depend on the choice of* $f^1,\dots,f^k$, *hence depends only on* I , *and is called* $\mathrm{red}_I f$.

(ii) $f \in I$ *if and only if* $\mathrm{red}_I f = 0$.

(iii) $\{f^1,\dots,f^k\}$ *is an ideal base of* I .

The proof is left to the reader; (i) depends on the fact that $\mathbb{N}^n = \coprod_j \Delta_j \amalg \bar{\Delta}$ is a disjoint decomposition, and (ii), (iii) are simple consequences.

Because of (iii), the following definiton makes sense:

Definition 2.4.2. *Let* $I \subseteq \mathbb{k}\langle X\rangle$. *Then* $\{f^1,\dots,f^k\} \subseteq I$ *is called a standard base of* I (with respect to a given linear ordering of the monomials) *if and only if* $\{LM(f^1),\dots,LM(f^k)\}$ *is a base of* $LM(I)$.

Since standard bases exist, we get

Corollary 2.4.3. *The rings* $\mathbb{k}\langle X\rangle$ *are noetherian*.

This clearly implies Theorem 1.3.2.

Remark 2.4.4. a) Not every ideal base is a standard base.

b) If $\{f^1,\ldots,f^k\}$ is a standard base, the initial forms $L(f^1),\ldots,L(f^k)$ generate the initial ideal $L(I)$, i.e. the ideal $L(I)$ having the property that $gr_{\mathfrak{m}_n}(\mathbb{k}\langle X\rangle/I) \cong \mathbb{k}\langle X\rangle/L(I)$. In particular, $L(f^1),\ldots,L(f^k)$ define the tangent cone at 0 of $\mathrm{Spec}(\mathbb{k}\langle X\rangle/I)$.

c) If $f^1,\ldots,f^k$ are polynomials defining an ideal $I \subseteq \mathbb{k}\langle X\rangle$, there is a constructive procedure for deriving a standard base from them using the division algorithm, which is based on the fact that $\{f^1,\ldots,f^k\}$ is a standard base if and only if each monomial syzygy of the leitmonomials lifts to a syzygy of $f^1,\ldots,f^k$. Dividing the $f := \sum g_i f^i$, where the $g_1,\ldots,g_k$ run through a generating system of the monomial syzygies of the leitmonomials, by $f^1,\ldots,f^k$ and adding the nonzero remainders leads eventually to a standard base. See [44], [55] and [62]. An implementation of the algorithm of [44] is available on the computer algebra system Macaulay [4]. This allows the computation of the Hilbert function of a homogeneous ideal (based on III, (1.3.4)), [53].

d) For $\mathbb{k}[X]$ one obtains, using maximal monomials instead of minimal ones, a proof of the Hilbert Basissatz along identical lines. In this case, a standard base is known as a Gröbner base ([3],[44], and [46]).

2.5. Applications of standard bases: The General Weierstrass Preparation Theorem and the Krull Intersection theorem

Any ideal $I \subseteq \mathbb{k}\langle X\rangle$ has a canonical standard base with respect to a given linear order in the following way: By Dickson's Lemma, $LM(I)$ has a unique base of monomials minimal with respect to divisibility, $\{X^{A_1},\ldots,X^{A_k}\}$ say. By Proposition 2.4.1, we have well-defined remainders $red_I X^{A_j}$. We thus obtain

Theorem 2.5.1 (The General Weierstrass Preparation Theorem). Let $I \subseteq \mathbb{k}\langle X\rangle$ be an ideal, Λ a linear order on $M(n)$, and $\mathcal{B} \subseteq LM_\Lambda(I)$ the canonical base consisting of the minimal elements with respect to the divisibility relation. Then $\{\omega_M \mid \omega_M := M - red_I M \text{ for } M \in \mathcal{B}\}$ is a standard base of I.

We refer to this base as the Weierstrass base of I (with respect to the given linear order).

As a further application of the Division Theorem we prove:

<u>Theorem 2.5.2</u> (Krull Intersection Theorem). <u>If</u> $R \in \underline{\underline{la/\mathbb{k}}}$,

$$\bigcap_{p=0}^{\infty} m_R^p = \{0\} \quad .$$

<u>Proof.</u> One has to show that, if $I \subseteq \mathbb{k}\{X_1,\ldots,X_n\}$ is any ideal, $\bigcap_{p=0}^{\infty} (I + m_n^p) = I$.

Choose a standard base $\{f^1,\ldots,f^k\}$ of I . Let $f \in \bigcap_{p=0}^{\infty} (I+m_n^p)$, and let p_0 be so large that all the $X^{A_j} := LM(f^j)$, $1 \leq j \leq k$, have degree less that p_0 . Let $p \geq p_0$. Fix the order $(X^{A_j},\ldots,X^{A_k})$ and then all X^A with degree $X^A = p$ in some order, and apply the Division Algorithm with respect to $f^1,\ldots,f^k$ and the X^A to f ; so f can be written

$$f = g_1^{(p)} f^1 + \ldots + g_k^{(p)} f^k + \sum_{\deg X^A = p} g_A^{(p)} \cdot X^A \quad .$$

But the Division Algorithm 2.3.1 shows that the $g_j^{(p)}$, $1 \leq j \leq k$, do not depend on p as soon as $p \geq p_0$. Hence the remainder $\sum_{\deg X^A = p} g_A^{(p)} \cdot X^A$ does not depend on p and so is in $\bigcap_{p \geq p_0} m_n^p$, which is zero by Corollary 1.2.9.

<u>2.6. The classical Weierstrass Theorems</u>

These are the classical cornerstones of Local Complex Analysis and direct consequences of the Division Theorem 2.3.2.

We introduce the notation $X' := (X_1,\ldots,X_{n-1})$, and so $X = (X',X_n)$.

<u>The Weierstrass Division Theorem 2.6.1</u> (Stickelberger-Späth; see the discussion in [26], p. 36). <u>Let</u> $f \in \mathbb{k}<X>$ <u>be such that</u> $f(0,X_n) = X_n^b \cdot u$ <u>for some integer</u> $b \geq 1$ <u>and</u> $u \in \mathbb{k}<X_n>$ <u>a unit</u> (we then say f is <u>regular in</u> X_n <u>of order</u> b.) <u>Then any</u> $e \in \mathbb{k}<X>$ <u>can be uniquely written as</u>

$$e = g\cdot f + h$$

with $g \in \mathbb{k}\langle X\rangle$ and $h \in \mathbb{k}\langle X'\rangle[X_n]$ of X_n-degree strictly less than b.

For this, just note that the condition on f ensures, after reversing the numbering of the coordinates, the existence of a positive linear form Λ on $\mathbb{R}^n$ making $LM_\Lambda(f) = X_n^b$; then apply the Division Theorem 2.3.2. Uniqueness, i.e. independence of g and h of the order, comes from the fact that $\Delta_1 = (0,b) + \mathbb{N}^n$, $\overline{\Delta} = \mathbb{N}^{n-1} \times [0,b-1]$ do not depend on the choice of order.

Corollary 2.6.2. $\mathbb{k}\langle X\rangle/(f) \cong \mathbb{k}\langle X'\rangle^b$ as a $\mathbb{k}\langle X'\rangle$-module.

Hence $\mathbb{k}\langle X\rangle/(f)$ is a finite $\mathbb{k}\langle X'\rangle$-module. This fact is the main reason why Local Complex Analysis is accessible to algebraic methods. It will be considerably generalized in the sequel to the extent that any local analytic algebra is finite over a convergent power series ring (see 6.2.4), leading in geometric terms to the Local Representation Theorem 6.3.1, which realizes any analytic space germ as a finite branched cover of a domain in some number space.

The Weierstrass Preparation Theorem 2.6.3. Let $f \in \mathbb{k}\langle X\rangle$ be regular in X_n of order b. Then f can be uniquely written as

$$f = e\cdot\omega ,$$

where e is unit in $\mathbb{k}\langle X\rangle$ and $\omega \in \mathbb{k}\langle X'\rangle[X_n]$ a Weierstrass polynomial, i.e. it is monic with coefficients in $\mathfrak{m}_{n-1}$.

Just apply Theorem 2.5.1. to $I = (f)$, and with linear order as above. The fact that the coefficients of ω are in $\mathfrak{m}_{n-1}$ follows from comparison of coefficients in the relation $X_n^b\cdot u = e(0,X_n)\cdot\omega(0,X_n)$.

§ 3. Complex spaces and the Equivalence Theorem

From now on, $\mathbb{k} = \mathbb{C}$, and $\underline{\mathrm{la}} := \underline{\mathrm{la}/\mathbb{C}}$. The standard coordinates on $\mathbb{C}^n$ are denoted $z_1,\dots,z_n$.

The main result of this section will be Grothendieck's Equivalence Theorem which states the equivalence of the "algebraic" category of local analytic $\mathbb{C}$-algebras and the "geometric" category of complex analytic spacegerms (or "singularities"), or rather its dual. This is a local analogue to the equivalence in Algebraic Geometry between the categories of rings and of affine schemes. Although well-known, proofs are not readily accessible; one is in [64], Exposé 13, and one in [40], the latter one, however, makes use of the machinery of coherence, which we will, following the viewpoint of Grothendieck ([64], p. 9 - 10) make no unnecessary use of.

3.1. Complex spaces

Higher dimensional complex manifolds and complex spaces with singular points arise naturally in the deformation and classification of varying complex structures on smooth complex curves. The systematic construction of these spaces by means of his philosophy of representable functors led Grothendieck to consider nilpotents in the structure sheaf (see his exposés 7 - 17 in [64]), and it is only when allowing arbitrary nonreduced spaces that phenomena as, for instance subspaces which have plenty of infinitesimal deformations but no actual one within the ambient space (corresponding to nonreduced isolated points of the Hilbert scheme), can be satisfactorily understood. At the same time Grauert [22], also led by the consideration of moduli problems, introduced nonreduced complex spaces.

I will assume that the notion of a ringed space is known and just fix some notation concerning them; full discussions are available in [28], [31], [40] and [64], Exposś 9.

As usual, a ringed space consists of a topological space X and a sheaf of (commutative, unitary) rings $\mathcal{O}_X$ on it and is denoted $(X,\mathcal{O}_X)$, or $\underline{X}$, if $\mathcal{O}_X$ is understood. We denote the stalk of $\mathcal{O}_X$ at $x \in X$ by $\mathcal{O}_{X,x}$, and, if it is a local ring, its maximal ideal by $\mathfrak{m}_x$. A morphism between ringed spaces is a pair $(f,f^0) : (X,\mathcal{O}_X) \longrightarrow (Y,\mathcal{O}_Y)$, where $f : X \longrightarrow Y$ is continuous and f^0 a sheaf morphism $\mathcal{O}_Y \longrightarrow f_*\mathcal{O}_X$; if no confusion is possible, we also denote the canonical adjoint by $f^0 : f^{-1}\mathcal{O}_Y \longrightarrow \mathcal{O}_X$ because $\mathrm{Hom}(\mathcal{O}_Y,f_*\mathcal{O}_X) = \mathrm{Hom}(f^{-1}\mathcal{O}_Y,\mathcal{O}_X)$ naturally. Again, we abbreviate by writing $\underline{f} : \underline{X} \longrightarrow \underline{Y}$.

I further assume the notions of an open subspace and an closed subspace defined by an ideal $\mathcal{J} \subseteq \mathcal{O}_X$ which we always will assume to be locally finitely generated or, as I will say, locally finite. A subspace will always mean a locally closed subspace, i.e. a closed subspace of an open subspace. Corresponding to these notions there are the notions of an open immersion, closed immersion and immersion.

For later use, we note the following simple Lemma:

Lemma 3.1.1. Let $(X,\mathcal{O}_X)$ be a ringed space, $\mathcal{I},\mathcal{J}\subseteq\mathcal{O}_X$ ideals, and $\mathcal{I}$ locally finite. Then any $x\in X$ such that $\mathcal{I}_x\subseteq\mathcal{J}_x$ has a neighbourhood U such that $\mathcal{I}|U\subseteq\mathcal{J}|U$.

The proof is left to the reader.

We make $\mathbb{C}^n$ into a ringed space by defining the structure sheaf $\mathcal{O}_{\mathbb{C}^n}$ to be the sheaf of germs of holomorphic functions, in other words, $\mathcal{O}_{\mathbb{C}^n}(U) := \{f\,|\,f : U \longrightarrow \mathbb{C} \text{ holomorphic}\}$ for any open $U\subseteq\mathbb{C}^n$. For any $a = (a_1,\dots,a_n)\in\mathbb{C}^n$ the stalk $\mathcal{O}_{\mathbb{C}^n,a}$ is then canonically isomorphic to the convergent power series ring $\mathbb{C}\{X_1-a_1,\dots,X_n-a_n\}$, and we will identify these two rings: in particular, $\mathcal{O}_{\mathbb{C}^n,0} = \mathbb{C}\{X_1,\dots,X_n\}$. Moreover, we will identify the indeterminates X_j with the standard coordinate functions z_j on $\mathbb{C}^n$. We can now define complex (analytic) spaces.

Definition 3.1.2.

(i) (Local model spaces). A local model space is a ringed space $(M,\mathcal{O}_M)$ given by the following data:

1) an open set $U\subseteq\mathbb{C}^n$,

2) elements $f^1,\dots,f^k\in\mathcal{O}_{\mathbb{C}^n}(U)$ ("equations"),

in the following way: If $\mathcal{I} := (f^1,\dots,f^k)\cdot\mathcal{O}_U$, then

$$\begin{aligned} M &:= \operatorname{supp}(\mathcal{O}_U|\mathcal{I}) \\ &= \left\{x\in U\,|\,\forall 1\leq j\leq k : f^j_x\in\mathfrak{m}_x\subseteq\mathcal{O}_{\mathbb{C}^n,x}\right\} \\ &= \left\{x\in U\,|\,\forall 1\leq j\leq k : f^j(x) = 0\right\} \quad, \end{aligned}$$

and $\mathcal{O}_M := (\mathcal{O}_X/I)|M$. We then write $M = N(n,U,(f^1,\dots,f^k))$ or $M = N(n,U,I)$; if $U \subseteq \mathbb{C}^n$ is understood we simply write $M = N(f^1,\dots,f^k)$ or $M = \mathbf{N}(I)$, and call it the null space of I .

(ii) (Morphisms of local models). A morphism between local models $M = N(m,U,(f^1,\dots,f^k))$ and $N = N(n,V,(g^1,\dots,g^\ell))$ is a morphism $(f,f^0) : (M;\mathcal{O}_M) \longrightarrow (N,\mathcal{O}_N)$ induced by a holomorphic map $F : U \longrightarrow V$ with the property $\forall 1 \leq j \leq \ell : g^j \circ F \in (f^1,\dots,f^k)\cdot\mathcal{O}_U$ in the following way:

1) $f := F|M$;

2) $f^0 : \mathcal{O}_N \longrightarrow f_*\mathcal{O}_M$ is induced by the mapping
$F^0_W : \mathcal{O}_V(W) \longrightarrow \mathcal{O}_U(F^{-1}W)$, $g \longmapsto g \circ F$, for all open $W \subseteq V$.

(iii) (The Category of complex spaces). A complex space is a ringed space which is locally isomorphic to a local model. A morphism of complex spaces, or holomorphic map, is morphism $(f,f^0) : (X,\mathcal{O}_X) \longrightarrow (Y,\mathcal{O}_Y)$ of the complex spaces $(X,\mathcal{O}_X)$, $(Y,\mathcal{O}_Y)$ within the category of ringed spaces which locally is isomorphic to a morphism of local models. This defines the category cpl of complex spaces.

In fact, any morphism between complex spaces within the category of ringed spaces turns out to be a holomorphic map; see Corollary 3.3.4.

If X is a complex space, an open or closed subspace in the category of ringed spaces, as defined before, is itself a complex space, and we can talk about open, closed, or arbitrary subspaces, and of open, closed, and arbitrary, immersions.

Example 3.1.3. Let $X = \{x\}$ be a one point space and $A \in$ la be artinian. Then $(\{x\},A)$ is a complex space. In fact the converse is true: any one point complex space arises in this way. This is astonishingly difficult to prove; it is a special case of the Rückert Nullstellensatz, and essentially equivalent to it; see § 5.

3.2. Constructions in cpl . It should be kept in mind that the following constructions are categorical; that means that the spaces and morphisms whose existence is asserted do not exist only settheoretically, but also the sheaves and sheaf maps have to be considered, and I urge the reader to convince himself of the details.

a) Glueing. Glueing data for a complex space consist of

(i) a family $(M_i, \mathcal{O}_{M_i})_{i \in I}$ of local models,

(ii) open subsets $U_{ij} \subseteq M_i$, $U_{ji} \subseteq M_j$ and isomorphisms

$$\underline{f}_{ij} : (U_{ij}, \mathcal{O}_{M_i}) | U_{ij} \xrightarrow{\cong} (U_{ji}, \mathcal{O}_{M_j} | U_{ji})$$

for all $i, j \in I$ such that the cocycle identity

$$\underline{f}_{jk} \circ \underline{f}_{ij} = \underline{f}_{ik}$$

holds for all $i, j, k \in I$.

Given glueing data, there is, up to isomorphism, a unique complex space $(X, \mathcal{O}_X)$ which has local models $(M_i, \mathcal{O}_{M_i})$.

In a similar way, a morphism $(f, f^0) : (X, \mathcal{O}_X) \longrightarrow (Y, \mathcal{O}_Y)$ can be given by glueing data which I will not write down explicitely.

b) Intersections. Let $\underline{X}, \underline{X}' \hookrightarrow \underline{Y}$ be closed complex subspaces of the complex space $\underline{Y}$, defined by the locally finite ideals $\mathcal{I}, \mathcal{I}' \subseteq \mathcal{O}_Y$. The intersection $\underline{X} \cap \underline{X}'$ is defined to be the largest complex subspace $\underline{X}'' \hookrightarrow \underline{Y}$ such that any morphism $\underline{Z} \longrightarrow \underline{Y}$ which factors through $\underline{X}$ and $\underline{X}'$ also factors through $\underline{X}''$; it is given by the locally finite ideal $\mathcal{I} + \mathcal{I}'$.

c) Inverse images. Let $\underline{f} : \underline{X} \longrightarrow \underline{Y}$ be a morphism in $\underline{\underline{cpl}}$. If $\underline{Z} \hookrightarrow \underline{Y}$ is a complex subspace, the inverse image $\underline{f}^{-1}(\underline{Z}) \hookrightarrow \underline{X}$ is the complex subspace with the universal property that if $\underline{f}' : \underline{X}' \longrightarrow \underline{X}$ is in $\underline{\underline{cpl}}$ and $f \circ f'$ factors through $\underline{Z}$, $\underline{f}'$ factors through $\underline{f}^{-1}\underline{Z}$. If $\underline{Z} \hookrightarrow \underline{Y}$ is a closed complex subspace defined by the locally finite ideal $\mathcal{I}$, $f^{-1}(\underline{Z}) \hookrightarrow \underline{X}$ is defined by $\underline{f}^{-1}\mathcal{I} := \mathcal{I} \cdot \mathcal{O}_X$, the ideal generated in $\mathcal{O}_X$ by $\mathcal{I}$ under $f^0 : f^{-1}\mathcal{O}_Y \longrightarrow \mathcal{O}_X$. A special case of this construction are the fibres $\underline{f}^{-1}(\underline{y}) \hookrightarrow \underline{X}$, $y \in Y$, of the morphism $\underline{f}$.

d) Products. In $\underline{\underline{cpl}}$, the categorical product

$$(3.2.1)\qquad \underline{X}\times\underline{Y} \begin{array}{l} \xrightarrow{\underline{pr}_X} \underline{X} \\ \xrightarrow[\underline{pr}_Y]{} \underline{Y} \end{array}$$

exists for $\underline{X},\underline{Y}\in \underline{\underline{cpl}}$. Locally, it is given as follows:
If U,V are open subsets of number spaces,

$$(3.2.2)\qquad \underline{U}\times\underline{V} \begin{array}{l} \xrightarrow{\underline{pr}_U} \underline{U} \\ \xrightarrow[\underline{pr}_V]{} \underline{V} \end{array}$$

is given by the usual product $U\times V$ with the canonical complex structure, and $\underline{pr}_U$, $\underline{pr}_V$ by the usual projections and, on the sheaf level, by lifting holomorphic functions via these. If $\underline{X} = (m,U,(f^1,\dots,f^k))$ and $\underline{V} = (n,V,(g^1,\dots,g^\ell))$ are local models, (3.2.1) is given by b) and c) as $\underline{X}\times\underline{Y} := \underline{pr}_U^{-1}(\underline{X}) \cap \underline{pr}_V^{-1}(\underline{Y})$ and $\underline{pr}_X := \underline{pr}_U|\underline{X}$, $\underline{pr}_Y := pr_V|\underline{Y}$; this means that $\underline{X}\times\underline{Y}$ is the local model $(m+n, U\times V, f^1\circ pr_U,\dots,f^k\circ pr_U, g^1\circ pr_V,\dots,g^\ell\circ pr_V)$. In the general case, cover $\underline{X}$ and $\underline{Y}$ by local models, form their products, and use the universal property of the product to obtain glueing data for (3.2.1) according to a).

e) Diagonals. If $\underline{X}\in\underline{\underline{cpl}}$, the diagonal $\underline{\Delta}_X \hookrightarrow \underline{X}$ is the complex subspace with the property that for any morphism $\underline{f} : \underline{Z} \longrightarrow \underline{X}$ in $\underline{\underline{cpl}}$, $\underline{f}\times\underline{f} : \underline{Z} \longrightarrow \underline{X}\times\underline{X}$ factors uniquely through $\underline{\Delta}_X$. For a local model $\underline{X}\subseteq\underline{U}$, where U is open in some $\mathbb{C}^n$, $\underline{\Delta}_X := (\underline{X}\times\underline{X}) \cap \underline{\Delta}_U$, and $\underline{\Delta}_U$ is the obvious diagonal of $\underline{U}$; for the general case, glue according to a).

f) Fibre products. In $\underline{\underline{cpl}}$, categorical fibre products exist. Given $\underline{f} : \underline{X} \longrightarrow \underline{Y}$, $g : \underline{Y}' \longrightarrow \underline{Y}$, the cartesian square

$$(3.2.3)\qquad \begin{array}{ccc} X' & \xrightarrow{\underline{g}'} & \underline{X} \\ \underline{f}'\downarrow & & \downarrow\underline{f} \\ \underline{Y}' & \xrightarrow[\underline{g}]{} & \underline{Y} \end{array}$$

is defined by putting $\underline{X}' := \underline{X} \times_{\underline{Y}} \underline{Y}' := ((\underline{f} \circ \underline{pr}_X) \times (\underline{g} \circ \underline{pr}_{Y'}))^{-1}(\underline{\Delta}_Y)$, and f', g' defined by the projections $\underline{pr}_X, \underline{pr}_Y : \underline{X} \times \underline{Y}' \longrightarrow \underline{X}, \underline{Y}'$. The universal property of the fibre product is implied by the universal properties of the inverse image and the diagonal.

g) Graph spaces. A special case of f) is the graph space $\underline{\Gamma}_f$ of a morphism $\underline{f} : \underline{X} \longrightarrow \underline{Y}$; it is defined by the cartesian square

(3.2.4)

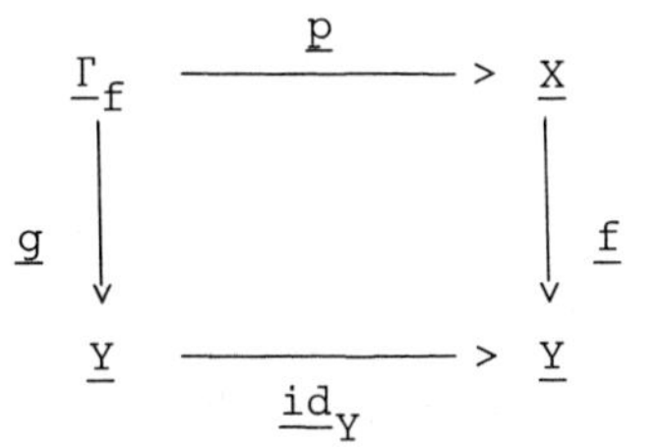

and is a complex subspace of $\underline{X} \times \underline{Y}$. By the universal property of the fibre product the morphisms $\underline{id}_X : \underline{X} \longrightarrow \underline{X}$ and $\underline{f} : \underline{X} \longrightarrow \underline{Y}$ define $\underline{i} := \underline{id}_X \times_Y \underline{f} : \underline{X} \longrightarrow \underline{\Gamma}_f$, and one gets the commutative diagram

(3.2.5)

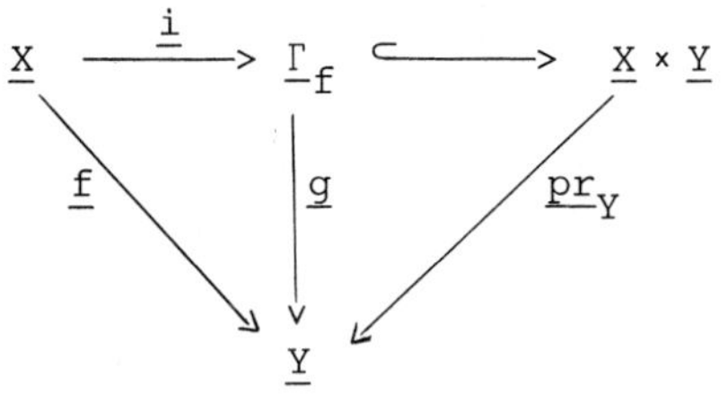

where $\underline{i}$ is an isomorphism, inverse to $\underline{p}$. Hence, we have:

Proposition 3.2.1. Any morphism $\underline{f} : \underline{X} \longrightarrow \underline{Y}$ is isomorphic to the restriction of a projection to a complex subspace.

If X and Y are Hausdorff, $\underline{\Gamma}_f$ is a closed complex subspace, and so $\underline{id} \times f : \underline{X} \longrightarrow \underline{X} \times \underline{Y}$ is a closed immersion with image $\underline{\Gamma}_f$.

The proposition will be important in the study of finite morphisms in the following paragraphs, since it allows to reduce locally to the situation of linear projections of number spaces restricted to closed complex subspaces.

h) Supports of modules.

Definition 3.2.2. Let $X \in \underline{\underline{cpl}}$, and M be an $\mathcal{O}_X$-module. M is called admissible if and only if it is locally of finite presentation, i.e. if and only if every $x \in X$ has an open neighbourhood such that there is a short exact sequence

$$(3.2.6) \qquad \mathcal{O}_X^q \mid U \xrightarrow{\varphi} \mathcal{O}_X^p \mid U \longrightarrow M \mid U \longrightarrow 0 \quad .$$

If M is admissible, the Fitting ideals $F_n(M)$ are defined as

$$(3.2.7) \qquad F_n(M) \mid U := \text{ideal generated in } \mathcal{O}_X \mid U \text{ by the } (p-n) \times (p-n)\text{-minors of the } p \times q\text{-matrix given by } \varphi \text{ in } (3.2.6).$$

A theorem of Fitting [15] implies that the $F_n(M)$ are globally well-defined. By construction, they are locally finite. We then define the support of M to be

$$(3.2.8) \qquad \underline{\text{supp}}\, M := \text{the closed complex subspace of } \underline{X} \text{ defined by } F_0(M) \; .$$

The underlying topological space of $\underline{\text{supp}}\, M$ is $\text{supp}\, M := \{x \in X \mid M_x \neq 0\}$; for this, just tensorize (3.2.6) at $x \in X$ with $\mathbb{C} \cong \mathcal{O}_{X,x} \mid m_x$.

Remark. If $\text{Ann}(M)$ is the annihilator ideal of M , then $F_0(M) \subseteq \text{Ann}(M) \subseteq \sqrt{F_0(M)}$. The first inclusion is by elementary linear algebra, whereas the second one lies considerably deeper and follows from the Rückert Nullstellensatz 5.3.1.

i) Image spaces. Let $\underline{f} : \underline{X} \longrightarrow \underline{Y}$ be a morphism in $\underline{\underline{cpl}}$. Then $\text{im}(f) = \text{supp}(f_* \mathcal{O}_X)$ settheoretically, so if $f_*\mathcal{O}_X$ happens to be an admissible $\mathcal{O}_Y$-module, $\text{supp}(f_* \mathcal{O}_X)$ has a natural structure as a closed complex subspace of $\underline{Y}$ via $F_0(f_*\mathcal{O}_X)$ in view of a). We call this space the complex image space of $\underline{f}$, denoted $\underline{\text{im}}(\underline{f})$ or $\underline{f}(\underline{X})$.

3.3. The Equivalence Theorem

The Equivalence Theorem asserts the equivalence of the "geometric" category of complex space singularities with the "algebraic" category of local analytic $\mathbb{C}$-algebras. Its explicit formulation seems to be due to Grothendieck ([64], Exposé 13).

We begin with describing the morphisms of a complex space $\underline{X}$ to $\underline{\mathbb{C}}^n$. If $R \in \underline{\underline{la}}$, $R/m_R \cong \mathbb{C}$ canonically via the augmentation mapping induced by the $\mathbb{C}$-algebra-structure; hence, if $\underline{X} \in \underline{\underline{cpl}}$, any section $f \in \mathcal{O}_X(X)$ defines a function $[f] : X \longrightarrow \mathbb{C}$ via

$$\forall x \in X : [f](x) := f_x \bmod m_x \quad . \tag{3.3.1}$$

Proposition 3.3.1. If $\underline{X} \in \underline{\underline{cpl}}$, we get a bijection

$$\begin{aligned} \mathrm{Hom}_{\underline{\underline{cpl}}}(\underline{X},\underline{\mathbb{C}}^n) &\longrightarrow \mathcal{O}_X(X)^n \\ \underline{f} &\longmapsto (f_X^0(z_1),\dots,f_X^0(z_n)) \quad , \end{aligned}$$

where $f_X^0 : (f^{-1}\mathcal{O}_{\mathbb{C}^n})(X) = \mathcal{O}_{\mathbb{C}^n}(\mathbb{C}^n) \longrightarrow \mathcal{O}_X(X)$.

Sketch of proof.

(i) Injectivity: Since $z_j \circ f = [f^0(z_j)]$, the $f^0(z_j)$ determine the settheoretic map $f : X \longrightarrow \mathbb{C}^n$. Now, if $\underline{f}, \underline{g} \in \mathrm{Hom}_{\underline{\underline{cpl}}}(\underline{X},\underline{\mathbb{C}}^n)$ have $f^0(z_j) = g^0(z_j)$ for $1 \leq j \leq n$, then $f = g$, and $f_x^0, g_x^0 : \mathcal{O}_{\mathbb{C}^n,y} \longrightarrow \mathcal{O}_{X,x}$, where $y := f(x) = g(x)$, agree on the z_j for $1 \leq j \leq n$. But then they agree on $\mathcal{O}_{\mathbb{C}^n,y}$, since $\mathcal{O}_{\mathbb{C}^n,y} \cong \mathbb{C}\{z_1,\dots,z_n\}$ is a free object in $\underline{\underline{la}}$ by Theorem 1.3.4.

(ii) Surjectivity: Let $(f_1,\dots,f_n) \in \mathcal{O}_X(X)^n$ be given. First suppose X is a local model space in some open $U \subseteq \mathbb{C}^n$, and the f_j are induced by holomorphic functions $F_j : U \longrightarrow \mathbb{C}$ for $1 \leq j \leq n$. Then $F := (F_1,\dots,F_n) : U \longrightarrow \mathbb{C}^n$ induces a morphism $\underline{f} : \underline{X} \longrightarrow \underline{\mathbb{C}}^n$ with $f^0(z_j) = f_j$ for $1 \leq j \leq n$. In the general case cover $\underline{X}$ with local models and glue the local morphisms obtained on the overlaps by means of (i).

Definition 3.3.2 (Germs of complex spaces).

(i) A complex spacegerm, or singularity, is a tuple $(\underline{X},x)$ with $X \in \underline{\underline{cpl}}$ and $x \in X$.

(ii) A morphism of complex spacegerms , or complex mapgerm, is a morphism $\underline{f} : \underline{U} \longrightarrow \underline{V} \in \underline{\underline{cpl}}$ of an open neighbourhood $\underline{U}$ of x into an open neighbourhood $\underline{V}$ of y with $f(x) = y$, where one identifies those morphisms which coincide after restriction to possibly smaller neighbourhoods.

The complex space germs with their morphisms form a category, which I will denote $\underline{\underline{cpl}}_0$. If $(\underline{X},x) \in \underline{\underline{cpl}}_0$, and U is any open neighbourhood of x in X , $(\underline{X},x) = (\underline{U},x)$ up to isomorphism in $\underline{\underline{cpl}}_0$, and I will refer to this as "possibly shrinking $\underline{X}$ " .

There is a canonical contrafunctor

$$\mathcal{O} : \underline{\underline{cpl}}_0 \longrightarrow \underline{\underline{la}}$$

mapping $(\underline{X},x) \in \underline{\underline{cpl}}_0$ to $\mathcal{O}_{X,x}$ and $\underline{f} : (\underline{X},x) \longrightarrow (\underline{Y},y)$ to $f_x^0 : \mathcal{O}_{Y,y} \longrightarrow \mathcal{O}_{X,x}$.

Theorem 3.3.3 (The Equivalence Theorem; Grothendieck [64], Exposé 13). $\mathcal{O} : \underline{\underline{cpl}}_0^{opp} \longrightarrow \underline{\underline{la}}$ is an equivalence of categories.

Sketch of proof. We have to show two properties:

(i) essential surjectivity on objects: For $R \in \underline{\underline{la}}$ there exists $(\underline{X},x) \in \underline{\underline{cpl}}_0$ with $\mathcal{O}_{X,x} \cong R$.

(ii) bijectivity on morphisms:

$$\mathrm{Hom}_{\underline{\underline{cpl}}_0}((\underline{X},x),(\underline{Y},y)) \longrightarrow \mathrm{Hom}_{\underline{\underline{la}}}(\mathcal{O}_{Y,y},\mathcal{O}_{X,x})$$

$$f \longmapsto f_x^0$$

is a bijection.

(i): is trivial from the constructions.

(ii): Since the question is local, we may assume, after possibly shrinking $\underline{X}$ and $\underline{Y}$, that $\underline{X} \hookrightarrow \underline{U} \subseteq \mathbb{C}^m$, $\underline{Y} \hookrightarrow \underline{V} \subseteq \mathbb{C}^n$ are local model where U and V are open, and $x = 0 \in \mathbb{C}^m$, $y = 0 \in \mathbb{C}^n$.

Injectivity: We may assume $\underline{Y} = \underline{\mathbb{C}}^n$; the claim then follows from Proposition 3.3.1.

Surjectivity: Let $\varphi : \mathcal{O}_{Y,y} \longrightarrow \mathcal{O}_{X,x} \in \underline{\underline{la}}$ be given. By Theorem 1.3.4 there is a commutative diagram

$$\begin{array}{ccc} \mathcal{O}_{\mathbb{C}^n,0} & \overset{\psi}{\dashrightarrow} & \mathcal{O}_{\mathbb{C}^m,0} \\ \downarrow & \circlearrowright & \downarrow \\ \mathcal{O}_{Y,y} & \underset{\varphi}{\longrightarrow} & \mathcal{O}_{X,x} \end{array} \qquad (3.3.2)$$

Let $(F_j)_0 := \psi(z_j) \in \mathcal{O}_{\mathbb{C}^m,0}$, $1 \leq j \leq n$; after possibly shrinking U, we may assume the $(F_j)_0$ have representatives $F_j : U \longrightarrow \mathbb{C}$, which together define the holomorphic map

$$F := (F_j,\ldots,F_n) : U \longrightarrow \mathbb{C}^n .$$

Let $\underline{X}$ be defined by $g^1,\ldots,g^k \in \mathcal{O}_{\mathbb{C}^m}(U)$ and $\underline{Y}$ by $h^1,\ldots,h^\ell \in \mathcal{O}_{\mathbb{C}^n}(V)$. Define the $\mathcal{O}_U$-ideals

$$I := (g^1,\ldots,g^h)\cdot\mathcal{O}_U$$

$$J := (h_0^1 \circ F,\ldots,h_0^\ell \circ F)\cdot\mathcal{O}_U .$$

Then $J_0 \subseteq I_0$ because of the commuative diagram (3.3.2). By Lemma 3.1.1. we may therefore assume $J \subseteq I$. But then F induces a morphism $\underline{f} : \underline{X} \longrightarrow \underline{Y}$ by Definition 3.1.2.(ii), and $f_x^0 = \varphi$ by construction.

Corollary 3.3.4. $\underline{\underline{cpl}}$ is a full subcategory of the category $\underline{\underline{lrsp}}$ of spaces locally ringed in $\mathbb{C}$-algebras.

For the same proof as in 3.3.3. shows the injectivity of $\mathrm{Hom}_{\underline{\underline{lrsp}}_0}((\underline{X},x),(\underline{Y},y)) \longrightarrow \mathrm{Hom}_{\underline{\underline{la}}}(\mathcal{O}_{Y,y},\mathcal{O}_{X,x})$.

Corollary 3.3.5. Morphisms $\underline{f} : (\underline{X},x) \longrightarrow (\mathbb{C}^n,0)$ correspond one-to-one to $\mathfrak{m}_x$-sequences $(f_1,\ldots,f_n)$ (i.e. sequences $(f_1,\ldots,f_n)$ with $f_j \in \mathfrak{m}_x$ for $1 \leq j \leq n$).

Remark 3.3.6. By Corollary 3.3.5, special morphisms of germs should correspond to $\mathfrak{m}_x$-sequences with special properties. We will see instances of this principle later on (4.4.2, 6.2.3., 6.3.1.).

3.4 The analytic spectrum

For later use, we shortly discuss a further application of Proposition 3.3.1.

Let A be a finitely generated $\mathbb{C}$-algebra. Picking generators $a_1,\ldots,a_n \in A$ gives an epimorphism

$$\varphi : \mathbb{C}[z_1,\ldots,z_n] \longrightarrow\!\!\!\!\!\!\rightarrow A \quad .$$

Let I be the kernel of φ , and $\mathcal{I} \subseteq \mathcal{O}_{\mathbb{C}^n}$ the ideal sheaf generated by I . $\mathcal{I}$ defines a closed complex subspace $\underline{Z} \hookrightarrow \underline{\mathbb{C}}^n$, and there is a canonical homomorphism $\zeta : A \longrightarrow \mathcal{O}_Z(Z)$, such that for given $a \in A$ the germ $\zeta(a)_z$ at a given $z \in Z$ is the germ induced by $f_z \in \mathcal{O}_{\mathbb{C}^n,z}$ where f is any preimage of a under φ . We then have the following generalization of Proposition 3.3.1.

Proposition 3.4.1. The pair $(\underline{Z},\zeta)$ represents the functor $\underline{\underline{\text{cpl}}}^{\text{opp}} \longrightarrow \underline{\underline{\text{sets}}}$ given by $\underline{X} \longmapsto \text{Hom}_{\underline{\underline{\text{cpl}}}}(\underline{X},\underline{Z})$, in other words, the canonical map

$$\begin{aligned} \text{Hom}_{\underline{\underline{\text{cpl}}}}(\underline{X},\underline{Z}) &\longrightarrow \text{Hom}_{\underline{\underline{\mathbb{C}\text{-alg}}}}(A,\mathcal{O}_X(X)) \\ f &\longmapsto f_X^0 \circ \zeta \end{aligned}$$

induces a natural equivalence of functors.

Here, f_X^0 is the homomorphism $\mathcal{O}_Z(Z) \longrightarrow \mathcal{O}_X(X) = (f_*\mathcal{O}_X)(Z)$ given by the sheaf map $f^0 : \mathcal{O}_Z \longrightarrow f_*\mathcal{O}_X$.

The proof of the Proposition is simple, using 3.3.1, and left to the reader. For the general formalism of representable functors see [64], Exposé 11, by Grothendieck.

It follows that the pair $(\underline{Z},\zeta)$ is unique up to unique isomorphism, and so the following definition makes sense:

Definition 3.4.2. If A is a finitely generated $\mathbb{C}$-algebra, the pair $(\underline{Z},\zeta)$, or the complex space $\underline{Z}$ alone when ζ is understood, constructed above is called the analytic spectrum of A and denoted Specan(A).

§ 4. Local Weierstrass Theory II: Finite morphisms

Classically finite maps arose naturally by solving systems of polynomial equations via Kronecker's elimination theory (see e.g. [51]); successively eliminating indeterminates by forming resultants of polynomials turns some indeterminates into free parameters, which can be varied arbitrarily and whose number should be thought of as the dimension of the solution variety; the rest of the indeterminates become algebraic functions of these parameters. Geometrically, this amounts to representing the solution variety as a finite branched cover of an affine space, and algebraically to the fact that the coordinate ring of the solution variety is a finite integral extension of a polynomial ring. This is nowadays known as "Noether normalization", and fairly easy to prove, without using elimination theory.

This picture remains true locally in the complex analytic case, but this is much harder to prove. As already mentioned before, the main reason for the applicability of local algebra to local complex analysis is the fact that, under the equivalence 3.3.3, finite mapgerms will correspond to finite, and hence integral, ring extensions of local analytic algebras, and so a kind of "relative algebraic situation" emerges. This will be the subject of the main result of this paragraph, the Integrality Theorem 4.4.1. Fundamental for it is the famous Finite Mapping Theorem 4.3.1. of Grauert and Remmert; in the proof of it, the elimination procedure of the algebraic case is mimicked geometrically by a sequence of linear projections along a line.

4.1. Finite morphisms

From now on, all topological spaces under consideration will be Hausdorff, locally compact, and paracompact. For general facts of topology quoted in the sequel see [7] , and also [14].

A continuous map $f : X \longrightarrow Y$ of topological spaces is called proper if the inverse image of a compact subset of Y is compact in X . This is equivalent to the requirement that f is closed (i.e. maps closed sets to closed sets) and has compact fibres. A proper map with finite fibres is called finite, so a map is finite iff it is closed with finite fibres. Finally, a morphism $\underline{f} : \underline{X} \longrightarrow \underline{Y}$ of complex spaces is called finite if the underlying map $f : X \longrightarrow Y$ of topological spaces is so. Elementary considerations from topology show that any $y \in Y$ has a neighbourhood basis consisting of open neighbourhoods V such that $f^{-1}V = \coprod_{x\in f^{-1}(y)} U_x$ for open neighbourhoods U_x of x in X and $f|U_x : U_x \longrightarrow V$ is finite. Thus, there are canonical homomorphisms for a sheaf $\mathcal{M}$ on X ,

$$(4.1.1) \qquad \varepsilon_y : (f_*\mathcal{M})_y \longrightarrow \bigoplus_{x\in f^{-1}(y)} \mathcal{M}_x \quad , \quad \text{for all } y \in Y ,$$

induced from $\mathcal{M}(f^{-1}V) \longrightarrow \bigoplus_{x\in f^{-1}(y)} \mathcal{M}(U_x)$ via $s \longmapsto \sum_{x\in f^{-1}(y)} s|U_x$, and one gets:

Theorem 4.1.1. Let $f : X \longrightarrow Y$ be a finite morphism of complex spaces. Let $\mathcal{O}_X$-mod and $\mathcal{O}_Y$-mod denote the category of $\mathcal{O}_X$-modules and $\mathcal{O}_Y$-modules respectively. Then:

(i) The homomorphisms ε_y in (4.1.1) are isomorphisms for all $\mathcal{M} \in \mathcal{O}_X$-mod ;

(ii) the functor $f_* : \mathcal{O}_X\text{-mod} \longrightarrow \mathcal{O}_Y\text{-mod}$ is exact.

4.2. Weierstrass maps (see [28]). These are the prototypes of finite morphisms in local complex analytic geometry and play a prominent rôle in what follows, since any finite morphism locally will embed in a Weierstrass map. So ultimatively basic properties of finite morphisms will be proved using Weierstrass maps.

Let $\omega^{(j)} \in \mathcal{O}_{\mathbb{C}^n,0}[w_j]$ be monic polynomials

$$(4.2.1) \qquad \omega^{(j)} = w_j^{b_j} + \sum_{\nu=0}^{b_j-1} a_\nu^{(j)}(z) w_j^\nu \quad , \quad 1 \leq j \leq k \ ,$$

$a_\nu^{(j)} \in \mathcal{O}_{\mathbb{C}^n,0}$, and $b_j \geq 1$, for $1 \leq j \leq k$. Let $B \subseteq \mathbb{C}^n$ be a domain containing $0 \in \mathbb{C}^n$ such that the $\omega^{(j)}$ have representatives, also called $\omega^{(j)}$, defined on B. We get the closed subspace $\underline{A} := \underline{N}(\omega^{(1)},\ldots,\omega^{(k)}) \hookrightarrow \underline{B} \times \underline{\mathbb{C}}^k$, and the projection $\underline{pr}_B : \underline{B} \times \mathbb{C}^k \longrightarrow \underline{B}$ defines

$$(4.2.2) \qquad \underline{\pi} := \underline{pr}_B \mid A : \underline{A} \longrightarrow \underline{B} \quad .$$

We call $\underline{\pi}$ a <u>Weierstrass map</u>.

Given $z \in B$, the equations (4.2.1) have only finitely many solutions. Moreover, if $\omega = w^b + \sum_{\nu=0}^{b-1} a_\nu(z) w^\nu \in \mathcal{O}_{\mathbb{C}^n,0}[w]$ and $\omega(z_0,w_0) = 0$, we have the simple estimate

$$|w_0| \leq \max\left(1, \sum_{\nu=0}^{b-1} |a_\nu(z_0)|\right) \quad ,$$

which shows that the inverse image of a bounded set is bounded. Hence:

<u>Proposition 4.2.1</u>. <u>A Weierstrass map is finite</u>.

Somewhat deeper lies:

<u>Proposition 4.2.2</u>. <u>A Weierstrass map is open</u>.

This is implied by the following easy but very useful consequence of the Weierstrass Preparation Theorem:

<u>Lemma 4.2.3</u> (Hensel's Lemma).

<u>Let</u> $\omega := \omega(z,w) = w^b + \sum_{\nu=0}^{b-1} a_\nu(z) w^\nu \in \mathcal{O}_{\mathbb{C}^n,0}[w]$ <u>be a monic polynomial of degree</u> $b \geq 1$. <u>Let</u> $\omega(0,\omega) = (w-c_1)^{b_1} \cdot \ldots \cdot (w-c_r)^{b_r}$. <u>Then there exist unique monic polynomials</u> $\omega_1,\ldots,\omega_r \in \mathcal{O}_{\mathbb{C}^n,0}[w]$, $\deg \omega_j = b_j$ <u>for</u> $1 \leq j \leq r$, <u>such that</u> $\omega = \omega_1 \cdot \ldots \cdot \omega_r$.

For the proof of 4.2.3, one just applies the Preparation Theorem successively in the rings $\mathcal{O}_{\mathbb{C}^n,0}[w-c_1]$, $\mathcal{O}_{\mathbb{C}^n,0}[w-c_2]$, and so on.

Now the Weierstrass map (4.2.2) clearly is open at $0 \in A$ since the equations (4.2.1) have a solution for any $z \in B$, but by Hensel's Lemma the germ $\underline{\pi} : (\underline{A},a) \longrightarrow (\underline{B},\pi(a))$ is locally around any $a \in A$ a Weierstrass map, so π is open at all $a \in A$, and so is open, which proves Proposition 4.2.2.

4.3. The Finite Mapping Theorem

The following theorem is the fundamental result in local complex analytic geometry, and is due to Grauert and Remmert ([24], Satz 27). Recall the notion of an admissible module (Definition 3.2.2.).

Theorem 4.3.1 (The Finite Mapping Theorem). Let $\underline{f} : \underline{X} \longrightarrow \underline{Y}$ be a finite morphism of complex spaces. Then, if M is an admissible $\mathcal{O}_X$-module, $f_* M$ is an admissible $\mathcal{O}_Y$-module.

Corollary 4.3.2. If $\underline{f} : \underline{X} \longrightarrow \underline{Y}$ is a finite morphism of complex spaces, the complex image space $\mathrm{im}(\underline{f})$ in the sense of 3.2.i) exists.

This Corollary is an obvious consequence of the Theorem.

The proof of this basic result is done in various steps. The details are in [28], Chapter 3, but since the full machine of coherence is employed there, I will give an outline, indicating the minor modifications which are necessary when not invoking the notion of coherence.

In the first step, one considers the special case where $\underline{f}$ is a Weierstrass map $\underline{\pi} : \underline{A} \longrightarrow \underline{B}$. Let the notation be as in 4.2. Let $\mathbb{N}^{n+k} = \overline{\Delta} \amalg \coprod_{j=1}^{k} \Delta_j$ be the decomposition given by the monomials $w_j^{b_j}$ according to Theorem 2.3.2; hence

$$(4.3.1) \qquad \overline{\Delta} = \mathbb{N}^n \times \overline{\Delta}_0 \quad \text{with} \quad \overline{\Delta}_0 := \left\{ B \in \mathbb{N}^k \mid 0 \leq B^i < b_i \quad \text{for} \quad 1 \leq i \leq k \right\} .$$

Let $\mathcal{O}_B^{\overline{\Delta}_0}$ be the $\mathcal{O}_B$-module defined by

(4.3.2) $\mathcal{O}_B^{\bar{\Delta}_0}(U) := \left\{ \sum_{B\in\bar{\Delta}_0} f_B w^B \mid f_B \in \mathcal{O}_{\mathbb{C}^n}(U) \right\}$

for $U \subseteq B$ open. There is a natural $\mathcal{O}_B$-module homomorphism

(4.3.3) $\overset{0}{\pi} : \mathcal{O}_B^{\bar{\Delta}_0} \longrightarrow \pi_* \mathcal{O}_A$

given as follows: If $U \subseteq B$ is open, $\sum_{B\in\bar{\Delta}_0} f_B w^B$ is defined on $\mathcal{O}_{B\times\mathbb{C}^k}(U\times\mathbb{C}^k) = \mathcal{O}_B(\pi^{-1}U)$; this defines

(4.3.4) $\overset{0}{\pi}_U : \mathcal{O}_B^{\bar{\Delta}_0}(U) \longrightarrow \mathcal{O}_{B\times\mathbb{C}^k}(\pi^{-1}U) \xrightarrow{\text{restriction}} \mathcal{O}_A(\pi^{-1}U)$,

and so (4.3.3). The following theorem substantially generalizes Corollary 2.6.2:

Theorem 4.3.3. *$\overset{0}{\pi}$ is an isomorphism of $\mathcal{O}_B$-modules.*

This in turn is an immediate consequence of the following parametrized generalization of the Division Theorem:

Theorem 4.3.4 (The Generalized Division Theorem). *Let the notation be as in 4.2. Let $y \in B$, and let, for all $x_j \in \pi^{-1}(y)$, germs $f_j \in \mathcal{O}_{\mathbb{C}^{n+k},x_j}$ be given. Then there exist unique germs $g_{\alpha j} \in \mathcal{O}_{\mathbb{C}^{n+k},x_j}$, $\alpha = 1,\dots,k$, and a unique polynomial $h \in \mathcal{O}_{\mathbb{C}^n}[w_1,\dots,w_k]$ of the form $h = \sum_{A\in\mathbb{N}^k} h_A w^A$ with $0 \leq A^i < b_i$ for $1 \leq i \leq k$ such that for all* $x_j \in \pi^{-1}(y)$

$$f_j = g_{1j}\omega_{x_j}^{(1)} + \dots + g_{kj}\omega_{x_j}^{(k)} + h_{x_j} \quad \text{in } \mathcal{O}_{\mathbb{C}^{n+k},x_j} .$$

The main point of this theorem is that one h works for all x_j. The proof is a formal consequence of the Division Theorem and Hensel's Lemma 4.2.3., and I refer to [28] for it.

Theorem 4.3.3. is then proved as follows: By Theorem 4.1.1. (i), $(\pi_*\mathcal{O}_A)_y \cong \bigoplus_{x_j\in\pi^{-1}(y)} \mathcal{O}_{A,x_j}$, so any element s_y of $(\pi_*\mathcal{O}_A)_y$ is represented by a family $(f_{x_j})_{x_j \in \pi^{-1}(y)}$, $f_{x_j} \in \mathcal{O}_{\mathbb{C}^{n+k},x_j}$. Dividing the f_{x_j} by $\omega_{x_j}^{(1)},\ldots,\omega_{x_j}^{(k)}$ via Theorem 4.3.4 shows there is an unique $h_y \in \overline{\mathcal{O}}^{\Delta 0}_{B,y}$ mapping to s_y , so (4.3.3) is bijective on stalks, and so bijective by Theorem 4.1.1. (ii).

The second step reduces the general case to the case of a linear projection. For this, one observes that the statement of Theorem 4.3.1 is local in the sense that any $x \in X$ has an open neighbourhood U such that $\pi|U : U \longrightarrow \pi(U)$ is again finite, and so we may assume that $\underline{X} \hookrightarrow \underline{B}'$, $\underline{Y} \hookrightarrow \underline{B}$, where $B' \subseteq \mathbb{C}^n$ and $B \subseteq \mathbb{C}^k$ are domains. One gets a commutative diagram

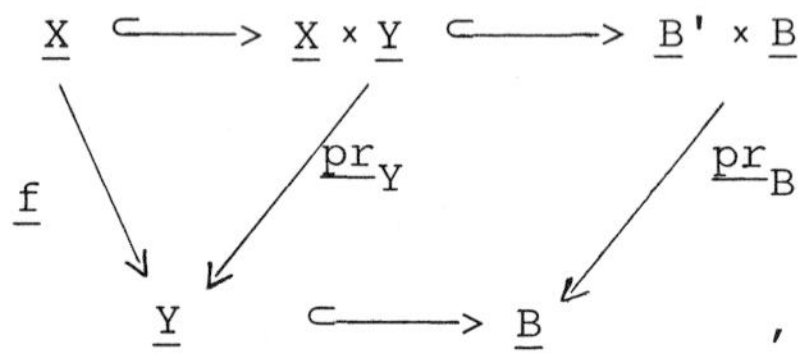

where the horizontal arrows in the upper row are closed immersions, the left hand triangle is defined by the graph construction (3.2.5), and the right hand square is defined by the closed immersions $\underline{X} \hookrightarrow \underline{B}'$, $\underline{Y} \hookrightarrow \underline{B}$. Identifying $\underline{X}$ with its image in $\underline{B}' \times \underline{B}$ we may assume we have a commutative diagram

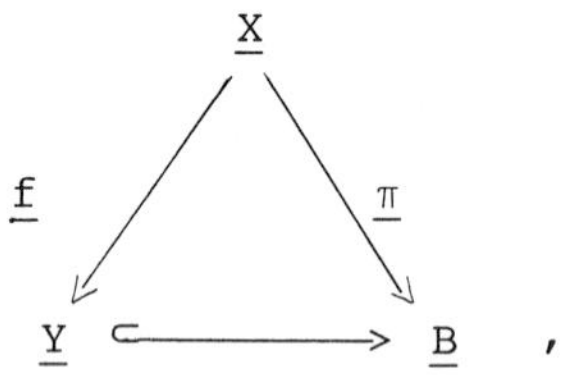

where π is given by the restriction of a linear projection to $\underline{X}$ which is finite, or, as I will say, where π is a finite linear projection. One now has the following lemma.

Lemma 4.3.5. Let $X \in \underline{\underline{cpl}}$, $\underline{Y} \overset{i}{\hookrightarrow} \underline{X}$ a closed complex subspace, and $\mathcal{M}$ an $\mathcal{O}_Y$-module. Then $\mathcal{M}$ is an admissible $\mathcal{O}_Y$-module if and only if $i_*\mathcal{M}$ is an admissible $\mathcal{O}_X$-module.

The proof is a simple diagram chase and left to the reader.

This lemma shows that it suffices to prove Theorem 4.3.1 for $\underline{\pi}$.

The last step reduces now everything to the first step. We may assume that $\underline{f}$ is a finite linear projection. We may even assume that $k = 1$, for we can factor $\underline{f}$ successively into a sequence of projections along lines, and Corollary 4.3.2 and Lemma 4.3.5 reduce everything to that case.Then choose a nonzero $g \in \mathcal{O}_{\mathbb{C}^{n+1},0}$ which vanishes on X near 0 ; after possibly shrinking $\underline{X}$ and $\underline{B}$ we may assume g is a Weierstrass polynomial by Theorem 2.6.3. We then have the commutative triangle

$$\begin{array}{ccc} \underline{X} & \overset{i}{\hookrightarrow} & \underline{N}(g) =: A \\ & {\scriptstyle \underline{f}} \searrow \quad \swarrow {\scriptstyle \underline{\pi}} & \\ & \underline{B} & \end{array}$$

and, again by Lemma 4.3.5, we are reduced to prove Theorem 4.3.1 for the Weierstrass map $\underline{\pi}$. Now let M be an admissible $\mathcal{O}_A$-module; after shrinking A and B, we may assume there is an exact sequence

$$\mathcal{O}_A^q \longrightarrow \mathcal{O}_A^p \longrightarrow M \longrightarrow 0 \quad ,$$

so there is an exact sequence, since π_* is exact by 4.1.1. (ii):

$$\left(\pi_* \mathcal{O}_A\right)^q \longrightarrow \left(\pi_* \mathcal{O}_A\right)^p \longrightarrow \pi_* M \longrightarrow 0 \quad ,$$

(note π_* commutes with direct sums). But $\pi_* \mathcal{O}_A \cong \mathcal{O}_B^b$ for some b by Theorem 4.3.3, hence Theorem 4.3.1 follows.

As a corollary of the proof we obtain:

<u>Corollary 4.3.6</u>. <u>Let $\underline{f} : \underline{X} \longrightarrow \underline{Y}$ be quasifinite at</u> $x \in X$ (i.e. x is an isolated point of the fibre $f^{-1}f(x)$) . <u>Then</u> x <u>has a neighbourhood</u> U <u>and</u> $f(x)$ <u>a neighbourhood</u> V <u>with</u> $f(U) \subseteq V$ <u>such that</u> $\underline{f}|\underline{U} : \underline{U} \longrightarrow \underline{V}$ <u>is finite</u>.

The proof is identical with the reduction procedure in the above proof, reducing it to the case of a Weierstrass map, which is finite by Proposition 4.2.1.

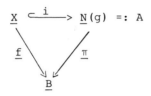

4.4. The Integrality Theorem

Recall the equivalence of categories

$$\mathcal{O} : \underline{\underline{\mathrm{cpl}}}_0^{\mathrm{opp}} \longrightarrow \underline{\underline{\mathrm{la}}}$$

given by the Equivalence Theorem 3.3.3. We are now in a position to describe which homomorphisms in $\underline{\underline{\mathrm{la}}}$ correspond to the finite mapgerms in $\underline{\underline{\mathrm{cpl}}}_0$, and this will finally allow to describe algebraic invariants of local analytic algebras in geometric terms of $\underline{\underline{\mathrm{cpl}}}_0$.

Theorem 4.4.1 (The Integrality Theorem). Let $f : (\underline{X},x) \longrightarrow (\underline{Y},y)$ be a holomorphic mapgerm; recall that by Theorem 3.3.3 this is equivalent to having a homomorphism $\varphi : \mathcal{O}_{Y,y} \longrightarrow \mathcal{O}_{X,x}$ of local analytic algebras. The following statements are equivalent:

(i) f is quasifinite, i.e. x is isolated in $f^{-1}f(x)$ for some (or any) representative of f .

(ii) f is finite, i.e. some representative of f is a finite morphism of complex spaces.

(iii) φ is quasifinite, i.e $\mathcal{O}_{X,x} / \mathfrak{m}_y \cdot \mathcal{O}_{X,x}$ is a finite dimensional complex vectorspace.

(iv) φ is finite, i.e. $\mathcal{O}_{X,x}$ is a finite $\mathcal{O}_{Y,y}$-module via φ .

We can visualize this situation by the following diagram:

(i) f quasifinite at x $\overset{}{\underset{\mathcal{O}}{\Longleftrightarrow}}$ (iii) $\dim_{\mathbb{C}} \mathcal{O}_{X,x}/\mathfrak{m}_y\mathcal{O}_{X,x} < \infty$

$\Updownarrow$ $\Updownarrow$

(ii) f finite near x $\overset{}{\underset{\mathcal{O}}{\Longleftrightarrow}}$ (iv) $\mathcal{O}_{X,x}$ finite over $\mathcal{O}_{Y,y}$

equivalence in $\underline{\underline{\mathrm{cpl}}}_0$ equivalence in $\underline{\underline{\mathrm{la}}}$

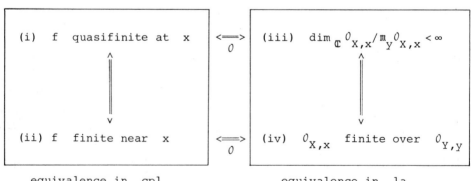

I will give a bare outline of the argument, following the diagram clockwise via (i) ⇒ (iii) ⇒ (iv) ⇒ (ii) ⇒ (i) . Arguing as in the last section, I may assume throughout $\underline{f}$ is represented by a finite linear projection, $\underline{Y} = \underline{B} \subseteq \mathbb{C}^n$ is a domain containing $y = 0 \in \mathbb{C}^n$; $\underline{X}$ is defined in $\underline{Y} \times \underline{V}$, $\underline{V}$ a domain in $\underline{\mathbb{C}}^k$, by a finitely generated ideal $I \subseteq \mathcal{O}_{\mathbb{C}^{n+k}}(Y \times V)$, $x = 0 \in \mathbb{C}^{n+k}$; and $\underline{f}$ is induced by the projection $\underline{pr}_Y^{\mathbb{C}} : \underline{Y} \times \underline{V} \longrightarrow \underline{Y}$. (See Figure 2). Let $R := \mathcal{O}_{X,x}/\mathfrak{m}_y\mathcal{O}_{X,x}$.

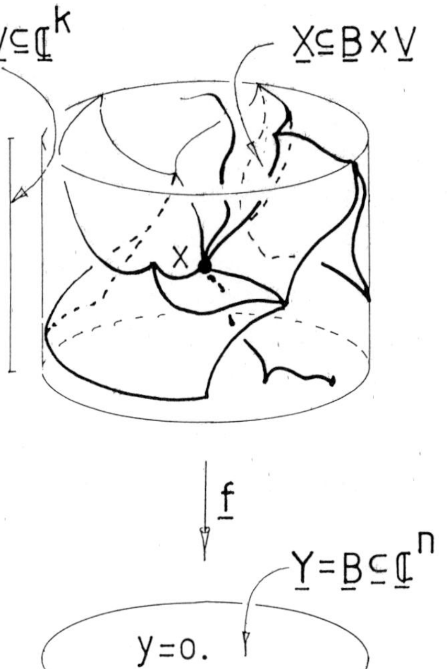

Fig. 2

(i) ⇒ (iii). The fibre $\underline{f}^{-1}(\underline{y})$ is defined by the ideal $\mathfrak{m}_y \cdot \mathcal{O}_X$, by 3.2.c). The Corollary 4.3.6 then shows $\mathcal{O}_{\underline{f}^{-1}(\underline{y}),x}$ is an admissible $\mathcal{O}_{\underline{Y},y}$-module by Theorem 4.3.1, where $\underline{y} = (\{y\}, \mathcal{O}_{\mathbb{C}^d,y}/\mathfrak{m}_d)$.

(iii) ⇒ (iv). (iii) means that there is an integer $b \geq 1$ with $\mathfrak{m}_R^b = 0$. This implies that, after possibly shrinking X and Y , there are integers b_j , $1 \leq j \leq k$, and $g_\nu^j \in \mathcal{O}_{\mathbb{C}^{n+k}}(Y \times V)$, $\nu = 1,\dots,n$, such that

$$(4.4.1) \qquad \omega^j(z,w) := w_j^{b_j} + \sum_{\nu=1}^{n} g_\nu^j(z,w) \cdot z_\nu \in I \quad \text{for} \quad 1 \leq j \leq k ,$$

where $I \subseteq \mathcal{O}_{\mathbb{C}^{n+k}}(Y \times V)$ defines X . One can then show that there is a positive linear form Λ with $w_1 <_\Lambda \ldots <_\Lambda w_k <_\Lambda z_1 <_\Lambda \ldots <_\Lambda z_n$ such that

$$(4.4.2) \qquad LM_\Lambda(\omega^j) = w_j^{b_j}$$

for $1 \leq j \leq k$. Given any $f \in \mathcal{O}_{\mathbb{C}^{n+k},0}$, divide it by $\omega^1, \ldots, \omega^k$ according to the Division Theorem 2.3.2:

$$(4.4.3) \qquad f = g_1\omega^1 + \ldots + g_k\omega^k + h$$

with $\operatorname{supp}(h) \subseteq \overline{\Delta}$. Because of (4.4.2), h can be written as

$$(4.4.4) \qquad h = \sum_{A \in \overline{\Delta}_0} h_A(z) w^A$$

with $\overline{\Delta}_0 := \left\{ A \in \mathbb{N}^k \mid \forall_j : 0 \leq A^j < b_j \right\}$, $h_A(z) \in \mathcal{O}_{\mathbb{C}^n,0} = \mathcal{O}_{Y,y}$. Taking (4.4.3) mod I , we see by (4.4.4) that the monomials w^A for $A \in \overline{\Delta}_0$ generate $\mathcal{O}_{X,x}$ over $\mathcal{O}_{Y,y}$.

<u>(iv) ⇒ (ii)</u>. Since $\mathcal{O}_{X,x}$ is finite over $\mathcal{O}_{Y,y}$, there are integral equations

$$(4.4.5) \qquad \omega^j(z,w) := w_j^{b_j} + \sum_{\nu=0}^{b_j-1} a_j^{(\nu)}(z) w_j^\nu \in I$$

for the w_j as elements in $\mathcal{O}_{X,x}$ over $\mathcal{O}_{Y,y}$. After possibliy shrinking X and Y , this gives the commutative diagram

$$(4.4.6) \qquad \begin{array}{ccc} \underline{X} & \stackrel{\underline{i}}{\hookrightarrow} & \underline{A} := \underline{N}(\omega^1, \ldots, \omega^k) \\ & \underline{f} \searrow \quad \swarrow \underline{\pi} & \\ & \underline{B} & \end{array}$$

where $\underline{i}$ is a closed immersion and $\underline{\pi}$ a Weierstrass map. $\underline{\pi}$ is finite by Proposition 4.2.1, hence so is $\underline{f}$.

<u>(ii) ⇒ (i)</u>. This is clear.

<u>Corollary 4.4.2.</u> (i) <u>Let</u> $\underline{f} : (\underline{X},x) \longrightarrow (\mathbb{C}^n,0)$ <u>be defined by the elements</u> $f_j \in \mathfrak{m}_x$, $j = 1,\ldots,n$. <u>Then</u>

a) $\underline{f}$ is finite if and only if $(f_1,\ldots,f_n)$ generates an $\mathfrak{m}_x$-primary ideal of $\mathcal{O}_{X,x}$.

b) $\underline{f}$ is a local immersion if and only if $(f_1,\ldots,f_n)$ generates $\mathfrak{m}_x$.

(ii) $(\underline{X},x) \in \underline{\underline{\mathrm{cpl}}}_0$ is smooth, i.e. $(\underline{X},x) \cong (\underline{\mathbb{C}}^n,0)$ for some n, if and only if $\mathcal{O}_{X,x}$ is a regular local ring.

Sketch of proof.

(i) a): is clear because of Theorem 4.4.1.

(i) b): If $(f_1,\ldots,f_n)$ generate $\mathfrak{m}_x$, $\dim_{\mathbb{C}}(\mathcal{O}_{X,x}/\mathfrak{m}_n\cdot\mathcal{O}_{X,x}) = 1 < \infty$; hence $\mathcal{O}_{X,x}$ is finite over $\mathcal{O}_{\mathbb{C}^n,0}$ via f_x^0, and Nakayama's Lemma tells us that $f_x^0 : \mathcal{O}_{\mathbb{C}^n,0} \longrightarrow \mathcal{O}_{X,x}$ is surjective. So it factors as $\mathcal{O}_{\mathbb{C}^n,0} \twoheadrightarrow \mathcal{O}_{Y,0} \cong \mathcal{O}_{X,x}$ with $(\underline{Y},0) \subseteq (\underline{\mathbb{C}}^n,0)$ defined by the ideal $I := \operatorname{Ker} f_x^0$. Conversely, if $\underline{f}$ is a local immersion, f_x^0 factors as $\mathcal{O}_{\mathbb{C}^n,0} \twoheadrightarrow \mathcal{O}_{Y,0} \cong \mathcal{O}_{X,x}$, hence is surjective, and so $f_x^0(\mathfrak{m}_n) = \mathfrak{m}_x$.

(ii). If $(\underline{X},x)$ is smooth, $\mathcal{O}_{X,x} \cong \mathcal{O}_{\mathbb{C}^n,0}$, which is regular. If $\mathcal{O}_{X,x}$ is regular, a regular system of parameters of $\mathcal{O}_{X,x}$ gives a homomorphism $\varphi : \mathcal{O}_{\mathbb{C}^n,0} \longrightarrow \mathcal{O}_{X,x}$ such that $\hat{\varphi} : \hat{\mathcal{O}}_{\mathbb{C}^n,0} \longrightarrow \hat{\mathcal{O}}_{X,x}$ is an isomorphism. This implies φ is injective and $\dim_{\mathbb{C}}(\mathcal{O}_{X,x}/\mathfrak{m}_n\,\mathcal{O}_{X,x}) = \dim_{\mathbb{C}}(\hat{\mathcal{O}}_{X,x}/\hat{\mathfrak{m}}_n\hat{\mathcal{O}}_{X,x}) = 1$, so φ is finite and hence surjective by Nakayama's Lemma again. Hence φ is also surjective, hence an isomorphism, which implies $(\underline{X},x) \cong (\underline{\mathbb{C}}^n,0)$ by the Equivalence Theorem 3.3.3.

Exercise. Prove 4.4.2. without passing to the completion (use 2.6.2).

§ 5. Dimension and Nullstellensatz

Pursueing the analogy with elimination theory further, it is shown that a complex spacegerm has a well-defined local dimension, given as the minimal number of free parameters such that in the system of holomorphic equations defining the germ the rest of the unknowns are algebraic functions of them (this will be geometrically and algebraically exploited in 6.1 and 6.2). This local dimension coincides with the Chevalley dimension of the corresponding local ring. We introduce active elements, providing good inductive proofs for the dimension, and give a short proof of the Rückert Nullstellensatz, from which we deduce that the decomposition of a complex space germ into irreducible analytic setgerms corresponds in a one-to-one fashion to the minimal primes of the corresponding local analytic algebra.

5.1. Local dimension

Recall that by Corollary 3.3.5 mapgerms $\underline{f} : (\underline{X},x) \longrightarrow (\mathbb{C}^n,0) \in \underline{\underline{cpl}}_0$ correspond in a one-to-one fashion to sequences $(f_1,\ldots,f_n)$ with $f_1,\ldots,f_n \in \mathfrak{m}_x$.

Proposition and Definition 5.1.1 (Local dimension). Let $(\underline{X},x) \in \underline{\underline{cpl}}_0$. The following integers are the same:

$$\min\left\{n \mid \exists\, f_1,\ldots,f_n \in \mathfrak{m}_x : x \text{ is isolated in } N(f_1,\ldots,f_k)\right\} ,$$

$$\min\left\{n \mid \exists \text{ finite mapgerm } \underline{f} : (\underline{X},x) \longrightarrow (\mathbb{C}^n,0)\right\} ;$$

their common value is called the (local) dimension of X at x and denoted $\dim_x \underline{X}$.

This is immediate from the Integrality Theorem 4.4.1. We list the following properties:

Proposition 5.1.2. The local dimension has the following properties:

(i) $\dim_x \underline{X} \leq n$ if and only if $(\underline{X},x)$ admits a finite holomorphic mapgerm $(\underline{X},x) \longrightarrow (\mathbb{C}^n,0)$.

(ii) $\underline{f} : (\underline{X},x) \longrightarrow (\underline{Y},x)$ finite $\Rightarrow \dim_x \underline{X} \leq \dim_y \underline{Y}$.

(iii) If $(\underline{X},x) \in \underline{\underline{cpl}}_0$, define $(\underline{X}_{red},x) \hookrightarrow (\underline{X},x)$ as the subgerm corresponding to the projection $\mathcal{O}_{X,x} \twoheadrightarrow \mathcal{O}_{X,x}|N_x$, where N_x is the nilradical of $\mathcal{O}_{X,x}$, via the Equivalence Theorem 3.3.3. Then $\dim_x \underline{X} = \dim_x \underline{X}_{red}$.

(iv) If $\underline{X} \in \underline{\underline{cpl}}$, $x \longmapsto \dim_x \underline{X}$ is upper semicontinuous.

(v) If $(\underline{Y},x) \subseteq (\underline{X},x)$ is a subgerm and $\dim_x \underline{Y} < \dim_x \underline{X}$, $(Y,x) \neq (X,x)$ as germs of sets.

(vi) x is isolated in X if and only if $\dim_x \underline{X} = 0$.

(vii) $\dim_x \underline{X} = \dim \mathcal{O}_{X,x}$, the Chevalley dimension of the local ring $\mathcal{O}_{X,x}$.

Of this, (i) - (vi) are immediate from the definitions, only (vii) deserves a comment. Recall that the Chevalley dimension of a noetherian local ring R is defined to be the minimal length of a system $\underline{f} := (f_1,\ldots,f_n)$ of elements which generate an $\mathfrak{m}_R$-primary ideal; the latter condition is in our case equivalent to
$\dim_{\mathbb{C}}(R/\underline{f}R) = \text{lenght}_R(R/\underline{f}R) < \infty$. Then the claim (vii) follows directly from the Integrality Theorem 4.4.1.

5.2. Active elements and the Active Lemma

Active elements generalize nonzerodivisors. The main result is the Active Lemma 5.2.2, which makes inductive proofs work. Since, as we will see, activity of an element of a local analytic algebra restricts only its behaviour on the irreducible components of the corresponding complex space germ and not its behaviour on the embedded ones, it is a more flexible notion than that of nonzerodivisors.

Proposition and Definiton 5.2.1. Let R be a noetherian local ring. Then $f \in R$ is called active iff it satisfies one of the following equivalent conditions:

(i) $\forall\, \mathfrak{p} \in \operatorname{Min}(R) : f \notin \mathfrak{p}$.

(ii) $\forall\, g \in R : f \cdot g \in \mathfrak{n}_R \Rightarrow g \in \mathfrak{n}_R$, where $\mathfrak{n}_R$ is the nilradical of R .

(iii) f is a nonzero-divisor in the reduction $R_{red} := R/\mathfrak{n}_R$.

Lemma 5.2.2 (The Active Lemma). Let $(\underline{X},x) \in \underline{\underline{cpl}}_0$ and $f \in \mathfrak{m}_x$ be active. Then

$$\dim_x \underline{N}(f) = \dim_x \underline{X} - 1 \ .$$

<u>Idea of proof</u>. It suffices to show $\dim_x \underline{N}(f) \leqq \dim_x \underline{X} - 1$. Let $d := \dim_x \underline{X}$ and $\underline{\pi} : (\underline{X}, x) \longrightarrow (\underline{\mathbb{C}}^d, 0)$ be a finite holomorphic mapgerm; then f satisfies an integral equation

$$f^k + a_{k-1} f^{k-1} + \ldots + a_1 f + a_0 = 0$$

with $a_j \in \mathcal{O}_{\mathbb{C}^d,0}$, $0 \leqq j \leqq k-1$, by the Integrality Theorem 4.4.1. By activity, we may assume $a_0 \neq 0$. Then (5.2.1) induces the commutative diagram of complex space germs

$$\begin{array}{ccc}
(\underline{N}(f), x) & \hookrightarrow & (\underline{X}, x) \\
{\scriptstyle \underline{\pi}|\underline{N}(f)} \downarrow & & \downarrow {\scriptstyle \underline{\pi}} \\
(\underline{N}(a_0), 0) & \hookrightarrow & (\underline{\mathbb{C}}^d, 0)
\end{array} ,$$

where the horizontal arrows are closed immersions, and so $\underline{\pi}|\underline{N}(f)$ is a finite holomorphic mapgerm. Hence $\dim_x \underline{N}(f) \leqq \dim_0 \underline{N}(a_0)$ by Proposition 5.1.2 (ii). But since $a_0 \neq 0$, there is a line $L \subseteq \mathbb{C}^d$ such that 0 is isolated in $N(a_0) \cap L$ by the Identity Theorem for holomorphic functions in one variable, which easily implies $\dim_0 \underline{N}(a_0) \leqq d - 1$. This proves the Active Lemma.

The Active Lemma has numerous consequences as we will see in the next sections. Immediate is the following one:

<u>Corollary 5.2.3</u>. $\dim_0 \mathbb{C}^n = n$.

<u>Remark 5.2.4</u>. If $\dim_x X > 0$, active elements do exist in $\mathcal{O}_{X,x}$ (see [28], p. 99).

<u>5.3. The Rückert Nullstellensatz</u>

If $\mathbb{k}$ is an algebraically closed field and A a finitely generated $\mathbb{k}$-algebra, elements $f \in A$ define regular functions $[f] : X \longrightarrow \mathbb{k}$

on the variety $X = \operatorname{Spec} A$ (we consider only the closed points). The famous Hilbert Nullstellensatz states that $[f]$ is zero as a function if and only if f is a nilpotent element of A, or, what is equivalent in this case, nilpotent in all local rings $\mathcal{O}_{X,x}$, $x \in X$. The proof of the Nullstellensatz is rather easy in this algebraic case: One proves (i) the "weak Nullstellsatz" that any ideal $I \neq 1$ in $\mathbb{k}[X_1,\ldots,X_n]$, $n \geq 1$, has a zero, and then (ii) applies the Rabinowitsch trick (see [71], § 121). Usually (i) is proven by means of Noether normalization, which is easy in the algebraic case but hard in the complex analytic case (in fact it is our final aim in this chapter to prove it as the Local Representation Theorem in § 6); there are even more elementary proofs using the Division Algorithm in polynomial rings (which is similar to Theorem 2.3.2, but much easier to establish), see [3] for the Divison Algorithm and [46] for the Division Algorithm and the Nullstellensatz.

Although the Nullstellensatz remains true in the complex analytic case, the above approach will not work because (ii) fails; the result lies considerably deeper in this case, and was first proved by Rückert in his fundamental paper [59], in which for the first time algebraic methods were systematically introduced into Local Complex Analytic Geometry. In the treatment here, it will be a consequence of the Active Lemma.

Theorem 5.3.1 (Rückert Nullstellensatz). Let $X \in \underline{\underline{\mathrm{cpl}}}$, $f \in \mathcal{O}_X(X)$, and $[f] : X \longrightarrow \mathbb{C}$ the function defined by f (see (3.3.1)). Then $[f] = 0$ if and only if $f_x \in \mathcal{O}_{X,x}$ is nilpotent for all $x \in X$.

Idea of proof.

The "if"-part is clear. For the "only if"-part, let $x \in \underline{X}$ be given; one decomposes the nilradical $\mathcal{N}_x \subseteq \mathcal{O}_{X,x}$:

$$\mathcal{N}_x = \bigcap_{\mathfrak{p} \in \operatorname{Min}(\mathcal{O}_{X,x})} \mathfrak{p} \quad . \tag{5.3.1}$$

For $\mathfrak{p} \in \operatorname{Min}(\mathcal{O}_{X,x})$, let the immersion $(\underline{X}_{\mathfrak{p}},x) \hookrightarrow (\underline{X},x)$ of germs correspond to the projection $\mathcal{O}_{X,x} \twoheadrightarrow \mathcal{O}_{X,x}/\mathfrak{p}$ via the Equivalence Theorem 3.3.3. Then $\mathcal{O}_{X_{\mathfrak{p}},x} = \mathcal{O}_{X,x}/\mathfrak{p}$ is an integral domain, and so $f_{\mathfrak{p}} := f|\underline{X}_{\mathfrak{p}}$ is either 0 or active in $\mathcal{O}_{X_{\mathfrak{p}},x}$. But it cannot be active, since then by the Active Lemma $\dim_{\underline{x}} N(f_{\mathfrak{p}}) < \dim_{\underline{x}} X_{\mathfrak{p}}$, and so it would not vanish near x on $X_{\mathfrak{p}}$, which it must since f vanishes on X by assumption. So $(f_{\mathfrak{p}})_x = 0$ in $\mathcal{O}_{X_{\mathfrak{p}},x}$, which means $f_x \in \mathfrak{p}$. Since this holds for all $\mathfrak{p} \in \mathrm{Min}\, \mathcal{O}_{X,x}$, $f_x \in N_x$ by (5.3.1).

There are other useful formulations of this result:

Corollary 5.3.2. The following statements are equivalent to Theorem 5.3.1 and do therefore hold:

(i) Let $X \in \underline{\underline{\mathrm{cpl}}}$, $\underline{Y} \hookrightarrow \underline{X}$ a closed complex subspace defined by the locally finite ideal $I \subseteq \mathcal{O}_X$. Let J_Y be the ideal defined as $J_Y(U) := \{f \in \mathcal{O}_X(U) \mid [f]|Y = 0\}$ for $U \subseteq X$ open. Then $J_Y = \sqrt{I}$ (this is the traditional formulation of the Nullstellensatz).

(ii) Let M be an admissible $\mathcal{O}_X$-module, and let $f \in \mathcal{O}_X(X)$ be such that it vanishes on $\mathrm{supp}(M)$ as a function, i.e. $[f] \mid \mathrm{supp}\, M = 0$. Then any $x \in X$ has an open neighbourhood such that $f^t \cdot M = 0$ for some integer $t \geq 1$.

For 5.3.1 ⇒ (ii) see [28] , p. 67, Corollary (use $F_0(M)$ instead of $\mathrm{Ann}(M)$ there and the fact $F_0(M) \subseteq \mathrm{Ann}(M)$) . The implications (ii) ⇒ (i) and (i) ⇒ 5.3.1 are easy.

5.4. Analytic sets and local decomposition

Let $\underline{X}$ be a complex space. A subset $A \subseteq X$ is called analytic iff it is locally around any $x \in X$ the null set of finitely many sections of $\mathcal{O}_X$ defined near x . The ideal $J_A \subseteq \mathcal{O}_X$ with $J_A(U) := \{f \in \mathcal{O}_X(U) \mid [f]|A = 0\}$ is called the vanishing ideal of A .

If $A \subseteq X$ is analytic, it has a well-defined local dimension at $a \in A$: Since $\mathcal{O}_{X,a}$ is noetherian by the Rückert Basissatz 1.3.2, a has an open neighbourhood U such that $A \cap U$ is, the underlying set of a closed complex subspace of U defined by a finitely generated $\mathcal{O}_U$-ideal I which is such that $I_a = J_{A,a}$, and two such ideals coincide locally near a by Lemma 3.1.1. So there is, up to isomorphy, a well-defined germ $(\underline{A},a) \in \underline{\underline{\mathrm{cpl}}}_0$ defined by any such I in U , and we put $\dim_a A := \dim_a \underline{A}$. Especially, X is an analytic set in $\underline{X}$, and

$J_{X,x} = N_x$, the nilradical of $O_{X,x}$, by the Rückert Nullstellensatz 5.3.1, and so $\dim_x \underline{X} = \dim_x X$ by Proposition 5.1.2 (iii). If $x \in X$, we have the usual notion of the germ of an analytic set at x, denoted (A,x), which is the equivalence class of an analytic set A defined in an open neighbourhood of x with respect to the equivalence relation which identifies two locally defined analytic sets when they coincide near x. We call such germs analytic setgerms. Unions of analytic germs are well-defined and so there is the notion of an irreducible germ, this being one which cannot be written as a nontrivial union. It is then an easy exercise to show that an analytic setgerm has a unique decomposition into irreducible ones which corresponds to the associated primes of their vanishing ideal; this is a consequence of the Rückert Basissatz (see [28], Chapter 4, § 1.). Together with the Nullstellensatz we get the following result:

Proposition 5.4.1 (Local decomposition). Let $(\underline{X},x) \in \underline{\underline{\mathrm{cpl}}}_0$. If $I \subseteq O_{X,x}$ is any ideal, let the inclusion $(\underline{X}_I,x) \hookrightarrow (\underline{X},x)$ of complex spacegerms be defined by the projection $O_{X,x} \twoheadrightarrow O_{X,x}/I$ via the Equivalence Theorem 3.3.3. Then:

(i) The complex space subgerms of $(\underline{X},x)$ correspond bijectively to the ideals of $O_{X,x}$ under $I \longmapsto (\underline{X}_I,x)$, and the analytic setgerms to the radical ideals of $O_{X,x}$ under $I \longmapsto (X_I,x)$.

(ii) $(X,x) = \bigcup_{\mathfrak{p} \in \mathrm{Min}(O_{X,x})} (X_{\mathfrak{p}},x)$ is the unique decomposition of the analytic setgerm (X,x) into irreducible ones.

I refer to the decomposition in (ii) as the local decomposition of $\underline{X}$ at x into irreducible components.

I call the $X_{\mathfrak{p}}$ the local irreducible components of $\underline{X}$ at x (they are called prime components in [28]). Germs with exactly one irreducible component are called irreducible.

Using the Active Lemma, one proves the following result (see [28], p. 103), which is a converse to Proposition 5.1.2. (v) and which will be needed in § 6.

Theorem 5.4.2. Let Y be a closed complex subspace of the complex space $\underline{X}$, $x \in \underline{Y}$, and suppose $\dim_x \underline{Y} = \dim_x \underline{X}$. Then $\underline{X}$ and $\underline{Y}$ have a common local irreducible component at x .

Corollary 5.4.3 (Lemma of Ritt). Let $\underline{X}$ be a complex space, $\underline{Y} \hookrightarrow \underline{X}$ a closed complex subspace. The following statements are equivalent:

(i) $\dim_y \underline{Y} < \dim_y \underline{X}$ for all $y \in Y$.

(ii) Y is nowhere dense in X .

The proof is left as an exercise (use 5.2.2 (v) and 5.4.2).

§ 6. The Local Representation Theorem for complex space germs (Noether normalization)

In this paragraph, we are finally in a position to interprete geometrically the concepts of dimension and of a system of parameters fo a local analytic algebra and to see that they give rise locally to a situation identical with Noether normalization in the algebraic case, as described at the beginning of § 4. The dimension turns out to be the unique integer d that the complex space germ corresponding to the given local analytic algebra lies spread out finitely over a germ $(\mathbb{C}^d,0)$, and these finite branched covering mapgerms are precisely those given by a system of parameters according to Corollary 3.3.5.

6.1. Openness and dimension

We now can give a geometric characterization of the local dimension. The geometric characterization in question is the openness of a map at a point; here, a continuous map $f : X \longrightarrow Y$ of topological spaces is said to be open at a point $x \in X$ iff it maps every neighbourhood of x in X onto a neighbourhood of $f(x)$ in Y .

Lemma 6.1.1 (Open Lemma I). Let $\underline{f} : (\underline{X},x) \longrightarrow (\underline{Y},y) \in \underline{\underline{\mathrm{cpl}}}_0$ be finite. Then f is open at x if and only if each element of $\mathrm{Ker}(f^0_x : \mathcal{O}_{Y,y} \longrightarrow \mathcal{O}_{X,x})$ is nilpotent.

Proof. Since f is finite, $f(X)$ is an analytic set in Y by Corollary 4.3.2. f is open at x if and only if $(f(X),y) = (Y,y)$ as germs of sets at $y \in Y$, which means $\mathcal{J}_{f(X),y} = \mathcal{J}_{Y,y}$, where

$J_{f(X)}$ and J_Y are the vanishing ideals of the analytic sets $f(X)$ and Y in $\underline{Y}$. But $J_{f(X),y} = (f_x^0)^{-1}N_{X,x}$ and $J_{Y,y} = N_y$ by the Nullstellensatz, and so

$$\begin{aligned} f \text{ open at } x &\iff (f_x^0)^{-1}N_x = N_y \\ &\iff \operatorname{Ker} f_x^0 \subseteq N_y \quad , \end{aligned}$$

which proves the claim.

Lemma 6.1.2 (Open Lemma II). Let $\underline{f} : (\underline{X},x) \longrightarrow (\underline{Y},y) \in \underline{\underline{\mathrm{cpl}}}_0$ be finite.

(i) f open at $x \Rightarrow \dim_x \underline{X} = \dim_y \underline{Y}$.

(ii) If Y is locally irreducible at y (i.e. Y has only one local irreducible component at y , see 5.4), then $\dim_x \underline{X} = \dim_y \underline{Y} \Rightarrow f$ open at x .

Proof. (i): We may assume $\dim_x \underline{X} > 0$. After possibly shrinking $\underline{X}$ and $\underline{Y}$, we may assume there is $g \in \mathcal{O}_Y(Y)$ which is active at y such that $f_x^0(g) =: g'$ is active at y by the so-called Lifting Lemma (see [28], p. 99; the proof there actually does not need the assumptions that X and Y are reduced). This gives the commutative diagram

$$\begin{array}{ccccc} \underline{N}(g') & =: & \underline{X}' & \hookrightarrow & \underline{X} \\ & \underline{f}|\underline{N}(g') =: \underline{f}' & \Big\downarrow & & \Big\downarrow \underline{f} \\ \underline{N}(g) & =: & \underline{Y}' & \hookrightarrow & \underline{Y} \end{array}$$

with $\underline{f}'$ finite and open, and this allows to induct over $\dim_x X$.

(ii) $\dim_x X \leq \dim_y f(X) \leq \dim_y Y$ by Proposition 5.1.2, hence $\dim_y f(X) = \dim_y Y$, and the claim follows from Theorem 5.4.2.

6.2. Geometric interpretation of the local dimension and of a system of parameters; algebraic Noether normalization

Combining the results in 6.1. gives immediately:

Theorem 6.2.1 (Dimension Theorem). Let $(\underline{X},x) \in \underline{\underline{cpl}}_0$. If $f : (\underline{X},x) \longrightarrow (\underline{\mathbb{C}}^n,0)$ is a finite holomorphic mapgerm, the following statements are equivalent:

(i) $f_x^0 : \mathcal{O}_{\mathbb{C}^n,0} \longrightarrow \mathcal{O}_{X,x}$ is injective,

(ii) f is open at x,

(iii) $n = \dim_x \underline{X}$.

Corollary 6.2.2. Let $(\underline{X},x) \in \underline{\underline{cpl}}_0$. Then $\dim_x \underline{X}$ is the unique integer n such that $(\underline{X},x)$ admits a finite mapgerm to $(\underline{\mathbb{C}}^n,0)$ which is open at x.

Corollary 6.2.3. Let $R \in \underline{\underline{la}}$, $f_1,\ldots,f_n \in \mathfrak{m}_R$. Then $(f_1,\ldots,f_n)$ is a system of parameters for R if and only if the mapgerm $f : (\underline{X},x) \longrightarrow (\underline{\mathbb{C}}^n,0) \in \underline{\underline{cpl}}_0$ corresponding to $(f_1,\ldots,f_n)$ via Corollary 3.3.5 is finite and open at x.

Corollary 6.2.4 (Algebraic Noether normalization). Let $R \in \underline{\underline{la}}$, and let $(f_1,\ldots,f_d)$ be a system of parameters for R. Then the analytic subring generated by $f_1,\ldots,f_d$ is isomorphic to $\underline{\mathbb{C}}\{X_1,\ldots,X_d\}$, and R is finite over it.

Proof. If $R \in \underline{\underline{la}}$ and $f_1,\ldots,f_k \in \mathfrak{m}_R$, the simplest way to define the analytic subring generated by them is to declare it to be the image of the homomorphism $\varphi : \mathcal{O}_{\mathbb{C}^k,0} \longrightarrow R$ defined by mapping z_i to f_i for $1 \leq i \leq k$ according to Theorem 1.3.4. By the way φ is defined, this subring should consist of the (in R) convergent infinite series $\sum_{A \in \mathbb{N}^k} c_A f^A$, $c_A \in \mathbb{C}$, and in fact one can put a topology on R, the topology of analytic convergence (see [26]) so that this statement makes sense and is true; this analytic subring then is just the closure of the subring generated by the f_i in the algebraic sense. The claim of 6.2.4 is immediate from 6.2.1 and Theorem 4.4.1 (iv).

6.3. The Local Representation Theorem; geometric Noether normalization

We now can more fully exploit the geometry of a system of parameters of a local analytic algebra R or, what is the same according to Corollary 6.2.3, of a finite open mapgerm of a complex space germ onto a number space germ. We have already proven the "algebraic Noether normalization", namely that the system of parameters generate a subalgebra which is a convergent power series ring over which R is finite. It now will turn out that locally this implies the same geometric situation that we have in the algebraic case, where the variety corresponding to a finite $\mathbb{k}$-algebra R is a branched covering over an affine space of dimension $\dim R$, but this time the proof is substantially more difficult and needs the whole machinery described up to now.

Anyway, the following local description of a complex space germ holds, which is a kind of geometric Noether normalization:

Theorem 6.3.1 (The Local Representation Theorem). Let $(\underline{X},x) \in \underline{\underline{\mathrm{cpl}}}_0$, $d = \dim_x \underline{X}$, and let $\underline{f} : (\underline{X},x) \longrightarrow (\underline{\mathbb{C}}^d,0)$ be a finite holomorphic mapgerm; such mapgerms exist by the definition of the local dimension, and they correspond to systems of parameters for $\mathcal{O}_{X,x}$. Then $\underline{f}$ has arbitrarily small representatives $\underline{f} : \underline{X} \longrightarrow \underline{B}$, where B is a domain in $\underline{\mathbb{C}}^d$, such that the following holds:

(i) There exists a closed complex subspace $\underline{\Delta} \hookrightarrow \underline{B}$ which is nowhere dense and has the property that $X - f^{-1}(\Delta)$ is dense in $X_0 := \{x' \in X \mid \dim_{x'} \underline{X} = d\}$. $\underline{\Delta}$ can be chosen to be a hypersurface, i.e. $\underline{\Delta} = \underline{N}(\delta)$ for a nonzero $\delta \in \mathcal{O}_{\mathbb{C}^d}(B)$.

(ii) $f|X - f^{-1}(\Delta) : X - f^{-1}(\Delta) \longrightarrow B - \Delta$ is a topological covering map.

(iii) If, in addition, $\underline{X}$ is reduced at x, i.e. the nilradical N_X of $\mathcal{O}_{X,x}$ is zero, $\underline{f}|\underline{X} - f^{-1}(\Delta)$. $\underline{X} - f^{-1}(\Delta) \longrightarrow \underline{B} - \Delta$ is a holomorphic covering of complex manifolds.

We call these representatives good representatives. $\underline{\Delta}$ is called a discriminant locus for $\underline{f}$.

I will not give the detailed proof here , but describe the main ingredients, so that the rest of it is a careful exploitation on the basis of the results described until now.

It is clear that it suffices to prove (i) and (iii) for a germ reduced at X , for we can pass from $(\underline{X},x)$ to $(\underline{X}_{red},x) \hookrightarrow (\underline{X},x)$ defined by $\mathcal{O}_{X,x} \twoheadrightarrow \mathcal{O}_{X,x}/\mathcal{N}_X$.

First, one treats the case of a Weierstrass map $\underline{\pi} : \underline{A} \longrightarrow \underline{B}$ (see 4.2.) with the additional property that the defining monic polynomials $\omega^{(j)} \in \mathcal{O}_{\mathbb{C}^d}(B)[w_j]$, $1 \leq j \leq k$, have no multiple factors.

Put $\tilde{\delta} := \prod_{j=1}^{k} \operatorname{discr}(\omega^{(j)})$, where $\operatorname{discr}(\omega^{(j)}) \in \mathcal{O}_{\mathbb{C}^d}(B)$ is the discriminant of $\omega^{(j)}$, and let $\underline{\Delta}(\pi) := \underline{N}(\tilde{\delta})$. Then Hensel's Lemma 4.2.3 tells us that around $z_0 \in B - \Delta(\pi)$ we can write

$$\omega^{(j)}(z,w_j) = \prod_{\nu=1}^{b_j}(w_j - c_\nu^{(j)}(z)) \ , \ 1 \leq j \leq k \tag{6.3.1}$$

for holomorphic functions $c_\nu^{(j)}$ defined near z_0 . If $a = (z_0,c) \in A - \pi^{-1}(\Delta)$ and, for $1 \leq j \leq k$, ν_j is such that $c_{\nu_j}^{(j)}(z_0) = c_j$, this forces

$$\mathcal{O}_{A,a} = \mathcal{O}_{\mathbb{C}^{d+k},a} / (w_1 - c_{\nu_1}^{(1)}(z),\ldots,w_k - c_{\nu_k}^{(k)}(z)) \ , \tag{6.3.2}$$

so that clearly $\pi_a^0 : \mathcal{O}_{\mathbb{C}^d,z_0} \longrightarrow \mathcal{O}_{A,a}$ is isomorphic. Hence $\underline{\pi}$ is locally isomorphic over $B - \Delta(\pi)$ by the Equivalence Theorem 3.3.3. This shows (iii).

(i) follows from the fact that $\tilde{\delta} \neq 0$ since the $\omega^{(j)}$ have no multiple factors, hence $\Delta(\pi)$ is nowhere dense in B by the identity theorem for holomorphic functions, and so $\pi^{-1}(\Delta(\ ))$ is nowhere dense in A , since π is open by Proposition 4.2.2.

For the general case of a reduced $(\underline{X},x)$ we may assume $\underline{f}$ is induced by a linear projection $pr : \mathbb{C}^{d+k} \longrightarrow \mathbb{C}^d$. With the notation of

4.4. we get the embedding (4.4.6) of $\underline{f}$ into a Weierstrass map, which, in addition, we may assume to be of the above type, since $\mathcal{O}_{X,x}$ has no nilpotents. Let

$$(6.3.3.) \qquad (X,x) = \bigcup_{\mathfrak{p}\in \mathrm{Min}(\mathcal{O}_{X,x})} (X_{\mathfrak{p}},x)$$

$$(6.3.4) \qquad (A,0) = \bigcup_{\mathfrak{p}\in \mathrm{Min}(\mathcal{O}_{A,0})} (A_{\mathfrak{q}},0)$$

be the decompositions into locally irreducible components according to 5.4.

Let $M_0 := \left\{\mathfrak{p} \in \mathrm{Min}(\mathcal{O}_{X,x}) \mid \dim_x X_{\mathfrak{p}} = d\right\} = \mathrm{Assh}(\mathcal{O}_{X,x})$ and $M_1 := \mathrm{Min}(\mathcal{O}_{X,x}) - M_0 = \left\{\mathfrak{p} \in \mathrm{Min}(\mathcal{O}_{X,x}) \mid \dim_x X_{\mathfrak{p}} < d\right\}$ (equality by Proposition 5.1.2 (ii)). Choosing X,A,B small enough one can achieve:

1) for each $\mathfrak{p} \in M_0$ there is exactly one $\mathfrak{q} =: \mathfrak{q}(\mathfrak{p}) \in \mathrm{Min}\,\mathcal{O}_{A,0}$ with $X_{\mathfrak{p}} = A_{\mathfrak{q}(\mathfrak{p})}$; this is by Theorem 5.4.2;

2) for all $\mathfrak{p} \in M_1$ and all $x' \in X_{\mathfrak{p}}$: $\dim_{x'} X_{\mathfrak{p}} < d$; this is by upper semicontinuity of dimension (Proposition 5.1.2 (iv));

3) $\Delta(\pi) \cup \bigcup_{\substack{\mathfrak{q},\mathfrak{q}'\in\mathrm{Min}(\mathcal{O}_{A,0})\\ \mathfrak{q}\neq\mathfrak{q}'}} \pi(A_{\mathfrak{q}} \cap A_{\mathfrak{q}'}) \cup \left(\bigcup_{\mathfrak{p}\in M_1} f(X_{\mathfrak{p}})\right) =: \Delta(f)$ is an analytic subset of B ; this is by Corollary 4.3.2.

4) $N(f) \subseteq N(\delta)$ for a nonzero $\delta \in \mathcal{O}_{\mathbb{C}^d}(B)$.

One checks that for this $\Delta := N(\delta)$ the conditions (i) and (iii) of the Local Representation Theorem hold; the main ingredient is the Open Lemma II,6.1.2.

Remark 6.3.2. For small enough representatives, $\Delta(f)$ can in fact be defined as a complex subspace since $\underline{\Delta}(\pi)$, $\underline{\pi}(\underline{A}_{\mathfrak{q}} \cap \underline{A}_{\mathfrak{q}'})$ and $\underline{f}(\underline{X}_{\mathfrak{p}})$ exist naturally as complex subspace germs at $0 \in B$, and so their union exists as a complex subspace germ defined by the intersection of the corresponding ideals in $\mathcal{O}_{B,0}$. Moreover

(6.3.5) $$\bigcup_{\substack{\mathfrak{q},\mathfrak{q}'\in \mathrm{Min}\left(\mathcal{O}_{A,0}\right)\\ \mathfrak{q}\neq\mathfrak{q}'}} \pi(A_{\mathfrak{q}} \cap A_{\mathfrak{q}'}) \subseteq \Delta(\pi) ,$$

(see II 2.2.1), so

(6.3.6) $$\Delta(f) = \Delta(\pi) \cup \bigcup_{\mathfrak{p}\in M_1} f(X_{\mathfrak{p}}) ,$$

the natural choice.

Remark 6.3.3. Finally, I have to make a short remark on prime germs, i.e. $(\underline{X},x) \in \underline{\underline{\mathrm{cpl}}}_0$ with $\mathcal{O}_{X,x}$ an integral domain, so that especially $(\underline{X},x)$ is locally irreducible (see 5.4.). Let $\underline{f} : (\underline{X},x) \longrightarrow (\underline{\mathbb{C}}^d,0)$ be as in the Local Representation Theorem 6.2.1, then $f_x^0 : \mathcal{O}_{\mathbb{C}^d,0} \hookrightarrow \mathcal{O}_{X,x}$ is an integral ring extension by the Integrality Theorem 4.4.1. Let $h \in \mathcal{O}_{X,x}$ be a primitive element for the corresponding field extension and form, for a suitably small representative $\underline{f} : \underline{X} \longrightarrow \underline{B}$:

$$\omega(z,t) := \prod_{x'\in f^{-1}(z)} (t - h(x')) \in \mathcal{O}_B(B - \Delta)[t] .$$

Then $\omega(z,t)$ extends over Δ since Δ is nowhere dense in B by the classical Riemann Extension Theorem (for a nice proof of the latter see [30],p.9),and gives a monic irreducible polynomial $\omega \in \mathcal{O}_B(B)[t]$. The homomorphism

$$v_x^0 := \varphi : \mathcal{O}_{B\times\mathbb{C},0} \longrightarrow \mathcal{O}_{X,x} ,$$

which maps z_i to $f_x^0(z_i)$ for $1 \leq i \leq n$ and t to h_x ,annihilates ω, and so defines a morphism, via the Equivalence Theorem 3.3.3,

$$\begin{array}{ccccc} (\underline{X},x) & \xrightarrow{\underline{v}} & (\underline{Y},y) := (\underline{N}(\omega),0) & \hookrightarrow & (\underline{B}\times\underline{\mathbb{C}},0) \\ & {\scriptstyle \underline{f}}\searrow & \swarrow{\scriptstyle \underline{\pi}} & & \\ & & (\underline{B},0) & & \end{array}$$

from $\underline{f}$ into the Weierstrass mapgerm $\underline{\pi}$ given by the irreducible monic polynomial ω . It can be shown that v is isomorphic outside a nowhere dense closed subspace of $\underline{B}$ for suitable representatives (exercise; for a direct proof not using 6.3.1 see [40], § 46). If we replace $\underline{\Delta}$ of 6.3.1 with this subspace, $Y - \pi^{-1}(\Delta)$ is connected since ω is irreducible, and so we get

<u>Corollary 6.3.4.</u> <u>If, in the situation of</u> 6.3.1, $(\underline{X},x)$ <u>is a prime germ, i.e. reduced and locally irreducible,</u> $X - f^{-1}(\Delta)$ <u>is connected</u>.

§ 7. Coherence

7.1. Coherent sheaves

Definition 7.1.1. (i). Let R be a ring. A finitely presentable R-module M is called coherent if all its finitely generated submodules are also finitely presentable. R is called coherent if it is coherent as a module over itself, i.e. if every finitely generated ideal is finitely presentable.

(ii). Let $(X,\mathcal{O}_X)$ be a ringed space. An admissible $\mathcal{O}_X$-module M is called coherent if all its locally finitely generated submodules are also admissible. $\mathcal{O}_X$ is called coherent if it is coherent as a module over itself, i.e. if every locally finitely generated $\mathcal{O}_X$-ideal is admissible.

I discuss the notion of coherence for sheaves; the discussion for modules over a ring is analogous. The coherent $\mathcal{O}_X$-modules over a ringed space $(X,\mathcal{O}_X)$ form a good category Coh/X in the sense that it is stable under various operations on sheaves (called the "yoga of coherent sheaves", see [28], Annex). From this yoga one infers:

Lemma 7.1.2. Let $(X,\mathcal{O}_X)$ be a ringed space, $\mathcal{O}_X$ a coherent sheaf of rings. Then an $\mathcal{O}_X$-module is coherent if and only if it is admissible.

So in this case the admissible modules are the right category to work with, and, given a ringed space, the question is basic whether its structure sheaf is coherent. For complex spaces, the answer is given by the following famous theorem.

Theorem 7.1.3 (Oka's Coherence Theorem). For every complex space $(X,\mathcal{O}_X)$, $\mathcal{O}_X$ is a coherent sheaf of rings.

For a nice proof, which deduces this from the Weierstrass isomorphism 4.3.3, see [28], 2.5. Other proofs are in [64], Exposé 18, and [40], where it is deduced immediately, but in a not very enlightening way, from the classical Weierstrass Preparation Theorem 2.6.3.

So from now on we identify admissible and coherent $\mathcal{O}_X$-modules on a complex space.

7.2. Nonzerodivisors

Oka's Coherence Theorem immediately entails:

Proposition 7.2.1. Let $\underline{X}$ be a complex space, $f \in \mathcal{O}_X(X)$. Then, if f is a nonzerodivisor at x, it is a nonzerodivisor near x.

See [28], p. 68, (or just look at the kernel of $\mathcal{O}_X \xrightarrow{.f} \mathcal{O}_X$).

7.3. Purity of dimension and local decomposition

Let $(\underline{X},x) \in \underline{\underline{\mathrm{cpl}}}_0$, and let

$$(X,x) = \bigcup_{\mathfrak{p}\in \mathrm{Min}(\mathcal{O}_{X,x})} (X_{\mathfrak{p}},x) \tag{7.3.1}$$

be its decomposition into local irreducible components according to Proposition 5.4.1.

Definition 7.3.1 (X,x) is called equidimensional (or pure dimensional) if and only if $\dim_x X_{\mathfrak{p}} = \dim_x X$ for all $\mathfrak{p} \in \mathrm{Min}(\mathcal{O}_{X,x})$.

In terms of local algebra this means $\mathrm{Assh}(\mathcal{O}_{X,x}) = \mathrm{Min}(\mathcal{O}_{X,x})$.

Theorem 7.3.2 (Purity of dimension). Let the complex space $\underline{X}$ be equidimensional at x. Then it is equidimensional near x.

The proof is left as an exercise. For it, assume $\underline{X}$ is reduced at x and represent $(\underline{X},x)$ via $\underline{f} : (\underline{X},x) \longrightarrow (\mathbb{C}^d,0)$ as in the Representation Theorem 6.3.1. Then $f_x^0(\delta)$ is a nonzerodivisor at x; apply 7.2.1 and Ritt's Lemma 5.4.4. to conclude $X_0 = X$ near x.

Corollary 7.3.3 (Open Mapping Lemma). Let $\underline{f} : \underline{X} \longrightarrow \underline{B}$ be a finite morphism from the complex space $\underline{X}$ to an open subspace $\underline{B} \subset \mathbb{C}^d$. If $\underline{f}$ is open at $x \in \underline{X}$, and $\underline{X}$ is equidimensional at x, $\underline{f}$ is open near x.

This follows from the Purity Theorem 7.3.2. and the Dimension Theorem 6.2.1.

Corollary 7.3.4. In the decomposition (7.3.1), for suitably small representatives, $X_{\mathfrak{p}} \cap X_{\mathfrak{p}'}$ is nowhere dense in $X_{\mathfrak{p}}$ and $X_{\mathfrak{p}'}$ for all $\mathfrak{p}, \mathfrak{p}' \in \mathrm{Min}(\mathcal{O}_{X,x})$ with $\mathfrak{p} \neq \mathfrak{p}'$.

Proof. Exercise; use 7.3.2 to conclude $\dim_{x'}(X_{\mathfrak{p}} \cap X_{\mathfrak{p}'}) < \dim_{x'}(X_{\mathfrak{p}})$ for x' near x.

7.4. Reduction.

The significance and importance of the notion of coherence cannot be described by a few words; they manifest themselves in the numerous results they imply. From this point on, coherence is indisputable for the further developments of the theory, which comprise coherence of the sheaf of nilpotents (Cartan's Coherence theorem), theory of reduction, analyticitiy of the singular locus, normalization . For this, see the book [28].

Theorem 7.4.1 (Cartan's Coherence Theorem). For every complex space $(X, \mathcal{O}_X)$, the nilradical $\mathcal{N}_X \subseteq \mathcal{O}_X$ is coherent.

For proofs see [28],[40], [64], and the sketch below.

Corollary 7.4.2. If A is an analytic set in the complex space X, the vanishing ideal $\mathcal{J}_A$ (see I, 5.4.) is coherent and endowes A with the canonical structure of a reduced complex space. Especially the analytic set X has a canonical structure as a reduced complex space and is called the reduction $\underline{X}_{red}$ of $\underline{X}$; one has $\mathcal{O}_{X_{red}} = \mathcal{O}_X/\mathcal{N}_X$ by the Rückert Nullstellensatz 5.3.1.

Here a complex space is called reduced if all its local rings have no nilpotents.

Sketch of proof of 7.4.1.

The assertion is local; so let $(\underline{X},x) \in \underline{\underline{\mathrm{cpl}}}_0$, and we must show that there is a representative $\underline{X}$ such that $\mathcal{N}_X$ is locally finite.

Assume first $\underline{X}$ is reduced at x . Choose a representative $\underline{X}$ and a finite map $\underline{f} : \underline{X} \longrightarrow \underline{B}$ as in the Local Representation Theorem 6.3.1. Let

$$I_0 := \bigcap_{\mathfrak{p} \in \mathrm{Assh}(\mathcal{O}_{X,x})} \mathfrak{p}$$

$$I_1 := \bigcap_{\mathfrak{p} \in \mathrm{Ass}(\mathcal{O}_{X,x}) - \mathrm{Assh}(\mathcal{O}_{X,x})} \mathfrak{p} \quad .$$

After possibly shrinking X , these define locally finite ideal sheaves $\mathcal{I}_j \subseteq \mathcal{O}_{X,x}$ and so two closed complex subspaces $\underline{X}_j \xrightarrow{g_j} \underline{X}$ for $j = 0,1$. Then, and here Oka's Coherence Theorem comes in, $\mathcal{I}_0 \cap \mathcal{I}_1$ is locally finite; hence, since $(\mathcal{I}_0 \cap \mathcal{I}_1)_x = I_0 \cap I_1 = \{0\}$, we may assume $\mathcal{I}_0 \cap \mathcal{I}_1 = 0$ after eventually shrinking $\underline{X}$, by Lemma 3.1.1. Further shrinking $\underline{X}$ we may assume $\dim_{x'}\underline{X}_0 = d$ for all $x' \in X_0$ and $\dim_{x'}X_1 < d$ for all $x' \in X_1$ by Theorem 7.3.2 and Proposition 5.2.2 (iv).

Let $\underline{\Delta} = \underline{N}(\delta)$ be as in 6.3.1, with $\delta \in \mathcal{O}_{\mathbb{C}^d}(B)$. We may choose $\underline{X}$ so small that $f^0(\delta)$ is a nonzerodivisor in $\mathcal{O}_{X_0,x'}$ at any $x' \in X_0$, because it is a nonzerodivisor in $\mathcal{O}_{X_0,x}$, and we then apply Proposition 7.2.1. I then propose to show $\mathcal{N}_X = 0$.

Let $x' \in X$. Choose, after possibly shrinking X , a locally finite ideal $\mathcal{J} \subseteq \mathcal{N}_X$ with $\mathcal{J}_{x'} = \mathcal{N}_{X,x'}$. Then $\mathrm{supp}\,\mathcal{J} \subseteq \mathrm{supp}\mathcal{N}_X \subseteq N(f^0(\delta))$, and so there is $t \in \mathbb{N}$, $t \geq 1$, such that $f^0(\delta)^t \cdot \mathcal{J} = 0$ near x' by the Rückert Nullstellensatz in the form of Corollary 5.3.2 (ii). Hence $\mathcal{N}_{X,x'} = \mathcal{J}_{x'}$ is contained in

$\mathcal{I}_{0,x'} = \mathrm{Ker}(\mathcal{O}_{X,x'} \xrightarrow{g^0_{0,x'}} \mathcal{O}_{X_0,x'})$, so $\mathcal{N}_X \subseteq \mathcal{I}_0$ near x' . Since

$\mathcal{I}_1 = \mathcal{K}er(\mathcal{O}_X \xrightarrow{g^0_1} \mathcal{O}_{X_1})$, $\mathcal{N}_X \cap \mathcal{K}er\, g^0_1 = \mathcal{N}_X \cap \mathcal{I}_1 \subseteq \mathcal{I}_0 \cap \mathcal{I}_1 = 0$, and so g^0_1 injects $\mathcal{N}_X$ into $\mathcal{N}_{X_1}$. Since $\dim_{x'}\underline{X}_1 < d$ for all $x' \in X_1$, $\mathcal{N}_{X_1} = 0$ by the induction assumption, and so $\mathcal{N}_X = 0$.

Finally, if $(\underline{X},x)$ is arbitrary, choose, after shrinking $\underline{X}$, a locally finite ideal $\mathcal{J} \subseteq \mathcal{N}_X$ with $\mathcal{J}_x = \mathcal{N}_{X,x}$. Let $\underline{Y}$ be the closed complex subspace of $\underline{X}$ defined by $\mathcal{J}$. Then $\underline{Y}$ is reduced at x and so $\mathcal{N}_Y = 0$ by what we proved above. But $\mathcal{N}_Y = \sqrt{\mathcal{J}}/\mathcal{J}$, and so $\mathcal{N}_X = \mathcal{J}$, which is locally finite.

II. GEOMETRIC MULTIPLICITY

The concept of multiplicity arises as a natural generalization of the multiplicity of a solution to a polynomial equation in one indeterminate. Consider a system

(1) $$f_j(z_1,\ldots,z_n) = 0 \quad , \quad j = 1,\ldots,k$$

of holomorphic equations, and suppose $0 \in \mathbb{C}^n$ is a solution. Heuristically, the multiplicity of 0 as a solution should be the number of solutions "concentrated near 0", i.e. the algebraic number m of distinct generic solutions arbitrarily near to 0 (cf. [51], p. 17, Definition). Symbolically:

(2) $$m = \varinjlim_{U} (\sup \# \{z \in U \mid f_j(z) = 0 \ , \ j = 1,\ldots,k \ \text{"distinct" solutions}\}) \ ,$$

where U runs over the neighbourhoods of 0 in $\mathbb{C}^n$, and the solutions are properly counted. In modern terms, the $f_1,\ldots,f_k$ define an ideal $I \in \mathcal{O}_{\mathbb{C}^n,0}$ and so a germ $(\underline{X},x) \in \underline{\underline{cpl}}_0$, and the multiplicity in question is called <u>the multiplicity of</u> x <u>on</u> $\underline{X}$, <u>denoted</u> $m(\underline{X},x)$.

To clarify what this means, consider the corresponding algebraic situation, where the f_j above are polynomials in $\mathbb{k}[z_1,\ldots,z_n]$ for some field $\mathbb{k}$. Kronecker's elimination theory ([43], [42], and [51], which is, in a sense, still quite readable and has become a classic) represents the solutions, after a general linear coordinate transformation, as algebraic functions of some of the coordinates, $z_1,\ldots,z_d$ say, which act as free parameters. The correct definition of the global multiplicity, i.e. the algebraic number of distinct generic solutions, was debated quite a time after Kronecker's 1882 paper [43] (see e.g. [42]) and found 30 years later by Macaulay [50]. In modern terms:

(3) $$\begin{aligned} M &:= \dim_K K \otimes_{\mathbb{k}} R \\ &= \sum_{\mathfrak{p} \in \mathrm{Assh}(R)} \mathrm{length}(R_{\mathfrak{p}}) \cdot [R/\mathfrak{p} : K] \end{aligned}$$

with $K := \mathbb{k}(z_1,\dots,z_d)$ and $R := \mathbb{k}[z_1,\dots,z_n]/(f_1,\dots,f_k)$, a natural generalization, after all, of the case of one variable. (It is interesting to look at the attempts in [42] to define the correct coefficient of $[R/\mathfrak{p}:K]$ via the degrees of the factors of the resolvent and Macaulay's criticism of it in [50]. This is a good lesson how painfully and slowly concepts developed which nowadays are considered to be utterly self-explanatory and trivial. This applies equally well to primary decomposition and the notion of local multiplicity below).

Geometrically, this corresponds to representing the solution variety $X \subseteq A^n$ as branched cover

(4) $$\pi : X \longrightarrow \mathbb{A}^d \quad , \quad d = \dim X = \dim R$$

with π induced by a generic projection, and putting

(5) $$M := \text{algebraic global mapping degree of } \pi$$
$$= \sum_\lambda \ell_\lambda \cdot \#(\pi^{-1}(z) \cap X_\lambda) \quad ,$$

where the X_λ are the irreducible components of X , $\ell_\lambda = \text{length}\, \mathcal{O}_{X,X_\lambda}$ and $z \in \mathbb{A}^d$ is any point outside the image of the branching locus (a "generic" z). (That (3) and (5) agree will be proved, in a local version, in 5.1.4 below).

The local multiplicity $m(X,x)$ of X at x , then, should be the <u>local</u> mapping degree of a generic projection. This means one wishes to take a small neighbourhood U around x such that $\pi(U)$ is open in $\mathbb{A}^d$ and $\pi^{-1}\pi(x) \cap U = \{x\}$; then $m(X,x)$ should be

(6) $$m(X,x) = \sum_\lambda \ell_\lambda \cdot \#(\pi^{-1}(z) \cap U_\lambda)$$

where the U_λ are the local branches of X at x and ℓ_λ the length of a maximal primary chain starting at the primary defining U_λ , which measures the multiplicity of the generic solution on U_λ .

Unfortunately, there are no small neighbourhoods in the algebraic situation, and so it took several decades to master the concept of multiplicity. There are three ways out of this difficulty:

(i) One tries to make sense out of the limit process in (1) algebraically, i.e. out of the concept of "solutions coming together at 0 ". This leads to the theory of specialization multiplicity of v.d. Waerden and Weil ([72], [73], and [74]). This will not be touched further upon here.

(ii) One passes to formal ("infinitesimal") neighbourhoods via completion; then the analogue of the local mapping degree makes sense. This leads to the definition of Chevalley ([9], [10]; see also Chapter 1, (6.7), and 5.1.5 and 5.1.8 below).

(iii) One uses the sophisticated approach to define multiplicity via the highest coefficient of the Hilbert function of the associated graded ring; this is the definitive and commonly accepted definition of Samuel [60]. It has the advantage of being concise, and it works very well in the practice of algebraic manipulations. (Ultimately, it leads via Serre's notes [67] and the paper of Auslander and Buchsbaum on codimension and multiplicity (Ann. of Math. 68 (1958), 625-657, esp. Theorem 4.2) to the definition presented in Chapter I, (1.2).) Although the geometric significance of this definition must have been known to the experts, it seems to have been rarely explicited (it was already known to Macaulay, see [50], footnotes on p.82 and 115, and [37], which makes quite a tense reading). It corresponds, geometrically, to approximating X at x by its tangent cone and taking the local multiplicity of the tangent cone at its vertex; for cones, the problem of small neighbourhoods does not pose itself, since the local and global mapping degree of a projection of a cone agree, due to the latter's homogeneous structure.

Fortunately, small neighbourhoods do exist in Complex Analytic Geometry, and so the definition of multiplicity as the local mapping degree of a generic projection makes perfect sense; this must have been, in the reduced case, folklore ever since (cf. [13], [38] and [75]). This formalism is set up in the first three paragraphs of this part II. To handle the nonreduced case, we make use of the properties of compact Stein neighbourhoods to relate the properties of nearby analytic local rings to those of one algebraic object, the coordinate ring of the compact Stein neighbourhood; this guarantees the constancy of the numbers ℓ_λ in (6) along the local brances U_λ . This is exposed in § 1. In

§ 2, we define the local mapping degree, and in § 3 the geometric multiplicity $m(\underline{X},x)$ of $(\underline{X},x) \in \underline{\underline{cpl}}_0$. In § 4, we explain the geometry of Samuel multiplicity alluded to above, and in the last paragraph we prove that the local mapping degree definition of the multiplicity of $(\underline{X},x) \in \underline{\underline{cpl}}_0$ coincides with the Samuel multiplicity $e(\mathcal{O}_{X,x})$ of the corresponding local ring.

This geometric description of multiplicity will then be put to work in the next chapter, since it is basic for geometric proofs of equimultiplicity results due to Hironaka, Lipman, Schickhoff, and Teissier.

§ 1. Compact Stein neighbourhoods

1.1. Coherent sheaves on closed subsets

Let $\underline{X}$ be a complex space and $A \subseteq X$ a closed set.

Definition 1.1.1. A coherent module on A is a sheaf of the form $\widetilde{M}|A$, where $\widetilde{M}$ is a coherent $\mathcal{O}_V$-module on some open neighbourhood V of A .

Here, $\widetilde{M}|A$ is the restriction in the sense of sheaves of abelian groups, in other words, for $U \subseteq A$ open in A , $(\widetilde{M}|A)(U)$ are the continuous sections of the "espace étalé" associated to $\widetilde{M}$ over A . It is not to be confused with the coherent $\mathcal{O}_A$-module $i^*\widetilde{M}$ if $\underline{i}$: $\underline{A} \hookrightarrow \underline{V}$ happens to be a closed complex subspace, so in this case one has to distinguish between "coherent modules on A " and "coherent $\mathcal{O}_A$-modules". Especially, we have to distinguish $\mathcal{O}|A := \mathcal{O}_X|A$ and $\mathcal{O}_A$ in this case.

Directly from the definitions and the "yoga of coherent sheaves" the following simple lemma follows:

Lemma 1.1.2. If M,N are coherent modules on A , and $\alpha : M \longrightarrow N$ is a homomorphism of $\mathcal{O}|A$-modules, then $\mathcal{K}er(\alpha)$ and $\mathcal{C}oker(\alpha)$ are coherent modules on A .

1.2. Stein subsets

In the following I assume known the simplest properties of sheaf cohomology groups for sheaves of abelian groups. They can be defined as the higher right derived functors of the section functor. On paracompact spaces they can be computed by the Čech procedure (based on alternating cochains), and on complex manifolds by the Dolbeault cohomology of (p,q)-forms (see [39], [27], [40], and [30] , at least in the locally free case).

The notion of Stein subsets is closely related to the following three statements, which have their traditional names. Let $A \subseteq X$ be a closed set in a complex space.

"Theorem A". Any coherent module on A is generated by its global sections.

"Theorem B". $H^q(A,\mathcal{M}) = 0$ for all coherent modules $\mathcal{M}$ on A and all $q \geq 1$.

"Theorem F". If $\alpha : \mathcal{M} \longrightarrow \mathcal{N}$ is a surjective homomorphism of coherent modules on A , $\alpha_A : \mathcal{M}(A) \longrightarrow \mathcal{N}(A)$ is surjective.

The long exact cohomology sequence gives immediately:

Proposition 1.2.1. Theorem B implies Theorem A and Theorem F.

Definition 1.2.2. Let X be a complex space. A closed subset $A \subseteq X$ is called a Stein subset if and only if Theorem B holds for A .

In a sense, a Stein subset should be thought of as the analogue of an affine set in the case of algebraic varieties, so there should be a correspondence between coherent modules on them and modules over the coordinate ring. For this however, we have to make an additional compactness assumption, which we do in the following section.

1.3. Compact Stein subsets and the Flatness Theorem

Let now $A = K \subseteq X$ be a compact subset. It is then easy to see that in this case the coherent modules on K are just the finitely presented $\mathcal{O}|K$-modules. Using this and standard arguments based on Proposition 1.2.1, one gets the following proposition, which states that compact Stein neighbourhoods are the appropriate analogues of the affine subsets in the algebraic case. Let $\mathcal{O}(K) := \Gamma(K,\mathcal{O}_X)$.

Proposition 1.3.1. Let X be a complex space, $K \subseteq X$ a compact Stein subset. Let $coh(K)$ be the category of coherent modules on K , and $adm(\mathcal{O}(K))$ the category of admissible, i.e. finitely presented,

$\mathcal{O}(K)$-modules. Then:

(i) $\mathcal{O}(K)$ is a coherent ring (cf. I 7.1.1. (i));

(ii) the section functor induces a natural equivalence:

(1.3.1) $$\Gamma : \mathrm{coh}(K) \longrightarrow \mathrm{adm}(\mathcal{O}(K))$$, which has

(1.3.2) $(-) \otimes_{\mathcal{O}(K)} (\mathcal{O}|K) : \mathrm{adm}(\mathcal{O}(K)) \longrightarrow \mathrm{coh}(K)$ as an inverse.

Theorem 1.3.2 (Flatness Theorem). Let K be a Stein compact subset in the complex space X. Then, for any $x \in K$, the natural morphism

(1.3.3) $$\lambda_x : \mathcal{O}(K) \longrightarrow \mathcal{O}_{X,x}$$
$$f \longmapsto f_x$$

is flat.

This follows from Proposition 1.3.1, because the section functor is exact by Theorem B, and hence so is $(-) \otimes_{\mathcal{O}(K)} (\mathcal{O}|K)$.

Remark 1.3.3. In the case where X is an algebraic variety (by this I mean an algebraic scheme of finite type over a field) and K is an affine set, the analogue of Theorem 1.3.2 is immediate, since λ_x is just the algebraic localization of $\mathcal{O}(K)$ with respect to the prime ideal corresponding to x. In this case, the local rings $\mathcal{O}_{X,x}$ are "semiglobal" in the sense that any element is a quotient of two sections defined on the whole of K. In the complex analytic case, λ_x does not arise by this simple construction, and, moreover, one has to work with compact Stein subsets, which makes the result much harder; we are going to show in the next section that sufficiently small compact Stein neighbourhoods always exist.

1.4. Existence of compact Stein neighbourhoods

The theory of Stein spaces is concerned with various criteria which characterize Stein subsets (or Stein spaces). The basic reference for this is the book [27], of which I will need only the first three chapters. Fundamental for the theory is the following Theorem 1.4.1, which goes back to Cartan and Serre; it directly implies the existence of compact Stein neighbourhoods (Corollary 1.4.2) needed for the applications of Theorem 1.3.2 in the sequel, e.g. for Definition 2.2.6 and for the proofs of Theorem 5.1.4 and Theorem 5.2.1.

A <u>compact stone</u> in $\mathbb{C}^n$ with coordinates $(z_1,\dots,z_n)$ will be a compact interval in the space $\mathbb{R}^{2n}$ with coordinates $(\operatorname{Re} z_1, \operatorname{Im} z_1, \dots, \operatorname{Re} z_n, \operatorname{Im} z_n)$.

<u>Theorem 1.4.1</u>. <u>A compact stone in</u> $\mathbb{C}^n$ <u>is a Stein subset</u>.

A detailed and clear proof of this is in Chapter III of [27]. Since the result is so basic, I give a short summary of the strategy of the proof. It is considerably more difficult than the proof of the corresponding statement for affine sets, which ultimately rests on localization of rings, a technique which one has not at its disposal in Complex Analytic Geometry, since the coherent sheaves on smaller open subsets of Stein subsets do not arise by localization. Complex analysis ultimately shows up by solving the $\bar\partial$-equation.

<u>1st Step</u>. There are two basic Vanishing Theorems for compact stones. One is elementary and uses simple combinatorical arguments on subdivisions of stones together with alternating Čech cochains to show that $\exists\, q_0 = q_0(n)$ with $H^q(Q,\mathcal{S}) = 0$ for $q \geq q_0$ and all sheaves $\mathcal{S}$ on Q. The other lies deeper and uses Dolbeault cohomology; by explicitely solving the $\bar\partial$-equation (in the so-called $\bar\partial$-Poincaré-Lemma due to Grothendieck, see [27], II, § 3) one shows that $H^q(Q,\mathcal{O}) = 0$ for $q \geq 1$. These two Vanishing Theorems show that Theorem A implies Theorem B for compact stones, and so it suffices to show Theorem A for compact stones. ([27], III, § 3.2).

<u>2nd Step</u>. Theorem A is proven by induction on the <u>real</u> dimension d of the compact stone Q. If A_d, B_d, and F_d are the statements of Theorem A, Theorem B, and Theorem F for compact stones of dimension $\leq d$, it suffices by the first step and Proposition 1.2.1. to prove

$$(1.4.1) \qquad A_{d-1} \text{ and } F_{d-1} \Rightarrow A_d .$$

3rd Step. Since sections of sheaves over a compact set extend over an open neighbourhood, one easily sees that by subdividing a one dimensional side of the d-dimensional stone Q into sufficiently small pieces the claim follows if we are able to deal with the following situation. Suppose $Q = Q^- \cup Q^+$ arises by cutting Q into two halves by a section orthogonal to a one-dimensional side (see Figure 3).

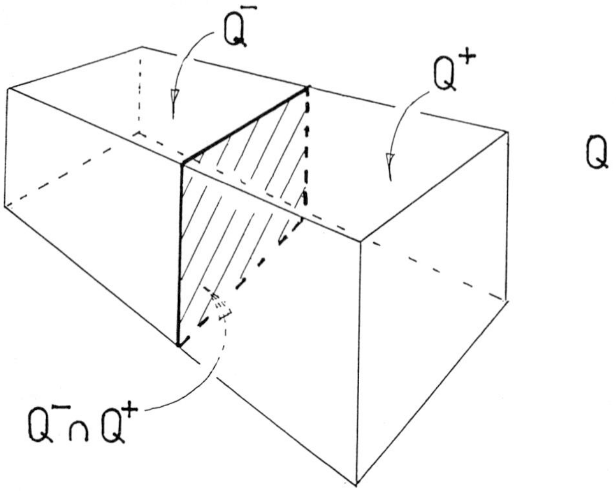

Fig. 3

Let $\mathcal{M}$ be a coherent module on Q, $\mathcal{O} := \mathcal{O}|Q$, and suppose there are given $\mathcal{O}$-module epimorphisms $h^- : \mathcal{O}^p|Q^- \twoheadrightarrow \mathcal{M}|Q^-$, $h^+ : \mathcal{O}^q|Q^+ \twoheadrightarrow \mathcal{M}|Q^+$ such that the images of h^- and h^+ generate the same subsheaf of $\mathcal{O}^p|Q^- \cap Q^+$. We then want to glue h^- and h^+ into an $\mathcal{O}$-module epimorphism $\mathcal{O}^{p+q} \twoheadrightarrow \mathcal{M}$; this will then complete step 2. Let $t_1^-,\dots,t_p^- \in \mathcal{M}(Q^-)$ and $t_1^+,\dots,t_q^+ \in \mathcal{M}(Q^+)$ be the sections defining h^- and h^+. Then one can write

$$(1.4.2) \qquad \begin{pmatrix} t_1^- \\ \cdot \\ \cdot \\ \cdot \\ t_p^- \end{pmatrix}^T \Bigg|_{Q^- \cap Q^+} = \begin{pmatrix} t_1^+ \\ \cdot \\ \cdot \\ \cdot \\ t_q^+ \end{pmatrix}^T \Bigg|_{Q^- \cap Q^+} \cdot A$$

with a matrix $A \in M(q \times p, \mathcal{O}|Q^- \cap Q^+)$. Now suppose we could find holomorphic invertible matrices $C^{\pm} \in GL(Q^{\pm}, \mathcal{O})$ such that

(1.4.3) $\quad \mathbb{1}_p \mid Q^- \cap Q^+ = (C^- \mid Q^- \cap Q^+) \cdot (C^+ \mid Q^+ \cap Q^+)$, where

$\mathbb{1}_p \in GL(p,\mathcal{O})$ is the identity matrix. This would imply

(1.4.4)
$$\begin{pmatrix} t_1^- \\ \cdot \\ \cdot \\ \cdot \\ t_p^- \end{pmatrix}^T \cdot C^- \Bigg|_{Q^- \cap Q^+} = \begin{pmatrix} t_1^+ \\ \cdot \\ \cdot \\ \cdot \\ t_q^+ \end{pmatrix}^T \Bigg|_{Q^{-1} \cap Q^+} \cdot A \cdot (C^+)^{-1} \Bigg|_{Q^- \cap Q^+}$$

So, if we then define new sections $\tilde{t}_1^-,\ldots,\tilde{t}_p^- \in M(Q^-)$ via

$$\begin{pmatrix} \tilde{t}_1^- \\ \cdot \\ \cdot \\ \cdot \\ \tilde{t}_p^- \end{pmatrix}^T := \begin{pmatrix} t_1^- \\ \cdot \\ \cdot \\ \cdot \\ t_p^- \end{pmatrix}^T \cdot \overline{C} \qquad ,$$

they still define an epimorphism $\tilde{h}^- : \mathcal{O}^p|Q^- \twoheadrightarrow M|Q^-$, since C^- is invertible. Now make the

(1.4.5) $\quad$ <u>assumption</u>: A extends over Q .

Then one could extend the sections $\tilde{t}_1^-,\ldots,\tilde{t}_p^-$ to sections $\tilde{t}_1,\ldots,\tilde{t}_p$ over Q by (1.4.4), and this would give an $\mathcal{O}$-homomorphism $\tilde{h}^- : \mathcal{O}^p \longrightarrow M$ which restricts to an epimorphism over Q^- . In the same way one would produce an $\mathcal{O}$-homomorphism $\tilde{h}^+ : \mathcal{O}^q \longrightarrow M$ which restricts to an epimorphism over Q^+ . Then $h := \tilde{h}^- \oplus \tilde{h}^+ : \mathcal{O}^{p+q} \twoheadrightarrow M$ would be the desired epimorphism.

<u>Last Step</u>. (1.4.5) does not hold in general. One has to approximate A by a holomorphic matrix $\hat{A}$ defined on Q, which can be done via an appproximation theorem of Runge; this then forces to have a decomposition (1.4.3) not only of 1_p, but of holomorphic $p \times p$ - matrices close to $\mathbb{1}_p$. That this can be done is the content of the famous Cartan Patching Lemma [27],III, § 1,3. This Lemma is, by a delicate interation procedure, reduced to an additive decomposition of holomorphic functions on an open polycylinder which itself is a union of two open polycylinders, the so-called Cousin Patching Lemma [27],III, § 1,1.

This Lemma, finally, is proven by explicitely solving the $\bar{\partial}$-equation. All details are in §§ 1 and 2 of Chapter III of [27].

Corollary 1.4.2. Let $\underline{X}$ be a complex space. Then any $x \in X$ has a neighbourhood basis consisting of compact Stein subsets. For this, one can take the compact sets in the inverse image of the system of compact stones 0 in $\mathbb{C}^n$ under any local immersion $(X,x) \overset{i}{\hookrightarrow} (\mathbb{C}^n,0)$.

Proof. Let $\underline{X} \hookrightarrow \underline{U}$ be a closed complex subspace of an open set $U \subseteq \mathbb{C}^n$, $x = 0 \in X \subseteq U$. Let K be a compact polydisc centered at 0 . Let M be a coherent module on $K \cap X$. After possibly shrinking U , we may assume M is the restriction of a coherent $\mathcal{O}_X$-module $\widetilde{M}$. Then $i_*\widetilde{M}$ is a coherent $\mathcal{O}_U$-module, and so $H^p(X \cap K, M) = H^p(K, i_*\widetilde{M}) = 0$ for $p \geq 1$, since K is Stein by Theorem 1.4.1.

§ 2. Local mapping degree

In this paragraph, I assign to each finite mapgerm $\underline{f} : (\underline{X},x) \longrightarrow (\underline{\mathbb{C}}^d,0)$ a local mapping degree $\deg_x \underline{f} \in \mathbb{N}_{>0}$, which counts the algebraic number of preimages of a "general" point of $\mathbb{C}^d$ close to 0 . This will be basic for the definition of multiplicity.

2.1. Local decomposition revisited

In order to count the number of preimages of such an $\underline{f}$ as above algebraically, I have to weight a preimage point lying on a local irreducible component where $\underline{X}$ is possibly not reduced by a certain positive number, which will appear as the value of some locally constant function along a generic subset of that component; here, I call a subset of a topological space generic if it contains an open dense subset. It is the purpose of this section to exhibit such generic subsets.

First I introduce some terminology. Let $\underline{X}$ be a complex space, $x \in X$. Define the germ $(\underline{X}_{red},x)$ as in I, 5.1.2 (iii). We then have the following loci:

(2.1.1) $$X_{reg} := \left\{ x \in X \mid (\underline{X}_{red}, x) \text{ is smooth} \right\} = \left\{ x \in X \mid \mathcal{O}_{X,x}/\mathcal{N}_x \text{ is regular} \right\}$$

(2.1.2) $$X_{ir} := \left\{ x \in X \mid (\underline{X}, x) \text{ is irreducible} \right\} = \left\{ x \in X \mid \mathcal{O}_{X,x}/\mathcal{N}_x \text{ is an integral domain} \right\} .$$

Obviously,

(2.1.3) $$X_{reg} \subseteq X_{ir} \quad .$$

Now let $(\underline{X},x) \in \underline{\underline{\mathrm{cpl}}}_0$, and let $\underline{X}$ be a <u>good representative,</u> i.e. there should be a finite map from $\underline{X}$ to $\underline{B}$, a domain in $\mathbb{C}^d$ satisfying the Local Representation Theorem I 6.3.1. Let

(2.1.4) $$X = \bigcup_{\lambda \in \Lambda} X_\lambda$$

be the local decomposition of $(\underline{X},x)$ into irreducible components as in I 5.4. This decomposition has the following properties:

<u>Proposition 2.1.1.</u> <u>There are arbitrarily small good representatives</u> $\underline{X}$ <u>such that the following statements hold</u>:

(i) $X_\lambda \cap X_\mu$ <u>is nowhere dense in</u> X_λ <u>for all</u> $\lambda \in \Lambda$ <u>and all</u> $\mu \in \Lambda$ <u>with</u> $\mu \neq \lambda$.

(ii) X <u>is locally reducible at all points of</u> $\bigcup_{\substack{\lambda,\mu\in\Lambda \\ \lambda\neq\mu}} (X_\lambda \cap X_\mu)$.

<u>Proof</u>.
(i) is just I 7.3.4., and (ii) follows from (i) and elementary properties of the local decomposition of analytic sets (see [28] , p. 108).

Corollary 2.1.2. Let the notations be as in Proposition 2.1.1. Put

(2.1.5) $$X_\lambda^0 := X - \bigcup_{\mu\neq\lambda} X_\mu .$$

Then, for all $\lambda \in \Lambda$:

(2.1.6) X_λ^0 is connected, open and dense in X_λ , and open in X ;

(2.1.7) $X_\lambda \cap X_{ir} = (X_\lambda^0)_{ir}$ is connected, and this set is generic in X_λ ;

(2.1.8) $$X_{ir} = \coprod_{\lambda\in\Lambda} (X_\lambda \cap X_{ir}) .$$

Proof. X_λ^0 is clearly open both in X and X , since $\bigcup_{\substack{\mu \\ \mu\neq\Lambda}} X_\mu$ is closed as a finite union of analytic sets. It is dense by Proposition 2.1.1 (i). Let $\underline{f}_\lambda : \underline{X}_\lambda \longrightarrow \underline{B}_\lambda$ satisfy the assumption of the Local Representation Theorem I 6.3.1. So, after possibly shrinking $\underline{f}_\lambda$, $\underline{f}_\lambda$ is open by the Open Mapping Lemma I 7.3.3, and therefore $f_\lambda^{-1}(\Delta_\lambda)$ is nowhere dense in X_λ , as Δ_λ is nowhere dense in B_λ . This shows that $X_\lambda - f_\lambda^{-1}(\Delta_\lambda)$ is open and dense in X_λ , and it is connected by I 6.3.4. Since $X_\lambda - f_\lambda^{-1}(\Delta_\lambda) \subseteq X_\lambda^0$ for some $\Delta_\lambda \subseteq X_\lambda^0$, this shows X_λ^0 is connected, and dense in X_λ . Finally, $X_\lambda \cap X_{ir} = (X_\lambda^0)_{ir}$ follows from Proposition 2.1.1 (ii), and so $X_\lambda \cap X_{ir}$, containing $X_\lambda - f_\lambda^{-1}(\Delta_\lambda)$, is generic in X_λ , and connected. (2.1.8) finally is obvious from $X = \bigcup_{\lambda\in\Lambda} X_\lambda$. •

Remark 2.1.3. One has, again by Proposition 2.1.1 (ii), that $X_\lambda \cap X_{reg} = (X_\lambda^0)_{reg}$, and that $(X_\lambda^0)_{reg}$, containing $X_\lambda - f_\lambda^{-1}(\Delta_\lambda)$, is generic in X_λ . Using the Jacobian criterion for regularity one may show it is the complement of a nowhere dense analytic set in X_λ . It follows that $X_{reg} = \coprod_{\lambda\in\Lambda} (X_\lambda^0)_{reg}$ is the complement of a nowhere dense analytic set in X . This implies that for any $\underline{X} \in \underline{\text{cpl}}$ the locus X_{reg} is also the complement of a nowhere dense analytic set.

Remark 2.1.4. Using the local results above, one can show the following. Let $\underline{X}$ be any complex space. Decompose X_{reg} into connected components:

$$X_{reg} = \coprod_{\lambda \in \Lambda} Z_\lambda$$

and put $X_\lambda := \overline{Z_\lambda}$. The decomposition

$$X = \bigcup_{\lambda \in \Lambda} X_\lambda$$

then will satisfy Corollary 2.1.2. Moreover, this decomposition is unique and characterized by the fact that it is a decomposition of X into irreducible analytic sets, i.e. analytic sets which cannot be written as a proper union of analytic sets. We call this decomposition the decomposition of X into (global) irreducible components. Locally this decomposition induces the decomposition given by the local decomposition into irreducible analytic setgerms. (See [40], § 49). So in the local situation above, the decomposition (2.1.4) is indeed the decomposition into global irreducible components and we will call it so, but we will make use only of the properties in Corollary 2.1.2.

2.2. Local mapping degree

We first introduce the weights with which to count preimage points. Let R be a noetherian ring, Ac(R) the set of active elements. Since

$$(2.2.1) \qquad \mathrm{Ac}(R) = \bigcap_{\mathfrak{p} \in \mathrm{Min}(R)} (R - \mathfrak{p})$$

by I 5.2.1, Ac(R) is a multiplicative subset, and we can form the localization of R with respect to Ac(R) .

Definition 2.2.1. $\widetilde{\mathrm{Quot}}(R) := (\mathrm{Ac})^{-1}R$ is called the modified ring of fractions of R .

Lemma 2.2.2. $\widetilde{\mathrm{Quot}}(R)$ has the following properties:

(i) $\widetilde{\text{Quot}}(R)$ is artinian, and $\text{length}(\widetilde{\text{Quot}}(R)) = \sum_{\mathfrak{p}\in\text{Min}(R)} \text{length}(R_{\mathfrak{p}})$;

(ii) if R has no embedded primes, $\widetilde{\text{Quot}}(R) = \text{Quot}(R)$, the usual total ring of fractions of R .

Proof. (i): All primes of $\widetilde{\text{Quot}}(R)$ are minimal by construction, so $\widetilde{\text{Quot}}(R)$ is artinian. By the well-known structure of artinian rings (see [6], Chapter IV, § 2.5, Corollary 1 of Proposition 9).

$$S := \widetilde{\text{Quot}}(R) \cong \prod_{\tilde{\mathfrak{p}}\in\text{Min}(S)} S_{\tilde{\mathfrak{p}}} = \prod_{\mathfrak{p}\in\text{Min}(R)} R_{\mathfrak{p}} ,$$

and so $\text{length}(\widetilde{\text{Quot}}(R)) = \sum_{\mathfrak{p}\in\text{Min}(R)} \text{length}(R_{\mathfrak{p}})$.

(ii): In this case, $\text{Ac}(R) = R - \bigcup_{\mathfrak{p}\in\text{Ass}(R)} \mathfrak{p}$ is the set of nonzero-divisors of R .

Proposition 2.2.3. Let X be a complex space, $X = \bigcup_{\lambda\in\Lambda} X_\lambda$ the decomposition into irreducible components. Then for any $x \in X_{ir}$ the modified ring of fractions $\widetilde{\text{Quot}}(\mathcal{O}_{X,x})$ is of finite length, and the function $x \longmapsto \text{length}(\widetilde{\text{Quot}}(\mathcal{O}_{X,x}))$ is, for each λ , constant along the generic subset $X_\lambda \cap X_{ir}$ of X_λ .

Proof. $\widetilde{\text{Quot}}(\mathcal{O}_{X,x})$ is artinian by Lemma 2.2.2, so is of finite lenght. Since $x \in X_{ir}$, $\widetilde{\text{Quot}}(\mathcal{O}_{X,x}) = (\mathcal{O}_{X,x})_{N_x}$. So, because of (2.1.7), it suffices to prove that the function $x \longmapsto \text{length}((\mathcal{O}_{X,x})_{N_x})$ is locally constant. Let $x \in X_{ir}$ and fix a compact Stein neighbourhood K of x according to Corollary 1.4.2. From the construction there one sees that one can take K so that it has a fundamental system of open neighbourhoods $(U_\alpha)_{\alpha\in A}$ such that each U_α is irreducible and $U_\alpha \subseteq X_\lambda$, where λ is the unique $\mu \in \Lambda$ such that $x \in X^0_\mu$ by (2.1.8). Since $x \in X^0_\lambda$, and X_λ is open in X , we may, replacing X by a small open subspace contained in X^0_λ , forget about λ and assume $X = X_\lambda$. Now, by I Corollary 7.4.2, X has the structure of a complex space $\underline{X}_{red}$ by putting $\mathcal{O}_{X_{red}} := \mathcal{O}_X/N_X$. Let N be the $\mathcal{O}(K)$-ideal $N_X(K) = \Gamma(K,N_X)$. I claim N is prime. Since the section functor is exact by Proposition 1.3.1 (ii) (or Theorem B),

$$\Gamma(K,\mathcal{O}_{X_{red}}) = \Gamma(K,\mathcal{O}_X) \,/\, \Gamma(K,\mathcal{N}_X) \quad .$$

But

$$\Gamma(K,\mathcal{O}_{X_{red}}) = \varinjlim_{\alpha\in A} \Gamma(U_\alpha,\mathcal{O}_{X_{red}}) \quad ,$$

and the $\Gamma(U_\alpha,\mathcal{O}_{X_{red}})$ are integral domains because the U_α are irreducible, so $\Gamma(K,\mathcal{O}_{X_{red}})$ is an integral domain, and N is indeed prime. Now the natural morphism

$$(2.2.2) \qquad \lambda_{x'} : \Gamma(K,\mathcal{O}_X) \longrightarrow \mathcal{O}_{X,x'}$$

is flat for all $x' \in K \cap X_{ir}$ by Theorem 1.3.2. The ideal N generates in $\mathcal{O}_{X,x'}$ the ideal $\mathcal{N}_{x'}$ via $\lambda_{x'}$ because of Proposition 1.3.1. Localizing (2.2.2) at N gives that

$$(2.2.3) \qquad (\lambda_{x'})_N : \Gamma(K,\mathcal{O}_X)_N \longrightarrow (\mathcal{O}_{X,x'})_{\mathcal{N}_{x'}}$$

is flat, since flatness localizes. Hence (2.2.3) is faithfully flat, being a flat local morphism of local rings. Pushing composition series of $\Gamma(K,\mathcal{O}_X)_N$ to $(\mathcal{O}_{X,x'})_{\mathcal{N}_{x'}}$ then shows by standard arguments

$$(2.2.4) \qquad \text{length}((\mathcal{O}_{X,x'})_{\mathcal{N}_{x'}}) = \text{length}(\Gamma(K,\mathcal{O}_X)_N)$$

(see the following Lemma 2.2.4). But the right hand side does not depend on x' , and this shows the Proposition.

From the literature, I cite the following lemma.

Lemma 2.2.4. ([31], Chapter O, Corollary (6.6.4)). *Let* $\rho : A \longrightarrow B$ *be a local flat homomorphism of local rings,* M *an A-module. Then*

$$\text{length}_B(M \otimes_A B) = \text{length}_A(M)\cdot\text{length}(B/m_A B)$$

<u>in the sense that the left side is finite if and only if the right hand side is finite, and then the equality holds</u>.

We now consider finite mapgerms $\underline{f} : (\underline{X},x) \longrightarrow (\mathbb{C}^d,0)$ and choose a good <u>representative</u> $\underline{f} : \underline{X} \longrightarrow \underline{B}$, which here is defined to mean

(i) B is a domain in $\mathbb{C}^d$;

(ii) if $\dim_x \underline{X} < d$, we choose $\underline{f} : \underline{X} \longrightarrow \underline{B}$ so small that $\dim_{x'} X < d$ for all $x' \in X$ (which can be done by I 5.1.2, (iv)); put $\underline{\Delta} := \underline{\text{im}}(\underline{f})$ (then Δ is nowhere dense in B);

(iii) if $\dim_x \underline{X} = d$, $\underline{f}$ should have the properties of the Local Representation Theorem I 6.3.1 ;

(iv) Proposition 2.1.1 and Corollary 2.1.2 hold for $\underline{X}$.

Note that always $\dim_x \underline{X} \leq d$ by I 5.1.2, (iv), and that we may take good representatives to be arbitrarily small, i.e. we are allowed to shrink them when necessary.

<u>Proposition 2.2.5. Let $\underline{f} : \underline{X} \longrightarrow \underline{B}$ be a good representative for the finite mapgerm $\underline{f} : (\underline{X},x) \longrightarrow (\mathbb{C}^d,0)$ in $\underline{\underline{\text{cpl}}}_0$ with discriminant locus $\underline{\Delta}$. Then the number $\sum_{x' \in f^{-1}(y)} \text{length}(\widetilde{\text{Quot}}(\mathcal{O}_{X,x'}))$ does not depend on the choice of $y \in B - \Delta$.</u>

<u>Proof</u>. Let $y \in B - \Delta$. Then $X - f^{-1}(\Delta) \subseteq X_{ir}$, and so all the $x' \in f^{-1}(y)$ are in X_{ir} . The claim then follows from the fact that $f : X - f^{-1}(\Delta) \longrightarrow B - \Delta$ is a covering map and from Proposition 2.2.3.

I can now make the main definition:

<u>Definition 2.2.6. Let $\underline{f} : (\underline{X},x) \longrightarrow (\mathbb{C}^d,0)$ be a finite mapgerm in $\underline{\underline{\text{cpl}}}_0$, and $\underline{f} : \underline{X} \longrightarrow \underline{B}$ be a good representative with discriminant</u>

locus Δ. Then the well-defined number,

$$\deg_x \underline{f} := \sum_{x' \in f^{-1}(y)} \operatorname{length}(\widetilde{\operatorname{Quot}}(\mathcal{O}_{X,x'})) \quad ,$$

y any point in $B - \Delta$, is called the local mapping degree of the germ $\underline{f}$.

Remark 2.2.7. Since $\operatorname{length}(\widetilde{\operatorname{Quot}}(\mathcal{O}_{X,x}))$ may be difficult to compute, one hopes for a nicer formula. In fact, one may show that, in the situation of Definition 2.2.6, one can find a nowhere dense subspace $\Delta' \subseteq B$ such that $\underline{X} - f^{-1}(\Delta')$ is Cohen-Macaulay at all x lying over $B - \Delta'$ (see Theorem 2.2.11); consequently

$$\deg_x \underline{f} = \sum_{x' \in f^{-1}(y)} \operatorname{length}(\operatorname{Quot}(\mathcal{O}_{X,x}))$$

$$= \sum_{x' \in f^{-1}(y)} \dim_{\mathbb{C}} (\mathcal{O}_{X,x'}/\mathfrak{m}_y \cdot \mathcal{O}_{X,x'}) \quad ,$$

for all $y \in B - \Delta'$, where $\mathfrak{m}_y$ is the maximal ideal of $\mathcal{O}_{\mathbb{C}^d,y}$.

We have the following simple but important fact:

Theorem 2.2.8 (Degree Formula). Let $\underline{f} : \underline{X} \longrightarrow \underline{B}$ be as in Definition 2.2.6. Then

$$\deg_x \underline{f} = \sum_{x' \in f^{-1}(y)} \deg_{x'} \underline{f}$$

for all $y \in B$.

This follows from the geometry of Definition 2.2.6. An algebraic proof will appear below, cf. 5.1.7. Theorem 2.2.8. has the important application that multiplicity will be upper semi-continuous along complex spaces, see Theorem 5.2.4.

Exercise 2.2.9. In the situation of Defintion 2.2.6

(2.2.5) $$\deg_x \underline{f} = \sum_{x' \in f^{-1}(y)} \dim_{\mathbb{C}} (\mathcal{O}_{X,x'}/m_y \cdot \mathcal{O}_{X,x'})$$

for $y \in B - \Delta$ and Δ a suitable nowhere dense analytic set in B.

For this, proceed as follows:

(i) Show by means of Fitting ideals that for an admissible module M on a reduced complex space Y the set $LF(M) := \{y \in Y | M$ is locally free at $y\}$ is the complement of a nowhere dense analytic set (cf.[28], Chapter 4, § 4).

(ii) Let now $\underline{f}$ be as in Definition 2.2.6; choose Δ in such a way that $f_* \mathcal{O}_X$ is locally free on $B - \Delta$.

Exercise 2.2.10. Use 2.2.9 (ii) to prove the following

Theorem 2.2.11. Let $\underline{X}$ be a complex space. Then the Cohen-Macaulay-locus $X_{CM} := \{x \in \underline{X} | \mathcal{O}_{X,x}$ is Cohen-Macaulay$\}$ is the complement of a nowhere dense analytic set.

What is with the smooth locus $X_{sm} := \{x \in X | \mathcal{O}_{X,x}$ is regular$\}$?

§ 3. Geometric multiplicity

We now use the notion of the local mapping degree of a finite mapgerm to define the geometric multiplicity $m(\underline{X},x)$ of a complex space germ $(\underline{X},x) \in \underline{\underline{cpl}}_0$.

Geometric multiplicity in the reduced case is discussed in [13], [38], [61], [70] and [75].

3.1. The tangent cone.

Let $(\underline{X},x) \in \underline{\underline{\mathrm{cpl}}}_0$, and $\mathrm{gr}_{\mathfrak{m}_x}(\mathcal{O}_{X,x}) := \bigoplus_{k\geq 0} \mathfrak{m}_x^k/\mathfrak{m}_x^{k+1}$, which is a finitely generated $\mathbb{C}$-algebra. Recall the notion of the analytic spectrum of a finitely generated $\mathbb{C}$-algebra in I 3.4.

Definition 3.1.1. $\underline{C}(\underline{X},x) := \underline{\mathrm{Specan}}(\mathrm{gr}_{\mathfrak{m}_x}(\mathcal{O}_{X,x}))$, the tangent cone of $(\underline{X},x) \in \underline{\underline{\mathrm{cpl}}}_0$.

To describe it in a more concrete way, choose generators $f_1,\ldots,f_n$ of $\mathfrak{m}_x$, i.e. an embedding $(\underline{X},x) \hookrightarrow (\mathbb{C}^n,0)$ by I 4.4.2. This gives a surjection

$$\varphi : \mathbb{C}[z_1,\ldots,z_n] = \mathrm{gr}_{\mathfrak{m}_n}(\mathcal{O}_{\mathbb{C}^n,0}) \twoheadrightarrow \mathrm{gr}_{\mathfrak{m}_x}(\mathcal{O}_{X,x}) \quad ,$$

and so $\underline{C}(\underline{X},x)$ is defined in $\underline{\mathbb{C}}^n$ by the homogeneous ideal $\mathrm{Ker}(\varphi)$, hence is a cone. If the ideal $I \subseteq \mathcal{O}_{\mathbb{C}^n,0}$ defines $(\underline{X},x)$, one can show that $\mathrm{Ker}(\varphi) = L(I)$, the ideal generated by the leitforms $L(f)$ of all the $f \in I$. So if I is generated by finitely many polynomials, the standard base algorithm discussed in I Remark 2.4.4, gives finitely many equations which define $\underline{C}(\underline{X},x)$.

Proposition 3.1.2. $\mathrm{Dim}_x\,\underline{C}(\underline{X},x) = \dim_x \underline{X} = \dim \mathrm{gr}_{\mathfrak{m}_x}(\mathcal{O}_{X,x})$.

Proof. A geometric proof is somewhat involved (see Proposition 3.1.3 (iii) below), so we use the elementary properties of dimension of local rings. Now $\mathrm{gr}_{M_x^+}(\mathcal{O}_{C(\underline{X},x),x}) = \mathrm{gr}_{\mathfrak{m}_x}(\mathcal{O}_{X,x})$, where M_x^+ is the irrelevant maximal ideal of $\mathrm{gr}_{\mathfrak{m}_x}(\mathcal{O}_{X,x})$. Since these two rings have the same Hilbert function, the result follows from the well-known main result of dimension theory of local rings (see e.g. [1], Theorem 11.14.) and the fact that this Hilbert function is just the Hilbert function of $\mathcal{O}_{X,x}$.

We now shortly touch upon another, more geometric description of the tangent cone, which puts it into a flat deformation of $(\underline{X},x)$; this appears in [45], [70], and is a special case of Fulton's and Macpherson's "deformation to the normal cone" (see [17] for the algebraic case; the analytic case is analogous):

Let $(\underline{X},x) \hookrightarrow (\mathbb{C}^n,0)$ be defined by the ideal $I \subseteq \mathcal{O}_{\mathbb{C}^n,0}$. For $f \in I$, let $f^* \in \mathcal{O}_{\mathbb{C}^n\times\mathbb{C},0}$ be defined by

$$f^*(z,t) := \frac{1}{t^{\nu(f)}} \cdot f(tz) \quad ,$$

where $\mathbb{C}^n$ has coordinates z and $\mathbb{C}$ has coordinate t , and $\nu(f)$ is the order of f (I(1.1.3)). Let $I^* \subseteq \mathcal{O}_{\mathbb{C}^n\times\mathbb{C},0}$ be the ideal generated by the f^* for $f \in I$. It defines a germ $(\underline{X},0) \hookrightarrow (\underline{\mathbb{C}}^n\times\underline{\mathbb{C}},0)$, and the projection $\underline{\mathbb{C}}^n\times\underline{\mathbb{C}} \longrightarrow \underline{\mathbb{C}}$ defines a morphism $\underline{p} : (\underline{X},0) \longrightarrow (\underline{\mathbb{C}},0)$ and so $\underline{p} : \underline{X} \longrightarrow \underline{B}$, where $\underline{B} \subseteq \underline{\mathbb{C}}$ is an open disk around 0 (in fact, it is easy to see that $\underline{p}$ is defined over $\underline{\mathbb{C}}$). Then the following statements do hold:

<u>Proposition 3.1.3</u> (Deformation to the tangent cone).

(i) $(\underline{p}^{-1}(t),(0,t)) \cong (\underline{X},x)$ <u>for all</u> $t \neq 0$.

(ii) $(\underline{p}^{-1}(0),(0,0)) \cong (\underline{C}(\underline{X},x),x)$.

(iii) $p_X^0(t-p(x))$ <u>is a nonzerodivisor in</u> $\mathcal{O}_{X,x}$ <u>for all</u> $x \in X$, <u>and so</u> $\underline{p}$ <u>is flat; especially</u> $\dim_x \underline{C}(\underline{X},x) = \dim_x \underline{X}$.

(iv) $\overline{X - p^{-1}(0)} = X$.

<u>Corollary 3.1.4.</u>

$$C(\underline{X},x) = \bigcup \left\{ \ell \mid \ell = \lim_{\substack{x\to x' \\ x\neq x'}} \overline{xx'} \right\} ,$$

where $\overline{xx'}$ is the complex line through x and x', and the limit is taken in $\mathbb{P}^{n-1}$.

In other words, settheoretically is $C(\underline{X},\underline{x})$ the union of limits of secants of $\underline{X}$ through x, whence the name "tangent cone".

3.2. Multiplicity

Let now $(\underline{X},x) \in \underline{\underline{cpl}}_0$, $d := \dim_x \underline{X}$. We fix generators $f_1,\ldots,f_n \in \mathfrak{m}_x$, so an embedding $(\underline{X},x) \hookrightarrow (\underline{\mathbb{C}}^n,0)$, and so an embedding $\underline{C}(\underline{X},x) \hookrightarrow \underline{\mathbb{C}}^n$ as in 3.1. Note that $d = n$ implies $(\underline{X},x) \cong (\underline{\mathbb{C}}^n,0)$ by I 4.4.2. We now consider finite linear projections of $(\underline{X},x)$ onto $(\underline{\mathbb{C}}^d,0)$.

Definition 3.2.1. Let $\mathrm{Grass}^d(\mathbb{C}^n)$ denote the Grassmannian of d-codimensional linear subspaces $L \subseteq \mathbb{C}^n$ (see e.g. [30], Chapter 1, Section 5). Let $(\underline{X},x) \in \underline{\underline{cpl}}_0$. Then $L \in \mathrm{Grass}^d(\mathbb{C}^n)$ is called good for $(\underline{X},x)$ if and only if x is isolated in $L \cap X$, and excellent for $(\underline{X},x)$ if and only if it is good for $(C(\underline{X},x),x)$, i.e. $L \cap C(\underline{X},x) = \{x\}$.

We put

$$P_g^d(\underline{X},x) := \left\{L \in \mathrm{Grass}^d(\mathbb{C}^n) \mid L \quad \text{good for} \quad (\underline{X},x)\right\} \quad , \tag{3.2.1}$$

$$P_e^d(\underline{X},x) := \left\{L \in \mathrm{Grass}^d(\mathbb{C}^n) \mid L \quad \text{excellent for} \quad (\underline{X},x)\right\} \quad , \tag{3.2.2}$$

and use the notations

$$L \pitchfork_x X \; :\Longleftrightarrow \; L \in P_g^d(\underline{X},x) \tag{3.2.3}$$

$$L \pitchfork_x C(\underline{X},x) :\Longleftrightarrow \; L \in P_e^d(\underline{X},x) \quad . \tag{3.2.4}$$

If $L \in \mathrm{Grass}^d(\mathbb{C}^n)$, choose coordinates $(z_1,\ldots,z_n)$ so that L is $\mathbb{C}^{n-d}$ with coordinates $(z_{d+1},\ldots,z_n)$; then the projection $\pi_L : \mathbb{C}^n \longrightarrow \mathbb{C}^d$ along L defines the linear projection $\underline{p}_L := \underline{\pi}_L \mid (\underline{X},x) : (\underline{X},x) \longrightarrow (\underline{\mathbb{C}}^d,0)$. Then Corollary I 4.3.6 immediately implies

Proposition 3.2.2. If $L \in P^d_g(X,x)$, p_L is finite.

We now show that there is an ample supply of these finite projections p_L .

For this, we exploit the transversality condition algebraically; the following observation seems to be due to Lipman [49], see also [69].

Let $\underline{f} : (\underline{X},x) \longrightarrow (\underline{Y},y)$ be a mapgerm; then $\underline{f}$ induces $\mathrm{gr}_{\mathfrak{m}}(f^0_x) : \mathrm{gr}_{\mathfrak{m}_y}(\mathcal{O}_{Y,y}) \longrightarrow \mathrm{gr}_{\mathfrak{m}_x}(\mathcal{O}_{X,x})$, so by localizing at the irrelevant maximal ideal a homomorphism $\mathcal{O}_{\underline{C}(\underline{Y},y),y} \longrightarrow \mathcal{O}_{\underline{C}(\underline{X},x),x}$, and hence a mapgerm

$$d_x\underline{f} : (\underline{C}(\underline{X},x),x) \longrightarrow (\underline{C}(\underline{Y},y),y)$$

called the differential of f at x .

Proposition 3.2.3. Let $\underline{f} : (\underline{X},x) \longrightarrow (\underline{\mathbb{C}}^d,0)$ be a mapgerm, $d = \dim_x \underline{X}$. The following conditions are equivalent:

(i) $d_x\underline{f} : (\underline{C}(\underline{X},x),x) \longrightarrow (\mathbb{C}^d,0)$ is finite ;

(ii) the ideal $\mathfrak{q}_x := f^0_x(\mathfrak{m}_d)\cdot\mathcal{O}_{X,x}$ is a minimal reduction of $\mathfrak{m}_x$.

In particular, then, $\underline{f} : (\underline{X},x) \longrightarrow (\underline{\mathbb{C}}^d,0)$ is finite.

Proof. Let $\underline{f}$ be defined by $f_1,\ldots,f_d \in \mathfrak{m}_x$. To simplify notation, let $G := \mathrm{gr}_{\mathfrak{m}_x}(\mathcal{O}_{X,x})$, and let $M^+ \subseteq G$ be the irrelevant maximal ideal, $M^+ := \bigoplus_{k>0} G_k$. Let f_j^* be the image of f_j in $G_1 = \mathfrak{m}_x/\mathfrak{m}_x^2$, $j = 1,\dots,d$, and $Q := (f_1^*,\dots,f_d^*)\cdot G$. Let $\mathfrak{q}_x^* := (f_1^*,\dots,f_d^*)\cdot \mathcal{O}_{C(\underline{X},x),x}$. Consider the injections

$$G/Q \underset{\varphi}{\hookrightarrow} (G/Q)_{M^+\cdot(G/Q)} \underset{\psi}{\hookrightarrow} \mathcal{O}_{C(\underline{X},x),x} / \mathfrak{q}_x^* \quad .$$

If $\dim_{\mathbb{C}}(\mathcal{O}_{C(\underline{X},x),x} / \mathfrak{q}_x^*) < \infty$, it follows that $\dim_{\mathbb{C}}(G/Q) < \infty$. Conversely, if $\dim_{\mathbb{C}}(G/Q) < \infty$, G/Q is artinian, and so, since it is graded, must be local, so that in fact φ is an isomorphism. Now ψ is faithfully flat by (4.1.3), and so by Lemma 2.2.4 we get $\dim_{\mathbb{C}}((G/Q)_{M^+\cdot(G/Q)}) = \dim_{\mathbb{C}}(\mathcal{O}_{C(\underline{X},x),x} / \mathfrak{q}_x^*)$. Consequently, $\dim_{\mathbb{C}}(\mathcal{O}_{C(\underline{X},x),x} / \mathfrak{q}_x^*) = \dim_{\mathbb{C}}(G/Q)$ hence is finite. It follows that $\dim_{\mathbb{C}}(\mathcal{O}_{C(\underline{X},x),x} / \mathfrak{q}_x^*) < \infty$ is equivalent to $\dim_{\mathbb{C}}(gr_{\mathfrak{m}}(\mathcal{O}_{X,x})/Q) < \infty$. But the first inequality means $d_x\underline{f}$ is finite by the Integrality Theorem I 4.4.1, and the second one that $\mathfrak{q}$ is a minimal reduction of $\mathfrak{m}_x$ by Chapter II, Theorem (10.14) and Corollary (10.15). Especially, $\mathfrak{q}_x$ is $\mathfrak{m}$-primary, and so $\dim_{\mathbb{C}}\mathcal{O}_{X,x} / \mathfrak{q}_x < \infty$, whence $\underline{f}$ is finite by the Integrality Theorem I 4.4.1.

We now get:

<u>Proposition 3.2.4.</u>

(i) <u>If</u> $L \in \mathrm{Grass}^d(\mathbb{C}^n)$, $L \pitchfork_x C(\underline{X},x)$ <u>implies</u> $L \pitchfork_x X$, <u>and so</u> $P_e^d(\underline{X},x) \subseteq P_g^d(\underline{X},x)$.

(ii) $P_e^d(\underline{X},x)$, <u>and so a fortiori</u> $P_g^d(\underline{X},x)$, <u>is generic in</u> $\mathrm{Grass}^d(\mathbb{C}^n)$.

<u>Proof</u>.

(i) is direct from Proposition 3.2.3.

(ii) We may assume $1 \leq d \leq n-1$. Put

$$R := \left\{ (L,\ell) \in \mathrm{Grass}^d(\mathbb{C}^n) \times \mathbb{P}^{n-1} \mid \ell \subseteq L \right\} ,$$

where $\mathbb{P}^k := \mathbb{P}^k(\mathbb{C})$ denotes complex projective k-space. We have the diagram

$$\begin{array}{ccc} R & \xrightarrow{\ q\ } & \mathbb{P}^{n-1} \\ {\scriptstyle p}\downarrow & & \\ \mathrm{Grass}^d(\mathbb{C}^n) & & \end{array} \qquad ,$$

where p,q are projections. We now use some elementary Algebraic Geometry, as e.g. in the first chapter of [56]. This is a diagram of algebraic varieties and algebraic morphisms, and p is proper. Let $\mathbb{P}C(\underline{X},x) \subseteq \mathbb{P}^{n-1}$ be the projective tangent cone. It follows that $p(q^{-1}(\mathbb{P}C(\underline{X},x)))$ is an analytic, even algebraic, set in $\mathrm{Grass}^d(\mathbb{C}^n)$, and so is either nowhere dense in $\mathrm{Grass}^d(\mathbb{C}^n)$ or coincides with it, since $\mathrm{Grass}^d(\mathbb{C}^n)$ is a connected, smooth, and hence irreducible, variety. But equality $p(q^{-1}(\mathbb{P}C(\underline{X},x))) = \mathrm{Grass}^d(\mathbb{C}^n)$ means that any d-codimensional plane in $\mathbb{P}^{n-1}$ hits $\mathbb{P}C(\underline{X},x) \subseteq \mathbb{P}^{n-1}$, which cannot be since it has projective dimension $d-1$ by Proposition 3.1.2. Finally, note that $P_e^d(\underline{X},x) = \mathrm{Grass}^d(\mathbb{C}^n) - p(q^{-1}(\mathbb{P}C(\underline{X},x)))$, which implies (i) .

Remark 3.2.5. The inclusion $P_e^d(\underline{X},x) \subseteq P_g^d(\underline{X},x)$ says that if $L \in \mathrm{Grass}^d(\mathbb{C}^n)$ has $\dim_x L \cap X \geq 1$ there should be a line $\ell \subseteq L \cap C(\underline{X},x)$, which is intuitively clear, since $\dim_x L \cap X \geq 1$ tells us there are secants $\overline{xx'} \subseteq L$ with $x' \neq x$ arbitrarily close to x . So a geometric proof could be based on Corollary 3.1.4, for which, however, I did not give a complete proof. The existing geometric proofs of Proposition 3.2.4 (i) in the literature ([13], [75]) are somewhat involved. Proposition 3.2.4 (ii) is also in [75] (Chapter 7, Lemma 7N).

We are now ready for the definition of multiplicity.

Definition 3.2.6 (Geometric multiplicity). Let $(\underline{X},x) \in \underline{\underline{\mathrm{cpl}}}_0$, $d := \dim_x X$. Fix generators $f_1,\ldots,f_n$ of m_x , i.e. an embedding $(\underline{X},x) \hookrightarrow (\mathbb{C}^n,0)$. The geometric multiplicity $m(\underline{X},x)$ of $(\underline{X},x)$ (with respect to this embedding) is defined to be

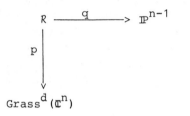

$$m(\underline{X},x) := \min_{L \in \mathcal{P}^d_g(\underline{X},x)} \left\{ \deg_x p_L \right\} .$$

Proposition 3.2.4 (ii) implies that this definition is not empty. $m(\underline{X},x)$ depends a priori on the embedding. It will be shown algebraically in Theorem 5.2.1 that this is not so.

Exercise 3.2.7. (i) Prove by geometric means that $m(\underline{X},x)$ depends only on the isomorphism class of $(\underline{X},x)$ in $\underline{\underline{cpl}}_0$.

Hints: First show that it suffices to compare embeddings of equal dimensions; here (2.2.5) might be of use. Then use Proposition I 3.2.1, to show that

$$(3.2.5) \qquad m(\underline{X},x) = \min_{\substack{\underline{f}:(\underline{X},x) \to (\mathbb{C}^d,0) \\ \underline{f} \text{ finite}}} \left\{ \deg_x \underline{f} \right\} .$$

(ii) Conclude that $m(\underline{X},x) = 1$ when $(\underline{X},x)$ is smooth. Show that conversely $(\underline{X},x)$ is smooth when $(\underline{X},x)$ is equidimensional and $m(\underline{X},x) = 1$ (Criterion of multiplicity one).

Hints: For the converse prove that a finite extension $\mathcal{O}_{\mathbb{C}^d,0} \hookrightarrow \mathcal{O}_{Y,y}$ of degree one, where $\mathcal{O}_{Y,y}$ is an integral domain, is surjective. For this, use the Local Representation Theorem I 6.3.1 and the classical Riemann Extension Theorem (see I Remark 6.3.3).

Example 3.2.8. If $L \notin \mathcal{P}^d_{e_2}(\underline{X},x)$, it can happen that $\deg_x \underline{p}_L \neq m(\underline{X},x)$. For instance, let $\underline{X} \hookrightarrow \underline{\mathbb{C}}^2$ be defined by $z_1 - z_2^2 = 0$, $L :=$ the z_2-axis, $x = 0$. Then $m(\underline{X},x) = 1$ by Exercise 3.2.7 (ii) above, but $\deg_x \underline{p}_L = 2$. However, $L \in \mathcal{P}^d_e(\underline{X},x)$ will imply $\deg_x \underline{p}_L = m(\underline{X},x)$. See Theorem 5.2.1.

§ 4. The geometry of Samuel multiplicity

The purpose of this paragraph is to give a geometric interpretation of the Samuel multiplicity $e(\mathfrak{q},\mathcal{O}_{X,x})$ of an $\mathfrak{m}_x$-primary ideal $\mathfrak{q}$ in the local analytic $\mathbb{C}$-algebra $\mathcal{O}_{X,x}$; it will turn out to be canonically the geometric multiplicity $m(\underline{C},0)$, where $\underline{C}$ is the geometric affine cone corresponding to $gr_{\mathfrak{q}}(\mathcal{O}_{X,x})$, and $0 \in C$ its vertex; see Theorem 4.4.2. This has, of course, to do with very classical Algebraic Geometry, namely the fact that the Hilbert function of a projective variety determines its degree, which is the number of intersection points with a generic complementary linear subspace. This should explain, or at least motivate, the usual abstract definition of $e(\mathfrak{q},\mathcal{O}_{X,x})$ by means of the Hilbert function of $gr_{\mathfrak{q}}(\mathcal{O}_{X,x})$. The reader who takes this definition of $e(\mathfrak{q},\mathcal{O}_{X,x})$ for granted may skip this paragraph.

4.1. Degree of a projective variety

Let $\underline{Z} \subseteq \mathbb{P}^{n-1}$ be a projective variety, i.e. an algebraic $\mathbb{C}$-scheme of finite type. We denote the structure sheaf of $\underline{Z}$, when $\underline{Z}$ is regarded as an algebraic variety, by $\mathcal{O}_Z^{alg}$; so $\underline{Z}$ is given by the ideal sheaf generated in $\mathcal{O}^{alg}_{\mathbb{P}^{n-1}}$ by a homogeneous ideal $I \subseteq \mathbb{C}[z_1,\ldots,z_n]$. Let $\underline{C} \subseteq \underline{\mathbb{C}}^n$ be the corresponding cone; as an algebraic variety, $\underline{C} = \underline{\mathrm{Spec}}(R)$, and as a complex space, $\underline{C} = \underline{\mathrm{Specan}}(R)$, where R is the graded ring $\mathbb{C}[z_1,\ldots,z_n] / I$.

Classically, the degree $\deg(\underline{Z})$ of $\underline{Z}$ is defined to be the number of intersection points of $\underline{Z}$ with a general (d-1)-codimensional projective plane $P \subseteq \mathbb{P}^{n-1}$, where $d-1$ is the projective dimension of $\underline{Z}$, and hence d is the affine dimension of $\underline{C}$. One has, however, to be a little careful what "general" means, and what "number of intersection points" means when $\underline{Z}$ is not reduced.

In analogy to Proposition 3.2.3 for cones one has

Proposition 4.1.1. The set $P_e^{d-1}(\underline{Z}) := \{P \in \mathrm{Grass}^{d-1}(\mathbb{P}^{n-1}) \mid P \cap Z$ is finite$\}$ is generic in $\mathrm{Grass}^{d-1}(\mathbb{P}^{n-1})$.

Proof. We use basic Projective Algebraic Geometry, see e.g. the first 72 pages of [56]. There is $P' \in \mathrm{Grass}^{d}(\mathbb{P}^{n-1})$ with $P' \cap Z = \emptyset$ since $\dim \underline{Z} = d-1$; then the linear projection $q_{P'} : \underline{Z} \longrightarrow \mathbb{P}^{d-1}$ along P' is finite, hence for $z \in Z$, the linear space through P' and z hits $\underline{Z}$ in only finitely many points. So $P_e^{d-1}(\underline{Z}) \neq \emptyset$. Now let $P := \{(z,L) \in \mathbb{P}^{n-1} \times \mathrm{Grass}^{d-1}(\mathbb{P}^{n-1}) \mid z \in L\}$; it has a canonical structure as an algebraic variety $\underline{P}$, and the projection gives a fibre bundle $\underline{P} \longrightarrow \mathbb{P}^{n-1}$, which, by pulling back via $\underline{Z} \hookrightarrow \mathbb{P}^{n-1}$ gives us a fibre bundle $\underline{Z} \xrightarrow{p} \underline{Z}$ with
$Z = \{(z,L) \in Z \times \mathrm{Grass}^{d-1}(\mathbb{P}^{n-1}) \mid z \in L\}$. The projection $\underline{Z} \xrightarrow{q} \underline{\mathrm{Grass}}^{d-1}(\mathbb{P}^{n-1})$ is proper and finite at some point, so it is finite over a nonempty Zariski-open subset of $\mathrm{Grass}^{d-1}(\mathbb{P}^{n-1})$, say over $\mathrm{Grass}^{d-1}(\mathbb{C}^{n-1}) - \Delta(\underline{Z})$, where $\Delta(\underline{Z})$ is a proper Zariski-closed subset. Q.e.d.

Remark 4.1.2. Since $\underline{q}$ is finite outside a nowhere dense analytic set of $\mathrm{Grass}^{d-1}(\mathbb{P}^{n-1})$, $q_* \mathcal{O}_{\underline{Z}}$ is locally free outside a nowhere dense analytic set. One may use this to prove that the set

$$P_{CM}^{d-1}(\underline{Z}) := \left\{ P \in P_e^{d-1}(\underline{Z}) \mid P \cap Z \subseteq Z_{CM} \right\} \qquad (4.1.1)$$

with $Z_{CM} := \{z \in Z \mid \mathcal{O}_{Z,z}$ is Cohen-Macaulay$\}$, is generic in $\mathrm{Grass}^{d-1}(\mathbb{P}^{n-1})$. Similarly, if $\underline{Z}$ is reduced, $\underline{q}$ is locally isomorphic outside a nowhere dense analytic set, and one can equally show that then

(4.1.2) $$P^{d-1}_{reg}(\underline{Z}) := \left\{ P \in P^{d-1}_{e}(\underline{Z}) \mid P \cap Z \subseteq Z_{reg} \text{ and } P \text{ is transversal to } Z_{reg} \text{ along } P \cap Z \right\}$$

is generic in $\mathrm{Grass}^{d-1}(\mathbb{P}^{n-1})$.

Definition 4.1.3. The degree $\deg(\underline{Z})$ of $\underline{Z} \hookrightarrow \mathbb{P}^{n-1}$ is defined to be

$$\deg(\underline{Z}) := \sum_{z \in \underline{Z} \cap P} \deg_{(z,P)} \underline{q}$$

where $\underline{q} : \underline{Z} \longrightarrow \mathrm{Grass}^{d-1}(\mathbb{P}^{n-1})$ and $\Delta(\underline{Z})$ are as above, and $P \in \mathrm{Grass}^{d-1}(\mathbb{P}^{n-1}) - \Delta(\underline{Z}) = P^{d-1}_{e}(\underline{Z})$.

That this number is independent of P can be proven as in Proposition 2.2.5, but it is simpler here, since we will see that we could have worked with the algebraic local rings, and then the local constancy of the $\deg_{(z,P)}\underline{q}$ along Z_{ir} follows without using compact Stein neighbourhoods; see Corollary 4.1.5 below.

Lemma 4.1.4. Let Z be an algebraic variety over $\mathbb{C}$. Let Z_{ir} be the locus of points where Z is locally irreducible as a complex space. Then, if $z \in Z_{ir}$, z lies on a unique irreducible component of Z as an algebraic variety, Z_λ say, and

$$\mathrm{length}(\widetilde{\mathrm{Quot}}(\mathcal{O}_{Z,z})) = \mathrm{length}(\widetilde{\mathrm{Quot}}(\hat{\mathcal{O}}_{Z,z})) = \mathrm{length}(\widetilde{\mathrm{Quot}}(\mathcal{O}^{alg}_{Z,z}))$$
$$= \mathrm{length}(\mathcal{O}^{alg}_{Z,Z_\lambda}) ,$$

<u>where</u> $\mathcal{O}^{alg}_{Z,Z_\lambda}$ <u>is the local ring of</u> Z <u>along</u> Z_λ . <u>In particular, it is constant along</u> $Z_\lambda \cap Z_{ir}$.

<u>Proof</u>. Consider the inclusions

$$(4.1.3) \qquad \mathcal{O}^{alg}_{Z,z} \underset{\varphi}{\hookrightarrow} \mathcal{O}_{Z,z} \underset{\psi}{\hookrightarrow} \hat{\mathcal{O}}_{Z,z} = \hat{\mathcal{O}}^{alg}_{Z,z} .$$

Then, since $(\mathcal{O}_{Z,z})_{red}$ is integral, so is $(\mathcal{O}^{alg}_{Z,z})_{red}$, and z is on a unique Z_λ . Moreover, ψ and $\psi \circ \varphi$ are faithfully flat as completion morphisms, and hence so is φ .

Now it is known (and this is a nontrivial result) that for an integral local analytic $\mathbb{C}$-algebra R the completion $\hat{R}$ is integral. For this see [64], Exposé 21, Théorème 3 on p. 21-13. Or use the fact that the normalization R' of R is again a local analytic algebra ([26], Satz 2 on p. 136); since R is excellent, the minimal primes of $\hat{R}$ correspond to the maximal ideals of R' ([12], Theorem 6.5), and so $\hat{R}$ is integral. Applying this to $R := (\mathcal{O}_{Z,z})_{red}$, one has $\hat{R} = \hat{\mathcal{O}}_{Z,z} / N_z \cdot \hat{\mathcal{O}}_{Z,z}$ is integral, so $N_z \cdot \hat{\mathcal{O}}_{Z,z}$ is prime and so equals $\hat{N}_z$, the nilradical of $\hat{\mathcal{O}}_{Z,z}$. We thus get $N^{alg}_z \cdot \mathcal{O}_{Z,z} = N_z$, $N_z \cdot \hat{\mathcal{O}}_{Z,z} = \hat{N}_z$. We now can localize and get morphisms

$$(4.1.4) \qquad \widetilde{Quot}(\mathcal{O}^{alg}_{Z,z}) \hookrightarrow \widetilde{Quot}(\mathcal{O}_{Z,z}) \hookrightarrow \widetilde{Quot}(\hat{\mathcal{O}}_{Z,z}) ,$$

which are faithfully flat, and Lemma 2.2.4 gives $\text{length}(\widetilde{Quot}(\mathcal{O}^{alg}_{Z,z}))) =$ $= \text{length}(\widetilde{Quot}(\hat{\mathcal{O}}_{Z,z})) = \text{length}(\widetilde{Quot}(\mathcal{O}_{Z,z}))$. Finally, assume $\underline{Z} = \underline{Spec}(A)$ affine, where A is a finitely generated $\mathbb{C}$-alebra, with Z_λ corresponding to $\mathfrak{p} \in Min(A)$, and z to a maximal ideal $\mathfrak{m}$ of $\underline{spec}(A)$. Then $\mathfrak{p} \subseteq \mathfrak{m}$, and so $\widetilde{Quot}(\mathcal{O}^{alg}_{Z,z}) = (A_\mathfrak{m})_\mathfrak{p} =$ $= A_\mathfrak{p} = \mathcal{O}^{alg}_{Z,Z_\lambda}$.

Corollary 4.1.5. $\deg(\underline{Z})$ does not depend on the choice of P. Especially, if Z is irreducible and reduced,

$$\deg(\underline{Z}) = \#(Z \cap P) \quad ,$$

where $P \in P^{d-1}_{reg}(\underline{Z})$ arbitrary (this is the classical definition).

Lemma 4.1.6. Let X be either an algebraic variety over $\mathbb{C}$ or a complex space, and let $\mathcal{O}$ denote either the algebraic or complex analytic structure. Then, for all $k \geq 0$, and $x \in X_{ir}$ (the irreducible locus with respect to the complex analytic structure), $(x,0) \in (X \times \mathbb{C}^k)_{ir}$, and

$$\mathrm{length}(\widetilde{\mathrm{Quot}}(\mathcal{O}_{X,x})) = \mathrm{length}(\widetilde{\mathrm{Quot}}(\mathcal{O}_{X\times\mathbb{C}^k,(x,0)})) \quad .$$

In particular, if $\underline{f} : (\underline{X},x) \longrightarrow (\underline{\mathbb{C}}^d,0)$ is finite, $\deg_x \underline{f} = \deg_{(x,0)}(\underline{f} \times \mathrm{id}_{\mathbb{C}^k})$ for all k.

Proof. We may assume $k=1$. Consider the faithfully flat extension

$$(4.1.5) \qquad \hat{\mathcal{O}}_{X,x} \longrightarrow \hat{\mathcal{O}}_{X\times\mathbb{C},(x,0)} = \hat{\mathcal{O}}_X[[t]] \quad .$$

The nilradical of $\hat{\mathcal{O}}_X[[t]]$ is $N_x \cdot \mathcal{O}_x[[t]]$, and so $((\hat{\mathcal{O}}_X[[t]])_{red} = (\hat{\mathcal{O}}_{X,x}/\hat{N_x\mathcal{O}}_{X,x})[[t]] = (\hat{\mathcal{O}}_{X,x})_{red}[[t]]$ by the proof of 4.1.4; so if N_x is prime, $\hat{N}_x = N_x \cdot \hat{\mathcal{O}}_{X,x}$ is prime, so $x \in X_{ir}$ implies $(x,0) \in (X \times \mathbb{C})_{ir}$. The claim now follows again by Lemma 2.2.4 and Lemma 4.1.4.

Proposition 4.1.7. Let $Z \hookrightarrow \mathbb{P}^{n-1}$ be a projective variety of dimension $d-1$ and with homogeneous coordinate ring R. Then for any $P \in P^{d-1}_e(\underline{Z})$ and P' a hyperplane in P with $Z \cap P' = \emptyset$:

$$\deg(\underline{Z}) = \sum_{z \in Z \cap P} \deg_z \underline{q}_{P'} \qquad ,$$

<u>where</u> $\underline{q}_{P'} : \underline{Z} \longrightarrow \underline{\mathbb{P}}^{d-1}$ <u>is the projection with centre</u> P' .

(cf. (5.3) and (5.4.) in Mumford's book [56]) .

<u>Outline of proof</u>.

Let the notations be as above. Fix P and P' . Let $\mathbb{P}^{n-2} \subseteq \mathbb{P}^{n-1}$ be a hyperplane containing P' and not meeting $Z \cap P$. Finally, let $\mathbb{P}^{d-1} \subseteq \mathbb{P}^{n-1}$ be such that $\mathbb{P}^{d-1} \cap P' = \emptyset$ and $\mathbb{P}^{d-1} \cap \mathbb{P}^{n-2}$ is a hyperplane in $\mathbb{P}^{d-1}$ (see Figure 4).

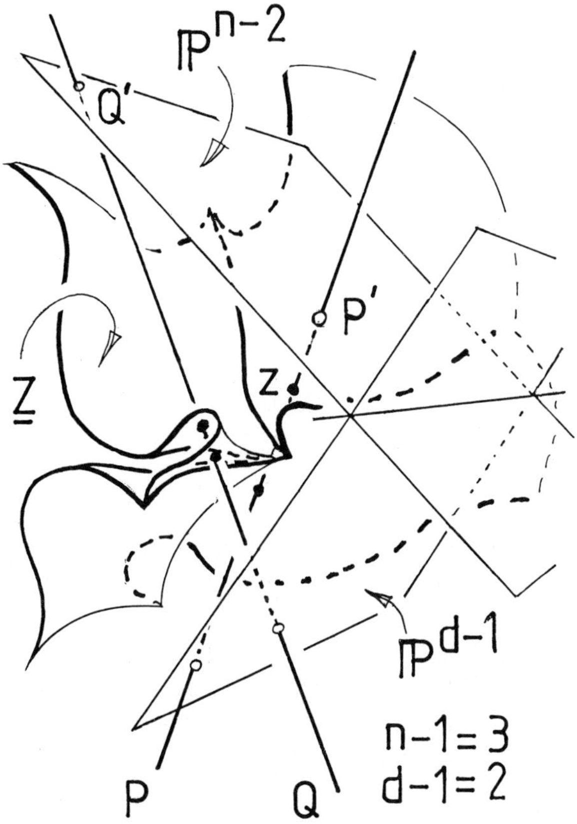

Fig. 4

We say two planes $L, L' \subseteq \mathbb{P}^{n-1}$ are <u>transversal</u>, denoted $L \pitchfork L'$, if $L \cap L'$ has minimal possible dimension. Put $\mathbb{P}_0^{d-1} := \mathbb{P}^{d-1} - (\mathbb{P}^{d-1} \cap \mathbb{P}^{n-2})$, $\mathrm{Grass}^{d-1}(\mathbb{P}^{n-1})_0 := \{Q \in \mathrm{Grass}^{d-1}(\mathbb{P}^{n-1}) \mid Q \pitchfork \mathbb{P}^{d-1}, Q \pitchfork \mathbb{P}^{n-2}, Q \pitchfork \mathbb{P}^{d-1} \cap \mathbb{P}^{n-2}\}$, and $\mathrm{Grass}^{d-1}(\mathbb{P}^{n-2})_0 := \{Q' \in \mathrm{Grass}^{d-1}(\mathbb{P}^{n-2}) \mid Q' \pitchfork \mathbb{P}^{d-1} \cap \mathbb{P}^{n-2}\}$. These are nowhere dense Zariski-open subsets. Finally put $\underline{Z}_0 := \underline{Z} - \mathbb{P}^{n-2}$, and $\underline{Z}_0 := \underline{p}^{-1}(\underline{Z}_0) \cap \underline{q}^{-1}(\underline{\mathrm{Grass}}^{d-1}(\mathbb{P}^{n-1})_0)$ (notations as in the proof of Proposition 4.1.1). One then gets the diagram

(4.1.6)
$$\begin{array}{ccc}
\underline{Z}_0 & \overset{\underline{f}}{\underset{\underline{g}}{\rightleftarrows}} & \underline{Z}_0 \times \underline{\mathrm{Grass}}^{d-1}(\mathbb{P}^{n-2})_0 \\
\Big\downarrow \underline{q} & & \Big\downarrow \underline{q}_{P'} \times \underline{\mathrm{id}} \\
\mathrm{Grass}^{d-1}(\mathbb{P}^{n-1})_0 & \overset{\underline{h}}{\underset{\underline{k}}{\rightleftarrows}} & \underline{\mathbb{P}}_0^{d-1} \times \underline{\mathrm{Grass}}^{d-1}(\mathbb{P}^{n-2})_0
\end{array},$$

where
$\underline{f} : (z,Q) \longmapsto (z, Q \cap \mathbb{P}^{n-2})$,

$\underline{g} : (z,Q') \longmapsto (z, Q' \vee z)$, where $Q' \vee z$ denotes the plane spanned by Q' and z,

$\underline{h} : Q \longmapsto (Q \cap \mathbb{P}^{d-1}, Q \cap \mathbb{P}^{n-2})$,

$\underline{k} : (z,Q') \longmapsto Q' \vee z$.

Then $\underline{f}$ and $\underline{g}$ are inverse to each other, and so are $\underline{h}$ and $\underline{k}$. Over $P \in \mathrm{Grass}^{d-1}(\mathbb{P}^{n-1})_0$, the diagram is commutative, and so for $z \in Z \cap P$:

$$\deg_{(z,P)} \underline{q} = \deg_{(z,P')} (\underline{q}_{P'} \times \underline{\mathrm{id}}) = \deg_z (q_{P'}),$$

the last equality from Lemma 4.1.6. This proves the Proposition.

Theorem 4.1.8. *Let* $\underline{Z} \hookrightarrow \mathbb{P}^{n-1}$ *be a projective variety of dimension* $d-1$ *with homogeneous coordinate ring* R , *and let* $\underline{C} \hookrightarrow \mathbb{C}^n$ *be the corresponding affine cone. Then*

$$\deg(\underline{Z}) = m(\underline{C},0) \quad ,$$

the geometric multiplicity of $\underline{C}$ *at its vertex.*

Proof. Let $\mathbb{C}^n$ have coordinates $(z_1,\ldots,z_n)$; we may assume $\mathbb{P}^{n-2} \subset \mathbb{P}^{n-1}$ in 4.1.7 is given by $z_n = 0$. Let $L' \in \mathrm{Grass}^{d-1}(\mathbb{C}^n)$ correspond to $P \in \mathrm{Grass}^{d-1}(\mathbb{P}^{n-1})$, and $\mathbb{C}^{n-1} \subseteq \mathbb{C}^n$ be the hyperplane corresponding to $\mathbb{P}^{n-2}$. Let $L := L' \cap \mathbb{C}^{n-1}$ and put $\underline{C}_0 := \underline{C} - \mathbb{C}^{n-1}$, where $\underline{C}$ is te affine cone corresponding to $\underline{Z}$. Let $H_1 \subseteq \mathbb{C}^n$ be the affine hyperplane given by $z_n = 1$, and put $\underline{C}_1 := \underline{C}_0 \cap \underline{H}_1$. Now consider the commutative diagram of morphisms of algebraic varieties

$$(4.1.7) \qquad \begin{array}{ccccccc}
\underline{C}_1 & \xrightarrow[\underline{j}_n]{} & \underline{C}_1 \times \underline{\mathbb{C}}^* & \xrightarrow[\underline{u}]{\cong} & \underline{C}_0 & \xrightarrow[\pi_n]{} & \underline{Z}_0 \\
\downarrow {\scriptstyle \underline{p}_L|H_1} & \circlearrowright & \downarrow {\scriptstyle \underline{p}_L} & \circlearrowright & \downarrow {\scriptstyle \underline{p}_L} & \circlearrowright & \downarrow {\scriptstyle \underline{q}_{P'}} \\
\underline{\mathbb{C}}^{d-1} & \xrightarrow{\underline{j}_d} & \underline{\mathbb{C}}^{d-1} \times \underline{\mathbb{C}}^* & = & \underline{\mathbb{C}}^d - \mathbb{C}^{d-1} & \xrightarrow{\pi_d} & \underline{\mathbb{P}}_0^{d-1}
\end{array}$$

Here, the left horizontal arrows are inclusions via $z' \mapsto (z',1)$, u is induced by $\mathbb{C}^{n-1} \times \mathbb{C}^* \longrightarrow \mathbb{C}^n - \mathbb{C}^{n-1}$ with $(z',\lambda) \mapsto (\lambda z',\lambda)$, and the right horizontal arrows are induced by the canonical projection $\pi_N : \mathbb{C}^N - \{0\} \twoheadrightarrow \mathbb{P}^{N-1}$. u is isomorphic, the inverse being induced by $\mathbb{C}^n - \mathbb{C}^{n-1} \longrightarrow \mathbb{C}^{n-1} \times \mathbb{C}^*$, $z = (z',z_n) \mapsto (z'/z_n, z_n)$ (see Figure 5).

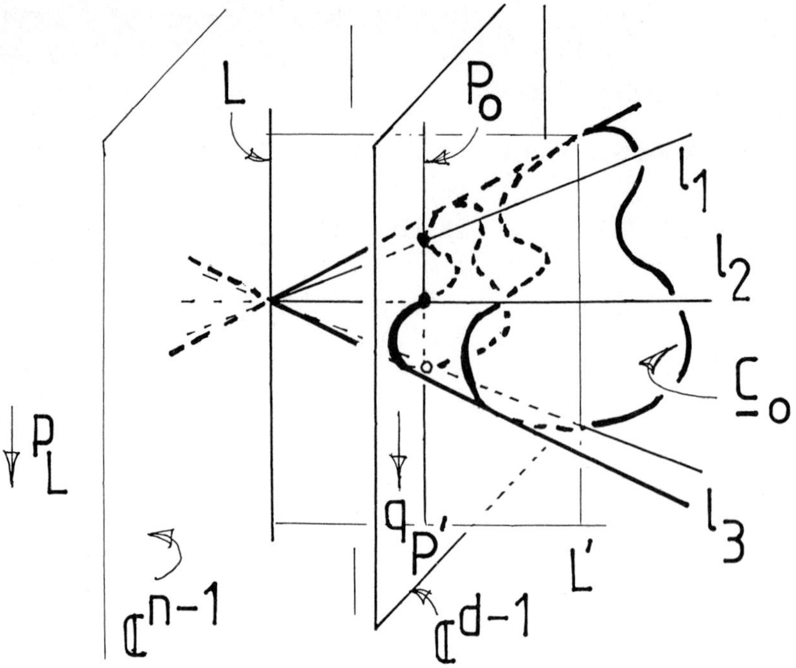

Fig. 5

From this figure, the result should be intuitively clear, since the intersection of P with Z corresponds to the intersection lines of L' with C, which in turn correspond to the intersection points of the affine plane $P_0 := P - \mathbb{C}^{n-1} = H_1 \cap L'$ with C; but we must check the multiplicities.

The composite horizontal arrows give isomorphisms, so, since $\mathbb{C}^{n-1}$ is disjoint from $Z \cap P$, $\deg(\underline{Z}) = \sum_{w \in Z \cap P} \deg_w \underline{q}_{P'} =$

$\sum_{z' \in \underline{C}_1 \cap P_0} \deg_{z'} (\underline{p}_L | H_1)$. But this equals $\deg_0 \underline{p}_L$ by Lemma 4.1.6 and the middle square in (4.1.7.). So $\deg(\underline{Z}) = \deg_0 \underline{p}_L$ for all $L \in P^d_g(\underline{C}, 0) = P^d_e(\underline{C}, 0)$, which proves the claim.

Corollary 4.1.9. Let $Z_{\mathfrak{p}}$, $\mathfrak{p} \in \mathrm{Assh}(R)$, be the irreducible components of Z of dimension $d-1$, given by a homogeneous primary decomposition of 0 in R. Then

$$(4.1.8) \qquad \deg(\underline{Z}) = \sum_{\mathfrak{p}\in \mathrm{Assh}(R)} \mathrm{length}(R_{\mathfrak{p}})\cdot \deg(Z_{\mathfrak{p}}) \quad .$$

Proof. As $P \in P_e^{d-1}(\underline{Z})$ hits $Z_{\mathfrak{p}}$ for $\mathfrak{p} \in \mathrm{Assh}(R)$, and these correspond to the maximal irreducible components of $\underline{C}_0$, it suffices to show $\mathrm{length}(\widetilde{\mathrm{Quot}}(\mathcal{O}_{C,z})) = \mathrm{length}(R_{\mathfrak{p}})$ for $z \in (C_0)_{ir}$ and $\mathfrak{p}$ corresponding to the irreducible component on which z lies. Now the affine coordinate ring of $\underline{C}_0$ is $R_{(z_n)}$, and $z_n \notin \mathfrak{p}$, since otherwise $\underline{Z} \cap \underline{P}$ would not be disjoint to $\mathbb{P}^{n-2}$. Then $\mathcal{O}^{alg}_{C,z} = (R_{(z_n)})_{\mathfrak{p}} = R_{\mathfrak{p}}$, and the claim follows from Lemma 4.1.4.

4.2. Hilbert functions

The following result is classical; it was, at least in the reduced irreducible case, known to Hilbert ([32], p. 244), and, in general, to Macaulay [50], footnotes on pp. 82 and 115).

Theorem 4.2.1. Let R be the coordinate ring of a projective variety $\underline{Z} \hookrightarrow \mathbb{P}^{n-1}$ of dimension d. Then the Hilbert function $H(R,k) := \dim_{\mathbb{C}} R_k$ has the form

$$(4.2.1) \qquad H(R,k) = \frac{\deg(\underline{Z})}{(d-1)!}\, k^{d-1} + \text{lower terms}$$

for $k >> 0$.

One way of geometric thinking about this goes as follows: For any projective variety $\underline{Z}$ and coherent $\mathcal{O}_Z$-module M put

$$(4.2.2) \qquad \chi(\underline{Z},M) := \sum_{i\geq 0} (-1)^i \dim_{\mathbb{C}} H^i(\underline{Z},M) \quad ,$$

where all $H^i(\underline{Z},M)$ are finite dimensional and 0 for $i > d-1$ ([65]), and one may either take analytic or algebraic sheaf cohomology ([66]).

Let M be a f.g. graded module over $\mathbb{C}[X_1,\ldots,X_n]$ and $\mathcal{M}$ the corresponding coherent $\mathcal{O}_{\mathbb{P}^{n-1}}$-module. By celebrated results of [65], $H^i(\mathbb{P}^{n-1},\mathcal{M}(k)) = 0$ for $i > 0$ and $k >> 0$, and $M_k \cong \Gamma(\mathbb{P}^{n-1},\mathcal{M}(k))$ for $k >> 0$, hence

(4.2.3) $$H(M,k) := \dim_{\mathbb{C}} M_k = \chi(\mathbb{P}^{n-1},\mathcal{M}(k)) \quad \text{for} \quad k >> 0 \quad .$$

Now take any hyperplane $H \overset{i}{\hookrightarrow} \mathbb{P}^{n-1}$, defined by a linear form F; then the exact sequence

(4.2.4) $$0 \longrightarrow \mathcal{O}_{\mathbb{P}^{n-1}}(-1) \xrightarrow{\cdot F} \mathcal{O}_{\mathbb{P}^{n-1}} \longrightarrow i_*\mathcal{O}_H \longrightarrow 0$$

induces (loc. cit. p. 277)

(4.2.5) $$0 \longrightarrow \mathcal{M}(k-1) \longrightarrow \mathcal{M}(k) \longrightarrow i_*(i^*\mathcal{M}(k)) \longrightarrow 0$$

for all k as soon as H is in general position with respect to $\underline{\text{supp}}\,\mathcal{M}$, namely F should not belong to any prime of the homogeneous primary decomposition of M, except the possibly present irrelevant maximal ideal.

By additivity of χ, then,

(4.2.6) $$\chi(\underline{\mathbb{P}}^{n-1},\mathcal{M}(k)) = \chi(\underline{\mathbb{P}}^{n-1},\mathcal{M}(k-1)) + \chi(\underline{H},i^*\mathcal{M}(k)) \quad .$$

Applying this to $M = R$ gives the recursion

(4.2.7) $$\chi(\underline{Z},\mathcal{O}_Z(k)) = \chi(\underline{Z},\mathcal{O}_Z(k-1) + \chi(\underline{Z} \cap \underline{H},\mathcal{O}_{Z\cap H}(k)) \quad ,$$

and by doubly inducting over k and d one gets

(4.2.8) $$\chi(\underline{Z},\mathcal{O}_Z(k)) = \sum_{j=0}^{d-1} (\underline{Z} \cap \underline{H}^{(j)},\mathcal{O}_{\underline{Z}\cap H^{(j)}}) \cdot \binom{j+k-1}{j} \quad ,$$

where $H_1,\ldots,H_{d-1}$ are hyperplanes in general position defined by linear forms $F_1,\ldots,F_k$, and $\underline{H}^{(j)} := \underline{H}_1 \cap \ldots \cap \underline{H}_j$. So

$$(4.2.9) \qquad H(R,k) = \chi(\underline{Z},\mathcal{O}_Z(k)) \quad \text{for} \quad k \gg 0$$

is indeed a polynomial of degree $d-1$ in k, whose leading coefficient is $\frac{1}{(d-1)!}\chi(\underline{Z}\cap\underline{P},\mathcal{O}_{\underline{Z}\cap\underline{P}})$, where P is a $(d-1)$-codimensional plane in general position, and $\underline{Z}\cap\underline{P}$ the scheme-theoretic intersection. But since $P_e^{d-1}(\underline{Z})$ is generic in $\text{Grass}^{d-1}(\mathbb{P}^{n-1})$, we then have that, for a general choice of $H_1,\ldots,H_{d-1}$, the intersection $Z\cap P$ consists of finitely many points. Then

$$(4.2.10) \qquad \mathcal{O}_{Z\cap P} = \bigoplus_{z\in Z\cap P} \mathcal{O}_{Z\cap P,z} \quad ,$$

a direct sum of artinian rings, and so

$$(4.2.11) \qquad \chi(\underline{Z}\cap\underline{P},\mathcal{O}_{Z\cap P}) = \sum_{z\in Z\cap P} \dim_{\mathbb{C}}(\mathcal{O}_{Z\cap P,z}) \quad .$$

Choosing $P'\subseteq P$ a hyperplane in P with $P'\cap Z=\emptyset$, $q_{P'}:\underline{Z}\longrightarrow\mathbb{P}^{d-1}$ will be finite ; so $(q_{P'})_*(\mathcal{O}_Z)$ being a coherent sheaf, will be generically finite. Moving the H_j we may assume that $\mathcal{O}_{Z,z}$ is locally free over $\mathcal{O}_{\mathbb{P}^{d-1},z'}\cong\mathcal{O}_{\mathbb{C}^{d-1},0}$ for all $z\in Z\cap P$ with $P\cap\mathbb{P}^{d-1}=\{z'\}$.
But then

$$(4.2.12) \qquad \begin{aligned} \text{length}(\widetilde{\text{Quot}}(\mathcal{O}_{Z,z})) &= \text{rank}_{\mathcal{O}_{\mathbb{C}^{d-1},0}}(\mathcal{O}_{Z,z}) \\ &= \dim_{\mathbb{C}}(\mathcal{O}_{Z,z}/m_{d-1}\mathcal{O}_{Z,z}) \\ &= \dim_{\mathbb{C}}(\mathcal{O}_{Z\cap P,z}) \quad , \end{aligned}$$

which implies $\deg(\underline{Z}) = \sum_{z\in Z\cap P} \deg_z q_{P'} = \chi(\underline{Z}\cap\underline{P},\mathcal{O}_{Z\cap P})$. Q.e.d.

For a more classical proof which does not use sheaf cohomology see [56], p. 112 ff, which works for the case $\underline{Z}$ reduced irreducible. Since $H(-,k)$ is additive on modules,

$$(4.2.13) \qquad H(R,k) = \sum_{\mathfrak{p}\in\text{Assh}(R)} \text{length}(R_{\mathfrak{p}})H(R/\mathfrak{p},k) \quad ,$$

and so the general case follows also from this because of Corollary 4.1.9.

4.3. A generalization

Let $A \in \underline{\underline{la}}$ be an artinian local $\mathbb{C}$-algebra corresponding to a one-point complex space $\underline{S} = (\{s\}, A) \in \underline{\underline{cpl}}$.

Definition 4.3.1.

(i) $\underline{\mathbb{P}}_A^{n-1} := \underline{S} \times \underline{\mathbb{P}}^{n-1}$, projective (n-1)-space over A .

(ii) A projective variety $\underline{Z}$ over A is a closed complex subspace $\underline{Z} \hookrightarrow \underline{\mathbb{P}}_A^{n-1}$ defined by a homogeneous ideal $I \subseteq A[Z_1,\dots,Z_n]$ for some n .

Remark 4.3.2. Projective varieties correspond to finitely generated graded A-algebras (positively graded, $B_0 = A$, generated by B_1). In fact if $\underline{Z}$ is as above, $R := A[Z_1,\dots,Z_n]/I$, $\underline{Z} = \underline{\text{Projan}}(R)$ (see III 1.2.8), the complex space associated to the projective scheme Proj(R) .

Corresponding to $\underline{\mathbb{P}}_A^{n-1}$ there is affine n-space $\underline{\mathbb{A}}_A^n := \underline{S} \times \underline{\mathbb{C}}^n$ over A . Corresponding to $\underline{Z} \hookrightarrow \underline{\mathbb{P}}_A^{n-1}$ there is an affine variety $\underline{C} \hookrightarrow \mathbb{A}_A^n$, in fact $\underline{C} = \underline{\text{Specan}}(R)$ as a complex space. We call again $\underline{C}$ the cone associated to $\underline{Z}$, and $\underline{Z}$ the projective cone $\underline{\mathbb{P}\,C}$ of $\underline{C}$.

Let $\underline{\mathbb{P}}^{n-1} \xrightarrow{\underline{r}} \underline{\mathbb{P}}_A^{n-1}$ be the morphism given by $A[Z_1,\dots,Z_n] \twoheadrightarrow (A/m_A)[Z_1,\dots,Z_n]$. If $\underline{Z} \hookrightarrow \underline{\mathbb{P}}_A^n$ is a projective variety over A , we put $\underline{Z}_0 =: \underline{r}^{-1}(\underline{Z})$ and

$$\deg(\underline{Z}) := (\dim_{\mathbb{C}} A)\cdot(\deg(\underline{Z}_0)) \quad . \tag{4.3.1}$$

Now let M be a finitely generated B-module. Define again the Hilbert function H(M,k) to be

$$H(M,k) := \dim_{\mathbb{C}} M_k \quad . \tag{4.3.2}$$

Then Theorem 4.1.8 and Theorem 4.2.1 still hold with the convention (4.3.1) for $\deg(\underline{Z})$.

4.4. Samuel multiplicity

Let now $(\underline{X},x) \in \underline{\underline{\mathrm{cpl}}}_0$, $\mathfrak{q}$ an $\mathfrak{m}_x$-primary ideal of $\mathcal{O}_{X,x}$, defining a zero dimensional complex subspace of $\underline{X}$ supported on x , which we call $\underline{x}(\mathfrak{q})$.

Definition 4.4.1 (Normal cone). The normal cone of $\underline{x}(\mathfrak{q})$ in $\underline{X}$ is defined to be

$$\underline{C}(\underline{X},\underline{x}(\mathfrak{q})) := \underline{\mathrm{Specan}}(\mathrm{gr}_{\mathfrak{q}}(\mathcal{O}_{X,x})) .$$

In case $\mathfrak{q} = \mathfrak{m}$, $\underline{C}(\underline{X},\underline{x}(\mathfrak{q})) = \underline{C}(\underline{X},x)$, the tangent cone.

The epimorphism $\mathrm{Sym}(\mathfrak{q}/\mathfrak{q}^2) \twoheadrightarrow \mathrm{gr}_{\mathfrak{q}}(\mathcal{O}_{X,x})$ gives an embedding $\underline{\mathbb{P}C}(\underline{X},\underline{x}(\mathfrak{q})) \hookrightarrow \underline{\mathbb{P}}_A^{d-1}$, where $d := \dim_{\mathbb{C}}(\mathfrak{q}/\mathfrak{q}^2)$ and $A := R/\mathfrak{q}$. Taking the Hilbert function of $\underline{\mathbb{P}C}(\underline{X},\underline{x}(\mathfrak{q}))$ with respect to this embedding we get from Theorem 4.1.8, Theorem 4.2.1 and the discussion in 4.3:

Theorem 4.4.2. $e(\mathfrak{q},\mathcal{O}_{X,x}) = m(\underline{C}(\underline{X},\underline{x}(\mathfrak{q})),x)$.

Remark 4.4.3. For an extension of this to the general scheme-theoretic context see the paper [57] of C.P. Ramanujam.

§ 5. Algebraic multiplicity

In this paragraph, I show the equality $m(\underline{X},x) = e(\mathfrak{m}_x,\mathcal{O}_{X,x})$ for a complex space germ $(\underline{X},x)$.

5.1. Algebraic degree

I now give some algebraic formulae for the local mapping degree, which relate it to Samuel multiplicity.

Proposition 5.1.1. Let X be a complex space, M a coherent $\mathcal{O}_X$-module, and Z be an irreducible component of the support $\operatorname{supp}(M)$ of M. Let the ideal $P \subseteq \mathcal{O}_X$ define Z in X. Then

(i) For $z \in Z_{ir}$, the localization $(M_z)_{P_z}$ is an artinian $(\mathcal{O}_{X,z})_{P_z}$-module;

(ii) the function $z \longmapsto \operatorname{length}(M_z)_{P_z}$ is locally constant on Z_{ir}.

Proof. The proof is analogous to the proof of Proposition 2.2.3, so the details are omitted. One proves

$$\operatorname{length}_{(\mathcal{O}_{X,z'})_{P_{z'}}}((M_{z'})_{P_{z'}}) = \operatorname{length}_{\mathcal{O}(K)_P} M_P \quad \text{for} \quad z' \in Z_{ir} \cap K ,$$

where, for given $z \in Z_{ir}$, K is a suitable compact Stein neighbourhood of z, P is the $\mathcal{O}(K)$-ideal $\Gamma(K,P)$, and M the $\mathcal{O}(K)$-module $\Gamma(K,M)$, by localizing the flat map

$$\lambda_{z'} : \mathcal{O}(K) \longrightarrow \mathcal{O}_{X,z'}$$

at P and again using Lemma 2.2.4.

We now apply this, with $f : X \longrightarrow B$ as in Definition 2.2.6, to the coherent $\mathcal{O}_B$-module $f_*\mathcal{O}_X$:

Corollary 5.1.2. The number

$$(5.1.1) \qquad \sum_{x' \in f^{-1}(y)} \dim_{\operatorname{Quot}(\mathcal{O}_{\mathbb{C}^d,y})} \left(\operatorname{Quot}(\mathcal{O}_{\mathbb{C}^d,y}) \otimes_{\mathcal{O}_{\mathbb{C}^d,y}} \mathcal{O}_{X,x'} \right)$$

is independent of $y \in B$.

Proof. By I Theorem 4.1.1,

$$(f_*\mathcal{O}_X)_y = \bigoplus_{x'\in f^{-1}(y)} \mathcal{O}_{X,x'} \quad \text{for all } y\in B$$

as an $\mathcal{O}_{\mathbb{C}^d,y}$-module. The claim now follows by Proposition 5.1.1.

Recall now Serre's notation: Let R be a local ring, $\mathfrak{q}$ an $\mathfrak{m}_R$-primary ideal, M an R-module, $d\in\mathbb{N}$ such that $\dim_R M \leq d$; then put

$$(5.1.2)\qquad e_{\mathfrak{q}}(M,d) := \begin{cases} e(\mathfrak{q},M) & \text{if } \dim_R M = d \\ 0 & \text{else} \end{cases}$$

(see [67], p. V-3). We then have the formula (loc.cit, or Chapter I, Theorem (1.8)):

$$(5.1.3)\qquad e_{\mathfrak{q}}(M,d) = \sum_{\dim(R/\mathfrak{p})=d} \operatorname{length}(M_{\mathfrak{p}})\cdot e_{\mathfrak{q}}(R/\mathfrak{p},d)$$

(because of additivity of length).

Corollary 5.1.3. In the situation of Corollary 5.1.2, the number

$$(5.1.4)\qquad \sum_{x'\in f^{-1}(y)} e_{\mathfrak{q}_{x'}}(\mathcal{O}_{X,x'},d)$$

is also independent of $y\in B$, where $\mathfrak{q}_{x'}$ is the ideal in $\mathcal{O}_{X,x'}$ generated by the maximal ideal $\mathfrak{m}_y$ of $\mathcal{O}_{\mathbb{C}^d,y}$; in fact it equals the number (5.1.1.).

Proof. The number in question is $e_{\mathfrak{m}_y}((f_*\mathcal{O}_X)_y,d)$, which by (5.1.3) is just $\operatorname{length}(\operatorname{Quot}(\mathcal{O}_{\mathbb{C}^d,y})\otimes_{\mathcal{O}_{\mathbb{C}^d,y}}(f_*\mathcal{O}_{X,x})_y)$, since $R=\mathcal{O}_{\mathbb{C}^d,y}$ is regular and so $e(\mathfrak{m}_d,R)=1$. And this number is (5.1.1).

We now can characterize the local mapping degree algebraically.

Theorem 5.1.4 (Multiplicity formula). Let f be as in Definition 2.2.6. Then the following numbers are equal:

(i) the local mapping degree $\deg_{\underline{x}} \underline{f}$;

(ii) $\dim_{\mathrm{Quot}\left(\mathcal{O}_{\mathbb{C}^d,0}\right)} \left(\mathrm{Quot}\left(\mathcal{O}_{\mathbb{C}^d,0}\right) \otimes_{\mathcal{O}_{\mathbb{C}^d,0}} \mathcal{O}_{X,x}\right)$;

(iii) the Samuel multiplicity $e_{\mathfrak{q}}(\mathcal{O}_{X,x},d)$ with $\mathfrak{q} = \mathfrak{m}_d \cdot \mathcal{O}_{X,x} = (f_1,\ldots,f_d)\cdot\mathcal{O}_{X,x}$, where $(f_1,\ldots,f_d)$ define $\underline{f}$ according to I, Corollary 3.3.5.

Remark 5.1.5.

a) For a complete local ring containing a field which is an integral domain, (ii) was Chevalley's original definition of the multiplicity $e(\mathfrak{q}_x,\mathcal{O}_{X,x})$ (up to multiplying with the degree of the residue field extension, which is 1 here) in [9], § IV. Somewhat later he extended it to quasi-unmixed local rings in [10], Definition 3 on p. 13, and his definition can be shown to be again the number (ii). In other words, the philosophy behind his definition was to mimic, by passing to the completion, the notion of local mapping degree by an algebraic construction. See also Remark 5.1.8.

b) The equality of (ii) and (iii) is a special case of the Projection Formula (Theorem (6.3) in Chapter I).

Proof of Theorem 5.1.4. We may assume $\dim_{\underline{x}} \underline{X} = d$, since otherwise all numbers are 0 . The equality of (ii) and (iii) has just been seen in the proof of Corollary 5.1.3.

To prove the equality of (i) and (ii), we are reduced, by Corollary 5.1.2, to prove the equality

$$(5.1.5) \qquad \mathrm{length}(\mathrm{Quot}(\mathcal{O}_{X,x'})) = \dim_{\mathrm{Quot}(\mathcal{O}_{\mathbb{C}^d,y})}(\mathrm{Quot}(\mathcal{O}_{\mathbb{C},y}) \otimes_{\mathcal{O}_{\mathbb{C}^d,y}} \mathcal{O}_{X,x'})$$

in the special case where in the diagram

$$\begin{array}{ccc} (\underline{X}_{red}, x') & \overset{\underline{i}}{\hookrightarrow} & (\underline{X}, x') \\ & \underline{g} := \underline{f} \circ \underline{i} \searrow \quad \swarrow \underline{f} & \\ & (\underline{\mathbb{C}}^d, y) & \end{array}$$

$\underline{g}$ is an isomorphism and where $\underline{i}$ is defined by $\mathcal{O}_{X,x'} \twoheadrightarrow \mathcal{O}_{X,x'} / \mathcal{N}_{x'}$. We thus have that in the situation

$$\mathcal{O}_{\mathbb{C}^d,y} \xrightarrow{f^0_{x'}} \mathcal{O}_{X,x'} \xrightarrow{i^0_{x'}} \mathcal{O}_{X,x'} / \mathcal{N}_{x'} \quad ,$$

$f^0_{x'}$ is injective by I Theorem 6.2.1 and $i^0_{x'} \circ f^0_{x'}$ is an isomorphism. The claim then follows from the following Lemma.

Lemma 5.1.6. Let $R \hookrightarrow S$ be a finite extension of local analytic $\mathbb{C}$-algebras such that R is an integral domain and the nilradical $\mathfrak{n}_S$ of S is prime. Then $\mathrm{Quot}(R) \otimes_R S \cong \widetilde{\mathrm{Quot}(S)}$.

Proof. Since $\mathfrak{n}_S$ is prime, any element of S is either nilpotent or active by (2.2.1). By the argument in the proof of the Active Lemma I 5.2.2 and $t \in \mathrm{Ac}(S) = S - \mathfrak{n}_S$ satisfies an integral equation

$$t^k + r_{k-1}t^{k-1} + \ldots + r_1 t + r_0 = 0 \tag{5.1.6}$$

with $k \geq 1$, $r_j \in R$ for $0 \leq j \leq k-1$, and $r_0 \neq 0$.

Now any element of $\mathrm{Quot}(R) \otimes_R S$ can be written as a fraction s/r with $s \in S$, $r \in R - \{0\}$. Since $R - \{0\} \hookrightarrow S - \mathfrak{n}_S$, we can consider this as an element of $\widetilde{\mathrm{Quot}(S)}$, and this gives a homomorphism

$$\varphi : \mathrm{Quot}(R) \otimes_R S \longrightarrow \widetilde{\mathrm{Quot}(S)} \ . \tag{5.1.7}$$

I claim φ is an isomorphism.

Injectivity of φ : Suppose $s/r \in \mathrm{Quot}(R) \otimes_R S$ maps to 0 in $\widetilde{\mathrm{Quot}(S)}$. This means there is $t \in \mathrm{Ac}(S)$ with $t \cdot s = 0$. Multiplying

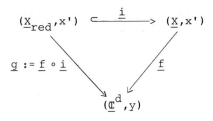

(5.1.6) with s shows $r_0 \cdot s = 0$, with $r_0 \in R - \{0\}$, hence $s/r = 0$ in $\mathrm{Quot}(R) \otimes_R S = (R - \{0\})^{-1} S$.

Surjectivity of φ: Let $s/t \in \widetilde{\mathrm{Quot}(S)}$; it suffices to produce $u \in S$ such that $tu = r \in R - \{0\}$, for then $s/t = su/r$.

Now $t \in \mathrm{Ac}(S)$, therefore (5.1.6) gives

$$t\left(t^{k-1} + r_{k-1}t^{k-2} + \ldots + r_1\right) = -r_0 ,$$

so it suffices to take $u := t^{k-1} + r_{k-1}t^{k-2} + \ldots + r_1$ and $r := -r_0$.

Remark 5.1.7. The degree formula 2.2.8. holds.

This is now immediate by 5.1.2 and 5.1.4.

Remark 5.1.8. Formula (3.2.5) can be written as

$$(5.1.8) \qquad m(\underline{X},x) = \min_{\substack{(f_1,\ldots,f_d)\text{s.o.p.} \\ \text{of } \mathcal{O}_{X,x}}} \left\{ \dim_{\mathrm{Quot}(\mathcal{O}_{\mathbb{C}^d,0})} \mathrm{Quot}(\mathcal{O}_{\mathbb{C}^d,0}) \otimes_{\mathcal{O}_{\mathbb{C}^d,0}} \mathcal{O}_{X,x} \right\} .$$

By the proof of Lemma 4.1.4, $\mathrm{Quot}(\mathcal{O}_{X,x}) \longrightarrow \mathrm{Quot}(\hat{\mathcal{O}}_{X,x})$ is a flat morphism of local rings with residue field extension of degree 1; from this one can show

$$\dim_{\mathrm{Quot}(\mathcal{O}_{\mathbb{C}^d,0})} (\mathrm{Quot}(\mathcal{O}_{\mathbb{C}^d,0}) \otimes_{\mathcal{O}_{\mathbb{C}^d,0}} \mathcal{O}_{X,x}) =$$

$$\dim_{\mathrm{Quot}(\hat{\mathcal{O}}_{\mathbb{C}^d,0})} (\mathrm{Quot}(\hat{\mathcal{O}}_{\mathbb{C}^d,0}) \otimes_{\hat{\mathcal{O}}_{\mathbb{C}^d,0}} \hat{\mathcal{O}}_{X,x})$$

which is just Chevalley's definition of his $e(\mathcal{O}_{X,x}; f_1,\ldots,f_d)$. So $m(\underline{X},x)$ corresponds to taking the minimal value of these multiplicities, as asserted in the Historical Remark Chapter I, (6.7),c).

5.2. Algebraic multiplicity

We now characterize the geometric multiplicity algebraically.

Theorem 5.2.1 (The Multiplicity Theorem). Let $(\underline{X},x) \hookrightarrow (\mathbb{C}^n,0)$ be an embedding of $(\underline{X},x) \in \underline{\underline{\mathrm{cpl}}}_0$, $d := \dim_x \underline{X}$, and $L \in P_g^d(\underline{X},x)$. Then

(i) $\deg_x \underline{p}_L \geqq e(\mathfrak{m}_x, \mathcal{O}_{X,x})$;

(ii) if $L \in P_e^d(\underline{X},x)$, $\deg_x \underline{p}_L = e(\mathfrak{m}_x, \mathcal{O}_{X,x})$, and if $(\underline{X},x)$ is pure dimensional, the converse holds;

(iii) $m(\underline{X},x) = e(\mathfrak{m}_x, \mathcal{O}_{X,x})$, i.e. the geometric multiplicity of $(\underline{X},x)$ equals the Samuel multiplicity of $\mathcal{O}_{X,x}$. Especially, $m(\underline{X},x)$ does not depend on the embedding $(\underline{X},x) \hookrightarrow (\mathbb{C}^n,0)$, but only on the isomorphism class of (X,x) in $\underline{\underline{\mathrm{cpl}}}_0$.

Proof.

(i). We have $\deg_x \underline{p}_L = e(\mathfrak{q}_x, \mathcal{O}_{X,x})$ by Theorem 5.1.4, where $\mathfrak{q}_x = \underline{p}^0_{L,x}(\mathfrak{m}_d) \cdot \mathcal{O}_{X,x}$. Since $\mathfrak{q}_x \subseteq \mathfrak{m}_x$ is $\mathfrak{m}_x$-primary, $e(\mathfrak{q}_x, \mathcal{O}_{X,x}) \geq e(\mathfrak{m}_x, \mathcal{O}_{X,x})$ by the definition $e(\mathfrak{q}_x, \mathcal{O}_{X,x})$

(ii). If $L \in P_e^d(\underline{X},x)$, $L \pitchfork_x C(\underline{X},x)$, which means $d_x \underline{p}_L$ is quasi-finite at $x \in C(\underline{X},x)$, and hence finite as a mapgerm $d_x \underline{p}_L : (\underline{C}(\underline{X},x),x) \longrightarrow (\underline{\mathbb{C}}^d,0)$ by I Corollary 4.3.6. So $\mathfrak{q}_x$ is a minimal reduction of $\mathfrak{m}_x$ by Proposition 3.2.3, and so $e(\mathfrak{q}_x, \mathcal{O}_{X,x}) = e(\mathfrak{m}_x, \mathcal{O}_{X,x})$ by Chapter I, Proposition (4.14.). The converse is just the Theorem of Rees, Chapter III, Theorem (19.3).

(iii). This is immediate from (i) and (ii). Q.e.d.

For geometric proofsof Rees's Theorem in the reduced case for the maximal ideal see [13], Th. 6.3 and [75], Chap. 7, Th. 7P. For the geometric interpretation of the general case of Rees' Theorem see III, 3.2.2.

Corollary 5.2.2. $m(\underline{X},x) = m(\underline{C}(\underline{X},x),x)$.

This gives a geometric proof of the following well-known fact:

Proposition 5.2.3. Let $(\underline{X},x) \in \underline{\underline{cpl}}_0$ be equidimensional. Then $m(\underline{X},x) = 1$ implies $(\underline{X},x)$ is smooth.

Proof. $m(\underline{X},x) = m(\underline{C}(\underline{X},x),x)$ by Corollary 5.2.2.

$= \deg(\mathbb{P}\,\underline{C}(\underline{X},x))$ by Theorem 4.1.8, where $\underline{C}(\underline{X},x) \hookrightarrow \mathbb{C}^n$ with $n = \dim_{\mathbb{C}}(\mathfrak{m}_x/\mathfrak{m}_x^2)$. But $\deg(\mathbb{P}\,\underline{C}(\underline{X},x)) = 1$ implies that $\mathbb{P}\,C(\underline{X},x)$ is a $(d-1)$-dimensional linear space (see Exercise) and so $d = n$, since otherwise $\mathfrak{m}_x$ could be generated by less than n elements which cannot be. This proves the claim.

Exercise: $\mathbb{P}\,C(X,x)$ is equidimensional (Hint: Consider 3.1.3. Or blow up $\underline{X}$ at x).

As an application of 5.2.1, we now prove:

Theorem 5.2.4. (Upper Semicontinuity of Multiplicity). Let $X \in \underline{\underline{cpl}}$. Then the function $x \longmapsto e(\mathfrak{m}_x,\mathcal{O}_{X,x})$ is upper semicontinuous, i.e. any $x \in X$ has a neighbourhood U such that $e(\mathfrak{m}_{x'},\mathcal{O}_{X,x'}) \leq e(\mathfrak{m}_x,\mathcal{O}_{X,x})$ for all $x' \in U$.

Proof. Since the claim is local, we may assume $(\underline{X},x) \hookrightarrow (\mathbb{C}^n,0)$ for some n . Let $L \in P^d_e(\underline{X},x)$ where $d = \dim_x X$, then $L \pitchfork_x X$ by Proposition 3.2.4 (i), and so $p_L : (\underline{X},x) \longrightarrow (\mathbb{C}^d,0)$ is quasifinite and hence finite by Proposition 3.2.2. So $L + x' \in P^d_g(\underline{X},x')$ for x' near x . Choosing U sufficiently small, we have

$$e(\mathfrak{m}_x,\mathcal{O}_{X,x}) = \deg_x \underline{p}_L \qquad \text{by Theorem 5.2.1, (ii)}$$

$$= \sum_{\tilde{x}\in \underline{p}^{-1}(x')} \deg_{\tilde{x}} \underline{p}_L \qquad \text{by Theorem 2.2.8}$$

$$\geq \deg_{x'} \underline{p}_L$$

$$\geq e(\mathfrak{m}_{x'},\mathcal{O}_{X,x'}) \qquad \text{by Theorem 5.2.1, (i)} .$$

III. GEOMETRIC EQUIMULTIPLICITY

As exposed in the preface of this book, one of the numerical conditions to be imposed on a subspace $\underline{Y}$ of a complex space $\underline{X}$ as to qualify for a suitable centre of blowing up is that $\underline{X}$ should have the same multiplicity along $\underline{Y}$. This condition has been studied algegraically in Chapter IV, and it is the purpose of this part to give a description of it from a geometric point of view.

The appropriate geometric property of the blowup $\tilde{\underline{X}} \xrightarrow{\pi} \underline{X}$ of $\underline{X}$ along $\underline{Y}$ which is controlled by the multiplicity in case $\underline{Y}$ is smooth is the equidimensionality of $\underline{\pi}$ restricted to the exceptional divisor. In terms of the normal cone, it is called normal pseudoflatness of $\underline{X}$ along $\underline{Y}$; in terms of local algebra, it is just the condition $ht(I) = s(I)$, where I defines $\underline{Y}$ in $\underline{X}$ locally. Normal pseudoflatness has been introduced by Hironaka in [34], and the name originates from the fact that it is just that weaker version of normal flatness which keeps the essential topological properties of the latter. The surprising fact that equimultiplicity is equivalent to normal pseudoflatness is due in the special case of a surface along a smooth curve to Zariski, and, in the general case,to Hironaka and Schickhoff.

In the first paragraph I introduce the notions of normal cone, blowup,and normal flatness and pseudoflatness for the complex analytic case. In the following section, I give a detailed account of the result of Hironaka and Schickhoff and related results of Lipman and Teissier. These results could have been,in principle, mostly derived from the corresponding algebraic results by the method of compact Stein neighbourhoods, but I have preferred to give a geometric proof more or less along the original lines. This was done partly to give an introduction to the geometric method, where multiplicity appears as a local mapping degree and which is used explicitely by the authors mentioned above, and partly to illustrate the geometric content of various other algebraic notions and methods; in particular, the relation of equimultiplicity with reduction and integral dependence, which is emphasized in the preface of this book, is commented on. The last paragraph, finally, describes more briefly the geometric content of equimultiplicity and normal flatness along a nonsmooth centre, where equimultiplicity in the former sense has to be modified to a general type of multiplicity, which however, can again be described geometrically by local mapping degrees.

My general contention is that the relation between equimultiplicity and normal pseudoflatness asserts, on the geometric level, that the local mapping degree of a linear projection of a complex spacegerm (embedded in a number space) is a measure for the contact of the kernel of the projection with the spacegerm at the intersection. In that sense, the requirement of equimultiplicity of a space X along a subspace Y puts a transversality condition on the intersection of the space with the family of projections defining the multiplicity of the space along the subspace. This transversality appears as growth conditions on the local coordinates of X in directions normal to Y, and so the relations with integral dependence and normal pseudoflatness emerge. From this the fundamental rôle of the Theorem of Rees-Böger should be apparent, and I have tried to indicate the connections with this theorem at the appropriate places.

§ 1. Normal flatness and pseudoflatness

Here I discuss the notions of normal flatness and normal pseudoflatness of a complex space along a closed complex subspace. Basic is the result that these notions are open, and generic when the subspace is reduced. It is derived from the algebraic case by the method of compact Stein neighbourhoods, and for this some technical preparations are needed, which are supplied in 1.1. In 1.2 the notions of the analytic and projective spectrum over an arbitrary base $\underline{S} \in \underline{\underline{cpl}}$ are discussed; these constructions are fundamental for the construction of the normal cone and of the blowup. Section 1.3 contains a proof that flatness is open, and generic along a reduced base. Finally, in 1.4, we define the normal cone, the blowup, and discuss normal flatness and normal pseudoflatness.

1.1. Generalities from Complex Analytic Geometry

In the sequel I need some general facts from Complex Analytic Geometry which I collect here.

First some notation. Let $\underline{X}$ be a complex space. If $x \in X$, $\mathfrak{p} \subseteq \mathcal{O}_{X,x}$ a prime, I put

$$(1.1.1) \qquad \mathbb{k}(\mathfrak{p}) := \mathrm{Quot}(\mathcal{O}_{X,x}/\mathfrak{p}) \quad ,$$

the residue field of the local ring $(\mathcal{O}_{X,x})_{\mathfrak{p}}$. Let M be a coherent $\mathcal{O}_X$-module, $x \in X$, then

$$M(x) := M_x/\mathfrak{m}_x M_x = M_x \otimes_{\mathcal{O}_{X,x}} \mathbb{C} \quad . \tag{1.1.2}$$

Proposition 1.1.1. *Let the notation be as above.*

(i) $\dim_{\mathbb{C}} M(x) \geq \dim_{\mathbb{k}(\mathfrak{p})}(M_x \otimes_{\mathcal{O}_{X,x}} \mathbb{k}(\mathfrak{p}))$ *for all* $\mathfrak{p} \in \mathrm{Spec}(\mathcal{O}_{X,x})$.

(ii) $\dim_{\mathbb{C}} M(x) \geq \dim_{\mathbb{C}} M(x')$ *for all* x' *near* x , *i.e. the function* $y \longmapsto \dim_{\mathbb{C}} M(y)$ *is upper semicontinuous.*

(iii) *The freenees locus* $LF(M) := \{x \in X \mid M$ *is locally free at* $x\}$ *is the complement of an analytic set* $\mathrm{Deg}(M)$.

(iv) *If* X *is reduced,* M *is locally free at* x *if and only if the function* $y \longmapsto \dim_{\mathbb{C}} M(y)$ *is constant near* x . *Further,* $\mathrm{Deg}(M)$ *is nowhere dense.*

Proof.

(i). Let $m := \dim_{\mathbb{C}} M(x)$. Then m generators of M_x over $\mathcal{O}_{X,x}$ give m generators of $(M_x)_{\mathfrak{p}}$ over $(\mathcal{O}_{X,x})_{\mathfrak{p}}$. Then apply Nakayama's Lemma.

(ii). Let $F_n(M)$ be the n-th Fitting ideal of M (cf. I 3.2.h)) and $\underline{X}_n(M)$ the closed complex subspace defined by it. Tensorizing the exact sequence of I (3.2.6) at x with $\mathbb{C}$ shows

$$X_n(M) = \{y \in X \mid \dim_{\mathbb{C}} M(y) > n\} \quad . \tag{1.1.3}$$

Now, with $m = \dim_{\mathbb{C}} M(x)$, $x \in X - X_m(M)$, which is open.

(iii). It is easy to see that

$$M_x \text{ is locally free of rank } n \iff F_n(M)_x = {}_{X,x} \text{ and } F_{n-1}(M)_x = 0 \ . \tag{1.1.4}$$

Hence,

$$LF(M) = X - \bigcap_{n\geq 0}(X_n(M) \cup \mathrm{supp}\, F_{n-1}(M)) \quad , \tag{1.1.5}$$

and $\bigcap_{n\geq 0}(X_n(M) \cup \mathrm{supp}\, F_{n-1}(M))$ is analytic since the family $(X_n(M) \cup \mathrm{supp}\, F_{n-1}(M))_{n\in\mathbb{N}}$ becomes locally stationary.

(iv). Let $r := r(M) := \min\{\dim_{\mathbb{C}} M(x) \mid x \in X\}$. Then $X(r) := X - X_r(M)$ is nonempty and open. Now all $x \in X(r)$ are in $X_{r-1}(M)$, so $F_{r-1}(M) \mid X(r) \subseteq N_X \mid X(r)$, which implies $F_{r-1}(M)_x = 0$ for $x \in X(r)$ since $\underline{X}$ is reduced. The claim now follows by replacing $\underline{X}$ with any open neighbourhood of a given $x \in X$ and applying (1.1.4).

Theorem 1.1.2 (Cartan). Let M be a coherent module on the complex space X and $M_0 \subseteq M_1 \subseteq M_2 \subseteq \dots \subseteq M$ an increasing chain of coherent submodules. Then this chain is locally stationary.

For a slick elementary proof see [28] , Chapter 5, § 6; see also [14], 0.40.

Next, we set up a formalism ([5],[29],[38],[41],[63]) by which results in Algebraic Geometry can often be transferred to Complex Analytic Geometry; we will use it in 1.4 to deduce the fact that normal flatness is generic from the Krull-Seidenberg-Theorem in Chapter IV, (24.4). This idea seems to have originated from footnote 18 on p. 136 of [33]. We partly follow the presentation of [38].

In the following, $\underline{X}$ is a local model in some open set $U \subseteq \mathbb{C}^n$.

Definition 1.1.3. A distinguished compact Stein set in $\underline{X}$ is a compact neighbourhood of some $x \in X$ of the form $Q \cap X$, where Q is a compact stone in U .

By II Corollary 1.4.2, any $x \in X$ has a neighbourhood basis consisting of distinguished compact Stein subsets.

We first need a noetherian property for distinguished compact Stein subsets. The following result is a special case of a theorem due to Frisch ([16], Théorème (I, 9)) and Siu ([68], Theorem 1).

Proposition 1.1.4. Let K be a distinguished compact Stein subset in a complex space $\underline{X}$. Then $\mathcal{O}(K) = \Gamma(K,\mathcal{O}_X)$ is a noetherian ring.

Proof. We may assume $\underline{X} \overset{i}{\hookrightarrow} \underline{U}$ is a local model, where $U \subseteq \mathbb{C}^n$. Let $Q \subseteq U$ be a compact stone which defines K , i.e. $K = X \cap Q$. The surjection $\mathcal{O}_U \twoheadrightarrow i_*\mathcal{O}_X$ induces the surjection $\Gamma(Q,\mathcal{O}_U) \twoheadrightarrow \Gamma(K,\mathcal{O}_X)$ by Theorem B. So it suffices to prove $\Gamma(Q,\mathcal{O}_U)$ is noetherian. For this we induct over the real dimension d of Q .

If $d = 0$, Q is a point, and the claim is just the Rückert Basissatz, I 1.3.2. Let $d \geq 1$, and suppose the proposition is true for (d-1)-dimensional compact stones. Suppose $I \subseteq \Gamma(Q,\mathcal{O}_X)$ were not finitely generated, so we can find a sequence $f_1, f_2, f_3, \dots$ of elements in I such that we get a strictly increasing sequence

$I_1 \subset I_2 \subset I_3 \subset \ldots$ with $I_j := (f_1,\ldots,f_j)\cdot\Gamma(Q,\mathcal{O}_X)$.

Now we may write

$$(1.1.6) \qquad Q = \bigcup_{\ell=1}^{2(d+1)} Q_\ell \amalg \overset{\circ}{Q} \quad ,$$

where the Q_ℓ are compact $(d-1)$-dimensional stones, and $\overset{\circ}{Q}$ is a stone which is open in the real vectorsubspace of $\mathbb{C}^n$ spanned by Q . By the induction assumption there are finitely many elements $g_1,\ldots,g_t \in \Gamma(Q,\mathcal{O}_X)$ such that $I\cdot\Gamma(Q_\ell,\mathcal{O}_X) = (g_1,\ldots,g_t)\cdot\Gamma(Q_\ell,\mathcal{O}_X)$ for $\ell = 1,\ldots,2(d+1)$. Let U be an open neighbourhood of Q in $\mathbb{C}^n$ such that $g_1,\ldots,g_t \in \Gamma(U,\mathcal{O}_X)$. Define ideal sheaves $\mathcal{I}_j \subseteq \mathcal{O}_U$ via

$$(1.1.7) \qquad \mathcal{I}_j(V) := \begin{cases} (g_1,\ldots,g_t)\cdot\mathcal{O}_V \ , & V \subseteq U - Q \ \text{ open} \\ (g_1,\ldots,g_t,f_1,\ldots,f_j)\cdot\mathcal{O}_V \ , & V \subseteq U \ \text{ open, } V \cap Q \neq \emptyset \ . \end{cases}$$

Then $\mathcal{I}_1 \subset \mathcal{I}_2 \subset \mathcal{I}_3 \subset \ldots$ is a strictly increasing sequence of coherent $\mathcal{O}_U$-ideals, so it cannot become eventually stationary on the compact set Q . This contradicts Theorem 1.1.2. Q.e.d.

A point $x \in K$ defines a character $\chi_x : \mathcal{O}(K) \longrightarrow \mathbb{C}$ via $\chi_x(f) := f(x)$, called a <u>point character</u>. Its kernel is a maximal ideal of $\mathcal{O}(K)$, denoted M_x . Let $\underline{K}$ be the ringed space $(K,\mathcal{O}|K)$, and $\underline{\mathrm{Spec}}(\mathcal{O}(K))$ be the usual prime spectrum as a ringed space. We get a map of ringed spaces

$$(1.1.8) \qquad \underline{\Phi}_K : \underline{K} \longrightarrow \underline{\mathrm{Spec}}(\mathcal{O}(K))$$

by putting

$$(1.1.9) \qquad \phi_K(x) := M_x = \mathrm{Ker}(\chi_x) \quad \text{for} \quad x \in K \ ,$$

and

$$(1.1.10) \qquad \begin{aligned} \phi^0_{K,D(f)} := \mathcal{O}(K)_{(f)} &\longrightarrow \Gamma(D(f),\mathcal{O}_X) \\ g/f^m &\longmapsto (x \longmapsto g(x)/f(x)^m) \end{aligned}$$

for $f \in \mathcal{O}(K)$.

We call a subset $A \subseteq K$ <u>analytic in</u> K if there is an analytic subset $\widetilde{A}$ of some open neighbourhood $V \supseteq K$ such that $A = \widetilde{A} \cap K$; this is the same as requiring that there is a finitely generated ideal sheaf $\mathcal{I} \subseteq \mathcal{O}|K$ such that $A = N(\mathcal{I})$. The following result is basic.

Proposition 1.1.5. If $B \subseteq \mathrm{Spec}(\mathcal{O}(K))$ is Zariski-closed, $\phi_K^{-1}(B) =: A$ is analytic in K , in fact $A = N(\mathcal{I})$, when $B = V(I)$ for $I \subseteq \mathcal{O}(K)$ an ideal and $\mathcal{I} = I \cdot \mathcal{O}|K$. In particular, ϕ_K is a morphism of ringed spaces.

Proof. Let $B = V(I)$; since $\mathcal{O}(K)$ is noetherian, $I = (f_1,\dots,f_k)\cdot\mathcal{O}(K)$ for some $f_1,\dots,f_k \in \mathcal{O}(K)$, and $\mathcal{I} = (f_1,\dots,f_k)\cdot\mathcal{O}|K$. Then $x \in \phi_K^{-1}(V(I)) \Longleftrightarrow M_x \supseteq I \Longleftrightarrow f_1,\dots,f_k \in \mathrm{Ker}(\mathcal{X}_x) \Longleftrightarrow x \in N(\mathcal{I})$.

Remark 1.1.6.

(i) The sheaf morphism ϕ_K^0 is regular on the stalks. From this one may deduce the openness of certain analytic loci, e.g. the regular locus, the Cohen-Macaulay locus, or the normal locus of a complex space, from the corresponding scheme-theoretic results, which, as a rule, are easier to prove; see [38].

(ii) One may use Proposition 1.1.5 to deduce the openness of the flatness locus of a coherent $\mathcal{O}_X$-module $\mathcal{M}$ with respect to a morphism $\underline{f} : \underline{X} \longrightarrow \underline{Y}$ of complex spaces from the corresponding algebraic result (Theorem of Frisch); see [41].

1.2. The analytic and projective analytic spectrum

This section generalizes I, 3.4 and II, 4.3 to the case of families of affine respectively projective varieties parametrized by a complex space.

Definition 1.2.1. Let $\underline{S} \in \underline{\underline{\mathrm{cpl}}}$, $\mathcal{A}$ a sheaf of $\mathcal{O}_S$-algebras. $\mathcal{A}$ is called an admissible $\mathcal{O}_S$-algebra, or an $\mathcal{O}_S$-algebra locally of finite presentation, if every $x \in S$ has an open neighbourhood U such that there are sections $g_1,\dots,g_\ell \in \mathcal{O}_S(U)[T_1,\dots,T_k]$ and an epimorphism

(1.2.1) $$\psi_U : \mathcal{O}_U[T_1,\dots,T_k] \twoheadrightarrow A|U$$

of $\mathcal{O}_U$-algebras such that $Ker(\psi_U)$ is the ideal generated by $g_1,\dots,g_\ell$.

Now consider the category $\underline{\underline{cpl/S}}$ of complex spaces over $\underline{S}$, whose objects are the morphisms $\underline{\varphi} : \underline{W} \longrightarrow \underline{S}$ in $\underline{\underline{cpl}}$ and whose morphisms are the commutative diagrams

(1.2.2) $$\begin{array}{ccc} \underline{W} & \xrightarrow{\underline{f}} & \underline{W}' \\ & \underline{\varphi} \searrow \quad \swarrow \underline{\varphi}' & \\ & \underline{S} & \end{array}$$

Then an $\mathcal{O}_S$-algebra A induces a contrafunctor

(1.2.3) $$\mathrm{Hom}_{\mathcal{O}_S\text{-}\underline{\underline{alg}}}(A,-) : \underline{\underline{cpl/S}} \longrightarrow \underline{\underline{sets}}$$

as follows: It assigns to an object $\underline{\varphi} : \underline{W} \longrightarrow \underline{S}$ in $\underline{\underline{cpl/S}}$ the set $\mathrm{Hom}_{\mathcal{O}_S\text{-}\underline{\underline{alg}}}(A,\underline{\varphi}_*\mathcal{O}_W)$, and to the commuative triangle (1.2.2) the map

(1.2.4) $$\begin{array}{ccc} \mathrm{Hom}_{\mathcal{O}_S\text{-}\underline{\underline{alg}}}(A,\underline{\varphi}'_*\mathcal{O}_{W'}) & \longrightarrow & \mathrm{Hom}_{\mathcal{O}_S\text{-}\underline{\underline{alg}}}(A,\underline{\varphi}_*\mathcal{O}_W) \\ \alpha & \longmapsto & \underline{\varphi}'_*(f^0) \circ \alpha \end{array} .$$

Theorem 1.2.2 (see [64], Exposé 19). If A is an admissible $\mathcal{O}_S$-algebra, the functor (1.2.3) is representable in $\underline{\underline{cpl/S}}$.

This means the following: There is an object $\underline{\pi}_X : \underline{X} \longrightarrow \underline{S}$ in $\underline{\underline{cpl/S}}$ and an element $\zeta_X \in \mathrm{Hom}_{\mathcal{O}_S\text{-}\underline{\underline{alg}}}(A,(\underline{\pi}_X)_*\mathcal{O}_X)$ such that the natural transformation

(1.2.5) $$\mathrm{Hom}_{\underline{\underline{cpl/S}}}(-,\pi_X) \longrightarrow \mathrm{Hom}_{\mathcal{O}_S\underline{\underline{alg}}}(A,-) \quad ,$$

which assigns to $\underline{\varphi} : \underline{W} \longrightarrow \underline{S} \in \underline{\underline{cpl/S}}$ the map

(1.2.6) $$\mathrm{Hom}_{\underline{\underline{\mathrm{cpl}/S}}}(\underline{\varphi},\underline{\pi}_X) \longrightarrow \mathrm{Hom}_{\mathcal{O}_S\text{-}\underline{\underline{\mathrm{alg}}}}(A,\underline{\varphi}_*\mathcal{O}_W)$$

$$\begin{array}{ccc} \underline{W} & \xrightarrow{\underline{f}} & \underline{X} \\ & \underline{\varphi}\searrow \quad \swarrow \underline{\pi}_X & \\ & \underline{S} & \end{array} \longmapsto (\underline{\pi}_X)_*(\underline{f}^0) \circ \zeta_X \quad ,$$

is a natural equivalence of functors.

As usual, the pair $(\underline{\pi}_X,\zeta_X)$ is unique up to unique isomorphism.

The universal property together with the glueing construction I 3.2 a) reduces the proof to the case $A = \mathcal{O}_S[T_1,\dots,T_k]/I$, where I is generated by sections $g_1,\dots,g_\ell \in \mathcal{O}_S(S)[T_1,\dots,T_k]$. Now there is a natural morphism $\mathcal{O}_S(S)[T_1,\dots,T_k] \longrightarrow \mathcal{O}_{S\times\mathbb{C}^k}(S\times\mathbb{C}^k)$, hence $g_1,\dots,g_\ell$ generate an ideal $J \subseteq \mathcal{O}_{S\times\mathbb{C}^k}$, and one defines $\underline{\pi}_X$ via

(1.2.7) $$\begin{array}{ccc} \underline{X} := \underline{N}(J) & \overset{\underline{i}}{\hookrightarrow} & \underline{S}\times\underline{\mathbb{C}}^k \\ \underline{\pi}_X \downarrow & \circlearrowleft & \swarrow \underline{pr}_S \\ \underline{S} & & \end{array}$$

The homomorphism $\zeta : \mathcal{O}_S(S)[T_1,\dots,T_k] \hookrightarrow \mathcal{O}_{S\times\mathbb{C}^k}(S\times\mathbb{C}^k) \xrightarrow{i_X^0} \mathcal{O}_X(\underline{\pi}^{-1}S)$,

factors through I and restricts over any open $U \subseteq S$, defining ζ_X . Details are left to the reader.

Definition 1.2.3. The pair $(\underline{\pi}_X,\zeta_X)$, or, if no confusion is possible, the complex space $\underline{X}$ over $\underline{S}$, is called the analytic spectrum of the admissible $\mathcal{O}_S$-algebra A and denoted $\underline{\mathrm{Specan}}(A)$.

We also write, 'par abus de languague', $\underline{\pi}_A$: $\underline{\mathrm{Specan}}(A) \longrightarrow \underline{S}$ for $\underline{\pi}_X : \underline{X} \longrightarrow \underline{S}$.

The analytic spectrum has the expected functional properties, see [64], Exposé 19. We mention here:

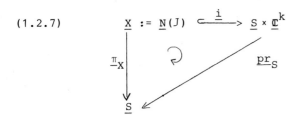

Proposition 1.2.4 (Base change). Let A be an admissible $\mathcal{O}_S$-algebra, $\underline{\varphi} : \underline{T} \longrightarrow \underline{S} \in \underline{\underline{cpl}}$. Let $\underline{\psi} : \underline{\text{Specan}}(\underline{\varphi}^*A) \longrightarrow \underline{\text{Specan}}(A) \in \underline{\underline{cpl}}$ correspond to the canonical morphism $A \longrightarrow \underline{\varphi}_*\underline{\varphi}^*A$ via (1.2.6). Then the diagram

$$(1.2.8)\qquad \begin{array}{ccc} \underline{\text{Specan}}(\varphi^*A) & \xrightarrow{\ \underline{\psi}\ } & \underline{\text{Specan}}(A) \\ \Big\downarrow{\scriptstyle \underline{\pi}_{\underline{\varphi}^*A}} & \circlearrowright & \Big\downarrow{\scriptstyle \underline{\pi}_A} \\ \underline{T} & \xrightarrow[\ \underline{\varphi}\ ]{} & \underline{S} \end{array}$$

is cartesian, i.e. $\underline{\text{Specan}}(\underline{\varphi}^*A) = \underline{\text{Specan}}(A) \times_{\underline{S}} \underline{T}$.

From this we see the following: Let A_s be the stalk of A at $s \in S$, $\mathfrak{m}_s \subseteq \mathcal{O}_{S,s}$ the maximal ideal, and put

$$(1.2.9)\qquad A(s) := A_s/\mathfrak{m}_s \cdot A_s = A_s \otimes_{\mathcal{O}_{S,s}} \mathbb{C} \quad ,$$

which is a finitely generated $\mathbb{C}$-algebra. Then in 1.2.7.

$$(1.2.10)\qquad \underline{X}_s := \underline{\pi}^{-1}(s) = \underline{\text{Specan}}(A(s))$$

by base change, i.e. we may think informally of $\underline{X} = \underline{\text{Specan}}(A)$ as a family of affine varieties (considered as complex spaces) parametrized by the points of the complex space $\underline{S}$ via $\underline{\pi} : \underline{X} \longrightarrow \underline{S}$. This motivates the following result, which I just quote:

Proposition 1.2.5 ([64], Exposé 19, Prop. 3 and 4).

(i) The points of $\underline{X}_s$ correspond bijectively to the elements of $Vm(\mathfrak{m}_s A_s) := \{\mathfrak{n} \in \text{Specm}(A_s) \mid \mathfrak{n} \supseteq \mathfrak{m}_s A_s\}$ under $x \in \underline{X}_s \longmapsto \text{Ker}(A_s \xrightarrow{\zeta^0_{X,s}} \mathcal{O}_{X,x})$

(ii) Let $\mathfrak{n} \in Vm(\mathfrak{m}_s A_s)$ correspond to $x \in \underline{X}_s$. Then $\zeta^0_{X,s}$ factors as $A_s \longrightarrow (A_s)_{\mathfrak{n}} \xrightarrow{\varphi_x} \mathcal{O}_{X,x}$, and

$$(1.2.11)\qquad \hat{\varphi}_x : (A_s)_{\mathfrak{n}} \longrightarrow \hat{\mathcal{O}}_{X,x}$$

is an isomorphism.

We now come to the projective analytic spectrum.

Definition 1.2.6. Let $S \in \underline{cpl}$. An admissible graded $\mathcal{O}_S$-algebra is an admissible $\mathcal{O}_S$-algebra such that

(i) A is positively graded, i.e. $A = \bigoplus_{n \geq 0} A_n$, and locally generated by A_1 as $\mathcal{O}_S$-algebra.

(ii) The local representations (1.2.1) can be so chosen that ψ_U is a graded homomorphism of degree zero, where $T_1, \ldots, T_k$ have degree one.

Proposition 1.2.7 ([47], 1.4). Let A be a graded $\mathcal{O}_S$-algebra which is locally finitely generated as $\mathcal{O}_S$-algebra. Then the following statements are equivalent:

(i) A is an admissible graded $\mathcal{O}_S$-algebra.

(ii) A_k is a coherent $\mathcal{O}_S$-module for all $k \geq 0$.

Since the reference may be not easily accessible, I give a short idea of the proof.

(i) ⇒ (ii): Consider (1.2.1); $\mathcal{K}er(\psi_U)$ is a locally finite $\mathcal{O}_U$-module, so $A_k|U \cong (\mathcal{O}_U[T_1, \ldots, T_n]_k)/\mathcal{K}er(\psi_U)_k$ is coherent.

(ii) ⇒ (i): The question is local, so we may assume we have an epimorphism

$$(1.2.12) \qquad \mathcal{O}_S[T_1, \ldots, T_k] \xrightarrow{\psi} A$$

of graded $\mathcal{O}_S$-algebras. Let $K := \mathcal{K}er(\psi)$, and put for $n \in \mathbb{N}$

$$(1.2.13) \qquad A^{(n)} := \mathcal{O}_S[T_1, \ldots, T_k] / \bigoplus_{k \leq n} K_k \quad .$$

Then $A^{(0)} \twoheadrightarrow A^{(1)} \twoheadrightarrow \ldots$ is a decreasing tower of admissible $\mathcal{O}_S$-algebras. This gives us an increasing chain of coherent $\mathcal{O}_{S \times \mathbb{C}^k}$-ideals

$I^{(0)} \subseteq I^{(1)} \subseteq \dots$, where $I^{(h)}$ defines $\underline{X}^{(n)} := \underline{\mathrm{Specan}}(A^{(n)}) \subseteq \underline{S} \times \underline{\mathbb{C}}^k$. The claim then follows from Theorem 1.1.2.

If A is an admissible graded $\mathcal{O}_S$-algebra, we have local representations (1.2.1) with $\mathcal{Ker}(\psi_U)$ homogeneous. Therefore, in the local construction of $\underline{\mathrm{Specan}}(A)$ in diagram (1.2.7), the $\mathcal{O}_S$-homogeneous ideal J defines a closed complex subspace $\underline{Z} \hookrightarrow \underline{S} \times \underline{\mathbb{P}}^{k-1}$, and we get the commutative diagram

(1.2.14)

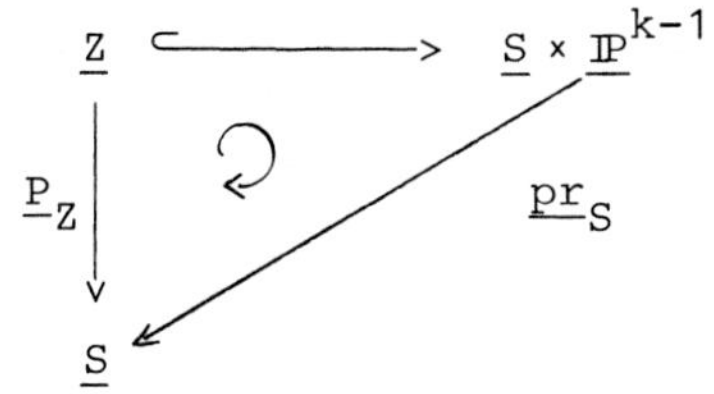

The $\underline{p}_Z$ glue well because of the functorial properties of the Specan-construction; so, for any admissible graded $\mathcal{O}_S$-algebra, we have constructed a complex space $p_{\underline{Z}} : \underline{Z} \longrightarrow \underline{S}$ over $\underline{S}$.

Definition 1.2.8. The space so obtained is called the projective analytic spectrum of A and denoted $\underline{p}_A : \underline{\mathrm{Projan}}(A) \longrightarrow \underline{S}$, or $\underline{\mathrm{Projan}}(A)$ for short.

Remark 1.2.9. As in 1.2.4, base change holds for the Projan-construction.

1.3. Flatness of admissible graded algebras

Definition 1.3.1. Let $\underline{S} \in \underline{\mathrm{cpl}}$, A an admissible $\mathcal{O}_S$-algebra. Then A is called flat along S at $s \in \underline{S}$ if and only if A_s is a flat $\mathcal{O}_{S,s}$-module. A is called flat along S if and only if it is flat along S at all $s \in S$.

Remark 1.3.2. If A is flat along S , $(A_s)_n$ is $\mathcal{O}_{S,s}$-flat for all

s and all $\mathfrak{n} \in \operatorname{Specan}(A_s)$, hence $\widehat{(A_s)_\mathfrak{n}} = \hat{\mathcal{O}}_{X,x}$ is $\hat{\mathcal{O}}_{S,s}$ flat for all $s \in S$ and $x \in \underline{\pi}_X^{-1}(s)$, where $\underline{\pi}_X : \underline{X} \longrightarrow \underline{S}$ is $\operatorname{Specan}(A)$, by Proposition 1.2.5. It follows that $\underline{\pi}_X : \underline{X} \longrightarrow \underline{S}$ is a flat morphism.

Proposition 1.3.3. Let $S \in \underline{\text{cpl}}$ be reduced, A an admissible graded $\mathcal{O}_S$-algebra. The following statements are equivalent:

(i) A is a flat $\mathcal{O}_S$-algebra.

(ii) The functions $s \longmapsto \dim_{\mathbb{C}} A_k(s)$ (see (1.2.9)) are locally constant for all k.

Proof. A is a flat $\mathcal{O}_S$-algebra if and only if A_k is a flat $\mathcal{O}_S$-module for all k. But each A_k is a coherent $\mathcal{O}_S$-module by Proposition 1.2.7. The claim then follows from Proposition 1.1.1 (iv), since over a local ring, to be flat means to be free.

We now have the following theorem, which has been stated by Hironaka in [33], p. 136, and proved by means of Proposition 1.1.3 in [38], and by other means in [47].

Theorem 1.3.4 (Flatness is generic). Let A be an admissible graded $\mathcal{O}_S$-algebra on the complex space S. Then the set $F(A) := \{s \in S \mid A_s$ is a flat $\mathcal{O}_{S,s}$-module$\}$ is the complement of an analytic set. If S is reduced, $S - F(A)$ is nowhere dense.

Proof. The question is local. Let $K \subseteq S$ be a distinguished compact Stein subset, and let $A_k := \Gamma(K, A_k)$, $A = \bigoplus_{k \geq 0} A_k$, $R := \Gamma(K, \mathcal{O}_X) = \mathcal{O}(K)$; R is noetherian by Proposition 1.1.4. Let $s \in K$. Then

(1.3.1) A_s is $\mathcal{O}_{S,s}$-flat $\iff \forall k \geq 0 : (A_k)_s$ is $\mathcal{O}_{S,s}$-flat

$\iff \forall k \geq 0 : (A_k)_{M_s}$ is R_{M_s}-flat, since $R_{M_s} \longrightarrow \mathcal{O}_{S,s}$ is faithfully flat by II 1.3.2.

$\iff A_M$ is R_{M_s}-flat .

Hence $K \cap F(A) = \phi_K^{-1}(F(A))$. The first claim now follows by the Krull-Seidenberg-Grothendieck - Theorem (Chapter IV, (24.4)) and by Proposition 1.1.5. The second claim follows from Proposition 1.1.1 (iv) and 1.3.3. (ii):

(1.3.2) $$S - F(A) = \bigcup_{k \geq 0} \mathrm{Deg}(A_k)$$

has empty interior as a countable union of nowhere dense analytic sets by the theorem of Baire.

Remark 1.3.5. Theorem 1.3.4 can be interpreted more concretely, without using the Krull-Seidenberg-Grothendieck-Theorem, as follows, using 1.3.3. instead. Let $\underline{S}$ be reduced. Then 1.3.4 would follow from 1.3.3, if one were able to show that the Hilbert functions $H(A(s),-)$ were constant for s near s_0 , i.e. if $k \longmapsto H(A(s),k)$ were independent of s near s_0 . Note that this is a priori stronger that the statement (ii) of 1.3.3, since the neighbourhoods of s_0 on which the functions $\dim_{\mathbb{C}} A_k(s)$ are constant might depend on k .

Now it is known that each Hilbert function $k \longmapsto H(A(s),k)$ becomes a polynomial, of degree $d_0(s)-1$, say, for k above some number $k_0 = k_0(s)$, and so is determined by any $d_0(s)$ values at numbers $k > k_0(s)$. So the constancy of finitely many functions $\dim_{\mathbb{C}} A_k(s)$ near s_0 would guarantee the constancy of all of them if we were able to bound $d_0(s)$ and $k_0(s)$ near s_0 ; this would then imply 1.3.4 because of 1.3.3 (ii). So what one wants to show is:

(1.3.3) For any $s_0 \in S$, there are a neighbourhood U of s_0 and natural numbers d_0 and k_0 such that $H(A(s),k)$ is a polynomial in k for all $k > k_0$ of degree $< d_0$ for $s \in U$.

There might be two ways to establish (1.3.3). For the first one, results of Grauert and Remmert for projective morphisms over a basis in cpl (concering the vanishing of the sheaves $(R^i \underline{p})_* \mathcal{M}(n)$ for $\underline{p} = \underline{p}_A$: Projan$(A) \longrightarrow \underline{S}$ and $\underline{\mathcal{M}}$ a coherent module on Projan(A) and generalizing well-known facts from the scheme-theoretic case; (see [25], [2] Chapter IV)) suggest that one should have: There is a neighbourhood U of s_0 and a number k_0 such that

$$H(A(s),k) = \chi(\underline{Z}_s, \mathcal{O}_{Z_s}(k)) \quad \text{for} \quad k \geq k_0\ , s \in U \quad ,$$

where $\underline{p} : \underline{Z} \longrightarrow \underline{S}$ is $\underline{\text{Projan}(A)}$, $\underline{Z}_s$ the fibre $\underline{p}^{-1}(s)$, and $\mathcal{O}_{Z_s}(k) = \mathcal{O}_Z(1)^{\otimes k}$, $\mathcal{O}_Z(1)$ the canonical linebundle on $\underline{Z}$. Then (1.3.3) holds with $d_0 = \max\{\dim \underline{Z}_s \mid s \in U\} + 1$.

The other approach might be based on a parametrized version of the division algorithm for rings of the form $\mathcal{O}_{S,s}[Z_1,\ldots,Z_\ell]$ (see [20], (1.2.7) and [62], 1.3). Applying this to the ideal $I_{s_0} \subseteq \mathcal{O}_S[Z_1,\ldots,Z_\ell]$, where $A \cong \mathcal{O}_S[T_1,\ldots,T_\ell]/I$ locally, should give a leitideal generated by monomials $\lambda_A Z^A$, where λ_A are germs in $\mathcal{O}_{S,s_0}$. Now the Hilbert function of a homogeneous ideal $I \subseteq \mathbb{C}[Z_1,\ldots,Z_\ell] =: R$ is the Hilbert function of the leitideal $LM(I)$, and so (see [53])

$$H(R/I,k) = \sum_{j=0}^{t} (-1)^k \sum_{1 \leq i_1 < \ldots < i_j \leq t} \binom{\ell-1+\deg \ell cm(M_{i_1},\ldots,M_{i_j})+k}{\ell-1}_+ ,$$

where the monomials $M_1,\ldots,M_t$ generate $LM(I)$. From this it may be possible to see that $H(A(s),k) = H(R/I_sR,k)$ is constant outside the subspace of (S,s_0) defined by the λ_A and can only increase over there, so that $s \longmapsto H(A(s),k)$ is upper semicontinuous, and that the $H(A(s),k)$ are polynomials for all s near s_0 for k above a fixed value k_0 . (Added in proof: By oral communication of J.L. Vicente this effective approach has been worked out in complete detail in a forthcoming book of Aroca, Hironaka, and Vicente on the resolution of singularities of complex spaces).

1.4. The normal cone, normal flatness, and normal pseudoflatness

Let $\underline{X}$ be a complex space, $\underline{Y} \overset{i}{\hookrightarrow} \underline{X}$ a closed complex subspace, defined by the locally finite ideal $I \subseteq \mathcal{O}_X$.

Lemma 1.4.1. The graded $\mathcal{O}_X$-algebra $\mathcal{B}(I,\mathcal{O}_X) := \bigoplus_{k\geq 0} I^k$ is an admissible graded $\mathcal{O}_X$-algebra.

Proof. Since $I = B_1(I, O_X)$ is locally finite and generates $B(I, O_X)$, the O_X-algebra $B(I, O_X)$ is locally finitely generated. Moreover, I is coherent, so all I^k, $k \geq 0$, are coherent, and the claim follows from Proposition 1.2.7.

Corollary 1.4.2. The graded O_Y-algebra

$$(1.4.1) \qquad G(I, O_X) := \bigoplus_{k \geq 0} I^k / I^{k+1}$$

is an admissible graded O_Y-algebra.

Proof. $G(I, O_X) = i^* B(I, O_X)$, and $B(I, O_X)$ is an admissible graded O_X-algebra.

Hence, the following definition makes sense:

Definition 1.4.3. $\pi_{G(I,O_X)} : \mathrm{Specan}(G(I, O_X)) \longrightarrow \underline{Y}$ is called the normal cone of $\underline{Y}$ in $\underline{X}$ and denoted $\underline{\nu} : \underline{C}(\underline{X}, \underline{Y}) \longrightarrow \underline{Y}$.

For geometric applications to equimultiplicity we need a geometric description of $C(\underline{X}, \underline{Y})$, which will explain the name 'normal cone'. Recall that a blowup $\underline{\pi} : \underline{\tilde{X}} \longrightarrow \underline{X}$ of $\underline{X}$ along $\underline{Y}$ is a morphism which is universal among the morphisms $\varphi : \underline{X}' \longrightarrow \underline{X}$ having the property that $\varphi^{-1}\underline{Y}$ is a hypersurface in $\underline{X}'$, i.e. locally generated by a nonzero-divisor. It is unique up to unique isomorphism.

Theorem 1.4.4. $\underline{p} : \mathrm{Projan}(B(I, O_X)) \longrightarrow \underline{X}$ is the blowup of $\underline{X}$ along $\underline{Y}$.

I will not prove Theorem 1.4.4, but make some remarks which I will use anyway. Let I be generated over the open subspace $\underline{U} \hookrightarrow \underline{X}$ by $g_1, \ldots, g_k \in O_X(U)$, and consider the morphism

$$(1.4.2) \qquad \underline{\gamma} : \underline{U} - Y \longrightarrow \mathbb{P}^{k-1}$$
$$x \longmapsto [g_1(x) : \ldots : g_k(x)] \quad .$$

It can then be shown that $\underline{p}|U$ above is given as

$$(1.4.3)\qquad \begin{array}{ccc} \overline{\underline{\Gamma}_\gamma} & \hookrightarrow & \underline{U} \times \underline{\mathbb{P}}^{k-1} \\ {\scriptstyle \underline{p}|U}\big\downarrow & \circlearrowleft & \swarrow {\scriptstyle \underline{pr}_U} \\ \underline{U} & & \end{array} ,$$

where $\underline{\Gamma}_\gamma \hookrightarrow (\underline{U} - Y) \times \underline{\mathbb{P}}^{k-1}$ is the graph space of γ according to I 3.2 g), and $\overline{\underline{\Gamma}_\gamma}$ is the idealtheoretic closure of $\underline{\Gamma}_\gamma$, i.e. the smallest closed complex subspace of $\underline{U} \times \underline{\mathbb{P}}^{k-1}$ containing $\underline{\Gamma}_\gamma$ as an open subspace (for this see [14], 0.44). It is then not difficult to show, using the factorization criterion for holomorphic maps through a closed complex subspace (see [28], Chapter I, § 2.3), that (1.4.3) constitutes the blowup locally, which proves 1.4.4 by universality. (The diagram (1.4.3) coincides with the local description given by Hironaka and Rossi in [37]; consult this paper for details).

<u>Corollary 1.4.5.</u> <u>If $\underline{\pi} : \underline{\tilde{X}} \longrightarrow \underline{X}$ blows up $\underline{Y}$, $\underline{\pi}^{-1}(\underline{Y}) \cong \mathbb{P}\,C(\underline{X},\underline{Y})$, the projectivized normal cone.</u>

<u>Proof.</u> $\mathbb{P}\,C(\underline{X},\underline{Y})$ is defined as $\underline{\text{Projan}}(G(I,\mathcal{O}_X))$. But $G(I,\mathcal{O}_X) = i^* B(I,\mathcal{O}_X)$, where $i : \underline{Y} \hookrightarrow \underline{X}$ is the inclusion, and the claim follows by base change for Projan (Remark 1.2.9).

This gives the following description of the fibre $\nu^{-1}(y)$ of the normal cone $\underline{\nu} : \underline{C}(\underline{X},\underline{Y}) \longrightarrow \underline{Y}$ at a point $y \in Y$. Choose generators $g_1,\dots,g_k \in \mathcal{O}_{X,y}$ of the stalk I_y, where the ideal $I \subseteq \mathcal{O}_X$ defines $\underline{Y} \hookrightarrow \underline{X}$, and add elements $h_1,\dots,h_f$ such that $h_1,\dots,h_f, g_1,\dots,g_k$ generate the maximal ideal. After possibly shrinking $\underline{X}$, we may assume these generators are in $\mathcal{O}_X(X)$, and they define, according to I 4.2.2, an embedding $\underline{X} \overset{\underline{i}}{\hookrightarrow} \underline{\mathbb{C}}^n$, $n := f + k$, as a locally closed subspace. Then $g_1,\dots,g_k$ are induced by the coordinates $z_{f+1},\dots,z_n$ of $\mathbb{C}^n$ via $\underline{i}$. Let $K := \mathbb{C}^k \times 0$, and let $\underline{p} : \underline{\mathbb{C}}^n \longrightarrow \underline{K}$ be the projection; then $\gamma(x) = p(\overline{yx}) \subseteq K$, and (1.4.3) gives, together with

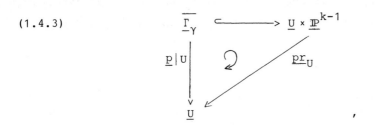

Corollary 1.4.5.

$$\nu^{-1}(y) = \cup\left\{\ell \mid \ell = \lim_{\substack{x\to y\\ x\in X-Y}} p(\overline{yx})\right\} \subseteq K \quad . \tag{1.4.4}$$

Corollary 1.4.6. $\dim_\xi \underline{C}(\underline{X},\underline{Y}) = \dim_{\nu(\xi)}\underline{X}$ for all $\xi \in C(\underline{X},\underline{Y})$. If $(\underline{X},\nu(\xi))$ is equidimensional, so is $(\underline{C}(\underline{X},\underline{Y}),\xi)$.

Proof. There is a canonical embedding $\underline{Y} \hookrightarrow \underline{C}(\underline{X},\underline{Y})$, corresponding to the augmentation homomorphism $G(I,\mathcal{O}_X) \twoheadrightarrow \mathcal{O}_X/I = i_*\mathcal{O}_Y$, where $I \subseteq \mathcal{O}_X$ defines $\underline{Y} \overset{i}{\hookrightarrow} \underline{X}$, via the universal property of the Specan-construction. In the sequel, therefore, we may view $\underline{Y}$ as being naturally embedded in $\underline{C}(\underline{X},\underline{Y})$.

Let $\xi \in C(\underline{X},\underline{Y})$. We may assume $(Y,\nu(\xi)) \neq (X,\nu(\xi))$. First, let $\xi \notin Y$, so it is not a vertex of a fibre of $\underline{\nu}$. Then ξ corresponds to a line on $C(\underline{X},\underline{Y})$, i.e. to a point $x' \in \tilde{X}$, where $\underline{\pi} : \underline{\tilde{X}} \to \underline{X}$ is the blowup of $\underline{X}$ along $\underline{Y}$. Now $\underline{\pi}|\underline{\tilde{X}} - \pi^{-1}(Y) : \underline{\tilde{X}} - \pi^{-1}(Y) \to \underline{X} - Y$ is isomorphic; so there are points on X_{reg} arbitrarily close to $\pi(x') =: x = \nu(\xi)$, hence $\dim_{x'}\tilde{X} = \dim_{\nu(\xi)}X$. Since $\underline{\pi}^{-1}(\underline{Y}) = \mathbb{P}\,\underline{C}(\underline{X},\underline{Y})$ is a hypersurface in $\underline{\tilde{X}}$, i.e. locally generated by a nonzerodivisor, $\dim_{x'}\tilde{X} = \dim_{x'}\mathbb{P}\,\underline{C}(\underline{X},\underline{Y}) + 1$ by the Active Lemma I 5.2.2. Thus we get $\dim_\xi C(\underline{X},\underline{Y}) = \dim_{x'}\tilde{X} = \dim_{\nu(\xi)}X$.

If ξ is a vertex, there are points ξ' arbitrarily close to ξ on $\underline{C}(\underline{X},\underline{Y}) - Y$, where $\dim_{\xi'}C(\underline{X},\underline{Y}) = \dim_{\nu(\xi')}X$ by the first case; this again implies $\dim_\xi C(\underline{X},\underline{Y}) = \dim_{\nu(\xi)}X$.
The last claim is obvious. Q.e.d.

Remark. For the algebraic proof, see Chapter II, Theorem (9.7).

Definition 1.4.7 (Hironaka). Let $X \in \underline{cpl}$, $\underline{Y} \overset{i}{\hookrightarrow} \underline{X}$ a closed complex subspace, $y \in Y$. Then $\underline{X}$ is called normally flat along Y at y if and only if $G(I,\mathcal{O}_X)_y$ is a flat $\mathcal{O}_{Y,y}$-module. $\underline{X}$ is called normally flat along Y if and only if it is normally flat along Y at all $y \in Y$.

The following theorem with an idea of proof was formulated by Hironaka ([33], p. 136) and proved in [38], Theorem 1.5, and in [46], Théorème 8.1.3.

Theorem 1.4.8. Let $X \in \underline{\underline{cpl}}$, $\underline{Y} \overset{i}{\hookrightarrow} \underline{X}$ a closed complex subspace, and let $F(X,Y) := \{y \in Y \mid X$ is normally flat along Y at $y\}$. Then $F(X,Y)$ is the complement of an analytic set in $\underline{Y}$. Moreover, when $\underline{Y}$ is reduced, $Y - F(X,Y)$ is nowhere dense.

Proof. This is immediate from Theorem 1.3.4.

We finally need the following weaker notion, whose importance was also discovered by Hironaka ([34], Definition (2.4) and Remark (2.5)). We use throughout $\dim_y \nu^{-1}(y) = \dim \nu^{-1}(y)$, cf. II, Proposition 3.1.2.

Proposition and Definition 1.4.9. Let $\underline{X} \in \underline{\underline{cpl}}$, $\underline{Y} \hookrightarrow \underline{X}$ a closed complex subspace, and $\underline{\nu} : \underline{C}(\underline{X},\underline{Y}) \longrightarrow \underline{Y}$ be the normal cone. Let $\underline{X}$ be equidimensional at $y \in Y$. The following statements are equivalent.

(i) $\underline{\nu}$ is universally open near y , i.e. there is an open neighbourhood U of y in Y such that, for any base change $\underline{U}' \longrightarrow \underline{U}$ in $\underline{\underline{cpl}}$, $(\underline{\nu}|U) \times_{\underline{U}} \underline{U}'$ is an open map;

(ii) $\dim \underline{\nu}^{-1}(\underline{z})$ does not depend on z near y ;

(iii) $\dim \nu^{-1}(z) = \dim_y X - \dim_y Y$.

We call X normally pseudoflat along Y $\underline{\underline{at}}$ y if and only if one of these statements holds true (this clearly is an open condition on y).

The statement (iii) just means $ht(I_y) = s(I_y)$, where $I_y \subseteq \mathcal{O}_{X,y}$ defines $(\underline{Y},y) \hookrightarrow (\underline{X},y)$; see Proposition 2.2.5 below.

Outline of proof. We may assume $\underline{U} = \underline{Y}$ and $\underline{Y}$ reduced. We have the following general facts for a morphism $\underline{f} : \underline{W} \longrightarrow \underline{Z}$ in $\underline{\underline{cpl}}$:

1) $z \mapsto \dim \underline{f}^{-1}(\underline{z})$ is upper semicontinuous ([14], 3.4).

2) $\forall w \in \underline{W} : \dim_w \underline{f}^{-1} f(w) + \dim_{f(w)} \underline{Z} \geqq \dim_w \underline{W}$ ([14], 3.9.) .

3) If $\underline{Z}$ is equidimensional at all points and $\underline{f}$ is open, equality holds in 2) for all $w \in \underline{W}$. Conversely, if $\underline{Z}$ is irreducible at $z = f(w)$ and equality holds in 2) for w , f is open at w ([14], 3.10 and 3.9).

We may assume $\underline{X}$ to be equidimensional of dimension d at all points of Y by I Theorem 7.3.2. Then $\underline{C}(\underline{X},\underline{Y})$ is equidimensional of dimensional d at all points by Corollary 1.4.6.

Let $(Y\ )_{\lambda \in \Lambda}$ be the irreducible components of $\underline{Y}$; we may assume Λ finite and the $\underline{Y}_\lambda$ given by the local decomposition of $(\underline{Y},y)$ by II Remark 2.1.4.

<u>(i) ⇒ (iii)</u>: Make the base change $\underline{Y}_\lambda \hookrightarrow \underline{Y}$ and get from 3)

$$\dim \nu^{-1}(y) + \dim_y Y_\lambda = d \qquad \text{for all } \lambda \quad .$$

<u>(iii) ⇒ (ii)</u>: This follows from 1) and 2).

<u>(ii) ⇒ (i)</u>: (cf. [34]) Since $\underline{C}(\underline{X},\underline{Y})$ is equidimensional, we may, through any given point $\xi \in S(\underline{X},\underline{Y})$ and for any λ find an irreducible subgerm $(\underline{W}_\lambda,\xi) \subseteq (\underline{C}(\underline{X},\underline{Y}),\xi)$ such that $\dim_\xi \underline{W}_\lambda = \dim_{\nu(\xi)} \underline{Y}_\lambda$ and $\nu|\underline{W}_\lambda : (\underline{W},\xi) \longrightarrow (\underline{Y}_\lambda,\nu(\xi))$ is finite. Then, for suitable representatives, $\nu|\underline{W}_\lambda : \underline{W}_\lambda \longrightarrow \underline{Y}_\lambda$ is universally open; for this, use the fundamental facts on open finite mappings of I, § 6. Since this holds for all λ and ξ , ν must be, after a possible shrinking, universally open.

<u>Remark 1.4.10.</u> A motivation for the definition is the following: If $\underline{X}$ is normally flat along $\underline{Y}$, the normal cone map $\underline{\nu} : \underline{C}(\underline{X},\underline{Y}) \longrightarrow \underline{Y}$ is a flat map of complex spaces by Remark 1.3.2. Now it is known that flatness is stable under base extension and that a flat map is open, hence a flat map is universally open (see [14], 3.15 and 3.19, and [36], p. 225).

This is in fact the main topological property of a flat map, which, in particular, implies that the fibres of a flat map have the expected minimal generic dimension. In this sense, normal pseudoflatness retains the topological essence of normal flatness.

Remark 1.4.11. Normal flatness of X along Y at y implies normal pseudoflatness at this point. Hence, in the situation of 1.4.9, if Y is reduced, the set $PF(\underline{X},\underline{Y}) := \{y \in Y \mid \underline{X}$ is normally pseudoflat along $\underline{Y}$ at $y\}$ is generic in $\underline{Y}$.

Proposition 1.4.12. Let the situation be as in 1.4.9. Let y be a smooth point on $\underline{Y}$. Then the following statements are equivalent:

(i) $\underline{X}$ is normally flat along $\underline{Y}$ at y.

(ii) The natural morphism

$$\underline{\nu}^{-1}(\underline{y}) \times \underline{C}(\underline{Y},y) \longrightarrow \underline{C}(\underline{X},y) \tag{1.4.5}$$

is an isomorphism.

Proof. Since (1.4.5) corresponds to an algebraic morphism of the corresponding projectivized cones, the celebrated results of [66] imply that (1.4.5) is an isomorphism of complex spaces if and only if it is an isomorphism of algebraic schemes. In view of this, the Proposition 1.4.12 is a mere restatement of Chapter IV, Corollary (21.11), in geometric form.

§ 2. Geometric equimultiplicity along a smooth subspace

In this paragraph we analyse the geometric significance of a complex space X having the same multiplicity along a subspace Y near a smooth point y of Y, and give various characterizations due to Hironaka, Schickhoff, Lipman, and Teissier (see Theorem 2.2.2 below). The motivation, of course, is to understand which restrictions this requirement puts on the blowup of X along Y; see the preface of this book. The result of Hironaka-Schickhoff is that equimultiplicity in the above sense is equivalent to normal pseudoflatness, so we have the noteworthy fact that the dimension of the normal cone fibres are controlled by the multiplicity. The underlying reason why this is so is that the requirement of equimultiplicity and of the normal cone fibre having the generic minimal dimension both put a transversality condition on X along Y relating the two properties. To be more precise, let us embed X locally around y in some $\mathbb{C}^n$ so that Y becomes a linear subspace. Let $L \in P_e^d(\underline{X},y)$ be a projection centre whose corresponding projection onto $\mathbb{C}^d$ has the multiplicity $m(\underline{X},y)$ as local mapping degree. It turns out that both requirements amount to the requirement that $Y \times L$ and X intersect transversally along Y in the sense that $Y \times L \cap C(\underline{X},\underline{Y}) = Y$. If X is normally pseudoflat along Y at y, this fact comes about by blowing up X and $Y \times L$ along Y, and the various projection centres in $Y \times L$ parametrized by points of Y yield projections whose local mapping degrees are constant and give the multiplicity of X along Y. The converse direction, starting from equimultiplicity and reaching transversality, is more delicate and is essentially the geometric version of the Theorem of Rees-Böger. Inherent is the principle that multiplicity was defined as a minimal mapping degree, and this minimality forces the projection centre defining the multiplicity to be generic and hence transversal. Archetypical for this situation is $(\underline{X},x) \hookrightarrow (\underline{\mathbb{C}}^n,0)$ given by a Weierstraß equation so that the z_n-axis L has 0 as isolated intersection point with X; it is then a challenging exercise to convince oneself that the projection along L has minimal mapping degree if and only if L is transversal to the tangent cone. We end by analysing some further geometric conditions and their relationship to various algebraic characterizations of equimultiplicity, especially to the notion of reduction and integral dependence, as exposed in the first four chapters of this book. It is instructive to return again to the above Weierstraß example and to convince oneself that the transversality of L to the tangent cone is, in this case, equivalent to z_n

being, as a function on X, integrally dependent on the ideal generated in $\mathcal{O}_{X,x}$ by $z_1,\dots,z_{n-1}$. In particular, it appears that the algebraic connection between reduction and integral dependence is reflected geometrically by the fact that the transversality condition stated above is equivalent to growth conditions on the coordinate functions of $\mathbb{C}^n$ along normal directions of Y in X.

2.1. Zariski-equimultiplicity

Throughout this section we employ the following notation. $\underline{X}$ is a complex space, $\underline{Y}$ a closed complex subspace, $y \in \underline{Y}$ a smooth point on $\underline{Y}$, $I \subseteq \mathcal{O}_Y$ the ideal defining $\underline{Y} \xhookrightarrow{i} \underline{X}$, and $\mathfrak{p}_z \in \operatorname{Spec}(\mathcal{O}_{X,z})$ the ideal defining the subgerm $(\underline{Y},z) \subseteq (\underline{X},z)$ for $z \in Y_{ir}$. If $(R,\mathfrak{m}_R)$ is a local noetherian ring, $e(R) := e(\mathfrak{m}_R,R)$.

<u>Definition 2.1.1</u> (Zariski-equimultiplicity). <u>Let</u> $(\underline{X},\underline{Y},y)$ <u>be as above. Then</u> X <u>is called Zariski-equimultiple along</u> Y <u>at</u> y <u>if and only if the function</u> $z \longmapsto m(\underline{X},z)$ <u>on</u> Y <u>is constant near</u> y.

The following result exploits this definition algebraically [38],[49]).

<u>Theorem 2.1.2</u> (algebraic characterization of equimultiplicity). <u>Let</u> $(\underline{X},\underline{Y},y)$ <u>be as stated above. The following conditions are equivalent</u>:

(i) $\underline{X}$ <u>is Zariski-equimultiple along</u> Y <u>near</u> y.

(ii) $e(\mathcal{O}_{X,y}) = e((\mathcal{O}_{X,y})_{\mathfrak{p}_y})$, <u>where</u> $\mathfrak{p}_y \in \operatorname{Spec}(\mathcal{O}_{X,y})$ <u>defines</u> $(\underline{Y},y) \hookrightarrow (\underline{X},y)$.

This will be an immediate consequence of the following proposition, which explains the geometric significance of the number $e((\mathcal{O}_{X,y})_{\mathfrak{p}_y})$.

<u>Proposition 2.1.3</u> <u>Let</u> $(\underline{W},w) \in \underline{\underline{\mathrm{cpl}}}_0$, $(\underline{Z},w) \hookrightarrow (\underline{W},w)$ <u>a prime subgerm. Then, after suitably shrinking</u> $\underline{W}$:

(i) $m(\underline{W},w) \geq e((\mathcal{O}_{\underline{W},w})_{\mathfrak{p}_w})$, where $\mathfrak{p}_w \in \mathrm{Spec}(\mathcal{O}_{W,w})$ defines $(\underline{Z},w)$.

(ii) There is a nowhere dense analytic set $A \subseteq Z$ such that $m(\underline{W},z) = e((\mathcal{O}_{W,w})_{\mathfrak{p}_w})$ for all $z \in Z - A$.

In other words, $e((\mathcal{O}_{W,w})_{\mathfrak{p}_w})$ is the generic multiplicity of W along the subspace $\underline{Z} \hookrightarrow \underline{W}$ defined locally by $\mathfrak{p}_w$.

Proof of 2.1.3. Since $\underline{Z}$ is reduced at w , we may assume, after possibly shrinking $\underline{W}$, that there is a nowhere dense analytic set A such that $\underline{Z} - A$ is reduced and smooth, and $\underline{W}$ is normally flat along $\underline{Z} - A$; this follows from I 6.3.1, and 1.4.8. Now consider the chain

$$(2.1.1) \qquad m(\underline{W},y) \overset{(1)}{\geq} m(\underline{W},z) \overset{(2)}{=} e((\mathcal{O}_{W,z})_{\mathfrak{p}_z}) \overset{(3)}{=} e((\mathcal{O}_{W,w})_{\mathfrak{p}_w}) , \ z \in Z - A .$$

(1): This is just the upper semicontinuity of multiplicity in II Theorem 5.2.4.

(2): This is II Theorem 5.2.1 (iii) and Corollary (21.12) of Chapter IV.

(3): This results from the following Lemma 2.1.4.

This proves the Proposition 2.1.3.

Proof of Theorem 2.1.2. After shrinking $\underline{X}$, let $A \subseteq Y$ be such that 2.1.3 (ii) holds, so $e((\mathcal{O}_{X,y})_{\mathfrak{p}_y})$ is the generic value of $m(\underline{X},z)$, and

$$(2.1.2) \qquad m(\underline{X},y) \geq m(\underline{X},z) \geq e((\mathcal{O}_{X,y})_{\mathfrak{p}_y}) ,$$

both inequalities by upper semicontinuity of multiplicity (II Theorem 5.2.4). Q.e.d.

Lemma 2.1.4 Let W be a complex space , $\mathcal{M}$ a coherent $\mathcal{O}_W$-module, and Z an irreducible component of supp $\mathcal{M}$. Then the function $z \mapsto e((\mathcal{M}_z)_{\mathfrak{p}_z})$ is locally constant on Z_{ir} .

Proof. This is done by the methods of compact Stein neighbourhoods and is similar to the proof of II 2.2.3, so I will be brief. Let $I \subseteq \mathcal{O}_X$ define $\underline{Z} \hookrightarrow \underline{W}$. Let $z_0 \in Z_{ir}$, and choose a compact Stein neighbourhood K of z_0 in W . Let $R := \Gamma(K,\mathcal{O}_W)$, $P := \Gamma(K,I)$, which is a prime ideal of R by II, proof of 2.2.3. Finally, put $M := \Gamma(K,M)$. If $z \in K \cap Z_{ir}$, the homomorphism

(2.1.2) $$(\lambda_z)_P \;:\; R_P \longrightarrow (\mathcal{O}_{W,z})_{\mathfrak{p}_z} ,$$

where $\mathfrak{p}_z \in \mathrm{Spec}(\mathcal{O}_{W,z})$, defines $(\underline{Z},z) \hookrightarrow (\underline{W},z)$,and is faithfully flat by II Theorem 1.3.2. Moreover,

(2.1.3) $$(M_z)_{\mathfrak{p}_z} = M_P \otimes_{R_P} (\mathcal{O}_{W,z})_{\mathfrak{p}_z} .$$

Then, for all $k \geqq 0$, we get by II Lemma 2.2.4:

(2.1.4) $$\mathrm{length}((M_z)_{\mathfrak{p}_z} / \mathfrak{p}_z^k (M_z)_{\mathfrak{p}_z}) = \mathrm{lenght}(M_P / P^k M_P) ,$$

this proves the lemma.

Remark 2.1.5 If one just wants Theorem 2.1.2 without the characterization in Proposition 2.1.3, one could use the chain

$$m(\underline{X},y) \overset{(1)}{\geqq} m(\underline{X},z) \overset{(2')}{\geqq} e((\mathcal{O}_{\underline{X},z})_{\mathfrak{p}_z}) \overset{(3)}{=} e((\mathcal{O}_{\underline{X},y})_{\mathfrak{p}_y}) \quad \text{for } z \text{ near } y ,$$

with (2') given by Proposition (30.1) of Chapter VI.

Corollary 2.1.6. Let $\underline{X}$ be a complex space, $\underline{Y}$ a smooth closed complex subspace. If $\underline{X}$ is normally flat along $\underline{Y}$, then $\underline{X}$ is Zariski-equimultiple along $\underline{Y}$.

Proof. Condition (ii) of Theorem 2.1.2 holds by Chapter IV, Corollary (21.12).

Remark 2.1.7. Corollary (21.12) in Chapter IV relates normal flatness to an equality of Hilbert functions. In fact, normal flatness can be characterized by this; this is the content of the following famous theorem.

Theorem of Bennett (complex analytic case). Let X be a complex space, $Y \hookrightarrow X$ a smooth connected closed complex subspace. The following statements are equivalent:

(i) X is normally flat along Y.

(ii) All local rings $\mathcal{O}_{X,y}$, $y \in Y$, have the same Hilbert function, i.e. $z \mapsto H^{(0)}(\mathcal{O}_{X,z},-)$ is constant for z near y.

The algebraic analogue, the original Theorem of Bennett, is Theorem (22.24) in Chapter IV. The complex analytic version above is proven in [48], Theorem (4.11).

Remark 2.1.8. Definition 2.1.1 makes sense for $(\underline{X},y)$ and $(\underline{Y},y) \hookrightarrow (\underline{X},y)$ arbitrary. I leave an appropriate statement of Theorem 2.1.2 in the general case to the reader.

2.2. The Hironaka-Schickhoff-Theorem

We have seen in Corollary 2.1.6 that normal flatness along a smooth subspace implies Zariski-equimultiplicity along this subspace. It is a remarkable discovery of Hironaka and Schickhoff that normal pseudoflatness along a smooth subspace is equivalent to Zariski-equimultiplicity (see Theorem 2.2.2 below). Recall that we employ the property (ii) of Proposition 1.4.9 as the definition of normal pseudoflatness, but it is property (i) which characterizes normal pseudoflatness as the notion carrying the topological essence of normal flatness, so it is this topological essence which 'interpretes' Zariski-equimultiplicity along a smooth subspace geometrically (for Zariski-equimultiplicity along a nonsmooth subspace see § 3). Hironaka proved that normal pseudoflatness along smooth centres implies equimulitiplicity in [34], Remark (3.2). Schickhoff proved the converse in [61], p. 49; in fact he proved the stronger statement below, which is analogous to Proposition 1.4.11, and shows how much from normal flatness is lost by normal pseudoflatness. Both proofs were geometric, and I will given the outlines in the sequel; the algebraic essence of the Hironaka-Schickhoff-Theorem is Theorem (20.9) in Chapter IV of this book; using the method of compact Stein neighbourhoods, it would be possible to derive the Hironka-Schickhoff-Theorem from this algebraic result.

Before formulating the main result, I fix some terminology. Let $(\underline{X},y) \in \underline{\underline{cpl}}_0$ be a complex spacegerm of dimension d, $(\underline{Y},y) \hookrightarrow (\underline{X},x)$ a complex subspacegerm. After possibly shrinking $\underline{X}$, we may assume:

(2.2.1) (i) $\underline{X} \hookrightarrow \underline{U}$ as a closed complex subspace, where $U \subseteq \mathbb{C}^n$ is open, such that $\underline{X}$ is equidimensional at all points if $(\underline{X},y)$ was equidimensional, and $y = 0 \in U$.

(ii) $\underline{Y} \hookrightarrow \underline{X}$ is a closed complex subspace, and $\underline{Y} = \underline{X} \cap \underline{G}$, where G is the linear subspace of $\mathbb{C}^n$ given by $z_{f+1} = \ldots = z_n = 0$.

This can always be achieved by choosing generators $g_1,\ldots,g_\ell \in \mathcal{O}_{X,y} =: R$ of the ideal $I \subseteq R$ defining $(\underline{Y},y) \hookrightarrow (\underline{X},y)$ and adding elements $h_1,\ldots,h_m \in R$ such that $h_1,\ldots,h_m$, $g_1,\ldots,g_\ell$ generate the maximal ideal of R. Then $n := m + \ell$, and we write points in $\mathbb{C}^n$ as pairs (u,t) with $u = (z_1,\ldots,z_m)$ and $t = (z_{m+1},\ldots,z_n)$; the $h_1,\ldots,h_m$ are induced by $z_1,\ldots,z_m$, and the $g_1,\ldots,g_\ell$ by $z_{m+1},\ldots,z_n$.

(iii) If $(\underline{Y},y)$ is smooth, $\underline{Y}$ is connected and smooth everwhere, and $m = \dim_y Y =: f$.

Since $\underline{Y} \hookrightarrow \underline{G}$, $\underline{Y} \times \underline{\mathbb{C}}^\ell \hookrightarrow \underline{\mathbb{C}}^n$. Any $h \in \mathcal{O}_{\mathbb{C}^n,0}$, considered as an element in $\mathcal{O}_{Y\times\mathbb{C}^\ell,0} = \mathcal{O}_{Y,y}\{t_1,\ldots,t_\ell\}$, can be written as

$$(2.2.2) \qquad h = \sum_{k=\nu_Y(h)}^{\infty} h_k \quad , \quad h_k \in \mathcal{O}_{Y,y}[t_1,\ldots,t_\ell]$$

with $\nu_Y(h)$ uniquely determined by requiring $\nu_Y(h) \neq 0$. We call $\nu_Y(h)$ <u>the order of</u> h <u>along</u> Y <u>at</u> y, and $h_{\nu_Y(h)} =: L_Y(h)$ the Y-<u>leitform of</u> h. The germ $(\underline{C}(\underline{X},\underline{Y}),y) \hookrightarrow (\underline{Y} \times \underline{\mathbb{C}}^k,0)$ is then defined by the ideal generated by all $L_Y(h)$ for $h \in J$, where the ideal $J \subseteq \mathcal{O}_{\mathbb{C}^n,0}$ defines $(\underline{X},y) \hookrightarrow (\underline{\mathbb{C}}^n,0)$. This ideal is called the Y-<u>leitideal of</u> J and denoted $L_Y(J)$. It is possible to find finitely many generators $f_1,\ldots,f_s$ of J such that $L_Y(f_1),\ldots,L_Y(f_s)$ generate $L_Y(J)$; we call $\{f_1,\ldots,f_s\}$ a Y-<u>standard-base of</u> J.

After possibly shrinking $\underline{X}$, we may assume that $L_Y(f_1),\dots,L_Y(f_s)$ are defined on $\underline{Y}\times\underline{\mathbb{C}}^{\ell}$; then $\underline{C}(\underline{X},\underline{Y})\hookrightarrow\underline{Y}\times\underline{\mathbb{C}}^{\ell}$, and $\underline{\nu}:\underline{C}(\underline{X},\underline{Y})\longrightarrow\underline{Y}$ is induced by the projection $\underline{Y}\times\underline{\mathbb{C}}^{\ell}\longrightarrow\underline{\mathbb{C}}^{\ell}$.

We make all these assumptions in the sequel of this section.

Example 2.2.1.

1) $\underline{X}\hookrightarrow\underline{\mathbb{C}}^3$ given by $g(x,y,z) = z^2 - x^2y = 0$, $\underline{Y}$ the x-axis, i.e. defined by $(y,z)\cdot\mathcal{O}_{\mathbb{C}^3,0}$. Then $\nu_Y(g) = 1$, and $g_\nu = -x^2y$ defines $\underline{C}(\underline{X},\underline{Y})$. See Figure 6 .

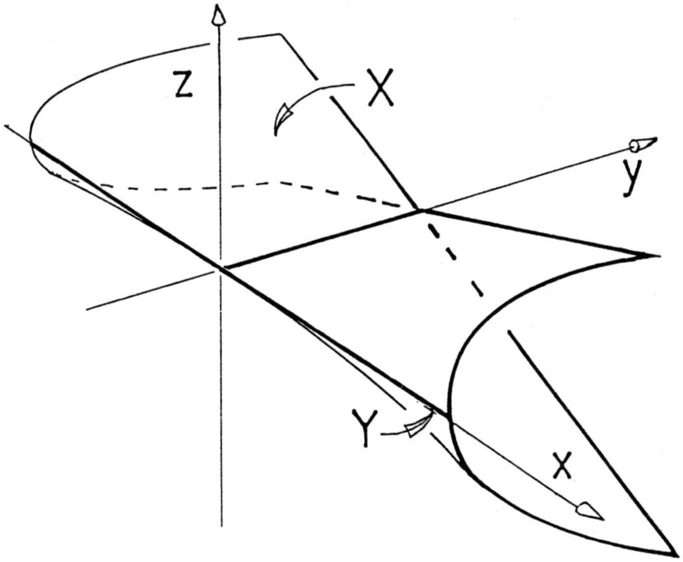

Fig. 6

2) $\underline{X}\hookrightarrow\underline{\mathbb{C}}^3$ given by $g(x,y,z) = z^2 - y^2(y+x^2) = 0$, $\underline{Y}$ again the x-axis defined by $(y,z)\cdot\mathcal{O}_{\mathbb{C}^3}$. Then $\nu_F(g) = 2$, and $g_{\nu_I(g)} = z^2 - y^2x^2$ defines $\underline{C}(\underline{X},\underline{Y})$. See Figure 7 .

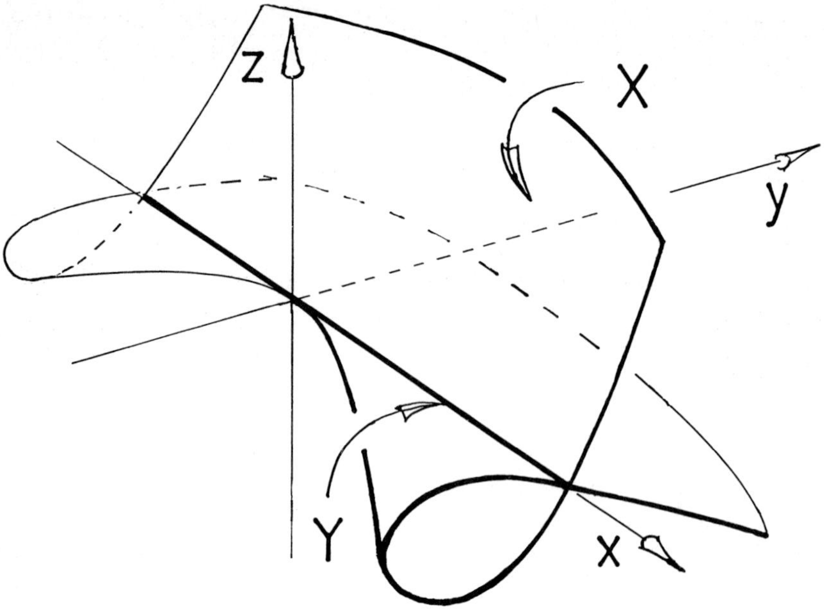

Fig. 7

The main result on the geometric significance of equimultiplicity is now the following theorem.

Theorem 2.2.2 (Geometric analysis of equimultiplicity; Hironaka-Lipman-Schickhoff-Teissier). Let $(\underline{Y},y) \hookrightarrow (\underline{X},y) \hookrightarrow (\mathbb{C}^n,0)$ be embeddings of complex spacegerms, $(\underline{X},x)$ equidimensional of dimension d, $(\underline{Y},y)$ smooth of dimension f, and let X, Y be chosen as stated above. The following statements are equivalent.

(i) X is Zariski-equimultiple along Y at y.

(ii) There is $L \in \mathrm{Grass}^d(\mathbb{C}^n)$ and a neighbourhood V of y in X such that $L_z \cap V = \{z\}$ and $L_z \in P_e^d(\underline{X},z)$ for all $z \in V \cap Y$, where $L_z := L + z$ ([61]).

(iii) There is a nonempty Zariski-open subset $\mathcal{U}$ of $\mathrm{Grass}^d(\mathbb{C}^n)$ such that for any $L \in \mathcal{U}$ there is a neighbourhood V of y in X such that $L_z \cap V = \{z\}$ for all $z \in V \cap Y$ ([69]).

(iv) X is normally pseudoflat along Y at y, i.e. $\dim \nu^{-1}(y) = d - f$, ([34], [61]).

<u>Moreover, if one of these condition holds, one may take</u> $V = P_e^d(\underline{X},x)$ <u>in (iii), and then</u> $L \in P_e^d(\underline{X},z)$ <u>for all</u> $L \in V$ <u>and</u> $z \in Y$ near y .

<u>Addendum to Theorem 2.2.2</u> (cf. Teissier [69], Chapter I, 5.5).
<u>The condition</u> (iii) <u>is equivalent with</u>

(iii') <u>There exists a nonempty Zariski-open subset</u> $U \subseteq \mathrm{Grass}^{d-f}(\mathbb{C}^n,Y)$ $:= \{H \in \mathrm{Grass}^{d-f}(\mathbb{C}^n) \mid H \supseteq Y\}$ <u>such that</u> $(Y,y) = (X \cap H,y)$ <u>as analytic setgerms for all</u> $H \in U$.

<u>Exercise 2.2.3</u>. <u>Analyse the given conditions in the two cases of Example 2.2.1</u>.

The rest of this section is devoted to an outline of the proof, which will be geometric.

Basic is a careful setup for a finite projection $\underline{h} : (\underline{X},y) \longrightarrow (\underline{\mathbb{C}}^d,0)$, which is to give $m(\underline{X},z)$ for all z on Y near y . For this, we collect the following facts, which hold after possibly shrinking $\underline{X}$.

<u>2.2.3</u>.

<u>(i)</u>. Let $(\underline{X},x) \hookrightarrow (\underline{\mathbb{C}}^n,0)$ be a complex subspacegerm, $d := \dim_x \underline{X}$, $f \in \mathbb{N}$ with $0 \leq f \leq d$. Let $K \in \mathrm{Grass}^f(\mathbb{C}^n)$. We say

(2.2.3) K <u>weakly transverse to</u> $\underline{X}$ <u>at</u> x : $\Longleftrightarrow$
$\dim_x X \cap K = d - f$, denoted $K \pitchfork_x \underline{X}$;

K <u>transverse to</u> $\underline{X}$ <u>at</u> x : $\Longleftrightarrow$
$\dim_x C(\underline{X},x) \cap K = d - f$, denoted $K \pitchfork \underline{C}(\underline{X},x)$;

and put

$$(2.2.4) \qquad P_g^f(\underline{X},x) := \{K \in \mathrm{Grass}^f(\mathbb{C}^n) \mid K \pitchfork_x \underline{X}\} ,$$

$$P_e^f(\underline{X},x) := \{K \in \mathrm{Grass}^f(\mathbb{C}^n) \mid K \pitchfork C(\underline{X},x)\} .$$

Then $P_e^f(\underline{X},x) \subseteq P_g^f(\underline{X},x)$. To see this, note that $C(\underline{X} \cap \underline{K},x) \subseteq C(\underline{X},x) \cap K$; so, if $\dim_x C(\underline{X},x) \cap K = d-f$, we have $\dim C(\underline{X} \cap \underline{K},x) = \dim_x \underline{X} \cap \underline{K} \leq d-f$; since always $\dim_x \underline{X} \cap \underline{K} \geq d-f$ (for instance by the Active Lemma, I 5.2.2), we get equality.

The set $P_e^f(\underline{X},x)$ is a nonempty Zariski-open subset of $\mathrm{Grass}^f(\mathbb{C}^n)$, so $P_g^f(\underline{X},x)$ is generic in $\mathrm{Grass}^f(\mathbb{C}^n)$. The proof is a straightforward generalization of the case $f = d$ in II, 4.1: If $\underline{Z} \hookrightarrow \mathbb{P}^{n-1}$ is a (d-1)dimensional variety, consider the fibre bundle given by the "incidence correspondence"

$$\begin{array}{l} \mathcal{Z} := \{(z,K) \in Z \times \mathrm{Grass}^f(\mathbb{P}^{n-1}) \mid z \in K\} \\ \quad \Big\downarrow q \\ \mathrm{Grass}^f(\mathbb{C}^n) \end{array} .$$

Then, by Elementary Algebraic Geometry, q has fibre dimension $(d-f)-1$ outside a proper Zariski-closed subset (see e.g. [56], Chapter 3, (3.15)). Now apply this to $\underline{Z} := \mathbb{P}C(\underline{X},x)$.

We finally define the notion of being <u>strongly transverse</u>, which is based on the following theorem.

<u>Theorem</u>. <u>Let $\underline{X} \in \underline{\mathrm{cpl}}$. Then the Cohen-Macaulay-locus $X_{CM} := \{x \in X \mid \mathcal{O}_{X,x}$ is Cohen-Macaulay$\}$ is the complement of a nowhere dense analytic set. Moreover, if $\mathfrak{p} \in \mathrm{Spec}(\mathcal{O}_{X,y})$ defines $(\underline{Y},y) \hookrightarrow (\underline{X},y)$, $(Y,y) \cap (X_{CM},y) \neq \emptyset$ if and only if $(\mathcal{O}_{X,y})_{\mathfrak{p}}$ is Cohen-Macaulay</u>.

This can be proved by the methods of distinguished compact Stein neighbourhoods, see Remark 1.1.6 (i). For the first statement, see also II Theorem 2.2.11; the second statement can also be proved by the methods of [64], Exposé 21. We will make use only of the first statement at the moment.

Further, if $(\underline{X},x) \hookrightarrow (\underline{\mathbb{C}}^n,0)$, and $(A,x) \subseteq (X,x)$ is an analytic set-germ with $(A,x) \neq (X,x)$, the set of $K \in \mathrm{Grass}^f(\mathbb{C}^n)$ with $(A \cap K,x) \neq (X \cap K,x)$ is generic in $\mathrm{Grass}^f(\mathbb{C}^n)$ for $0 \leq f < d$ (for this, one may assume A being defined by one equation, and then the proof is left to the reader). We define

$$(2.2.5)\qquad P^f_{gs}(\underline{X},x) := \begin{cases} \{K \in P^f_g(X,x) \mid ((X\cap K)_\lambda,x)\cap(X_{CM},x) \neq \emptyset \text{ for all irreducible components } ((X\cap K)_\lambda,x) \text{ of } (X\cap K,x) \text{ of dimension } d\}, \text{ if } f<d\,; \\ \{K\in P^f_g(\underline{X},x) \mid K \pitchfork C(\underline{X},\underline{Y}\cap\underline{K}) \text{ if } f=d\}; \end{cases}$$

$$P^f_{es}(\underline{X},x) := P^f_e(\underline{X},x)\cap P^f_{gs}(\underline{X},x) \quad .$$

These are generic sets in $\mathrm{Grass}^f(\mathbb{C}^n)$. If $K\in P^f_{es}(\underline{X},x)$, we say K is <u>strongly transverse to</u> $(\underline{X},x)$.

We have the following lemma:

<u>Lemma</u>. <u>Let</u> $(\underline{X},x) \overset{i}{\hookrightarrow} (\mathbb{C}^n,0)$ <u>be in</u> $\underline{\underline{\mathrm{cpl}}}_0$, $L\in P^d_g(\underline{X},x)$, <u>and</u> $K\in P^f_{gs}(\underline{X},x)$ <u>with</u> $K\supseteq L$. <u>Let</u> $\underline{h} : (\underline{X},x) \longrightarrow (\underline{\mathbb{C}}^d,0)$ <u>be the projection along</u> L , <u>and</u> $\underline{h}_x : (\underline{X}\cap\underline{K},x) \longrightarrow (\underline{\mathbb{C}}^d\cap\underline{K},0)$ <u>be the restriction of</u> $\underline{h}$ <u>to</u> $\underline{X}\cap\underline{K}$. <u>Then</u>

$$\deg_x \underline{h} = \deg_x \underline{h}_x$$

<u>Proof</u>. There is $b\in\mathbb{C}^d\cap K$ such that $h^{-1}(b) = h_x^{-1}(b) \subseteq X_{CM}\cap X_{reg}$. Then II Remark 2.2.7 shows that we get the same contributions to $\deg_x\underline{h}$ and $\deg_x\underline{h}_x$.

<u>Corollary</u>. $m(\underline{X},x) = m(\underline{X}\cap\underline{K},x)$.

<u>Proof</u>. Choose $L\in P^d_e(\underline{X},x)$; since $C(\underline{X}\cap\underline{K},x)\subseteq C(\underline{X},x)$, $L\in P^{d-f}_e(\underline{X}\cap\underline{K},x;K) := \{L_0\in \mathrm{Grass}^{d-f}(K) \mid L_0 \pitchfork C(\underline{X}\cap\underline{K},x)\}$. Then $\deg_x\underline{h} = m(\underline{X},x)$, and $\deg_x\underline{h}_x = m(\underline{X}\cap\underline{K},x)$. Q.e.d.

If $K\in P^d_g(\underline{X},x)$, $K_z := K+z\in P^d_g(\underline{X},z)$ for z near x . If $K\in P^d_{es}(\underline{X},x)$, $K_z\in P^d_{gs}(\underline{X},x)$ for z near x . This follows, because the $\underline{X}_z := \underline{X}\cap\underline{K}_z$

are the fibres of the projection $\underline{P}_K \mid \underline{X}$, where $\underline{P}_K : \underline{\mathbb{C}}^n \longrightarrow \underline{\mathbb{C}}^n$ is the projection along K , and from the openness of X_{CM} .

(iii) Let $(\underline{X},x) \hookrightarrow (\underline{\mathbb{C}}^n,0)$ be equidimensional of dimension d , and $K \in P_e^f(\underline{X},x)$. Then

$$C(\underline{X} \cap \underline{K},x) = C(\underline{X},x) \cap K \quad .$$

Proof ([61], 2.9). First remark that for any $(\underline{W},w) \in \underline{\underline{\mathrm{cpl}}}_0$, $\dim_w \underline{W} = \dim_z W_{reg}$ for z near w , the dimension of the manifold of regular points on W ; this follows from the local representation Theorem I 6.3.1 (iii), since $\dim_w \underline{W} = \dim_w \underline{W}_{red}$.

Consider the deformation $\underline{p} : (\underline{X},(0,0)) \subseteq (\underline{\mathbb{C}}^n \times \underline{\mathbb{C}},(0,0)) \longrightarrow (\underline{\mathbb{C}},0)$ to the tangent cone $C(\underline{X},y) \cong X_0 := \underline{p}^{-1}(\underline{0})$ in II Proposition 3.1.3 and the resulting description of $C(\underline{X},y)$ by II Corollary 3.1.4. From this the inclusion $C(\underline{X} \cap \underline{K},y) \subseteq C(\underline{X},y) \cap K$ is obvious. For the converse, note that X_0 is nowhere dense in X by II 3.1.3 (iv) and so $(X,(0,0))$ is equidimensional of dimension $d+1$ by the introductory remark. So $\dim_{(z,t)}(X \cap (K \times \mathbb{C})) \geq d+1-f$ for all (z,t) close to $(0,0)$, but $\dim_{(0,0)}(X_0 \cap (K \times \{0\}) = d-f$ by assumption. Hence there is the strict inclusion $(X \cap (K \times \mathbb{C}),(0,0)) \supset (X_0 \cap (K \times \{0\}),(0,0))$ of analytic setgerms, and this proves $C(\underline{X} \cap \underline{K},y) \supseteq C(\underline{X},y) \cap K$. Q.e.d.

(iv) Let $(\underline{Y},y) \hookrightarrow (\underline{X},y) \hookrightarrow (\underline{\mathbb{C}}^n,0)$ be as in Theorem 2.2.2. Consider the diagram of projections

$$R := \{(K,L) \in \mathrm{Grass}^f(\mathbb{C}^n) \times \mathrm{Grass}^d(\mathbb{C}^n) \mid L \subseteq K\}$$

$$\mathrm{Grass}^f(\mathbb{C}^n) \xleftarrow{\ q\ } R \xrightarrow{\ r\ } \mathrm{Grass}^d(\mathbb{C}^n) \quad .$$

We then define various sets:

$$P_{\lambda\mu\nu}(\underline{X},\underline{Y},y) := q^{-1}(P^f_{\lambda\mu}(\underline{X},y) \cap P^f_g(Y,y)) \cap r^{-1}(P^d_\nu(\underline{X},y))$$

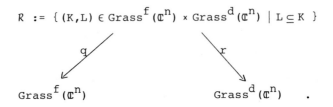

where λ, ν are the letters "g","e", and μ is the blank or "s". These are generic subsets of $\mathcal{R}$.

Moreover, given $L \in \mathcal{P}^d_\mu(\underline{X},y)$, the sets $\mathcal{P}_{\mu\mu\nu}(\underline{X},\underline{Y},y) \cap r^{-1}(L)$ are generic in $r^{-1}(L)$; so, for given $L \in \mathcal{P}^d_\mu(\underline{X},y)$, there is $K \in \mathcal{P}^f_{\mu\nu}(\underline{X},y)$ (for both values of ν) such that $K \supseteq L$.

Elements $(K,L) \in \mathcal{P}_{x\mu\nu}(\underline{X},\underline{Y},y)$ now allow to perform the basic construction for the proof of Theorem 2.2.2:

Let $(K,L) \in \mathcal{P}_{\lambda\mu\nu}(\underline{X},\underline{Y},y)$ be given. Let the coordinates $(z_1,\ldots,z_n)$ on $\mathbb{C}^n$ be such that K is defined by $z_1 = \ldots = z_f = 0$. We use the following notations:

(2.2.6) $\underline{P}_K : (\underline{\mathbb{C}}^n,0) \longrightarrow (\underline{Y},y)$ the projection along K;

$\underline{\rho} : \underline{X} \longrightarrow \underline{Y}$ the restriction $\underline{P}_K \mid \underline{X}$;

$\underline{X}_z := \underline{\rho}^{-1}(z) = \underline{X} \cap \underline{K}_z$ for $z \in Y$ near y, with K_z the affine plane $K + z$ parallel to K through z;

$\underline{E}$: a d-dimensional plane containing Y complementary to L;

$\underline{P}_L : \underline{\mathbb{C}}^n \longrightarrow \underline{E}$ the projection along L;

$\underline{h} : \underline{X} \longrightarrow \underline{E}$ the restriction $\underline{P}_L \mid \underline{X}$;

$\underline{h}_z : \underline{X}_z \longrightarrow \underline{E}_z := \underline{E}\ \underline{K}_z$ the restriction of $\underline{h}$ to $\underline{K}_z$ and hence the projection along L_z;

$\underline{P} := \underline{P}_y : \underline{\mathbb{C}}^n \longrightarrow \underline{K}$ the projection along $\underline{Y}$.

The following figure may illustrate the situation.

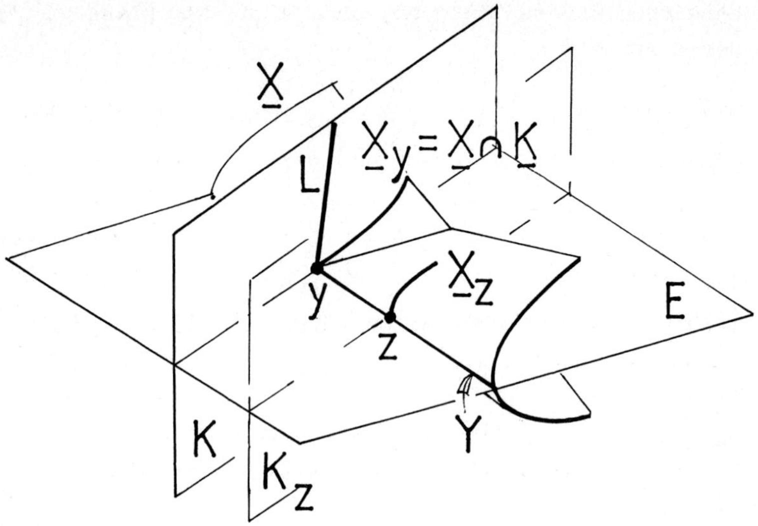

Fig. 8

We now come to the actual proof of Theorem 2.2.2. We use the notations of (2.2.6) throughout. Further, if $K \in \mathrm{Grass}^f(\mathbb{C}^n)$ is given, it defines an embedding $\underline{C}(\underline{X},\underline{Y}) \hookrightarrow \underline{Y} \times \mathbb{C}^{n-f} \hookrightarrow \mathbb{C}^n$ of the normal cone, with $\underline{\nu} : \underline{C}(\underline{X},\underline{Y}) \longrightarrow \underline{Y}$ induced by the projection $\underline{Y} \times \mathbb{C}^{n-f} \longrightarrow \underline{Y}$ according to the description given in 1.4. If $K \in P^r_g(\underline{X},x)$, we have the settheoretic inclusions

(2.2.7) (i) $C(\underline{X} \cap \underline{K},y) \subseteq \nu^{-1}(y) \subseteq K$

(ii) $C(\underline{X} \cap \underline{K},y) \subseteq C(\underline{X},x) \cap K$,

and, if $K \in P^f_e(\underline{X},x)$,

(iii) $C(\underline{X} \cap \underline{K},y) = C(\underline{X},x) \cap K$;

this will be used without further comment.

We proceed according to the pattern

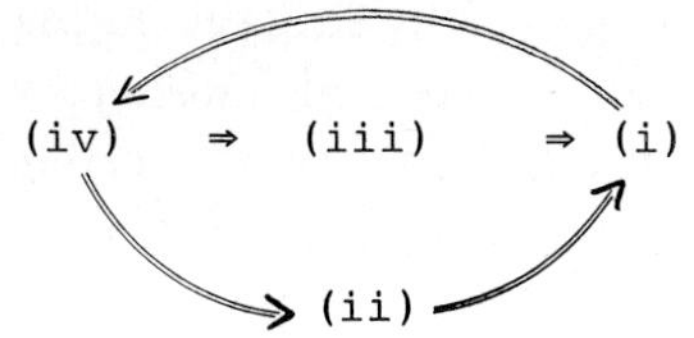

<u>(iv) ⇒ (iii)</u> (cf. [34], [69]) Choose $K \in P_g^f(\underline{X},y)$. With the conventions above, K is given by $z_1 = \ldots = z_f = 0$, $\underline{p} : \mathbb{C}^n \longrightarrow \underline{K}$ denotes the projection along Y, and $\underline{\nu}^{-1}(\underline{y})$ may be regarded as a subvariety of $\underline{K}$, which is of dimension $d-f$ by assumption. So $P_e^{d-f}(\underline{\nu}^{-1}(\underline{y}),y;K) :=$ $:= \{L \in \mathrm{Grass}^{d-f}(K) \mid L \pitchfork \nu^{-1}(y)\}$ is a nonempty Zariski-open set of $\mathrm{Grass}^{d-f}(K)$. Put $V_0(y) := \{L \in \mathrm{Grass}^d(\mathbb{C}^n) \mid p(L) \in P_e^{d-f}(\underline{\nu}^{-1}(\underline{y}),y;K\}$. This is a nonempty Zariski-open subset of $\mathrm{Grass}^d(\mathbb{C}^n)$, and the claim is that (iii) holds for $V := V_0(y)$. Suppose this were not so. We could then find an $L \in V_0(y)$ and a sequence $(x^{(j)})_{j \in \mathbb{N}}$ such that $x^{(j)} \in$ $(X-Y) \cap (L+x^{(j)}), h(x^{(j)}) \in Y$, and $x^{(j)} \longrightarrow y$. After selecting a suitable subsequence we may assume $p(\overline{x^{(j)}y})$ converges to a line ℓ in $\mathbb{P}(K)$, since $\mathbb{P}(K)$ is compact. But then $\ell \subseteq \nu^{-1}(y)$ by (1.4.4), and $\ell \subseteq p(L)$ by construction, which contradicts the fact that $p(L) \in P_e^{d-f}(\underline{\nu}^{-1}(\underline{y}),y;K)$. So we have (iii).

Before showing (iii) ⇒ (i), one shows the following consequence of (iv):

(2.2.8) <u>Assume</u> (iv) <u>holds. Let</u> $K \in P_g^f(\underline{X},y)$ <u>and</u> $L \subseteq K$ <u>be in</u> $P_e^{d-f}(\underline{\nu}^{-1}(\underline{y}),y;K)$. <u>Then</u> $L_z \in P_e^{d-f}(\underline{\nu}^{-1}(\underline{z}),z;K_z)$ <u>for all</u> z <u>outside a nowhere dense analytic subset of</u> Y.

For this, let $\pi : \widetilde{\mathbb{C}}^n \longrightarrow \mathbb{C}^n$ be the blowup of $\mathbb{C}^n$ along $\mathbb{C}^f \times \underline{0}$. The strict transforms of $\underline{X}$ and $\underline{Y} \times \underline{L}$ under π give the blowups $\underline{\widetilde{X}}$ and $(\underline{Y} \times \underline{L})^{\sim}$ along $\underline{Y}$. Their exceptional divisors $\underline{\mathbb{P}C}(\underline{X},\underline{Y})$ and $\underline{Y} \times \underline{\mathbb{P}}(L)$ are subvarieties of $\underline{Y} \times \underline{\mathbb{P}}(K)$, and so meet in a subvariety of $\underline{Y} \times \underline{\mathbb{P}}(K)$, whose image under $\underline{Y} \times \underline{\mathbb{P}}(K) \longrightarrow \underline{Y}$ is a subvariety of $\underline{Y}$ since this map is proper. This shows (2.2.8).

<u>(iii) ⇒ (i)</u> (cf. loc. cit.) By Proposition 2.1.3, the function $z \longmapsto m(\underline{X},z)$ has a generic value, m say, outside a nowhere dense analytic set A in Y. By Theorem 1.4.8, we may assume $Y - A \subseteq F(X,Y)$,

the flatness locus of X along Y. So (iv) holds at all points of $Y-A$. We choose $K \in P^f_{gs}(\underline{X},y)$; after shrinking Y, we may assume $K_z \in P^f_{gs}(\underline{X},z)$ for all $z \in Y$ by (2.2.3) (ii). Choose a $w \in Y-A$ and an L in the generic set $V \cap V_0(w) \cap \mathrm{Grass}^{d-f}(K)$ of $\mathrm{Grass}^{d-f}(K)$.

Since $L \in V_0(w)$, we know by (2.2.8) that $L_z \in P^{d-f}_e(\underline{\nu}^{-1}(\underline{z}),z;K_z)$ outside a nowhere dense analytic set B ; we may assume $B \supseteq A$. Since $C(\underline{X}_z,z) \subseteq \nu^{-1}(z)$ always, we have $L_z \in P^{d-f}_e(\underline{X}_z,z;K_z)$. The Lemma and Corollary of 2.2.3 (ii) imply:

$$\text{(2.2.9)} \qquad \deg_z \underline{h} = \deg_z \underline{h}_z = m(\underline{X}_z,z) = m(\underline{X},z) \qquad ,$$

so $\deg_z \underline{h}$ must have the generic value m on $Y-B$.

On the other hand, we have $L \in V$. Now the degree formula II Theorem (2.2.8), applied to $\underline{h}$, gives

$$\text{(2.2.10)} \qquad \deg_y \underline{h} = \sum_{z' \in h^{-1}h(z)} \deg_{z'} \underline{h} \qquad ,$$

for z near y. But the assumption (iii) forces $h^{-1}h(z) = \{z\}$ near y, so

$$\text{(2.2.11)} \qquad \deg_y \underline{h} = \deg_z \underline{h}$$

for z near y. This implies $\deg_y \underline{h} = m$ by (2.2.9) so we have equimultiplicity by upper semicontinuity of multiplicity (II Theorem 5.2.4).

<u>(i) ⇒ (iv)</u> (cf. [61]). Let $\underline{X}$ be equimultiple along $\underline{Y}$ at y. Let $L \in P^d_e(\underline{X},x)$ and $\underline{h} : \underline{X} \longrightarrow \underline{E}$ be the corresponding projection as in (2.2.10). Then $\deg_y \underline{h} = m(\underline{X},y)$, and so by (2.2.11), $\deg_y h \geq \deg_z \underline{h} \geq m(\underline{X},z)$ for z near y, hence we have $\deg_y h = m(\underline{X},z)$ for $z \in Y$ near y by equimultiplicity.

We will now show: If $L \in P^d_g(\underline{X},x)$ is such that for the corresponding projection we have $\deg_y h = m(\underline{X},z)$ for $z \in Y$ near y, then $L \pitchfork \nu^{-1}(y)$; this will obviously establish (i) ⇒ (iv). One proves this first in case $(\underline{X},y)$ is a hypersurface, and then for general $(\underline{X},y)$ by the classical device of reducing it to the hypersurface case via a finite projection. We let $K \in P_e^f(\underline{X},x)$ be the plane given by $z_1 = \ldots = z_f = 0$ and define the normal cone $\underline{C}(\underline{X},\underline{Y}) \hookrightarrow \underline{\mathbb{C}}^n$ by this K.

So let $\underline{X}$ be a hypersurface in $\underline{\mathbb{C}}^n = \underline{\mathbb{C}}^{d+1}$. We choose coordinates $z_1,\ldots,z_n$ in such a way that Y is given by $z_{f+1} = \ldots = z_n = 0$. We decompose $\mathbb{C}^n = \mathbb{C}^f \times \mathbb{C}^k$ and write points in $\mathbb{C}^n$ as (z,t) with $z = (z_1,\ldots,z_f)$ and $t = (z_{f+1},\ldots,z_n)$. Let $g \in \mathcal{O}_{\mathbb{C}^n}(U)$ be an equation for $\underline{X}$; one can write

$$g(z,t) = \sum_{A \in \mathbb{N}^k} g_A(z) \cdot t^A \tag{2.2.12}$$

(notation as in I, §§ 1-2)), where the $g_A(z)$ are holomorphic functions on $Y = (\mathbb{C}^f \times 0) \cap U$. The Y-leitform of g (as defined in (2.2.2)) is

$$L_Y(g) = \sum_{\substack{A \in \mathbb{N}^k \\ |A| = \nu}} g_A(z) \cdot t^A \quad , \tag{2.2.13}$$

where $\nu = \nu_Y(g)$ is the degree of the first nonzero monominal t^A appearing in (2.2.13) with respect to the lexicographic degree order.

Now the equimultiplicity assumption on X along Y at y implies that the $g_A(z)$ with $|A| = \nu$ cannot simultaneously vanish at $y = 0$. For suppose this were the case. The analytic set defined by the simultaneous vanishing of the $g_A(z)$ with $|A| = \nu$ is nowhere dense in Y because $L_Y(g)$ does not vanish identically on K since $K \in P_e^f(\underline{X},y)$. So there are, arbitrarily close to y, points $z_0 \in Y$ such that $g_A(z_0) \neq 0$ for at least one A with $|A| = \nu$. But then all monomials in the development of $g(z,t) \in \mathbb{C}\{z,t\}$ of (2.2.13) would have degree $> \nu$, whereas in the corresponding development of $g(z,t) \in \mathbb{C}\{z-z_0,t\}$ there would appear monomials of degree ν, and the multiplicity $m(\underline{X},y) > \nu$ would drop to $m(\underline{X},z_0) = \nu$ which cannot be by assumption (here we agree on $m(\underline{X},z_0) = 0$ if $z_0 \notin X$). Note that this argument establishes, in particular:

(2.2.14) $\nu_Y(g)$ = generic multiplicity $m(\underline{X},z)$ for $z \in Y$ near y.

It follows that $\nu_Y(g) = m(\underline{X},y)$, hence the leitform $L(g)$ is

(2.2.15) $$L(g) = \sum_{\substack{A \\ |A|=\nu}} g_A(0) \cdot t^A \quad .$$

(2.2.13) and (2.2.15) show:

(2.2.16) $$\underline{C}(\underline{X},y) = \underline{\nu}^{-1}(\underline{y}) \times \underline{C}(\underline{Y},y) \quad ,$$

and so X <u>is normally flat along</u> Y <u>at</u> y. In particular, we get

(2.2.17) $$\nu^{-1}(y) = C(\underline{X},y) \cap K = C(\underline{X} \cap \underline{K},y) \quad .$$

We now turn to $L \in P^d_g(\underline{X},y)$. In suitable coordinates $v = (v',v_n)$ of $\mathbb{C}^n$, we may assume g is a Weierstraß polynomial $g(v',v_n) = v_n^b + a_{b-1}(v')v_n^{b-1} + \ldots + a_1(v')v_n + a_0(v')$, and L is given by $v' = 0$. Then $\deg_y \underline{h} = b$, and, by assumption, $b = m(\underline{X},y) = \nu(g)$. So v_n^b appears in $L(g)$, which means $L \in P^d_e(\underline{X},y)$. So we can choose $K \in P^f_e(\underline{X},y)$ with $K \supseteq L$, and then (2.2.17) shows $L \pitchfork \nu^{-1}(y)$.

We now treat the general case. So let $\underline{Y} \hookrightarrow \underline{X} \hookrightarrow \underline{U}$ be as in Theorem 2.2.2, and let $L \in P^d_g(X,x)$ be such that

(2.2.18) $$\deg_y \underline{h} = m(\underline{X},z)$$

for all $z \in Y$ near y, $\underline{h}$ the projection along L. We want to show $L \pitchfork \nu^{-1}(y)$, where $\underline{\nu} : \underline{C}(\underline{X},Y) \longrightarrow \underline{Y}$ is the normal cone. For this, it suffices to show $L^\lambda \cap \nu^{-1}(y) = \{y\}$ for each line $L^\lambda \subseteq L$.

We may assume $\underline{X}$ is reduced. Namely, by the degree formula (II Theorem 2.2.8), we have

(2.2.19) $$\deg_y \underline{h} = \sum_{z' \in h^{-1}h(z)} \deg_{z'} \underline{h} \geq \deg_z \underline{h} \quad ,$$

so our assumption forces $h^{-1}h(z) = \{z\}$ and $\deg_z \underline{h} = \deg_y \underline{h}$ for $z \in Y$ near y. But then $\deg_z \underline{h}_{red} = \deg_y \underline{h}_{red}$ for $z \in Y$ near y. Moreover, $\deg_z h_{red} = m(\underline{X}_{red}, z)$, and so we have our assumption on L with respect to $\underline{X}_{red}$. By the limit description (1.4.4), $\nu^{-1}(y)$ depends only on $\underline{X}_{red}$, and so it suffices to consider the case $\underline{X} = \underline{X}_{red}$.

We describe lines in L by linear forms $\lambda \in \check{L} - \{0\}$, where $\check{L} := \mathrm{Hom}(L, \mathbb{C})$ is the dual of L, in the following way: We fix $\lambda \in L - \{0\}$ and choose L^λ to be a complementary line to $\mathrm{Ker}(\lambda)$.

This gives us the following situation.

(2.2.20)
$$\begin{array}{ccc} \underline{X} & \xrightarrow{\underline{p}_\lambda} & \underline{X}_\lambda \\ & {}_{\underline{h}}\searrow \;\circlearrowleft\; \swarrow_{\underline{h}_\lambda} & \\ & U_E & \end{array}$$

Here, we have assumed $U = U_E \times U_L$ with U_E open in E, U_L open in L. The maps are finite projections; $\underline{p}_\lambda := \underline{\pi}_\lambda \mid X$ with $\underline{\pi}_\lambda : \underline{\mathbb{C}}^n \longrightarrow \underline{E} \oplus \underline{L}^\lambda \cong \underline{\mathbb{C}}^{d+1}$ the projection along $\mathrm{Ker}(\lambda)$, $\underline{X}_\lambda := \underline{\mathrm{im}}(\pi_\lambda)$, and $\underline{h}_\lambda$ the projection along L^λ. $\underline{X}_\lambda \hookrightarrow \underline{U}_E \times \underline{L}^\lambda$ is a hypersurface, given be the equation

(2.2.21)
$$\omega^\lambda(z,t) := \prod_{x \in h^{-1}h(z) = L+z} (t - \lambda(x-z))^{\deg_x \underline{h}} \in \mathcal{O}(U_E)[t] ,$$

where we regard $\mathcal{O}(U_E)[t] \hookrightarrow \mathcal{O}(U_E \times L^\lambda)$ under $t \longmapsto \lambda$. This follows because π_λ is given by

(2.2.22)
$$\pi_\lambda(v) = (p_L(v), \rho_\lambda(v)) ,$$

where $\rho_\lambda : \mathbb{C}^n \longrightarrow L^\lambda$ is the projection along $E \oplus \mathrm{Ker}(\lambda)$, and from the classical arguments involving the elementary symmetric functions in the $\lambda(x-z)$ for $x \in h^{-1}(z)$. We have

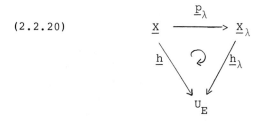

$$(2.2.23) \qquad X = \bigcap_{\lambda \in L-\{0\}} \pi_\lambda^{-1}(X^\lambda) \qquad ;$$

namely, $X \subseteq \pi_\lambda^{-1}(X^\lambda)$ for all λ since X is equidimensional, and on the other hand, for any $v \in \mathbb{C}^n - X$, there is $\lambda \in \check{L} - \{0\}$ with $\lambda(x-v) \neq 0$ for all $x \in h^{-1}p_L(v)$, and so $\pi_\lambda(v) \notin X^\lambda$ by (2.2.21) and (2.2.22).

From (2.2.21), we see $\deg_z \underline{h}_\lambda = \sum_{x \in h^{-1}h(z)} \deg_x \underline{h} = \deg_z \underline{h}$ and so, putting $z = y$, in particular $m(\underline{X}^\lambda, y) = m(\underline{X}, x)$. Let $C(\underline{X},y)^\lambda := \pi_\lambda(C(\underline{X},x))$; then, since π_λ is proper, one may show, by the limit description of tangent cones,

$$(2.2.24) \qquad C(\underline{X},y)^\lambda = C(\underline{X}^\lambda, y) \qquad .$$

So $\pi_\lambda : C(\underline{X},y) \longrightarrow C(\underline{X}^\lambda, y)$ is finite, and, in particular, if $K \in P_e^{f\lambda}(\underline{X},y)$, we have $K^\lambda := \pi_{\lambda_\lambda}(K) \in P_e^f(\underline{X}^\lambda, y)$. If we define the normal cones of $\underline{Y}$ in $\underline{X}^\lambda$ by the K^λ, we get, by the hypersurface case proved above, that $L^\lambda \pitchfork \nu_\lambda^{-1}(y)$. Again by the properness of π_λ and the limit description of normal cones, there results

$$(2.2.25) \qquad (\nu^{-1}(y))^\lambda = \nu_\lambda^{-1}(y) \qquad ,$$

where $(\nu^{-1}(y))^\lambda := \pi_\lambda(\nu^{-1}(y))$. Hence $L^\lambda \cap \nu^{-1}(y) = \{y\}$, as we wanted to show. So (i) $\Rightarrow$ (iv) is established.

Note that this proof shows, in addition ,

$$(2.2.26) \qquad C(\underline{X},y) = \nu^{-1}(y) \times C(\underline{Y},y) \qquad .$$

This follows, because, by (2.2.16), we have

$$(2.2.27) \qquad C(\underline{X}^\lambda,y) = \nu_\lambda^{-1}(y) \times C(\underline{Y},y) \quad \text{for all} \quad \lambda \in \check{L} - \{0\} \qquad ;$$

then, by (2.2.24) and (2.2.25), we get (2.2.26) by intersecting (2.2.27) over all λ and using (2.2.23) (for X^λ, $(\nu^{-1}(y))^\lambda$, and $(C(\underline{X},y))^\lambda$). In particular, we get

$$(2.2.28) \qquad \nu^{-1}(y) = C(\underline{X},y) \cap K = C(\underline{X} \cap \underline{K}, y)$$

for $K \in \mathcal{P}^f_e(\underline{X},y)$ under the condition (i). This is in fact equivalent to (i) and hence to (2.2.26), because it clearly implies $\dim \nu^{-1}(y) = d - f$, so (iv) holds, and we have already (iv) ⇒ (i).

(iv) ⇒ (ii) ⇒ (i): By the proof of (iv) ⇒ (iii) and (2.2.8) we even know that (ii) holds for all $L \in \mathcal{V}_0(y)$. The implication (ii) ⇒ (i) follows because we have (2.2.11) for the projection $\underline{h}$ along L by the same reasoning as in the step (iii) ⇒ (i); by assumption, we have $\deg_z \underline{h} = m(\underline{X},z)$ for all z near y in addition, and this shows (i).

This establishes the equivalence of (i) - (iv). For the additional statements, note that the step (iv) ⇒ (iii) showed we may take $\mathcal{V} = \mathcal{V}_0(y) := \{L \in \mathrm{Grass}^d(\mathbb{C}^n) \mid p(L) \in \mathcal{P}^{d-f}_e(\underline{\nu}^{-1}(\underline{y});K)\}$. If one of the statements of Theorem 2.2.2 holds, we know all of them hold for all $z \in Y$ near y, and then (2.2.28) and (2.2.8) applied to z, show $L_z \in \mathcal{P}^d_e(\underline{X},z)$ for all $L \in \mathcal{V}$ and $z \in Y$ near y. This concludes the proof of Theorem 2.2.2.

The proof of the Addendum is left to the reader.

Before commenting further on the significance of the various characterizations of normal pseudoflatness, let us remark that the proof of (i) ⇒ (iv) gave further important characterizations. Recall, for $g \in \mathcal{O}_{X,y}$, the notions of the order $\nu(g)$ (I, (1.1.3)) and the order $\nu_Y(g)$ of g along Y ((2.2.2)).

<u>Theorem 2.2.2 (cont.).</u> <u>Let $\underline{Y} \hookrightarrow \underline{X} \hookrightarrow \underline{U}$ be as in Theorem 2.2.2. Then the following statements are equivalent to (i) - (iv) of Theorem 2.2.2</u>:

(v) <u>Let $\mathcal{I} \subseteq \mathcal{O}_U$ be the ideal defining $\underline{X} \hookrightarrow \underline{U}$. There are finitely many equations $g_\lambda \in \mathcal{I}(U)$ with the following properties. Let $\underline{X}_\lambda := \underline{N}(g_\lambda)$, and $\underline{\nu}_\lambda : \underline{C}(\underline{X}_\lambda,\underline{Y}) \longrightarrow \underline{Y}$ be the normal cones for all</u> λ. <u>Then</u>:

1) $\nu(g_\lambda) = \nu_Y(g_\lambda)$ for all λ ;

2) $C(\underline{X},y) = \bigcap_\lambda C(\underline{X}_\lambda,y)$;

3) $\nu^{-1}(y) = \bigcap_\lambda \nu_\lambda^{-1}(y)$,

where $\nu^{-1}(y)$, $\nu_\lambda^{-1}(y)$ are defined in $\mathbb{C}^n$ with respect to some $K \in P_e^f(\underline{X},y)$.

(vi) $C(\underline{X},y) = \nu^{-1}(y) \times C(\underline{Y},y)$ with respect to some $K \in P_g^f(\underline{X},y)$.

(vii) $\nu^{-1}(y) = C(\underline{X} \cap \underline{K},y)$ with respect to some $K \in P_g^f(\underline{X},y)$.

If one of the conditions (i) - (vii) holds, (vii) holds for all $K \in P_g^f(\underline{X},y)$.

Moreover, if X is a hypersurface, the following condition is also equivalent to (i) - (vii):

(iv') $\underline{X}$ is normally flat along $\underline{Y}$ at y .

I leave it to the reader to show (i) $\Rightarrow$ (v) $\Rightarrow$ (vi); all the other implications have been mentioned above.

Conditions (v) and (vi) are particularly interesting for the relation between normal flatness and normal pseudoflatness; (v) shows algebraically, and (vi) geometrically, how much is lost when passing from normal flatness to normal pseudoflatness. For normal flatness, condition (v) would require, in addition to $\nu_Y(g_\lambda) = \nu(g_\lambda)$, that the $L_Y(g_\lambda)$ generate the normal cone $C(\underline{X},\underline{Y})$ (note that this implies that the g_λ generate the ideal defining $\underline{X} \longrightarrow \underline{U}$, so $C(\underline{X},x) = \bigcap_\lambda C(X_\lambda,x)$ Condition (vi) would require $\underline{C}(\underline{X},y) = \underline{\nu}^{-1}(\underline{y}) \times \underline{C}(\underline{Y},y)$ so normal pseudoflatness keeps the geometric content of normal flatness, but looses the possibly nonreduced structure.

In order to connect Theorem 2.2.2 with the algebraic equimultiplicity results of Chapter IV of this book, we formulate the following result.

Proposition 2.2.3. Let $(\underline{X},y) \in \underline{\mathrm{cpl}}_0$, $(\underline{Y},y) \hookrightarrow (\underline{X},y)$ a complex subspacegerm defined by the ideal $I \subseteq R := \mathcal{O}_{X,y}$. Then:

(2.2.29) $$\operatorname{codim}_{\underline{y}} \underline{Y} = \operatorname{ht}(I) \quad ;$$

(2.2.30) $$\dim \nu^{-1}(y) = s(I) \; .$$

Proof. A local analytic algebra is catenary (e.g. by the Active Lemma I 5.2.2). This gives (2.2.29) by the Dimension Formula, Chapter III (18.6.1). Further, by base change for Specan, Proposition 1.2.4, $\underline{\nu}^{-1}(\underline{y}) = \underline{\mathrm{Specan}}(\bigoplus_{k\geq 0} I^k/m_x I^k)$. This gives (2.2.30).

By 2.1.2 and 2.2.3, then, we see that the equivalence (i) $\Longleftrightarrow$ (iv) of Theorem 2.2.2 is, for local analytic $\mathbb{C}$-algebras, equivalent to Theorem (20.9) of Chapter IV of this book, thus elucidating its geometric content in this case. Conversely, (20.9) gives an algebraic proof of the Hironaka-Schickhoff-Theorem, based on 2.1.2, which used compact Stein neighbourhoods to interprete invariants of localizations of local analytic $\mathbb{C}$-algebras geometrically (note that the localization of $R \in \underline{\underline{la}}$ is no longer in $\underline{\underline{la}}$, so does not correspond directly to a geometric object via the Equivalence Theorem I 3.3.3). This is a particular case of the general principle that distinguished compact Stein neighbourhoods provide a systematic way of translating results from local complex analytic geometry into local algebra and vice versa. In this vein, the equivalence (iv) $\Longleftrightarrow$ (vi) $\Longleftrightarrow$ (vii) of Theorem 2.2.2 is the geometric content of Proposition (23.15) of Chapter IV (see also the discussion in [49], § 5), and we will deduce geometric properties in $\underline{\underline{cpl}}_0$ from local algebra in 3.2. below.

Exercise 2.2.4. Try to express the statement (ii) of Theorem 2.2.2 in terms of local algebra and to show its being equivalent to the equimultiplicity condition $e(R) = e(R_{\mathfrak{p}})$ algebraically.

(ii) Try to translate the proof of Theorem 2.2.2 into an algebraic proof of Theorem (20.9) of Chapter IV. What do the choices of the f- and d-codimensional planes K and L mean algebraically?

I close this section by some comments on the geometric and algebraic significance of the various conditions in Theorem 2.2.2 and 2.2.2 (cont.); these will be partly, within this limited account, informal.

The equivalence (i) $\Longleftrightarrow$ (v), i.e. that the size of the normal cone is controlled by equimultiplicity, is geometrically a transversality statement, as we will see now. This should be, in a sense, not too surprising, since multiplicity was defined as a generic mapping degree, and we have already seen in II Theorem 5.2.1, that a projection has generic mapping degree if its kernel is transverse to the tangent cone.

The appropriate generalization of this is the following theorem, which we actually proved in the course of establishing (i) $\Leftrightarrow$ (iv) of Theorem 2.2.2.

Theorem 2.2.5. Let $\underline{Y} \hookrightarrow \underline{X} \hookrightarrow \underline{U}$ be as in Theorem 2.2.2, and let $L \in P_g^d(\underline{X},y)$. The following conditions are equivalent:

(i) $\deg_y \underline{h} = m(\underline{X},z)$ for $z \in Y$ near y , where $\underline{h}$ is the projection along L .

(ii) $Y \times L \pitchfork_Y X$, i.e. $Y \times L$ intersects X transversally along Y in the sense that $Y \times L \cap C(\underline{X},\underline{Y}) = Y$ ($C(\underline{X},\underline{Y})$ defined with respect to any (n-f)-dimensional plane $K \supseteq L$) .

Remark 2.2.6.

1) If we put $Y = \{y\}$, we get the statement (ii) of II Theorem 5.2.1 which is the geometric form of the Theorem of Rees in Chapter III, Theorem (19.3), for reductions of the maximal ideal. For primary ideals, see Proposition 3.2.2 (ii) below. In fact, Theorem 2.2.5 is a variant of the geometric form of the Theorem of Böger (Chapter III, Theorem (19.6)) for the case of a regular prime ideal. The transversality condition in (ii) just means that the ideal generated in $\mathcal{O}_X$ via the projection $\underline{X} \longrightarrow \underline{\mathbb{C}}^{d-f}$ along $Y \times L$ is a minimal reduction of the ideal generating Y . This gives a geometric picture of the meaning of a minimal reduction in this case. For the general case, see Theorem 3.2.7 below.

2) We did not use the Theorem of Rees (i.e. the important direction (i) $\Rightarrow$ (ii) in Chapter III, Theorem (19.3)) to establish (i) $\Rightarrow$ (ii) above, so we really gave it a geometric proof. The direction (ii) $\Rightarrow$ (i) was also established in a geometric way, although one may object that I made use of the fact that, if $L \in P_e^d(\underline{X},y)$, one has $\deg_y \underline{h} = m(\underline{X},y)$, which was established in an algebraic way in II Theorem 5.2.1 using the theory of reduction. We will see in the proof of Theorem 2.2.8 below how to interprete this more geometrically.

I now turn to a discussion of condition (vi). Note that the equivalence (iv) $\Leftrightarrow$ (vii) means, in particular:

Either $\nu^{-1}(y) = C(\underline{X} \cap \underline{K},y)$ or $\dim \nu^{-1}(y) > d-f$ (where $K \in P_e^f(\underline{X},x)$), which is, at first sight, rather surprising. Trying to understand this sheds some more light on the geometry of equimultiplicity, so I give an informal account. For this, we have to take a closer look how normal directions arise geometrically.

Proposition 2.2.7 (Existence of testarcs). Let $(\underline{X},x) \in \underline{\underline{cpl}}_0$, and $(A,x) \subsetneq (X,x)$ be an analytic setgerm. Then there exists a morphism $\underline{\alpha} : (\mathbb{D},0) \longrightarrow (\underline{X},x)$, where $\mathbb{D} \subseteq \mathbb{C}$ is the open unit disc, such that $\alpha(\mathbb{D}-0) \subseteq X-A$ and $\alpha(0) = x$. We call $\underline{\alpha}$ a testarc for (A,x).

Sketch of proof. If $(\underline{C},c) \in \underline{\underline{cpl}}_0$ is onedimensional, we get $\underline{\alpha} : (\mathbb{D},0) \longrightarrow (\underline{C},c)$ by parametrizing an irreducible component. This reduces the proof to the case $(\underline{X},x) = (\underline{\mathbb{C}}^d,0)$ via the Local Representation Theorem I 6.3.1. Then just parametrize a complex line transverse to A at x. Q.e.d.

Applying this to the blowup $\pi : \underline{X} \longrightarrow \underline{X}$ of $\underline{X}$ along $\underline{Y}$, with x being a point in $\pi^{-1}(Y)$ and $A := \pi^{-1}(Y)$, we see that in the limit description (1.4.4) of $\nu^{-1}(y)$ we can restrict the limit process to testarcs for (Y,y):

(2.2.31) ℓ is a line in $\nu^{-1}(y) \iff \ell = \lim_{t\to 0} p(\overline{y\alpha(t)})$ for some testarc $\underline{\alpha} : (\underline{\mathbb{D}},0) \longrightarrow (\underline{X},y)$ for (Y,y).

Here, it is understood we have choosen $K \in P_g^f(\underline{X},y)$, and $\underline{p} : \underline{\mathbb{C}}^n \longrightarrow \underline{K}$ is the projection along $\underline{Y}$. The normals at y now fall into two classes: Those that belong to $C(\underline{X} \cap \underline{K},y)$, which I call ordinary normals, and those that do not, which I call excess normals. The equivalence of (vi) and (vii) says that the failure of normal pseudoflatness is due to the existence of excess normals. These are characterized as follows:

(2.2.32) $\ell \subseteq K$ is an excess normal $\iff \ell = (p \circ \alpha)^{\cdot}(0)$, where $\underline{\alpha} : (\underline{\mathbb{D}},0) \longrightarrow (\underline{X},y)$ is a testarc for (Y,y) such that $\alpha^{\cdot}(0)$ is a tangent line of $\underline{X}$ at x, but $(p \circ \alpha)^{\cdot}(0)$ is not a tangent line of $\underline{X} \cap \underline{K}$ at y.

Here I have put $\dot{\beta}(0) := \lim_{t\to 0} \overline{y\beta(t)}$ for a testarc $\underline{\beta}$.

The following picture may illustrate the situation.

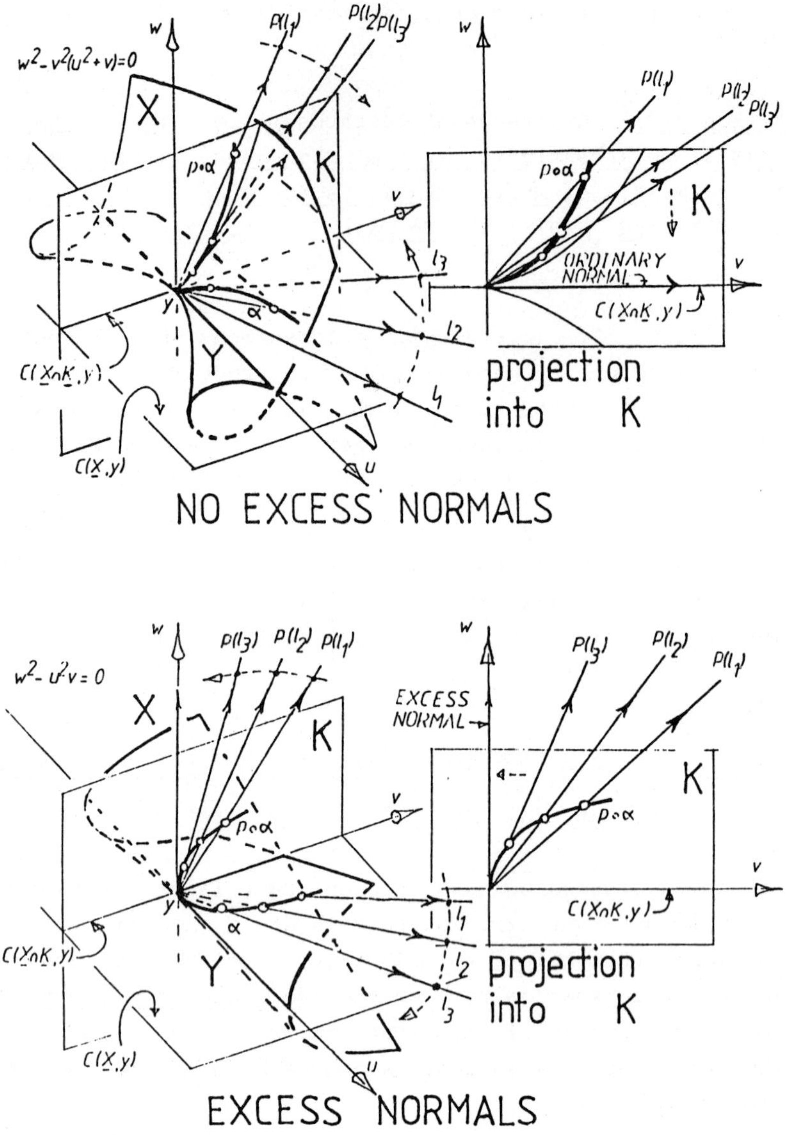

Fig. 9

So we have to analyse what it means, in terms of testarcs, that a line $\ell \subseteq \mathbb{C}^p$ is not tangent to a given $(\underline{W},w) \hookrightarrow (\underline{\mathbb{C}}^p,0)$. Clearly

$$\text{(2.2.33)} \qquad \ell \not\subseteq C(\underline{W},w) \iff \text{for all testarcs } \underline{\alpha} : (\mathbb{D},0) \longrightarrow (\underline{W},w) : \ell \neq \dot{\alpha}(0)$$

Choose coordinates $(z_1,\ldots,z_p)$ such that ℓ is given by $z_1 = \ldots = z_{p-1} = 0$. It is conceivable that the requirement $\ell \neq \dot{\alpha}(0)$ puts growth conditions on the coordinate functions $z_1,\ldots,z_d$ restricted to α , as the following picture suggests:

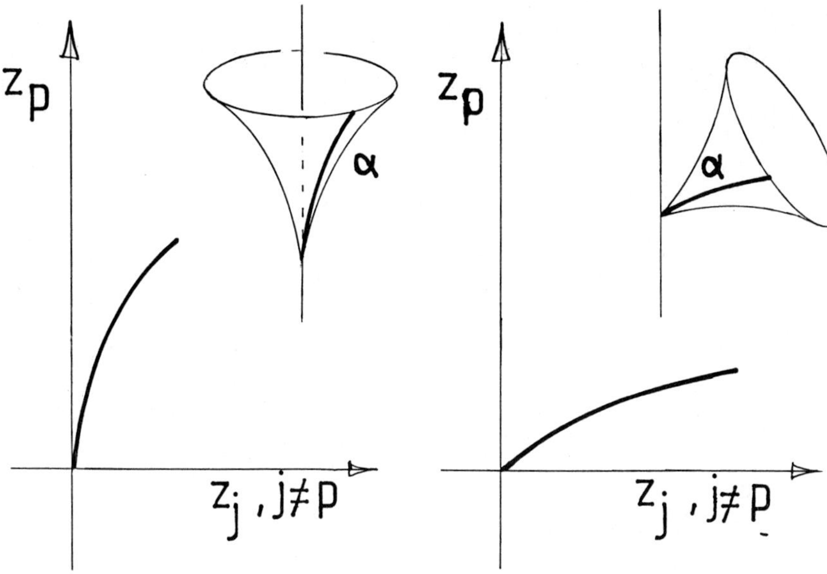

Fig. 10

It turns out that the appropriate growth conditions are:

(2.2.34) $\ell \neq \dot{\alpha}(0) \iff$ <u>there is a neighbourhood</u> V <u>of</u> $0 \in \mathbb{C}$ <u>and</u> $C \in \mathbb{R}_{>0}$ <u>such that</u>

$$|z_p \circ \alpha(t)| \leq C \sup_{1 \leq j \leq p-1} |z_j \circ \alpha(t)|$$

<u>for all</u> $t \in V$.

Now testarcs $\underline{\alpha}$ define valuations v_α on $R := \mathcal{O}_{W,w}$ in the sense of

Chapter I , Definition (4.18), via

$$(2.2.35) \qquad v_\alpha(f) := \nu(\alpha^0(f)) = \mathrm{ord}_0(\alpha^0(f))$$

where $f \in \mathcal{O}_{W,w}$, $\alpha^0 : \mathcal{O}_{W,w} \longrightarrow \mathcal{O}_{\mathbb{C},0}$ is given by $\underline{\alpha} : (\underline{\mathbb{D}},0) \longrightarrow (\underline{W},w)$, and ord_0 denotes the order of vanishing at $0 \in \mathbb{C}$. Then the condition (2.2.3) reads

$$(2.2.3\ ) \qquad \ell \neq \dot{\alpha}(0) \iff v_\alpha(z_p) \geqq v_\alpha((z_1,\ldots,z_{p-1})\mathcal{O}_{W,w}) \quad \text{for all } \alpha ,$$

and so the valuation criterion of integral dependence of Chapter I, (4.20) strongly suggests that $\ell \neq \dot{\alpha}(0)$ <u>is equivalent to</u> z_p , <u>regarded as a function on</u> W , <u>being integrally dependent on the ideal</u> $(z_1,\ldots,z_{p-1})\cdot\mathcal{O}_{W,w}$.

In fact, there is the following proposition:

<u>Proposition 2.2.8</u> ([69]). <u>Let</u> $(\underline{X},x) \in \underline{\underline{\mathrm{cpl}}}_0$, $I \subseteq \mathcal{O}_{X,x}$ <u>an ideal</u> , $f \in \mathcal{O}_{X,x}$. <u>The following statements are equivalent</u>:

(i) <u>For all testarcs</u> $\underline{\alpha} : (\underline{\mathbb{D}},0) \longrightarrow (\underline{X},x)$, $v_\alpha(f) \geqq v_\alpha(I)$.

(ii) <u>For all systems of generators</u> $(g_1,\ldots,g_\ell)$ <u>of</u> I <u>there is a neighbourhood</u> V <u>of</u> x <u>in</u> X <u>and</u> $C \in \mathbb{R}_{>0}$ <u>such that</u>

$$|f(y)| \leqq C \cdot \sup_{1\leqq j\leqq \ell} |g_j(y)|$$

<u>for</u> $y \in V$.

(iii) $f \in \overline{I}$.

(i) ⇒ (iii) depends on the fact that in the proof of (ii) ⇒ (i) of Proposition (4.20) of Chapter I the valuations v_α suffice, see the argument in the proof of Chapter I, 1.3.4 of [69]. (iii) ⇒ (ii) follows because the equation of integral dependence gives the necessary estimates, and (ii) ⇒ (i) is immediate. For the complex analytic proof see [69], Chapter I, 1.3.1 and 1.3.4.

From this results we see:

Theorem 2.2.9. Let $(\underline{W},w) \hookrightarrow (\mathbb{C}^p,0)$ be of dimension d , L the d-codimensional plane given by $z_1 = \dots = z_d = 0$. Then $L \pitchfork C(\underline{W},w)$ if and only if $z_{d+1},\dots,z_n \in \overline{(z_1,\dots,z_d)\cdot \mathcal{O}_{W,w}}$.

It is in this way how the algebraic notion of integral dependence comes in when describing the geometric notion of transversality.

We can now translate the condition (vii) into algebra. We formulate (2.2.32) in the following way:

(2.2.37) There are no excess normals, i.e. (vi) holds $\Longleftrightarrow$ for all testarcs α such that $(p \circ \alpha)^{\cdot}(0)$ is not a tangent line of $\underline{X} \cap \underline{K}$ at y , $\dot{\alpha}(0)$ is not a tangent line of $\underline{X}$ at y .

This can be exploited as follows.

We first get the generalization of Theorem 2.2.9:

Theorem 2.2.10. Let $(\underline{Y},y) \hookrightarrow (\underline{X},y) \hookrightarrow (\mathbb{C}^n,0)$ be as in Theorem 2.2.2, $L \in P^d_g(\underline{X},y)$. Choose any $(n-f)$-dimensional plane $K \supseteq L$, thus $K \in P^f_g(\underline{X},y)$ (defining an embedding $\underline{C}(\underline{X},\underline{Y}) \hookrightarrow \mathbb{C}^n$). Let the coordinates $z_1,\dots,z_d = 0$ be such that L is defined by $z_1 = \dots = z_d$, and K be $z_1 = \dots = z_f = 0$. Then $Y \times L \pitchfork C(\underline{X},\underline{Y})$ if and only if

$$\overline{(z_{f+1},\dots,z_n)\cdot \mathcal{O}_{X,x}} = \overline{(z_{f+1},\dots,z_d)\cdot \mathcal{O}_{X,x}} .$$

This follows by applying Theorem 2.2.9 to (2.2.32), since there are no excess normals if and only if (vii) holds, i.e. we have equimultiplicity, and so (vii) is equivalent to $Y \times L \pitchfork C(\underline{X},\underline{Y})$ by Theorem 2.2.5. The geometric content of this is that transversality is equivalent to growth conditions on the coordinates of X along directions normal to Y , and this is the geometric interpretation of the fact that a (minimal) reduction is characterized by integral dependence.

Further, it is now easy to see that we have, using Theorem 2.2.5:

(2.2.38) $\underline{X}$ is equimultiple along $\underline{Y}$ at y if and only if for all $L \in P^d_g(\underline{X},y)$ we have $L \pitchfork \nu^{-1}(y) \Longleftrightarrow L_z \pitchfork \nu^{-1}(z)$ for all $z \in Y$ near y outside some nowhere dense analytic subset.

Since normal pseudoflatness holds outside a nowhere dense analytic set, so that we can apply Theorem 2.2.10 there, we get, putting together our achievments, the following theorem.

Theorem 2.2.2 (cont.) Let $Y \overset{i}{\hookrightarrow} X \hookrightarrow U$ be as in Theorem 2.2.2, and let the ideal $I \subseteq \mathcal{O}_X$ define i. The following statements are equivalent to the statements of Theorem 2.2.2.

(viii) ("Principle of specialization of (minimal) reduction"). Let $J \subseteq I$. Then J_y is a (minimal) reduction of I_y if and only if J_z is a (minimal) reduction of I_z for all $z \in Y$ near y outside a nowhere dense analytic set in Y.

(ix) ("Principle of specialization of integral dependence"; cf. [69], Chapter I, 5.1). Let $f \in \mathcal{O}_X(X)$. Then $f_y \in \overline{I}_y$ if and only if $f_z \in \overline{I}_z$ for all $z \in Y$ near y outside a nowhere dense analytic set in Y.

The discussion of (ix) is similar to that of (viii) by embedding $X \hookrightarrow \mathbb{C}^n$ in such a way that f is a coordinate on K. One can also show (viii) $\Longleftrightarrow$ (ix) directly.

§ 3. Geometric equimultiplicity along a general subspace

If a complex space X has the same multiplicity along a smooth subspace Y, the results of the last paragraph show that this numerical condition gives control over the blowup $\pi : \tilde{X} \longrightarrow X$ of X along Y to the extent that $\pi | D : D \longrightarrow Y$ is equidimensional, where $D \hookrightarrow \tilde{X}$ is the exceptional divisor (which is the same as saying that X is normally pseudoflat along Y). This is no longer so when Y becomes singular, and it turns out that the "naive" equimultiplicity condition above has to be replaced by a more refined equimultiplicity condition in order to guarantee normal pseudoflatness. The algebraic formulation of this result is Theorem (20.5) of Chapter IV, and it is the purpose of this paragraph to survey the geometric significance of these and related results in that case.

In general, these two notions of equimultiplicty are not related. To visualize this, I give in the first section a short description of the geometric significance of the first one, a result due to Lipman. In the subsequent section I give a somewhat more detailed description of the geometric meaning of the refined equimultiplicity condition and various other equivalent geometric and algebraic conditions, including normal pseudoflatness. These are the appropriate analogues of the smooth case, formulated in Theorem 2.2.2 above, and correspond to the algebraic results (20.5) and (23.15) of Chapter IV. I also describe the relation with the reduction of ideals and integral dependence. The main difference to the smooth case is that one has to replace the tangent cones by the normal cones to possibly nonreduced one-point-subspaces induced in $\underline{X}$ along $\underline{Y}$ by a suitable projection, and to change the multiplicities accordingly. These are also local mapping degrees.

The underlying geometric principle is again that the local mapping degree of a projection measures the order of contact of the kernel of this projection with the spacegerm on which it is defined. Hence, the equimultiplicity condition of a space along a subspace controls the intersection behaviour of the family of this projection centres along the subspace with the space under consideration and so represents a transverality condition on the normal cone. The algebraic notion corresponding to transversality is that of the reduction of an ideal (or integral dependence), and so it is not surprising that the Theorem of Rees-Böger is fundamental to equimultiplicity considerations and contains, in a sense, the essence of it; I have made some comments on this at the end.

3.1. Zariski-equimultiplicity

The following result shows that the geometric description of Zariski-equimultiplicity in Theorem 2.2.2 (ii) can be maintained. It will, however, no longer control the dimension of the normal cone fibres, which makes this notion therefore not very interesting for the study of the blowup along a nonsmooth centre. The main reason for this is that along a general subspace the tangent cones to the ambient space are not related to the fibres of the normal cone and to the normal cones of a transverse plane section, which was the case in the smooth situation.

For the definition of Zariski-equimultiplicity see Remark 2.1.8.

Theorem 3.1.1 (Geometric analysis of Zariski-equimultiplicity; [49], Proposition (4.3)). Let $(\underline{X},y) \hookrightarrow (\mathbb{C}^n,0)$ be an equidimensional spacegerm of dimension d, $(\underline{Y},y) \hookrightarrow (\underline{X},y)$ a complex subspacegerm. The following statements are equivalent.

(i) $\underline{X}$ is Zariski-equimultiple along $\underline{Y}$ at y .

(ii) There is $L \in \mathrm{Grass}^d(\mathbb{C}^n)$ and a neighbourhood V of y in $\underline{X}$ such that $L_z \cap V = \{z\}$ and $L_z \in P_e^d(\underline{X},z))$ for all $z \in V \cap Y$.

(iii) For all $L \in P_e^d(\underline{X},y)$ there is a neighbourhood V of y in $\underline{X}$ such that $L_z \cap V = \{z\}$ and $L_z \in P_e^d(\underline{X},z)$ for all $z \in V \cap Y$.

Proof. For $L \in P_e^d(\underline{X},y)$, let $\underline{h} := \underline{p}_L : (\underline{X},y) \longrightarrow (\underline{E},0)$ be the projection along L to a d-dimensional plane $E \subseteq \mathbb{C}^n$ complementary to L . We have

$$\begin{array}{ccccc} \deg_y \underline{h} & = & \sum\limits_{z' \in h^{-1}h(z)} \deg_{z'} \underline{h} & \geq & \deg_z \underline{h} \\ \vee\!| & & & & \vee\!| \\ m(\underline{X},y) & & & & m(\underline{X},z) \end{array} \tag{3.1.1}$$

for z near y on Y .

(i) ⇒ (iii) If $L \in P_e^d(\underline{X},y)$, (3.1.1) implies $h^{-1}h(z) = \{z\}$ and $\deg_z \underline{h} = m(\underline{X},z)$ for z near y on Y . Then $L_z \in P_e^d(\underline{X},z)$ by the geometric form of the Theorem of Rees, Remark 2.2.6,1).

(iii) ⇒ (ii) This is obvious.

(ii) ⇒ (i) By (3.1.1), $m(\underline{X},z) = \deg_z \underline{h} = \deg_y \underline{h}$ for z near y on Y .

3.2. Normal pseudoflatness

As mentioned before, if we have $(\underline{Y},y) \hookrightarrow (\underline{X},y)$, the tangent cone $\underline{C}(\underline{X},y)$ will in general not be related to the fibre $\underline{\nu}^{-1}(\underline{y})$ of the normal cone $\underline{\nu} : \underline{C}(\underline{X},\underline{Y}) \longrightarrow \underline{Y}$, and so it cannot be expected that its dimension is controlled by the multiplicity of X along Y near y .

Recall that the geometric analysis of equimultiplicity along a smooth subspace in 2.2. depended heavily on the use of a finite projection, $\underline{h}$. It turns out that the correct cones which to replace the tangent cones with are the normal cones $\underline{C}(\underline{X},\underline{y})$, where $\underline{y} \hookrightarrow \underline{X}$ is the one-point-space defined in $\underline{X}$ be the primary ideal of $\mathcal{O}_{X,y}$ generated via the finite projection, and that the correct multiplicities are the sums of the multiplicities corresponding to these cones in the fibres of the projection restricted to Y . This will be described now. Since the results are a natural generalization of the smooth case, which has been exposed in detail in § 2, arguments are only sketched, or omitted. The corresponding algebraic results are (20.5) and (23.15) of Chapter IV of which the exposition here describes the geometric content.

<u>Definition 3.2.1.</u> <u>Let</u> $(\underline{X},x) \in \underline{\underline{cpl}}_0$, $\dim_x X =: d$, $L' \in \mathrm{Grass}^k(\mathbb{C}^n)$ <u>such that</u> x <u>is isolated in</u> $X \cap L'$ <u>with</u> $d \leqq k \leqq n$, <u>and</u> $\mathfrak{q}' := p^0_{L'}(m_k) \subseteq m_x$ <u>the</u> m_x<u>-primary ideal generated via the projection</u> $\underline{p}_{L'}(\underline{X},x) \longrightarrow (\underline{\mathbb{C}}^k,0)$ <u>along</u> L' . <u>Let</u> $\underline{x} \hookrightarrow \underline{X}$ <u>be the one-point complex spacegerm defined by</u> $\mathfrak{q}'$.

(i) $$P^d_g(\underline{X},\underline{x}) := \{L \in P^d_g(\underline{X},x) \mid L \supseteq L'\}$$

$$P^d_e(\underline{X},\underline{x}) := \{L \in P^d_g(\underline{X},\underline{x}) \mid L \pitchfork C(\underline{X},\underline{x})\}$$

$$= P^d_g(C(\underline{X},\underline{x}),\underline{x}) \quad ,$$

<u>where</u> $\underline{C}(\underline{X},\underline{x})$ <u>is the normal cone of</u> $\underline{x} \hookrightarrow \underline{X}$. (These are both generic subspaces of the grassmannian of d-codimensional planes in $\mathbb{C}^n$ containing L'.)

(ii) $$m(\underline{X},\underline{x}) := \min\{\deg_x \underline{p}_L \mid L \in P^d_g(\underline{X},\underline{x})\} \quad .$$

In generalization of II, Theorem 5.2.1, one has

<u>Proposition 3.2.2.</u> <u>Let the notation be as in Definition 3.2.1; in particular,</u> L' , <u>or</u> $\mathfrak{q}'$, <u>is fixed.</u>

(i) $\deg_x \underline{p}_L \geqq e(\mathfrak{q}',\mathcal{O}_{X,x})$ <u>for all</u> $L \in {}^d_g(\underline{X},\underline{x})$.

(ii) (Theorem of Rees). <u>If</u> $L \in P^d_e(\underline{X},\underline{x})$, $\deg_x \underline{p}_L = e(\mathfrak{q}',\mathcal{O}_{X,x})$. <u>Conversely, if</u> $(\underline{X},x)$ <u>is equidimensional and</u> $\deg_x \underline{p}_L = e(\mathfrak{q}',\mathcal{O}_{X,x})$, <u>then</u> $L \in P^d_e(\underline{X},x)$.

<u>Notation 3.2.3</u>. We consider $(\underline{Y},y) \hookrightarrow (\underline{X},y) \hookrightarrow (\underline{\mathbb{C}}^n,0)$, $\dim_y Y =: f$, and $(\underline{X},x)$ equidimensional of dimension d . We assume the conventions (2.2.1) (i),(ii), and (iii) made at the beginning of 2.2; so we assume $\underline{Y} \hookrightarrow \underline{X} \hookrightarrow \underline{U}$ with U a domain in $\mathbb{C}^n$, and $\underline{Y} = \underline{X} \cap \underline{G}$, where G is an m-codimensional plane in $\mathbb{C}^n$ such that $\underline{Y} = \underline{X} \cap \underline{G}$, called a <u>generating plane for</u> Y . Let $\mathcal{I} \subseteq \mathcal{O}_X$ define $\underline{Y} \hookrightarrow \underline{X}$. Further, let $K \in \mathcal{P}^f_g(\underline{Y},y)$ (cf. (2.2.4)). We let the coordinates on $\mathbb{C}^n$ be chosen in such a way that K is given by $z_1 = \ldots = z_f = 0$ and G by $z_{f+1} = \ldots = z_{f+m} = 0$. Let $L' := G \cap K$. Then L' $\mathrm{Grass}^{f+m}(\mathbb{C}^n)$. The projection along L' defines a finite map $\underline{h}' : \underline{X} \longrightarrow \underline{\mathbb{C}}^{f+m}$, and we will use the multiplicities induced by h' in $\underline{X}$ along $\underline{Y}$ to control the fibres of the normal cone (see Figure 11).

For this, put $\underline{y} := \underline{y}(K) := \underline{Y} \cap \underline{K} = (\underline{h}')^{-1}(0)$; the multiplicity in question is $m(\underline{X},\underline{y})$, the behaviour of which along $\underline{Y}$ is relevant for normal pseudoflatness. One has $m(\underline{X},\underline{y}) = e(\mathcal{I}_y(\underline{x}),\mathcal{O}_{X,y})$, $\underline{x} := (z_1,\ldots,z_f)$ the set of parameters of $\mathcal{O}_{X,y}$ defining K (cf. Chap.I, (3.6)). Put

(3.2.1) $$\underline{h}_K : \underline{X} \longrightarrow \underline{F}$$

to be the projection along K , where $\underline{F} = \underline{\mathbb{C}}^f \times \underline{0} \hookrightarrow \underline{\mathbb{C}}^n$. We get the commutative diagram

(3.2.2)

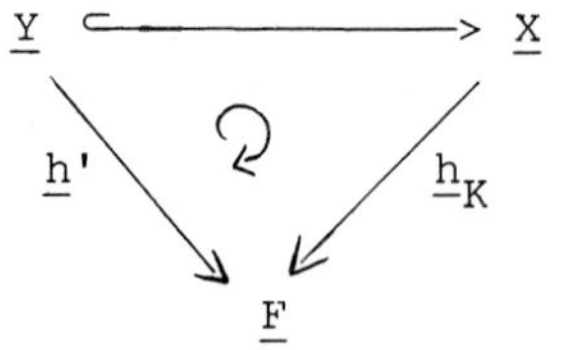

and, for $z \in F$ near y , $(\underline{h}')^{-1}(z) = \underline{Y} \cap \underline{K}_z$. The behaviour of $m(\underline{X},\underline{y})$ along $\underline{Y}$ is as follows.

<u>Proposition 3.2.4</u>. <u>Put, for</u> $z \in F$ <u>near</u> y ,

(3.2.3.) $$m(\underline{X},\underline{Y} \cap \underline{K}_z) := \sum_{z' \in Y \cap K_z} m(\underline{X},\underline{z}') \quad .$$

<u>Then</u>

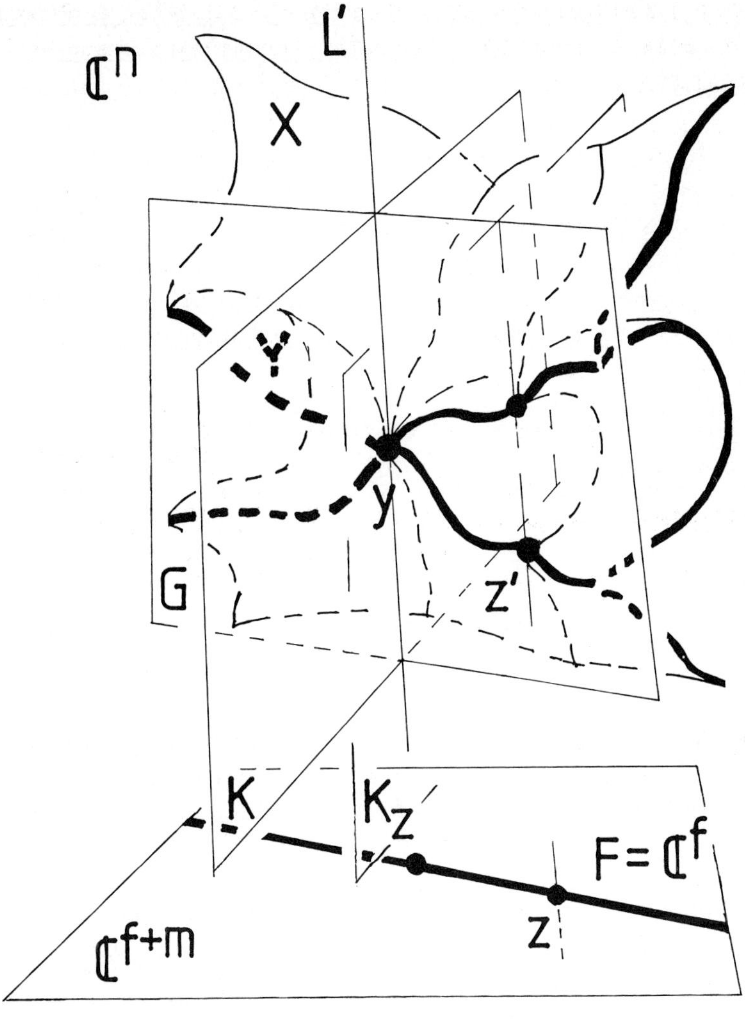

Fig. 11

(i) $m(\underline{X},\underline{Y}\cap K_z) \leq m(\underline{X},\underline{y})$ for all z near y, and has a constant value for z near y outside a nowhere dense analytic subset $A \subseteq Y$, denoted $m(\underline{X},\underline{Y},\underline{K})$.

(ii) If $I \subseteq \mathcal{O}_X$ defines $\underline{Y} \hookrightarrow \underline{X}$,

$$m(\underline{X},\underline{Y},\underline{K}) = e(\underline{x},I_y,\mathcal{O}_{X,y}),$$

where $\underline{x}$ is the set $(z_1,\ldots,z_f)$ of parameters of $\mathcal{O}_{X,y}$ defining K, and $e(\underline{x},I_y,\mathcal{O}_{X,y})$ is the generalized multiplicity of Chapter I, (3.9).

The proof is similar to the proof of Theorem 2.1.2 ; one considers the admissible graded $\mathcal{O}_F$-algebra $G((\underline{h}')_* I,\mathcal{O}_F) = \bigoplus_{k\geq 0} (\underline{h}')_*(I^k/I^{k+1})$ and uses the fact that normal flatness is generic, i.e. Theorem 1.4.8. See [54]

This leads to the following definition:

Definition 3.2.5. Let $y \in \underline{Y} \hookrightarrow \underline{X} \hookrightarrow \underline{U}$ be as in 3.2.3. If $K \in P^f_g(\underline{Y},y)$, X is said to be K-equimultiple along $\underline{Y}$ at y if and only if the function $z \longmapsto m(\underline{X},\underline{Y}\cap \underline{K}_z)$ is constant for all $z \in F$ near y.

For equimultiplicity considerations, one wants to proceed as in the smooth case and choose an $L \in P^d_e(\underline{X},\underline{y})$ with $L \subseteq K$, in order to use the local mapping degree of the projection $\underline{h} =: \underline{h}_{K,L} : \underline{X} \longrightarrow \underline{E} := \mathbb{C}^d \times \underline{0}$ along $\underline{Y}$ to compare the various $m(\underline{X},\underline{Y}\cap \underline{K}_z)$. For this, one may show there is an open neighbourhood V of K in $\mathrm{Grass}^f(\mathbb{C}^n)$ such that $m(\underline{X},\underline{Y}\cap \underline{K}'_z)$ does not depend on K' for $K' \in V$ and z near y (this are grassmannian arguments similar to those employed in II, 4.1.). So, since $P^f_{es}(\underline{X},y)$ is generic in $\mathrm{Grass}^f(\mathbb{C}^n)$, we may replace K with some $K' \in P^f_{es}(\underline{Y},y)$ without affecting $m(\underline{X},\underline{Y}\cap \underline{K}_z)$ (this is the geometric content of (20.3) and (20.4) in Chapter IV). So we may always assume, for questions concerning $m(\underline{X},\underline{Y}\cap \underline{K}_z)$, that $K \in P^f_{es}(\underline{Y},y)$. Then $C(\underline{X},\underline{y}) \cap K = C(\underline{X}\cap \underline{K},\underline{y})$, and the set $P^{d-f}_e(\underline{X}\cap \underline{K},\underline{y};K)$ $:= \{L \in \mathrm{Grass}^{d-f}(K) \mid L' \subseteq L$ and $L \pitchfork C(\underline{X},\underline{x})\}$ is generic in $\mathrm{Grass}^{d-f}(K)$, so we can always choose an $L \in P^{d-f}_e(\underline{X}\cap \underline{K},\underline{y};K)$. Then $L \in P^d_e(\underline{X},\underline{y})$, and if $\underline{h} : \underline{X} \longrightarrow \underline{E} := \mathbb{C}^d \times \underline{0}$ is the projection along L, there is the fundamental chain of inequalities for $z \in F$ near y :

$$(3.2.4) \qquad m(\underline{X},\underline{y}) = \deg_y \underline{h} = \sum_{z' \in h^{-1}(z)} \deg_{z'} \underline{h} \overset{(1)}{\geq} \sum_{z' \in (h')^{-1}(z)} \deg_{z'} \underline{h}$$

$$\overset{(2)}{\geq} \sum_{z' \in (h')^{-1}(z)} m(\underline{X},\underline{z}') = m(\underline{X},\underline{Y} \cap \underline{K}_z) \quad .$$

The inequality (1) holds because $(h')^{-1}(z) \subseteq h^{-1}(z)$, and (2) holds because $\deg_{z'} \underline{h} = e(\mathfrak{q}, \mathcal{O}_{X,z'}) \geq e(\mathfrak{q}', \mathcal{O}_{X,z'}) = \deg_{z'} \underline{h}' =: m(\underline{X},z')$, where $\mathfrak{q}' \supseteq \mathfrak{q}$ are the primary ideals induced by $\underline{h}'$ and $\underline{h}$ from the maximal ideal of $\mathcal{O}_{E,z}$.

The various aspects of K-equimultiplicity of $\underline{X}$ along $\underline{Y}$ at y are now summarized in the following theorem.

Theorem 3.2.6 (Geometric analysis of equimultiplicity). Let $y \in \underline{Y} \hookrightarrow \underline{X} \hookrightarrow \underline{U} \hookrightarrow \mathbb{C}^n$ be as described in 3.2.3. Let $K \in P^f_{es}(\underline{Y},y)$. The following conditions are equivalent:

(i) $\underline{X}$ is K-equimultiple along Y at y .

(ii) There is $L \in \mathrm{Grass}^d(\mathbb{C}^n)$ and a neighbourhood V of y in X such that, for all $z \in \mathbb{C}^f \times \{0\} \cap V$, $V \cap L_z = Y \cap K_z$ and $L_{z'} \in P^d_e(\underline{X},\underline{z}')$ for all $z' \in Y \cap K_z$.

(iii) For all $L \in P^d_e(\underline{X},\underline{y})$ there is a neighbourhood V of y in X such that $V \cap L_z = Y \cap K_z$ for all $z \in \mathbb{C}^f \times \{0\} \cap V$.

(iv) $\underline{X}$ is normally pseudoflat along $\underline{Y}$ at y , i.e. $\dim \nu^{-1}(y) = d - f$, where $\underline{\nu} : \underline{C}(\underline{X},\underline{Y}) \longrightarrow \underline{Y}$ is the normal cone.

(v) There is $L \in \mathrm{Grass}^d(\mathbb{C}\ )$ such that $(G+L) \cap C(\underline{X},\underline{Y}) = Y$.

(vi) $C(\underline{X},\underline{y}) = \nu^{-1}(y) \times (\mathbb{C}^f \cap \{0\})$.

(vii) $C(\underline{X} \cap \underline{K},\underline{y}) = \nu^{-1}(y)$.

(viii) ("Principle of specialization of minimal reduction"). Let $I \subseteq \mathcal{O}_X$ define $\underline{Y} \hookrightarrow \underline{X}$. Let $J \subseteq I$. Then J_y is

<u>a (minimal) reduction of</u> I_y <u>if and only if</u> J_z <u>is a (minimal) reductio</u> <u>of</u> I_z <u>for all</u> $z \in Y$ <u>near</u> y <u>outside a nowhere dense analytic set</u>.

(ix) ("Principle of specialization of integral dependence", cf.[69],). <u>Let</u> $f \in \mathcal{O}_X(X)$. <u>Then</u> $f_y \in \overline{I_y}$ <u>if and only if</u> $f_z \in \overline{I_z}$ <u>for all</u> $z \in Y$ <u>near</u> y <u>outside a nowhere dense analytic set</u>.

<u>If one of these conditions holds</u>, (i) <u>and</u> (vii) <u>hold for all</u> $K \in P_g^f(\underline{Y},y)$.

The implications (iv) ⇒ (iii) ⇒ (i) follow , analogously to 2.2, by blowing up $\mathbb{C}^n$ along $\underline{G}$ and using (3.2.4).(i) ⟺ (ii) follows from (3.2.4 and the Theorem of Rees (Proposition 3.2.2. (ii)). I do not know of a geometric proof of (i) ⇒ (iv), but (i) ⟺ (iv) follows from the corresponding algebraic result, namely Theorem (20.5) of Chapter IV in view of Proposition 3.2.4 (cf. [54]), and it is a useful exercise to visualize the proof of that Theorem geometrically using the geometric form of Böger's Theorem below. The equivalence (iv) ⟺ (vi) ⟺ (vii) follows from the corresponding algebraic result Proposition (23.15) of Chapter IV. The implications (vi) ⇒ (vii) ⇒ (iv) are also direct geometrically. The equivalence (v) ⟺ (viii) can be treated as in the smooth case, and (viii) ⟺ (ix) is left to the reader. One may also derive the equivalence (iv) ⟺ (viii) as a direct consequence of Böger's Theorem:

<u>Theorem 3.2.7</u> (Theorem of Böger). <u>Let</u> $(\underline{Y},y) \hookrightarrow (\underline{X},y) \hookrightarrow (\underline{\mathbb{C}}^n,0)$ <u>be as in Theorem 3.2.6. Let</u> $L \in P_g^d(\underline{X},\underline{y})$, <u>and</u> $K \in P_g^f(\underline{Y},y)$ <u>containing</u> L . <u>Let</u> $G \in \mathrm{Grass}^m(\mathbb{C}^n)$ <u>be such that</u> $\underline{Y} = \underline{X} \cap \underline{G}$, <u>and let</u> $\underline{h} : \underline{X} \to \mathbb{C}^d$ <u>be the projection along</u> L . <u>The following statements are equivalent</u>.

(i) $\deg_z \underline{h} = m(\underline{X}, \underline{Y} \cap \underline{K}_z)$ <u>for all</u> z <u>near</u> Y <u>outside a nowhere dense analytic set, and</u> $(G+L) \cap X = Y$ <u>near</u> Y .

(ii) $G+L$ <u>intersects</u> X <u>transversally along</u> Y , <u>i.e.</u> $(G+L) \cap C(\underline{X},\underline{Y}) = Y$.

<u>Exercise 3.2.8</u> (i) Derive this theorem from Böger's Theorem (19.6) in Chapter IV, and show the equivalence (I) ⟺ (v) of Theorem 3.2.6.

We end our survey of Theorem 3.2.6 by establishing (iv) $\Longleftrightarrow$ (viii). The implication (viii) $\Rightarrow$ (iv) is left to the reader. For (iv) $\Rightarrow$ (viii), the "only if" statement is obvious, because I is locally finitely generated at y. For the "if"-statement, let $J \subseteq I$ be a minimal reduction, $J = (g_1,\dots,g_\ell)\cdot \mathcal{O}_X$. We may assume $\underline{X}$ is so embedded in $\mathbb{C}^n$ that $(g_1,\dots,g_\ell) = (z_{f+1},\dots,z_d)$. The assumptions then imply that condition (i) of Theorem 3.2.7 holds, and the conclusion follows from (ii) of the theorem.

An interpretation of this is that the content of Böger's Theorem, beyond the content of Rees' Theorem, is essentially the statement of the principle of specialization of integral dependence. This is also apparent from the proof of (19.6) in Chapter III.

Finally, as an application of Theorem 3.2.6 we mention the following geometric variant of proof of the result (31.1) (b) in Chapter VI.

<u>Theorem. Let $(\underline{X},y) \hookrightarrow (\mathbb{C}^n,0)$ be in $\underline{\underline{\mathrm{cpl}}}_0$, $(\underline{Y},y) \hookrightarrow (\underline{X},y)$ a complex spacegerm, and let the notation be as in 3.2.3. Let $K \in P^f_g(\underline{Y},y)$ and suppose $\underline{X}$ is K-equimultiple along $\underline{Y}$ at y. Let $\underline{\pi} : \underline{\widetilde{X}} \longrightarrow \underline{X}$ be the blowup of $\underline{X}$ along $\underline{Y}$ and let $\widetilde{y} \in \pi^{-1}(y)$. Then</u>

$$m(\underline{\widetilde{X}},\widetilde{y}) \leqq m(\underline{X},\underline{Y},\underline{K}) \quad .$$

<u>Idea of proof</u>. If $(\underline{C},0) \hookrightarrow (\mathbb{C}^p,0)$ is a cone, $m(\underline{C},c) \leqq m(\underline{C},0)$ for all $c \in C$ by the Degree Formula II 2.2.8. Now let the line $\ell \subseteq C(\underline{X},\underline{Y})$ correspond to $\widetilde{y} \in \pi^{-1}(y)$ and let $\xi \in \ell - \{0\}$. By Theorem 3.2.6 (vii) we may assume $\xi \in \nu^{-1}(y)$. We have the chain of inequalities:

$$\begin{aligned} m(\underline{\widetilde{X}},\widetilde{y}) &\leqq m(\underline{\pi}^{-1}(\underline{y}),\widetilde{y}) = m(\underline{C}(\underline{X},\underline{Y}),\xi) \\ &\leqq m(\underline{\nu}^{-1}(\underline{y}),y) = m(\underline{X},y) \leqq m(\underline{X},\underline{y}) \\ &= m(\underline{X},\underline{Y},\underline{K}) \quad , \end{aligned}$$

which proves the claim.

BIBLIOGRAPHY

[1] ATIYAH, M.F. and MACDONALD, I.G.: Introduction to Commutative Algebra. Addison-Wesley,1969.

[2] BĂNIČA, C. and STĂNAŠILĂ, O.: Algebraic methods in the global theory of complex spaces. John Wiley and Sons, 1976.

[3] BAYER, D.: The division algorithm and the Hilbert scheme. Ph. D. Dissertation, Harvard University, 1982.

[4] ———— and STILLMAN, M.: The Macaulay System. Discette and manual Version 1.2, July 1986. Design and implementation David Bayer, Columbia University, and Michael Stillman, Brandeis University.

[5] BINGENER, J.: Schemata über Steinschen Algebren. Schriftenreihe des Math. Inst. der Universität Münster, 2. Serie, Heft 10, Januar 1976.

[6] BOURBAKI, N.: Commutative Algebra. Hermann and Addison-Wesley, 1972.

[7] ————: General Topology, Part I. Hermann and Addison-Wesley, 1966.

[8] BRIANCON, J.: Weierstraß préparé à la Hironaka. Astérisque N° 7 et 8 (1973), 67-76.

[9] CHEVALLEY, C.: On the theory of local rings. Ann. of Math. 44 (1943), 690-708.

[10] ———— : Intersections of algebraic and algebroid varieties. Trans. AMS 57 (1945), 1-85.

[11] DICKSON, L.E.: Finiteness of the odd perfect and primitive abundant numbers with n distinct prime factors. Am. J. of Math. 35 (1913), 413-426.

[12] DIEUDONNÉ, J.: Topics in local algebra. Notre Dame Mathematical Lectures Nr. 10, Notre Dame, Indiana, 1967.

[13] DRAPER, R.N.: Intersection Theory in Analytic Geometry. Math. Ann. 180 (1969), 175-204.

[14] FISCHER, G.: Complex Analytic Geometry. SLN 538.

[15] FITTING, H.: Über den Zusammenhang zwischen dem Begriff der Gleichartigkeit zweier Ideale und dem Äquivalenzbegriff der Elementarteilertheorie. Math. Ann. 112 (1936), 572-582.

[16] FRISCH, J.: Points de platitude d'un morphisme d'espaces analytiques complexes. Inv. Math. 4 (1967), 118-138.

[17] FULTON, W.: Intersection Theory. Ergebnisse der Mathematik und ihrer Grenzgebiete 3. Folge, Band 2, Springer Verlag 1984.

[18] GALLIGO, A.: Sur le théorème de preparation de Weierstraß pour un idéal de $k\{x_1,\dots,x_n\}$. Astérisque N° 7 et 8 (1973), 165-169.

[19] GALLIGO, A.: A propos du théorème de préparation de Weierstraß. SLN 409 (1973), 543-579.

[20] ―――――― : Théorème de division et stabilité en géométrie analytique locale. Ann. Inst. Fourier Tome XXIX (1979), 107-184.

[21] ―――――― et HOUZEL, C.: Module des singularités isolées d'apres Verdier et Grauert (Appendice). Astérisque N° 7 et 8 (1973), 139-163.

[22] GRAUERT, H.: Ein Theorem der analytischen Garbentheorie und die Modulräume komplexer Strukturen. Publ. IHES N° 5 (1960), 233-292.

[23] ―――――― : Über die Deformation isolierter Singularitäten analytischer Mengen. Inv. Math. 15 (1971), 171-198.

[24] ―――――― und REMMERT, R.: Komplexe Räume. Math. Ann. 136 (1958), 245-318.

[25] ――――――――――――――: Bilder und Urbilder analytischer Garben. Ann. of Math. 68 (1964), 393-443.

[26] ――――――――――――――: Analytische Stellenalgebren. Springer Verlag, 1971.

[27] ――――――――――――――: Theorie der Steinschen Räume. Springer Verlag, 1977.

[28] ――――――――――――――: Coherent Analytic Sheaves. Springer Verlag, 1984.

[29] GRECO, S. and TRAVERSO, C.: On seminormal schemes. Comp. Math. 40 (1980), 325-365.

[30] GRIFFITHS, P. and HARRIS. J.: Principles of Algebraic Geometry. John Wiley and Sons 1978.

[31] GROTHENDIECK, A. et DIEUDONNÉ, J.: Éléments de Géométrie Algébrique. Springer Verlag, 1971.

[32] HILBERT, D.: Über die Theorie der algebraischen Formen. Math. Ann. 36 (1890), 473-534. In: Gesammelte Abhandlungen Band II, 199-263, Springer 1970.

[33] HIRONAKA, H.: Resolution of singularities of an algebraic variety over a field of characteristic zero: I, II. Ann. of Math. 79 (1964), 109-326.

[34] ―――――― : Normal cones in analytic Whitney stratifications. Publ. IHES 36 (1969), 127-138.

[35] ―――――― : Bimeromorphic smoothing of a complex analytic space. University of Warwick notes 1971.

[36] ―――――― : Stratification and flatness. In: Real and complex singularities Oslo 1976. Ed. P. Holm Sythoff and Nordhoff, 1977.

[37] ―――――― and ROSSI, H.: On the Equivalence of Imbeddings of Exceptional Complex Spaces. Math. Ann. 156 (1964), 313-333.

[38] IDÀ, M. and MANARESI, M.: Some remarks on normal flatness and multiplicitiy in complex spaces. In: Commutative Algebra. Proceedings of the Trento Conference, ed. by S. Greco and G. Valla. Lecture notes in pure and applied mathematics Vol. 84, 171-182, Marcel Dekker 1983.

[39] IVERSEN, B.: Cohomology of sheaves. Springer 1986.

[40] KAUP, B. and KAUP, L.: Holomorphic Functions of Several Variables, Walter de Gruyter 1983.

[41] KIEHL, R.: Note zu der Arbeit von J. Frisch "Points de platitude d'un morphisme d'espaces analytiques complexes". Inv. Math. 4 (1967), 139-141.

[42] KÖNIG, J.: Einleitung in die allgemeine Theorie der algebraischen Größen. B.G. Teubner, Leipzig, 1903.

[43] KRONECKER, L.: Grundzüge einer arithmetischen Theorie der algebraischen Größen. J. reine u. ang. Math. 92 (1882), 1-122. In: Werke, herausg. von K. Hensel. Zweiter Band, Chelsea Reprint 1968.

[44] LAZARD, D.: Gröbner bases, Gaussian elimination, and resolution of systems of algebraic equations. Proc. EUROCAL 1983, SLN in Comp. Sci. 162, 146-156.

[45] LÊ, D.T.: Limites d'espaces tangents sur les surfaces. Centre de Math. de l'École Polytechnique. M 361.0678, Juin 1978.

[46] LEJEUNE-JALABERT, M.: Effectivité de calculs polynomiaux. Cours de D.E.A, Université de Grenoble I, 1984 - 85.

[47] ———————— et TEISSIER, B.: Contributions à l'étude des singularités du point de vue du polygone de Newton. Thèse, Université Paris VII, 1973.

[48] ————————————: Normal cones and sheaves of relative jets. Comp. Math. 281 (1974), 305-331.

[49] LIPMAN, J.: Equimultiplicity, reduction, and blowing up. In: Commutative Algebra - analytic methods, ed. by R.N. Draper. Lecture notes in pure and applied mathematics, Vol. 68, 111-147, Marcel Dekker 1982.

[50] MACAULAY, F.S.: On the Resolution of a given Modular System into Primary Systems including some Properties of Hilbert Numbers. Math. Ann. 74 (1913), 66-121.

[51] ——————: The algebraic theory of modular systems. Stechert-Hafner Service Agency, New York and London, 1964 (Reprint from the Cambridge University Press Edition 1916).

[52] ——————: Some properties of enumeration in the theory of modular systems. Proc. London Math. Soc. 26 (1927), 531-555.

[53] MOELLER, H.H. and MORA, F.: The computation of the Hilbert function. Proc. EUROCAL 83, SLN in Comp. Sci. 162 (1983), 157-167.

[54] MOONEN, B.: Transverse Equimultiplicity. To appear.

[55] MORA, F.: An algorithm to compute the equations of tangent cones. Proc. EUROCAM 82, SLN in Comp. Sci. 144 (1982), 158-165.

[56] MUMFORD, D.: Algebraic Geometry I. Complex Projective Varieties. Springer Verlag 1976.

[57] RAMANUJAM, C.P.: On a Geometric Interpretation of Multiplicity. Inv. Math. 22 (1973), 63-67.

[58] ROBBIANO, L.: Term orderings on the polynomial ring. Proc. EUROCAL 85, II, SLN in Comp. Sci. 204 (1985), 513-517.

[59] RÜCKERT, W.: Zum Eliminationsproblem der Potenzreihenideale. Math. Ann. 107 (1933), 259-281.

[60] SAMUEL, P.: La notion de multiplicité en Algèbre et en Géométrie algébrique, Journal de Math., tome XXX (1951), 159-274.

[61] SCHICKHOFF, W.: Whitneysche Tangentenkegel, Multiplizitätsverhalten, Normal-Pseudoflachheit und Äquisingularitätstheorie für Ramissche Räume. Schriftenreihe des Math. Inst. der Universität Münser, 2. Serie, Heft 12, 1977.

[62] SCHREYER, F.-O.: Die Berechnung von Syzygien mit dem verallgemeinerten Weierstraßschen Divisionssatz und eine Anwendung auf analytische Cohen-Macaulay Stellenalgebren minimaler Multiplizität. Diplomarbeit Hamburg, Oktober 1980.

[63] SELDER, E.: Eine algebraische Definition lokaler analytischer Schnittmultiplizitäten. Rev. Roumaine Math. Pures Appl. 29 (1984), 417-432.

[64] SÉMINAIRE CARTAN 1960/61: Familles d'espaces complexes et fondements de la géométrie analytique. Sécrétariat mathematique, 11 rue Pierre Curie, PARIS 5e, 1962, Reprint W.A. Benjamin 1967.

[65] SERRE, J.-P.: Faisceaux algébriques cohérents. Ann. of Math. 61 (1955), 197-278.

[66] ——————: Géometrie algébrique et géométrie analytique. Ann. Inst. Fourier 6 (1956), 1-42.

[67] ——————: Algèbre locale. Multiplicités, SLN 11, seconde édition 1965.

[68] SIU, Y.T.: Noetherianness of rings of holomorphic functions on Stein compact subsets. Proc. AMS 31 (1969), 483-489.

[69] TEISSIER, B.: Variétés polaires II. Multiplicités polaires, sections planes, et conditions de Whitney. In: Algebraic Geometry, Proc. La Rábida 1981, SLN 961, 314-491.

[70] THIE, P.R.: The Lelong Number of a Point of a Complex Analytic Set, Math. Ann. 172 (1967), 269-312.

[71] v.d. WAERDEN, B.L.: Algebra II. Vierte Auflage. Springer Verlag 1959.

[72] v.d. WAERDEN, B.L.: Zur algebraischen Geometrie 20. Der Zusammenhangssatz und der Multiplizitätsbegriff. Math. Ann. 193 (1971), 89-108.

[73] ——————— : The foundations of Algebraic Geometry from Severi to André Weil. Arch Hist. of Exact Sci. 7 (1971), 171-180.

[74] WEIL, A.: Foundations of algebraic geometry. Am. Math. Soc. Publ. 29, 1946.

[75] WHITNEY, H.: Complex Analytic Varieties. Addison-Wesley 1972.

[76] ZARISKI, O. and SAMUEL, P.: Commutative Algebra, Vol. I, II. Graduate Texts in Mathematics 28, 29, Springer Verlag, 1975.

GENERAL INDEX